INFORMATION C

AF838519

... on or before the date ...

The UK Pesticide Guide 2004

Editor: R. Whitehead BA, MSc

BCPC

CABI Publishing

© 2004 CAB International and BCPC (British Crop Protection Council). All rights reserved. No part of this publication may be reproduced in any form or by any means, electronically, mechanically, by photocopying, recording or otherwise, without the prior permission of the copyright owners.

CABI Publishing (a division of CAB International) is one of the world's foremost publishers of databases, books and journals in agriculture, and applied life sciences. It has a worldwide reputation for producing high quality, value-added information, drawing on its links with the scientific community. From its Headquarters in Wallingford, UK, CABI Publishing runs a worldwide operation, distributing books, journals and electronic products to customers in over 150 countries, and selling its products through a network of international agents. For further information, please contact CABI Publishing, CAB International, Wallingford, Oxon OX10 8DE, UK.

Telephone: (01491) 832111
Fax: (01491) 833508
e-mail: publishing@cabi.org
Web: www.cabi-publishing.org

BCPC (British Crop Protection Council) is a self-supporting limited company with charity status, which was formed in 1968 to promote the knowledge and understanding of the science and practice of crop protection. The corporate members include government departments and research councils; advisory services; associations concerned with the farming industry; agrochemical manufacturers; agricultural engineering, agricultural contracting and distribution services; universities; scientific societies; organisations concerned with the environment; and some experienced independent members. Further details available from BCPC, 7 Omni Business Centre, Omega Park, Alton, Hampshire GU34 2QD, UK.

Telephone: (01420) 593200
Fax: (01420) 593209
e-mail: md@bcpc.org
Web: www.bcpc.org

ISBN 0 85199 737 6

INFORMATION CENTRE
CENTRAL SCIENCE LABORATORY
SAND HUTTON
YORK
YO4 1LZ
Tel 01904 462272

CALL NO	632.95 CAB
DYNIX BARCODE	-141841
DYNIX BIB NO	98642
DATE:	26-5-04

Typeset and printed by Page Bros, Norwich, UK

Contents

Disclaimer

Every effort has been made to ensure that the information in this book is correct at the time of going to press, but the Editor and the publishers do not accept liability for any error or omission in the content, or for any loss, damage or other accident arising from the use of the products listed herein. Omission of a product does not necessarily mean that it is not approved or that it is unavailable.

It is essential to follow the instructions on the approved label when handling, storing or using any crop protection product. Approved 'off-label' extensions of use are undertaken entirely at the risk of the user.

The Dangerous Preparations Directive

The Dangerous Preparations Directive (1999/45/EEC) is due to be implemented in the UK on 30 July 2004. The Directive will affect the classification, packaging, labelling and safety data sheets of crop protection products. In most cases it will result in additional hazard symbols, and associated risk and safety phrases relating to environmental and health hazards, appearing on the label.

At the time of going to press the conversion of labels of products included in this edition, to comply with the Directive, was in progress. The result is that some products are shown with their new environmental hazard classification and risk phrases, while other similar or identical products are not yet so designated. No disadvantage should be inferred for those products whose labels have already been changed to show these additional hazards and risks.

It is hoped that these discrepancies will disappear from the 2005 edition of the *Guide*, by which time all product labels must be compliant with the Directive.

Editor's Note

This seventeenth edition comes at a time of more major changes in the crop protection industry. The ever-sharpening focus on the control of use of crop protection chemicals through legislation and voluntary measures makes it increasingly important that everyone involved remains fully up to date and informed. *The UK Pesticide Guide* seeks to do just that.

During the currency of this edition the revised Dangerous Preparations Directive comes into force in the UK - in July 2004 (page v). This will necessitate substantial changes to packaging and labelling of crop protection products. The process of conversion was only partially complete at the time this edition went to press, but where the changes have been made, they are included here. The remainder will be included in the 2005 edition. Similarly, the project under the Voluntary Initiative to review and improve the clarity of product labels is in progress (page vii), and we have amended the presentation of information in the Guide to reflect this.

About 30 active substances and all products containing them, whose approvals were lost in July 2003 as a result of the EU Pesticide Review Programme, have been deleted, apart from those few for which 'Essential Use' derogations have been granted until 2007. The list of the latter has been updated. The list of Products Also Registered (Section 5) has been brought up to date and, when combined with the actively marketed products in shown in Section 4, represents a comprehensive listing of products that have extant approval in 2004 and may legally be used.

There are over 130 new product entries in this edition. Many of these reflect further changes in the commercial world, such as the formation of Bayer CropScience. Re-registered products in the new company name are duplicated if they remain in the supply chain with their previous packaging and registration number. Nevertheless, there are 13 new active substances, or mixtures of active substances, in this edition. These include six new chemicals listed for the first time: four fungicides (dimoxystrobin, boscalid, fenamidone and flutolanil); one insecticide (methoxyfenozide); and one herbicide (iodosulfuron).

The crop protection market place is dynamic and ever-changing. To provide a better service to users of the *Guide*, any important developments or new products notified to the Editor during the year will be shown on updates at www.ukpesticideguide.co.uk

As always, criticisms and suggestions are welcome and, in particular, notification of any errors or omissions.

R. Whitehead
Editor

The information in this publication has been selected from official sources and from suppliers' product labels and product manuals of pesticides approved for use in the UK under the Control of Pesticides Regulations 1986 or the Plant Protection Products Regulations 1995.

The content is based on information received by the Editor up to October 2003.

For developments since publication of this edition, and new products notified to the Editor during the year, see:

www.ukpesticideguide.co.uk

The Voluntary Initiative

The Voluntary Initiative is a five-year programme of measures, agreed by the crop protection industry and related organisations with Government, to minimise the environmental impacts of crop protection products. By protecting water quality and enhancing farmland biodiversity, the programme aims to reduce the risk of a pesticide tax. Anyone involved with the products in this Guide, whether as a user or advisor, is likely to be affected directly or indirectly by the projects in the programme.

This edition of the *Guide* can help in two specific ways.

- It is illegal to use, or even store, products whose approval has expired. While mere storage is unlikely to have any environmental impact, their incorrect or inappropriate disposal could undoubtedly do so. The Obsolete Pesticides Disposal Campaign project in the Voluntary Initiative is designed to address this problem by encouraging farmers and growers to identify any obsolete products in their possession, and providing a mechanism for relatively inexpensive disposal. All the products listed in Sections 4 and 5 of this Guide have extant approval for all or part of 2004. A date is shown for any whose approval will expire before the end of the year. Therefore any products not listed here are likely to be obsolete, although before disposal a check should be made with the manufacturer or supplier.
- Product labels contain much information that is often too crowded and complicated. Another project in the Voluntary Initiative, undertaken by the Crop Protection Association (CPA) together with the Pesticides Safety Directorate and the Health and Safety Executive, is a review of product labels and the production of Best Practice Labelling Guidance. Member companies of the CPA are gradually adapting their labels to be compliant with these guidelines, at the same time as making the mandatory changes required by the Dangerous Preparations Directive. In this edition of the Guide we have amended the presentation of information to reflect this improved clarity as far as possible. In the sister publication, *The e-UK Pesticide Guide*, we have gone further by using the new symbols for environmental hazard, product activity and LERAP category, as well as providing links to company websites to enable access to product environmental information sheets (also being produced as a Voluntary Initiative project).

Further information on the Voluntary Initiative is available at www.voluntaryinitiative.org.uk

Changes Since 2003 Edition

Pesticides and adjuvants added or deleted since the 2003 edition are listed below. Every effort has been made to ensure the accuracy of this list, but late notifications to the Editor are not included.

Products added

Products new to this edition are shown here. In addition, products that were listed in the previous edition whose PSD/HSE registration number *and* supplier have changed are included. Where only the PSD/HSE registration number has changed, the product is not listed.

Product	Reg'n No	Supplier
Agriguard Bentazone 480	11569	AgriGuard
Agriguard Cerusite	11494	AgriGuard
Agriguard Isoxaben	11652	AgriGuard
Aliette 80 WG	11213	Certis
Amber	ADJ0367	Interagro
Antec Durakil 1.5 SC	H7560	Antec
Antec Durakil 6SC	H7559	Antec
Arelon 500	11639	Nufarm UK
Arizona	11722	Makhteshim
ASAP	11485	Me2
Atlantis	11679	Bayer CropScience
Aurora 50 WG	11613	Belchim
BackRow	ADJ0472	Interagro
Barclay Piza 400FL	11495	Barclay
Barclay Rebel	11648	Barclay
Barramundi	ADJ0376	Interagro
Baytan Flowable	11714	Makhteshim
Bio-Flex	ADJ0481	Greenaway
BioPower	ADJ0454	Bayer CropScience
C-Cure	ADJ0467	Interagro
Centium 360 CS	11607	Belchim
Ceres Platinum	ADJ0445	Interagro
Chipko Green	11211	Bayer Environ.
Cirrus	11751	Belchim
Companion PCT12	ADJ0482	Ciba Specialty
Compass	11740	BASF
Contact Plus	ADJ0418	Interagro
Cosavet DF	11477	Headland
Croptex Bronze	11329	Certis
Crotale	11593	Sipcam
Dacthal W-75	11323	Certis
Danadim	11550	Headland
Decis Protech	11502	Bayer CropScience
Diamant	11557	BASF
Diurex 50SC	11472	Makhteshim
Elvaron Multi	11422	Bayer CropScience
Emblem	11559	Nufarm UK
Equity	11520	Dow
Esterol	ADJ0330	Interagro
Ethosat 500	11501	Makhteshim
Euro	11511	Me2
Fernpath Graminite	11646	AgriGuard
Fernpath Haptol Flo	11481	AgriGuard
Fernpath Hatchet	11510	AgriGuard
Fernpath Ipex	11391	AgriGuard
Filan	11449	BASF
Flame	11583	Dow
Flamenco	11699	BASF
Foil	11700	BASF
Force ST	11671	Bayer CropScience
Foundation	11708	Headland
Fungazil 100 SL	11762	BASF
Fury 10 EW	11608	Belchim
Fusilade Max	11519	Syngenta
Galmano	11650	Bayer CropScience
Galmano Plus	11645	Bayer CropScience
Gex 1664	ADJ0366	Headland
Gharda Bonanza	11571	Nufarm UK
Glydate	10999	Nufarm UK
Glyper	11095	Interfarm
Glyphosate 360	11579	Cardel
Goltix 90	11578	Makhteshim
Goltix Flowable	11540	Makhteshim

Product	Reg'n No	Supplier
Goltix WG	11539	Makhteshim
Greencrop Cajole	11402	Greencrop
Greencrop Champion	11755	Greencrop
Greencrop Malin	11787	Greencrop
Greencrop Pomeroy	11769	Greencrop
Greengard	11715	SumiAgro Amenity
Halo	11546	Headland
Herbasan Flow	11448	Nufarm UK
Heritage	11383	Scotts
Hussar	11205	Bayer CropScience
HY-D Super	11618	Agrichem
Impact XL	11547	Headland
Intro Plus	11630	Interfarm
Jockey	11689	BASF
Jockey F	11690	BASF
Katalyst	ADJ0450	Interagro
Linuron 500	11590	Nufarm UK
Magnate 100 SL	11705	Makhteshim
Marnoch Clodim	11460	Me2
Marshal 10G	11682	Belchim
Marshal Soil Insecticide suSCon CR Sachets	11754	Fargro
Me2 Notin	11655	Me2
Midstream GSR	11674	Scotts
Minuet EW	11610	Belchim
Monceren IM	11426	Bayer CropScience
n2n Diuron 80	11821	n2n
n2n Diuron Flowable	11822	n2n
n2n Garlon 4	11806	n2n
n2n Mosskiller	11823	n2n
Nettle	ADJ0466	Interagro
Novak	11693	BASF
Nuturf Carbendazim	11469	Nufarm UK
Pan Oasis	ADJ0411	Pan Agriculture
Pan Panorama	ADJ0412	Pan Agriculture
Platform S	11611	Belchim
Pointer	11522	Headland
Poraz	11701	BASF
Posse 10G	11640	Belchim
Premis	11742	BASF
Primer	11782	AgriGuard
Priori Xtra	11518	Syngenta
Prompt	11535	Headland
QuikSilver	ADJ0471	Interagro

Product	Reg'n No	Supplier
Ranman Twinpack	20628	Belchim
Repulse	11328	Certis
Rhino	11802	Certis
Rhythm	11669	Interfarm
Rizolex Flowable	11399	Certis
Robust	11807	BASF
Romany	11000	Greencrop
Ronstar Liquid	11215	Certis
Rovral Flo	11702	BASF
Rovral Liquid FS	11703	BASF
Rovral WP	11694	BASF
Runner	11470	Bayer CropScience
Sakarat D Wax Bait	H7489	Killgerm
Samurai	10334	Monsanto
Scala	11695	BASF
Seradix 1	11330	Certis
Seradix 2	11331	Certis
Seradix 3	11332	Certis
Shark	11616	Belchim
SHL Granular Feed and Weed	10970	Sinclair
SHL Granular Feed, Weed & Mosskiller	10972	Sinclair
SHL Turf Feed and Weed	10963	Sinclair
SHL Turf Feed, Weed & Mosskiller	10973	Sinclair
Signum	11450	BASF
Snooker	11744	BASF
Sonata	11570	Bayer CropScience
Spartan	ADJ0375	Interagro
Spotlight 24 EC	11617	Belchim
Spread and Seal	ADJ0449	Mandops
Standon FFA60 Plus	11649	Standon
Standon Fluroxypyr 200	11534	Standon
Starion	11641	Belchim
Stealth	11514	Syngenta
Super Nova	ADJ0364	Interagro
Swing Gold	11658	BASF
Talstar	11325	Certis
Terbine	11407	Nufarm UK
Thistlex	11533	Dow
Trimaran	11400	Nufarm UK
UPL Diuron 80	07619	United Phosphorus

Product	Reg'n No	Supplier
Uranus	11637	Makhteshim
Vizzler	11409	Interfarm
Weedazol-TL	11430	Bayer CropScience
Wetta Plus	ADJ0480	Interagro

Products deleted

The appearance of a product name in the following list may not necessarily mean that it is no longer available. In cases of doubt, refer to the supplier.

Product	Reg'n No	Supplier
Actellic 2% Dust	10508	Syngenta
Actellic D	06930	Zeneca
Actellic Smoke Generator No 20	06627	Zeneca
Actipron	ADJ0013	Batsons
Adagio	10057	Interfarm
Afalon	00100	Aventis
Agrichem Flowable Thiram	06245	Agrichem
Aitken's Lawn Sand	05253	Aitken
Aitken's Lawn Sand Plus	04542	Aitken
Alistell	10510	Syngenta
Alpha Captan 83 WP	04806	Makhteshim
Alpha Terbutryn 50 EC	04809	Makhteshim
Alphabird	10146	Killgerm
Amistar	08517	Zeneca
Amistar Pro	08871	Zeneca
Apache	10514	Syngenta
Aphox	06633	Zeneca
Aquator	ADJ0425	Syngenta
Array	00352	Bayer
Array	11238	Bayer CropScience
Arsenal	05537	Nomix Chipman
Arsenal 50	05587	Nomix Chipman
Ashlade Carbendazim Flowable	06213	Nufarm UK
Ashlade Linuron FL	06221	Nufarm UK
Ashlade Trimaran	06228	Nufarm UK
Asset	07243	Aventis
Atlas 460:46	07705	Atlas
Atlas Atrazine	07702	Nufarm UK
Atlas Chlormequat 700	10706	Nufarm UK
Atlas Red	07724	Nufarm UK
Atlas Simazine	07725	Nufarm UK
Atlas Terbine	07709	Nufarm UK
Auger	10896	Headland
Avadex Excel 15G	10299	Monsanto
Avenge 2	10210	BASF
Axiom	ADJ0136	Batsons
Barclay Dart	05129	Barclay
Barclay Fentin Flow	07914	Barclay
Barclay Fentin Flow 532	09434	Barclay
Barclay Haybob II	08532	Barclay
Barclay Meadowman II	08525	Barclay
Barclay Pirimisect	09057	Barclay
Bayer UK 413	08277	Bayer
Bayfidan	02672	Bayer
Bayleton	00221	Bayer
Benazalox	07246	Aventis
Betosip	06787	Sipcam
Betosip Combi	08630	Sipcam
Borocil K	09924	Aventis Environ.
Brestan 60 SP	07305	Aventis
Bromomethane 100%	09244	Brian Jones
Bullet	10654	Feinchemie
Butoxone Plus	10499	Headland
Cadence	00570	Barclay
Carbetamex	10265	Feinchemie
Centium 360 CS	10591	FMC
Charger	09986	Aventis
Checkmate	09122	Certis
Chipko Diuron 80	00497	Nomix Chipman
Chipman Diuron 80	08054	Nomix Chipman
Chipman Garlon 4	06016	Nomix Chipman

Product	Reg'n No	Supplier
Choir	09778	Nufarm UK
Chryzoplus Grey	07984	Fargro
Chryzopon Rose	07982	Fargro
Chryzosan White	07983	Fargro
Chryzotek Beige	07125	Fargro
Chryzotop Green	07129	Fargro
Cirrus	11751	Belchim
Commando	10215	BASF
Comodor 600	08398	Agrichem
Companion	ADJ0239	Ciba Specialty
Comulin	ADJ0164	Batsons
Condox	10522	Syngenta
Contact	ADJ0191	Interagro
Cultar	06649	Zeneca
Cutonic Foliar Booster	ADJ0228	Lambson
Danadim	09583	Cheminova
Deosan Chlorate Weedkiller (Fire Suppressed)	08521	Johnson Diversey
Deosan Fly Spray	H5246	Johnson Diversey
Dithane NT Dry Flowable	09868	Interfarm
Divora	08960	Syngenta
Dorado	06657	Zeneca
Dorado	10524	Syngenta
Dovetail	07973	Zeneca
Draza	00765	Bayer
Dursban 4	07815	Dow
Elvaron WG	04855	Bayer
Eminent	10504	Monsanto
Endeavour	07385	Bayer
Fairy Ring Destroyer	05541	Vitax
Farmatin 560	09877	Aventis
Favour 600 SC	08405	Syngenta
Folio	08547	Syngenta
Fongarid	08407	Syngenta
Fortrol	10666	Feinchemie
Fubol 75 WP	08409	Novartis
Fubol Gold WP	08812	Syngenta
Fungazil 100 SL	10270	Aventis
Fusilade 250 EW	06531	Zeneca
Fyfanon	11014	Headland
Galtak 50 SC	07258	Aventis
Garlon 2	06616	Zeneca
Garnet	06391	Bayer
Glyper	07968	PBI
Goltix Flowable	11284	Bayer CropScience
Gramoxone 100	06674	Zeneca

Product	Reg'n No	Supplier
Grasp	06675	Zeneca
Greencrop Mooncoin	09392	Greencrop
Greencrop Rosette	09648	Greencrop
Greenshield	07988	Scotts
Hallmark with Zeon technology	09809	Zeneca
Headland Addstem DF	08904	Headland
Headland Archer	08814	Headland
Headland Guard Pro	ADJ0423	Headland
Headland Inflo XL	ADJ0329	Headland
Headland Regain	08675	Headland
Headland Relay Turf	08935	Headland
Headland Rocket	10998	Headland
Headland Swift	10437	Headland
Hydon	01088	Nomix Chipman
Hyvar X	01105	DuPont
Jupital DG	10529	Syngenta
Kandu	ADJ0424	Syngenta
Karate Ready-to-Use Rat and Mouse Bait	08658	Johnson Diversey
Karate Ready-to-Use Rodenticide Sachets	08659	Johnson Diversey
Karmex	11045	Headland
Klerat	06869	Sorex
Klerat Wax Blocks	06827	Sorex
Krenite	01165	DuPont
Landgold Aldicarb 10G	06036	Landgold
Landgold Cyproconazole 100	06463	Landgold
Landgold DFF 625	06274	Landgold
Landgold Flupyrsulfuron	10777	Landgold
Landgold Iprodione 250	06465	Landgold
Landgold Lambda-C	09205	Landgold
Landgold Mancozeb 80 W	06507	Landgold
Landgold Propiconazole	06291	Landgold
Landgold Propyzamide 50	09753	Landgold

Product	Reg'n No	Supplier
Landgold Trifloxystrobin 125	10653	Landgold
Legumex Extra	08676	Aventis
Littac	H5176	Sorex
Luxan Oil-H	ADJ0321	Luxan
Lyrex	ADJ0177	Batsons
Lyrol	ADJ0165	Batsons
Malathion 60	08018	United Phosphorus
Mallard	08662	Syngenta
Manzate 75 WG	11046	Headland
Marshal 10G	09804	FMC
Mascot Selective-P	06105	Rigby Taylor
Mascot Super Selective-P	06106	Rigby Taylor
Maton	10366	Headland
Maxicrop Moss Killer & Conditioner	04635	Maxicrop
MCPA 25%	07998	Nufarm UK
Me2 Aldee	09781	Me2
Me2 Lambda	09712	Me2
Midstream	09267	Scotts
Mildothane Liquid	10002	Aventis
Monceren Flowable	04907	Bayer
Monicle	07375	Bayer
MSS 2,4-D Amine	10183	Nufarm UK
MSS 2,4-DB + MCPA	01392	Nufarm UK
MSS Glyfield	11117	Nufarm UK
MSS MCPA 50	11144	Nufarm UK
MSS MCPB + MCPA	01413	Nufarm UK
MTM CIPC 40	05895	MTM Agrochem.
Nemathorin 10G	08915	Zeneca
Neosorexa	07756	Sorex
Neosorexa Ratpacks	04653	Sorex
Nimrod	10530	Syngenta
Nimrod-T	09268	Scotts
Nomix Chipman Flowable Diuron	10837	Nomix Chipman
Nomix-Chipman Mosskiller	06271	Nomix Chipman
Panacide M21	H4923	Coalite
Panaclean 730	H5075	Coalite
Pancil T	09261	Landseer
Parade	10568	Nomix Chipman
Pasturol Plus	08581	FCC
Patafol	07397	Nufarm UK

Product	Reg'n No	Supplier
PDQ	06518	Zeneca
PDQ	10241	Zeneca
Permasect 25 EC	01576	Mitchell Cotts
Phase	ADJ0279	Loveland
Phorate 10G	08007	United Phosphorus
Plenum	09816	Novartis
Plenum	09816	Syngenta
Powertwin	10442	Feinchemie
Premiere	10866	Scotts
Prophet	08433	Syngenta
Pynosect 30 Water Miscible	01653	Mitchell Cotts
Python	09791	Feinchemie
Racumin Bait	H6754	Bayer Environ.
Ranman TwinPack	10628	BASF
Rat & Mouse Bait	00764	B H & B
Recoil	08483	Syngenta
Redipon Extra	07518	United Phosphorus
Reglone	09646	Zeneca
Rezist	08576	Barrettine
Rhizopon A Powder	07131	Fargro
Rhizopon A Tablets	07132	Fargro
Rhizopon AA Powder (0.5%)	07126	Fargro
Rhizopon AA Powder (1%)	07127	Fargro
Rhizopon AA Powder (2%)	07128	Fargro
Rhizopon AA Tablets	07130	Fargro
Rhizopon B Powder (0.1%)	07133	Fargro
Rhizopon B Powder (0.2%)	07134	Fargro
Rhizopon B Tablets	07135	Fargro
Rhythm	09909	Interfarm
Ridomil mbc 60 WP	08437	Syngenta
Ridomil Plus	08353	Syngenta
Riposte Pepite	08485	Syngenta
Rival	10316	Monsanto
Roundup	10317	Monsanto
Roundup Greenscape	10699	Monsanto
Roundup Rapide	10331	Monsanto
Sanuron	09236	Headland
Satellite	08969	Cyanamid
Satellite	10395	BASF

Product	Reg'n No	Supplier
Setter 33	05623	Dow
Sheen	08442	Syngenta
Shirlan	07091	Zeneca
Shirlan Programme	10574	Syngenta
SHL Granular Feed and Weed	09926	Sinclair
SHL Granular Feed, Weed and Mosskiller	09925	Sinclair
SHL Turf Feed & Weed & Mosskiller	04438	Sinclair
SHL Turf Feed and Weed	04437	Sinclair
Sistan	01957	Unicrop
Skater	09790	Feinchemie
Sorex Warfarin 250 ppm Rat Bait	07371	Sorex
Sorex Warfarin 500 ppm Rat Bait	07372	Sorex
Sorex Warfarin Sewer Bait	07373	Sorex
Sorexa CD Mouse Bait	08207	Sorex
Sorexa Checkatube	09474	Sorex
Sorexa Gel	08315	Sorex
Sphinx	10561	Aventis
Sponsor	08674	Aventis
Standon Alpha-C10	08823	Standon
Standon Cyanazine 50	09230	Standon
Standon IPU	08671	Standon
Standon Propyzamide 50	09054	Standon
Standon Trifloxystrobin	10431	Standon
Sterilite Hop Defoliant	05060	Coventry Chemicals
Stride	10537	Syngenta
Stryper	08310	Crompton
Super-Tin 80 WP	07605	SumiAgro
Swipe P	08452	Syngenta
Teal	08453	Syngenta
Tedion V-18 EC	03820	Certis
Thiovit	08493	Syngenta
Thyram Plus	10421	Agrichem
Tilt	08456	Syngenta
Topas C	08459	Syngenta
Torero	10250	Feinchemie
Touchdown	06326	Zeneca
Touchdown LA	09270	Scotts
Tribunil	02169	Bayer
Tricur	10377	Bayer
Triflur	10750	Nufarm UK
Triptic 48 EC	09294	United Phosphorus
Trustan WDG	05050	DuPont
Turfclear	07506	Scotts
Twin	09875	Feinchemie
Twincarb	08777	Vitax
UPL Carbendazim 500 FL	07472	United Phosphorus
Vista CT	09368	Aventis
Warfarin 0.5% Concentrate	02325	B H & B
Warfarin Ready Mixed Bait	02333	B H & B
Zulu	08464	Syngenta

SECTION 1
INTRODUCTORY INFORMATION

About *The UK Pesticide Guide*

Purpose

The primary aim of this book is to provide a practical handbook of pesticides, plant growth regulators and adjuvants that the farmer or grower can realistically and legally obtain in the UK, and to indicate the purposes for which they may be used. It is designed to help in the identification of products appropriate for a particular problem, and a Crop/Pest Guide is included to facilitate this. In addition to uses recommended on product labels, details are provided of those uses that do not appear on labels but which have been granted specific off-label approval (SOLA). Such uses are not endorsed by the manufacturer, and are undertaken entirely at the risk of the user.

As well as identifying the products available, the book provides guidance on how to use them safely and effectively, but without giving details of doses, volumes, spray schedules or approved tank mixtures. Other sections provide essential background information on a wide range of pesticide-related issues, including legislation, codes of practice, poisons and treatment of poisoning, products for use in special situations, and weed and crop growth stage keys.

While we have tried to cover all other important factors, this book does **not** provide a full statement of product recommendations. **Before using any pesticide product it is essential that the user should read the label carefully and comply strictly with the instructions it contains**.

Scope

Individual profiles are provided of some 350 individual active ingredients, and a further 150 profiles of mixtures containing these active ingredients. Each profile has a list of available approved products for use in arable agriculture, horticulture (including amenity horticulture), forestry and areas near water, together with the name of the supplier from whom each may be obtained. Within these fields of use, all types of pesticide covered by the Control of Pesticides Regulations 1986 or the Plant Protection Products Regulations are included. This embraces acaricides, algicides, fungicides, herbicides, insecticides, lumbricides, molluscicides, nematicides and rodenticides, together with plant growth regulators used as straw shorteners, sprout inhibitors and for various horticultural purposes. The total number of pesticides covered in the *Guide* is about 1400.

Products are included in the *Guide* only if requested by the supplier and supported by evidence of approval. This is intended to ensure that only marketed products are listed.

In addition, the *Guide* gives information on more than 100 authorised adjuvants which, although not themselves pesticides, may be added to pesticides to improve their effectiveness in use.

The *Guide* does **not** include products approved solely for amateur, home garden, domestic, food storage, public health or animal health uses.

Sources of Information

The information for this edition has been drawn from these authoritative sources:

- approved labels and product manuals received from suppliers of pesticides up to October 2003
- websites of the Pesticides Safety Directorate (PSD, www.pesticides.gov.uk) and the Health and Safety Executive (HSE, www.hse.gov.uk)
- *The Pesticides Monitor* (incorporating *The Pesticides Register*), published monthly by PSD and HSE, listing new UK approvals (including off-label approvals) up to and including the issues for August/September 2003
- PSD lists of approval expiries.

Criteria for inclusion

To be included in the *Guide*, a product must meet the following conditions:

- it must have extant approval under UK pesticides legislation
- information on the approved uses must have been provided by the supplier
- the product must be expected to be on the UK market during the currency of this edition.

When a company changes its name, whether by merger, take-over or joint venture, it is obliged to re-register its products in the name of the new company, and new Ministry registration numbers are normally assigned. Where stocks of the previously registered products remain legally available, both old and new identities are included in the *Guide* and remain until approval for the former lapses, or stocks are notified as exhausted.

Products that have been withdrawn from the market and whose approval will finally lapse during 2004 are identified in each profile. After the indicated date, sale, storage or use of the product bearing that approval number becomes illegal. Where there is a direct replacement product, this is indicated.

How To Use The *Guide*

The book consists of six main sections:

1 Introductory Information
2 Adjuvants
3 Crop/Pest Guide
4 Pesticide Profiles
5 Products Also Registered
6 Appendices

1 Introductory Information

This section summarises legislation covering approval, storage, sale and use of pesticides in the UK. There are brief articles on the broader issues of resistance management and environmental protection. Chemicals subject to the Poisons Laws are listed, and there is a summary of first aid measures if pesticide poisoning should be suspected. The section also provides lists of products approved for use in or near water, in forestry, as seed treatments, and for aerial application.

2 Adjuvants

Adjuvants are listed in a table ordered alphabetically by product name. For each product, details are shown of the main supplier, the authorisation number and the mode of action (e.g. wetter, sticker, etc.) as shown on the label. A brief statement of the uses of the adjuvant is given. Protective clothing requirements and label precautions are listed using the codes from Appendix 4 (Section 6).

3 Crop/Pest Guide

This section enables the user to identify which active ingredients are approved for a particular crop/pest combination. The crops are grouped as shown in the Crop Index. For convenience, some crops and targets have been grouped into generic units (e.g. 'cereals' are handled as a group, not individually). Therefore indications from the Crop/Pest Guide must always be followed up by reference to the specific entry in the pesticide profiles section, because a product may not be approved on all crops in the group. Chemicals indicated as having uses in cereals, for example, may be approved for use only on winter wheat and barley, not for other cereals. Because of differences in the wording of product labels, it may sometimes be necessary to refer to broader categories of organism in addition to specific organisms (e.g. annual grasses and annual weeds in addition to annual meadow grass).

4 Pesticide Profiles

Each active ingredient has a separate, numbered profile entry, as does each mixture of active ingredients. The entries are arranged in alphabetical order using the common names approved by the British Standards Institution, and used in *The Pesticide Manual* (BCPC, 2003). Where an active ingredient is available only in mixtures, this is stated. The ingredients of the mixtures are themselves ordered alphabetically, and the entries appear at the correct point for the first named ingredient.

Within each profile entry, a table lists the products approved and available on the market, in the following style:

Product name	Main supplier	Active ingredient content	Formulation type	Registration number
1 Gremlin	Sipcam	285:176 g/l	SC	09468
2 Magnum	BASF	275:170 g/l	SC	08635
3 Spectron	Aventis	211:200 g/l	SC	07284

Many of the **product names** are registered trademarks, but no special indication of this is given. Individual products are indexed under their entry number in the Index of Proprietary Names of Products at the back of the book. The **main supplier** indicates the marketing outlet through which the product may be purchased. Full addresses and telephone/fax numbers of suppliers are listed in Appendix 1. Some website and e-mail addresses are also given. For mixtures, the **active ingredient contents** are given in the same order as they appear in the profile heading. The **formulation types** are listed in full in the Key to Abbreviations and Acronyms (Appendix 5). The **registration number** normally refers to registration with the Pesticides Safety Directorate. In cases where products registered with the Health and Safety Executive are included, HSE numbers are quoted, e.g. H0462.

Below the product table, a **Uses** section lists all approved uses (on-label and off-label) notified by suppliers to the Editor by October 2003, giving both the principal target organisms and the recommended crops or situations (identified in bold italics). Where there is an important condition of use, or the approval is off-label, this is shown in parentheses, e.g. (off-label). Numbers in square brackets refer to the numbered products in the table above. Thus, in the example shown above, a use approved for Spectron (product 3) but not for Gremlin or Magnum (products 1 and 2) appears as:

Annual dicotyledons in ***mangels*** [3]

Below the Uses paragraph, **Notes** are listed under the following headings. Unless otherwise stated, any reference to dose is made in terms of product rather than active ingredient. Where notes refer to particular products, rather than to the entry generally, this is indicated by numbers in square brackets as described above.

Specific off-label approvals (SOLAs)	Where SOLAs exist for any of the products in the profile, these are listed. Each SOLA has a separate entry and shows the crops to which it applies, the Notice of Approval number, the expiry date (if any), and the product reference number in square brackets.
Approval information	Information of a general nature about the approval status of the active ingredient or products in the profile is given here. Notes on approval for aerial or ULV application are given. Where a product approval will finally expire in 2004, the expiry date is shown here together with the registration number of its direct replacement, if any.
Efficacy	Factors important in making the most effective use of the treatment. Certain factors, such as the need to apply chemicals uniformly, the need to spray at appropriate volume rates, and the need to prevent settling out of active ingredient in spray tanks, are assumed to be important for all treatments and are not emphasised in individual profiles.
Restrictions	Notes are included in this section where products are subject to the Poisons Law, and where the label warns about organophosphorus and/or anticholinesterase compounds. Factors important in optimising performance and minimising the risk of crop damage, including statutory conditions relating to the maximum permitted number of applications, are listed. Any restrictions on crop varieties that may not be sprayed are mentioned here.
Environmental safety	Where the label specifies an environmental hazard classification it is noted, together with the associated risk phrases. Any other special operator precautions, and any conditions concerning withholding livestock from treated areas, are specified. Where any of the products in the profile are subject to Category A, Category B or broadcast air-assisted buffer zone restrictions under the Local Environment Risk Assessment for Pesticides (LERAP) scheme, the relevant classification is shown. Other environmental hazards are also noted here, including potential dangers to livestock, game, wildlife, bees and fish. The need to avoid drift onto

neighbouring crops, and to wash out equipment thoroughly after use, is important with all pesticide treatments but may receive special mention here if of particular significance.

Crop-specific information

Instructions about timing of application or cultivations that are specific to a particular crop, rather than generally applicable, are listed here. The latest permitted use and harvest intervals (if any) are shown for individual crops.

Following crops guidance

Any specific label instructions about what may be grown after a treated crop, whether harvested normally or after failure, are shown here.

Hazard classification and safety precautions

The label hazard classification(s) and precautions are shown using a series of letter or number codes which are explained in Appendix 4. The codes are listed under a number of sub-headings:

- Hazard
- Risk phrases
- Operator protection
- Environmental protection
- Consumer protection
- Storage and disposal
- Treated seed
- Vertebrate/rodent control products
- Medical advice

This section is given for information only and should not be used for the purpose of making a COSHH assessment without reference to the actual label of the product to be used.

5 Products Also Registered

Products with extant approval for all or part of the year of the edition in which they appear are listed in this section, if they have not been requested by their supplier or manufacturer for inclusion in Section 4. Details shown are the same (apart from formulation) as those in the product tables in Section 4 and, where relevant, an approval expiry date is shown. Not all the products listed here will be available in the market, but this list, together with the products in Section 4, comprises a comprehensive listing of all approved products for uses within the scope of the *Guide* for the year in question.

6 Appendices

Appendix 1

Gives names, addresses, telephone and fax numbers of all companies listed as main suppliers in the pesticide profiles. E-mail and website addresses are also listed where available.

Where a supplier is no longer trading under that name (usually following a merger or takeover), but products in that name are still listed in the *Guide* (because they are still available in the supply chain), a cross-reference indicates the new 'parent' company from which technical or commercial information can be obtained.

Appendix 2

Gives names, addresses, telephone and fax numbers of useful contacts, including the National Poisons Information Service. Website addresses are also included where available.

Appendix 3

Gives details of the keys to crop and weed growth stages, including the publication reference for each. The numeric codes are used in the descriptive sections of the pesticide profiles (Section 4).

Appendix 4	Shows the full text for code letters and numbers used in the pesticide profiles (Section 4) to indicate personal protective equipment requirements and label precautions.
Appendix 5	Shows the full text for the formulation abbreviations used in the pesticide profiles (Section 4). Other abbreviations and acronyms used in the *Guide* are explained here.

Definitions

The descriptions used in this *Guide* for the crops or situations in which products are approved for use are those used on the approved product labels. These are now standardised in a Crop Hierarchy published by the Pesticides Safety Directorate, in which definitions are given. To assist users of this Guide, the definitions of some of the terminology where misunderstandings can occur are reproduced below.

Amenity grassland. Areas of semi-natural or planted grassland subject to minimal management. It includes areas that may be accessed by the public (*e.g. railway and motorway embankments, airfields, and grassland nature reserves*). These areas may be managed for their botanical interest, and the relevant authority should be contacted before using pesticides in such locations.

Amenity vegetation. Areas of semi-natural or ornamental vegetation, including trees, or bare soil around ornamental plants, or soil intended for ornamental planting. It includes areas to which the public have access. It does NOT include hedgerows around arable fields.

Farm forestry. Groups of trees established on arable land or improved grassland including those planted for short rotation coppicing. It includes mature hedgerows around arable fields.

Forest nursery. Areas where young trees are raised outside for subsequent forest planting.

Forest. Groups of trees being grown in their final positions. Covers all woodland grown for whatever objective, including commercial timber production, amenity and recreation, conservation and landscaping, ancient traditional coppice and farm forestry, and trees from natural regeneration, colonisation or coppicing. Also includes restocking of established woodlands and new planting on both improved and unimproved land.

Green cover on land temporarily removed from production. Includes fields covered by natural regeneration or by a planted green cover crop that will not be harvested (*e.g. green cover on setaside*). It does NOT include industrial crops.

Hard surfaces. Man-made impermeable surfaces that are not intended to bear vegetation (*e.g. pavements, tennis courts, industrial areas, railway ballast*).

Indoors (for rodenticide use). Situations where the bait is placed within a building or other enclosed structure, and where the target is living or feeding predominantly within that building or structure.

Managed amenity turf. Areas of frequently mown, intensively managed, turf that is not intended to flower and set seed. It includes areas that may be for intensive public use (*e.g. all types of sports turf*).

Natural surfaces not intended to bear vegetation. Areas of soil or natural outcroppings of rock that are not intended to bear vegetation, including areas such as sterile strips around fields. It may include areas to which the public have access. It does not include the land between rows of crops.

Ornamental plant production. All ornamental plants that are grown for sale or are produced for replanting into their final growing position (*e.g. flowers, house plants, nursery stock, bulbs grown in containers or in the ground*).

Permanent grassland. Grazed areas that are intended to be permanent in nature (*e.g. permanent pasture and moorland that can be grazed*).

Permeable surfaces overlying soil. Any man-made permeable surface (excluding railway ballast) such as gravel that overlies soil and is not intended to bear vegetation.

Rotational grass. Short-term grass crops grown on land that is likely to be growing different crops in future years (*e.g. short-term intensively managed leys for one to three years that may include clover*).

Pesticide Legislation

Anyone who advertises, sells, supplies, stores or uses a pesticide is bound by legislation, including those who use pesticides in their own homes, gardens or allotments. There are numerous UK statutory controls, but the major legal instruments are outlined below.

The Food and Environment Protection Act 1985 (FEPA) and Control of Pesticides Regulations 1986 (COPR)

FEPA introduced statutory powers to control pesticides, with the aims of protecting human beings, creatures and plants, safeguarding the environment, ensuring safe, effective and humane methods of controlling pests, and making pesticide information available to the public.

Control of pesticides is achieved by COPR. These Regulations lay down the Approvals required before any pesticide may be sold, stored, supplied, advertised or used, subject to conditions which are contained in a series of Schedules. Schedule 1 relates to advertisement; Schedule 2 to sale, supply and storage; Schedule 3 to use; and Schedule 4 to aerial application of pesticides. The Schedules may be changed at any time, following Parliamentary approval. Details are given on the websites of the Pesticides Safety Directorate (PSD, www.pesticides.gov.uk) and the Health and Safety Executive (HSE, www.hse.gov.uk), under 'Guide to Pesticides' and 'Publications', respectively, and are updated in the monthly publication *The Pesticides Monitor.*

The controls currently in force include the following.

- Only approved products may be sold, supplied, stored, advertised or used.
- No advertisement may contain any claim for safety beyond that which is permitted in the approved label text.
- Only products specifically approved for the purpose may be applied from the air.
- A recognised Storeman's Certificate of Competence is required by anyone who stores for sale or supply pesticides approved for agricultural use.
- A recognised Certificate of Competence is required by anyone who gives advice when selling or supplying pesticides approved for agricultural use.
- Users of pesticides must comply with the Conditions of Approval relating to use.
- A recognised Certificate of Competence is required for all contractors and for persons born after 31 December 1964 applying pesticides approved for agricultural use (unless working under direct supervision of a certificate holder). Proposals are now in place that every user of agricultural pesticides will have to hold a Certificate of Competence regardless of age or supervision. A suitable transition period will allow this requirement to be enacted.
- Only those adjuvants authorised by PSD may be used.
- Regarding tank-mixes, 'no person shall combine or mix for use two or more pesticides which are anti-cholinesterase compounds unless such a mixture is expressly permitted by the conditions of approval given in relation to at least one of those pesticides', and 'no person shall combine or mix for use two or more pesticides unless all the conditions of Approval given in relation to each of those pesticides can be complied with'.

The Plant Protection Products Directive

European Council Directive 91/414/EEC is intended to harmonise national arrangements for the authorisation of plant protection products within the European Community. It became effective in the UK on 25 July 1993. Under the provisions of the Directive, individual Member States are responsible for authorisation within their own territory of products containing active substances that appear in a list agreed at Community level. This list, known as Annex I, is being created over a period of time by review of existing active ingredients (to ensure they meet present safety standards) and authorisation of new ones. About 60 active substances had achieved Annex I listing by mid-2003.

Individual Member States are required to amend their national arrangements and legislation in order to meet the requirements of Directive 91/414/EEC. In the UK this has been achieved by a series of Plant Protection Products Regulations (PPPR) under which, over a period of time, all agricultural and horticultural pesticides will come to be regulated. Meanwhile, existing product approvals are being maintained under COPR, and new ones are granted for products containing active ingredients that were already on the market by 25 July 1993. Products containing new active substances that were not on the market at this date are being granted provisional approval in advance of Annex I listing of their active ingredients. As active ingredients are placed on Annex I of the Directive, products containing only those active ingredients will be regulated solely under PPPR. Active ingredients in this *Guide* that have been included in Annex I are identified in the 'Approval' section of the profile.

The Directive also provides for a system of 'mutual recognition' of products registered in other Member States. Annex I listing, and a relevant approval in the Member State on which the mutual recognition is to be based, are essential prerequisites. Authorisation may be subject to conditions set to take account of differences such as climate and agricultural practice.

The Review Programme

The process of reviewing existing active ingredients is taking considerably longer than originally anticipated. The programme is designed to ensure that all available plant protection products are supported by up-to-date information on safety and efficacy. At the start of the programme it was envisaged that all active substances that were on the market on 25 July 1993 would have been reviewed by 25 July 2003, but the process is now not scheduled for completion until December 2008. Approximately 850 substances on the EU market are being handled in four phases but, because of the cost of providing this information, many are not being supported and their approvals are being revoked. By the end of 2003 about half the original list had been lost. For the remainder, the complex packages of data have to be evaluated, and further losses are likely following scientific risk assessments.

A few substances are temporarily reprieved by derogations for 'Essential Uses' granted by the European Commission (see below). However, even allowing for this temporary relief, the effect on the horticulture industry will be serious, with fewer products, especially herbicides, available and little prospect of new developments for vegetable crops coming forward.

Derogations for 'Essential Uses' permitted until 31 December 2007

Because of the scale of the likely loss of compounds and products, and the impact this would have on some sectors, requests have been made to the European Commission by Member States to allow the use of a few active substances after 2003 for key uses where no alternative exists. UK requests were co-ordinated by BCPC (British Crop Protection Council) and the Horticultural Development Council. The Commission decided in June 2002 to allow the use of 49 active substances for such uses, until December 2007, in order to provide time for alternatives to be researched and developed. In the UK this includes nine herbicides for vegetables, three fungicides, one acaricide and one insecticide (Table 1.1). These 'Essential Uses' are only for the crops specified, and only apply to SOLAs or products already registered for the use. Extrapolations to other crops are not permitted. Affected products must be re-labelled for 2004.

It is possible that there will be further losses of products as a result of product rationalisation by companies, or failure of some active substances to achieve Annex 1 listing. Even where active substances are supported in EC Review rounds 2 and 3, not all uses will be supported, and growers need to be aware of these so that requests for SOLAs can be made.

Table 1.1 Active substances and products permitted, for the following 'Essential Uses' only, until 31 Dec 2007 (unless approval expiry precedes this date)

Active substance	Products containing the active substance	Permitted crop use(s)
Cyanazine	Bullet*	Combining peas Field beans (spring sown)
	Fortrol (11174) Barclay Canter Cleancrop Vectro Clayton Gazette (exp 31 7 04) Standon Cyanazine 50 (exp 31 7 04) I T Cyanazine	Combining peas Vining peas
	Fortrol (10666 and 11174) Cleancrop Vectro	Calabrese (SOLA) Cauliflowers (SOLA) Cabbage (SOLA) Kale (SOLA) Collards (LTAEU)
	Fortrol (onion) (07009 exp 30 6 04), (10222 exp 30 6 04) Barclay Canter (onion) I T Cyanazine Cleancrop Vectro Clayton Gazette (onion) exp 31 7 04 Standon Cyanazine 50 (onion) exp 31 7 04)	Bulb onions Salad onions Leeks (LTAEU from salad onions)
	Fortrol (10666 exp 31 7 05), (11174)	Bulb onions only
	Fortrol (11174) Barclay Canter I T Cyanazine Cleancrop Vectro Clayton Gazette exp 31 7 04 Standon Cyanazine 50 exp 31 7 04	Winter oilseed rape Narcissi
	Fortrol	Forestry (SOLA)
Fenuron	Croptex Chrome*	Spinach Runner beans (SOLA)
Fomesafen	Reflex T*	Combining peas (spring sown) Vining peas (spring sown) Broad beans (spring sown) Field beans (spring sown) Lupins (LTAEU)
	Flex	Dwarf French beans Soya beans (SOLA) Runner beans (LTAEU)
Metoxuron	Dosaflo	Carrots Parsnips (LTAEU)
Pentanochlor	Atlas Brown* Atlas Solan	Celeriac Celery

Active substance	Products containing the active substance	Permitted crop use(s)
Pentanochlor *cont.*	Croptex Bronze Atlas Solan 40 Atlas Brown*	Carrots Parsnips Hardy ornamentals Ornamentals (LTAEU)
	Croptex Bronze Atlas Solan Atlas Brown*	Parsley Protected parsley Herbs (LTAEU)
	Atlas Brown*	Herbs (SOLA)
Prometryn	Gesagard Alpha Prometryne 50WP	Bulb onions (SOLA) Salad onions (SOLA)
	Pennine 50 Alpha Prometryne 50WP	Leeks (transplanted)
	Gesagard	Leeks (direct drilled and transplanted) (SOLA)
	Gesagard Alpha Prometryne 50WP Pennine 50	Carrots Parsnips (LTAEU)
	Gesagard Alpha Prometryne 50WP Pennine 50	Celery
	Alpha Prometryne 50WP Gesagard	Protected celery
	Gesagard Alpha Prometryne 50WP Pennine 50	Parsley Protected parsley Herbs (LTAEU)
	Gesagard Alpha Prometryne 50WP	Herbs (SOLA) Protected herbs (SOLA)
Sodium mono-chloroacetate	Croptex Steel Atlas Somon	Bulb onions Salad onions Leeks Cabbages Brussels sprouts Kale Collards (LTAEU)
	Croptex Steel	Calabrese (SOLA) Cauliflowers (SOLA)
	Croptex Steel	Hops (SOLA)
	Croptex Steel	Raspberries (SOLA) Blackberries (SOLA) Loganberries (SOLA) Hybrid berries (SOLA)
Terbacil	Sinbar	Herbs (SOLA) Protected herbs (SOLA)
Terbutryn	Opogard*	Combining peas Vining peas Broad beans Field beans Lupins (spring sown)

Active substance	Products containing the active substance	Permitted crop use(s)
Terbutryn *cont.*	Batallion*	Combining peas Vining peas Edible podded peas Broad beans Field beans Lupins (LTAEU)
2-aminobutane	Certis 2-aminobutane	Stored seed potatoes
Azaconozole	Nectec Paste*	Ornamentals
Oxycarboxin	Plantvax 75	Ornamentals Protected ornamentals
Fenpropathrin	Meothrin	Blackcurrants
Resmethrin	Pynosect 30 Water Miscible*	Mushrooms (SOLA)

*Product in mixture with a supported active substance.
SOLA, specific off-label approval; LTAEU, long-term arrangements for extension of use.

The Control of Substances Hazardous to Health Regulations 1988 (COSHH)

The COSHH regulations, which came into force on 1 October 1989, were made under the Health and Safety at Work Act 1974, and are also important as a means of regulating the use of pesticides. The regulations cover virtually all substances hazardous to health, including those pesticides classed as Very toxic, Toxic, Harmful, Irritant or Corrosive; other chemicals used in farming or industry; and substances with occupational exposure limits. They also cover harmful micro-organisms, dusts and any other material, mixture, or compound used at work which can harm people's health.

The original Regulations, together with all subsequent amendments, have been consolidated into a single set of regulations: The Control of Substances Hazardous to Health Regulations 1994 (COSHH 1994).

The basic principle underlying the COSHH regulations is that the risks associated with the use of any substance hazardous to health must be assessed before it is used, and the appropriate measures taken to control the risk. The emphasis is changed from that pertaining under the Poisonous Substances in Agriculture Regulations 1984 (now repealed), whereby the principal method of ensuring safety was the use of protective clothing, to preventing or controlling exposure to hazardous substances by a combination of measures. In order of preference, the measures should be:

1 substitution with a less hazardous chemical or product
2 technical or engineering controls (e.g. the use of closed handling systems, etc.)
3 operational controls (e.g. operators located in cabs fitted with air-filtration systems, etc.)
4 use of personal protective equipment, which includes protective clothing.

Consideration must be given to whether it is necessary to use a pesticide at all in a given situation, and if so, the product posing the least risk to humans, animals and the environment must be selected. Where other measures do not provide adequate control of exposure and the use of personal protective equipment is necessary, the items stipulated on the product label must be used as a minimum. It is essential that equipment is properly maintained and the correct procedures adopted. Where necessary, the exposure of workers must be monitored, health checks carried out, and employees must be instructed and trained in precautionary techniques. Adequate records of all operations involving pesticide application must be made and retained for at least 3 years.

Certificates of Competence - the roles of BASIS and the National Proficiency Tests Council

COPR, COSHH and other legislation places certain obligations on those who handle and use pesticides. Minimum standards are laid down for the transport, storage and use of pesticides, and the law requires those who act as storekeepers, sellers and advisors to hold recognised Certificates of Competence.

BASIS is an independent registration scheme for the pesticide industry, recognised under COPR. It is responsible for organising training courses and examinations to enable staff to obtain a Certificate of Competence.

In addition, BASIS undertakes annual assessments of pesticide supply stores, enabling distributors, contractors and seedsmen to meet their obligations under the Code of Practice for Suppliers of Pesticides. Further information can be obtained from BASIS (www.basis-reg.co.uk; see Appendix 2).

Certain spray operators also require Certificates of Competence under the Control of Pesticides Regulations. These certificates are awarded by the National Proficiency Tests Council (NPTC) to candidates who pass an assessment carried by an approved NPTC or Scottish Skills Testing Service Assessor in one or more of a series of competence modules. Holders are required to produce their certificate on demand for inspection to any person authorised to enforce COPR. Further information can be obtained from NPTC (www.nptc.org.uk; see Appendix 2).

Maximum Residue Levels

A small number of pesticides are liable to leave residues in foodstuffs, even when used correctly. Where residues can occur, statutory limits, known as Maximum Residue Levels (MRLs), have been established. MRLs provide a check that products have been used as directed; they are not safety limits. However, they do take account of consumer safety because they are set at levels that ensure normal dietary intake of residues presents no risk to health. Wide safety margins are built in, and eating food containing residues above the MRL does not automatically imply a risk to health. Nevertheless, it is an offence to put into circulation any produce where the MRL is exceeded.

The surrounding legislation is complex. MRLs may be specified by several different bodies. The UK has set statutory MRLs since 1988. The European Union intends eventually to introduce MRLs for all pesticide/commodity combinations. These are being introduced initially by a series of priority lists, but will subsequently be covered by the review programme under Directive 91/414. However, in cases where no information is available, EU Directive 2000/42/EC requires many MRLs to be set at the limit of determination (LOD). As a result of this, certain approvals are being withdrawn where such use would leave residues above the MRL set in the Directive.

MRLs apply to imported as well as to home-produced foodstuffs. Details of those that have been set have been published in *The Pesticides (Maximum Residue Levels in Crops, Food and Feeding Stuffs) Regulations 1994*, and successive amendments to these Regulations. These Statutory Instruments are available from The Stationery Office.

MRLs are set for many chemicals not currently marketed in Britain. Because of this, and the ever-changing information on MRLs, no quantitative information is given in this Guide. Instead, users of the sister CD-ROM publication *The e-UK Pesticide Guide 2004* may gain direct weblink access to the comprehensive and authoritative MRL database compiled and maintained by Leatherhead Food International. This online database sets out in table form those levels specified by UK Regulations, EC Directives, and the Codex Alimentarius for each commodity.

The e-UK Pesticide Guide (2004 edition) may be purchased from BCPC Publication Sales, 01420 593200, or online via www.bcpc.org/bookshop

Approval (On-label and Off-label)

Only officially approved pesticides may be marketed and used in the UK. Approvals are granted by UK Government Ministers in response to applications that are supported by satisfactory data on safety, efficacy and, where relevant, humaneness. The Pesticides Safety Directorate (PSD, www.pesticides.gov.uk) is the UK Government Agency for regulating agricultural pesticides and plant protection products. The Health and Safety Executive (HSE, www.hse.gov.uk) fulfils the same role for other pesticides. The main focus of the regulatory process is the protection of human health and the environment.

Approvals are normally granted only in relation to individual products and for specified uses. It is an offence to use non-approved products or to use approved products in a manner that does not comply with the statutory conditions of use, except where the crop or situation is the subject of an off-label extension of use (see below).

Statutory Conditions of Use

Statutory conditions have been laid down for the use of individual products and may include:

- field of use (e.g. agriculture, horticulture etc.)
- crop or situations for which treatment is permitted
- maximum individual dose
- maximum number of treatments or maximum total dose
- maximum area or quantity which may be treated
- latest time of application or harvest interval
- operator protection or training requirements
- any other specific restrictions relating to particular pesticides.

All products must display these statutory conditions of use in a 'statutory box' on the label.

Types of approval

There are three categories of approval that may be granted, under the Control of Pesticides Regulations, to products containing active ingredients that were already on the market by 25 July 1993:

- **Full** (granted for an unstipulated period)
- **Provisional** (granted for a stipulated period)
- **Experimental Permit** (granted for the purposes of testing and developing new products, formulations or uses). Products with only an Experimental Permit may not be advertised or sold and do not appear in this *Guide*.

Products containing new active substances, and those containing older ingredients once they have been listed in Annex I to Directive 91/414 (see previous section), are granted approval under the Plant Protection Products Regulations. Again, there are three categories, similar to those listed above:

- **Standard approval** (granted for a period not exceeding 10 years). Only applicable to products whose active substances are listed in Annex I.
- **Provisional approval** (granted for a period not exceeding three years, but renewable). Applicable where Annex I listing of the active substance is awaited.
- **Approval for research and development** (to enable field experiments or tests to be carried out with active substances not otherwise approved).

The official lists of approved products, including all the above categories except those for experimental purposes, are shown on the websites of PSD (www.pesticides.gov.uk) and HSE (www.hse.gov.uk) under 'Guide to Pesticides' and 'Publications' respectively.
Details of new approvals, amendments to existing approvals and off-label approvals (see below) are published monthly in *The Pesticides Register.*

Withdrawal of approval

Product approvals may be reviewed, amended, suspended or revoked at any time. Revocation may occur for various reasons, such as commercial withdrawal or failure by the approval holder to meet data requirements. Where there are no safety concerns, a phased approval revocation is implemented, specifying a 'wind-down' period to allow the using up of existing stocks by persons other than the approval holder.

Until April 2003 the wind-down period has normally been set by PSD at 2 or 3 years, but it has become apparent that the European Commission considers 18 months is a reasonable period for the withdrawal or use of stocks in the supply chain.

From 1 April 2003 phased revocations have been as follows.

- A **maximum** of six months for the approval holder to advertise, sell, supply and use the product.
- A further **maximum** of 12 months for sale, supply and use by persons other than the approval holder. This allows a total of 18 months from the date of revocation for stocks to pass through the supply chain and be used. Anyone may store the product during this 18-month period.

Shorter periods will continue to be applied where necessary.

The expiry dates shown in this *Guide* continue to be the final date of legal use.

Approval of commodity substances

Commodity substances are compounds that have a variety of alternative and often widespread non-pesticidal uses, but also have potential for use as a pesticide. For a commodity substance to be used as a pesticide, it requires approval under the Control of Pesticides Regulations. Approval is given only for the use of the substance; approval is not given for sale, storage, supply or advertisement. There is no approval holder, and no pesticide product label, but the approval and associated conditions of use are published in the Pesticides Guide on the PSD website. Nineteen such substances have been approved for certain specified uses, as laid down in Annex D of the Guide to Pesticides. Of these, carbon dioxide, formaldehyde, methyl bromide, sodium chloride, sodium hypochlorite, strychnine hydrochloride and sulphuric acid are approved for agricultural or horticultural use, and are included in this *Guide*.

Off-label extension of use

Products may legally be used in a manner not covered by the printed label in several ways:

- in accordance with the 'Long-Term Arrangements for Extension of Use' (see below)
- in accordance with a specific off-label approval (SOLA). SOLAs are uses for which approval has been sought by individuals or organisations other than the manufacturer. The Notices of Approval are published by Defra and are widely available from ADAS or National Farmers' Union offices. Users of SOLAs must first obtain a copy of the relevant Notice of Approval and comply strictly with the conditions laid down therein
- in tank mixture with other approved pesticides in accordance with Consent C(i) made under FEPA. Full details of Consent C(i) are given in Annex A of the Guide to Pesticides on the PSD website, but there are two essential requirements for tank mixes. First, all the conditions of approval of all components of a mixture must be complied with. Second, no person may mix or combine pesticides which are cholinesterase compounds unless allowed by the label of at least one of the pesticides in the mixture
- in conjunction with authorised adjuvants
- in reduced spray volume under certain conditions
- the use of certain herbicides on specified set-aside areas subject to restrictions which differ between Scotland and the rest of the UK
- by mutual recognition of a use fully approved in another Member State of the European Union and authorised by PSD.

Although approved, off-label uses are not endorsed by manufacturers and such treatments are made entirely at the risk of the user.

Long-Term Arrangements for Extension of Use

Since 1 January 1990, arrangements have been in place allowing many approved products to be used for additional minor uses, without the need for specific off-label approvals (SOLAs). These arrangements - the Long-Term Arrangements for Extension of Use (LTAEU) - are regularly reviewed to ensure they stay in line with current science and regulations. They were revised in December 1994, and again in December 1999. They must be reviewed again before the end of December 2004. However, rapid changes are taking place within the agriculture/horticulture sector, and novel crops are being developed. It is recognised that this creates a need for additions to the list of permitted extrapolations to be made on a more regular basis, and the first such revision took place at the end of 2001. As a result some new extrapolations have been added, subject to certain restrictions.

The scheme is founded on the principle that the risks to the operator, the consumer, wildlife and the environment arising from the extended (off-label) use are acceptable, and no greater than those which arise from the approved uses of a product, which appear on its label. All off-label approvals, including those covered by these arrangements, undergo a full safety assessment by the PSD, in consultation with the Advisory Committee on Pesticides. There is, however, no test of efficacy - **use of these extensions of use, as with any off-label use, is made at the user's choosing, and the commercial risk is entirely theirs**.

The current arrangements are set out in full at Appendix II of the Guide to Pesticides on the PSD website. The following is a **summary** of the main points and uses.

Specific restrictions for extension of use

Certain restrictions are necessary to ensure that the extension of use does not increase the risk to the operator, the consumer or the environment.

- The arrangements apply to label and SOLA recommendations, but ONLY for the use of products approved for use as Agricultural and/or Horticultural pesticides.
- Safety precautions and statutory conditions relating to use (shown in the statutory box on the label) MUST be observed. If extrapolation from a SOLA is to be used, all conditions specified on the SOLA Notice of Approval MUST be observed in addition to the safety precautions and statutory conditions on the specified product's label.
- Pesticides must only be used in the same situation (i.e. outdoor or protected) as that specified on the product label or SOLA Notice. Approvals for use on tomatoes, cucumbers, lettuces, chrysanthemums and mushrooms include protected crops unless otherwise stated. Use on other protected crops may only occur when the product label specifically allows use under protection on the crop on which the extrapolation is based. Similarly, pesticides approved only for use in protected situations must not be used outdoors. **If a label or SOLA Notice does not specify a situation, then extrapolation only to an outdoor use is permitted**.
- The application method must be as stated on the product label and in accordance with relevant Codes of Practice and COSHH requirements. Where application of a pesticide under these arrangements is planned to be made by hand-held equipment users MUST ensure that hand-held use is appropriate for the current label or SOLA Notice conditions.
- **Note:** unless otherwise stated, spray applications to protected crops include hand-held uses.

In cases where hand-held application is not appropriate to the use on which extrapolation is to be made, hand-held application may nevertheless be permitted provided certain conditions specified in the arrangements are met (consult Appendix II of the Guide to Pesticides on the PSD website, www.pesticides.gov.uk). When planning to apply a pesticide under these arrangements by broadcast air-assisted sprayer, only those products with a specific label or SOLA recommendation for such use may be used. Any associated buffer zone restrictions must also be observed.

- Use in or near water (including coastal waters), or by aerial application, is not permitted under the arrangements.
- Rodenticides and other vertebrate control agents are not included, nor is use on land not intended for cropping.

- Pesticides classed as harmful, dangerous, extremely dangerous or high risk to bees must not be used off-label during flowering unless otherwise permitted. Applications must also not be made when flowering weeds are present or when bees are actively foraging. If the label or SOLA Notice specifies an aquatic buffer zone, users must conduct a Local Environment Risk Assessment for Pesticides (LERAP) assessment for the extension of use.
- The user of a pesticide under these arrangements must take all reasonable precautions to safeguard wildlife and the environment.

The arrangements apply as follows.

Non-edible crops and plants

Subject to the specific restrictions set out above:

- Pesticides approved for use on any growing crop may be used on commercial agricultural and horticultural holdings and forest nurseries on: (i) ornamental crops including hardy nursery stock, plants, bulbs, flowers and seed crops where neither the seed nor any part of the plant is to be consumed by humans or animals; (ii) forest nursery crops prior to final planting out.
- In addition, pesticides approved for use on any growing edible crop (except seed treatments) may be used on non-ornamental crops grown for seed subject to the same consumption restrictions as above (but not including seed crops of potatoes, cereals, oilseeds, peas, beans and other pulses).
- Pesticides (except seed treatments) approved for use on oilseed rape may be used on hemp grown for fibre and woad (*Isatis tinctoria* and *I. indigotica*).
- Pesticides (except seed treatments) approved for use on cereals, grass and forage maize may be used on commercial agricultural and horticultural holdings on *Miscanthus* spp. (Elephant grass), but not after the crop is 1 m high. The crops and its products must not be used for food or feed.
- Herbicides approved for use on oilseed rape may be used on sunflower, quinoa, fodder radish and sweet clover grown as game cover.
- Herbicides approved for use on cereals and forage maize may be used on canary grass, tanka millet, white millet and sorghum grown as game cover.

Farm forestry and rotational cropping

Subject to the specific restrictions set out above:

- Herbicides approved for use on cereals may be used in the first 5 years of establishment in farm forestry on land previously under arable cultivation or improved grassland, and reclaimed brownfield sites.
- Herbicides approved for use on cereals, oilseed rape, sugar beet, potatoes, peas and beans may be used in the first year of regrowth after cutting in coppices established on land previously under arable cultivation or improved grassland, or reclaimed brownfield sites.

Nursery fruit crops

Subject to the specific restrictions set out above:

- Pesticides approved for use on any crop for human or animal consumption may be used on commercial agricultural and horticultural holdings on nursery fruit trees, nursery grape vines prior to final planting out, bushes, canes and non-fruiting strawberry plants provided any fruit harvested within 12 months is destroyed. Applications must NOT be made if fruit is present.

Hops

Subject to the specific restrictions set out above:

- On commercial and horticultural holdings pesticides may be used on mature stock or mother plants kept specifically for propagation, hop propagules prior to final planting out, and first year 'nursery hops' in their final planting position but not taken to harvest in that year. Treated hops must not be harvested for human or animal consumption (including idling) within 12 months of treatment.

Crops used partly or wholly for consumption by humans or livestock

Subject to the specific restrictions set out above:

- Pesticides may be used on commercial agricultural and horticultural holdings on crops in the first column of Table 1.2 and 1.3 below if they have been approved for use on the crop(s) opposite them in the second column. **However, it is the responsibility of the user to ensure that the proposed use does not result in any statutory UK MRL being exceeded. These extrapolations may NOT be used where the MRL for the crop in column 1 is lower than the MRL for the crop in column 2, or where an MRL has been established for the crop in column 1 but not for the crop in column 2.**

Table 1.2 Crops used partly or wholly for consumption by humans or livestock

Minor use	Crops on which use is approved	Additional special conditions
Arable crops		
Poppy (grown for oilseed), sesame	Sunflower	
Mustard, linseed, evening primrose, honesty	Oilseed rape	
Borage (grown for oilseed), canary flower e.g. *Echium vulgare, E. plantagineum* (grown for oilseed)	Oilseed rape	Seed treatments are not permitted
Rye, triticale	Barley	Treatments applied before first spikelet of inflorescence just visible
Rye, triticale	Wheat	
Grass seed crop	Grass for grazing or fodder	
Grass seed crop	Wheat, barley, oats, rye, triticale	Treated crops must not be grazed or cut for fodder until 90 days after treatment Seed treatments are not permitted Use of chlormequat-containing products is not permitted
Lupins	Combining peas or field beans	
Fruit crops		
Almond, chestnut, hazelnut, walnut	Apple, cherry or plum	For herbicides used on the orchard floor ONLY
Quince, crab apple	Apple or pear	

Minor use	Crops on which use is approved	Additional special conditions
Almond, chestnut, hazelnut, walnut	Any two of the following: almond, chestnut, hazelnut, walnut	
Nectarine, apricot	Peach	
Blackberry, dewberry, *Rubus* species (e.g. tayberry, loganberry)	Raspberry	
Whitecurrant, bilberry, cranberry, blueberry, other *Vaccinium* species	Blackcurrant or redcurrant	
Gooseberry	Blackcurrant	
Redcurrant	Blackcurrant	
Vegetable crops		
Parsley root	Carrot or radish	
Fodder beet, mangel	Sugar beet	
Horseradish	Carrot or radish	
Parsnip	Carrot	
Salsify	Carrot or celeriac	
Swede	Turnip	
Turnip	Swede	
Garlic, shallot	Bulb onion	
Leek	Salad onion	Herbicides only Latest timing up to two true leaves of the crop
Cayenne pepper	Pepper	
Aubergine	Tomato	
Squash, pumpkin, marrow, watermelon,	Melon	
Broccoli	Calabrese	
Calabrese	Broccoli	
Roscoff cauliflower	Cauliflower	
Collards	Kale	

Minor use	Crops on which use is approved	Additional special conditions
Chinese cabbage, pak choi, choi sum	Kale	
Lamb's lettuce, frisée/ frise, radicchio, cress, scarole	Lettuce	
Leaf herbs and edible flowers*	Lettuce or spinach or parsley or sage or mint or tarragon	
Leafy brassicas grown for baby leaf production	Lettuce	Seed treatments are not permitted Pre-crop emergence use not permitted
Spinach	Lettuce	Seed treatments are not permitted
Beet leaves, red chard, white chard, yellow chard	Spinach	
Edible podded peas (e.g. mange-tout, sugar snap)	Edible podded beans	
Runner beans	Dwarf French beans	
Broad beans	Vining pea	Applications must NOT be made during flowering or where bees are actively foraging Seed treatments are not permitted
Rhubarb, cardoon	Celery	
Edible fungi other than mushroom (e.g. oyster mushroom)	Mushroom	

*This extension of use applies to the following leaf herbs and edible flowers: *Agastache* spp., angelica, balm, basil, bay, borage, burnet (salad), caraway, catnip, camomile, chervil, chives, clary, coriander, curry plant, dill, dragonhead, fennel, fenugreek, feverfew, hyssop, land cress, lavender, lavandin, lovage, marjoram, marigold, mint, nasturtium, nettle, oregano, *Origanum heracleoticum*, parsley, rocket, rosemary, rue, sage, savory, sorrel, spike lavender, tarragon, thyme, verbena (lemon), woodruff, violet.

For applications in store on crops **partly or wholly for human or animal consumption**, the extensions of use are set out in Table 1.3.

Table 1.3 Applications in store on crops used partly or wholly for consumption by humans or animals

Minor use	Crops on which use is approved
Rye, barley, oats, buckwheat, millet, sorghum, triticale	Wheat
Dried peas	Dried beans
Dried beans	Dried peas
Mustard, sunflower, honesty, sesame, evening primrose, poppy (grown for oilseed), borage (grown for oilseed), canary flower e.g. *Echium vulgare/E. plantaginium* (grown for oilseed)	Oilseed rape

Clarifications

Under these arrangements the crops listed in Table 1.4 are considered synonymous or equivalent and, as such, uses on crops in column 1 can be read across to uses in column 2.

Table 1.4 Crops considered synonymous or equivalent

Column 1	Column 2 (equivalent)
Hazelnut	Cobnuts, filberts
French bean	Navy bean
Vining pea	Picking pea, shelling pea, non-edible podded pea
Linseed	Linola, flax
Wheat	Durum wheat

Guidance on the Use of Pesticides

The information given in *The UK Pesticide Guide* provides some of the answers needed to assess health risks, including the hazard classification and the level of operator protection required. However, the *Guide* cannot provide all the details needed for a complete hazard assessment, which must be based on the product label itself and, where necessary, the Health and Safety Data Sheet and other official literature.

Detailed guidance on how to comply with the Regulations is available from several sources.

Pesticides: Code of Practice

Pesticides: Code of Practice for the Safe Use of Pesticides on Farms and Holdings, 1998 (ISBN 0 11 242892 4)

The *Code of Practice* is available from Defra Publications and has been substantially revised and updated from the previous (1990) edition. It gives comprehensive guidance on how to comply with COPR and the COSHH regulations under the following headings: User training and certification; Planning and preparation; Working with pesticides; Disposal of pesticide waste; Keeping records. In particular it details what is involved in making a COSHH assessment. The principal source of information for making such an assessment is the approved product label. In most cases the label provides all the necessary information but in certain circumstances other sources must be consulted, and these are listed in the *Code of Practice*.

Other codes of practice

- *Code of Practice for the Use of Approved Pesticides in Amenity and Industrial Areas* (National Association of Agricultural Contractors/British Agrochemicals Association, ISBN 1 871140 12 9)
- *The Safe Use of Pesticides for Non-Agricultural Purposes - Approved Code of Practice* (Defra Booklet PB 3529)
- *Code of Practice for Suppliers of Pesticides to Agriculture, Horticulture and Forestry* (Defra Booklet PB 0091)
- *Code of Good Agricultural Practice for the Protection of Soil* (Defra Booklet PB 0617)
- *Code of Good Agricultural Practice for the Protection of Water* (Defra Booklet PB 0587)
- *Code of Good Agricultural Practice for the Protection of Air* (Defra Booklet PB 0618)

Other guidance and practical advice

Health and Safety Executive (from HSE Books; see Appendix 2)

- *A Step by Step Guide to COSHH Assessment, HS(G) 97* (ISBN 0 11 886 379 7)
- *Food and Environment Protection Act 1985/Control of Pesticides Regulations 1986. An Open Learning Course* (ISBN 0 11 8857 43 6)
- *COSHH in Agriculture*, AS 28
- *COSHH in Forestry*, AS 30
- *Occupational Exposure Limits 1995, Guidance Note EH 40/95* (ISBN 0 7176 0876 X)
- *Biological Monitoring of Workers Exposed to Organophosphorus Pesticides, Guidance Note MS 17* (ISBN 0 11 883951 9)

Defra (from The Stationery Office; see Appendix 2)

- 'Guide to Pesticides', www.pesticides.gov.uk
- *The Pesticides Monitor* (monthly) (ISSN 1466-5670)
- *Local Environment Risk Assessments for Pesticides - A Practical Guide* (PB4168)

Crop Protection Association (see Appendix 2)

- *The Plain Man's Guide to Pesticides and the COSHH Assessment*
- *Amenity Handbook: A Guide to the Selection and Use of Amenity Pesticides*
- *Pesticides and Water Quality*
- *Sprayer Cleaning*
- *Container Cleaning*
- *Agrochemical Disposal*
- *Hand Protection*
- *Good Agrochemical Storage*
- *Avoiding Drift*

BCPC (British Crop Protection Council) (see Appendix 2)

- *The Pesticide Manual* (13th edition) (ISBN 1 901396 13 4)
- *The e-Pesticide Manual* PC CD-ROM (Version 3.0) (ISBN 1 901396 34 7)
- *The BioPesticide Manual* (2nd edition) (ISBN 1 901396 29 0)
- *IdentiPest* PC CD-ROM (ISBN 1 901396 05 3)
- *Garden Detective* PC CD-ROM (ISBN 1 901396 32 0)
- *Hand-held and Amenity Sprayers Handbook* (ISBN 1 901396 03 7)
- *Boom and Fruit Sprayers Handbook* (ISBN 1 901396 02 9)
- *Using Pesticides: A Complete Guide to Safe Effective Spraying* (ISBN 1 901396 01 0)
- *Safety Equipment Handbook* (ISBN 1 901396 06 1)

The Environment Agency (see Appendix 2)

- *The Prevention of Pollution of Controlled Waters by Pesticides* (Leaflet PPG9)
- *The Use of Herbicides in or near Water*
- *Introduction of the Groundwater Regulations*

Resistance Management

Pest species are, by definition, adaptable organisms. The development of resistance to some crop protection chemicals is just one example of this adaptability. Repeated use of products with the same mode of action will clearly favour those individuals in the pest population able to tolerate the treatment. This leads to a situation where the tolerant (or resistant) individuals can dominate the population and the product is ineffective. In general, the more rapidly the pest species reproduces and the more mobile it is, the faster is the emergence of resistant populations, although some weeds seem able to evolve resistance more quickly than would be expected.

The speed of appearance of resistance depends on the mode of action of the crop protection chemicals, as well as the manner in which they are used. Resistance among insects and fungal diseases has been evident for much longer than weed resistance to herbicides, but examples in all three categories are now widespread and increasing.

This has created a need for agreement on the advice given for the use of crop protection chemicals in order to reduce the likelihood of the development of resistance and to avoid the loss of potentially valuable products in the chemical armoury.

In the UK, key independent research organisations, chemical manufacturers and other organisations have collaborated to share knowledge and expertise on resistance issues through three action groups. Participants include ADAS, the Pesticides Safety Directorate, universities, colleges and the Home Grown Cereals Authority. The groups have a common aim of monitoring resistance in UK and devising and publishing management strategies designed to combat it where it occurs. The groups are:

Weed Resistance Action Group (WRAG), formed in 1989 (Secretary, Dr Stephen Moss, IACR-Rothamsted, Harpenden, Herts AL5 2JQ *Tel: 01582 7631330 ext. 2521*)

Fungicide Resistance Action Group (FRAG), formed in 1995 (Secretary, Mr Oliver Macdonald, Pesticides Safety Directorate, Mallard House, King's Pool, 3 Peasholme Green, York YO1 2PX *Tel: 01904 455864*)

Insecticide Resistance Action Group (IRAG), formed in 1997 (Secretary, Dr Ian Denholm, IACR-Rothamsted, Harpenden, Herts AL5 2JQ *Tel: 01582 763133 ext. 2324*)

The above groups publish detailed advice on resistance management relevant for each sector and, in some cases, specific to a pest problem. This information, together with further details about the function of each group, can be obtained from the Pesticides Safety Directorate web site: www.pesticides.gov.uk/committees/resistance

The general guidelines for resistance management are similar for all three problem areas.

Preparation in advance

- Be aware of the factors that favour the development of resistance, such as repeated annual use of the same product, and assess the risk.
- Plan ahead and aim to integrate all possible means of control.
- Use cultural measures such as rotations, stubble hygiene, variety selection and, for fungicides, removal of primary inoculum sources, to reduce reliance on chemical control.
- Monitor crops regularly.
- Keep aware of local resistance problems.
- Take professional advice.

Using crop protection products

- Optimise product efficacy by using it as directed, at the right time, in good conditions.
- Treat pest problems early.
- Mix or alternate chemicals with different modes of action.
- Avoid repeated applications of very low doses.
- Keep accurate field records.

Label guidance depends on the appropriate strategy for the product. Most frequently it consists of a warning of the possibility of poor performance due to resistance, and a restriction on the number of treatments that should be applied in order to minimise the development of resistance. This information is summarised in the profiles in this Guide, but detailed guidance must always be obtained by reading the label itself, before use.

International Action Committees

Resistance to crop protection products is an international problem. Agrochemical industry collaboration on a global scale is via three action committees whose aims are to support a co-ordinated industry approach to the management of resistance worldwide. In particular they produce lists of crop protection chemicals classified according to their mode of action. These lists, and other information, can be obtained from the respective websites:

Herbicide Resistance Action Committee (HRAC): www.plantprotection.org/hrac

Fungicide Resistance Action Committee (FRAC): www.frac.info

Insecticide Resistance Action Committee (IRAC): www.irac-online.org

Protection of Bees

Honey bees

Honey bees are a source of income for their owners, and important to farmers and growers as pollinators of their crops. It is irresponsible and unnecessary to use pesticides in such a way that may endanger them. Pesticides vary in their toxicity to bees, but those that present a special hazard are classed as 'harmful', 'dangerous' or 'extremely dangerous' to bees. Products so classified carry a specific warning in the precautions section of the label. These are indicated in this *Guide* in the **personal protective equipment/label precautions** section of the pesticide profile by the numbers E12b, E12c, E12d, E12e, depending on which phrase applies. The phrases are stated in full in Appendix 4.

In July 1996 a new classification was introduced that more accurately reflects the actual risk to honey bees when a product is used. Products assessed by the Pesticides Safety Directorate as posing such a risk are classified as 'High risk to bees'. Such products carry a new warning in the precautions section of the label, indicated in this *Guide* by the number E12a.

Product labels indicate the necessary environmental precautions to take, but where use of an insecticide on a flowering crop is contemplated, the British Beekeepers Association (see Appendix 2) has produced the following guidelines for growers:

- target insect pests with the most appropriate product
- choose a product that will cause minimal harm to beneficial species
- follow the manufacturer's instructions carefully
- inspect and monitor crops regularly
- avoid spraying crops in flower or where bees are actively foraging
- keep down flowering weeds
- spray late in the day, in still conditions
- avoid excessive spray volume and run-off
- adjust sprayer pressure to reduce production of fine droplets and drift
- give local beekeepers as much warning of your intention as possible.

Wild bees

Wild bees also play an important role. Bumblebees are useful pollinators of spring-flowering crops and fruit trees because they forage in cool, dull weather when honey bees are inactive. They play a particularly important part in pollinating field beans, red and white clover, lucerne and borage. Bumblebees nest and overwinter in field margins and woodland edges. Avoidance of direct or indirect spray contamination of these areas, in addition to the creation of hedgerows and field margins, and late cutting or grazing of meadows, all help the survival of these valuable insects.

The Campaign Against Illegal Poisoning of Wildlife

The Campaign Against Illegal Poisoning of Wildlife, aimed at protecting some of Britain's rarest birds of prey and wildlife while also safeguarding domestic animals, was launched in March 1991 by the (then) Ministry of Agriculture, Fisheries and Food and the Department of the Environment, Transport and the Regions. The main objective is to deter those who may be considering using pesticides illegally.

The Campaign is supported by a range of organisations associated with animal welfare, nature preservation, field sports and game keeping, including the Royal Society for the Protection of Birds, English Nature, the British Field Sports Society and the Game Conservancy Trust.

The three objectives are:

- to advise farmers, gamekeepers and other land managers on legal ways of controlling pests
- to advise the public on how to report illegal poisoning incidents and to respect the need for legal alternatives
- to investigate incidents and prosecute offenders.

A freephone number (0800 321 600) is available to make it easier for the public to report incidents, and numerous leaflets, posters, postcards, coasters and stickers have been created to publicise the existence of the Campaign.

The Campaign arose from the results of the Wildlife Incident Investigation Scheme for the investigation of possible cases of illegal poisoning. Under this scheme, all reported incidents are considered and thoroughly investigated where appropriate. Enforcement action is taken wherever sufficient evidence of an offence can be obtained, and numerous prosecutions have been made since the start of the Campaign.

Further information about the Campaign is available from:

Helena Cooke
Pesticides Safety Directorate
Room 317
Mallard House
3 Peasholme Green
York YO1 7PX
E-mail: helena.cooke@psd.defra.gsi.gov.uk
Website: www.pesticides.gov.uk/citizen/caip.htm

Chemicals Subject to the Poisons Law

Certain products in this book are subject to the provisions of the Poisons Act 1972, the Poisons List Order 1982, and the Poisons Rules 1982 (copies are obtainable from The Stationery Office; see Appendix 2). These Rules include general and specific provisions for the storage, sale and supply of listed non-medicine poisons. Full details can be accessed in Annex C of the Guide to Pesticides on the PSD website. The nature of the formulation and the concentration of the active ingredient allow some products to be exempted from the rules (see below).

The chemicals approved for use in the UK and included in this book are specified under Parts I and II of the Poisons List, as follows.

Part I Poisons

Sale restricted to registered retail pharmacists and to registered non-pharmacy businesses provided sales do not take place on retail premises:

aluminium phosphide
chloropicrin
magnesium phosphide
methyl bromide
strychnine

Part II Poisons

Sale restricted to registered retail pharmacists and listed sellers registered with a local authority:

aldicarb
alphachloralose
dichlorvos (a,d)
formaldehyde
nicotine (b)
oxamyl (a)
paraquat (c)
sulphuric acid
zinc phosphide

(a) Granular formulations containing up to 12% w/w of this, or a combination of similarly flagged poisons, are exempt.
(b) Formulations containing not more than 7.5% of nicotine are exempt.
(c) Pellets containing not more than 5% paraquat ion are exempt.
(d) Preparations in aerosol dispensers containing not more than 1% a.i. are exempt. Materials impregnated with dichlorvos for slow release are exempt.

Occupational Exposure Limits

A fundamental requirement of the COSHH Regulations is that exposure of employees to substances hazardous to health should be prevented or adequately controlled. Exposure by inhalation is usually the main hazard, and in order to measure the adequacy of control of exposure by this route, various substances have been assigned occupational exposure limits.

There are two types of occupational exposure limits defined under COSHH: Occupational Exposure Standards (OES); and Maximum Exposure Limits (MEL). The key difference is that an OES is set at a level at which there is no indication of risk to health; for an MEL a residual risk may exist, and the level takes socio-economic factors into account. In practice, MELs have most often been allocated to carcinogens and to other substances for which no threshold of effect can be identified, and for which there is no doubt about the seriousness of the effects of exposure.

OESs and MELs are set on the recommendation of the Advisory Committee on Toxic Substances. Full details are published by the Health and Safety Executive (HSE) in *Occupational Exposure Limits 1995*, EH 40/95 (ISBN 0 7176 0876 X).

As far as pesticides are concerned, OESs and MELs have been set for relatively few active ingredients. This is because pesticide products usually contain other substances in their formulation, including solvents, which may have their own OES/MEL. In practice, inhalation of solvent may be at least, or more, important than that of the active ingredient. These factors are taken into account in the approval process of the pesticide product under the Control of Pesticide Regulations. This indicates one of the reasons why a change of pesticide formulation usually necessitates a new approval assessment under COPR.

Poisoning by Pesticides - First Aid Measures

If pesticides are handled in accordance with the required safety precautions, as given on the container label, poisoning should not occur. It is difficult, however, to guard completely against the occasional accidental exposure. Thus if a person handling, or exposed to, pesticides becomes ill, it is a wise precaution to apply first aid measures appropriate to pesticide poisoning even though the cause of illness may eventually prove to have been quite different. An employer has a legal duty to make adequate first aid provision for employees. Regular pesticide users should consider appointing a trained first aider even if numbers of employees are not large, as there is a specific hazard.

The first essential in a case of suspected poisoning is for the person involved to stop work, to be moved away from any area of possible contamination, and for a doctor to be called at once. If no doctor is available the patient should be taken to hospital as quickly as possible. In either event it is most important that the name of the chemical being used should be recorded, and preferably the whole product label or leaflet should be shown to the doctor or hospital concerned.

Some pesticides which are unlikely to cause poisoning in normal use are extremely toxic if swallowed accidentally or deliberately. In such cases get the patient to hospital as quickly as possible, with all the information you have.

General measures

Measures appropriate in all cases of suspected poisoning include:

- remove any protective or other contaminated clothing (taking care to avoid personal contamination)
- wash any contaminated areas carefully with water or with soap and water if available
- in cases of eye contamination, flush with plenty of clean water for at least 15 minutes
- lay the patient down, keep at rest and under shelter; cover with one clean blanket or coat, etc. and avoid overheating
- monitor level of consciousness, breathing and pulse rate
- if consciousness is lost, place the casualty in the recovery position (on his/her side with head down and tongue forward to prevent inhalation of vomit)
- if breathing ceases or weakens, commence mouth-to-mouth resuscitation: ensure that the mouth is clear of obstructions such as false teeth, that the breathing passages are clear, and that tight clothing around the neck, chest and waist has been loosened; if a poisonous chemical has been swallowed, it is essential that the first aider is protected by the use of a resuscitation device (several types are available on the market).

Specific measures

In case of poisoning with particular chemical groups, the following measures may be taken before transfer to hospital.

Dinitro compounds[1]

Keep the patient at rest and cool.

Organophosphorus and carbamate insecticides

Keep the patient at rest. The patient may suddenly stop breathing, so be ready to give artificial respiration.

[1] Approvals for use of dinoseb, dinoseb acetate, dinoseb amine, dinoterb and binapacryl were revoked on 22 January 1988. Approval for storage of these materials was withdrawn on 30 June 1988. Approvals for supply of DNOC were revoked in December 1988 and for its use in December 1989.

Organochlorine compounds

If convulsions occur, do not interfere unless the patient is in danger of injury; if so any restraint must be gentle. When convulsions cease, place the casualty on his or her side with head down, tongue forward (recovery position).

Paraquat, diquat

Irrigate skin and eye splashes copiously with water. If any chemical has been swallowed, take to hospital for tests.

Cyanide (including sodium cyanide)

Send for medical aid. Remove casualty to fresh air, if necessary using breathing apparatus and protective clothing. Remove casualty's contaminated clothing. Gently brush solid particles from the skin, making sure you protect your own skin from contamination. Wash the skin and eyes copiously with water. Transfer casualty to nearest accident and emergency hospital by the quickest possible means together with the first aid cyanide antidote, if held on the premises.

Reporting pesticide poisoning

Any cases of poisoning by pesticides must be reported without delay to an HM Agricultural Inspector of the Health and Safety Executive. In addition, any cases of poisoning by substances named in Schedule 2 of The Reporting of Injuries, Diseases and Dangerous Occurrences Regulations 1985 must also be reported to HM Agricultural Inspectorate (this includes organophosphorus chemicals, mercury and some fumigants).

Cases of pesticide poisoning should also be reported to the manufacturer concerned.

Additional information

General advice on the safe use of pesticides is given in a range of Health and Safety Executive leaflets available from HSE Books (see Appendix 2).

A useful booklet, *Guidelines for Emergency Measures in Cases of Pesticide Poisoning*, is available from CropLife International (see Appendix 2).

The major agrochemical companies are able to provide authoritative medical advice about their own pesticide products. Detailed advice is also available to doctors from the National Poisons Information Service, New Cross Hospital, London SE14 5ER (020 7635 9191) and from regional centres (see Appendix 2).

Agricultural Pesticides and Water

The Food and Environment Protection Act 1985 (FEPA) places a special obligation on users of pesticides to 'safeguard the environment and in particular avoid the pollution of water'. Under the Water Resources Act 1991 it is an offence to pollute any controlled waters (watercourses or groundwater), either deliberately or accidentally. Protection of controlled waters from pollution is the responsibility of the Environment Agency.

Users of pesticides therefore have a duty to adopt responsible working practices and, unless they are applying herbicides in or near water, to prevent them getting into water. Guidance on how to achieve this is given in the Defra *Code of Good Agricultural Practice for the Protection of Water* (available from The Stationery Office; see Appendix 2).

The duty of care covers not only the way in which a pesticide is sprayed, but also its storage, preparation and disposal of surplus, sprayer washings and the container. Products in this *Guide* that are a major hazard to fish, other aquatic life or aquatic higher plants carry one of several specific label precautions in their profile, depending on the assessed hazard level.

Advice on any water pollution problems is available from the Environment Agency and the Crop Protection Association (see Appendix 2).

Protecting surface waters

Surface waters are particularly vulnerable to contamination. One of the best ways of preventing those pesticides that carry the greatest risk to aquatic wildlife from reaching surface waters is to prohibit their application within a boundary adjacent to the water. Such areas are known as no-spray, or buffer, zones. Certain products in this *Guide* are restricted in this way, and have a legally binding label precaution to make sure the potential exposure of aquatic organisms to pesticides that might harm them is minimised.

Before 1999 the protected zones were measured from the edge of the water. The distances were 2 m for hand-held or knapsack sprayers, 6 m for ground crop sprayers, and a variable distance (but often 18 m) for broadcast air-assisted applications, such as in orchards. The introduction of LERAPs (see below) has changed the method of measuring buffer zones.

'Surface water' includes lakes, ponds, reservoirs, streams, rivers and watercourses (natural or artificial). It also includes temporarily or seasonally dry ditches, which have the potential to carry water at different times of the year.

Buffer zone restrictions do not necessarily apply to all products containing the same active ingredient. Those in formulations that are not likely to contaminate surface water through spray drift do not pose the same risk to aquatic life, and are not subject to the restrictions.

Local Environmental Risk Assessments for Pesticides

Local Environment Risk Assessments for Pesticides (LERAPs) were introduced in March 1999, and revised guidelines were issued in January 2002. They give users of most products currently subject to a buffer zone restriction the option of continuing to comply with the existing buffer zone restriction (using the new method of measurement), or carrying out a LERAP and possibly reducing the size of the buffer zone as a result. In either case, there is a new legal obligation for the user to record his decision, including the results of the LERAP.

The scheme has changed the method of measuring the buffer zone. Previously the zone was measured from the edge of the water, but it is now the distance from the top of the bank of the watercourse to the edge of the spray area.

The LERAP provides a mechanism for taking into account other factors that may reduce the risk, such as dose reduction, the use of low drift nozzles, and whether the watercourse is dry or flowing. The previous arrangements applied the same restriction regardless of whether there was actually water present. Now there is a standard zone of 1 m from the top of a dry ditch bank.

Other factors to include in a LERAP that may allow a reduction in the buffer zone are:

- size of the watercourse: the wider it is, the greater the dilution factor and the lower the risk of serious pollution
- dose applied: the lower the dose, the less is the risk
- application equipment: sprayers and nozzles are star-rated according to their ability to reduce spray drift fallout - equipment offering the greatest reductions achieves the highest rating of three stars. The scheme was initially restricted to ground crop sprayers; new, more flexible rules introduced in February 2002 included broadcast air-assisted orchard and hop sprayers.

Other changes introduced in 2002 allow the reduction of a buffer zone if there is an appropriate living windbreak between the sprayed area and a watercourse.

Not all products that had a label buffer zone restriction are included in the LERAP scheme. The option to reduce the buffer zone does not apply to organophosphorus or synthetic pyrethroid insecticides. These groups are classified as Category A products. All other products that had a label buffer zone restriction are classified as Category B. The wording of the buffer zone label precautions has been amended for all products to take account of the new method of measurement, and whether or not the particular product qualifies for inclusion in the LERAP scheme.

Products in this *Guide* that are in Category A or B, or have a broadcast air-assisted LERAP, are identified in the **Special Precautions/Environmental Safety** section of the profile. Updates to the list are published regularly by the Pesticides Safety Directorate, and details can be obtained from www.pesticides.gov.uk

The introduction of LERAPs is an important step forward because it demonstrates a willingness to reduce the impact of regulation on users of pesticides by allowing flexibility where local conditions make it safe to do so. This places a legal responsibility on the user to ensure the risk assessment is done either by himself or by the spray operator, or by a professional consultant or advisor. It is compulsory to record the LERAP and make it available for inspection by enforcement authorities.

Groundwater regulations

New Groundwater Regulations were introduced in 1999 to complete the implementation in the UK of the EU Groundwater Directive (Protection of Groundwater Against Pollution Caused by Certain Dangerous Substances - 80/68/EEC). These Regulations help prevent the pollution of groundwater by controlling discharges or disposal of certain substances, including all pesticides.

Groundwater is defined under the Regulations as any water contained in the ground below the water table. Pesticides must not enter groundwater unless it is deemed by the appropriate Agency to be permanently unsuitable for other uses.

The Regulations make it a criminal offence to dispose of pesticides onto land without official authorisation from the Environment Agency (in England and Wales) or the Scottish Environmental Protection Agency. Normal use of a pesticide in accordance with product approval does not require authorisation. This includes spraying the washings and rinsings back on the crop, provided that, in so doing, the maximum approved dose for that product on that crop is not exceeded.

The Agencies will review all authorisations within a 4-year interval. They may also grant authorisations for a limited period. The Agencies can serve notice at any time to modify the conditions of an authorisation where necessary to prevent pollution of groundwater.

In practice, the best advice to farmers and growers is to plan to use all diluted spray within the crop and to dispose of all washings via the same route, making sure they stay within the conditions of approval of the product.

Use of Herbicides In or Near Water

Products in this *Guide* approved for use in or near water are listed in Table 1.5. Before use of any product in or near water, the appropriate water regulatory body (Environment Agency/Local Rivers Purification Authority; or in Scotland the Scottish Environment Protection Agency) must be consulted. Guidance and definitions of the situation covered by approved labels are given in the Defra publication *Guidelines for the Use of Herbicides on Weeds in or near Watercourses and Lakes.* Always read the label before use.

Table 1.5 Products approved for use in or near water

Chemical	Product
2,4-D	Dormone, Ragox
dichlobenil	Casoron G, Luxan Dichlobenil Granules, Midstream GSR, Sierraron G
diquat	Agriguard Diquat
glyphosate (fruit, horticulture, forestry, amenity etc.)	Asteroid, Barclay Barbarian, Barclay Gallup 360, Barclay Gallup Amenity, Barclay Gallup Biograde 360, Barclay Gallup Biograde Amenity, Barclay Gallup Hi-Aktiv, Buggy SG, Clinic, Envision, Flame, Glydate, Glyfos, Glyfos Gold, Glyfos ProActive, Glyper, Glyphogan, Glyphosate 360, Greenaway Gly-490, Habitat, Kernel, Manifest, Nufosate, Romany, Roundup Amenity, Roundup Biactive, Roundup Pro Biactive, Spasor, Spasor Biactive, Tumbleweed Pro
maleic hydrazide	Regulox K
terbutryn	Clarosan

Use of Pesticides in Forestry

Table 1.6 lists those products in this *Guide* that have approved uses in forestry.

Table 1.6 Products approved for use in forestry

Chemical	Product	Use
2,4-D	Dicotox Extra, Easel	Herbicide
2,4-D + dicamba + triclopyr	Greengard, Nu-Shot	Herbicide
4-indol-3-yl-butyric acid + 2-(1-naphthyl)acetic acid with dichlorophen	Synergol	Growth regulator/stimulator/sprout suppressant
aluminium ammonium sulphate	Liquid Curb Crop Spray	Animal deterrent/repellent
ammonium sulphamate	Amcide, Root-Out	Herbicide, Tree killer
asulam	Asulox, I T Asulam	Herbicide
atrazine	Atrazol	Herbicide
azaconazole + imazalil	Nectec Paste	Fungicide
carbosulfan	Marshal Soil Insecticide suSCon CR granules, Marshal Soil Insecticide suSCon CR Sachets	Insecticide
chlorpyrifos	Agriguard Chlorpyrifos, Barclay Clinch II, Cyren, Dursban WG, Equity, Greencrop Pontoon, Lorsban T	Acaricide, Insecticide
copper hydroxide	Spin Out	Root acting plant growth regulator
cycloxydim	Greencrop Pomeroy, Greencrop Valentia, Marnoch Clodim, Standon Cycloxydim	Herbicide
dicamba	I T Dicamba	Herbicide
diquat + paraquat	ASAP, PDQ	Desiccant, Herbicide
fluazifop-P-butyl	Fusilade 250 EW, Fusilade Max	Herbicide
glufosinate-ammonium	Challenge, Harvest, Kaspar	Herbicide

Chemical	Product	Use
glyphosate (fruit, horticulture, forestry, amenity etc.)	Barclay Barbarian, Barclay Gallup, Barclay Gallup 360, Barclay Gallup Amenity, Barclay Gallup Biograde 360, Barclay Gallup Biograde Amenity, Barclay Gallup Hi-Aktiv, Barclay Garryowen, Buggy SG, Clinic, Envision, Glydate, Glyfos, Glyfos Gold, Glyfos ProActive, Glyper, Glyphogan, Glyphosate 360, Greenaway Gly-490, Greencrop Gypsy, Habitat, Hilite, Kernel, Manifest, Nufosate, Romany, Roundup Amenity, Roundup Pro Biactive, Stirrup, Tumbleweed Pro	Herbicide
isoxaben	Agriguard Isoxaben, Flexidor 125	Herbicide
metazachlor	Alpha Metazachlor 50 SC, Me2 Booty, Sultan 50 SC	Herbicide
paraquat	Agriguard Paraquat, Barclay Total, Fernpath Graminite, Gramoxone 100	Herbicide
pirimicarb	Phantom	Insecticide
propaquizafop	Barclay Rebel, Falcon, Greencrop Satchmo, Landgold PQF 100, Standon Propaquizafop	Herbicide
propyzamide	Agriguard Propyzamide 50 WP, Agriguard Propyzamide Flo, Base 50 W, Bulwark Flo, Kerb 50 W, Kerb 80 EDF, Kerb Granules, Kerb Pro Flo, Kerb Pro Granules, Menace 80 EDF, Mithras 80 EDF, Propose	Herbicide
simazine	Alpha Simazine 50 SC, Gesatop, Sipcam Simazine Flowable	Herbicide
triclopyr	Agriguard Triclopyr, Garlon 4, n2n Garlon 4, Timbrel	Herbicide
ziram	AAprotect	Animal deterrent/repellent

Pesticides Used as Seed Treatments

Information on the target pests for these products can be found in the relevant pesticide profile in Section 4.

Table 1.7 Products used as seed treatments (including treatments on seed potatoes)

Chemical	Product	Formulation	Crop(s)
aluminium ammonium sulphate	Guardsman STP Seed Dressing Powder	DS	Corms, Flower bulbs, Seeds
beta-cyfluthrin + imidacloprid	Chinook Blue	LS	Winter oilseed rape
	Chinook Colourless	LS	Winter oilseed rape
bitertanol + fuberidazole	Sibutol	FS	Spring oats, Spring rye, Spring wheat, Triticale, Winter oats, Winter rye, Winter wheat
bitertanol + fuberidazole + imidacloprid	Sibutol Secur	LS	Winter oats, Winter wheat
carboxin + thiram	Anchor	FS	Spring barley, Spring oats, Spring rye, Spring wheat, Triticale, Winter barley, Winter oats, Winter rye, Winter wheat
cymoxanil + fludioxonil + metalaxyl-M	Wakil XL	WS	Broad beans, Carrots, Combining peas, Parsnips, Spring field beans, Vining peas, Winter field beans
fludioxonil	Beret Gold	FS	Spring barley, Spring oats, Spring wheat, Winter barley, Winter oats, Winter wheat
fluquinconazole	Galmano	FS	Winter wheat
	Jockey F	FS	Winter wheat
fluquinconazole + prochloraz	Galmano Plus	FS	Winter wheat
	Jockey	FS	Winter wheat
flutolanil	Rhino	FS	Potatoes
fuberidazole + imidacloprid + triadimenol	Baytan Secur	FS	Winter barley, Winter oats, Winter wheat

Chemical	Product	Formulation	Crop(s)
fuberidazole + triadimenol	Baytan Flowable	FS	Spring barley, Spring oats, Spring rye, Spring wheat, Triticale, Winter barley, Winter oats, Winter rye, Winter wheat
guazatine	Panoctine	LS	Spring barley, Spring oats, Winter barley, Winter oats
	Ravine	LS	Spring barley, Spring oats, Spring wheat, Winter barley, Winter oats, Winter wheat
guazatine + triticonazole	Premis	FS	Spring wheat, Winter wheat
imazalil	Fungazil 100 SL	LS	Seed potatoes
imazalil + pencycuron	Monceren IM	DS	Potatoes
imazalil + thiabendazole	Extratect Flowable	FS	Potatoes, Seed potatoes
imazalil + triticonazole	Robust	FS	Spring barley, Winter barley
imidacloprid	Gaucho	WS	Broccoli, Brussels sprouts, Cabbages, Calabrese, Cauliflowers, Collards, Kale, Lettuce, Sugar beet
imidacloprid + tebuconazole + triazoxide	Raxil Secur	LS	Winter barley
iprodione	Rovral Liquid FS	FS	Linseed, Seed potatoes, Spring oilseed rape, Winter oilseed rape
pencycuron	Me2 Penny	DS	Potatoes
	Monceren DS	DS	Potatoes
	Standon Pencycuron DP	DS	Potatoes
prochloraz	Prelude 20LF	LS	Flax, Linseed
silthiofam	Latitude	FS	Spring wheat, Winter barley, Winter wheat
tebuconazole + triazoxide	Raxil S	FS	Spring barley, Winter barley
thiabendazole	Hykeep	DS	Ware potatoes
	Storite Clear Liquid	LS	Narcissi, Seed potatoes, Ware potatoes
	Storite Flowable	FS	Seed potatoes, Ware potatoes

Chemical	Product	Formulation	Crop(s)
thiabendazole + thiram	Hy-TL	FS	Broad beans, Bulb onions, Peas, Spring field beans, Winter field beans
	sHYlin	FS	Carrots, Flax, Linseed, Lupins, Parsnips
thiram	Agrichem Flowable Thiram	LS	Broad beans, Cabbages, Carrots, Cauliflowers, Celery, Dwarf beans, Fodder beet, Grass seed, Leeks, Lettuce, Maize, Mangels, Onions, Parsley, Peas, Radishes, Red beet, Runner beans, Salad onions, Spring field beans, Spring oilseed rape, Sugar beet, Turnips, Winter field beans, Winter oilseed rape
	Thiraflo	FS	Broad beans, Combining peas, Dwarf beans, Grass seed, Maize, Runner beans, Spring field beans, Spring oilseed rape, Vining peas, Winter field beans, Winter oilseed rape
	Thyram Plus	LS	Broad beans, Cabbages, Carrots, Cauliflowers, Dwarf beans, Grass seed, Leeks, Lettuce, Maize, Onions, Peas, Radishes, Runner beans, Salad onions, Spring field beans, Spring oilseed rape, Turnips, Winter field beans, Winter oilseed rape
tolclofos-methyl	Rizolex	DS	Potatoes
	Rizolex Flowable	FS	Potatoes

Aerial Application of Pesticides

Only those products specifically approved for aerial application may be so applied, and they may only be applied to specific crops or for specified uses. A complete list of products approved for application from the air can be accessed in Guide to Pesticides on the website of the Pesticides Safety Directorate (PSD, www.pesticides.gov.uk). The list in Table 1.8 is taken mainly from this source, with updating from issues of *The Pesticides Monitor* and product labels.

It is emphasised that the list is for guidance only – reference must be made to the product labels for detailed conditions of use, which must be complied with. The list does not include those products which have been granted restricted aerial application approval limiting the area which may be treated.

Detailed rules are imposed on aerial application regarding prior notification of the Nature Conservancy, water authorities, bee keepers, Environmental Health Officers, neighbours, hospitals, schools, etc. and the conditions under which application may be made. The full conditions are available from Defra and must be consulted before any aerial application is made.

Table 1.8 Products approved for aerial application

Chemical	Product	Crops
2-chloroethylphosphonic acid	Agriguard Cerusite	Winter barley
	Barclay Coolmore	Winter barley
	Cerone	Winter barley
asulam	Asulox	Amenity grass, Forestry, Permanent pasture, Rough grazing
	I T Asulam	Amenity grass, Forestry, Permanent pasture, Rough grazing
benalaxyl + mancozeb	Galben M	Early potatoes, Maincrop potatoes
	Intro Plus	Early potatoes, Maincrop potatoes
	Tairel	Early potatoes, Maincrop potatoes
carbendazim	Tripart Defensor FL	Spring oilseed rape, Winter oilseed rape, Winter wheat
chlormequat	Agriguard Chlormequat 700	Spring oats, Spring wheat, Winter oats, Winter wheat
	Agriguard Chlormequat 760	Spring oats, Spring wheat, Winter oats, Winter wheat
	Barclay Holdup	Spring oats, Spring wheat, Winter barley, Winter oats, Winter wheat
	Barclay Holdup 600	Spring oats, Spring wheat, Winter barley, Winter oats, Winter wheat

Chemical	Product	Crops
	Barclay Holdup 640	Spring oats, Spring wheat, Winter barley, Winter oats, Winter wheat
	BASF 3C Chlormequat 720	Spring oats, Spring rye, Spring wheat, Triticale, Winter barley, Winter oats, Winter rye, Winter wheat
	Greencrop Carna	Spring oats, Spring wheat, Winter barley, Winter oats, Winter wheat
	Greencrop Coolfin	Spring oats, Spring wheat, Winter oats, Winter wheat
	Hive	Spring oats, Spring wheat, Winter oats, Winter wheat
	Mandops Chlormequat 700	Spring oats, Spring wheat, Winter oats, Winter wheat
	Mirquat	Spring oats, Spring wheat, Winter oats, Winter wheat
	Sigma PCT	Spring wheat, Winter barley, Winter wheat
	Stabilan 700	Spring oats, Spring wheat, Winter oats, Winter wheat
	Terbine	Spring oats, Spring wheat, Winter oats, Winter wheat
	Tripart Brevis	Spring oats, Spring wheat, Winter oats, Winter wheat
	Uplift	Spring oats, Spring wheat, Winter barley, Winter oats, Winter wheat
chlormequat with choline chloride	Atlas 5C Quintacel	Spring oats, Spring wheat, Winter oats, Winter wheat
	Barclay Take 5	Spring oats, Spring wheat, Triticale, Winter barley, Winter oats, Winter rye, Winter wheat
	New 5C Cycocel	Spring oats, Spring rye, Spring wheat, Triticale, Winter barley, Winter oats, Winter rye, Winter wheat
chlorothalonil	Agriguard Chlorothalonil	Potatoes
	Barclay Corrib 500	Potatoes
	Bombardier FL	Potatoes

Chemical	Product	Crops
	Bravo 500	Potatoes
	Clortosip 500	Potatoes
	Flute	Potatoes
	Greencrop Orchid	Potatoes
	Repulse	Potatoes
	Rover	Potatoes
	Sipcam Echo 75	Potatoes
	Standon Chlorothalonil 500	Potatoes
	Tripart Faber	Potatoes
	Ultrafaber	Potatoes
chlorotoluron	Lentipur CL 500	Durum wheat, Triticale, Winter barley, Winter wheat
	Luxan Chlorotoluron 500 Flowable	Durum wheat, Triticale, Winter barley, Winter wheat
copper oxychloride	Cuprokylt	Potatoes
diflubenzuron	Dimilin Flo	Forestry
dimethoate	BASF Dimethoate 40	Spring rye, Spring wheat, Sugar beet, Sugar beet seed crops, Triticale, Winter rye, Winter wheat
	Danadim	Triticale, Winter rye, Winter wheat
	Rogor L40	Peas, Potatoes, Spring wheat, Sugar beet, Winter wheat
mancozeb	Dithane 945	Potatoes
	Dithane NT Dry Flowable	Potatoes
	Dithane Superflo	Potatoes
	Laminator DG	Potatoes
	Laminator FL	Potatoes
	Laminator WP	Potatoes
	Micene 80	Potatoes

Chemical	Product	Crops
	Micene DF	Potatoes
	Penncozeb WDG	Potatoes
	Quell Flo	Potatoes
maneb	Maneb 80	Potatoes
	Trimanzone	Potatoes
	X-Spor SC	Potatoes
metaldehyde	Brits	All edible crops, All non-edible crops
	Dixie 6	All edible crops, All non-edible crops
	Doff Horticultural Slug Killer Blue Mini Pellets	All edible crops, All non-edible crops
	Luxan Deal	All edible crops, All non-edible crops
	Superflor 6% Metaldehyde Slug Killer Mini Pellets	All edible crops, All non-edible crops
pirimicarb	Agriguard Pirimicarb	Spring barley, Spring oats, Spring wheat, Winter barley, Winter oats, Winter wheat
	Aphox	Durum wheat, Spring barley, Spring oats, Spring rye, Spring wheat, Triticale, Winter barley, Winter oats, Winter rye, Winter wheat
	Greencrop Glenroe	Durum wheat, Spring barley, Spring oats, Spring rye, Spring wheat, Triticale, Winter barley, Winter oats, Winter rye, Winter wheat
	Phantom	Durum wheat, Spring barley, Spring oats, Spring rye, Spring wheat, Triticale, Winter barley, Winter oats, Winter rye, Winter wheat
propiconazole	Barclay Bolt	Spring barley, Spring oats, Spring rye, Spring wheat, Sugar beet, Winter barley, Winter oats, Winter rye, Winter wheat
sulphur	Cosavet DF	Spring barley, Spring oilseed rape, Spring wheat, Winter barley, Winter oilseed rape, Winter wheat

Chemical	Product	Crops
	Kumulus DF	Spring barley, Spring oilseed rape, Spring wheat, Winter barley, Winter oilseed rape, Winter wheat

SECTION 2
ADJUVANTS

Adjuvants

Adjuvants are not themselves classed as pesticides and there is considerable misunderstanding over the extent to which they are legally controlled under the Food and Environment Protectiion Act. An adjuvant is a substance other than water which enhances the effectiveness of a pesticide with which it is mixed. Consent C(i)5 under the Control of Pesticides Regulations allows that an adjuvant can be used with a pesticide only if that adjuvant is authorised and on a list published from time to time in *The Pesticides Register*. An authorised adjuvant has an *adjuvant number* and may have specific requirements about the circumstances in which it may be used.

Adjuvant product labels must be consulted for full details of authorised use, but the table below provides a summary of the label information to indicate the area of use of the adjuvant. Label precautions refer to the keys given in Appendix 4, and may include warnings about products harmful or dangerous to fish. The table includes all adjuvants notified by suppliers as available in 2004.

Product	Supplier	Adj. No.	Type
Abacus	Loveland	A0434	adjuvant/wetter
Contains	505.2 g/l esterified rapeseed oil, 300 g/l alkylphenyl hydroxypolyoxyethylene and 60 g/l natural fatty acids		
Use with	All approved pesticides on edible and non-edible crops when used at half their approved dose or less. Also with approved pesticides on specified crops, up to specified growth stages, at up to their full approved dose		
Precautions	U02, U04a, U05a, U10, U19, U20b, E26, E30a, E31b, E34		
Activator 90	Loveland	A0337	non-ionic surfactant/spreader/wetter
Contains	750 g/l alkylphenyl hydroxypolyoxyethylene and 150 g/l natural fatty acids		
Use with	Any spray for which additional wetter is approved and recommended		
Protective clothing	A, C		
Precautions	R04a, R04b, U02, U04a, U05a, U10, U11, U19, U20b, E01, E13c, C03, E30a, E31a, E34, M03		
Admix-P	Loveland	A0301	wetter
Contains	80% w/w polyalkylene oxide modified heptamethyltrisiloxane and a maximum of 20 % w/w allyloxypolyethylene glycol methyl ether		
Use with	A wide range of pesticides applied as corm, tuber, onion and other bulb treatments in seed production and in seed potato treatment		
Protective clothing	A, C, H		
Precautions	R03a, R03c, R04a, R04b, R04e, U02, U05a, U08, U19, U20b, E01, E13c, C03, E30a, E31a, E34, M03		
Agral	Syngenta	A0421	non-ionic surfactant/spreader/wetter
Contains	948 g/l alkyl phenol ethylene oxide		
Use with	Any spray for which additional wetter is approved and recommended		
Protective clothing	A, C		
Precautions	R04a, R04b, U05a, U08, U20c, E01, E15, C03, E26, E30a, E31a, E34, R07d		

Product	Supplier	Adj. No.	Type
Agropen	Intracrop	A0143	vegetable oil
Contains	95% refined rapeseed oil		
Use with	Pesticides which have a recommendation for the addition of a wetter/ spreader		
Protective clothing	A, C		
Precautions	U02, U05a, U08, U19, U20b, E01, E13c, C03, E30a, E31a, E34		
Amber	Interagro	A0367	vegetable oil
Contains	95% w/w methylated rapeseed oil		
Use with	Sugar beet herbicides, oilseed rape herbicides, cereal graminicides and a wide range of other pesticides that have a label recommendation for use with authorised adjuvant oils on specified crops. See label for details		
Precautions	U05a, E01, E15, C03, E26, E30a, E31a		
Arma	Interagro	A0306	buffering agent/penetrant/wetter
Contains	500 g/l alkoxylated fatty amine + 500 g/l polyoxyethylene monolaurate		
Use with	Cereal growth regulators, cereal herbicides, cereal fungicides, oilseed rape fungicides and a wide range of other pesticides on specified crops		
Precautions	U05a, E01, E13c, C03, E26, E30a, E34		
AS Elite	Monsanto	A0406	extender
Contains	40% w/w ammonium sulphate		
Use with	Roundup Express, Roundup Rapide		
Precautions	U20a, E30a, E31b		
Atlas Adjuvant Oil	Nufarm UK	A0021	mineral oil
Contains	95% w/w highly refined mineral oil and surfactants		
Use with	Recommended rates of approved pesticides up to specified growth stages of a wide range of agricultural arable crops, and in non-crop situations. Use on edible crops beyond specified growth stages should only be with half recommended rates of the pesticide or less. See label for details of growth stage restrictions and specific recommendations for use with herbicides on beet crops.		
Protective clothing	A, C		
Precautions	R04a, R04b, R04c, U05a, U08, U19, U20b, E01, E13c, E37, C03, E26, E30a, E31b		
BackRow	Interagro	A0472	mineral oil
Contains	60% refined paraffinic petroleum oil		
Use with	Pre-emergence herbicides		
Protective clothing	A		
Precautions	R04b, R04g, U02, U05a, U08, U20b, E01, E13c, C03, E26, E30a, E31a, M05, R03		

Product	Supplier	Adj. No.	Type
Bandrift Plus	Ciba Specialty	xx	drift retardant
Contains	A non-ionic polyamide dispersed in oil		
Use with	A wide range of pesticides. See label for restrictions on use with wettable powders, suspension concentrates and water dispersible granules, and other usage limitations		
Protective clothing	A, C, M		
Precautions	U05a, U08, U20a, E01, E13c, C03, E30a, E34		
Banka	Interagro	A0245	penetrant/spreader/sticker/wetter
Contains	29.2% w/w alkyl pyrrolidones		
Use with	Potato fungicides and a wide range of other pesticides on specified crops		
Protective clothing	A, C		
Precautions	R04b, R04d, U02, U05a, U08, U11, U19, U20a, E01, E13c, E37, C03, E26, E30a, E34, R04, R07d		
Barclay Actol	Barclay	A0126	mineral oil
Contains	99% highly refined paraffinic oil		
Use with	Any approved pesticide for which the addition of a spraying oil is recommended		
Precautions	U19, U20b, E13c, E30a, E31b		
Barclay Clinger	Barclay	A0198	sticker/wetter
Contains	96% w/v poly-1-p-menthene		
Use with	Barclay Gallup, Barclay Gallup Amenity and a wide range of approved fungicides and insecticides. See label for details		
Precautions	U19, U20b, E13c, E30a, E31a		
Barramundi	Interagro	A0376	mineral oil
Contains	95% w/w mineral oil		
Use with	Sugar beet herbicides, oilseed rape herbicides, cereal graminicides and a wide range of other pesticides that have a label recommendation for use with authorised adjuvant oils on specified crops. See label for details		
Protective clothing	A		
Precautions	R04b, R04g, U02, U05a, U08, U20b, E01, E13c, C03, E26, E30a, E31a, M05, R03		
Bioduo	Intracrop	A0299	wetter
Contains	85% alkyl polyglycol ether and fatty acids		
Use with	A wide range of pesticides used in grassland, agriculture and horticulture and with pesticides used in non-crop situations		
Protective clothing	A, C		
Precautions	R03c, R04a, R04b, U05a, U08, U20c, E01, E13c, C03, E30a, E31a, E34, M03, R07d		
Biofilm	Intracrop	A0359	sticker/wetter
Contains	96% poly-1-p-menthene		
Use with	All approved fungicides and insecticides on edible crops and all approved formulations of glyphosate used pre-harvest on wheat, barley, oilseed rape, stubble, and in non-crop situations and grassland destruction		
Precautions	U19, U20c, E13c, E30a, E32a		

Product	Supplier	Adj. No.	Type
Bio-Flex	Greenaway	A0481	sticker/wetter
Contains	95 g/l vegetable oil		
Use with	GLY 490 (MAPP 11064)		
Precautions	U05a, U08, U20b, E37, C03, E26, E31a		
BioPower	Bayer CropScience	A0454	wetter
Contains	6.7% w/w 3,6-dioxaeicosylsulphate sodium salt and 20.2% w/w 3,6-dioxaoctadecylsulphate sodium salt		
Use with	Atlantis and all other approved cereal herbicides		
Protective clothing	A, C		
Precautions	R04a, R04b, U02, U05a, U08, U13, U19, U20b, E01, E13c, C03, E26, E30a, E31a, E34, R04		
Biothene	Intracrop	A0360	sticker/wetter
Contains	96% poly-1-p-menthene		
Use with	All approved fungicides and insecticides on edible crops and all approved formulations of glyphosate used pre-harvest on wheat, barley, oilseed rape, stubble, and in non-crop situations and grassland destruction		
Precautions	U19, U20c, E13c, E30a, E32a		
Bond	Loveland	A0184	drift retardant/extender/sticker/wetter
Contains	450 g/l synthetic latex and 100 g/l alkylphenylhydroxy-polyoxyethylene		
Use with	Blight sprays on potatoes and other pesticides on a wide range of crops. See label or contact supplier for details		
Protective clothing	A, C		
Precautions	R04a, R04b, U05a, U08, U19, U20a, E01, E13c, C03, E30a, E31a, M03		
C-Cure	Interagro	A0467	mineral oil
Contains	60% w/w refined mineral oil		
Use with	Pre-emergence herbicides		
Protective clothing	A		
Precautions	R04b, R04g, U02, U05a, U08, U20b, E01, E13c, C03, E26, E30a, E31a, M05, R03		
Celect	Fargro	A0334	extender/sticker/wetter
Contains	95% emulsified vegetable oil		
Use with	Approved pesticides in horticulture (but see label for certain active ingredient exclusions)		
Protective clothing	A, C		
Precautions	U19, U20b, E30a, E31b		
Celect	Microcide	A0413	extender/sticker/wetter
Contains	95% emulsified vegetable oil		
Use with	Approved pesticides in horticulture (but see label for certain active ingredient exclusions)		
Protective clothing	A, C		
Precautions	U19, U20b, E30a, E31b		

Product	Supplier	Adj. No.	Type
Ceres Platinum	Interagro	A0445	penetrant/spreader/wetter
Contains	500 g/l alkoxylated fatty amine + 500 g/l polyoxyethylene monolaurate		
Use with	Cereal growth regulators, cereal herbicides, cereal fungicides, oilseed rape fungicides and a wide range of other pesticides on specified crops		
Precautions	U05a, E01, E13c, C03, E26, E30a, E34		
Chum	Greenhill	A0420	spreader/wetter
Contains	85% w/w rapeseed oil		
Use with	All approved pesticides that recommend the use of a wetter/spreader. In addition, may be use with recommended rates of approved pesticides up to specified growth stages of a wide range of agricultural arable and horticultural crops, and in non-crop situations. Use on edible crops beyond specified growth stages should only be with half recommended rates of the pesticide or less. See label for details of growth stage restrictions.		
Protective clothing	A, C		
Precautions	R04a, U05a, U08, U19, U20b, E01, E13c, C03, E30a, E31a, E34		
Codacide Oil	Microcide	A0011	vegetable oil
Contains	95% emulsified vegetable oil		
Use with	All approved pesticides and tank mixes. See label for details		
Protective clothing	A, C		
Precautions	U20b, E30a, E31b		
Companion PCT12	Ciba Specialty	A0482	spreader/sticker/wetter
Contains	25% w/w polyacryalmide		
Use with	Approved herbicides on cereals and any pesticide on non-food crops. See label for restrictions on use with wettable powders, suspension concentrates and water dispersible granules, and other usage limitations		
Precautions	U05a, U08, E01, E13c, C03, E30a		
Contact Plus	Interagro	A0418	mineral oil
Contains	95% w/w mineral oil		
Use with	Sugar beet herbicides, oilseed rape herbicides, cereal graminicides and a wide range of other pesticides that have a label recommendation for use with authorised adjuvant oils on specified crops. See label for details		
Protective clothing	A		
Precautions	R04b, R04g, U02, U05a, U08, U20b, E01, E13c, C03, E26, E30a, E31a, M05, R03		
Cutinol	Greenhill	A0281	vegetable oil
Contains	95% emulsifiable rapeseed oil		
Use with	All approved pesticides that recommend the use of a wetter/spreader. In addition, may be used with recommended rates of approved pesticides up to specified growth stages of a wide range of agricultural arable and horticultural crops, and in non-crop situations. Use on edible crops beyond specified growth stages should only be with half recommended rates of the pesticide or less. See label for details of growth stage restrictions.		
Precautions	R04a, R04b, U05a, U08, U20a, E01, E13c, C03, E30a, E31a		

Product	Supplier	Adj. No.	Type
Cutinol Plus	Greenhill	A0395	spreader/wetter
Contains	85% w/w rapeseed oil		
Use with	All approved pesticides that recommend the use of a wetter/spreader. In addition, may be use with recommended rates of approved pesticides up to specified growth stages of a wide range of agricultural arable and horticultural crops, and in non-crop situations. Use on edible crops beyond specified growth stages should only be with half recommended rates of the pesticide or less. See label for details of growth stage restrictions.		
Protective clothing	A, C		
Precautions	R04a, U05a, U08, U19, U20b, E01, E13c, C03, E30a, E31a, E34		
Designer	Loveland	A0322	drift retardant/extender/sticker/wetter
Contains	8.44% w/w organosilicone wetter and 50% w/w synthetic latex		
Use with	A wide range of fungicides, insecticides and trace elements for cereals and specified agricultural and horticultural crops		
Protective clothing	A, C		
Precautions	R04a, R04b, R04e, U02, U05a, U08, U19, U20b, E01, E13c, C03, E30a, E31a, E34, M03		
Drill	Loveland	A0409	adjuvant/wetter
Contains	589.4 g/l esterified rapeseed oil, 225 g/l alkylphenyl hydroxypolyoxyethylene and 45 g/l natural fatty acids		
Use with	All approved pesticides on edible and non-edible crops when used at half their approved dose or less. Also with approved pesticides on specified crops, up to specified growth stages, at up to their full approved dose		
Precautions	U02, U04a, U05a, U10, U19, U20b, E26, E30a, E31b, E34		
Du Pont Adjuvant	DuPont	A0119	non-ionic surfactant/wetter
Contains	900 g/l ethylene oxide condensate		
Use with	Any spray for which a non-ionic wetter is recommended		
Protective clothing	A, C		
Precautions	R04a, R04b, U05a, U08, U20c, E01, E13c, C03, E30a, E31a, R07d		
Elan	Intracrop	A0392	spreader/sticker/wetter
Contains	20% w/w polyoxyethylen-alpha-methyl-omega-[3-(1,3,3,3-tetramethyl-3-trimethylsiloxy)-disiloxanyl] propylether		
Use with	Approved pesticides used at full dose on a range of specified fruit and vegetable crops, and on managed amenity turf		
Protective clothing	A, C, H		
Precautions	U05a, U08, U20c, E01, E13c, E15, E37, C03, E30a, E31a		
Emerald	Intracrop	A0031	anti-transpirant/extender
Contains	96% di-1-p-menthene		
Use with	Alone or with most approved insecticides, fungicides or growth regulators See label for details. Do not use in mixture with adjuvant oils or surfactants		
Precautions	U19, U20b, E13c, E30a, E31a		

Product	Supplier	Adj. No.	Type
Enhance	Greenhill	A0147	non-ionic surfactant/spreader/wetter
Contains	900 g/l phenol ethylene oxide condensate		
Use with	Recommended rates of approved pesticides up to specified growth stages of a wide range of agricultural arable and horticultural crops. Use on edible crops beyond specified growth stages should only be with half recommended rates of the pesticide or less. See label for details of growth stage restrictions.		
Protective clothing	A, C		
Precautions	R04a, R04b, U05a, U08, U19, U20a, E01, E13c, E37, C03, E30a, E31a, E34		
Enhance Low Foam	Greenhill	A0148	non-ionic surfactant/spreader/wetter
Contains	900 g/l alkyl phenol ethylene oxide condensate with silicone anti-foaming agent		
Use with	Recommended rates of approved pesticides up to specified growth stages of a wide range of agricultural arable and horticultural crops. Use on edible crops beyond specified growth stages should only be with half recommended rates of the pesticide or less. See label for details of growth stage restrictions.		
Protective clothing	A, C		
Precautions	R04a, R04b, U05a, U08, U19, U20a, E01, E13c, E37, C03, E30a, E31a, E34		
Esterol	Interagro	A0330	vegetable oil
Contains	95% w/w methylated rapeseed oil		
Use with	Plant protection products that have a label recommendation for use with adjuvant oils. See label for details		
Precautions	U05a, E01, E15, C03, E26, E30a, E31a		
Ethokem	Greenhill	A0353	cationic surfactant
Contains	870 g/l polyoxyethylene tallow amine		
Use with	Glyphosate and a wide range of foliar acting herbicides in agriculture, horticulture, amenity and forestry. See leaflet for details of specific crop/product recommendations		
Protective clothing	A, C		
Precautions	R03c, R04a, R04b, U02, U05a, U08, U13, U19, U20b, E01, E13b, E37, C03, E31a, E34, M03, R07d		
Euroagkem Pace	Euroagkem	A0305	spreader/wetter
Contains	42% w/w propionic acid		
Use with	Chlormequat, glyphosate and a wide range of herbicides for use in cereals and other agricultural and horticultural crops. Must not be used with any pesticide to be applied in or near water. See label for details		
Protective clothing	A, C		
Precautions	R05b, U02, U04a, U05a, U11, U12, U13, U14, U15, U19, U20b, E01, E15, C03, E30a, E31a, M05		

Product	Supplier	Adj. No.	Type
Felix	Intracrop	A0178	spreader/wetter
Contains	60% ethylene oxide condensate		
Use with	Mecoprop, 2,4-D in cereals and amenity turf, and a range of grass weedkillers in agriculture. See label for details		
Protective clothing	A, C		
Precautions	R04a, R04b, U05a, U08, U20a, E01, E15, C03, E30a, E31a, R07d		
FMC Oil	Belchim	A0435	adjuvant/vegetable oil
Contains	842 g/l esterified rapeseed oil		
Use with	Spotlight 24 EC (MAPP 10702 and 11617)		
Precautions	E13c, E37		
Frigate	Unicrop	A0325	sticker/wetter
Contains	800 g/l tallow amine ethoxylate		
Use with	Roundup formulations		
Protective clothing	A, C		
Precautions	R03c, R04a, R04b, R04c, U05a, U09a, U19, U20a, E01, E13b, C03, E30a, E31a, M03, R07d		
Galion	Intracrop	A0162	spreader/wetter
Contains	60% polyoxyalkylene glycol		
Use with	Herbicides in grassland and amenity turf and with grass weedkillers in cereals, oilseed rape, sugar beet and other crops		
Protective clothing	A, C		
Precautions	R04a, R04b, U05a, U08, U20b, E01, E15, C03, E30a, E31a, R07d		
Gex 1664	Headland	A0366	spreader/wetter
Contains	900 g/l ethylene oxide concentrate		
Use with	All approved pesticides on edible and non-edible crops when used at half their approved dose or less. Also with approved pesticides on specified crops, up to specified growth stages, at up to their full approved dose		
Protective clothing	A, C		
Precautions	R04a, R04b, U05a, U08, U20b, E01, E13c, E37, C03, E30a, E31b, E34		
Gly-Flex	Greenaway	A0370	spreader/wetter
Contains	140 g/l refined mineral oil		
Use with	All approved formulations of 360 g/l glyphosate		
Protective clothing	A, C, D, H, M		
Precautions	R04a, R04b, U08, U20b, E13c, E15, E37, C03, E26, E31a		
Greencrop Astra	Greencrop	A0417	adjuvant/penetrant/spreader
Contains	500 g/l alkoxylated fatty amine and 500 g/l polyoxylene monolaurate		
Use with	Approved pesticides on a range of specified arable crops, and in forestry, managed amenity turf, stubbles and non-crop areas and situations		
Precautions	U05a, E01, E13c, C03, E26, E30a, E31b, E34		

Product	Supplier	Adj. No.	Type
Greencrop Dolmen	Greencrop	A0419	adjuvant/spreader
Contains	64% w/w polyalkeneoxide modified heptamethyltrisiloxane		
Use with	Approved pesticides on a range of specified arable crops, and in forestry, managed amenity turf, stubbles and non-crop areas and situations		
Protective clothing	A, C, H		
Precautions	R03a, R03c, R04b, R04d, R04e, U02, U05a, U08, U11, E01, E13c, C03, E30a, E31b, E34, M03		
Greenhill Spreader	Greenhill	A0196	spreader/wetter
Contains	23% w/w (235 gm/litre) alkyl phenol ethoxylate		
Use with	Pesticides which have a recommendation for the use of a wetter/spreader		
Protective clothing	A, C		
Precautions	R04a, R04b, U02, U05a, U08, U19, U20b, E01, E13c, C03, E30a, E31a, E34		
Grip	Loveland	A0211	drift retardant/extender/sticker/wetter
Contains	450 g/l synthetic latex and 100 g/l alkylphenyl hydroxypolyoxyethylene		
Use with	Blight sprays on potatoes and other pesticides on a wide range of crops. See label or contact supplier for details		
Protective clothing	A, C		
Precautions	R04a, R04b, U02, U05a, U08, U19, U20b, E01, E13c, C03, E30a, E31a, E34, M03		
GS 800	Greenhill	A0354	cationic surfactant
Contains	800 g/l polyoxyethylene tallow amine		
Use with	Glyphosate and a wide range of foliar acting herbicides in agriculture, horticulture, amenity and forestry. See leaflet for details of specific crop/product recommendations		
Protective clothing	A, C		
Precautions	R03c, R04a, R04b, U02, U05a, U08, U13, U19, U20b, E01, E13b, E37, C03, E31a, E34, M03, R07d		
Guide	Loveland	A0208	acidifier/drift retardant/penetrant
Contains	350 g/l modified soya lecithin, 100 g/l alkylphenyl hydroxypolyoxyethylene and 350 g/l propionic acid		
Use with	All plant growth regulators for cereals and oilseed rape, dimethoate and chlorpyrifos for wheat bulb fly control. For other uses see label or contact supplier for details		
Protective clothing	A, C		
Precautions	R04b, R04d, U02, U04a, U05a, U10, U11, U19, U20a, E01, E13c, C03, E30a, E31a, M03		
Headland Fortune	Headland	A0277	penetrant/spreader/vegetable oil/wetter
Contains	75% w/w mixed methylated fatty acid esters of seed oil and N-butanol		
Use with	Herbicides and fungicides in a wide range of crops. See label for details		
Protective clothing	A, C		
Precautions	R04e, U02, U05a, U14, U20a, E01, E13c, C03, E26, E30a, E31b, E34, R04		

Product	Supplier	Adj. No.	Type
Headland Guard 2000	Headland	A0369	extender/sticker/wetter
Contains	52% w/w synthetic latex solution		
Use with	All approved pesticides that recommend the use of a sticking, wetting or extending agent; also with all approved pesticides for use on crops not destined for human or animal consumption; also with all approved pesticides on beans, peas, edible podded peas, oilseed rape, linseed, sugar beet, cereals, maize, Brussels sprouts, potatoes, cauliflowers		
Precautions	U05a, U08, U19, U20b, E01, E13b, E37, C03, E26, E31b, E34		
Headland Intake	Headland	A0074	penetrant
Contains	450 g/l propionic acid		
Use with	All approved pesticides on any crop not intended for human or animal consumption, and with all approved pesticides on beans, peas, edible podded peas, oilseed rape, linseed, sugar beet, cereals (except triazole fungicides), maize, Brussels sprouts, potatoes, cauliflowers. See label for detailed advice on timing on these crops		
Protective clothing	A, C		
Precautions	R05b, U02, U05a, U10, U11, U14, U15, U19, E01, C03, E26, E30a, E31b, E34, R05		
Headland Rhino	Headland	A0328	wetter
Contains	85% w/w polyalkylene oxide modified heptamethyl siloxane		
Use with	Any herbicide, systemic fungicide, systemic insecticide or plant growth regulator (except any product applied in or near water) where the use of a wetting, spreading and penetrating surfactant is recommended to improve foliar coverage		
Protective clothing	A, C		
Precautions	R03a, R03c, R04d, R04e, U02, U05a, U08, U19, U20b, E01, E13c, C03, E26, E30a, E31b, E34, M03		
Hyspray	Fine	A0020	cationic surfactant
Contains	800 g/l polyethoxylated tallow amine		
Use with	All approved formulations of glyphosate		
Protective clothing	A, C		
Precautions	R03c, R04a, R04b, U02, U05a, U08, U13, U19, U20a, E01, E13b, C03, E30a, E31b, E34, R07d		
Intracrop Bla-Tex	Intracrop	A0173	extender/sticker
Contains	450 g/l synthetic latex		
Use with	Any approved pesticide where the addition of a sticker/extender is required and recommended		
Protective clothing	A, C		
Precautions	R04a, R04b, U02, U05a, U08, U19, U20a, E01, E13c, C03, E30a, E31a, M03		

Product	Supplier	Adj. No.	Type
Intracrop Green Oil	Intracrop	A0262	wetter
Contains	950 g/l refined rapeseed oil		
Use with	All approved pesticides on grassland and edible crops when used at half or less than the approved dose, and with approved pesticides up to their full approved dose on a range of specified fruit and vegetable crops		
Precautions	U08, U20b, E13c, E37, E30a, E31b		
Intracrop Non-Ionic Wetter	Intracrop	A0009	spreader/wetter
Contains	900 g/l alkyl phenol ethoxylate		
Use with	Pesticides which have a recommendation for use with a wetter or emulsified oil		
Protective clothing	A, C		
Precautions	R04a, R04b, U02, U05a, U08, U19, U20a, E01, E13c, C03, E30a, E31a, R07d		
Intracrop Rapeze	Intracrop	A0390	adjuvant/vegetable oil
Contains	842 g/l methylated rapeseed oil		
Use with	All approved pesticides on edible and non-edible crops when used at half or less than the approved dose, and with approved pesticides up to their full approved dose on a range of specified fruit and vegetable crops		
Precautions	U19, U20b, E13b, E30a, E31b		
Intracrop Rapide Beta	Intracrop	A0175	acidifier/non-ionic surfactant
Contains	350 g/l propionic acid and 100 g/l alkylphenyl-hydroxypolyoxyethylene		
Use with	Any spray where the addition of a wetter/spreader or self-emulsifying oil is required and recommended on the pesticide label		
Protective clothing	A, C		
Precautions	R05b, U02, U04a, U05a, U08, U10, U11, U10, U20a, E01, E13c, C03, E30a, E31a, E34, M03, R05		
Iona	Intracrop	A0232	spreader/wetter
Contains	921 g/l alkyl phenol ethoxylate		
Use with	Any approved pesticide for which the addition of a wetter/spreader is recommended		
Protective clothing	A, C		
Precautions	R04a, R04b, U05a, U08, U19, U20a, E01, E13c, C03, E30a, E32a, E34		
Iona Low Foam	Intracrop	A0255	spreader/wetter
Contains	921 g/l alkyl phenol ethoxylate and silicone antifoam agent		
Use with	Pesticides that have a recommendation for the use of a wetter/spreader		
Protective clothing	A, C		
Precautions	R04a, R04b, U05a, U08, U19, U20b, E01, E13c, C03, E30a, E31a, E34		
Jogral	Intracrop	A0226	cationic surfactant
Contains	800 g/l tallow amine ethoxylate		
Use with	Glyphosate		
Protective clothing	A, C		
Precautions	R03c, U08, U19, U20a, E01, E13b, C03, E28, E31b, E32a, E34, M03, R07d		

Product	Supplier	Adj. No.	Type
Katalyst	Interagro	A0450	penetrant/wetter
Contains	90% w/w alkoxylated fatty amine		
Use with	Glyphosate and a wide range of other pesticides on specified crops. See label for details		
Protective clothing	A, C		
Precautions	R03c, R04a, R04b, U02, U05a, U08, U19, U20b, E01, E13b, C03, E26, E30a, E31a, E34, M03, R03		
Klipper	Amega	A0260	spreader/wetter
Contains	600 g/l ethylene oxide condensate		
Use with	Approved salt formulations of mecoprop alone or in mixtures with 2,4-D on managed amenity turf		
Protective clothing	A, C		
Precautions	R04a, R04b, U05a, U08, U20c, E01, E15, C03, E30a, E31a, R07d		
LI-700	Loveland	A0176	acidifier/drift retardant/penetrant
Contains	350 g/l modified soya lecithin, 100 g/l alkylphenyl hydroxypolyoxyethylene and 350 g/l propionic acid		
Use with	All plant growth regulators for cereals and oilseed rape, dimethoate and chlorpyrifos for wheat bulb fly control. For other uses see label or contact supplier for details		
Protective clothing	A, C		
Precautions	R04b, R04d, U02, U04a, U05a, U10, U11, U19, U20a, E01, E13c, C03, E30a, E31a, M03		
Libsorb	Ciba Specialty	A0438	spreader/wetter
Contains	alkyl alcohol ethoxylate		
Use with	Any spray for which additional wetter is approved and recommended		
Protective clothing	A, C		
Precautions	R04a, R04b, U05a, U08, U20a, E01, E13c, C03, E30a, E31a		
Lo Dose G	Reabrook	A0349	wetter
Contains	431 g/l polyoxyethylated tallow amine		
Use with	All approved formulations of glyphosate used pre-harvest in wheat, barley and oats, post-harvest on stubbles and in non-crop areas. In addition may be used with all approved herbicides up to their full approved dose on a range of specified fruit and vegetable crops		
Protective clothing	A, C		
Precautions	R04a, R04b, U05a, U08, U13, U19, U20b, E01, E13c, E37, C03, E30a, E31a, E34, M03		
Logic	Microcide	A0288	vegetable oil
Contains	95% emulsified vegetable oil		
Use with	All approved pesticides on edible and non-edible crops for ground or aerial application		
Protective clothing	A, C		
Precautions	U19, U20b, E30a, E31b		

Product	Supplier	Adj. No.	Type
Logic Oil	Microcide	A0293	vegetable oil
Contains	95% emulsified vegetable oil		
Use with	All approved pesticides on edible and non-edible crops for ground or aerial application		
Protective clothing	A, C		
Precautions	U19, U20b, E30a, E31b		
Luxan Non-Ionic Wetter	Luxan	A0139	non-ionic surfactant/spreader/wetter
Contains	900 g/l alkyl phenol ethylene oxide condensate		
Use with	Any spray for which additional wetter/spreader is approved and recommended		
Protective clothing	A, C		
Precautions	R04a, R04b, U05a, U08, U20b, E01, E13c, C03, E30a, E31b, E34, R07d		
Mangard	Mandops	A0132	adjuvant/anti-transpirant
Contains	96% w/w di-1-p-menthene		
Use with	Glyphosate and a wide range of other herbicides, fungicides and insecticides. See label for details		
Precautions	U20c, E15, E30a, E31a		
Mixture B	Amega	A0161	non-ionic surfactant/spreader/wetter
Contains	500 g/l nonyl phenol ethylene oxide condensate and 500 g/l primary alcohol ethylene oxide condensate		
Use with	Approved pesticides applied in amenity and forestry situations and on a range of specified arable and vegetable crops		
Protective clothing	A, C		
Precautions	R03a, R03c, R04a, R04b, U05a, U08, U20b, E01, E13c, E37, C03, E30a, E31a, E34, M03		
Nettle	Interagro	A0466	mineral oil
Contains	60% w/w mineral oil		
Use with	Cereal graminicides and a wide range of other pesticides that have a label recommendation for use with authorised adjuvant oils on specified crops. See label for details		
Protective clothing	A		
Precautions	R04b, R04g, U02, U05a, U08, U20b, E01, E13c, C03, E26, E30a, E31a, M05, R03		
Newman Cropspray 11E	Loveland	A0195	mineral oil
Contains	99% highly refined mineral oil		
Use with	Approved herbicides for which the addition of an adjuvant oil is recommended		
Precautions	R04g, U19, U20a, E13c, E26, E30a, E31b, R03		

Product	Supplier	Adj. No.	Type
Newman's T-80	Loveland	A0192	spreader/wetter
Contains	78% w/w polyoxyethylene tallow amine		
Use with	Glyphosate		
Protective clothing	A, C		
Precautions	R03c, R04a, R04b, U05a, U08, U13, U19, U20a, E01, E13b, E17, E37, C03, E30a, E31a, E34, M04		
Nion	Amega	A0415	spreader/wetter
Contains	90% (900 g/l) ethylene oxide condensate		
Use with	Recommended rates of approved pesticides up to specified growth stages of a wide range of agricultural arable and horticultural crops. Use on edible crops beyond specified growth stages should only be with half recommended rates of the pesticide or less. See label for details of growth stage restrictions.		
Protective clothing	A, C		
Precautions	R04a, R04b, U05a, U08, U19, U20b, E01, E13c, E37, C03, E30a, E31a		
Nu Film 17	Intracrop	A0055	sticker/wetter
Contains	96% di-1-p-menthene		
Use with	A range of fungicides, insecticides and foliar nutrients. See label for details		
Precautions	U19, U20c, E13c, C02, E30a, E32a		
Nu Film P	Intracrop	A0039	sticker/wetter
Contains	96% poly-1-p-menthene		
Use with	Glyphosate and many other pesticides and growth regulators for which a protectant is recommended. Do not use in mixture with adjuvant oils or surfactants		
Precautions	U19, U20b, E13c, E30a, E31a		
Nufarm Cropoil	Nufarm UK	A0447	mineral oil
Contains	99% highly refined mineral oil		
Use with	Approved pesticides for use on certain specified crops (cereals, combinable break crops, field and dwarf beans, peas, oilseed rape, brassicas, potatoes, carrots, parsnips, sugar beet, fodder beet, mangels, red beet, maize, sweetcorn, onions, leeks, horticultural crops, forestry, amenity, grassland)		
Protective clothing	A, C		
Precautions	R04e, U19, U20b, E13c, E26, E30a, E31b, R04		
Nufarm Cropoil Gold	Nufarm UK	A0446	adjuvant/vegetable oil
Contains	97.5% w/w methylated rapeseed oil		
Use with	A wide range of pesticides used in agriculture, horticulture, forestry and amenity situations. See label for details of crops and dose rates.		
Protective clothing	A, H		
Precautions	R04e, U05a, U19, U20b, E01, E13c, C03, E26, E30a, E31b		

Product	Supplier	Adj. No.	Type
Output	Syngenta	A0429	mineral oil/surfactant
Contains	60% mineral oil and 40% surfactants		
Use with	Grasp		
Protective clothing	A, C		
Precautions	R04b, U05a, U08, U19, U20a, E01, E13c, C03, E30a, E31b		
Pal	Greenhill	A0396	vegetable oil
Contains	85% w/w rapeseed oil		
Use with	All approved pesticides that recommend the use of a wetter/spreader. In addition, may be use with recommended rates of approved pesticides up to specified growth stages of a wide range of agricultural arable and horticultural crops, and in non-crop situations. Use on edible crops beyond specified growth stages should only be with half recommended rates of the pesticide or less. See label for details of growth stage restrictions.		
Protective clothing	A, C		
Precautions	R04a, U05a, U08, U19, U20b, E01, E13c, C03, E30a, E31a, E34		
Pan Oasis	Pan Agriculture	A0411	vegetable oil
Contains	95% w/w methylate rapeseed oil		
Use with	A wide range of pesticides that have a label recommendation for use with authorised adjuvant oils. See label for details		
Protective clothing	A		
Precautions	R04a, U02, U05a, U08, U20b, E01, E15, C03, E26, E30a, E31a, R04		
Pan Panorama	Pan Agriculture	A0412	mineral oil
Contains	95% w/w mineral oil		
Use with	A wide range of pesticides that have a label recommendation for use with adjuvant oils. See label for details		
Protective clothing	A		
Precautions	R04b, R04g, U02, U05a, U08, U20b, E01, E13c, C03, E26, E30a, E31a, M05, R03		
Partna	Syngenta	A0428	mineral oil/surfactant
Contains	A blend of surfactants and mineral oils		
Use with	Formulations of fluazifop-P-butyl		
Protective clothing	A		
Precautions	U05a, U08, U19, U20b, E01, E13c, C03, E30a, E31c		
Perm-E8	Intracrop	A0304	penetrant/surfactant
Contains	42% w/w propionic acid		
Use with	Approved chlormequat formulations and other pesticides on a range of agricultural and horticultural crops		
Protective clothing	A, C		
Precautions	R05b, U02, U04a, U05a, U10, U11, U13, U14, U15, U19, U20b, E01, C03, E30a, E31a, M05, R05		

Product	Supplier	Adj. No.	Type
Phase II	Loveland	A0331	vegetable oil
Contains	95% w/w esterified rapeseed oil		
Use with	Pesticides approved for use in sugar beet, oilseed rape, cereals (for grass weed control), and other specified agricultural and horticultural crops		
Precautions	U19, U20b, E13c, E26, E30a, E31b		
Pin-o-Film	Intracrop	A0209	sticker/wetter
Contains	96% di-1-p-menthene		
Use with	Approved pesticides on agricultural horticultural and forestry crops. Do not use in tank mixture with adjuvant oils or surfactants		
Precautions	U19, U20b, E13c, C02, E30a, E32a		
Planet	Intracrop	A0224	non-ionic surfactant/spreader/wetter
Contains	85% alkyl polyglycol ether and fatty acid		
Use with	Any spray for which additional wetter is recommended		
Precautions	R03c, R04a, U05a, U08, U19, U20b, E01, E13c, E30a, E31a, E32a, R07d		
Prima	Loveland	A0310	mineral oil
Contains	99% highly refined mineral oil		
Use with	All pesticides on all crops when used up to 50% of maximum approved dose for that use (mixtures with Roundup must only be used for treatment of stubbles); all approved pesticides at full dose on listed crops (see label). For other uses see label or contact supplier for details		
Precautions	R04g, U19, U20b, E13c, E26, E30a, E31b, R03		
QuikSilver	Interagro	A0471	non-ionic surfactant/spreader/wetter
Contains	75% w/w nonylphenol ethylene oxide condensate + fatty acids		
Use with	Pesticides that have a label recommendation for the use of an authorised non-ionic spreader or wetter		
Protective clothing	A, C		
Precautions	R03c, R04a, R04b, U02, U05a, U08, U11, U19, U20a, E01, E13c, C03, E26, E30a, E34, M03, R03		
Ranger	Loveland	A0296	acidifier/drift retardant/penetrant
Contains	350 g/l modified soya lecithin, 100 g/l alkylphenyl hydroxypolyoxyethylene and 350 g/l propionic acid		
Use with	All plant growth regulators for cereals and oilseed rape, dimethoate and chlorpyrifos for wheat bulb fly control. For other uses see label or contact supplier for details		
Protective clothing	A, C		
Precautions	R04b, R04d, U02, U04a, U05a, U10, U11, U19, U20a, E01, E13c, C03, E30a, E31a, M03		
Rapide	Intracrop	A0116	penetrant/surfactant
Contains	40% propionic acid		
Use with	A wide range of pesticides and growth regulators, especially chlormequat		
Protective clothing	A, C		
Precautions	R04a, R04b, R05b, U02, U05a, U08, U19, U20a, E01, C03, E30a, E31a, R05		

Product	Supplier	Adj. No.	Type
Ryda	Interagro	A0168	cationic surfactant/wetter
Contains	800 g/l polyethoxylated tallow amine		
Use with	Glyphosate		
Protective clothing	A, C		
Precautions	R03b, R03c, R04d, R14b, U02, U05a, U08, U13, U19, U20a, E01, E37, C03, E26, E30a, E31b, E34, M03, R03, R07d, R14		
SAS 90	Intracrop	A0311	spreader/wetter
Contains	100% polyoxyethylen-alpha-methyl-omega-[3-(1,3,3,3-tetramethyl-3-trimethylsiloxy)-disiloxanyl] propylether		
Use with	A wide range of pesticides on specified crops in agriculture and horticulture, and on amenity vegetation and land not intended for cropping		
Precautions	U02, U08, U19, U20b, E13c, E30a, E31a, E34		
Scout	Loveland	A0207	acidifier/drift retardant/non-ionic surfactant/penetrant
Contains	350 g/l modified soya lecithin, 100 g/l alkylphenyl hydroxypolyoxyethylene and 350 g/l propionic acid		
Use with	All plant growth regulators for cereals and oilseed rape, dimethoate and chlorpyrifos for wheat bulb fly control. For other uses see label or contact supplier for details		
Protective clothing	A, C		
Precautions	R04b, R04d, U02, U04a, U05a, U10, U11, U19, U20b, E01, E13c, C03, E30a, E31a, M03		
Signal	Intracrop	A0308	extender/sticker
Contains	450 g/l synthetic latex		
Use with	All approved pesticides where addition of a sticker/extender is required and recommended		
Protective clothing	A, C		
Precautions	R04a, R04b, U02, U05a, U08, U19, U20b, E01, E13c, C03, E30a, E31a, M03		
Siltex	Intracrop	A0398	spreader/sticker/wetter
Contains	27% w/w polyoxyethylen-alpha-methyl-omega-[3-(1,3,3,3-tetramethyl-3-trimethylsiloxy)-disiloxanyl] propylether and 7.3% w/w cocoiminodipropionate		
Use with	Approved pesticides used at full dose on a range of specified fruit and vegetable crops, and on managed amenity turf		
Protective clothing	A, C		
Precautions	U05a, U08, U20c, E01, E13c, E37, C03, E30a, E31a		
Silwet L-77	Loveland	A0193	drift retardant/spreader/wetter
Contains	80% w/w polyalkylene oxide modified heptamethyltrisiloxane and a maximum of 20 % w/w allyloxypolyethylene glycol methyl ether		
Use with	All approved fungicides on winter and spring sown cereals; all approved pesticides applied at 50% or less of their full approved dose. A wide range of other uses. See label or contact supplier for details		
Protective clothing	A, C		
Precautions	U02, U05a, U08, U19, U20a, E01, E13c, C03, E30a, E31a, E34, M03		

Product	Supplier	Adj. No.	Type
Slippa	Interagro	A0206	spreader/wetter
Contains	655 g/l polyalkyleneoxide modified heptamethyltrisiloxane + non-ionic wetters		
Use with	Cereal fungicides and a wide range of other pesticides and trace elements on specified crops		
Protective clothing	A, C		
Precautions	R03a, R03c, R04b, R04d, R04e, U02, U05a, U08, U11, U19, U20a, E01, E13c, E37, C03, E26, E30a, E31a, E34, M03, R03		
SM99	Loveland	A0134	mineral oil
Contains	99% w/w highly refined paraffinic oil		
Use with	Approved herbicides for which the addition of an adjuvant oil is recommended		
Precautions	R04g, U19, U20a, E13c, E15, E30a, E31b, R03		
Solar	Intracrop	A0225	activator/non-ionic surfactant/spreader
Contains	75% polypropoxypropanol and 15% alkyl polyglycol ether		
Use with	Foliar applied plant growth regulators		
Protective clothing	A, C		
Precautions	R04a, R04b, U04a, U05a, U08, U20a, E01, E13c, C03, E30a, E31b, E32a, E34, M03		
Spartan	Interagro	A0375	buffering agent/non-ionic surfactant/ penetrant/wetter
Contains	500 g/l alkoxylated fatty amine + 400 g/l polyoxyethylene monolaurate		
Use with	Cereal growth regulators, cereal herbicides, oilseed rape fungicides and a wide range of other pesticides on specified crops. See label for details		
Protective clothing	A, C		
Precautions	R03c, R04b, U02, U05a, U08, U19, U20b, E01, E13c, C03, E26, E30a, E31a, E34, M03, R03		
Sprayfast	Mandops	A0131	extender/sticker/wetter
Contains	334 g/l di-1-p-menthene and nonyl phenol ethylene oxide condensate		
Use with	Glyphosate and other pesticides, growth regulators or nutrients for which a coating agent is approved and recommended		
Precautions	U20b, E13c, E30a, E31a		
Sprayfix	Loveland	A0145	drift retardant/extender/sticker/wetter
Contains	450 g/l synthetic latex and 100 g/l alkylphenyl hydroxypolyoxyethylene		
Use with	Blight sprays on potatoes and other pesticides on a wide range of crops. See label or contact supplier for details		
Protective clothing	A, C		
Precautions	R04a, R04b, U02, U05a, U08, U19, U20a, E01, E13c, C03, E30a, E31a, M03		

Product	Supplier	Adj. No.	Type
Spraygard	Mandops	A0394	anti-transpirant/extender/sticker/wetter
Contains	400 g/l di-1-p-menthene		
Use with	Approved pesticides on all edible crops (except herbicides on peas) and as an anti-transpirant on vegetables before transplanting, evergreens, deciduous trees, shrubs, bushes and on turf		
Precautions	U20b, E13c, E37, E30a, E31a, M03		
Spraymac	Loveland	A0144	acidifier/non-ionic surfactant
Contains	350 g/l propionic acid and 100 g/l alkylphenyl hydroxypolyoxyethylene		
Use with	All plant growth regulators for cereals and oilseed rape and any spray for which the addition of a wetter/spreader is recommended. See label or contact supplier for details		
Protective clothing	A, C		
Precautions	R04a, R04b, U02, U04a, U05a, U10, U11, U19, U20a, E01, E13c, C03, E30a, E32a, M03		
Sprayprover	Fine	A0238	mineral oil
Contains	95% highly refined mineral oil		
Use with	Any spray for which the addition of a mineral oil is approved and recommended		
Protective clothing	A, C		
Precautions	R04a, R04b, U05a, U08, U19, U20a, E01, E13c, C03, E30a, E31b		
Spread and Seal	Mandops	A0449	extender/sticker/wetter
Contains	334 g/l d-1-p-menthene		
Use with	All pesticides on all edible crops when used at up to 50% of their approved dose, and all approved forulations of glyphosate when used on cereals. Must not be used in cobination with herbicides applied to pea crops		
Precautions	U20c, E13c, E15, E30a, E31a		
Stamina	Interagro	A0202	penetrant/wetter
Contains	100% w/w alkoxylated fatty amine		
Use with	Glyphosate and a wide range of other pesticides on specified crops		
Protective clothing	A, C		
Precautions	R03c, R04b, U02, U05a, U20a, E01, E13b, C03, E26, E30a, E34, M03, R03		
Super Nova	Interagro	A0364	penetrant/wetter
Contains	500 g/l alkoxylated fatty amine + 500 g/l polyoxyethylene monolaurate		
Use with	Cereal growth regulators, cereal herbicides, cereal fungicides, oilseed rape fungicides and a wide range of other pesticides on specified crops		
Precautions	U05a, E01, E13c, C03, E26, E30a, E34		
Sward	Amega	A0215	spreader/wetter
Contains	15.2% w/w polyalkylene oxide modified heptamethyltrisiloxane		
Use with	Specified pesticides in amenity situations		
Protective clothing	A, C		
Precautions	R04d, R04e, U05a, U08, U20c, E01, E13c, C03, E30a, E31a, R04		

Product	Supplier	Adj. No.	Type
T 25	Reabrook	A0352	spreader/wetter
Contains	800 g/l polyoxyethylated tallow amine		
Use with	Specified herbicides when used at half or less than their approved dose in a wide range of arable crops, forestry and amenity situations (see label for details) and also with approved formulations of diquat and glyphosate in specified situations up to the full approved dose of the herbicide		
Protective clothing	A, C		
Precautions	R03c, R04a, R04b, U02, U05a, U08, U13, U19, U20b, E01, E13b, E16e, C03, E30a, E31a, E34, M03, R07d		
Talzene	Intracrop	A0233	wetter
Contains	800 g/l polyoxyethylene tallow amine		
Use with	Glyphosate		
Protective clothing	A, C		
Precautions	R03c, R04a, R04b, U05a, U08, U13, U19, U20b, E01, E13c, C03, E30a, E31a, E34, M03		
Tech E	Greenhill	A0346	wetter
Contains	471 g/l polyethoxylated tallow amine		
Use with	All approved formulations of glyphosate in stubble and pre-harvest treatments of cereals and in non-crop areas; also with approved foliar acting herbicides on a wide range of agricultural and horticultural crops up to specified growth stages. Use on edible crops beyond specified growth stages should only be with half recommended rates of the pesticide or less. See label for details of growth stage restrictions.		
Protective clothing	A, C		
Precautions	R04a, R04b, U05a, U08, U13, U19, U20b, E01, E13c, E37, C03, E31a, E34, M03		
Tech G	Greenhill	A0345	wetter
Contains	431 g/l polyethoxylated tallow amine		
Use with	All approved formulations of glyphosate in stubble and pre-harvest treatments of cereals and in non-crop areas; also with approved foliar acting herbicides on a wide range of agricultural and horticultural crops up to specified growth stages. Use on edible crops beyond specified growth stages should only be with half recommended rates of the pesticide or less. See label for details of growth stage restrictions.		
Protective clothing	A, C		
Precautions	R04a, R04b, U05a, U08, U13, U19, U20b, E01, E13c, E37, C03, E31a, E34, M03		
Toil	Interagro	A0248	vegetable oil
Contains	95% w/w methylated rapeseed oil		
Use with	Sugar beet herbicides, oilseed rape herbicides, cereal graminicides and a wide range of other pesticides that have a label recommendation for use with authorised adjuvant oils on specified crops. See label for details		
Protective clothing	A		
Precautions	U05a, E01, E15, C03, E26, E30a, E31a		

Product	Supplier	Adj. No.	Type
Torpedo	Loveland	A0336	penetrant/wetter
Contains	400 g/l alkoxylated tallow amine, 375 g/l alkylphenylhydroxypolyoxyethylene and 75 g/l natural fatty acids		
Use with	Sulfonylurea herbicides and their approved tank mixes, cereal fungicides, oilseed rape fungicides and a wide range of other pesticides on specified crops		
Protective clothing	A, C		
Precautions	R03c, R04a, R04b, R04d, U02, U04a, U05a, U10, U11, U19, U20b, E01, E13b, E17, E37, C03, E30a, E31a, E34, M03		
Torpedo II	Loveland	A0393	penetrant/wetter
Contains	215.5 g/litre alkoxylated tallow amine, 375 g/litre alkylphenyl hydroxypolyoxyethylene and 75 g/litre natural fatty acids		
Use with	All pesticides on edible and non-edible crops when used at half or less than the full approved dose and with products up to their full approved dose on a wide range of specified fruit and vegetable crops.		
Protective clothing	A, C		
Precautions	R04a, R04b, U05a, U10, U11, U12, U20c, E01, E13b, E37, C03, E30a, E31a		
Tripart Acer	Tripart	A0097	acidifier/penetrant
Contains	750 g/l soyal phospholipids		
Use with	Chlormequat and a wide range of systemic pesticides and trace elements. Contact supplier for details		
Protective clothing	A, C		
Precautions	R04a, R04b, U02, U05a, U08, U19, U20b, E01, E13c, C03, E30a, E31a, M03		
Tripart Lentus	Tripart	A0117	extender/sticker
Contains	450 g/l synthetic latex		
Use with	A wide range of contact herbicides, fungicides and insecticides. Contact supplier for further details		
Protective clothing	A, C		
Precautions	R04a, R04b, U05a, U08, U19, U20a, E01, E13c, C03, E30a, E31a		
Tripart Minax	Tripart	A0108	non-ionic surfactant/spreader/wetter
Contains	900 g/l alkyl alcohol ethoxylate		
Use with	Any spray for which additional wetter or spreader is approved and recommended		
Precautions	E30a, E31a		
Tripart Orbis	Tripart	A0204	mineral oil
Contains	99% w/w highly refined paraffinic oil		
Use with	A wide range of approved pesticides on sugar beet, cereals, oilseed rape, maize, sweetcorn, fodder beet, mangels and red beet (including aerial use where recommended). See label for details		
Precautions	U19, U20a, E07, E15, E26, E30a, E31b		

Product	Supplier	Adj. No.	Type
Wayfarer	Certis	A0045	cationic surfactant/spreader/wetter
Contains	80% tallow amine ethoxylate		
Use with	Glyphosate in agriculture		
Protective clothing	A, C		
Precautions	R04a, R04b, U05a, U08, U20b, E01, E15, C03, E30a, E31a, R07d		
Wetta Plus	Interagro	A0480	non-ionic surfactant/spreader/wetter
Contains	75% w/w nonylphenol ethylene oxide condensate		
Use with	Pesticides that recommend the use of an authorised non-ionic spreader or wetter		
Protective clothing	A, C		
Precautions	R03c, R04a, R04b, U02, U05a, U08, U11, U19, U20a, E01, E13c, C03, E26, E30a, E34, M03, R03		

SECTION 3
CROP/PEST GUIDE

Crop/Pest Guide Index

Important Note: The Crop/Pest Guide Index refers to pages on which the subject can be located.

Crop/Pest Guide

Important note: For convenience, some crops and pests or targets have been brought together into generic groups in this guide, e.g. 'cereals', 'annual grasses'. It is essential to check the profile entry in Section 4 *and* the label to ensure that a product is approved for a specific crop/pest combination, e.g. winter wheat/blackgrass.

Arable and vegetable crops

Brassicas - Brassica seed crops

Diseases	Alternaria	iprodione
	Botrytis	iprodione
Weeds	Broad-leaved weeds	carbetamide, propyzamide, trifluralin (*off-label*)
	Crops as weeds	carbetamide
	Grass weeds	carbetamide, propyzamide, trifluralin (*off-label*)

Brassicas - Brassicas, general

Diseases	Alternaria	iprodione (*seed treatment*)
	Downy mildew	chlorothalonil
	Ring spot	chlorothalonil
Pests	Aphids	chlorpyrifos
	Caterpillars	chlorpyrifos
	Cutworms	chlorpyrifos
	Flies	chlorpyrifos
	Leatherjackets	chlorpyrifos

Brassicas - Fodder brassicas

Diseases	Alternaria	iprodione, iprodione (*seed treatment*)
	Black rot	copper oxychloride (*off-label*)
	Botrytis	chlorothalonil
	Damping off	fosetyl aluminium (*off label*)
	Downy mildew	chlorothalonil, fosetyl-aluminium (*off-label*), fosetyl-aluminium (*off-label - seedlings*)
	Ring spot	chlorothalonil, tebuconazole (*off-label*)
	Spear rot	copper oxychloride (*off-label*)
	White blister	boscalid + pyraclostrobin (*off-label*)
Pests	Aphids	carbosulfan, deltamethrin + pirimicarb, imidacloprid (*off-label - seed treatment*), nicotine, pirimicarb
	Beetles	alpha-cypermethrin, carbosulfan, deltamethrin
	Caterpillars	alpha-cypermethrin, cypermethrin, deltamethrin, deltamethrin + pirimicarb, nicotine
	Flies	carbosulfan, chlorpyrifos (*off-label*)
	Leaf miners	nicotine
	Pests, miscellaneous	dimethoate (*off-label*)
	Weevils	carbosulfan

Weeds	Broad-leaved weeds	chlorthal-dimethyl, chlorthal-dimethyl + propachlor, clopyralid, cyanazine (*off-label*), propachlor, sodium monochloroacetate, trifluralin
	Crops as weeds	fluazifop-P-butyl (*stockfeed only*)
	Grass weeds	chlorthal-dimethyl + propachlor, fluazifop-P-butyl (*off-label*), fluazifop-P-butyl (*stockfeed only*), propachlor, trifluralin
	Mayweeds	clopyralid

Brassicas - Leaf and flowerhead brassicas

Diseases	Alternaria	boscalid + pyraclostrobin, chlorothalonil, chlorothalonil + metalaxyl-M (*moderate control*), difenoconazole, tebuconazole
	Black rot	azoxystrobin (*off-label*), copper oxychloride (*off-label*)
	Botrytis	chlorothalonil
	Damping off	fosetyl-aluminium (*off-label*), thiram (*seed treatment*)
	Damping off and foot rot	etridiazole, etridiazole (*seeds and seedlings*)
	Damping off and wirestem	tolclofos-methyl
	Downy mildew	chlorothalonil, chlorothalonil + metalaxyl-M, fosetyl-aluminium (*off-label*), fosetyl-aluminium (*off-label - seedlings*), propamocarb hydrochloride
	Light leaf spot	benomyl, tebuconazole
	Phytophthora	etridiazole, etridiazole (*transplants*), propamocarb hydrochloride
	Powdery mildew	tebuconazole
	Pythium	propamocarb hydrochloride
	Rhizoctonia	chlorothalonil
	Ring spot	benomyl, boscalid + pyraclostrobin, chlorothalonil, chlorothalonil + metalaxyl-M (*reduction*), difenoconazole, tebuconazole, tebuconazole (*off-label*)
	Spear rot	copper oxychloride (*off-label*)
	White blister	boscalid + pyraclostrobin, boscalid + pyraclostrobin (*off-label*), boscalid + pyraclostrobin (*qualified minor use*), chlorothalonil + metalaxyl-M, mancozeb + metalaxyl-M (*off-label*)
Pests	Aphids	bifenthrin, carbosulfan, chlorpyrifos, cypermethrin, deltamethrin + pirimicarb, dimethoate, fatty acids, imidacloprid (*off-label - seed treatment*), lambda-cyhalothrin + pirimicarb, nicotine, pirimicarb, pymetrozine (*off-label*), triazamate
	Beetles	alpha-cypermethrin, carbosulfan, deltamethrin, lambda-cyhalothrin + pirimicarb
	Birds/mammals	aluminium ammonium sulphate
	Caterpillars	alpha-cypermethrin, Bacillus thuringiensis, bifenthrin, chlorpyrifos, cypermethrin, deltamethrin, deltamethrin + pirimicarb, diflubenzuron, lambda-cyhalothrin, lambda-cyhalothrin + pirimicarb, nicotine
	Cutworms	chlorpyrifos
	Flies	carbosulfan, chlorpyrifos, chlorpyrifos (*off-label*)
	Leaf miners	nicotine
	Leatherjackets	chlorpyrifos

	Mealybugs	fatty acids
	Pests, miscellaneous	dimethoate (*off-label*)
	Scale insects	fatty acids
	Slugs/snails	methiocarb, thiodicarb
	Spider mites	fatty acids
	Weevils	carbosulfan
	Whiteflies	bifenthrin, chlorpyrifos, cypermethrin, fatty acids, lambda-cyhalothrin, lambda-cyhalothrin + pirimicarb
Weeds	Broad-leaved weeds	carbetamide, chlorthal-dimethyl, chlorthal-dimethyl + propachlor, clopyralid, cyanazine (*off-label*), metazachlor, pendimethalin, propachlor, pyridate, sodium monochloroacetate, sodium monochloroacetate (*off-label*), trifluralin
	Crops as weeds	carbetamide, cycloxydim, pendimethalin, tepraloxydim
	Grass weeds	carbetamide, chlorthal-dimethyl + propachlor, cycloxydim, fluazifop-P-butyl (*off-label*), metazachlor, pendimethalin, propachlor, tepraloxydim, trifluralin
	Mayweeds	clopyralid

Brassicas - Mustard

Crop control	Pre-harvest desiccation	diquat (*off-label*), glyphosate (agricultural uses)
Diseases	Alternaria	iprodione, iprodione (*seed treatment*)
	Botrytis	iprodione
Pests	Beetles	deltamethrin
	Midges	deltamethrin
	Weevils	deltamethrin
Weeds	Broad-leaved weeds	chlorthal-dimethyl, chlorthal-dimethyl + propachlor, diquat (*off-label*), glyphosate (agricultural uses), propachlor, trifluralin
	Crops as weeds	glyphosate (agricultural uses)
	Grass weeds	chlorthal-dimethyl + propachlor, glyphosate (agricultural uses), propachlor, trifluralin
	Weeds, miscellaneous	glyphosate (agricultural uses)

Brassicas - Root brassicas

Diseases	Alternaria	fenpropimorph (*off-label*), iprodione (*seed treatment*), iprodione + thiophanate-methyl (*off-label*), tebuconazole
	Canker	iprodione (*off-label*)
	Crown rot	fenpropimorph (*off-label*), iprodione + thiophanate-methyl (*off-label*)
	Damping off	thiram (*seed treatment*)
	Downy mildew	propamocarb hydrochloride (*off-label*)
	Fungus diseases	tebuconazole (*off-label*)
	Powdery mildew	fenpropimorph (*off-label*), sulphur, tebuconazole
	Rhizoctonia	azoxystrobin (*off-label*), tolclofos-methyl (*off-label - with fleece covers*), tolclofos-methyl (*off-label - without covers*)

	White blister	mancozeb + metalaxyl-M (*off-label*), propamocarb hydrochloride (*off-label*)
Pests	Aphids	deltamethrin + pirimicarb, nicotine, pirimicarb, pirimicarb (*off-label*)
	Beetles	deltamethrin
	Caterpillars	deltamethrin, deltamethrin + pirimicarb, nicotine
	Flies	chlorpyrifos (*off-label*)
	Leaf miners	nicotine
	Pests, miscellaneous	dimethoate (*off-label*)
	Weevils	lambda-cyhalothrin (*off-label*)
Plant growth regulation	Quality/yield control	sulphur
Weeds	Broad-leaved weeds	chlorthal-dimethyl, chlorthal-dimethyl + propachlor, clopyralid, ethofumesate (*off-label*), glyphosate (agricultural uses), metazachlor, propachlor, propachlor (*off-label*), trifluralin, trifluralin (*off-label*)
	Crops as weeds	cycloxydim, fluazifop-P-butyl (*stockfeed only*), glyphosate (agricultural uses)
	Grass weeds	chlorthal-dimethyl + propachlor, cycloxydim, ethofumesate (*off-label*), fluazifop-P-butyl (*off-label*), fluazifop-P-butyl (*stockfeed only*), glyphosate (agricultural uses), metazachlor, propachlor, propachlor (*off-label*), propaquizafop, trifluralin, trifluralin (*off-label*)
	Mayweeds	clopyralid
	Weeds, miscellaneous	ethofumesate (*off-label*), glyphosate (agricultural uses)

Brassicas - Salad brassicas

Diseases	Alternaria	azoxystrobin (*off-label - for baby leaf production*), iprodione, iprodione (*seed treatment*)
	Black rot	copper oxychloride (*off-label*)
	Botrytis	chlorothalonil
	Damping off	fosetyl-aluminium (*off-label*)
	Damping off and wirestem	tolclofos-methyl
	Downy mildew	azoxystrobin (*off-label - for baby leaf production*), chlorothalonil, fosetyl-aluminium (*off-label*), fosetyl-aluminium (*off-label - seedlings*), propamocarb hydrochloride
	Phytophthora	propamocarb hydrochloride
	Pythium	propamocarb hydrochloride
	Rhizoctonia	azoxystrobin (*off-label - for baby leaf production*)
	Ring spot	chlorothalonil, tebuconazole (*off-label*)
	Spear rot	copper oxychloride (*off-label*)
	White blister	boscalid + pyraclostrobin (*off-label*), boscalid + pyraclostrobin (*off-label - baby leaf production*)
Pests	Aphids	chlorpyrifos, cypermethrin (*off-label*), deltamethrin (*off-label*), nicotine, pirimicarb, pirimicarb (*off-label - for baby leaf production*), pymetrozine (*off-label - for baby leaf production*)
	Beetles	deltamethrin (*off-label*), deltamethrin (*off-label - for baby leaf production*)

	Birds/mammals	aluminium ammonium sulphate
	Caterpillars	alpha-cypermethrin (*off-label - for baby leaf production*), chlorpyrifos, cypermethrin (*off-label*), deltamethrin (*off-label*), deltamethrin (*off-label - for baby leaf production*), nicotine
	Cutworms	chlorpyrifos
	Flies	chlorpyrifos
	Leaf miners	nicotine
	Leatherjackets	chlorpyrifos
	Pests, miscellaneous	deltamethrin (*off-label*), dimethoate (*off-label*)
	Whiteflies	chlorpyrifos
Weeds	Broad-leaved weeds	propachlor (*off-label*), propachlor (*off-label - for baby leaf production*)
	Grass weeds	fluazifop-P-butyl (*off-label*), propachlor (*off-label*)

SECTION 3

Cereals - Barley

Crop control	Pre-harvest desiccation	diquat (*stockfeed only*), glufosinate-ammonium, glyphosate (agricultural uses)
Diseases	Brown foot rot and ear blight	carboxin + thiram (*seed treatment*), fludioxonil (*seed treatment*), fuberidazole + imidacloprid + triadimenol (*seed treatment*), fuberidazole + imidacloprid + triadimenol (*seed treatment - reduction*), fuberidazole + triadimenol (*seed treatment*), guazatine (*seed treatment - reduction*), imazalil + triticonazole (*moderate control*)
	Covered smut	carboxin + thiram (*seed treatment*), fludioxonil (*seed treatment*), fuberidazole + imidacloprid + triadimenol (*seed treatment*), fuberidazole + triadimenol (*seed treatment*), imazalil + triticonazole
	Eyespot	azoxystrobin + cyproconazole (*reduction*), bromuconazole (*reduction*), carbendazim, carbendazim + flusilazole, chlorothalonil + cyproconazole, cyproconazole, cyproconazole + cyprodinil, cyproconazole + prochloraz, cyprodinil, epoxiconazole (*reduction*), epoxiconazole + fenpropimorph (*reduction*), epoxiconazole + fenpropimorph + kresoxim-methyl (*reduction*), epoxiconazole + kresoxim-methyl (*reduction*), flusilazole, maneb, prochloraz, prochloraz + propiconazole, prochloraz + tebuconazole, propiconazole (*low levels only*)
	Foot rot	guazatine (*seed treatment*)
	Leaf stripe	carboxin + thiram (*seed treatment*), carboxin + thiram (*seed treatment - reduction*), fludioxonil (*seed treatment - reduction*), fuberidazole + triadimenol (*seed treatment*), guazatine (*seed treatment*), imazalil + triticonazole, imidacloprid + tebuconazole + triazoxide (*seed treatment*), tebuconazole + triazoxide (*seed treatment*)
	Loose smut	carboxin + thiram (*seed treatment - reduction*), fuberidazole + imidacloprid + triadimenol (*seed treatment*), fuberidazole + triadimenol (*seed treatment*), imazalil + triticonazole, imidacloprid + tebuconazole + triazoxide (*seed treatment*), tebuconazole + triazoxide (*seed treatment*)
	Net blotch	azoxystrobin, azoxystrobin + cyproconazole, azoxystrobin + fenpropimorph, bromuconazole,

	carboxin + thiram (*seed treatment*), chlorothalonil + cyproconazole, chlorothalonil + tetraconazole, chlorothalonil + tetraconazole (*autumn sown*), cyproconazole, cyproconazole + cyprodinil, cyproconazole + prochloraz, cyproconazole + propiconazole, cyproconazole + trifloxystrobin, cyprodinil, epoxiconazole, epoxiconazole + fenpropimorph, epoxiconazole + fenpropimorph + kresoxim-methyl, epoxiconazole + fenpropimorph + pyraclostrobin, epoxiconazole + kresoxim-methyl, epoxiconazole + kresoxim-methyl + pyraclostrobin, epoxiconazole + pyraclostrobin, famoxadone + flusilazole, fenpropimorph + flusilazole, flusilazole, fuberidazole + imidacloprid + triadimenol (*seed treatment - seed-borne only*), guazatine (*seed treatment*), imazalil + triticonazole (*moderate control*), imidacloprid + tebuconazole + triazoxide, iprodione, mancozeb, maneb, metconazole (*reduction*), picoxystrobin, prochloraz, prochloraz + tebuconazole, propiconazole, propiconazole + trifloxystrobin, pyraclostrobin, spiroxamine + tebuconazole, tebuconazole, tebuconazole + triadimenol, tebuconazole + triazoxide (*seed treatment*), trifloxystrobin
Powdery mildew	azoxystrobin, azoxystrobin + cyproconazole, azoxystrobin + fenpropimorph, bromuconazole, carbendazim + flusilazole, chlorothalonil + cyproconazole, chlorothalonil + tetraconazole, chlorothalonil + tetraconazole (*autumn sown*), cyproconazole, cyproconazole + cyprodinil, cyproconazole + prochloraz, cyproconazole + propiconazole, cyproconazole + trifloxystrobin, cyprodinil, epoxiconazole, epoxiconazole + fenpropimorph, epoxiconazole + fenpropimorph + kresoxim-methyl, epoxiconazole + fenpropimorph + pyraclostrobin, epoxiconazole + kresoxim-methyl, fenpropidin, fenpropimorph, fenpropimorph + flusilazole, fenpropimorph + kresoxim-methyl, fenpropimorph + quinoxyfen, fluquinconazole + prochloraz, flusilazole, flutriafol, fuberidazole + imidacloprid + triadimenol (*seed treatment*), fuberidazole + triadimenol (*seed treatment*), metconazole, metconazole (*moderate control*), picoxystrobin, prochloraz, prochloraz + tebuconazole, propiconazole, propiconazole + trifloxystrobin, quinoxyfen, spiroxamine, spiroxamine + tebuconazole, sulphur, tebuconazole, tebuconazole + triadimenol, tetraconazole, trifloxystrobin
Rhynchosporium	azoxystrobin, azoxystrobin + cyproconazole (*moderate control*), azoxystrobin + fenpropimorph, bromuconazole, carbendazim, chlorothalonil, chlorothalonil + cyproconazole, chlorothalonil + tetraconazole, chlorothalonil + tetraconazole (*autumn sown*), cyproconazole, cyproconazole + cyprodinil, cyproconazole + prochloraz, cyproconazole + propiconazole, cyproconazole + trifloxystrobin, cyprodinil, epoxiconazole, epoxiconazole + fenpropimorph, epoxiconazole + fenpropimorph + kresoxim-methyl, epoxiconazole + fenpropimorph + pyraclostrobin, epoxiconazole + kresoxim-methyl, epoxiconazole + kresoxim-methyl + pyraclostrobin,

		epoxiconazole + pyraclostrobin, famoxadone + flusilazole, fenbuconazole, fenpropidin, fenpropimorph, fenpropimorph + flusilazole, fenpropimorph + kresoxim-methyl, fluquinconazole + prochloraz, flusilazole, flutriafol, mancozeb, maneb, metconazole, picoxystrobin, prochloraz, prochloraz + propiconazole, prochloraz + tebuconazole, propiconazole, propiconazole + trifloxystrobin, pyraclostrobin (*moderate*), spiroxamine (*reduction*), spiroxamine + tebuconazole, tebuconazole, tebuconazole + triadimenol, trifloxystrobin
	Rust	azoxystrobin, azoxystrobin + cyproconazole, azoxystrobin + fenpropimorph, bromuconazole, carbendazim + flusilazole, chlorothalonil + cyproconazole, chlorothalonil + tetraconazole, chlorothalonil + tetraconazole (*autumn sown*), cyproconazole, cyproconazole + prochloraz, cyproconazole + propiconazole, cyproconazole + trifloxystrobin, epoxiconazole, epoxiconazole + fenpropimorph, epoxiconazole + fenpropimorph + kresoxim-methyl, epoxiconazole + fenpropimorph + pyraclostrobin, epoxiconazole + kresoxim-methyl, epoxiconazole + kresoxim-methyl + pyraclostrobin, epoxiconazole + pyraclostrobin, famoxadone + flusilazole, fenbuconazole, fenpropidin, fenpropimorph, fenpropimorph + flusilazole, fluquinconazole + prochloraz, flusilazole, flutriafol, fuberidazole + imidacloprid + triadimenol (*seed treatment*), fuberidazole + triadimenol (*seed treatment*), mancozeb, maneb, metconazole, picoxystrobin, prochloraz + tebuconazole, propiconazole, propiconazole + trifloxystrobin, pyraclostrobin, spiroxamine, spiroxamine + tebuconazole, tebuconazole, tebuconazole + triadimenol, tetraconazole, trifloxystrobin
	Septoria diseases	flusilazole
	Snow mould	fludioxonil (*seed treatment*)
	Snow rot	fuberidazole + imidacloprid + triadimenol (*seed treatment - reduction*)
	Sooty moulds	mancozeb
	Take-all	silthiofam (*seed treatment*)
Pests	Aphids	alpha-cypermethrin, bifenthrin, chlorpyrifos, cypermethrin, cypermethrin (*autumn sown*), deltamethrin, deltamethrin + pirimicarb (*on ears*), esfenvalerate, fuberidazole + imidacloprid + triadimenol (*seed treatment*), imidacloprid + tebuconazole + triazoxide (*seed treatment*), lambda-cyhalothrin, lambda-cyhalothrin + pirimicarb, pirimicarb, tau-fluvalinate, zeta-cypermethrin
	Birds/mammals	aluminium ammonium sulphate
	Flies	alpha-cypermethrin, chlorpyrifos, cypermethrin, cypermethrin (*autumn sown*), deltamethrin, tefluthrin
	Leatherjackets	chlorpyrifos
	Slugs/snails	thiodicarb
	Thrips	chlorpyrifos
	Wireworms	fuberidazole + imidacloprid + triadimenol (*reduction of damage*), imidacloprid + tebuconazole + triazoxide (*reduction of damage*), tefluthrin

SECTION 3

Plant growth regulation	Growth control	2-chloroethylphosphonic acid, 2-chloroethylphosphonic acid + mepiquat chloride, chlormequat, chlormequat + 2-chloroethylphosphonic acid, chlormequat + 2-chloroethylphosphonic acid + mepiquat chloride, chlormequat with choline chloride, trinexapac-ethyl
	Quality/yield control	2-chloroethylphosphonic acid + mepiquat chloride (*low lodging situations*), chlormequat, chlormequat with choline chloride, sulphur
Weeds	Broad-leaved weeds	2,4-D, 2,4-D + MCPA, 2,4-DB + MCPA, amidosulfuron, amidosulfuron + iodosulfuron-methyl-sodium, bromoxynil, bromoxynil + diflufenican + ioxynil, bromoxynil + fluroxypyr + ioxynil, bromoxynil + ioxynil, carfentrazone-ethyl, carfentrazone-ethyl + mecoprop-P, carfentrazone-ethyl + metsulfuron-methyl, carfentrazone-ethyl + thifensulfuron-methyl, chlorotoluron, chlorotoluron + isoproturon, cinidon-ethyl, clopyralid, dicamba + MCPA + mecoprop-P, dicamba + mecoprop-P, dichlorprop-P, dichlorprop-P + MCPA + mecoprop-P, diflufenican + flurtamone, diflufenican + flurtamone + isoproturon, diflufenican + isoproturon, diflufenican + trifluralin, florasulam, florasulam + fluroxypyr, flufenacet + pendimethalin, fluroxypyr, glufosinate-ammonium, glyphosate (agricultural uses), imazamethabenz-methyl, isoproturon, isoproturon + pendimethalin, isoproturon + simazine, linuron, linuron + trifluralin, MCPA, MCPA + MCPB, MCPB, mecoprop-P, mecoprop-P + metsulfuron-methyl, metsulfuron-methyl, metsulfuron-methyl + thifensulfuron-methyl, pendimethalin, pendimethalin + picolinafen, pendimethalin + simazine, thifensulfuron-methyl + tribenuron-methyl, tribenuron-methyl, trifluralin
	Crops as weeds	amidosulfuron + iodosulfuron-methyl-sodium, diclofop-methyl + fenoxaprop-P-ethyl, diflufenican + flurtamone, diflufenican + flurtamone + isoproturon, florasulam, fluroxypyr, glyphosate (agricultural uses), imazamethabenz-methyl, pendimethalin, tralkoxydim, tralkoxydim (*autumn sown*)
	General weed control	diquat (*stockfeed only*)
	Grass weeds	chlorotoluron, chlorotoluron + isoproturon, diclofop-methyl + fenoxaprop-P-ethyl, diflufenican + flurtamone, diflufenican + flurtamone + isoproturon, diflufenican + isoproturon, diflufenican + trifluralin, flufenacet + pendimethalin, glufosinate-ammonium, glyphosate (agricultural uses), imazamethabenz-methyl, isoproturon, isoproturon (*off-label*), isoproturon + pendimethalin, isoproturon + simazine, linuron, linuron + trifluralin, pendimethalin, pendimethalin + picolinafen, pendimethalin + simazine, tralkoxydim, tralkoxydim (*autumn sown*), tralkoxydim (*autumn sown - qualified minor use*), tralkoxydim (*qualified minor use*), trifluralin
	Mayweeds	amidosulfuron + iodosulfuron-methyl-sodium, bromoxynil + diflufenican + ioxynil, carfentrazone-ethyl + thifensulfuron-methyl, clopyralid, dicamba + MCPA + mecoprop-P, dicamba + mecoprop-P, dichlorprop-P + MCPA + mecoprop-P, florasulam + fluroxypyr, mecoprop-P + metsulfuron-methyl, metsulfuron-methyl, thifensulfuron-methyl + tribenuron-methyl, tribenuron-methyl

	Polygonums	dicamba + MCPA + mecoprop-P, dicamba + mecoprop-P, metsulfuron-methyl + thifensulfuron-methyl
	Speedwells	bromoxynil + diflufenican + ioxynil, bromoxynil + fluroxypyr + ioxynil, carfentrazone-ethyl + mecoprop-P, carfentrazone-ethyl + thifensulfuron-methyl, cinidon-ethyl, flufenacet + pendimethalin, mecoprop-P + metsulfuron-methyl, metsulfuron-methyl + thifensulfuron-methyl, pendimethalin + picolinafen
	Weeds, miscellaneous	glyphosate (agricultural uses)

Cereals - Cereals, general

Pests	Leatherjackets	methiocarb (*reduction*)
	Slugs/snails	methiocarb
Weeds	Broad-leaved weeds	2,4-DB + MCPA, isoxaben (*off-label*), MCPA + MCPB

Cereals - Maize/sweetcorn

Diseases	Damping off	thiram (*seed treatment*)
Pests	Aphids	nicotine, pirimicarb, pirimicarb (*off-label*)
	Caterpillars	nicotine
	Flies	chlorpyrifos, lambda-cyhalothrin (*off-label*)
	Leaf miners	nicotine
	Slugs/snails	methiocarb
Weeds	Broad-leaved weeds	atrazine, bromoxynil, bromoxynil (*off-label*), bromoxynil + prosulfuron, clopyralid, fluroxypyr, isoxaben (*off-label - grown for game cover*), pendimethalin, pendimethalin (*off-label - grown under crop covers and mulches*), pendimethalin (*off-label - under crop covers and mulches*), pyridate, rimsulfuron, simazine
	Crops as weeds	fluroxypyr, pendimethalin, rimsulfuron
	Grass weeds	atrazine, pendimethalin, pendimethalin (*off-label - under crop covers and mulches*), simazine
	Mayweeds	bromoxynil + prosulfuron, clopyralid

Cereals - Oats

Crop control	Pre-harvest desiccation	diquat (*stockfeed only*), glyphosate (agricultural uses)
Diseases	Brown foot rot and ear blight	carboxin + thiram (*seed treatment*), fludioxonil (*seed treatment*), fuberidazole + imidacloprid + triadimenol (*seed treatment - reduction*), fuberidazole + triadimenol (*seed treatment*), guazatine (*seed treatment - reduction*)
	Covered smut	carboxin + thiram (*seed treatment*)
	Crown rust	azoxystrobin, azoxystrobin + cyproconazole, chlorothalonil + tetraconazole, cyproconazole, epoxiconazole + fenpropimorph + pyraclostrobin, epoxiconazole + kresoxim-methyl + pyraclostrobin, epoxiconazole + pyraclostrobin, fuberidazole + triadimenol (*seed treatment*), picoxystrobin, propiconazole, pyraclostrobin, tebuconazole (*reduction*), tebuconazole + triadimenol, tetraconazole

	Eyespot	epoxiconazole + fenpropimorph + kresoxim-methyl (*reduction*), epoxiconazole + kresoxim-methyl (*reduction*)
	Fusarium root rot	bitertanol + fuberidazole (*seed treatment*), bitertanol + fuberidazole + imidacloprid (*seed treatment*)
	Loose smut	carboxin + thiram (*seed treatment - reduction*), fuberidazole + imidacloprid + triadimenol (*seed treatment*), fuberidazole + triadimenol (*seed treatment*)
	Powdery mildew	azoxystrobin, azoxystrobin + cyproconazole, cyproconazole, epoxiconazole, epoxiconazole + fenpropimorph, epoxiconazole + fenpropimorph + kresoxim-methyl, epoxiconazole + fenpropimorph + pyraclostrobin, epoxiconazole + kresoxim-methyl, fenpropimorph, fenpropimorph + kresoxim-methyl, fenpropimorph + quinoxyfen, fuberidazole + imidacloprid + triadimenol (*seed treatment*), picoxystrobin, propiconazole, quinoxyfen, sulphur, tebuconazole, tebuconazole + triadimenol
	Pyrenophora leaf spot	fludioxonil (*seed treatment*), fuberidazole + imidacloprid + triadimenol (*seed treatment*), fuberidazole + triadimenol (*seed treatment*), guazatine (*seed treatment*)
Pests	Aphids	bifenthrin, bitertanol + fuberidazole + imidacloprid (*seed treatment*), chlorpyrifos, cypermethrin, deltamethrin, deltamethrin + pirimicarb (*on ears*), fuberidazole + imidacloprid + triadimenol (*seed treatment*), lambda-cyhalothrin, lambda-cyhalothrin + pirimicarb, pirimicarb, zeta-cypermethrin
	Birds/mammals	aluminium ammonium sulphate
	Flies	chlorpyrifos, cypermethrin, deltamethrin
	Leatherjackets	chlorpyrifos
	Midges	chlorpyrifos
	Slugs/snails	thiodicarb
	Thrips	chlorpyrifos
	Wireworms	bitertanol + fuberidazole + imidacloprid (*reduction of damage*), fuberidazole + imidacloprid + triadimenol (*reduction of damage*), tefluthrin
Plant growth regulation	Growth control	chlormequat, chlormequat with choline chloride, trinexapac-ethyl
	Quality/yield control	sulphur
Weeds	Broad-leaved weeds	2,4-D, 2,4-D + MCPA, 2,4-DB + MCPA, amidosulfuron, bromoxynil, bromoxynil + ioxynil, carfentrazone-ethyl, carfentrazone-ethyl + flupyrsulfuron-methyl, carfentrazone-ethyl + mecoprop-P, carfentrazone-ethyl + metsulfuron-methyl, carfentrazone-ethyl + thifensulfuron-methyl, clopyralid, dicamba + MCPA + mecoprop-P, dicamba + mecoprop-P, dichlorprop-P, florasulam, florasulam + fluroxypyr, fluroxypyr, glyphosate (agricultural uses), linuron, MCPA, MCPA + MCPB, MCPB, mecoprop-P, metsulfuron-methyl, tribenuron-methyl
	Crops as weeds	florasulam, fluroxypyr, glyphosate (agricultural uses)
	General weed control	diquat (*stockfeed only*)
	Grass weeds	carfentrazone-ethyl + flupyrsulfuron-methyl, glyphosate (agricultural uses), linuron

	Mayweeds	carfentrazone-ethyl + thifensulfuron-methyl, clopyralid, dicamba + MCPA + mecoprop-P, dicamba + mecoprop-P, florasulam + fluroxypyr, metsulfuron-methyl, tribenuron-methyl
	Polygonums	dicamba + MCPA + mecoprop-P, dicamba + mecoprop-P
	Speedwells	carfentrazone-ethyl + mecoprop-P, carfentrazone-ethyl + thifensulfuron-methyl
	Weeds, miscellaneous	glyphosate (agricultural uses)

Cereals - Rye/triticale

Diseases	Blue mould	fuberidazole + triadimenol (*seed treatment*)
	Brown foot rot and ear blight	carboxin + thiram (*seed treatment*), fuberidazole + triadimenol (*seed treatment*)
	Eyespot	carbendazim, epoxiconazole + fenpropimorph + kresoxim-methyl (*reduction*), epoxiconazole + kresoxim-methyl (*reduction*), prochloraz, prochloraz + tebuconazole
	Fusarium root rot	bitertanol + fuberidazole (*seed treatment*)
	Powdery mildew	azoxystrobin, azoxystrobin + cyproconazole, cyproconazole, epoxiconazole, epoxiconazole + fenpropimorph, epoxiconazole + fenpropimorph + kresoxim-methyl, epoxiconazole + kresoxim-methyl, fenpropimorph, fenpropimorph + kresoxim-methyl, fenpropimorph + quinoxyfen, prochloraz, prochloraz + tebuconazole, propiconazole, quinoxyfen, spiroxamine, spiroxamine + tebuconazole, sulphur, tebuconazole, tebuconazole + triadimenol
	Rhynchosporium	azoxystrobin, azoxystrobin (*off-label*), azoxystrobin + cyproconazole (*moderate control*), epoxiconazole, epoxiconazole + fenpropimorph, epoxiconazole + fenpropimorph + kresoxim-methyl, epoxiconazole + kresoxim-methyl, fenpropimorph + kresoxim-methyl, prochloraz, prochloraz + tebuconazole, propiconazole, spiroxamine (*reduction*), tebuconazole
	Rust	azoxystrobin, azoxystrobin (*off-label*), azoxystrobin + cyproconazole, cyproconazole, epoxiconazole, epoxiconazole + fenpropimorph, epoxiconazole + fenpropimorph + kresoxim-methyl, epoxiconazole + kresoxim-methyl, fenpropimorph, fuberidazole + triadimenol (*seed treatment*), prochloraz + tebuconazole, propiconazole, spiroxamine, spiroxamine + tebuconazole, tebuconazole, tebuconazole + triadimenol
	Septoria diseases	epoxiconazole, epoxiconazole + fenpropimorph, epoxiconazole + fenpropimorph + kresoxim-methyl, epoxiconazole + kresoxim-methyl, fenpropimorph + kresoxim-methyl (*reduction*), prochloraz, propiconazole
Pests	Aphids	cypermethrin, dimethoate, lambda-cyhalothrin + pirimicarb, pirimicarb
	Flies	cypermethrin
	Slugs/snails	thiodicarb
Plant growth regulation	Growth control	2-chloroethylphosphonic acid, 2-chloroethylphosphonic acid + mepiquat chloride, chlormequat, chlormequat with choline chloride, trinexapac-ethyl

Weeds	Broad-leaved weeds	2,4-D, amidosulfuron, amidosulfuron + iodosulfuron-methyl-sodium, bromoxynil + diflufenican + ioxynil, bromoxynil + ioxynil, carfentrazone-ethyl, carfentrazone-ethyl + flupyrsulfuron-methyl, carfentrazone-ethyl + metsulfuron-methyl, chlorotoluron, cinidon-ethyl, clodinafop-propargyl + diflufenican, dicamba + MCPA + mecoprop-P, dicamba + mecoprop-P, diflufenican + isoproturon, diflufenican + trifluralin, fluroxypyr, iodosulfuron-methyl-sodium, isoproturon, isoproturon + pendimethalin, linuron + trifluralin, MCPA, metsulfuron-methyl, pendimethalin, tribenuron-methyl
	Crops as weeds	amidosulfuron + iodosulfuron-methyl-sodium, fluroxypyr, pendimethalin, tralkoxydim
	Grass weeds	carfentrazone-ethyl + flupyrsulfuron-methyl, chlorotoluron, clodinafop-propargyl, clodinafop-propargyl + diflufenican, diflufenican + isoproturon, iodosulfuron-methyl-sodium (*from seed*), isoproturon, isoproturon + pendimethalin, linuron + trifluralin, pendimethalin, tralkoxydim, tralkoxydim (*qualified minor use*)
	Mayweeds	amidosulfuron + iodosulfuron-methyl-sodium, bromoxynil + diflufenican + ioxynil, dicamba + MCPA + mecoprop-P, dicamba + mecoprop-P, iodosulfuron-methyl-sodium, metsulfuron-methyl, tribenuron-methyl
	Polygonums	dicamba + MCPA + mecoprop-P, dicamba + mecoprop-P
	Speedwells	bromoxynil + diflufenican + ioxynil, cinidon-ethyl, iodosulfuron-methyl-sodium

Cereals - Undersown cereals

Weeds	Broad-leaved weeds	2,4-DB (*undersown with lucerne*), 2,4-DB + MCPA, 2,4-DB + MCPA (*red or white clover*), bentazone + MCPA + MCPB, bromoxynil + ioxynil, dicamba + MCPA + mecoprop-P, dicamba + MCPA + mecoprop-P (*grass only*), dicamba + mecoprop-P, MCPA, MCPA (*grass only*), MCPA (*red clover or grass*), MCPA + MCPB, MCPB
	Mayweeds	dicamba + MCPA + mecoprop-P, dicamba + MCPA + mecoprop-P (*grass only*), dicamba + mecoprop-P
	Polygonums	2,4-DB + MCPA, dicamba + MCPA + mecoprop-P, dicamba + MCPA + mecoprop-P (*grass only*), dicamba + mecoprop-P

Cereals - Wheat

Crop control	Pre-harvest desiccation	glufosinate-ammonium, glyphosate (agricultural uses)
Diseases	Alternaria	iprodione
	Blue mould	fuberidazole + imidacloprid + triadimenol (*seed treatment*), fuberidazole + triadimenol (*seed treatment*)
	Botrytis	iprodione
	Brown foot rot and ear blight	bitertanol + fuberidazole + imidacloprid (*seed treatment - reduction*), bromuconazole, carboxin + thiram (*seed treatment*), dimoxystrobin + epoxiconazole, epoxiconazole, epoxiconazole (*reduction*),

	epoxiconazole + fenpropimorph, epoxiconazole + fenpropimorph (*reduction*), epoxiconazole + fenpropimorph + kresoxim-methyl (*reduction*), epoxiconazole + fenpropimorph + pyraclostrobin (*reduction*), epoxiconazole + kresoxim-methyl (*reduction*), fludioxonil (*seed treatment*), fuberidazole + imidacloprid + triadimenol (*seed treatment - reduction*), fuberidazole + triadimenol (*seed treatment*), guazatine (*seed treatment - reduction*), guazatine + triticonazole, metconazole (*reduction*), spiroxamine + tebuconazole, tebuconazole, tebuconazole + triadimenol
Drechslera leaf spot	dimoxystrobin + epoxiconazole
Eyespot	azoxystrobin + cyproconazole (*reduction*), bromuconazole (*reduction*), carbendazim, carbendazim + flusilazole, carbendazim + prochloraz, chlorothalonil + cyproconazole, cyproconazole, cyproconazole + cyprodinil, cyproconazole + prochloraz, cyprodinil, epoxiconazole (*reduction*), epoxiconazole + fenpropimorph (*reduction*), epoxiconazole + fenpropimorph + kresoxim-methyl (*reduction*), epoxiconazole + kresoxim-methyl (*reduction*), flusilazole, maneb, picoxystrobin (*reduction*), prochloraz, prochloraz + propiconazole, prochloraz + tebuconazole, propiconazole (*low levels only*)
Fusarium root rot	bitertanol + fuberidazole (*seed treatment*), bitertanol + fuberidazole + imidacloprid (*seed treatment*), fluquinconazole + prochloraz (*seed treatment*)
Late ear diseases	azoxystrobin, azoxystrobin + fenpropimorph, carbendazim, chlorothalonil, chlorothalonil + flutriafol, cyproconazole, cyproconazole + propiconazole, picoxystrobin, prochloraz + tebuconazole
Loose smut	bitertanol + fuberidazole (*seed treatment - reduction*), bitertanol + fuberidazole + imidacloprid (*seed treatment - reduction*), fuberidazole + imidacloprid + triadimenol (*seed treatment*), fuberidazole + triadimenol (*seed treatment*)
Powdery mildew	azoxystrobin, azoxystrobin + cyproconazole, bromuconazole, carbendazim + flusilazole, carbendazim + prochloraz, chlorothalonil + cyproconazole, chlorothalonil + flutriafol, chlorothalonil + tetraconazole, chlorothalonil + tetraconazole (*autumn sown*), cyproconazole, cyproconazole + cyprodinil, cyproconazole + prochloraz, cyproconazole + trifloxystrobin, cyprodinil, epoxiconazole, epoxiconazole + fenpropimorph, fenbuconazole (*reduction*), fenpropidin, fenpropimorph, fenpropimorph + flusilazole, fenpropimorph + quinoxyfen, fluquinconazole, fluquinconazole + prochloraz (*moderate control*), flusilazole, flutriafol, fuberidazole + imidacloprid + triadimenol (*seed treatment*), fuberidazole + triadimenol (*seed treatment*), maneb, metconazole (*moderate control*), prochloraz, prochloraz (*protection*), prochloraz + tebuconazole, propiconazole, quinoxyfen, spiroxamine, spiroxamine + tebuconazole, sulphur, tebuconazole, tebuconazole + triadimenol, tetraconazole
Rust	azoxystrobin, azoxystrobin + cyproconazole, azoxystrobin + fenpropimorph, bromuconazole, carbendazim + flusilazole, chlorothalonil + cyproconazole, chlorothalonil + flutriafol, chlorothalonil

SECTION 3

	+ tetraconazole, chlorothalonil + tetraconazole (*autumn sown*), cyproconazole, cyproconazole + cyprodinil, cyproconazole + prochloraz, cyproconazole + propiconazole, cyproconazole + trifloxystrobin, difenoconazole, dimoxystrobin + epoxiconazole, epoxiconazole, epoxiconazole + fenpropimorph, epoxiconazole + fenpropimorph + kresoxim-methyl, epoxiconazole + fenpropimorph + pyraclostrobin, epoxiconazole + kresoxim-methyl, epoxiconazole + kresoxim-methyl + pyraclostrobin, epoxiconazole + pyraclostrobin, famoxadone + flusilazole, fenbuconazole, fenpropidin, fenpropimorph, fenpropimorph + flusilazole, fluquinconazole, fluquinconazole (*seed treatment*), fluquinconazole + prochloraz, flusilazole, flutriafol, fuberidazole + imidacloprid + triadimenol (*seed treatment*), fuberidazole + triadimenol (*seed treatment*), mancozeb, maneb, metconazole, picoxystrobin, prochloraz + tebuconazole, propiconazole, propiconazole + trifloxystrobin, pyraclostrobin, spiroxamine, spiroxamine + tebuconazole, tebuconazole, tebuconazole + triadimenol, tetraconazole, trifloxystrobin
Septoria diseases	azoxystrobin, azoxystrobin + cyproconazole, azoxystrobin + fenpropimorph, bitertanol + fuberidazole (*seed treatment - reduction*), bromuconazole, carbendazim + flusilazole, carbendazim + prochloraz, carboxin + thiram (*seed treatment*), chlorothalonil, chlorothalonil + cyproconazole, chlorothalonil + flutriafol, chlorothalonil + mancozeb, chlorothalonil + tetraconazole, chlorothalonil + tetraconazole (*autumn sown*), cyproconazole, cyproconazole + cyprodinil, cyproconazole + prochloraz, cyproconazole + propiconazole, cyproconazole + trifloxystrobin, difenoconazole, dimoxystrobin + epoxiconazole, epoxiconazole, epoxiconazole + fenpropimorph, epoxiconazole + fenpropimorph + kresoxim-methyl, epoxiconazole + fenpropimorph + pyraclostrobin, epoxiconazole + kresoxim-methyl, epoxiconazole + kresoxim-methyl + pyraclostrobin, epoxiconazole + pyraclostrobin, famoxadone + flusilazole, fenbuconazole, fenpropidin, fenpropimorph + flusilazole, fenpropimorph + kresoxim-methyl (*reduction*), fluquinconazole, fluquinconazole (*seed treatment*), fluquinconazole + prochloraz, fluquinconazole + prochloraz (*seed treatment*), flusilazole, flutriafol, fuberidazole + imidacloprid + triadimenol (*seed treatment*), fuberidazole + imidacloprid + triadimenol (*seed treatment - reduction*), fuberidazole + triadimenol (*seed treatment*), guazatine (*seed treatment*), iprodione, mancozeb, maneb, metconazole, picoxystrobin, prochloraz, prochloraz + propiconazole, prochloraz + tebuconazole, propiconazole, propiconazole + trifloxystrobin, pyraclostrobin, spiroxamine + tebuconazole, tebuconazole, tebuconazole + triadimenol, tetraconazole, trifloxystrobin
Snow mould	fludioxonil (*seed treatment*)
Sooty moulds	bromuconazole, carbendazim, chlorothalonil + tetraconazole, epoxiconazole (*reduction*), epoxiconazole + fenpropimorph (*reduction*), epoxiconazole + fenpropimorph + kresoxim-methyl (*reduction*),

		epoxiconazole + kresoxim-methyl (*reduction*), mancozeb, maneb, propiconazole, spiroxamine + tebuconazole, tebuconazole, tebuconazole + triadimenol, tetraconazole
	Stinking smut	bitertanol + fuberidazole (*seed treatment*), bitertanol + fuberidazole + imidacloprid (*seed treatment*), carboxin + thiram (*seed treatment*), fludioxonil (*seed treatment*), fluquinconazole (*seed treatment*), fluquinconazole + prochloraz (*seed treatment*), fuberidazole + imidacloprid + triadimenol (*seed treatment*), fuberidazole + triadimenol (*seed treatment*), guazatine + triticonazole
	Take-all	fluquinconazole (*seed treatment - reduction*), fluquinconazole + prochloraz (*seed treatment - reduction*), silthiofam (*seed treatment*)
Pests	Aphids	alpha-cypermethrin, bifenthrin, bitertanol + fuberidazole + imidacloprid (*seed treatment*), chlorpyrifos, cypermethrin, cypermethrin (*autumn sown*), deltamethrin, deltamethrin + pirimicarb (*on ears*), dimethoate, esfenvalerate, fuberidazole + imidacloprid + triadimenol (*seed treatment*), lambda-cyhalothrin, lambda-cyhalothrin + pirimicarb, pirimicarb, tau-fluvalinate, zeta-cypermethrin
	Birds/mammals	aluminium ammonium sulphate
	Flies	alpha-cypermethrin, chlorpyrifos, cypermethrin, cypermethrin (*autumn sown*), deltamethrin, dimethoate, lambda-cyhalothrin, tefluthrin
	Leatherjackets	chlorpyrifos
	Midges	chlorpyrifos
	Slugs/snails	thiodicarb
	Thrips	chlorpyrifos
	Wireworms	bitertanol + fuberidazole + imidacloprid (*reduction of damage*), fuberidazole + imidacloprid + triadimenol (*reduction of damage*), tefluthrin
Plant growth regulation	Growth control	2-chloroethylphosphonic acid, 2-chloroethylphosphonic acid + mepiquat chloride, chlormequat, chlormequat + 2-chloroethylphosphonic acid, chlormequat + 2-chloroethylphosphonic acid + mepiquat chloride, chlormequat + choline chloride + imazaquin, chlormequat + mepiquat chloride, chlormequat with choline chloride, chlormequat with choline chloride (*autumn sown*), fuberidazole + imidacloprid + triadimenol (*seed treatment - reduction*), trinexapac-ethyl
	Quality/yield control	chlormequat + choline chloride + imazaquin, sulphur
Weeds	Broad-leaved weeds	2,4-D, 2,4-D + MCPA, 2,4-DB + MCPA, amidosulfuron, amidosulfuron + iodosulfuron-methyl-sodium, bromoxynil, bromoxynil + diflufenican + ioxynil, bromoxynil + fluroxypyr + ioxynil, bromoxynil + ioxynil, carfentrazone-ethyl, carfentrazone-ethyl + flupyrsulfuron-methyl, carfentrazone-ethyl + mecoprop-P, carfentrazone-ethyl + metsulfuron-methyl, carfentrazone-ethyl + thifensulfuron-methyl, chlorotoluron, chlorotoluron + isoproturon, cinidon-ethyl, clodinafop-propargyl + diflufenican, clodinafop-propargyl + trifluralin, clopyralid, dicamba + MCPA + mecoprop-P, dicamba + mecoprop-P, dichlorprop-P, dichlorprop-P + MCPA + mecoprop-P, diflufenican + flurtamone, diflufenican + flurtamone + isoproturon,

SECTION 3

	diflufenican + isoproturon, diflufenican + trifluralin, florasulam, florasulam + fluroxypyr, flufenacet + pendimethalin, flupyrsulfuron-methyl, flupyrsulfuron-methyl + metsulfuron-methyl, flupyrsulfuron-methyl + thifensulfuron-methyl, fluroxypyr, glufosinate-ammonium, glyphosate (agricultural uses), imazamethabenz-methyl, iodosulfuron-methyl-sodium, iodosulfuron-methyl-sodium + mesosulfuron-methyl, isoproturon, isoproturon (*autumn sown*), isoproturon + pendimethalin, isoproturon + pendimethalin (*autumn sown*), isoproturon + simazine, linuron, linuron + trifluralin, MCPA, MCPA + MCPB, MCPB, mecoprop-P, mecoprop-P + metsulfuron-methyl, metsulfuron-methyl, metsulfuron-methyl + thifensulfuron-methyl, pendimethalin, pendimethalin + picolinafen, pendimethalin + simazine, sulfosulfuron, thifensulfuron-methyl + tribenuron-methyl, tribenuron-methyl, trifluralin
Crops as weeds	amidosulfuron + iodosulfuron-methyl-sodium, diclofop-methyl + fenoxaprop-P-ethyl, diflufenican + flurtamone, diflufenican + flurtamone + isoproturon, florasulam, fluroxypyr, glyphosate (agricultural uses), imazamethabenz-methyl, pendimethalin, tralkoxydim, tralkoxydim (*autumn sown*)
Grass weeds	amitrole (*direct drilling*), carfentrazone-ethyl + flupyrsulfuron-methyl, chlorotoluron, chlorotoluron + isoproturon, clodinafop-propargyl, clodinafop-propargyl + diflufenican, clodinafop-propargyl + trifluralin, diclofop-methyl + fenoxaprop-P-ethyl, diflufenican + flurtamone, diflufenican + flurtamone + isoproturon, diflufenican + isoproturon, diflufenican + trifluralin, fenoxaprop-P-ethyl, flufenacet + pendimethalin, flupyrsulfuron-methyl, flupyrsulfuron-methyl + metsulfuron-methyl, flupyrsulfuron-methyl + thifensulfuron-methyl, glufosinate-ammonium, glyphosate (agricultural uses), imazamethabenz-methyl, iodosulfuron-methyl-sodium (*from seed*), iodosulfuron-methyl-sodium + mesosulfuron-methyl, isoproturon, isoproturon (*autumn sown*), isoproturon + pendimethalin, isoproturon + pendimethalin (*autumn sown*), isoproturon + simazine, linuron, linuron + trifluralin, pendimethalin, pendimethalin + picolinafen, pendimethalin + simazine, propoxycarbazone-sodium, sulfosulfuron, tralkoxydim, tralkoxydim (*autumn sown*), tralkoxydim (*autumn sown - qualified minor use*), tralkoxydim (*qualified minor use*), trifluralin
Mayweeds	amidosulfuron + iodosulfuron-methyl-sodium, bromoxynil + diflufenican + ioxynil, carfentrazone-ethyl + thifensulfuron-methyl, clopyralid, dicamba + MCPA + mecoprop-P, dicamba + mecoprop-P, dichlorprop-P + MCPA + mecoprop-P, florasulam + fluroxypyr, iodosulfuron-methyl-sodium, iodosulfuron-methyl-sodium + mesosulfuron-methyl, mecoprop-P + metsulfuron-methyl, metsulfuron-methyl, sulfosulfuron, thifensulfuron-methyl + tribenuron-methyl, tribenuron-methyl
Polygonums	dicamba + MCPA + mecoprop-P, dicamba + mecoprop-P, metsulfuron-methyl + thifensulfuron-methyl
Speedwells	bromoxynil + diflufenican + ioxynil, bromoxynil + fluroxypyr + ioxynil, carfentrazone-ethyl + mecoprop-P,

		carfentrazone-ethyl + thifensulfuron-methyl, cinidon-ethyl, flufenacet + pendimethalin, iodosulfuron-methyl-sodium, mecoprop-P + metsulfuron-methyl, metsulfuron-methyl + thifensulfuron-methyl, pendimethalin + picolinafen
	Weeds, miscellaneous	glyphosate (agricultural uses)

Edible fungi - Mushrooms

Diseases	Bacterial blotch	sodium hypochlorite (commodity substance)
	Dry bubble	chlorothalonil
	Fungus diseases	formaldehyde (commodity substance) (*spray or fumigant*)
	Mycogone	carbendazim
	Trichoderma	carbendazim (*off-label*)
	Wet bubble	chlorothalonil
Pests	Flies	deltamethrin (*off-label*), diflubenzuron, methoprene

Fruiting vegetables - Aubergines

Diseases	Blight	azoxystrobin (*off-label*)
	Botrytis	azoxystrobin (*off-label*), pyrimethanil (*off-label*)
	Didymella stem rot	azoxystrobin (*off-label*)
	Phytophthora	propamocarb hydrochloride
	Powdery mildew	azoxystrobin (*off-label*), sulphur (*off-label*)
	Pythium	propamocarb hydrochloride
Pests	Aphids	nicotine, pymetrozine (*off-label*), Verticillium lecanii
	Leaf miners	deltamethrin (*off-label*), dichlorvos (*off-label*), oxamyl (*off-label*)
	Leafhoppers	nicotine
	Thrips	deltamethrin (*off-label*), dichlorvos (*off-label*), nicotine
	Whiteflies	buprofezin, nicotine, Verticillium lecanii
Weeds	Broad-leaved weeds	paraquat (*off-label - inter-row directed treatment*)
	Grass weeds	paraquat (*off-label - inter-row directed treatment*)

Fruiting vegetables - Cucurbits

Diseases	Black spot	benomyl (*off-label*)
	Botrytis	iprodione
	Damping off and foot rot	etridiazole (*seeds and seedlings*)
	Gummosis	benomyl (*off-label*)
	Phytophthora	etridiazole (*transplants*), propamocarb hydrochloride
	Powdery mildew	bupirimate, bupirimate (*off-label*), copper ammonium carbonate
	Pythium	propamocarb hydrochloride
Pests	Aphids	fatty acids, nicotine, pirimicarb, pirimicarb (*off-label*)
	Caterpillars	nicotine
	Leaf miners	nicotine
	Mealybugs	fatty acids

	Scale insects	fatty acids
	Spider mites	fatty acids
	Whiteflies	fatty acids
Weeds	Broad-leaved weeds	isoxaben (*off-label*)
	Grass weeds	propyzamide (*off-label*)

Fruiting vegetables - Peppers

Diseases	Damping off	propamocarb hydrochloride (*off-label*)
	Phytophthora	propamocarb hydrochloride
	Pythium	propamocarb hydrochloride
	Root rot	propamocarb hydrochloride (*off-label*)
Pests	Aphids	fatty acids, nicotine, pirimicarb
	Caterpillars	nicotine
	Leaf miners	nicotine
	Mealybugs	fatty acids
	Scale insects	fatty acids
	Spider mites	fatty acids
	Whiteflies	fatty acids

Fruiting vegetables - Tomatoes

Diseases	Blight	Bordeaux mixture, copper ammonium carbonate, copper oxychloride, propamocarb hydrochloride
	Botrytis	iprodione, pyrimethanil (*off-label*), thiram
	Damping off	copper oxychloride
	Damping off and foot rot	etridiazole (*seeds and seedlings*)
	Foot rot	copper oxychloride
	Phytophthora	copper oxychloride, etridiazole (*transplants*), propamocarb hydrochloride
	Pythium	propamocarb hydrochloride
	Soil-borne diseases	metam-sodium (*Jersey only*)
Pests	Aphids	fatty acids, nicotine, pirimicarb
	Caterpillars	nicotine
	Free-living nematodes	metam-sodium (*Jersey only*)
	Leaf miners	nicotine
	Mealybugs	fatty acids
	Scale insects	fatty acids
	Soil pests	metam-sodium (*Jersey only*)
	Spider mites	fatty acids
	Whiteflies	fatty acids
Plant growth regulation	Fruiting control	(2-naphthyloxy)acetic acid, 2-chloroethylphosphonic acid
Weeds	Weeds, miscellaneous	metam-sodium (*Jersey only*)

Herb crops - Herbs

Crop control	Pre-harvest desiccation	diquat (*off-label*)
Diseases	Alternaria	fenpropimorph (*off-label*)
	Botrytis	propamocarb hydrochloride (*off-label*)
	Crown rot	fenpropimorph (*off-label*)
	Downy mildew	fosetyl-aluminium (*off-label*), propamocarb hydrochloride (*off-label*)
	Fungus diseases	mancozeb + metalaxyl-M (*off-label*)
	Powdery mildew	azoxystrobin (*off-label*), fenpropimorph (*off-label*), tebuconazole (*off-label*)
	Ring spot	azoxystrobin (*off-label*)
	Rust	azoxystrobin (*off-label*), tebuconazole (*off-label*)
	Seed-borne diseases	thiram (*seed soak*)
Pests	Aphids	nicotine, pirimicarb (*off-label*), pymetrozine (*off-label*)
	Beetles	deltamethrin (*off-label*)
	Caterpillars	Bacillus thuringiensis (*off-label*), deltamethrin (*off-label*), nicotine
	Cutworms	lambda-cyhalothrin (*off-label*)
	Flies	lambda-cyhalothrin (*off-label*)
	Leaf miners	dichlorvos (*off-label*), nicotine
	Leafhoppers	deltamethrin (*off-label*)
	Thrips	dichlorvos (*off-label*), pirimicarb (*off-label*)
Weeds	Broad-leaved weeds	asulam (*off-label*), chlorpropham, chlorpropham + pentanochlor, chlorthal-dimethyl, chlorthal-dimethyl (*off-label*), clopyralid (*off-label*), diquat (*off-label*), linuron, metamitron (*off-label*), paraquat (*off-label*), pendimethalin, pendimethalin (*off-label*), pentanochlor, prometryn, prometryn (*off-label*), propachlor, propachlor (*off-label*), propachlor (*off-label - salad*), simazine (*off-label*), terbacil (*off-label*), trifluralin, trifluralin (*off-label*)
	Crops as weeds	pendimethalin, propaquizafop (*off-label*)
	Grass weeds	chlorpropham, chlorpropham + pentanochlor, linuron, paraquat (*off-label*), pendimethalin, pendimethalin (*off-label*), pentanochlor, prometryn, prometryn (*off-label*), propachlor, propachlor (*off-label*), propachlor (*off-label - salad*), propaquizafop (*off-label*), propyzamide (*off-label*), simazine (*off-label*), terbacil (*off-label*), trifluralin, trifluralin (*off-label*)
	Mayweeds	clopyralid (*off-label*)
	Polygonums	chlorpropham

Leafy vegetables - Endives

Diseases	Botrytis	iprodione (*off-label*), propamocarb hydrochloride (*off-label*)
	Downy mildew	fosetyl-aluminium (*off-label*), propamocarb hydrochloride (*off-label*)
	Powdery mildew	azoxystrobin (*off-label*)
	Ring spot	azoxystrobin (*off-label*)
	Rust	azoxystrobin (*off-label*)
	Sclerotinia	iprodione (*off-label*)

Pests	Aphids	nicotine, pirimicarb (*off-label*)
	Caterpillars	nicotine
	Leaf miners	cypermethrin (*off-label*), nicotine
	Thrips	pirimicarb (*off-label*)
Weeds	Broad-leaved weeds	pendimethalin (*off-label*), pendimethalin (*off-label - under covers*), propachlor (*off-label*), propachlor (*off-label - under crop covers*)
	Grass weeds	pendimethalin (*off-label*), pendimethalin (*off-label - under covers*), propachlor (*off-label*), propachlor (*off-label - under crop covers*)

Leafy vegetables - Lettuce

Diseases	Big vein	carbendazim (*off-label*)
	Botrytis	iprodione (*off-label*), iprodione (*outdoor crops*), propamocarb hydrochloride (*off-label*), thiram (*outdoor crops*)
	Damping off	thiram (*seed treatment*)
	Downy mildew	fosetyl-aluminium (*off-label*), mancozeb, propamocarb hydrochloride (*off-label*)
	Fungus diseases	mancozeb + metalaxyl-M (*off-label*)
	Powdery mildew	azoxystrobin (*off-label*)
	Ring spot	azoxystrobin (*off-label*)
	Rust	azoxystrobin (*off-label*)
	Sclerotinia	azoxystrobin (*off-label*), iprodione (*off-label*)
Pests	Aphids	cypermethrin, cypermethrin (*outdoor crops*), deltamethrin (*off-label*), dimethoate, fatty acids, imidacloprid (*off-label*), imidacloprid (*off-label - seed treatment*), lambda-cyhalothrin + pirimicarb, nicotine, pirimicarb (*outdoor crops*), pymetrozine (*off-label*)
	Caterpillars	Bacillus thuringiensis (*off-label*), cypermethrin, cypermethrin (*outdoor crops*), deltamethrin (*off-label*), nicotine
	Cutworms	cypermethrin, cypermethrin (*outdoor crops*), deltamethrin, lambda-cyhalothrin, lambda-cyhalothrin + pirimicarb
	Leaf miners	cypermethrin (*off-label*), nicotine
	Mealybugs	fatty acids
	Scale insects	fatty acids
	Slugs/snails	methiocarb
	Spider mites	fatty acids
	Whiteflies	fatty acids
Weeds	Broad-leaved weeds	chlorpropham, paraquat (*off-label - inter-row directed treatment*), pendimethalin (*off-label*), pendimethalin (*off-label - under covers*), propachlor (*off-label*), propachlor (*off-label - under crop covers*), propyzamide, propyzamide (*outdoor crops*), trifluralin
	Grass weeds	chlorpropham, paraquat (*off-label - inter-row directed treatment*), pendimethalin (*off-label*), pendimethalin (*off-label - under covers*), propachlor (*off-label*), propachlor (*off-label - under crop covers*), propyzamide, propyzamide (*outdoor crops*), trifluralin
	Polygonums	chlorpropham

Leafy vegetables - Spinach

Diseases	Downy mildew	fosetyl-aluminium (*off-label*), metalaxyl-M (*off-label*)
Pests	Aphids	cypermethrin (*off-label*), dimethoate, nicotine, pirimicarb (*off-label*)
	Caterpillars	cypermethrin (*off-label*), nicotine
	Leaf miners	dimethoate, nicotine
	Slugs/snails	methiocarb
Weeds	Broad-leaved weeds	chlorpropham + fenuron, clopyralid (*off-label*)
	Crops as weeds	propaquizafop (*off-label*)
	Grass weeds	chlorpropham + fenuron, propaquizafop (*off-label*)
	Mayweeds	clopyralid (*off-label*)

Leafy vegetables - Watercress

Diseases	Downy mildew	fosetyl-aluminium (*off-label - during propagation*), propamocarb hydrochloride (*off-label - during propagation*), propamocarb hydrochloride (*off-label - under protection*)
	Phytophthora	etridiazole (*off-label*), fosetyl-aluminium (*off-label - during propagation*), propamocarb hydrochloride (*off-label - during propagation*), propamocarb hydrochloride (*off-label - under protection*)
	Pythium	copper oxychloride (*off-label*), copper oxychloride (*off-label - during propagation*), etridiazole (*off-label*), fosetyl-aluminium (*off-label - during propagation*), propamocarb hydrochloride (*off-label - during propagation*), propamocarb hydrochloride (*off-label - under protection*)
	Rhizoctonia	benomyl (*off-label*), copper oxychloride (*off-label*), copper oxychloride (*off-label - during propagation*)
Pests	Beetles	deltamethrin (*off-label*)
	Caterpillars	Bacillus thuringiensis (*off-label*)
	Leafhoppers	deltamethrin (*off-label*)

Legumes - Beans (Phaseolus)

Crop control	Pre-harvest desiccation	diquat (*off-label*)
Diseases	Anthracnose	carbendazim
	Botrytis	azoxystrobin (*off-label*), carbendazim, iprodione (*off-label*), vinclozolin
	Damping off	thiram (*seed treatment*)
	Rust	tebuconazole (*off-label*)
	Sclerotinia	iprodione (*off-label*)
Pests	Aphids	fatty acids, nicotine, pirimicarb
	Caterpillars	lambda-cyhalothrin (*off-label*), nicotine
	Flies	chlorpyrifos (*off-label - harvested as a dry pulse*)
	Leaf miners	nicotine
	Mealybugs	fatty acids
	Pests, miscellaneous	nicotine (*off-label*)
	Scale insects	fatty acids

	Spider mites	fatty acids
	Whiteflies	fatty acids
Weeds	Broad-leaved weeds	bentazone, chlorpropham + fenuron (*off-label*), chlorthal-dimethyl, chlorthal-dimethyl (*off-label*), fomesafen, pendimethalin (*off-label*), simazine (*off-label*), trifluralin
	Crops as weeds	chlorthal-dimethyl (*off-label*), cycloxydim, fomesafen
	Grass weeds	cycloxydim, fluazifop-P-butyl (*off-label*), pendimethalin (*off-label*), simazine (*off-label*), trifluralin

Legumes - Beans (Soya)

Crop control	Pre-harvest desiccation	diquat (*off-label*)
Diseases	Sclerotinia	carbendazim + vinclozolin (*off-label*)
Weeds	Broad-leaved weeds	fomesafen (*off-label*), linuron (*off-label*), trifluralin (*off-label*)
	Crops as weeds	fomesafen (*off-label*)
	Grass weeds	linuron (*off-label*), trifluralin (*off-label*)

Legumes - Beans (Vicia)

Crop control	Pre-harvest desiccation	diquat (*stock or pigeon feed only*), glufosinate-ammonium, glyphosate (agricultural uses)
Diseases	Ascochyta	azoxystrobin (*off-label*), benomyl, thiabendazole + thiram (*seed treatment*)
	Chocolate spot	benomyl, chlorothalonil, chlorothalonil + cyproconazole, cyproconazole (*with chlorothalonil*), iprodione, iprodione + thiophanate-methyl, tebuconazole, vinclozolin
	Damping off	thiabendazole + thiram (*seed treatment*), thiram (*seed treatment*)
	Downy mildew	chlorothalonil + metalaxyl-M, cymoxanil + fludioxonil + metalaxyl-M (*off-label - seed treatment*), fosetyl-aluminium
	Rust	chlorothalonil + cyproconazole, cyproconazole, fenpropimorph, tebuconazole, tebuconazole (*off-label*)
Pests	Aphids	fatty acids, nicotine, pirimicarb
	Birds/mammals	aluminium ammonium sulphate
	Caterpillars	nicotine
	Leaf miners	nicotine
	Mealybugs	fatty acids
	Scale insects	fatty acids
	Spider mites	fatty acids
	Weevils	alpha-cypermethrin, cypermethrin, deltamethrin, lambda-cyhalothrin, lambda-cyhalothrin + pirimicarb, zeta-cypermethrin
	Whiteflies	fatty acids
Plant growth regulation	Quality/yield control	chlormequat with di-1-p-menthene
Weeds	Broad-leaved weeds	bentazone, carbetamide, clomazone, cyanazine + pendimethalin, fomesafen + terbutryn, fomesafen + terbutryn (*spring sown*), glufosinate-ammonium, glyphosate (agricultural uses), pendimethalin (*off-label*),

		propyzamide, simazine, terbuthylazine + terbutryn, trifluralin
	Crops as weeds	carbetamide, cycloxydim, fluazifop-P-butyl, glyphosate (agricultural uses), propyzamide, quizalofop-P-ethyl, tepraloxydim
	Grass weeds	carbetamide, cyanazine + pendimethalin, cycloxydim, fluazifop-P-butyl, fluazifop-P-butyl (*off-label*), glufosinate-ammonium, glyphosate (agricultural uses), propaquizafop, propyzamide, quizalofop-P-ethyl, simazine, tepraloxydim, terbuthylazine + terbutryn, trifluralin
	Weeds, miscellaneous	glyphosate (agricultural uses)

Legumes - Clovers

Crop control	Pre-harvest desiccation	diquat
Weeds	Broad-leaved weeds	asulam (*off-label*), carbetamide, isoxaben (*off-label*), MCPA + MCPB, MCPB, MCPB (*seed crops*), propyzamide
	Crops as weeds	carbetamide
	Grass weeds	carbetamide, propyzamide

Legumes - Forage legumes, general

Weeds	Broad-leaved weeds	2,4-DB, carbetamide, chlorpropham, MCPA + MCPB, paraquat (*off-label*), propyzamide
	Crops as weeds	carbetamide, fluazifop-P-butyl (*off-label*)
	Grass weeds	carbetamide, chlorpropham, fluazifop-P-butyl (*off-label*), paraquat (*off-label*), propyzamide
	Polygonums	chlorpropham

Legumes - Peas

Crop control	Pre-harvest desiccation	diquat, glufosinate-ammonium, glyphosate (agricultural uses), sulphuric acid (commodity substance)
Diseases	Ascochyta	azoxystrobin, chlorothalonil, cymoxanil + fludioxonil + metalaxyl-M, thiabendazole + thiram (*seed treatment*), vinclozolin
	Botrytis	chlorothalonil, chlorothalonil + cyproconazole, iprodione (*off-label*), vinclozolin
	Damping off	cymoxanil + fludioxonil + metalaxyl-M, thiabendazole + thiram (*seed treatment*), thiram (*seed treatment*)
	Downy mildew	cymoxanil + fludioxonil + metalaxyl-M, fosetyl-aluminium (*off-label*), fosetyl-aluminium (*off-label - seed treatment*)
	Mycosphaerella	chlorothalonil, vinclozolin
	Sclerotinia	iprodione (*off-label*)
Pests	Aphids	alpha-cypermethrin, alpha-cypermethrin (*reduction*), cypermethrin, deltamethrin, deltamethrin + pirimicarb, dimethoate, fatty acids, lambda-cyhalothrin, lambda-cyhalothrin + pirimicarb, nicotine, pirimicarb, triazamate, zeta-cypermethrin
	Birds/mammals	aluminium ammonium sulphate

	Caterpillars	alpha-cypermethrin, cypermethrin, deltamethrin, lambda-cyhalothrin, lambda-cyhalothrin + pirimicarb, nicotine, zeta-cypermethrin
	Leaf miners	nicotine
	Mealybugs	fatty acids
	Midges	dimethoate, lambda-cyhalothrin + pirimicarb
	Scale insects	fatty acids
	Spider mites	fatty acids
	Thrips	dimethoate
	Weevils	alpha-cypermethrin, cypermethrin, deltamethrin, lambda-cyhalothrin, lambda-cyhalothrin + pirimicarb, zeta-cypermethrin
	Whiteflies	fatty acids
Plant growth regulation	Quality/yield control	chlormequat with di-1-p-menthene
Weeds	Broad-leaved weeds	bentazone, bentazone + MCPB, bentazone + pendimethalin, clomazone, cyanazine, cyanazine + pendimethalin, fomesafen + terbutryn (*spring sown*), glufosinate-ammonium, glyphosate (agricultural uses), MCPA + MCPB, MCPB, pendimethalin, simazine (*off-label*), terbuthylazine + terbutryn, trifluralin (*off-label*)
	Crops as weeds	cycloxydim, fluazifop-P-butyl, glyphosate (agricultural uses), pendimethalin, quizalofop-P-ethyl, tepraloxydim
	Grass weeds	cyanazine, cyanazine + pendimethalin, cycloxydim, fluazifop-P-butyl, glufosinate-ammonium, glyphosate (agricultural uses), pendimethalin, propaquizafop, quizalofop-P-ethyl, simazine (*off-label*), tepraloxydim, terbuthylazine + terbutryn, trifluralin (*off-label*)
	Mayweeds	bentazone + pendimethalin
	Weeds, miscellaneous	glyphosate (agricultural uses)

Miscellaneous arable - Miscellaneous arable crops

Crop control	Pre-harvest desiccation	diquat (*off-label*)
Diseases	Botrytis	iprodione + thiophanate-methyl (*off-label*)
Pests	Slugs/snails	metaldehyde
Weeds	Broad-leaved weeds	diquat (*between row treatment*), diquat + paraquat, isoxaben (*off-label*), paraquat, paraquat (*stale seedbed/ inter-row*), trifluralin (*off-label*)
	Crops as weeds	diquat + paraquat, paraquat, paraquat (*stale seedbed/ inter-row*)
	Grass weeds	diquat + paraquat, paraquat, paraquat (*stale seedbed/ inter-row*), trifluralin (*off-label*)
	Weeds, miscellaneous	fluazifop-P-butyl

Miscellaneous arable - Miscellaneous arable situations

Diseases	Soil-borne diseases	dazomet (*soil fumigation*)
Pests	Birds/mammals	aluminium ammonium sulphate (*seed treatment*), ziram
	Free-living nematodes	dazomet (*soil fumigation*)

	Slugs/snails	metaldehyde
	Soil pests	dazomet (*soil fumigation*)
Weeds	Broad-leaved weeds	amitrole, citronella oil, diquat + paraquat, glufosinate-ammonium, glufosinate-ammonium (*uncropped*), glyphosate (agricultural uses), paraquat, paraquat (*minimum cultivation*), thifensulfuron-methyl (*in green cover*)
	Crops as weeds	amitrole (*barley stubble*), diquat + paraquat, fluazifop-P-butyl, glyphosate (agricultural uses), paraquat, paraquat (*minimum cultivation*), tepraloxydim
	Grass weeds	amitrole, diquat + paraquat, fluazifop-P-butyl, glufosinate-ammonium, glufosinate-ammonium (*uncropped*), glyphosate (agricultural uses), paraquat, paraquat (*minimum cultivation*), paraquat (*stubble treatment*), tepraloxydim
	Weeds, miscellaneous	amitrole, cycloxydim, dazomet (*soil fumigation*), diquat + paraquat, fluazifop-P-butyl, glufosinate-ammonium, glyphosate (agricultural uses), glyphosate (agricultural uses) (*wiper application*), metsulfuron-methyl, paraquat, tepraloxydim

Miscellaneous field vegetables - All vegetables

Diseases	Soil-borne diseases	dazomet (*soil fumigation*)
Pests	Aphids	natural plant extracts, nicotine, rotenone
	Capsid bugs	nicotine
	Free-living nematodes	dazomet (*soil fumigation*)
	Leafhoppers	natural plant extracts, nicotine
	Mites	natural plant extracts
	Sawflies	nicotine
	Soil pests	dazomet (*soil fumigation*)
	Thrips	natural plant extracts, nicotine
	Whiteflies	natural plant extracts
Weeds	Broad-leaved weeds	glufosinate-ammonium
	Grass weeds	glufosinate-ammonium
	Weeds, miscellaneous	ammonium sulphamate (*pre-planting*), dazomet (*soil fumigation*)

Oilseed crops - Linseed/flax

Crop control	Pre-harvest desiccation	diquat, glufosinate-ammonium, glyphosate (agricultural uses)
Diseases	Alternaria	iprodione (*seed treatment*)
	Botrytis	tebuconazole (*reduction*)
	Damping off	thiabendazole + thiram (*seed treatment*)
	Fusarium root rot	thiabendazole + thiram (*seed treatment*)
	Grey mould	thiabendazole + thiram (*seed treatment*)
	Powdery mildew	tebuconazole
	Seed-borne diseases	prochloraz (*seed treatment*)
Pests	Beetles	zeta-cypermethrin
Weeds	Broad-leaved weeds	amidosulfuron, bentazone, bromoxynil, bromoxynil + clopyralid, clopyralid, glufosinate-ammonium,

		glyphosate (agricultural uses), MCPA, metsulfuron-methyl, trifluralin, trifluralin (*off-label*)
	Crops as weeds	cycloxydim, fluazifop-P-butyl, glyphosate (agricultural uses), quizalofop-P-ethyl, tepraloxydim
	Grass weeds	cycloxydim, fluazifop-P-butyl, glufosinate-ammonium, glyphosate (agricultural uses), propaquizafop, quizalofop-P-ethyl, tepraloxydim, trifluralin, trifluralin (*off-label*)
	Mayweeds	clopyralid, metsulfuron-methyl
	Weeds, miscellaneous	glyphosate (agricultural uses)

Oilseed crops - Lupins

Diseases	Damping off	thiabendazole + thiram (*seed treatment*)
Weeds	Broad-leaved weeds	terbuthylazine + terbutryn
	Grass weeds	terbuthylazine + terbutryn

Oilseed crops - Miscellaneous oilseeds

Crop control	Pre-harvest desiccation	diquat (*off-label*), diquat (*off-label - for morphine production*), glyphosate (agricultural uses) (*off-label - for morphine production*)
Diseases	Alternaria	iprodione (*off-label - for morphine production*)
	Botrytis	iprodione (*off-label - for morphine production*)
	Fire	iprodione (*off-label - for morphine production*)
	Powdery mildew	azoxystrobin (*off-label - for morphine production*), tebuconazole (*off-label*)
	Rust	tebuconazole (*off-label*)
	Sclerotinia	iprodione (*off-label - for morphine production*)
	White blister	propiconazole (*off-label*)
Pests	Aphids	deltamethrin (*off-label*), iprodione (*off-label - for morphine production*), pirimicarb (*off-label*)
	Beetles	deltamethrin (*off-label*), lambda-cyhalothrin (*off-label - for morphine production*)
Weeds	Broad-leaved weeds	bentazone (*off-label*), clopyralid (*off-label*), metazachlor (*off-label*), trifluralin (*off-label*)
	Grass weeds	metazachlor (*off-label*), propyzamide (*off-label*), trifluralin (*off-label*)
	Mayweeds	clopyralid (*off-label*)

Oilseed crops - Oilseed rape

Crop control	Pre-harvest desiccation	diquat, glufosinate-ammonium, glyphosate (agricultural uses)
Diseases	Alternaria	carbendazim + prochloraz, carbendazim + vinclozolin (*reduction*), difenoconazole, iprodione, iprodione (*seed treatment*), iprodione + thiophanate-methyl, metconazole, prochloraz, tebuconazole, vinclozolin
	Black scurf and stem canker	difenoconazole, iprodione + thiophanate-methyl, tebuconazole

	Botrytis	carbendazim + prochloraz, carbendazim + vinclozolin (*reduction*), chlorothalonil, iprodione, iprodione + thiophanate-methyl, prochloraz, vinclozolin
	Canker	carbendazim + flusilazole (*reduction*), carbendazim + prochloraz, metconazole (*foliar disease - reduction*), metconazole (*reduction*), prochloraz, prochloraz + propiconazole, tebuconazole
	Damping off	thiram (*seed treatment*)
	Downy mildew	chlorothalonil, mancozeb
	Light leaf spot	carbendazim, carbendazim (*reduction*), carbendazim + flusilazole, carbendazim + prochloraz, carbendazim + vinclozolin (*reduction*), cyproconazole, difenoconazole, flusilazole, iprodione + thiophanate-methyl, prochloraz, prochloraz + propiconazole, propiconazole, propiconazole (*reduction*), tebuconazole
	Phoma leaf spot	carbendazim + flusilazole, cyproconazole
	Powdery mildew	sulphur
	Ring spot	tebuconazole (*reduction*)
	Sclerotinia stem rot	boscalid, carbendazim + prochloraz, carbendazim + vinclozolin, iprodione, iprodione + thiophanate-methyl, metconazole (*reduction*), prochloraz, prochloraz + tebuconazole, tebuconazole, vinclozolin
	White leaf spot	carbendazim + prochloraz, prochloraz
Pests	Aphids	deltamethrin, deltamethrin + pirimicarb, lambda-cyhalothrin, lambda-cyhalothrin + pirimicarb, nicotine (*off-label*), pirimicarb, tau-fluvalinate
	Beetles	alpha-cypermethrin, beta-cyfluthrin + imidacloprid (*seed treatment*), bifenthrin, cypermethrin, deltamethrin, lambda-cyhalothrin, lambda-cyhalothrin + pirimicarb, tau-fluvalinate, zeta-cypermethrin
	Birds/mammals	aluminium ammonium sulphate
	Midges	alpha-cypermethrin, cypermethrin, deltamethrin, lambda-cyhalothrin, lambda-cyhalothrin + pirimicarb, zeta-cypermethrin
	Slugs/snails	methiocarb, thiodicarb
	Weevils	alpha-cypermethrin, bifenthrin, cypermethrin, deltamethrin, lambda-cyhalothrin, lambda-cyhalothrin + pirimicarb, zeta-cypermethrin
Plant growth regulation	Growth control	tebuconazole
	Quality/yield control	sulphur
Weeds	Broad-leaved weeds	carbetamide, chlorthal-dimethyl, clomazone, clopyralid, clopyralid + propyzamide, cyanazine, glufosinate-ammonium, glyphosate (agricultural uses), metazachlor, metazachlor + quinmerac, napropamide, propachlor, propyzamide, pyridate (*off-label*), trifluralin
	Crops as weeds	carbetamide, cycloxydim, fluazifop-P-butyl, glyphosate (agricultural uses), propyzamide, quizalofop-P-ethyl, tepraloxydim
	Grass weeds	carbetamide, clopyralid + propyzamide, cyanazine, cycloxydim, fluazifop-P-butyl, glufosinate-ammonium, glyphosate (agricultural uses), metazachlor, metazachlor + quinmerac, napropamide, propachlor, propaquizafop, propyzamide, quizalofop-P-ethyl, tepraloxydim, trifluralin
	Mayweeds	clopyralid, clopyralid + propyzamide

	Weeds, miscellaneous	glyphosate (agricultural uses)

Oilseed crops - Sunflowers

Crop control	Pre-harvest desiccation	diquat (*off-label*)
Pests	Slugs/snails	methiocarb
Weeds	Broad-leaved weeds	diquat (*off-label*), isoxaben (*off-label*), pendimethalin, trifluralin (*off-label*)
	Crops as weeds	pendimethalin
	Grass weeds	pendimethalin, trifluralin (*off-label*)

Root and tuber crops - Beet crops

Diseases	Black leg	hymexazol (*seed treatment*)
	Botrytis	iprodione (*off-label*)
	Canker	iprodione (*off-label*)
	Powdery mildew	carbendazim + flusilazole, cyproconazole, flusilazole, propiconazole, quinoxyfen, sulphur
	Ramularia leaf spots	cyproconazole, propiconazole, propiconazole (*reduction*)
	Rhizoctonia	azoxystrobin (*off-label*), propamocarb hydrochloride (*off-label*)
	Rust	carbendazim + flusilazole, cyproconazole, cyproconazole (*off-label*), fenpropimorph (*off-label*), flusilazole, propiconazole
	Seed-borne diseases	thiram (*seed soak*)
Pests	Aphids	carbosulfan, deltamethrin + pirimicarb, dimethoate, dimethoate (*excluding Myzus persicae*), imidacloprid (*seed treatment*), lambda-cyhalothrin + pirimicarb, nicotine, oxamyl, pirimicarb, pirimicarb (*off-label*), triazamate (*including Myzus persicae*)
	Beetles	benfuracarb, carbosulfan, chlorpyrifos, deltamethrin, imidacloprid (*seed treatment*), lambda-cyhalothrin, lambda-cyhalothrin + pirimicarb, oxamyl, tefluthrin (*seed treatment*)
	Birds/mammals	aluminium ammonium sulphate
	Caterpillars	cypermethrin, lambda-cyhalothrin (*off-label*), nicotine
	Cutworms	cypermethrin, lambda-cyhalothrin, lambda-cyhalothrin (*off-label*), lambda-cyhalothrin + pirimicarb, methiocarb (*reduction*), zeta-cypermethrin
	Flies	carbosulfan, dimethoate, imidacloprid (*seed treatment*), oxamyl
	Free-living nematodes	benfuracarb, carbosulfan, oxamyl
	Leaf miners	benfuracarb, dimethoate, lambda-cyhalothrin, lambda-cyhalothrin + pirimicarb, nicotine
	Leatherjackets	chlorpyrifos, methiocarb (*reduction*)
	Millipedes	benfuracarb, carbosulfan, imidacloprid (*seed treatment*), methiocarb (*reduction*), oxamyl, tefluthrin (*seed treatment*)
	Slugs/snails	methiocarb
	Springtails	benfuracarb, carbosulfan, imidacloprid (*seed treatment*), tefluthrin (*seed treatment*)

	Symphylids	benfuracarb, carbosulfan, imidacloprid (*seed treatment*), tefluthrin (*seed treatment*)
	Wireworms	carbosulfan
Plant growth regulation	Quality/yield control	sulphur
Weeds	Broad-leaved weeds	carbetamide, chloridazon, chloridazon + chlorpropham + metamitron, chloridazon + ethofumesate, chloridazon + metamitron, chloridazon + quinmerac, chlorpropham + metamitron, clopyralid, desmedipham + ethofumesate + phenmedipham, desmedipham + phenmedipham, diquat, ethofumesate, ethofumesate + metamitron, ethofumesate + phenmedipham, glufosinate-ammonium, glyphosate (agricultural uses), lenacil, lenacil + phenmedipham, metamitron, paraquat, phenmedipham, propyzamide, trifluralin, triflusulfuron-methyl, triflusulfuron-methyl (*off-label*)
	Crops as weeds	carbetamide, cycloxydim, fluazifop-P-butyl, glyphosate (agricultural uses), glyphosate (agricultural uses) (*wiper application*), paraquat, propaquizafop (*off-label*), propyzamide, quizalofop-P-ethyl, sodium chloride (commodity substance), tepraloxydim
	Grass weeds	carbetamide, chloridazon, chloridazon + chlorpropham + metamitron, chloridazon + ethofumesate, chloridazon + metamitron, chloridazon + quinmerac, chlorpropham + metamitron, cycloxydim, desmedipham + ethofumesate + phenmedipham, ethofumesate, ethofumesate + metamitron, ethofumesate + phenmedipham, fluazifop-P-butyl, fluazifop-P-butyl (*off-label*), glufosinate-ammonium, glyphosate (agricultural uses), lenacil, metamitron, paraquat, propaquizafop, propaquizafop (*off-label*), propyzamide, quizalofop-P-ethyl, tepraloxydim, trifluralin
	Mayweeds	chloridazon + quinmerac, clopyralid
	Polygonums	sodium chloride (commodity substance)
	Speedwells	chloridazon + quinmerac
	Weeds, miscellaneous	glyphosate (agricultural uses)

Root and tuber crops - Carrots/parsnips/celeriac

Diseases	Alternaria	azoxystrobin, fenpropimorph (*off-label*), iprodione + thiophanate-methyl (*off-label*), tebuconazole
	Blight	difenoconazole (*off-label*)
	Cavity spot	metalaxyl-M
	Crown rot	fenpropimorph (*off-label*), iprodione + thiophanate-methyl (*off-label*)
	Damping off	thiabendazole + thiram (*seed treatment*), thiram (*seed treatment*)
	Powdery mildew	azoxystrobin, fenpropimorph (*off-label*), sulphur (*off-label*), tebuconazole
	Pythium	cymoxanil + fludioxonil + metalaxyl-M (*off-label*)
	Sclerotinia	azoxystrobin (*off-label*), tebuconazole
	Seed-borne diseases	thiram (*seed soak*)
	Septoria diseases	chlorothalonil (*off-label*)

Pests	Aphids	aldicarb, carbosulfan, deltamethrin (*off-label*), lambda-cyhalothrin + pirimicarb, nicotine, pirimicarb, pirimicarb (*off-label*)
	Birds/mammals	aluminium ammonium sulphate
	Caterpillars	deltamethrin (*off-label*), nicotine
	Cutworms	chlorpyrifos, cypermethrin (*off-label*), lambda-cyhalothrin, lambda-cyhalothrin (*off-label*), lambda-cyhalothrin + pirimicarb
	Flies	lambda-cyhalothrin (*off-label*), tefluthrin (*off-label*)
	Free-living nematodes	aldicarb, carbosulfan
	Leaf miners	nicotine
	Pests, miscellaneous	deltamethrin (*off-label*), dimethoate (*off-label*)
Plant growth regulation	Growth control	maleic hydrazide (*off-label*)
Weeds	Broad-leaved weeds	chlorpropham, chlorpropham + pentanochlor, clomazone, ioxynil (*off-label*), isoxaben (*off-label*), linuron, linuron (*off-label*), metoxuron, metribuzin (*off-label*), pendimethalin, pentanochlor, prometryn, trifluralin, trifluralin (*off-label*)
	Crops as weeds	cycloxydim, fluazifop-P-butyl, pendimethalin, tepraloxydim
	Grass weeds	chlorpropham, chlorpropham + pentanochlor, cycloxydim, fluazifop-P-butyl, fluazifop-P-butyl (*off-label*), linuron, metribuzin (*off-label*), pendimethalin, pentanochlor, prometryn, propaquizafop, tepraloxydim, trifluralin, trifluralin (*off-label*)
	Mayweeds	metoxuron
	Polygonums	chlorpropham

Root and tuber crops - Potatoes

Crop control	Pre-harvest desiccation	carfentrazone-ethyl, diquat, glufosinate-ammonium, sulphuric acid (commodity substance)
Diseases	Black dot	azoxystrobin (*off-label - for baby tuber production*)
	Black scurf and stem canker	flutolanil (*tuber treatment*), imazalil + pencycuron (*tuber treatment - reduction*), imazalil + thiabendazole (*tuber treatment - reduction*), pencycuron (*tuber treatment*)
	Blight	benalaxyl + mancozeb, Bordeaux mixture, chlorothalonil, chlorothalonil + mancozeb, chlorothalonil + propamocarb hydrochloride, copper oxychloride, cyazofamid, cymoxanil, cymoxanil + famoxadone, cymoxanil + mancozeb, dimethomorph + mancozeb, fenamidone + mancozeb, fluazinam, fluazinam + metalaxyl-M, mancozeb, mancozeb + metalaxyl-M, mancozeb + propamocarb hydrochloride, mancozeb + zoxamide, maneb
	Dry rot	imazalil, imazalil + thiabendazole (*tuber treatment - reduction*), thiabendazole (*tuber treatment - post-harvest*)
	Gangrene	2-aminobutane, imazalil, imazalil + thiabendazole (*tuber treatment - reduction*), thiabendazole (*tuber treatment - post-harvest*)
	Rhizoctonia	flutolanil (*tuber treatment*), imazalil + pencycuron (*tuber treatment*), iprodione (*seed treatment*), pencycuron

		(*tuber treatment*), tolclofos-methyl (*off-label - chitted seed treatment*), tolclofos-methyl (*tuber treatment*)
	Silver scurf	2-aminobutane, imazalil, imazalil + pencycuron (*tuber treatment - reduction*), imazalil + thiabendazole (*tuber treatment - reduction*), thiabendazole (*tuber treatment - post-harvest*)
	Skin spot	2-aminobutane, imazalil, imazalil + thiabendazole (*tuber treatment - reduction*), thiabendazole (*tuber treatment - post-harvest*)
Pests	Aphids	aldicarb, deltamethrin + pirimicarb, dimethoate (*excluding Myzus persicae*), lambda-cyhalothrin, lambda-cyhalothrin + pirimicarb, nicotine, oxamyl, pirimicarb, pymetrozine
	Beetles	deltamethrin (*off-label*)
	Caterpillars	cypermethrin, nicotine
	Cutworms	chlorpyrifos, cypermethrin, lambda-cyhalothrin + pirimicarb, zeta-cypermethrin
	Cyst nematodes	1,3-dichloropropene, ethoprophos, fosthiazate, oxamyl
	Free-living nematodes	1,3-dichloropropene, aldicarb, oxamyl
	Leaf miners	nicotine
	Leatherjackets	methiocarb (*reduction*)
	Slugs/snails	methiocarb, thiodicarb
	Wireworms	ethoprophos, fosthiazate (*reduction*)
Plant growth regulation	Growth control	chlorpropham, chlorpropham (*fog*), maleic hydrazide
	Plant growth regulation, miscellaneous	maleic hydrazide
Weeds	Broad-leaved weeds	bentazone, carfentrazone-ethyl, clomazone, diquat, diquat + paraquat, flufenacet + metribuzin, glufosinate-ammonium, linuron, metribuzin, paraquat, pendimethalin, rimsulfuron
	Crops as weeds	carfentrazone-ethyl, cycloxydim, diquat + paraquat, metribuzin, paraquat, pendimethalin, rimsulfuron
	Grass weeds	cycloxydim, diquat + paraquat, flufenacet + metribuzin, glufosinate-ammonium, linuron, metribuzin, paraquat, pendimethalin, propaquizafop
	Weeds, miscellaneous	diquat

Stem and bulb vegetables - Asparagus

Crop control	Pre-harvest desiccation	diquat (*off-label*)
Diseases	Rust	difenoconazole (*off-label*)
	Stemphylium	iprodione (*off-label*)
Pests	Aphids	nicotine
	Beetles	cypermethrin (*off-label*)
	Caterpillars	nicotine
	Leaf miners	nicotine
Weeds	Broad-leaved weeds	clopyralid (*off-label*), diquat (*off-label*), diuron (*off-label*), isoxaben (*off-label*), metamitron (*off-label*), metribuzin (*off-label*), simazine, simazine (*off-label*)

SECTION 3

	Grass weeds	diuron (*off-label*), metamitron (*off-label*), metribuzin (*off-label*), simazine, simazine (*off-label*)
	Weeds, miscellaneous	diuron (*off-label*), glyphosate (fruit, horticulture, forestry, amenity etc.) (*off-label*), metribuzin (*off-label*)

Stem and bulb vegetables - Celery/chicory

Diseases	Blight	difenoconazole (*off-label*)
	Botrytis	azoxystrobin (*off-label*), iprodione (*off-label*)
	Damping off and foot rot	etridiazole (*seeds and seedlings*)
	Phytophthora	etridiazole (*transplants*), fosetyl-aluminium (*off-label - for forcing*), fosetyl-aluminium (*off-label - in forcing sheds*)
	Powdery mildew	azoxystrobin (*off-label*)
	Rhizoctonia	azoxystrobin (*off-label*)
	Ring spot	azoxystrobin (*off-label*)
	Rust	azoxystrobin (*off-label*)
	Sclerotinia	azoxystrobin (*off-label*), iprodione (*off-label*)
	Seed-borne diseases	thiram (*seed soak*)
	Septoria diseases	Bordeaux mixture, chlorothalonil (*qualified minor use*), copper ammonium carbonate, copper oxychloride
Pests	Aphids	deltamethrin (*off-label*), nicotine, pirimicarb (*off-label - for forcing*)
	Beetles	deltamethrin (*off-label*)
	Caterpillars	deltamethrin (*off-label*), lambda-cyhalothrin (*off-label*), nicotine
	Cutworms	lambda-cyhalothrin (*off-label - for forcing*)
	Flies	lambda-cyhalothrin (*off-label*)
	Leaf miners	cypermethrin (*off-label*), nicotine
	Leafhoppers	deltamethrin (*off-label*)
	Pests, miscellaneous	deltamethrin (*off-label*)
Weeds	Broad-leaved weeds	chlorpropham, chlorpropham + pentanochlor, chlorthal-dimethyl (*off-label*), paraquat (*off-label*), pendimethalin (*off-label*), pendimethalin (*off-label - under covers*), pentanochlor, prometryn, propachlor (*off-label*), propachlor (*off-label - under crop covers*)
	Crops as weeds	propaquizafop (*off-label*)
	Grass weeds	chlorpropham, chlorpropham + pentanochlor, paraquat (*off-label*), pendimethalin (*off-label*), pendimethalin (*off-label - under covers*), pentanochlor, prometryn, propachlor (*off-label*), propachlor (*off-label - under crop covers*), propaquizafop (*off-label*), propyzamide (*off-label*)
	Polygonums	chlorpropham

Stem and bulb vegetables - Globe artichokes/cardoons

Pests	Aphids	nicotine
	Caterpillars	nicotine
	Leaf miners	nicotine

Stem and bulb vegetables - Onions/leeks/garlic

Diseases	Bacterial blight	copper oxychloride (*off-label*)
	Botrytis	chlorothalonil, iprodione, propamocarb hydrochloride (*off-label*), thiabendazole + thiram (*off-label - seed*)
	Collar rot	iprodione
	Damping off	thiram (*seed treatment*)
	Downy mildew	azoxystrobin, azoxystrobin (*off-label*), chlorothalonil + metalaxyl-M (*qualified minor use*), dimethomorph + mancozeb (*off-label*), fosetyl-aluminium (*off-label*), mancozeb (*off-label*), propamocarb hydrochloride (*off-label*)
	Fungus diseases	mancozeb + metalaxyl-M (*off-label*)
	Fusarium diseases	thiabendazole + thiram (*off-label - seed*)
	Phytophthora	propamocarb hydrochloride
	Powdery mildew	azoxystrobin (*off-label*), tebuconazole (*off-label*)
	Purple blotch	azoxystrobin
	Pythium	propamocarb hydrochloride
	Rhynchosporium	chlorothalonil, propiconazole (*off-label*)
	Ring spot	azoxystrobin (*off-label*)
	Rust	azoxystrobin, azoxystrobin (*off-label*), cyproconazole, fenpropimorph, propiconazole, propiconazole (*off-label*), propiconazole (*qualified minor use*), tebuconazole, tebuconazole (*off-label*)
	Storage rots	copper oxychloride (*off-label*)
	White rot	tebuconazole (*off-label*)
	White tip	chlorothalonil + metalaxyl-M (*qualified minor use*), maneb (*off-label*), propamocarb hydrochloride (*off-label*)
Pests	Aphids	deltamethrin (*off-label*), nicotine, pymetrozine (*off-label*)
	Caterpillars	Bacillus thuringiensis (*off-label*), deltamethrin (*off-label*), nicotine
	Cutworms	chlorpyrifos
	Flies	tefluthrin (*off-label*)
	Leaf miners	dichlorvos (*off-label*), nicotine
	Pests, miscellaneous	deltamethrin (*off-label*), dimethoate (*off-label*), dimethoate (*off-label - seedlings*)
	Stem nematodes	aldicarb, oxamyl (*off-label*)
	Thrips	dichlorvos (*off-label*)
Plant growth regulation	Growth control	maleic hydrazide
Weeds	Broad leaved weeds	bentazone (*off-label*), chloridazon (*off-label*), chloridazon + propachlor, chloridazon + propachlor (*off-label - post-emergence*), chlorpropham, chlorthal-dimethyl, chlorthal-dimethyl (*off-label*), chlorthal-dimethyl + propachlor, clopyralid, clopyralid (*off-label*), cyanazine, cyanazine (*fen soils only*), ethofumesate (*off-label*), fluroxypyr (*off-label*), glyphosate (agricultural uses), ioxynil, ioxynil (*off-label*), linuron (*off-label*), metamitron (*off-label*), paraquat (*off-label*), pendimethalin, pendimethalin (*off-label*), pendimethalin (*off-label - drilled and transplanted*), pendimethalin (*off-label - drilled and transplanted onion sets*), prometryn, prometryn (*off-label*), propachlor, propachlor (*off-label*), pyridate, simazine (*off-label*), sodium

SECTION 3

		monochloroacetate, terbacil (*off-label*), trifluralin (*off-label*)
	Crops as weeds	cycloxydim, fluazifop-P-butyl, fluroxypyr (*off-label*), glyphosate (agricultural uses), pendimethalin, propaquizafop (*off-label*), tepraloxydim
	Grass weeds	chloridazon (*off-label*), chloridazon + propachlor, chlorpropham, chlorthal-dimethyl + propachlor, cyanazine, cyanazine (*fen soils only*), cycloxydim, ethofumesate (*off-label*), fluazifop-P-butyl, fluazifop-P-butyl (*off-label*), glyphosate (agricultural uses), paraquat (*off-label*), pendimethalin, pendimethalin (*off-label*), prometryn, prometryn (*off-label*), propachlor, propachlor (*off-label*), propaquizafop, propaquizafop (*off-label*), simazine (*off-label*), tepraloxydim, terbacil (*off-label*), trifluralin (*off-label*)
	Mayweeds	clopyralid, clopyralid (*off-label*)
	Polygonums	chlorpropham
	Weeds, miscellaneous	glyphosate (agricultural uses)

Flowers and ornamentals

Flowers - Bedding plants, general

Diseases	Alternaria	iprodione (*seed treatment*)
	Black root rot	carbendazim (*off-label*)
	Botrytis	thiram
	Petal blight	mancozeb
	Phytophthora	propamocarb hydrochloride
	Powdery mildew	bupirimate, copper ammonium carbonate, dinocap
	Pythium	propamocarb hydrochloride
	Rust	azoxystrobin (*off-label - in pots*), mancozeb, propiconazole (*off-label*), thiram
	Soil-borne diseases	metam-sodium
Pests	Aphids	imidacloprid, malathion, pirimicarb
	Flies	imidacloprid
	Free-living nematodes	metam-sodium
	Pests, miscellaneous	aldicarb (*off-label*)
	Soil pests	metam-sodium
	Stem nematodes	aldicarb (*off-label*)
	Weevils	imidacloprid
	Whiteflies	imidacloprid
Plant growth regulation	Flowering control	paclobutrazol
	Growth control	chlormequat, chlormequat with choline chloride, daminozide, paclobutrazol
Weeds	Broad-leaved weeds	chlorpropham, chlorpropham + pentanochlor, chlorthal-dimethyl, pentanochlor
	Grass weeds	chlorpropham, chlorpropham + pentanochlor, pentanochlor
	Polygonums	chlorpropham

	Weeds, miscellaneous	metam-sodium

Flowers - Bulbs/corms

Crop control	Pre-harvest desiccation	sulphuric acid (commodity substance)
Diseases	Black root rot	carbendazim (*off-label*)
	Botrytis	thiram
	Crown rot	metalaxyl-M (*off-label*)
	Fire	mancozeb, thiram
	Fungus diseases	formaldehyde (commodity substance) (*dip*)
	Fusarium diseases	thiabendazole
	Ink disease	chlorothalonil (*qualified minor use*)
	Phytophthora	etridiazole, propamocarb hydrochloride
	Powdery mildew	bupirimate
	Pythium	propamocarb hydrochloride
Pests	Aphids	malathion, nicotine
	Birds/mammals	aluminium ammonium sulphate (*bulb treatment*), aluminium ammonium sulphate (*corm treatment*)
	Capsid bugs	nicotine
	Flies	chlorpyrifos (*off-label*)
	Leaf miners	nicotine
	Leafhoppers	nicotine
	Sawflies	nicotine
	Stem nematodes	1,3-dichloropropene
	Thrips	nicotine
Plant growth regulation	Flowering control	paclobutrazol
	Growth control	2-chloroethylphosphonic acid, chlormequat, chlormequat with choline chloride, paclobutrazol
Weeds	Broad-leaved weeds	bentazone, chlorpropham, chlorpropham + pentanochlor, cyanazine, diquat, diquat + paraquat, paraquat, pentanochlor
	Crops as weeds	cycloxydim
	Grass weeds	chlorpropham, chlorpropham + pentanochlor, cyanazine, cycloxydim, diquat + paraquat, paraquat, pentanochlor
	Polygonums	chlorpropham

Flowers - Miscellaneous flowers

Diseases	Black spot	captan, kresoxim-methyl, mancozeb, myclobutanil
	Botrytis	propamocarb hydrochloride (*off-label*)
	Downy mildew	propamocarb hydrochloride (*off-label*)
	Powdery mildew	bupirimate, carbendazim, dinocap, dodemorph, kresoxim-methyl, myclobutanil
	Rust	mancozeb, myclobutanil
Pests	Aphids	imidacloprid, malathion, pirimicarb, pymetrozine (*off-label - lemon) and woodruff*), rotenone
	Flies	imidacloprid

	Leaf miners	abamectin
	Pests, miscellaneous	aldicarb (*off-label*)
	Sawflies	rotenone
	Spider mites	abamectin
	Stem nematodes	aldicarb (*off-label*)
	Thrips	abamectin
	Weevils	imidacloprid
	Whiteflies	imidacloprid
Plant growth regulation	Flowering control	paclobutrazol
	Growth control	paclobutrazol
Weeds	Broad-leaved weeds	chlorthal-dimethyl, dichlobenil, lenacil, pentanochlor, propyzamide, simazine
	Grass weeds	dichlobenil, fluazifop-P-butyl (*off-label*), lenacil, pentanochlor, propyzamide, simazine
	Weeds, miscellaneous	dichlobenil

Flowers - Pot plants

Diseases	Black root rot	carbendazim (*off-label*)
	Botrytis	iprodione
	Phytophthora	fosetyl-aluminium, propamocarb hydrochloride
	Pythium	propamocarb hydrochloride
	Root rot	fosetyl-aluminium
	Rust	mancozeb
Pests	Aphids	deltamethrin, imidacloprid, pirimicarb
	Caterpillars	deltamethrin
	Flies	imidacloprid
	Mealybugs	deltamethrin, petroleum oil
	Pests, miscellaneous	aldicarb (*off-label*)
	Scale insects	deltamethrin, petroleum oil
	Spider mites	petroleum oil
	Stem nematodes	aldicarb (*off-label*)
	Weevils	imidacloprid
	Whiteflies	deltamethrin, imidacloprid
Plant growth regulation	Flowering control	2-chloroethylphosphonic acid, paclobutrazol
	Fruiting control	paclobutrazol
	Growth control	2-chloroethylphosphonic acid, chlormequat, chlormequat with choline chloride, daminozide, paclobutrazol

Flowers - Protected bulbs/corms

Pests	Aphids	malathion, pirimicarb

Flowers - Protected flowers

Diseases	Crown rot	metalaxyl-M (*off-label*)

	Fungus diseases	difenoconazole (*off-label - ground grown*), difenoconazole (*off-label - pot grown*)
	Powdery mildew	imazalil
	Rust	azoxystrobin (*off-label*), propiconazole (*off-label*)
	Soil-borne diseases	metam-sodium
Pests	Aphids	malathion, nicotine, pirimicarb, Verticillium lecanii
	Free-living nematodes	metam-sodium
	Leaf miners	abamectin
	Pests, miscellaneous	aldicarb (*off-label*)
	Soil pests	metam-sodium
	Spider mites	abamectin, tebufenpyrad
	Stem nematodes	aldicarb (*off-label*)
	Thrips	abamectin, imidacloprid (*off-label*)
Plant growth regulation	Growth control	2-chloroethylphosphonic acid
Weeds	Weeds, miscellaneous	metam-sodium

Miscellaneous flowers and ornamentals - Miscellaneous uses

Pests	Birds/mammals	bone oil
	Caterpillars	diflubenzuron
	Leaf miners	deltamethrin (*off-label*)
	Slugs/snails	metaldehyde, methiocarb (*outdoor only*)
	Thrips	deltamethrin (*off-label*)
Plant growth regulation	Growth control	maleic hydrazide
Weeds	Broad-leaved weeds	amitrole, diquat + paraquat
	Grass weeds	amitrole, diquat + paraquat
	Weeds, miscellaneous	amitrole, glyphosate (fruit, horticulture, forestry, amenity etc.), glyphosate (fruit, horticulture, forestry, amenity etc.) (*wiper application*)

Miscellaneous flowers and ornamentals - Soils

Diseases	Soil-borne diseases	metam-sodium
Pests	Free-living nematodes	metam-sodium
	Leaf miners	deltamethrin (*off-label*)
	Soil pests	metam-sodium
	Thrips	deltamethrin (*off-label*)
Weeds	Weeds, miscellaneous	metam-sodium

Ornamentals - Nursery stock

Diseases	Alternaria	iprodione (*seed treatment*)
	Black root rot	carbendazim (*off-label*), carbendazim (*off-label - container grown*), carbendazim (*off-label - NFT*), carbendazim (*off-label - soil grown*)
	Botrytis	chlorothalonil, thiram (*except Hydrangea*)

	Canker	azaconazole + imazalil
	Damping off	copper ammonium carbonate, tolclofos-methyl
	Damping off and foot rot	etridiazole (*seeds and seedlings*)
	Foot rot	tolclofos-methyl
	Fungus diseases	prochloraz
	Phytophthora	etridiazole, etridiazole (*rooted cuttings and transplants*), fosetyl-aluminium, propamocarb hydrochloride
	Powdery mildew	imazalil
	Pythium	propamocarb hydrochloride
	Root rot	tolclofos-methyl
	Rust	oxycarboxin
	Silver leaf	azaconazole + imazalil
	Soil-borne diseases	metam-sodium
Pests	Ants	pirimiphos-methyl
	Aphids	cypermethrin, deltamethrin, dimethoate, imidacloprid, natural plant extracts, nicotine, pirimicarb, pirimiphos-methyl, pymetrozine, rotenone, Verticillium lecanii
	Birds/mammals	ziram
	Capsid bugs	cypermethrin, deltamethrin, pirimiphos-methyl
	Caterpillars	Bacillus thuringiensis, cypermethrin, deltamethrin, diflubenzuron, nicotine, teflubenzuron
	Cutworms	cypermethrin
	Earwigs	pirimiphos-methyl
	Flies	chlorpyrifos, imidacloprid
	Free-living nematodes	metam-sodium
	Leaf miners	abamectin, cypermethrin (*off-label*), deltamethrin (*off-label*), dichlorvos (*off-label*), dimethoate, nicotine, oxamyl (*off-label*), pirimiphos-methyl
	Leafhoppers	natural plant extracts, nicotine
	Mealybugs	deltamethrin
	Mites	natural plant extracts
	Pests, miscellaneous	aldicarb (*off-label*)
	Sawflies	pirimiphos-methyl
	Scale insects	deltamethrin
	Slugs/snails	metaldehyde
	Soil pests	metam-sodium
	Spider mites	abamectin, bifenthrin, dimethoate, fenbutatin oxide, pirimiphos-methyl
	Stem nematodes	aldicarb (*off-label*)
	Thrips	abamectin, cypermethrin, deltamethrin, deltamethrin (*off-label*), dichlorvos (*off-label*), natural plant extracts, nicotine, pirimiphos-methyl, spinosad
	Weevils	chlorpyrifos, fipronil, imidacloprid
	Whiteflies	buprofezin, cypermethrin, deltamethrin, imidacloprid, natural plant extracts, nicotine, pirimiphos-methyl, teflubenzuron, Verticillium lecanii
Plant growth regulation	Growth control	2-(1-naphthyl)acetic acid, 4-indol-3-yl-butyric acid, 4-indol-3-yl-butyric acid + 2-(1-naphthyl)acetic acid with dichlorophen, chlormequat, chlormequat with choline chloride, copper hydroxide, daminozide, indol-3-ylacetic acid

Weeds	Broad-leaved weeds	atrazine, clopyralid, cyanazine, diquat + paraquat, diuron, diuron (*not container grown*), diuron + paraquat, glufosinate-ammonium, isoxaben, isoxaben (*off-label*), isoxaben + trifluralin (*outdoor stock only*), napropamide, oxadiazon, paraquat (*off-label - inter-row directed treatment*), pentanochlor, propachlor, propyzamide, simazine (*for resale only*), trifluralin (*off-label*)
	Crops as weeds	cycloxydim
	Grass weeds	atrazine, cyanazine, cycloxydim, diquat + paraquat, diuron, diuron + paraquat, fluazifop-P-butyl (*off-label*), glufosinate-ammonium, isoxaben + trifluralin (*outdoor stock only*), napropamide, oxadiazon, paraquat (*off-label - inter-row directed treatment*), pentanochlor, propachlor, propyzamide, simazine (*for resale only*), trifluralin (*off-label*)
	Mayweeds	clopyralid
	Weeds, miscellaneous	ammonium sulphamate, ammonium sulphamate (*pre-planting*), metam-sodium

Ornamentals - Trees and shrubs

Diseases	Fungus diseases	prochloraz
	Powdery mildew	penconazole
	Scab	penconazole
	Verticillium wilt	chloropicrin (*soil fumigation*)
Pests	Aphids	deltamethrin
	Capsid bugs	deltamethrin
	Caterpillars	Bacillus thuringiensis, deltamethrin, diflubenzuron
	Mealybugs	deltamethrin
	Pests, miscellaneous	aldicarb (*off-label*)
	Scale insects	deltamethrin
	Stem nematodes	aldicarb (*off-label*)
	Thrips	deltamethrin
	Whiteflies	deltamethrin
Plant growth regulation	Growth control	maleic hydrazide
Weeds	Broad-leaved weeds	ammonium sulphamate, carbetamide + diflufenican + oxadiazon, chlorthal-dimethyl, dichlobenil, diuron, diuron + glyphosate, glufosinate-ammonium, isoxaben, isoxaben + trifluralin, metazachlor, paraquat, propyzamide, simazine
	Grass weeds	carbetamide + diflufenican + oxadiazon, cycloxydim (*off-label*), dichlobenil, diuron, diuron + glyphosate, glufosinate-ammonium, isoxaben + trifluralin, metazachlor, paraquat, propyzamide, simazine
	Weeds, miscellaneous	ammonium sulphamate, dichlobenil, diuron, glyphosate (fruit, horticulture, forestry, amenity etc.), glyphosate (fruit, horticulture, forestry, amenity etc.) (*wiper application*)
	Woody weeds/scrub	asulam, glyphosate (fruit, horticulture, forestry, amenity etc.)

SECTION 3

Ornamentals - Woody ornamentals

Diseases	Fungus diseases	prochloraz
Pests	Aphids	fatty acids
	Mealybugs	fatty acids
	Scale insects	fatty acids
	Spider mites	fatty acids
	Whiteflies	fatty acids
Weeds	Aquatic weeds	propyzamide
	Bindweeds	oxadiazon
	Broad-leaved weeds	dichlobenil, diuron, diuron + paraquat, glufosinate-ammonium, lenacil, oxadiazon, paraquat, propyzamide, simazine
	Grass weeds	dichlobenil, diuron, diuron + paraquat, glufosinate-ammonium, lenacil, oxadiazon, paraquat, propyzamide, simazine
	Weeds, miscellaneous	dichlobenil, glyphosate (fruit, horticulture, forestry, amenity etc.), propyzamide

Forestry

Forest nurseries, general - Forest nurseries

Pests	Aphids	pirimicarb
	Birds/mammals	aluminium ammonium sulphate
	Weevils	carbosulfan (*off-label*), chlorpyrifos
Weeds	Broad-leaved weeds	cyanazine (*off-label*), diquat + paraquat, napropamide, paraquat, paraquat (*stale seedbed*), simazine
	Grass weeds	cyanazine (*off-label*), cycloxydim, diquat + paraquat, napropamide, paraquat (*stale seedbed*), simazine

Forestry plantations, general - Forestry plantations

Crop control	Chemical stripping/ thinning	glyphosate (fruit, horticulture, forestry, amenity etc.)
Pests	Beetles	chlorpyrifos
	Birds/mammals	aluminium ammonium sulphate, warfarin, ziram
	Caterpillars	diflubenzuron
	Weevils	carbosulfan, chlorpyrifos
Weeds	Aquatic weeds	glyphosate (fruit, horticulture, forestry, amenity etc.), propyzamide
	Broad-leaved weeds	2,4-D, 2,4-D + dicamba + triclopyr, ammonium sulphamate, atrazine, clopyralid (*off-label*), diquat + paraquat, glufosinate-ammonium, glyphosate (fruit, horticulture, forestry, amenity etc.), glyphosate (fruit, horticulture, forestry, amenity etc.) (*wiper application*), isoxaben, metamitron (*off-label*), metazachlor, paraquat, propyzamide, simazine, triclopyr
	Crops as weeds	diquat + paraquat
	Grass weeds	ammonium sulphamate, atrazine, cycloxydim, cycloxydim (*off-label*), diquat + paraquat, glufosinate-ammonium, glyphosate (fruit, horticulture, forestry,

		amenity etc.), metamitron (*off-label*), metazachlor, paraquat, propyzamide, simazine
	Weeds, miscellaneous	ammonium sulphamate, glyphosate (fruit, horticulture, forestry, amenity etc.), paraquat, propyzamide
	Woody weeds/scrub	2,4-D, 2,4-D + dicamba + triclopyr, ammonium sulphamate, asulam, asulam (*off-label*), dicamba, glyphosate (fruit, horticulture, forestry, amenity etc.), triclopyr

Miscellaneous forestry situations - Cut logs/timber

Pests	Beetles	chlorpyrifos

Woodland on farms - Woodland

Pests	Aphids	lambda-cyhalothrin
	Beetles	lambda-cyhalothrin
	Birds/mammals	aluminium phosphide
	Sawflies	lambda-cyhalothrin
Weeds	Broad leaved weeds	2,4-D + dicamba + triclopyr, atrazine, cyanazine (*off-label*), lenacil (*off-label*), metazachlor, metazachlor (*off-label*)
	Crops as weeds	fluazifop-P-butyl
	Grass weeds	atrazine, cyanazine (*off-label*), cycloxydim, fluazifop-P-butyl, lenacil (*off-label*), metazachlor, metazachlor (*off-label*), propaquizafop
	Weeds, miscellaneous	fluazifop-P-butyl (*off-label*)
	Woody weeds/scrub	2,4-D + dicamba + triclopyr

Fruit and hops

Bush fruit - Bilberries/blueberries/cranberries

Diseases	Botrytis	pyrimethanil (*off-label*)
	Powdery mildew	fenpropimorph (*off-label*)
Pests	Gall, rust and leaf & bud mites	tebufenpyrad (*off-label*)
Weeds	Broad-leaved weeds	asulam (*off-label*), propachlor (*off-label*)
	Grass weeds	propachlor (*off-label*)
	Weeds, miscellaneous	glyphosate (fruit, horticulture, forestry, amenity etc.) (*off-label*)

Bush fruit - Currants

Diseases	Botrytis	chlorothalonil, chlorothalonil (*qualified minor use*), fenhexamid, fenhexamid + tolylfluanid, pyrimethanil (*off-label*), tolylfluanid
	Leaf spots	Bordeaux mixture, chlorothalonil, chlorothalonil (*qualified minor use*), copper ammonium carbonate, dodine, mancozeb

SECTION 3

	Powdery mildew	bupirimate, chlorothalonil, chlorothalonil (*qualified minor use*), fenpropimorph (*off-label*), kresoxim-methyl, myclobutanil, penconazole
	Rust	copper oxychloride, thiram
Pests	Aphids	chlorpyrifos, pirimicarb, tar oils
	Capsid bugs	chlorpyrifos, fenpropathrin
	Caterpillars	chlorpyrifos, diflubenzuron, tar oils
	Gall, rust and leaf & bud mites	fenpropathrin, sulphur, tebufenpyrad (*off-label*)
	Midges	fenpropathrin
	Sawflies	fenpropathrin
	Scale insects	tar oils
	Spider mites	chlorpyrifos, clofentezine (*off-label*), fenpropathrin
Weeds	Bindweeds	oxadiazon
	Broad-leaved weeds	asulam, asulam (*off-label*), chlorpropham, chlorthal-dimethyl, dichlobenil, diuron (*off-label*), glufosinate-ammonium, isoxaben, lenacil, MCPB, napropamide, oxadiazon, paraquat, pendimethalin, propachlor (*off-label*), propyzamide, simazine
	Crops as weeds	fluazifop-P-butyl, pendimethalin
	Grass weeds	chlorpropham, dichlobenil, fluazifop-P-butyl, glufosinate-ammonium, lenacil, napropamide, oxadiazon, paraquat, pendimethalin, propachlor (*off-label*), propyzamide, simazine
	Polygonums	chlorpropham
	Weeds, miscellaneous	dichlobenil, diuron (*off-label*), glyphosate (fruit, horticulture, forestry, amenity etc.) (*off-label*)

Bush fruit - Gooseberries

Diseases	Botrytis	chlorothalonil (*qualified minor use*), fenhexamid, fenhexamid + tolylfluanid, pyrimethanil (*off-label*), tolylfluanid
	Leaf spots	chlorothalonil (*qualified minor use*), mancozeb
	Powdery mildew	bupirimate, chlorothalonil (*qualified minor use*), fenpropimorph (*off-label*), myclobutanil, sulphur
Pests	Aphids	chlorpyrifos, pirimicarb, tar oils
	Capsid bugs	chlorpyrifos
	Caterpillars	chlorpyrifos, tar oils
	Gall, rust and leaf & bud mites	tebufenpyrad (*off-label*)
	Sawflies	nicotine, rotenone
	Scale insects	tar oils
	Spider mites	chlorpyrifos
Weeds	Bindweeds	oxadiazon
	Broad-leaved weeds	asulam (*off-label*), chlorpropham, chlorthal-dimethyl, dichlobenil, isoxaben, lenacil, MCPB, napropamide, oxadiazon, paraquat, pendimethalin, propachlor (*off-label*), propyzamide, simazine
	Crops as weeds	fluazifop-P-butyl, pendimethalin
	Grass weeds	chlorpropham, dichlobenil, fluazifop-P-butyl, lenacil, napropamide, oxadiazon, paraquat, pendimethalin, propachlor (*off-label*), propyzamide, simazine

	Polygonums	chlorpropham
	Weeds, miscellaneous	dichlobenil

Bush fruit - Miscellaneous bush fruit

Pests	Aphids	nicotine
	Birds/mammals	aluminium ammonium sulphate
	Capsid bugs	nicotine
	Leafhoppers	nicotine
	Sawflies	nicotine
Weeds	Broad-leaved weeds	diquat + paraquat, glufosinate-ammonium
	Crops as weeds	diquat + paraquat
	Grass weeds	diquat + paraquat, glufosinate-ammonium

Bush fruit - Protected bush fruit

Pests	Aphids	pymetrozine (*off-label*)

Cane fruit - Outdoor cane fruit

Crop control	Sucker/shoot control	sodium monochloroacetate (*off-label*)
Diseases	Botrytis	chlorothalonil, fenhexamid, fenhexamid + tolylfluanid, iprodione, thiram, tolylfluanid
	Cane blight	carbendazim (*off-label*)
	Cane spot	Bordeaux mixture, chlorothalonil, copper ammonium carbonate, copper oxychloride, thiram
	Powdery mildew	azoxystrobin (*off-label*), bupirimate, chlorothalonil, fenpropimorph (*off-label*)
	Purple blotch	copper oxychloride
	Spur blight	Bordeaux mixture, thiram
Pests	Aphids	1,3-dichloropropene, chlorpyrifos, pirimicarb, tar oils
	Beetles	chlorpyrifos, deltamethrin, rotenone
	Birds/mammals	aluminium ammonium sulphate
	Caterpillars	Bacillus thuringiensis, tar oils
	Free-living nematodes	1,3-dichloropropene
	Midges	chlorpyrifos
	Scale insects	tar oils
	Spider mites	bifenthrin (*off-label*), chlorpyrifos, clofentezine (*off-label*)
	Weevils	bifenthrin (*off-label*)
Weeds	Bindweeds	oxadiazon
	Broad-leaved weeds	asulam (*off-label*), atrazine, chlorthal-dimethyl, dichlobenil, diquat + paraquat, glufosinate-ammonium, isoxaben, lenacil, MCPB, napropamide, oxadiazon, paraquat, pendimethalin, propachlor (*off-label*), propyzamide, propyzamide (*England only*), simazine, trifluralin
	Crops as weeds	diquat + paraquat, fluazifop-P-butyl, pendimethalin
	Grass weeds	atrazine, dichlobenil, diquat + paraquat, fluazifop-P-butyl, glufosinate-ammonium, lenacil, napropamide, oxadiazon, paraquat, pendimethalin, propachlor (*off-*

		label), propyzamide, propyzamide (*England only*), simazine, trifluralin
	Weeds, miscellaneous	dichlobenil

Cane fruit - Protected cane fruit

Diseases	Botrytis	pyrimethanil (*off-label*)
	Powdery mildew	myclobutanil (*off-label*)
	Root rot	fluazinam (*off-label*)
Pests	Aphids	pymetrozine (*off-label*)
	Spider mites	bifenthrin (*off-label*), clofentezine (*off-label*)
	Weevils	bifenthrin (*off-label*)

Hops, general - Hops

Crop control	Chemical stripping/ thinning	diquat, diquat + paraquat, paraquat, sodium monochloroacetate (*off-label*)
	Pre-harvest desiccation	pymetrozine (*off-label*)
Diseases	Downy mildew	Bordeaux mixture, chlorothalonil, copper oxychloride, fosetyl-aluminium
	Powdery mildew	bupirimate, fenpropimorph (*off-label*), myclobutanil (*off-label*), penconazole, sulphur
Pests	Aphids	1,3-dichloropropene, bifenthrin, cypermethrin, deltamethrin, imidacloprid, tebufenpyrad
	Free-living nematodes	1,3-dichloropropene
	Spider mites	bifenthrin, tebufenpyrad
Weeds	Bindweeds	oxadiazon
	Broad-leaved weeds	asulam, diquat + paraquat, isoxaben, oxadiazon, paraquat, pendimethalin, simazine
	Crops as weeds	diquat + paraquat, fluazifop-P-butyl, pendimethalin
	Grass weeds	diquat + paraquat, fluazifop-P-butyl, oxadiazon, paraquat, pendimethalin, propyzamide (*off-label*), simazine

Miscellaneous fruit situations - Fruit crops, general

Pests	Aphids	fatty acids, natural plant extracts
	Leafhoppers	natural plant extracts
	Mealybugs	fatty acids
	Mites	natural plant extracts
	Scale insects	fatty acids
	Spider mites	fatty acids
	Thrips	natural plant extracts
	Whiteflies	fatty acids, natural plant extracts
Weeds	Broad-leaved weeds	glufosinate-ammonium
	Grass weeds	glufosinate-ammonium

Miscellaneous fruit situations - Fruit nursery stock

Weeds	Broad-leaved weeds	metazachlor, trifluralin (*off-label*)
	Grass weeds	metazachlor, trifluralin (*off-label*)

Miscellaneous fruit situations - Orchards

Diseases	Soil-borne diseases	methyl bromide with chloropicrin
Pests	Free-living nematodes	methyl bromide with chloropicrin
	Soil pests	methyl bromide with chloropicrin
Weeds	Broad-leaved weeds	paraquat
	Grass weeds	paraquat
	Weeds, miscellaneous	methyl bromide with chloropicrin

Miscellaneous fruit situations - Protected miscellaneous fruit

Diseases	Botrytis	benomyl (*off-label*), iprodione (*off-label*)
	Fungus diseases	benomyl (*off-label*)
Pests	Aphids	natural plant extracts, nicotine
	Leafhoppers	natural plant extracts
	Mites	natural plant extracts
	Thrips	natural plant extracts
	Whiteflies	natural plant extracts

Miscellaneous nuts - Nuts

Diseases	Bacterial canker	copper oxychloride (*off-label*)
	Blight	copper oxychloride (*off-label*)
Pests	Aphids	lambda-cyhalothrin (*off-label*)
	Caterpillars	lambda-cyhalothrin (*off-label*)
Weeds	Broad-leaved weeds	glufosinate-ammonium
	Grass weeds	glufosinate-ammonium
	Weeds, miscellaneous	glyphosate (fruit, horticulture, forestry, amenity etc.) (*off-label*)

Other fruit - Rhubarb

Diseases	Downy mildew	mancozeb + metalaxyl-M (*off-label*)
Weeds	Broad-leaved weeds	dichlobenil (*off-label - established*), paraquat (*off-label*), propyzamide, propyzamide (*outdoor*), simazine, simazine (*off-label*)
	Grass weeds	dichlobenil (*off-label - established*), paraquat (*off-label*), propyzamide, propyzamide (*outdoor*), simazine, simazine (*off-label*)
	Weeds, miscellaneous	glyphosate (fruit, horticulture, forestry, amenity etc.) (*off-label*)

Other fruit - Vines

Diseases	Botrytis	fenhexamid, iprodione (*off-label*), pyrimethanil (*off-label*)

	Bunch rots	iprodione (*off-label*)
	Downy mildew	copper oxychloride, fosetyl-aluminium (*off-label*), mancozeb + zoxamide, metalaxyl-M (*off-label*)
	Fungus diseases	mancozeb (*off-label*)
	Powdery mildew	dinocap (*off-label*), fenbuconazole (*off-label*), myclobutanil (*off-label*), sulphur
Pests	Aphids	nicotine
	Mealybugs	petroleum oil
	Scale insects	petroleum oil, tar oils
	Spider mites	petroleum oil
Weeds	Bindweeds	oxadiazon
	Broad-leaved weeds	dichlobenil (*off-label*), glufosinate-ammonium, isoxaben, oxadiazon, paraquat
	Grass weeds	glufosinate-ammonium, oxadiazon, paraquat, propyzamide (*off-label*)
	Weeds, miscellaneous	glyphosate (fruit, horticulture, forestry, amenity etc.) (*off-label*)

Soft fruit - Miscellaneous soft fruit

Pests	Aphids	nicotine, rotenone
	Capsid bugs	nicotine
	Leafhoppers	nicotine
	Sawflies	nicotine

Soft fruit - Strawberries

Crop control	Sucker/shoot control	paraquat
Diseases	Black spot	azoxystrobin (*off-label*)
	Botrytis	boscalid + pyraclostrobin, captan, chlorothalonil, chlorothalonil (*off-label*), chlorothalonil (*outdoor crops only*), fenhexamid, fenhexamid + tolylfluanid, iprodione, mepanipyrim, pyrimethanil, thiram, tolylfluanid
	Crown rot	chloropicrin (*soil fumigation*), fosetyl-aluminium (*off-label*)
	Powdery mildew	bupirimate, dinocap, fenpropimorph (*off-label*), kresoxim-methyl, myclobutanil, sulphur
	Red core	chloropicrin (*soil fumigation*), fosetyl-aluminium, fosetyl-aluminium (*off-label*)
	Verticillium wilt	chloropicrin (*soil fumigation*)
Pests	Aphids	1,3-dichloropropene, chlorpyrifos, nicotine, pirimicarb, pymetrozine (*off-label*)
	Beetles	methiocarb
	Birds/mammals	aluminium ammonium sulphate
	Caterpillars	Bacillus thuringiensis, chlorpyrifos
	Free-living nematodes	1,3-dichloropropene, chloropicrin (*soil fumigation*)
	Slugs/snails	methiocarb
	Spider mites	abamectin (*off-label*), bifenthrin, chlorpyrifos, clofentezine (*off-label*), fenbutatin oxide, tebufenpyrad
	Stem nematodes	1,3-dichloropropene
	Weevils	chlorpyrifos

Weeds	Broad-leaved weeds	asulam (*off-label*), chlorpropham, chlorthal-dimethyl, clopyralid, diquat + paraquat, ethofumesate (*off-label*), glufosinate-ammonium, isoxaben, lenacil, napropamide, pendimethalin, phenmedipham, propachlor, propachlor (*off-label*), propyzamide, simazine, trifluralin
	Crops as weeds	cycloxydim, diquat + paraquat, ethofumesate (*off-label*), fluazifop-P-butyl, pendimethalin
	Grass weeds	chlorpropham, cycloxydim, diquat + paraquat, ethofumesate (*off-label*), fluazifop-P-butyl, glufosinate-ammonium, lenacil, napropamide, pendimethalin, propachlor, propachlor (*off-label*), propyzamide, simazine, trifluralin
	Mayweeds	clopyralid
	Polygonums	chlorpropham

Tree fruit - Miscellaneous tree fruit

Diseases	Verticillium wilt	chloropicrin (*soil fumigation*)
Pests	Aphids	nicotine, rotenone
	Birds/mammals	aluminium ammonium sulphate, ziram
	Capsid bugs	nicotine
	Sawflies	nicotine
Weeds	Broad-leaved weeds	paraquat
	Grass weeds	paraquat
	Weeds, miscellaneous	glyphosate (fruit, horticulture, forestry, amenity etc.)

Tree fruit - Pome fruit

Crop control	Sucker/shoot control	glyphosate (fruit, horticulture, forestry, amenity etc.)
Diseases	Blossom wilt	vinclozolin
	Botrytis	iprodione (*off-label*)
	Botrytis fruit rot	thiram
	Canker	Bordeaux mixture, carbendazim, copper oxychloride
	Collar rot	copper oxychloride (*off-label*), fosetyl-aluminium
	Crown rot	fosetyl-aluminium
	Phytophthora	mancozeb + metalaxyl-M (*off-label - applied to orchard floor*)
	Powdery mildew	bupirimate, captan + penconazole, carbendazim, dinocap, fenbuconazole (*reduction*), kresoxim-methyl (*reduction*), myclobutanil, penconazole, sulphur
	Scab	Bordeaux mixture, captan, captan + penconazole, carbendazim, carbendazim (*off-label*), dithianon, dodine, fenbuconazole, kresoxim-methyl, mancozeb, maneb, myclobutanil, pyrimethanil, sulphur, thiram
	Storage rots	captan, carbendazim, carbendazim (*off-label*), thiram
Pests	Aphids	chlorpyrifos, cypermethrin, deltamethrin, nicotine, pirimicarb, tar oils, thiacloprid, triazamate
	Capsid bugs	chlorpyrifos, cypermethrin, deltamethrin, nicotine
	Caterpillars	Bacillus thuringiensis (*off-label*), chlorpyrifos, cypermethrin, deltamethrin, diflubenzuron, fenoxycarb, methoxyfenozide, tar oils

	Gall, rust and leaf & bud mites	amitraz, diflubenzuron
	Sawflies	chlorpyrifos, cypermethrin, deltamethrin, rotenone
	Scale insects	tar oils
	Spider mites	amitraz, bifenthrin, chlorpyrifos, clofentezine, fenpyroximate, tebufenpyrad
	Suckers	amitraz, chlorpyrifos, cypermethrin, deltamethrin, diflubenzuron, lambda-cyhalothrin, tar oils
	Weevils	chlorpyrifos
Plant growth regulation	Fruiting control	2-chloroethylphosphonic acid (*off-label - for cider making*), paclobutrazol
	Growth control	gibberellins (*off-label*)
	Plant growth regulation, miscellaneous	paclobutrazol
	Quality/yield control	gibberellins
Weeds	Bindweeds	oxadiazon
	Broad-leaved weeds	2,4-D, 2,4-D + dichlorprop-P + MCPA + mecoprop-P, asulam, asulam (*off-label*), clopyralid (*off-label*), dicamba + MCPA + mecoprop-P, dichlobenil, diquat + paraquat, diuron, fluroxypyr (*off-label*), glufosinate-ammonium, isoxaben, napropamide, oxadiazon, pendimethalin, propyzamide, simazine
	Crops as weeds	diquat + paraquat, pendimethalin
	Grass weeds	amitrole, dichlobenil, diquat + paraquat, diuron, glufosinate-ammonium, napropamide, oxadiazon, pendimethalin, propyzamide, simazine
	Mayweeds	dicamba + MCPA + mecoprop-P
	Polygonums	dicamba + MCPA + mecoprop-P
	Weeds, miscellaneous	amitrole, amitrole (*off-label*), dichlobenil, glyphosate (fruit, horticulture, forestry, amenity etc.)

Tree fruit - Stone fruit

Crop control	Sucker/shoot control	glyphosate (fruit, horticulture, forestry, amenity etc.)
Diseases	Bacterial canker	Bordeaux mixture, copper oxychloride
	Blossom wilt	fenbuconazole (*off-label*), myclobutanil (*off-label*)
	Botrytis	fenhexamid (*off-label*)
	Leaf curl	Bordeaux mixture, copper ammonium carbonate, copper oxychloride
	Rust	myclobutanil (*off-label*)
	Sclerotinia	fenbuconazole (*off-label*)
Pests	Aphids	chlorpyrifos, cypermethrin, deltamethrin, nicotine, pirimicarb, pirimicarb (*off-label*), tar oils, tebufenpyrad (*off-label*)
	Caterpillars	Bacillus thuringiensis (*off-label*), chlorpyrifos, cypermethrin, deltamethrin, diflubenzuron, tar oils
	Gall, rust and leaf & bud mites	diflubenzuron
	Sawflies	deltamethrin
	Scale insects	tar oils
	Spider mites	chlorpyrifos, clofentezine, fenpyroximate (*off-label*)

Plant growth regulation	Growth control	paclobutrazol (*off-label*)
Weeds	Broad-leaved weeds	asulam, asulam (*off-label*), diquat + paraquat, glufosinate-ammonium, isoxaben, napropamide, pendimethalin, propyzamide
	Crops as weeds	diquat + paraquat, pendimethalin
	Grass weeds	diquat + paraquat, glufosinate-ammonium, napropamide, pendimethalin, propyzamide, propyzamide (*off-label*)
	Weeds, miscellaneous	amitrole (*off-label*), glyphosate (fruit, horticulture, forestry, amenity etc.)

Grain/crop store uses

Stored produce - Food/produce storage

Diseases	Alternaria	iprodione
	Botrytis	iprodione
Pests	Ants	fenitrothion
	Bedbugs	fenitrothion
	Beetles	fenitrothion
	Caterpillars	fenitrothion
	Cockroaches	fenitrothion
	Crickets	fenitrothion
	Earwigs	fenitrothion
	Fleas	fenitrothion
	Flies	fenitrothion
	Food/grain storage pests	methyl bromide, methyl bromide with amyl acetate
	Mites	fenitrothion
	Pests, miscellaneous	methyl bromide
	Silverfish	fenitrothion

Stored seed - Stored grain/rape/linseed

Pests	Food/grain storage pests	chlorpyrifos-methyl, d-phenothrin + tetramethrin, pirimiphos-methyl
	Pests, miscellaneous	aluminium phosphide, magnesium phosphide

Grass

Grass seed - Grass seed crops

Diseases	Crown rust	propiconazole
	Damping off	thiram (*seed treatment*)
	Drechslera leaf spot	propiconazole
	Powdery mildew	propiconazole
	Rhynchosporium	propiconazole
Pests	Aphids	deltamethrin (*off-label*), dimethoate

Plant growth regulation	Growth control	trinexapac-ethyl
Weeds	Broad-leaved weeds	2,4-D, 2,4-D + mecoprop-P, bromoxynil + ethofumesate + ioxynil, clopyralid (*off-label*), dicamba + MCPA + mecoprop-P, ethofumesate, MCPA, mecoprop-P
	Crops as weeds	ethofumesate
	Grass weeds	bromoxynil + ethofumesate + ioxynil, ethofumesate
	Mayweeds	dicamba + MCPA + mecoprop-P
	Polygonums	dicamba + MCPA + mecoprop-P

Grassland - Leys

Diseases	Crown rust	propiconazole
	Drechslera leaf spot	propiconazole
	Powdery mildew	propiconazole
	Rhynchosporium	propiconazole
	Stinking smut	dicamba + mecoprop-P
Pests	Flies	chlorpyrifos, cypermethrin
	Leatherjackets	chlorpyrifos, methiocarb (*seed admixture*)
	Slugs/snails	methiocarb (*seed admixture*)
Weeds	Broad-leaved weeds	2,4-D + dicamba + triclopyr, 2,4-DB + MCPA, amidosulfuron, bentazone + MCPA + MCPB, bromoxynil + ethofumesate + ioxynil, bromoxynil + ioxynil, clopyralid, clopyralid + fluroxypyr + triclopyr, clopyralid + triclopyr, dicamba + MCPA + mecoprop-P, dicamba + mecoprop-P, ethofumesate, fluroxypyr, fluroxypyr + triclopyr, MCPA, MCPA + MCPB, MCPB, mecoprop-P
	Crops as weeds	ethofumesate, fluroxypyr
	Grass weeds	bromoxynil + ethofumesate + ioxynil, ethofumesate
	Mayweeds	clopyralid, dicamba + MCPA + mecoprop-P
	Polygonums	2,4-DB + MCPA, dicamba + MCPA + mecoprop-P
	Weeds, miscellaneous	diquat + paraquat

Grassland - Permanent pasture

Diseases	Stinking smut	dicamba + mecoprop-P
Pests	Aphids	pirimicarb
	Birds/mammals	aluminium ammonium sulphate, aluminium phosphide, strychnine hydrochloride (commodity substance) (*areas of restricted public access*)
	Flies	chlorpyrifos
	Leatherjackets	chlorpyrifos
Plant growth regulation	Quality/yield control	sulphur
Weeds	Aquatic weeds	glyphosate (agricultural uses), MCPA, triclopyr
	Broad-leaved weeds	2,4-D, 2,4-D + dicamba + triclopyr, 2,4-D + MCPA, 2,4-D + mecoprop-P, amidosulfuron, asulam, citronella oil, clopyralid, clopyralid + 2,4-D + MCPA, clopyralid + fluroxypyr + triclopyr, clopyralid + triclopyr, clopyralid + triclopyr (*off-label*), dicamba (*wiper application*), dicamba

		+ MCPA + mecoprop-P, dicamba + mecoprop-P, ethofumesate, fluroxypyr, fluroxypyr + triclopyr, glyphosate (agricultural uses) (*wiper application*), MCPA, MCPA + MCPB, MCPB, mecoprop-P, paraquat (*sward destruction/direct drilling*), thifensulfuron-methyl, triclopyr
	Crops as weeds	ethofumesate, fluroxypyr
	Grass weeds	ethofumesate, glyphosate (agricultural uses), paraquat (*sward destruction/direct drilling*)
	Mayweeds	clopyralid, dicamba + MCPA + mecoprop-P, dicamba + mecoprop-P
	Polygonums	dicamba + MCPA + mecoprop-P, dicamba + mecoprop-P
	Weeds, miscellaneous	diquat + paraquat, glufosinate-ammonium, glyphosate (agricultural uses)
	Woody weeds/scrub	asulam, clopyralid + triclopyr, dicamba, glyphosate (agricultural uses), triclopyr

SECTION 3

Turf/amenity grass - Amenity grassland

Diseases	Brown patch	iprodione
	Dollar spot	iprodione
	Melting out	iprodione
	Red thread	iprodione
	Snow mould	azoxystrobin, iprodione
Pests	Flies	chlorpyrifos
	Leatherjackets	chlorpyrifos
	Slugs/snails	metaldehyde
Plant growth regulation	Growth control	dicamba + maleic hydrazide + MCPA, maleic hydrazide, mefluidide, trinexapac-ethyl
Weeds	Broad-leaved weeds	2,4-D, 2,4-D + dicamba + triclopyr, 2,4-D + picloram, asulam, asulam (*not fine turf*), clopyralid + triclopyr, dicamba (*wiper application*), dicamba + maleic hydrazide + MCPA, dicamba + MCPA + mecoprop-P, fluroxypyr + triclopyr, isoxaben, MCPA, picloram
	Weeds, miscellaneous	diuron, glyphosate (fruit, horticulture, forestry, amenity etc.), glyphosate (fruit, horticulture, forestry, amenity etc.) (*wiper application*)
	Woody weeds/scrub	2,4-D + picloram, clopyralid + triclopyr, dicamba, fluroxypyr + triclopyr, picloram

Turf/amenity grass - Managed amenity turf

Crop control	Miscellaneous non-selective situations	glufosinate-ammonium
Diseases	Anthracnose	carbendazim + iprodione, chlorothalonil
	Brown patch	iprodione
	Cladosporium leaf spot	carbendazim + iprodione
	Dollar spot	carbendazim, chlorothalonil, fenarimol, iprodione, thiophanate-methyl
	Fungus diseases	dichlorophen
	Melting out	iprodione

	Red thread	carbendazim + iprodione, chlorothalonil, dichlorophen, fenarimol, iprodione, thiophanate-methyl
	Snow mould	azoxystrobin, carbendazim, carbendazim + epoxiconazole, carbendazim + iprodione, chlorothalonil, fenarimol, iprodione, thiophanate-methyl
Pests	Birds/mammals	aluminium phosphide
	Earthworms	carbendazim, thiophanate-methyl
	Flies	chlorpyrifos
	Leatherjackets	chlorpyrifos
	Slugs/snails	metaldehyde
Plant growth regulation	Growth control	maleic hydrazide, mefluidide, trinexapac-ethyl
Weeds	Broad-leaved weeds	2,4-D, 2,4-D + dicamba, 2,4-D + dicamba + fluroxypyr, 2,4-D + MCPA, 2,4-D + mecoprop-P, chlorthal-dimethyl, clopyralid + diflufenican + MCPA, clopyralid + fluroxypyr + MCPA, dicamba + dichlorprop-P + ferrous sulphate + MCPA, dicamba + dichlorprop-P + MCPA, dicamba + MCPA + mecoprop-P, dichlorprop-P + ferrous sulphate + MCPA, dichlorprop-P + MCPA, diquat + paraquat (*sward destruction/turf renovation*), ethofumesate, fluroxypyr + mecoprop-P, MCPA, MCPA + mecoprop-P, mecoprop-P
	Crops as weeds	2,4-D + dicamba + fluroxypyr, dichlorprop-P + MCPA, ethofumesate, mecoprop-P
	Grass weeds	diquat + paraquat (*sward destruction/turf renovation*), ethofumesate
	Mayweeds	dicamba + MCPA + mecoprop-P
	Polygonums	dicamba + MCPA + mecoprop-P
	Weeds, miscellaneous	dicamba + dichlorprop-P + ferrous sulphate + MCPA, dichlorophen, dichlorophen + ferrous sulphate, dichlorprop-P + ferrous sulphate + MCPA, ferrous sulphate

Non-crop pest control

Farm buildings/yards - Farm buildings

Diseases	Fungus diseases	formaldehyde (commodity substance)
Pests	Ants	fenitrothion
	Bedbugs	fenitrothion
	Beetles	dichlorvos, fenitrothion
	Birds/mammals	alphachloralose (*indoor use*), bone oil, brodifacoum, brodifacoum (*indoor use only*), bromadiolone, chlorophacinone, cholecalciferol, cholecalciferol + difenacoum, coumatetralyl, difenacoum, warfarin, zinc phosphide
	Caterpillars	fenitrothion
	Cockroaches	fenitrothion
	Crickets	fenitrothion
	Earwigs	fenitrothion
	Ectoparasites	dichlorvos
	Fleas	fenitrothion

	Flies	azamethiphos, cyromazine, dichlorvos, diflubenzuron, d-phenothrin + tetramethrin, fenitrothion, pyrethrins, tetramethrin
	Food/grain storage pests	alpha-cypermethrin
	Mites	fenitrothion
	Pests, miscellaneous	alpha-cypermethrin, fenitrothion
	Silverfish	fenitrothion
	Wasps	d-phenothrin + tetramethrin
Weeds	Weeds, miscellaneous	diuron, glyphosate (fruit, horticulture, forestry, amenity etc.)

Farm buildings/yards - Farmyards

Pests	Birds/mammals	bromadiolone, difenacoum

Farmland pest control - Farmland situations

Pests	Birds/mammals	aluminium phosphide, bone oil
Weeds	Crops as weeds	dichlobenil (*blight prevention*)

Miscellaneous non-crop pest control - Manure/rubbish

Pests	Flies	cyromazine, diflubenzuron, fenitrothion

Miscellaneous non-crop pest control - Miscellaneous pest control situations

Pests	Birds/mammals	carbon dioxide (commodity substance), paraffin oil (commodity substance) (*egg treatment*), strychnine hydrochloride (commodity substance) (*areas of restricted public access*)
	Wasps	resmethrin + tetramethrin

Protected salad and vegetable crops

Protected brassicas - Protected brassica vegetables

Diseases	Black rot	copper oxychloride (*off-label*)
	Damping off	fosetyl-aluminium (*off-label*)
	Downy mildew	fosetyl-aluminium (*off-label*), fosetyl-aluminium (*off-label - seedlings*)
	Spear rot	copper oxychloride (*off-label*)
Pests	Leaf miners	dichlorvos (*off-label*)
	Pests, miscellaneous	dimethoate (*off-label*)
	Thrips	dichlorvos (*off-label*)

Protected brassicas - Protected salad brassicas

Diseases	Black rot	copper oxychloride (*off-label*)

	Damping off	fosetyl-aluminium (*off-label*)
	Downy mildew	fosetyl-aluminium (*off-label*), fosetyl-aluminium (*off-label - seedlings*)
	Rhizoctonia	azoxystrobin (*off-label - for baby leaf production*)
	Spear rot	copper oxychloride (*off-label*)
Pests	Aphids	cypermethrin (*off-label*), deltamethrin (*off-label*), pymetrozine (*off-label - for baby leaf production*)
	Beetles	deltamethrin (*off-label - for baby leaf production*)
	Caterpillars	cypermethrin (*off-label*), deltamethrin (*off-label*), deltamethrin (*off-label - for baby leaf production*)
	Pests, miscellaneous	deltamethrin (*off-label*), dimethoate (*off-label*)

Protected crops, general - All protected crops

Diseases	Soil-borne diseases	dazomet (*soil fumigation*), metam-sodium
Pests	Aphids	nicotine, rotenone
	Capsid bugs	nicotine
	Free-living nematodes	dazomet (*soil fumigation*), metam-sodium
	Leaf miners	nicotine
	Leafhoppers	nicotine
	Sawflies	nicotine
	Slugs/snails	metaldehyde
	Soil pests	dazomet (*soil fumigation*), metam-sodium
	Thrips	nicotine
Weeds	Weeds, miscellaneous	dazomet (*soil fumigation*), metam-sodium

Protected crops, general - Glasshouses

Diseases	Fungus diseases	formaldehyde (commodity substance) (*spray, dip or fumigant*)
	Phytophthora	fosetyl-aluminium
	Root rot	fosetyl-aluminium
	Soil-borne diseases	metam-sodium, methyl bromide with amyl acetate, methyl bromide with chloropicrin
Pests	Free-living nematodes	metam-sodium, methyl bromide with amyl acetate, methyl bromide with chloropicrin
	Soil pests	metam-sodium, methyl bromide with amyl acetate, methyl bromide with chloropicrin
Weeds	Weeds, miscellaneous	metam-sodium, methyl bromide with amyl acetate, methyl bromide with chloropicrin

Protected crops, general - Protected vegetables, general

Diseases	White blister	propamocarb hydrochloride (*off-label*)
Pests	Aphids	natural plant extracts
	Leafhoppers	natural plant extracts
	Mites	natural plant extracts
	Thrips	natural plant extracts
	Whiteflies	natural plant extracts

Protected crops, general - Soils and compost

Diseases	Soil-borne diseases	formaldehyde (commodity substance) (*drench*), metam-sodium, methyl bromide with amyl acetate, methyl bromide with chloropicrin
Pests	Free-living nematodes	metam-sodium, methyl bromide with amyl acetate, methyl bromide with chloropicrin
	Soil pests	metam-sodium, methyl bromide with amyl acetate, methyl bromide with chloropicrin
Weeds	Weeds, miscellaneous	metam-sodium, methyl bromide with amyl acetate, methyl bromide with chloropicrin

Protected fruiting vegetables - Protected aubergines

Pests	Aphids	pymetrozine (*off-label*)
	Leaf miners	deltamethrin (*off-label*)
	Thrips	deltamethrin (*off-label*)

Protected fruiting vegetables - Protected cucurbits

Diseases	Botrytis	chlorothalonil, iprodione (*off-label*)
	Damping off	propamocarb hydrochloride (*off-label*)
	Downy mildew	azoxystrobin (*off-label*)
	Phytophthora	propamocarb hydrochloride
	Powdery mildew	azoxystrobin (*off-label*), chlorothalonil, imazalil, imazalil (*off-label*), sulphur (*off-label*)
	Pythium	propamocarb hydrochloride
	Root diseases	carbendazim (*off-label*)
	Root rot	propamocarb hydrochloride (*off-label*)
Pests	Ants	pirimiphos-methyl
	Aphids	deltamethrin, nicotine, pirimicarb (*off-label*), pirimiphos-methyl, pymetrozine, pymetrozine (*off-label*), Verticillium lecanii
	Capsid bugs	pirimiphos-methyl
	Caterpillars	Bacillus thuringiensis, deltamethrin
	Earwigs	pirimiphos-methyl
	Leaf miners	cypermethrin (*off-label*), deltamethrin (*off-label*), dichlorvos (*off-label*), oxamyl (*off-label*), pirimiphos-methyl
	Leafhoppers	nicotine
	Mealybugs	deltamethrin, petroleum oil
	Sawflies	pirimiphos-methyl
	Scale insects	deltamethrin, petroleum oil
	Spider mites	abamectin, abamectin (*off-label*), clofentezine (*off-label*), fenbutatin oxide, fenbutatin oxide (*off-label*), petroleum oil, pirimiphos-methyl
	Thrips	abamectin, deltamethrin, deltamethrin (*off-label*), dichlorvos, dichlorvos (*off-label*), nicotine, pirimiphos-methyl
	Whiteflies	buprofezin, cypermethrin, deltamethrin, nicotine, pirimiphos-methyl, Verticillium lecanii

Weeds	Broad-leaved weeds	isoxaben (*off-label*), paraquat (*off-label - inter-row directed treatment*)
	Grass weeds	paraquat (*off-label - inter-row directed treatment*), propyzamide (*off-label*)

Protected fruiting vegetables - Protected tomatoes

Diseases	Blight	azoxystrobin (*off-label*), chlorothalonil, copper oxychloride, propamocarb hydrochloride
	Botrytis	azoxystrobin (*off-label*), chlorothalonil, iprodione, pyrimethanil (*off-label*), thiram
	Damping off	copper oxychloride, propamocarb hydrochloride (*off-label*)
	Damping off and foot rot	etridiazole (*seeds and seedlings*)
	Didymella stem rot	azoxystrobin (*off-label*), carbendazim (*off-label*)
	Foot rot	copper oxychloride
	Leaf mould	chlorothalonil, copper ammonium carbonate
	Phytophthora	copper oxychloride, etridiazole (*transplants*), propamocarb hydrochloride
	Powdery mildew	azoxystrobin (*off-label*), bupirimate (*off-label*), sulphur (*off-label*)
	Pythium	propamocarb hydrochloride
	Root diseases	etridiazole (*off-label*)
	Root rot	propamocarb hydrochloride (*off-label*)
	Soil-borne diseases	metam-sodium
Pests	Ants	pirimiphos-methyl
	Aphids	deltamethrin, fatty acids, nicotine, pirimicarb, pirimiphos-methyl, Verticillium lecanii
	Capsid bugs	pirimiphos-methyl
	Caterpillars	Bacillus thuringiensis, deltamethrin, deltamethrin (*off-label*), nicotine
	Earwigs	pirimiphos-methyl
	Free-living nematodes	metam-sodium
	Leaf miners	abamectin, abamectin (*off-label*), deltamethrin (*off-label*), dichlorvos (*off-label*), nicotine, oxamyl (*off-label*), pirimiphos-methyl
	Leafhoppers	nicotine
	Mealybugs	deltamethrin, fatty acids, petroleum oil
	Sawflies	pirimiphos-methyl
	Scale insects	deltamethrin, fatty acids, petroleum oil
	Soil pests	metam-sodium
	Spider mites	abamectin, fatty acids, fenbutatin oxide, petroleum oil, pirimiphos-methyl
	Thrips	abamectin, deltamethrin (*off-label*), dichlorvos (*off-label*), nicotine, pirimiphos-methyl
	Whiteflies	buprofezin, deltamethrin, fatty acids, nicotine, pirimiphos-methyl, pymetrozine (*off-label*), Verticillium lecanii
Plant growth regulation	Fruiting control	(2-naphthyloxy)acetic acid, 2-chloroethylphosphonic acid
Weeds	Broad-leaved weeds	paraquat (*off-label - inter-row directed treatment*)

	Grass weeds	paraquat (*off-label - inter-row directed treatment*)
	Weeds, miscellaneous	metam-sodium

Protected herb crops - Protected herbs

Diseases	Botrytis	propamocarb hydrochloride (*off-label*)
	Downy mildew	fosetyl-aluminium (*off-label*), propamocarb hydrochloride (*off-label*)
	Fungus diseases	mancozeb + metalaxyl-M (*off-label*)
	Powdery mildew	sulphur (*off-label*)
	Rhizoctonia	azoxystrobin (*off-label*)
Pests	Aphids	nicotine, pymetrozine (*off-label*)
	Beetles	deltamethrin, deltamethrin (*off-label*)
	Caterpillars	Bacillus thuringiensis (*off-label*)
	Leafhoppers	deltamethrin, deltamethrin (*off-label*), nicotine
	Thrips	nicotine
	Whiteflies	nicotine
Weeds	Broad-leaved weeds	prometryn, prometryn (*off-label*), terbacil (*off-label*)
	Grass weeds	prometryn, prometryn (*off-label*), propyzamide (*off-label*), terbacil (*off-label*)

SECTION 3

Protected leafy vegetables - Mustard and cress

Diseases	Botrytis	iprodione (*off-label*), propamocarb hydrochloride (*off-label*)
	Damping off and foot rot	etridiazole
	Downy mildew	propamocarb hydrochloride (*off-label*)
	Powdery mildew	azoxystrobin (*off-label*)
	Ring spot	azoxystrobin (*off-label*)
	Rust	azoxystrobin (*off-label*)
	Sclerotinia	iprodione (*off-label*)
Pests	Aphids	nicotine
	Beetles	deltamethrin (*off-label*)
	Caterpillars	nicotine
	Leaf miners	nicotine
	Leafhoppers	deltamethrin (*off-label*)
Weeds	Broad-leaved weeds	pendimethalin (*off-label*), pendimethalin (*off-label - under covers*), propachlor (*off-label*)
	Grass weeds	pendimethalin (*off-label*), pendimethalin (*off-label - under covers*), propachlor (*off-label*)

Protected leafy vegetables - Protected leafy vegetables

Diseases	Big vein	carbendazim (*off-label*)
	Botrytis	dicloran, iprodione, propamocarb hydrochloride (*off-label*), pyrimethanil (*off-label*), thiram
	Downy mildew	fosetyl-aluminium, fosetyl-aluminium (*off-label*), mancozeb, propamocarb hydrochloride (*off-label*), thiram

	Fungus diseases	mancozeb + metalaxyl-M (*off-label*)
	Powdery mildew	sulphur (*off-label*)
	Rhizoctonia	azoxystrobin (*off-label*), dicloran, tolclofos-methyl
Pests	Aphids	nicotine, pirimicarb, pirimicarb (*off-label*), pymetrozine (*off-label*), Verticillium lecanii
	Beetles	deltamethrin (*off-label*)
	Caterpillars	Bacillus thuringiensis (*off-label*)
	Leaf miners	abamectin, dichlorvos (*off-label*)
	Leafhoppers	deltamethrin (*off-label*), nicotine
	Thrips	abamectin, dichlorvos (*off-label*), nicotine, pirimicarb (*off-label*)
	Whiteflies	cypermethrin, nicotine, Verticillium lecanii
Weeds	Broad-leaved weeds	chlorpropham with cetrimide, paraquat (*off-label - inter-row directed treatment*)
	Grass weeds	chlorpropham with cetrimide, paraquat (*off-label - inter-row directed treatment*), propyzamide (*off-label*)
	Polygonums	chlorpropham with cetrimide

Protected leafy vegetables - Protected spinach

Diseases	Downy mildew	metalaxyl-M (*off-label*)
Pests	Aphids	cypermethrin (*off-label*), pirimicarb (*off-label*)
	Caterpillars	cypermethrin (*off-label*)
	Pests, miscellaneous	dimethoate (*off-label*)
	Thrips	dichlorvos (*off-label*)

Protected legumes - Protected peas and beans

Pests	Aphids	Verticillium lecanii
	Leaf miners	dichlorvos (*off-label*), oxamyl (*off-label*)
	Thrips	dichlorvos (*off-label*)
	Whiteflies	Verticillium lecanii

Protected root and tuber vegetables - Protected carrots/parsnips/celeriac

Pests	Pests, miscellaneous	dimethoate (*off-label*)

Protected root and tuber vegetables - Protected potatoes

Diseases	Blight	benalaxyl + mancozeb
Pests	Aphids	nicotine
	Leafhoppers	nicotine
	Thrips	nicotine
	Whiteflies	nicotine

Protected root and tuber vegetables - Protected root brassicas

Diseases	Downy mildew	propamocarb hydrochloride (*off-label*)

	Rhizoctonia	tolclofos-methyl (*off-label*)
	White blister	mancozeb + metalaxyl-M (*off-label*)
Pests	Aphids	pirimicarb (*off-label*)

Protected stem and bulb vegetables - Protected asparagus

Pests	Aphids	nicotine

Protected stem and bulb vegetables - Protected celery/chicory

Diseases	Botrytis	azoxystrobin (*off-label*), carbendazim (*off-label*)
	Rhizoctonia	azoxystrobin (*off-label*), tolclofos-methyl (*off-label*)
	Sclerotinia	azoxystrobin (*off-label*)
Pests	Aphids	deltamethrin (*off-label*), nicotine
	Beetles	deltamethrin (*off-label*)
	Caterpillars	deltamethrin (*off-label*)
	Leaf miners	dichlorvos (*off-label*)
	Leafhoppers	deltamethrin (*off-label*)
	Pests, miscellaneous	deltamethrin (*off-label*)
	Thrips	dichlorvos (*off-label*)
	Whiteflies	cypermethrin
Weeds	Broad-leaved weeds	paraquat (*off-label - inter-row directed treatment*), prometryn
	Grass weeds	paraquat (*off-label - inter-row directed treatment*), prometryn, propyzamide (*off-label*)

Protected stem and bulb vegetables - Protected onions/leeks/garlic

Pests	Aphids	deltamethrin (*off-label*), nicotine
	Caterpillars	deltamethrin (*off-label*)
	Leafhoppers	nicotine
	Pests, miscellaneous	deltamethrin (*off-label*), dimethoate (*off-label - seedlings*)
	Stem nematodes	oxamyl (*off-label*)
	Thrips	dichlorvos (*off-label*), nicotine
	Whiteflies	nicotine
Weeds	Broad-leaved weeds	paraquat (*off-label - inter-row directed treatment*)
	Grass weeds	paraquat (*off-label - inter-row directed treatment*)

Total vegetation control

Aquatic situations, general - Aquatic situations

Plant growth regulation	Growth control	maleic hydrazide
Weeds	Aquatic weeds	2,4-D, dichlobenil, diquat, glyphosate (fruit, horticulture, forestry, amenity etc.), terbutryn
	Broad-leaved weeds	2,4-D

	Grass weeds	glyphosate (fruit, horticulture, forestry, amenity etc.)
	Weeds, miscellaneous	glyphosate (fruit, horticulture, forestry, amenity etc.)

Non-crop areas, general - Farm buildings/yards

Plant growth regulation	Growth control	maleic hydrazide

Non-crop areas, general - Miscellaneous non-crop situations

Pests	Birds/mammals	warfarin
Plant growth regulation	Growth control	maleic hydrazide
Weeds	Broad-leaved weeds	amitrole, asulam, MCPA, triclopyr
	Grass weeds	amitrole
	Weeds, miscellaneous	amitrole, diuron, glyphosate (fruit, horticulture, forestry, amenity etc.)
	Woody weeds/scrub	glyphosate (fruit, horticulture, forestry, amenity etc.), triclopyr

Non-crop areas, general - Non-crop farm areas

Pests	Slugs/snails	metaldehyde
Weeds	Aquatic weeds	triclopyr
	Broad-leaved weeds	amitrole, diquat + paraquat, diuron, diuron + glyphosate, glufosinate-ammonium, glyphosate (agricultural uses), glyphosate (fruit, horticulture, forestry, amenity etc.), MCPA, paraquat, picloram, triclopyr
	Crops as weeds	diquat + paraquat
	Grass weeds	amitrole, diquat + paraquat, diuron, diuron + glyphosate, glufosinate-ammonium, glyphosate (agricultural uses), glyphosate (fruit, horticulture, forestry, amenity etc.), paraquat
	Weeds, miscellaneous	amitrole, amitrole + 2,4-D + diuron, dichlobenil, diuron, diuron + paraquat, glyphosate (agricultural uses), glyphosate (fruit, horticulture, forestry, amenity etc.), glyphosate + oxadiazon (*soil surface treatment*), sodium chlorate
	Woody weeds/scrub	asulam, dicamba, glyphosate (fruit, horticulture, forestry, amenity etc.), picloram, triclopyr, triclopyr (*directed treatment*)

Non-crop areas, general - Paths/roads etc

Diseases	Fungus diseases	dichlorophen

Weeds	Broad-leaved weeds	2,4-D + dicamba + triclopyr, diquat + paraquat, diuron, MCPA
	Grass weeds	diquat + paraquat, diuron
	Weeds, miscellaneous	dichlorophen, diuron, diuron + paraquat, glyphosate (fruit, horticulture, forestry, amenity etc.), sodium chlorate
	Woody weeds/scrub	2,4-D + dicamba + triclopyr, glyphosate (fruit, horticulture, forestry, amenity etc.)

SECTION 4
PESTICIDE PROFILES

1 abamectin

A selective acaricide and insecticide for use in ornamentals

Products

Dynamec	Syngenta Bioline	18 g/l	EC	08701

Uses

Leaf miner in ***miscellaneous flowers***, ***ornamental specimens***, ***protected cherry tomatoes*** *(off-label)*, ***protected flowers***, ***protected lettuce***, ***protected ornamentals***, ***protected tomatoes***. Red spider mites in ***protected peppers*** *(off-label)*, ***protected strawberries*** *(off-label)*. Two-spotted spider mite in ***miscellaneous flowers***, ***ornamental specimens***, ***protected cucumbers***, ***protected flowers***, ***protected ornamentals***, ***protected tomatoes***. Western flower thrips in ***miscellaneous flowers***, ***ornamental specimens***, ***protected cucumbers***, ***protected flowers***, ***protected lettuce***, ***protected ornamentals***, ***protected tomatoes***.

Specific Off-Label Approvals (SOLAs)

- ***protected cherry tomatoes***, ***protected peppers***, ***protected strawberries*** *(OLA 031128) Dec 2008* [1]

Efficacy guidance

- Treat at first sign of infestation. Repeat sprays may be required
- For effective control total cover of all plant surfaces is essential, but avoid run-off
- Target pests quickly become immobilised but 3-5 d may be required for maximum mortality

Restrictions

- Number of treatments 6 on protected tomatoes and cucumbers (only 4 of which can be made when flowers or fruit present); 4 on protected lettuce; not restricted on flowers but rotation with other products advised
- Maximum concentration must not exceed 50 ml per 100 l water
- Do not mix with wetters, stickers or other adjuvants
- Do not use on ferns (*Adiantum* spp) or Shasta daisies
- Consult manufacturer for list of plant varieties tested for safety
- There is insufficient evidence to support product compatibility with integrated and biological pest control programmes
- Unprotected persons must be kept out of treated areas until the spray has dried

Crop-specific information

- On tomato or cucumber crops that are in flower, or have started to set fruit, treat only between 1 Mar and 31 Oct. Seedling tomatoes or cucumbers that have not started to flower or set fruit may be treated at any time. Do not treat cherry tomatoes
- Apply to lettuce only between 1 Mar and 31 Oct
- Some spotting or staining may occur on carnation, kalanchoe and begonia foliage
- HI 14 d for protected lettuce; 3 d for other protected edible crops

Environmental safety

- Extremely dangerous to fish or other aquatic life. Do not contaminate surface waters or ditches with chemical or used container
- High risk to bees. Do not apply to crops in flower or to those in which bees are actively foraging. Do not apply when flowering weeds are present
- Where bumble bees are used in tomatoes as pollinators, keep them out for 24 h after treatment
- Permissible in organic systems

Hazard classification and safety precautions

Hazard R03, R04
Risk phrases R03a, R03b, R03c, R04a
Operator protection A, C, D, H, K, M; U02, U05a, U09a, U19, U20a
Environmental protection E01, E12a, E13a, E15
Consumer protection C03
Storage and disposal E26, E30b, E31b, E34
Medical advice M04

SECTION 4

2 aldicarb

A soil-applied, systemic carbamate insecticide and nematicide

Products

1	Agriguard Aldicarb	AgriGuard	10% w/w	GR	10481
2	Me2 New Aldee	Me2	10% w/w	GR	10434
3	Standon Aldicarb 10G	Standon	10% w/w	GR	11024
4	Temik 10G	Bayer CropScience	10% w/w	GR	10021

Uses

Aphids in ***carrots, parsnips*** [1-4]; ***early potatoes, maincrop potatoes*** [1-3]; ***potatoes*** [4]. Free-living nematodes in ***carrots, parsnips*** [1-4]; ***early potatoes, maincrop potatoes*** [1-3]; ***potatoes*** [4]. Insect pests in ***amenity trees and shrubs*** *(off-label)*, ***bedding plants*** *(off-label)*, ***chrysanthemums*** *(off-label)*, ***dahlias*** *(off-label)*, ***herbaceous perennials*** *(off-label)*, ***nursery stock*** *(off-label)*, ***ornamental specimens*** *(off-label)*, ***pot plants*** *(off-label)*, ***protected roses*** *(off-label)* [4]. Spraing vectors in ***early potatoes, maincrop potatoes*** [1-3]; ***potatoes*** [4]. Stem and bulb nematodes in ***bulb onions*** [1-4]. Stem nematodes in ***amenity trees and shrubs*** *(off-label)*, ***bedding plants*** *(off-label)*, ***chrysanthemums*** *(off-label)*, ***dahlias*** *(off-label)*, ***herbaceous perennials*** *(off-label)*, ***nursery stock*** *(off-label)*, ***ornamental specimens*** *(off-label)*, ***pot plants*** *(off-label)*, ***protected roses*** *(off-label)* [4]; ***bulb onions*** [2, 4].

Specific Off-Label Approvals (SOLAs)

- ***amenity trees and shrubs, bedding plants, chrysanthemums, dahlias, herbaceous perennials, nursery stock, ornamental specimens, pot plants, protected roses*** *(OLA 001932) Dec 2008* [4]

Approval information

- In March 2003 the E U Council decided not to include aldicarb in Annex I under Directive 91/414. This meant that all approved products containing aldicarb have to be withdrawn. However for certain uses it was decided that this withdrawal can take place over a longer period than is usually provided, in order to allow time for the development of alternatives.
- In the UK, for uses on sugar beet, approvals for sale and supply were revoked in September 2003. Off-label approval for use on outdoor and protected sweetcorn seedlings and outdoor leeks will be revoked on 18 September 2004.
- For uses on potatoes, carrots, parsnips and onions approvals for sale and supply will be revoked on 30 June 2007 and stocks must be used by 31 December 2007. Off-label approval for use in ornamental plant production will also be revoked on 31 December 2007

Efficacy guidance

- Persistence and activity may be reduced in very wet soils or where pH exceeds 8.0. Do not use within 14 d of liming
- Use in potatoes reduces incidence of spraing disease

Restrictions

- Aldicarb is subject to the Poisons Rules 1982 and the Poisons Act 1972. See notes in Section 1
- This product contains an anticholinesterase carbamate compound. Do not use if under medical advice not to work with such compounds
- Maximum number of treatments 1 per crop
- Must be incorporated into soil by physical means. See label for details of application rates, timing, suitable applicators and techniques of incorporation
- No edible crops other than those listed (see label) should be planted into treated soil for at least 8 wk after application
- Keep unprotected persons out of treated glasshouses for at least 1 d

Crop-specific information

- Do not harvest spring sown bulb onions until mature bulb stage
- Latest use: at planting, sowing, drilling or transplanting
- HI potatoes 8 wk; carrots, parsnips 12 wk

Environmental safety

- Keep in original container, tightly closed, in a safe place, under lock and key
- Dangerous to game, wild birds and animals. Cover granules completely and immediately after application. Bury spillages. Failure to bury granules immediately and completely is hazardous to wildlife

- Dangerous to fish or other aquatic life. Do not contaminate surface waters or ditches with chemical or used container
- Option to return empty container as instructed by supplier [4]
- Keep all livestock out of treated areas for at least 13 wk. Bury or remove spillages

Hazard classification and safety precautions

Hazard R02

Risk phrases R02a, R02b, R02c

Operator protection A [2, 4]; B, H, K, M [1-4]; C [1]; C [2] (or D+E); C [3]; C [4] (or D+E); D, E [1, 3]; U02, U04a, U05a, U13, U19, U20a [1-4]; U09a [1, 3, 4]

Environmental protection E01, E10a, E13b; E06a (13 wk)

Consumer protection C03

Storage and disposal E30b, E32a, E34 [1-4]; E33 [4]

Medical advice M02 [2, 4]; M04 [1-4]

3 alphachloralose

A narcotic glucofuranose rodenticide used to kill mice

Products

Alphamouse	Killgerm	100%	CB	H6692

Uses

Mice in ***farm buildings*** *(indoor use).*

Efficacy guidance

- Place baits where mice are active, in runs or harbourages. Best results obtained where temperature is below 15°C
- Leave in position for 7-10 days only. If mouse activity remains continue treatment with a rodenticide containing a different active ingredient

Restrictions

- Alphachloralose is subject to the Poisons Rules 1982 and Poisons Act 1972. See notes in Section 1
- May be used only by persons instructed or trained in the use of alphachloralose
- Harmful by inhalation and if swallowed
- Do not use outdoors. Products must be used in situations where baits are placed within a building or other enclosed structure, and the target is living or feeding predominantly within that building or structure
- Wash out all mixing equipment thoroughly at the end of every operation
- Prevent access to baits by children, birds and non-target animals
- Do not prepare or lay baits where food, feed or water could become contaminated
- A suitable warning dye must always be used

Environmental safety

- Remove all remains of bait and bait containers after treatment and burn or bury. Do not dispose of in refuse sacks or on open rubbish tips
- Search for and burn or bury all rodent bodies. Do not place in refuse bins or on rubbish tips

Hazard classification and safety precautions

Hazard R03

Risk phrases R03b, R03c

Operator protection A, D, H; U02, U05a, U13, U20b

Environmental protection E01, E15

Consumer protection C03

Storage and disposal E30b, E32a, E34

Vertebrate/rodent control products V01b, V02, V03a, V04a

Medical advice M04

4 alpha-cypermethrin

A contact and ingested pyrethroid insecticide for use in arable crops

Products

1	Agriguard Alpha-Cyper	AgriGuard	100 g/l	EC	09920
2	Alphathrin	Nufarm UK	100 g/l	EC	11163
3	Antec Durakil 1.5 SC	Antec	15 g/l	SC	H7560
4	Antec Durakil 6SC	Antec	60 g/l	SC	H7559
5	Contest	BASF	15% w/w	WG	10216
6	I T Alpha-Cyper	I T Agro	100 g/l	EC	10483

Uses

Brassica pod midge in ***winter oilseed rape*** [1, 2, 5, 6]. Cabbage seed weevil in ***spring oilseed rape***, ***winter oilseed rape*** [1, 2, 5, 6]. Cabbage stem flea beetle in ***winter oilseed rape*** [1, 2, 5, 6]. Cabbage white butterfly in ***salad brassicas*** *(off-label - for baby leaf production)* [5]. Caterpillars in ***broccoli***, ***brussels sprouts***, ***cabbages***, ***calabrese***, ***cauliflowers***, ***kale*** [1, 2, 5, 6]. Cereal aphid in ***spring barley***, ***spring wheat***, ***winter barley***, ***winter wheat*** [1, 2, 5, 6]. Diamond-back moth in ***salad brassicas*** *(off-label - for baby leaf production)* [5]. Flea beetle in ***broccoli***, ***brussels sprouts***, ***cabbages***, ***calabrese***, ***cauliflowers***, ***kale*** [1, 2, 5, 6]. Lesser mealworm in ***poultry houses*** [3, 4]. Pea and bean weevil in ***broad beans***, ***combining peas***, ***spring field beans***, ***vining peas*** [5, 6]; ***winter field beans*** [1, 5, 6]. Pea aphid in ***combining peas***, ***vining peas*** [6]; ***combining peas*** *(reduction)*, ***vining peas*** *(reduction)* [5]. Pea moth in ***combining peas***, ***vining peas*** [5, 6]. Pollen beetle in ***spring oilseed rape***, ***winter oilseed rape*** [1, 2, 5, 6]. Poultry red mite in ***poultry houses*** [3, 4]. Rape winter stem weevil in ***winter oilseed rape*** [1, 2, 5, 6]. Small white butterfly in ***salad brassicas*** *(off-label - for baby leaf production)* [5]. Yellow cereal fly in ***winter barley***, ***winter wheat*** [5].

Specific Off-Label Approvals (SOLAs)

- ***salad brassicas*** *(for baby leaf production) (OLA 011221) Dec 2008* [5]

Approval information

- Approval expiry 31 Mar 2004 [1]

Efficacy guidance

- For cabbage stem flea beetle control spray oilseed rape when adult or larval damage first seen and about 1 mth later
- For flowering pests on oilseed rape apply at any time during flowering, on pollen beetle best results achieved at green to yellow bud stage (GS 3,3-3,7), on seed weevil between 20 pods set stage and 80% petal fall (GS 4,7-5,8)
- Spray cereals in autumn for control of cereal aphids, in spring/summer for grain aphids. (See label for details)
- For flea beetle, caterpillar and cabbage aphid control on brassicas apply when the pest or damage first seen or as a preventive spray. Repeat if necessary
- For pea and bean weevil control in peas and beans apply when pest attack first seen and repeat as necessary
- For lesser mealworm control in poultry houses apply a coarse, low-pressure spray as routine treatment after clean-out and before each new crop. Spray vertical surfaces and ensure an overlap onto ceilings. It is not necessary to treat the floor [3, 4]
- Use the higher recommended concentration where extreme residual action is required or surfaces are dirty or highly absorbent [3, 4]

Restrictions

- Maximum number of treatments 4 per crop on edible brassicas, 3 per crop on winter oilseed rape, peas, 2 on beans, spring oilseed rape (only 1 after yellow bud stage - GS 3,7)
- Maximum number of applications in animal husbandry use 2 per crop when used in premises that are occupied by poultry [3, 4]
- Apply up to 2 sprays on cereals in autumn and spring, 1 in summer between 1 Apr and 31 Aug. See label for details of rates
- Only 1 aphicide treatment may be applied in cereals between 1 Apr and 31 Aug in any one year and spray volume must not be reduced in this period

Crop-specific information

- For summer cereal application do not spray within 6 m from edge of crop and do not reduce volume when used after 31 Mar
- Do not apply to a cereal crop if any product containing a pyrethroid or dimethoate has been applied after the start of ear emergence (GS 51)
- Latest use: before the end of flowering for oilseed rape; before 31 Mar in year of harvest for cereals (autumn and spring application); before early dough stage (GS 83) for cereals (summer application).
- HI vining peas 1 d, brassicas 7 d, combining peas, broad beans, field beans 11 d

Environmental safety

- Dangerous for the environment [5]
- Dangerous to bees. Do not apply to crops in flower or to those in which bees are actively foraging. Do not apply when flowering weeds are present [2, 6]
- Extremely dangerous to fish or other aquatic life. Do not contaminate surface waters or ditches with chemical or used container
- LERAP Category A (except [3, 4])
- Recommended treatment presents low hazard to foraging bees in oilseed rape and has minimal effect on beneficial insects
- Do not apply directly to poultry; collect eggs before application [3, 4]

Hazard classification and safety precautions

Hazard R03 [1, 2, 5, 6]; R04, R07d [1, 2, 6]; R14 [5]
Risk phrases R03a, R04b, R04d [1, 2, 6]; R03c [1, 2, 5, 6]
Operator protection A, H [1-6]; C [1, 2, 6]; U02, U19 [1-4, 6]; U04a, U11 [1, 2, 6]; U05a, U10, U20b [1, 2, 5, 6]
Environmental protection E01, E16c, E16d [1, 2, 5, 6]; E02, E04, E05, E12d, E15 [3, 4]; E12c [2, 6]; E13a [1-6]
Consumer protection C03 [1, 2, 5, 6]; C09 [3, 4]
Storage and disposal E26, E30a, E31b [1, 2, 5, 6]; E32a [3, 4]; E34 [1, 2, 6]
Medical advice M03 [1, 2, 6]; M05 [2, 6]

5 aluminium ammonium sulphate

An inorganic bird and animal repellent

Products

1	Guardsman STP Seed Dressing Powder	Sphere	88% w/w	DS	03000
2	Liquid Curb Crop Spray	Sphere	83 g/l	SC	03164

Uses

Birds in ***all top fruit***, ***broad beans***, ***bush fruit***, ***cane fruit***, ***carrots***, ***flowerhead brassicas***, ***forest nursery beds***, ***forestry plantations***, ***grassland***, ***leaf brassicas***, ***peas***, ***spring field beans***, ***spring oilseed rape***, ***strawberries***, ***sugar beet***, ***winter barley***, ***winter field beans***, ***winter oats***, ***winter oilseed rape***, ***winter wheat*** [2]; ***corms*** *(corm treatment)*, ***flower bulbs*** *(bulb treatment)*, ***seeds*** *(seed treatment)* [1]. Damaging mammals in ***all top fruit***, ***broad beans***, ***bush fruit***, ***cane fruit***, ***carrots***, ***flowerhead brassicas***, ***forest nursery beds***, ***forestry plantations***, ***grassland***, ***leaf brassicas***, ***peas***, ***spring field beans***, ***spring oilseed rape***, ***strawberries***, ***sugar beet***, ***winter barley***, ***winter field beans***, ***winter oats***, ***winter oilseed rape***, ***winter wheat*** [2]; ***corms*** *(corm treatment)*, ***flower bulbs*** *(bulb treatment)*, ***seeds*** *(seed treatment)* [1].

Efficacy guidance

- Apply as overall spray to growing crops before damage starts or mix powder with seed depending on type of protection required
- Spray deposit protects growth present at spraying but gives little protection to new growth
- Product must be sprayed onto dry foliage to be effective and must dry completely before dew or frost forms. In winter this may require some wind

Restrictions

- Maximum number of treatments 1 per batch for corms, flower bulbs, seeds [1]

Crop-specific information

- HI fruit crops 6 wk

Hazard classification and safety precautions

Operator protection U04b, U10, U20a [2]; U20b [1]

Environmental protection E15

Storage and disposal E30a, E32a [1, 2]; E31b [2]

Treated seed S02, S05

Medical advice M05 [2]

6 aluminium phosphide

A phosphine generating compound used against vertebrates and grain store pests

Products

1	Degesch Fumigation Tablets	Rentokil	56% w/w	GE	09313
2	Luxan Talunex	Luxan	57% w/w	GE	06563
3	Phostoxin	Rentokil	56% w/w	GE	09315

Uses

Insect pests in ***stored grain*** [1]. Moles in ***farm woodland, grassland, lawns, managed amenity turf*** [2]; ***farmland*** [2, 3]. Rabbits in ***farm woodland, grassland, lawns, managed amenity turf*** [2]; ***farmland*** [2, 3]. Rats in ***farm woodland, grassland, lawns, managed amenity turf*** [2]; ***farmland*** [2, 3].

Efficacy guidance

- Product releases poisonous hydrogen phosphide gas in contact with moisture
- Place fumigation tablets in grain stores as directed [1]
- Place pellets in burrows or runs and seal hole by heeling in or covering with turf. Do not cover pellets with soil. Inspect daily and treat any new or re-opened holes [2, 3]
- Apply pellets by means of Luxan Topex Applicator [2]

Restrictions

- Aluminium phosphide is subject to the Poisons Rules 1982 and the Poisons Act 1972. See notes in Section 1
- Only to be used by operators instructed or trained in the use of aluminium phosphide and familiar with the precautionary measures to be taken. See label and HSE Guidance Notes for full precautions
- Only open container outdoors [2, 3] and for immediate use. Keep away from liquid or water as this causes immediate release of gas. Do not use in wet weather
- Do not use within 3 m of human or animal habitation. Before application ensure that no humans or domestic animals are in adjacent buildings or structures. Allow a minimum airing-off period of 4 h before re-admission

Environmental safety

- Product liberates very toxic, highly flammable gas
- Keep in original container, tightly closed, in a safe place, under lock and key
- Dangerous to fish or other aquatic life. Do not contaminate surface waters or ditches with chemical or used container [1-3]
- Pellets must never be placed or allowed to remain on ground surface
- Do not use adjacent to watercourses
- Take particular care to avoid gassing non-target animals, especially those protected under the Wildlife and Countryside Act (e.g. badgers. polecat, reptiles, natterjack toads, most birds). Do not use in burrows where there is evidence of badger or fox activity, or when burrows might be occupied by birds
- Dust remaining after decomposition is harmless and of no environmental hazard
- Dispose of empty containers as directed on label
- Keep livestock out of treated areas

Hazard classification and safety precautions

Hazard R01, R03, R07c [1-3]; R04 [2]

Risk phrases R01b, R01c, R03a [1-3]; R04a, R04c [2]

Operator protection A, H [1-3]; D [1, 3]; G [2]; U01, U07, U13, U19, U20a [1-3]; U05a [3]; U05b [1, 2]

Environmental protection E01, E13b [1-3]; E02 [1] (4 h min); E07 [1]

Consumer protection C03

Storage and disposal E26 [1]; E27 [2]; E29 [1, 3]; E30b, E32b, E34 [1-3]
Vertebrate/rodent control products V03a [2]; V04a [2, 3]
Medical advice M04

7 amidosulfuron

A post-emergence sulfonylurea herbicide for cleavers and other broad-leaved weed control in cereals

Products

1	Barclay Cleave	Barclay	75% w/w	WG	11340
2	Druid	Aventis	50% w/w	WG	08714
3	Eagle	Aventis	75% w/w	WG	07318
4	Landgold Amidosulfuron	Landgold	75% w/w	WG	09021
5	Pursuit	Aventis	75% w/w	WG	07333
6	Squire	Bayer CropScience	50% w/w	WG	08715

Uses

Annual dicotyledons in ***durum wheat*, *spring barley*, *spring oats*, *spring rye*, *spring wheat*, *triticale*, *winter barley*, *winter oats*, *winter rye*, *winter wheat*** [1-5]; ***linseed*** [1-3, 5]. Charlock in ***permanent pasture*, *rotational grassland*** [6]. Cleavers in ***durum wheat*, *spring barley*, *spring oats*, *spring rye*, *spring wheat*, *triticale*, *winter barley*, *winter oats*, *winter rye*, *winter wheat*** [1-5]; ***linseed*** [1-3, 5]; ***permanent pasture*, *rotational grassland*** [6]. Docks in ***permanent pasture*, *rotational grassland*** [6]. Forget-me-not in ***permanent pasture*, *rotational grassland*** [6]. Shepherd's purse in ***permanent pasture*, *rotational grassland*** [6].

Efficacy guidance

- For best results apply in spring (from 1 Feb) in warm weather when soil moist and weeds growing actively. When used in grassland following cutting or grazing, docks should be allowed to regrow before treatment
- Weed kill is slow, especially under cool, dry conditions. Weeds may sometimes only be stunted but will have little or no competitive effect on crop
- May be used on all soil types unless certain sequences are used on linseed. See label
- Spray is rainfast after 1 h
- Cleavers controlled from emergence to flower bud stage. If present at application charlock (up to flower bud), shepherds purse (up to flower bud) and field forget-me-not (up to 6 leaves) will also be controlled
- Amidosulfuron is a member of the ALS-inhibitor group of herbicides and products should be used in a planned Resistance Management strategy. See Section 1 for more information

Restrictions

- Maximum number of treatments 1 per crop, or 1 per yr when used on grass
- Do not apply to crops undersown or due to be undersown with clover or lucerne
- Do not use on swards containing red clover but may be used on swards with white clover. Where clover has been newly established apply from 3-leaf stage of grass and 1-2 trifoliate leaves of clover [6]
- Do not spray crops under stress, suffering drought, waterlogged, grazed, lacking nutrients or if soil compacted
- Do not spray if frost expected
- Do not roll or harrow within 1 wk of spraying
- Certain mixtures or sequences with other sulfonylurea products are permitted under explicitly detailed conditions. See label for details. There are no recommendations for mixtures with metsulfuron methyl products on linseed
- Certain mixtures with fungicides are expressly forbidden. See label for details

Crop-specific information

- Latest use: before first spikelets just visible (GS 51) for cereals; before flower buds visible for linseed
- Broadcast crops should be sprayed post-emergence after plants have a well established root system
- Hay or silage from treated grass crops must not be cut for at least 21 d after treatment [6]

Following crops guidance

- Only cereals, winter oilseed rape, mustard, winter field beans, vetches may be sown in the same year as treatment
- Cereals or potatoes must be sown as the following crop after use of permitted mixtures or sequences with other sulfonyl urea herbicides in cereals. Only cereals may be sown after the use of such sequences in linseed
- If a treated crop fails cereals may be sown after 15 d and thorough cultivation

Environmental safety

- Dangerous to fish or other aquatic life. Do not contaminate surface waters or ditches with chemical or used container
- Take care to wash out sprayers thoroughly. See label for details
- Avoid drift onto neighbouring broad-leaved plants or onto surface waters or ditches
- Keep livestock out of treated areas for at least 7 d and until foliage of any poisonous weeds such as ragwort has died and become unpalatable [6]

Hazard classification and safety precautions

Hazard R04 [2, 6]
Risk phrases R04a [2, 6]
Operator protection C [2, 6]; U05a [2, 6]; U20a [1, 3, 4]; U20b [2, 5, 6]
Environmental protection E01 [2, 6]; E07 [6] (7 d); E13b [1-6]
Consumer protection C03 [2, 6]
Storage and disposal E29 [6]; E31a [1-6]

8 amidosulfuron + iodosulfuron-methyl-sodium

A post-emergence sulfonylurea herbicide mixture for cereals

Products

Chekker	Bayer CropScience	12.5:1.25% w/w	WG	10955

Uses

Annual dicotyledons in ***spring barley, spring rye, spring wheat, triticale, winter barley, winter rye, winter wheat***. Chickweed in ***spring barley, spring rye, spring wheat, triticale, winter barley, winter rye, winter wheat***. Cleavers in ***spring barley, spring rye, spring wheat, triticale, winter barley, winter rye, winter wheat***. Mayweeds in ***spring barley, spring rye, spring wheat, triticale, winter barley, winter rye, winter wheat***. Volunteer oilseed rape in ***spring barley, spring rye, spring wheat, triticale, winter barley, winter rye, winter wheat***.

Efficacy guidance

- Best results obtained from treatment in warm weather when soil is moist and the weeds are growing actively
- Weeds must be present at application to be controlled
- Dry conditions resulting in moisture stress may reduce effectiveness
- Weed control is slow especially under cool dry conditions
- Occasionally weeds may only be stunted but they will normally have little or no competitive effect on the crop
- Amidosulfuron and iodosulfuron are members of the ALS-inhibitor group of herbicides and products should be used in a planned Resistance Management strategy. See Section 1 for more information

Restrictions

- Maximum number of treatments 1 per crop
- Must only be applied between 1 Feb in yr of harvest and specified latest time of application
- Do not roll or harrow within 1 wk of spraying
- Do not spray crops under stress from any cause or if the soil is compacted
- Do not spray if frost expected
- Do not apply in mixture or in sequence with any other ALS inhibitor

Crop-specific information

- Treat broadcast crops after the plants have a well-established root system
- Latest use: before first spikelet of inflorescence just visible (GS 51)

Following crops guidance

- Cereals, winter oilseed rape and winter field beans may be sown in the same yr as treatment provided they are preceded by ploughing or thorough cultivation
- A minimum of 3 mth must elapse between treatment and sowing winter oilseed rape

Environmental safety

- Extremely dangerous to fish or other aquatic life. Do not contaminate surface waters or ditches with chemical or used container
- LERAP Category B
- Take extreme care to avoid damage by drift onto broad-leaved plants outside the target area or onto ponds, waterways and ditches
- Observe carefully label instructions for sprayer cleaning

Hazard classification and safety precautions

Hazard R04
Risk phrases R04a
Operator protection A, C, H; U05a, U08, U11, U14, U15, U20b
Environmental protection E01, E13a, E16a, E16b
Consumer protection C03
Storage and disposal E31a

9 2-aminobutane

A fumigant alkylamine fungicide permitted for use only on stored seed potatoes

Products

Certis 2-Aminobutane	Certis	720 g/l	VP	11182

Uses

Gangrene in ***seed potatoes***. Silver scurf in ***seed potatoes***. Skin spot in ***seed potatoes***.

Approval information

- Products containing this active ingredient have been granted derogations for specified 'Essential Uses' for use until 31 December 2007. Sale and supply must cease by 30 June 2007 but growers have no guarantee that the products will continue to be available until then.
  For more information see 'The Review Programme' under 'Pesticide Legislation' in Section 1

Efficacy guidance

- Product used for treatment of seed potato tubers by fumigation in appropriate premises (see label)

Restrictions

- Maximum number of treatments 1 per batch. Maximum quantity to be fumigated in a single stack must not exceed 2000 tonnes
- Do not treat immature tubers. Allow period of healing before treating damaged tubers
- Treatment must only be carried out by trained operators in suitable fumigation chambers under licence from the British Technology Group
- Fumigate within 21 d of lifting
- This product must not be used on any crops other than those listed, including any extrapolations that would normally be permissible under the Long Term Arrangements for Extension of Use (see Section 1)

Environmental safety

- Keep in original container, tightly closed, in a safe place, under lock and key
- Do not empty into drains
- Do not contaminate surface waters or ditches with chemical or used container
- Do not supply treated potatoes for consumption by humans or lactating dairy cows
- Use must be in accordance with approved Code of Practice for the Control of Substances Hazardous to Health: Fumigation Operations

Hazard classification and safety precautions

Hazard R03, R04, R07c
Risk phrases R03b, R04a, R04b, R04c
Operator protection A, C; U02, U04a, U05b, U11, U13, U19, U20b
Environmental protection E01, E15
Consumer protection C03

Storage and disposal E30b, E31a, E34
Medical advice M03

10 amitraz

An amidine acaricide and insecticide for use in top fruit

Products

Mitac HF	Aventis	200 g/l	EC	07358

Uses

Pear sucker in ***pears***. Red spider mites in ***apples, pears***. Rust mite in ***apples***.

Efficacy guidance

- Best results achieved in dry conditions, do not spray if rain imminent

Restrictions

- Maximum total dose equivalent to two full dose treatments

Crop-specific information

- For red spider mites on apples and pears spray at 60-80% egg hatch and repeat 3 wk later
- For pear sucker control spray when significant numbers of nymphs have hatched but before there is significant contamination of the fruit with honeydew, normally Jun/Jul
- HI apples, pears 2 wk

Environmental safety

- Harmful to fish. Do not contaminate surface waters or ditches with chemical or used container
- Keep in original container, tightly closed, in a safe place, under lock and key

Hazard classification and safety precautions

Hazard R03
Risk phrases R03a, R03c
Operator protection A, C, H; U02, U04a, U05a, U08, U13, U19, U20a
Environmental protection E01, E13c
Consumer protection C02 (2 wk); C03
Storage and disposal E26, E30b, E31b, E34
Medical advice M03

11 amitrole

A translocated, foliar-acting, non-selective triazole herbicide

Products

1	Aminotriazole Technical	Nufarm UK	92.5% w/w	TC	11137
2	Weedazol-TL	Bayer	225 g/l	SL	02979
3	Weedazol-TL	Bayer CropScience	225 g/l	SL	11430

Uses

Annual and perennial weeds in ***apricots*** *(off-label)*, ***cherries*** *(off-label)*, ***peaches*** *(off-label)*, ***plums*** *(off-label)*, ***quinces*** *(off-label)* [2, 3]. Annual weeds in ***fallows, field margins, headlands, stubbles*** [2, 3]. Barren brome in ***apple orchards, pear orchards*** [2, 3]. Couch in ***amenity areas, industrial sites, non-crop areas*** [1]; ***apple orchards, fallows, field margins, headlands, pear orchards, stubbles, winter wheat*** *(direct drilling)* [2, 3]. Creeping bent in ***amenity areas, industrial sites, non-crop areas*** [1]. Creeping thistle in ***amenity areas, industrial sites, non-crop areas*** [1]; ***fallows, stubbles*** [2, 3]. Docks in ***amenity areas, industrial sites, non-crop areas*** [1]; ***fallows, field margins, headlands, stubbles*** [2, 3]. General weed control in ***apple orchards, pear orchards*** [2, 3]. Perennial weeds in ***apple orchards, fallows, field margins, headlands, pear orchards*** [2, 3]. Total vegetation control in ***amenity areas, industrial sites, non-crop areas*** [1]. Volunteer potatoes in ***stubbles*** *(barley stubble)* [2, 3].

Specific Off-Label Approvals (SOLAs)

- ***apricots, cherries, peaches, plums, quinces*** *(OLA 031869) Dec 2008* [3]
- ***apricots, cherries, peaches, plums, quinces*** *(OLA 920333) Dec 2008* [2]

Approval information

- Amitrole included in Annex I under EC Directive 91/414

Efficacy guidance

- In non-crop land may be applied at any time from Apr to Oct. Best results achieved in spring or early summer when weeds growing actively. For coltsfoot, hogweed and horsetail summer and autumn applications are preferred [1]
- In cropland apply when couch is in active growth and foliage at least 7.5 cm high [2, 3]
- In fallows and stubble plough 3-6 wk after application to depth of 20 cm, taking care to seal the furrow [2, 3]
- Uptake is via foliage and heavy rain immediately after application will reduce efficacy. Amitrole is less affected by drought than some residual herbicides and remains effective for up to 2 mth

Restrictions

- Maximum number of treatments 1 per crop for winter wheat, 1 per yr for stone fruit, 2 per yr for amenity turf
- Keep off suckers or foliage of desirable trees or shrubs [1]
- Do not spray areas into which the roots of adjacent trees or shrubs extend [1]
- Do not spray on sloping ground when rain imminent and run-off may occur [1]
- Apply round base of established fruit trees taking care to avoid contaminating trees, especially where bark is damaged. Keep off trees [2, 3]

Crop-specific information

- Latest use: at least 2 wk before direct drilling for winter wheat; before drilling for field crops; before end Jun or after harvest for apples, pears, stone fruit; end Oct for fallows [2]

Following crops guidance

- Apply in autumn on land to be used for spring barley [2, 3]
- Do not sow direct-drilled winter wheat less than 2 wk after application [2, 3]

Environmental safety

- Harmful to fish or other aquatic life. Do not contaminate surface waters or ditches with chemical or used container

Hazard classification and safety precautions

Operator protection A [1-3]; C [2, 3]; U08, U20a [1]; U09a [2, 3]; U19 [1-3]
Environmental protection E13c [1]; E15 [2, 3]
Consumer protection C03 [2, 3]
Storage and disposal E30a, E31b

SECTION 4

12 amitrole + 2,4-D + diuron

A total herbicide mixture of translocated and residual chemicals

Products

Trik	Nufarm UK	26.6:11.2:46.4% w/w	WP	11441

Uses

Total vegetation control in ***land not intended to bear vegetation***.

Approval information

- Amitrole and 2,4-D included in Annex I under EC Directive 91/414

Efficacy guidance

- Apply in spring or late summer/early autumn when weeds are growing actively and have sufficient leaf area to absorb chemical
- Apply maintenance treatment if necessary at lower rate when weeds 7-10 cm high
- Increase dose rate on areas of peat or high carbon content

Restrictions

- Do not use on ground under which roots of valuable trees or shrubs are growing

Environmental safety

- Harmful to fish. Do not contaminate surface waters or ditches with chemical or used container
- Keep livestock out of treated areas until foliage of poisonous weeds such as ragwort has died and become unpalatable

Hazard classification and safety precautions

Hazard R04
Risk phrases R04a, R04b, R04c

Operator protection A, C, D, H, M; U05a, U08, U19, U20a
Environmental protection E01, E07, E13c
Consumer protection C03
Storage and disposal E30a, E32a

13 ammonium sulphamate

A non-selective, inorganic, general purpose herbicide and tree-killer

Products

1	Amcide	B H & B	99.5% w/w	CR	04246
2	Root-Out	Dax	98.5% w/w	CR	03510

Uses

Annual weeds in ***forestry***, ***ornamental specimens*** *(pre-planting)*, ***vegetables*** *(pre-planting)* [1]; ***ornamental specimens***, ***trees and shrubs*** [2]. Perennial dicotyledons in ***forestry*** [1]; ***trees and shrubs*** [2]. Perennial grasses in ***forestry*** [1, 2]. Perennial weeds in ***ornamental specimens*** [2]; ***ornamental specimens*** *(pre-planting)*, ***vegetables*** *(pre-planting)* [1]. Rhododendrons in ***forestry*** [1, 2]. Woody weeds in ***forestry*** [1, 2].

Efficacy guidance

- Apply as spray to low scrub and herbaceous weeds from Apr to Sep in dry weather when rain unlikely and cultivate after 3-8 wk
- Apply as crystals in frills or notches in trunks of standing trees at any time of year
- Apply as concentrated solution or crystals to stump surfaces within 48 h of cutting. Rhododendrons must be cut level with ground and sprayed to cover cut surface, bark and immediate root area
- Stainless steel or plastic sprayers are recommended. Solutions are corrosive to mild steel, galvanised iron, brass and copper

Restrictions

- Keep spray at least 30 cm from growing plants. Low doses may be used under mature trees with undamaged bark

Following crops guidance

- Allow 8-12 wk after treatment before replanting

Environmental safety

- Harmful to fish. Do not contaminate surface waters or ditches with chemical or used container

Hazard classification and safety precautions

Operator protection U09b, U20b [1]; U11, U14, U19 [1, 2]; U15, U20a [2]
Environmental protection E01, E13c
Storage and disposal E30a [1, 2]; E32a [1]

14 asulam

A translocated carbamate herbicide for control of docks and bracken

Products

1	Asulox	Aventis	400 g/l	SL	09969
2	I T Asulam	I T Agro	400 g/l	SL	10186

Uses

Bracken in ***amenity vegetation*** [2]; ***forestry***, ***permanent pasture***, ***rough grazing*** [1, 2]; ***forestry*** *(off-label)*, ***non-crop areas*** [1]. Docks in ***amenity grass*** *(not fine turf)*, ***apple orchards***, ***blackcurrants***, ***cherries***, ***hops***, ***pear orchards***, ***permanent pasture***, ***plums*** [1, 2]; ***blackberries*** *(off-label)*, ***blueberries*** *(off-label)*, ***clover seed crops*** *(off-label)*, ***cranberries*** *(off-label)*, ***damsons*** *(off-label)*, ***gooseberries*** *(off-label)*, ***loganberries*** *(off-label)*, ***mint*** *(off-label)*, ***nectarines*** *(off-label)*, ***parsley*** *(off-label)*, ***quinces*** *(off-label)*, ***raspberries*** *(off-label)*, ***redcurrants*** *(off-label)*, ***road verges***, ***strawberries*** *(off-label)*, ***tarragon*** *(off-label)*, ***waste ground***, ***whitecurrants*** *(off-label)* [1].

Specific Off-Label Approvals (SOLAs)

- ***blackberries**, **blueberries**, **cranberries**, **damsons**, **gooseberries**, **loganberries**, **nectarines**, **quinces**, **raspberries**, **redcurrants**, **whitecurrants*** *(OLA 001892) Dec 2008* [1]
- ***clover seed crops*** *(OLA 001894) Dec 2008* [1]
- ***forestry*** *(OLA 001898) Dec 2008* [1]
- ***mint**, **parsley**, **tarragon*** *(OLA 001900) Dec 2008* [1]
- ***strawberries*** *(OLA 001896) Dec 2008* [1]

Approval information

- May be applied through CDA equipment
- Approved for aerial application on bracken in agricultural grassland, amenity grassland, forestry and rough upland intended for grazing [1, 2]. See notes in Section 1
- Approved for use near surface waters. See notes in Section 1

Efficacy guidance

- Spray bracken when fronds fully expanded but not senescent, usually Jul-Aug; docks in full leaf before flower stem emergence
- Bracken fronds must not be damaged by stock, frost or cutting before treatment
- Uptake and reliability of bracken control may be improved by use of specified additives - see label. Additives not recommended on forestry land
- To allow adequate translocation do not cut or admit stock for 14 d after spraying bracken or 7 d after spraying docks. Preferably leave undisturbed until late autumn
- Complete bracken control rarely achieved by one treatment. Survivors should be sprayed when they recover to full green frond, which may be in the ensuing year but more likely in the second year following initial application

Restrictions

- Maximum number of treatments 1 per crop or 1 per yr
- Do not apply in drought or hot, dry conditions
- Do not use in pasture before mowing for hay

Crop-specific information

- In forestry areas some young trees may be checked if sprayed directly (see label)
- In fruit crops apply as a directed spray
- Do not treat blackcurrant cuttings, hop sets or weak hills
- Some grasses and herbs will be damaged by full dose. Most sensitive are cocksfoot, Yorkshire fog, timothy, bents, annual meadow-grass, daisies, docks, plantains, saxifrage
- Apply as spot treatment in parsley, mint and tarragon, not directly to crop

Following crops guidance

- Allow at least 6 wk between spraying and planting any crop

Environmental safety

- Product is approved for use near surface waters. Whilst every care should be taken to avoid contamination of water, any that does occur from normal use should not offer harm to users, consumers of the water, or the environment
- The appropriate water regulatory body must be consulted before application near surface waters
- Keep livestock out of treated areas for at least 14 d and until foliage of poisonous weeds such as ragwort has died and become unpalatable

Hazard classification and safety precautions

Operator protection A, C, D, H [1, 2]; M [1] (for ULV application); M [2]; U08, U19, U20b [2]; U19, U20b [1] (ULV); U20c [1]

Environmental protection E07 [1] (14 d); E07 [2]; E15 [1, 2]

Storage and disposal E26, E31b [2]; E30a [1, 2]; E31a [1]

15 atrazine

A triazine herbicide with residual and foliar activity, with restricted permitted uses

Products

1	Aconite 50	DAPT	500 g/l	SC	10642
2	Alpha Atrazine 50 SC	Makhteshim	500 g/l	SC	04877
3	Atrazol	Sipcam	500 g/l	SC	07598

SECTION 4

Products – continued

4	DG90	Sipcam	90% w/w	WG	11200
5	Gesaprim	Syngenta	500 g/l	SC	08411

Uses

Annual dicotyledons in ***farm forestry***, ***forestry***, ***ornamental plant production***, ***raspberries*** [2]; ***maize***, ***sweetcorn*** [1-5]. Annual grasses in ***farm forestry***, ***forestry***, ***ornamental plant production***, ***raspberries*** [2]; ***maize***, ***sweetcorn*** [1-3, 5]. Perennial grasses in ***farm forestry***, ***forestry*** [2].

Approval information

- Atrazine was reviewed in 1992 and approvals for aerial use, and use on non-crop land (excluding home garden use) revoked. Restrictions were also placed on the number of applications that could be made to crops

Efficacy guidance

- Root activity enhanced by rainfall soon after application and reduced on high organic soils. Foliar activity effective on weeds up to 3 cm high
- Resistant weed strains may develop with repeated use of atrazine or other triazines

Restrictions

- Maximum number of applications (including other atrazine/simazine products) 1 per crop (or lower doses to 3.0 l/ha total) for maize, sweetcorn; 1 per season for conifer plantations, raspberries, roses
- Maximum total dose equivalent to one full dose treatment [2]
- Not recommended for use on soils with more than 10% organic matter
- Do not apply to raspberries in season of planting
- Do not use on Christmas trees

Crop-specific information

- May be used pre- or early post-weed emergence in maize and sweetcorn
- In conifers apply as overall spray in Feb-Apr. May be used in first spring after planting
- Apply to raspberries in spring before new cane emergence, not in season of planting
- Application rates vary with crop, soil type and weed problem. See label for details

Following crops guidance

- Latest use: 7 mth before a succeeding crop for maize, sweetcorn; Apr for conifer plantations; before cane emergence for raspberries; before weeds at 3 cm for roses

Environmental safety

- Dangerous for the environment [5]
- Very toxic to aquatic organisms. May cause long-term adverse effects in the aquatic environment [5]
- Dangerous to fish or other aquatic life and aquatic higher plants. Do not contaminate surface waters or ditches with chemical or used container
- LERAP Category B
- Use must be restricted to one product containing atrazine or simazine, and either to a single application at the maximum approved rate or (subject to any existing maximum permitted number of treatments) to several applications at lower doses up to the maximum approved rate for a single application
- On slopes heavy rainfall soon after application may cause surface run-off
- To reduce soil run-off, especially from forest plantations, users are advised to plant grass strips 6 m wide between treated areas and surface waters

Hazard classification and safety precautions

Hazard R03, R06, R14 [5]; R04 [4]
Risk phrases R14c [5]
Operator protection A, C, D, H, M [1-5]; B [3-5]; U05a [4]; U20b [2]; U20c [1, 3-5]
Environmental protection E01 [4, 5]; E13b [1-4]; E14b [1]; E15, E38 [5]; E16a, E16b [1-5]
Consumer protection C03 [4, 5]
Storage and disposal E26 [1-4]; E30a [1-5]; E31a [3, 4]; E31b [1, 2]; E31c, E32d [5]

16 azaconazole

A conazole available only in mixture

17 azaconazole + imazalil

A fungicide mixture for use in horticulture

Products

Nectec Paste	Certis	1:2% w/w	PA	08510

Uses

Canker in ***ornamental plant production***. Silver leaf in ***ornamental plant production***.

Approval information

- Imazalil included in Annex I under EC Directive 91/414
- Products containing azaconazole have been granted derogations for specified 'Essential Uses' for use until 31 December 2007. Sale and supply must cease by 30 June 2007 but growers have no guarantee that the products will continue to be available until then.
  For more information see 'The Review Programme' under 'Pesticide Legislation' in Section 1

Efficacy guidance

- Paint pruning cuts immediately. If necessary clean wounds and cut back any loose bark
- Ensure whole of cut area is fully covered beyond wound to surrounding healthy bark
- Cut back established cankers to sound healthy wood before treatment. Work paste into all crevices

Restrictions

- Maximum number of treatments 1 per wound per yr
- Treat only during dry weather and not in frosty conditions
- Use only during dormant periods
- Do not apply to grafting cuts

Environmental safety

- Harmful to fish or other aquatic life. Do not contaminate surface waters or ditches with chemical or used container

Hazard classification and safety precautions

Operator protection A; U05a, U20c
Environmental protection E01, E13c
Consumer protection C03
Storage and disposal E26, E30a

18 azamethiphos

A residual organophosphorus insecticide for fly control

Products

Alfacron 10 WP	Novartis A H	10% w/w	WP	09439

Uses

Flies in ***livestock houses***.

Approval information

- Approvals for azamethiphos allowed to continue, subject to additional operator and worker protection requirements, following the UK Review programme of cholinesterase compounds

Efficacy guidance

- Add water as directed to produce paint consistency and paint onto 2.5% of total wall and ceiling surface area at a minimum of 5 points in building or apply as spray to 30% of surface area

Restrictions

- This product contains an anticholinesterase organophosphorus compound. Do not use if under medical advice not to work with such compounds
- Only apply to areas out of reach of children and animals

SECTION 4

Environmental safety

- Dangerous to fish. Do not contaminate surface waters or ditches with chemical or used container
- Do not apply directly to livestock and poultry
- Do not apply to surfaces on which food or feed is stored, prepared or eaten. Cover feedstuffs and remove exposed milk and eggs before application

Hazard classification and safety precautions

Hazard R04
Risk phrases R04a, R04b, R04e
Operator protection A, C; U05a, U10, U13, U14, U15, U19, U20b
Environmental protection E01, E05, E13b
Consumer protection C03, C04, C05, C06, C07
Storage and disposal E30a, E32a
Medical advice M01

19 azoxystrobin

A systemic translaminar and protectant strobilurin fungicide for cereals

Products

1	Amistar	Syngenta	250 g/l	SC	10443
2	Barclay ZX	Barclay	250 g/l	SC	11336
3	Heritage	Scotts	50% w/w	WG	11383
4	Landgold Strobilurin 250	Landgold	250 g/l	SC	09595
5	Me2 Azoxystrobin	Me2	250 g/l	SC	09654
6	Standon Azoxystrobin	Standon	250 g/l	SC	09515

Uses

Alternaria in ***carrots*** [1]. Ascochyta in ***combining peas***, ***spring field beans*** *(off-label)*, ***vining peas***, ***winter field beans*** *(off-label)* [1]. Black dot in ***potatoes*** *(off-label - for baby tuber production)* [1]. Black rot in ***broccoli*** *(off-label)*, ***brussels sprouts*** *(off-label)*, ***cabbages*** *(off-label)*, ***calabrese*** *(off-label)*, ***cauliflowers*** *(off-label)* [1]. Black spot in ***protected strawberries*** *(off-label)*, ***strawberries*** *(off-label)* [1]. Botrytis in ***celery*** *(off-label)*, ***dwarf beans*** *(off-label)*, ***navy beans*** *(off-label)*, ***protected celery*** *(off-label)*, ***runner beans*** *(off-label)* [1]. Brown rust in ***spring barley***, ***spring wheat***, ***winter barley***, ***winter wheat*** [1, 2, 4-6]; ***spring rye*** *(off-label)*, ***triticale***, ***winter rye*** [1]. Crown rust in ***spring oats***, ***winter oats*** [1]. Dark leaf spot in ***salad brassicas*** *(off-label - for baby leaf production)* [1]. Didymella in ***inert substrate aubergines*** *(off-label)*, ***inert substrate tomatoes*** *(off-label)* [1]. Didymella stem rot in ***aubergines*** *(off-label)*, ***protected tomatoes*** *(off-label)* [1]. Downy mildew in ***bulb onions***, ***inert substrate courgettes*** *(off-label)*, ***inert substrate cucumbers*** *(off-label)*, ***inert substrate gherkins*** *(off-label)*, ***protected courgettes*** *(off-label)*, ***protected cucumbers*** *(off-label)*, ***protected gherkins*** *(off-label)*, ***salad brassicas*** *(off-label - for baby leaf production)*, ***salad onions*** *(off-label)* [1]. Fusarium patch in ***amenity grass***, ***managed amenity turf*** [3]. Glume blotch in ***spring wheat***, ***winter wheat*** [1, 2, 4-6]. Grey mould in ***aubergines*** *(off-label)*, ***inert substrate aubergines*** *(off-label)*, ***inert substrate tomatoes*** *(off-label)*, ***protected tomatoes*** *(off-label)* [1]. Late blight in ***aubergines*** *(off-label)*, ***inert substrate aubergines*** *(off-label)*, ***inert substrate tomatoes*** *(off-label)*, ***protected tomatoes*** *(off-label)* [1]. Late ear diseases in ***spring wheat***, ***winter wheat*** [1, 2, 4-6]. Net blotch in ***spring barley***, ***winter barley*** [1, 2, 4-6]. Powdery mildew in ***agastache spp.*** *(off-label)*, ***angelica*** *(off-label)*, ***aubergines*** *(off-label)*, ***balm*** *(off-label)*, ***basil*** *(off-label)*, ***bay*** *(off-label)*, ***borage*** *(off-label)*, ***camomile*** *(off-label)*, ***caraway*** *(off-label)*, ***carrots***, ***catnip*** *(off-label)*, ***chervil*** *(off-label)*, ***chives*** *(off-label)*, ***clary*** *(off-label)*, ***coriander*** *(off-label)*, ***cress*** *(off-label)*, ***curry plant*** *(off-label)*, ***dill*** *(off-label)*, ***dragonhead*** *(off-label)*, ***fennel*** *(off-label)*, ***fenugreek*** *(off-label)*, ***feverfew*** *(off-label)*, ***french lavender*** *(off-label)*, ***frise*** *(off-label)*, ***hyssop*** *(off-label)*, ***inert substrate aubergines*** *(off-label)*, ***inert substrate courgettes*** *(off-label)*, ***inert substrate cucumbers*** *(off-label)*, ***inert substrate gherkins*** *(off-label)*, ***inert substrate tomatoes*** *(off-label)*, ***korean mint*** *(off-label)*, ***lamb's lettuce*** *(off-label)*, ***land cress*** *(off-label)*, ***lavandin*** *(off-label)*, ***lavender*** *(off-label)*, ***lemon verbena*** *(off-label)*, ***lovage*** *(off-label)*, ***marigold*** *(off-label)*, ***marjoram*** *(off-label)*, ***mint*** *(off-label)*, ***nasturtium*** *(off-label)*, ***nettle*** *(off-label)*, ***oregano*** *(off-label)*, ***origanum heracleoticum*** *(off-label)*, ***parsley*** *(off-label)*, ***poppies*** *(off-label - for morphine production)*, ***protected courgettes*** *(off-label)*, ***protected cucumbers*** *(off-label)*, ***protected gherkins*** *(off-label)*, ***protected peppers*** *(off-label)*, ***protected tomatoes*** *(off-label)*, ***radicchio*** *(off-label)*, ***raspberries*** *(off-label)*, ***rocket*** *(off-label)*, ***rosemary*** *(off-label)*, ***rue*** *(off-label)*, ***sage*** *(off-label)*, ***salad burnet*** *(off-label)*, ***savory*** *(off-label)*, ***scarole*** *(off-label)*, ***sorrel*** *(off-label)*, ***spike lavender*** *(off-label)*, ***spring oats***, ***tarragon*** *(off-label)*, ***thyme*** *(off-label)*, ***triticale***, ***violet*** *(off-label)*, ***winter oats***, ***winter rye***, ***woodruff*** *(off-label)* [1]; ***spring barley***, ***winter barley*** [1, 2, 4-6]; ***spring wheat*** [1, 6]; ***winter wheat*** [6]. Purple blotch in ***leeks*** [1]. Rhizoctonia in ***celery*** *(off-label)*, ***protected agastache spp.*** *(off-label)*, ***protected angelica*** *(off-label)*, ***protected balm*** *(off-label)*, ***protected basil*** *(off-label)*, ***protected bay*** *(off-label)*, ***protected borage*** *(off-label)*, ***protected camomile*** *(off-label)*, ***protected caraway*** *(off-label)*, ***protected catnip*** *(off-label)*, ***protected celery*** *(off-label)*, ***protected chervil*** *(off-label)*, ***protected chives*** *(off-label)*, ***protected clary*** *(off-label)*, ***protected coriander*** *(off-label)*, ***protected curry plant*** *(off-label)*, ***protected dill*** *(off-label)*, ***protected dragonhead*** *(off-label)*, ***protected fennel*** *(off-label)*, ***protected fenugreek*** *(off-label)*, ***protected feverfew*** *(off-label)*, ***protected horehound*** *(off-label)*, ***protected hyssop*** *(off-label)*, ***protected land cress*** *(off-label)*, ***protected lavandin*** *(off-label)*, ***protected lavender*** *(off-label)*, ***protected lemon verbena*** *(off-label)*, ***protected lettuce*** *(off-label)*, ***protected lovage*** *(off-label)*, ***protected marigold*** *(off-label)*, ***protected marjoram*** *(off-label)*, ***protected mint*** *(off-label)*, ***protected nasturtium*** *(off-label)*, ***protected nettle*** *(off-label)*, ***protected oregano*** *(off-label)*, ***protected origanum heracleoticum*** *(off-label)*, ***protected parsley*** *(off-label)*, ***protected plantain*** *(off-label)*, ***protected rocket*** *(off-label)*, ***protected rosemary*** *(off-label)*, ***protected rue*** *(off-label)*, ***protected sage*** *(off-label)*, ***protected salad brassicas*** *(off-label - for baby leaf production)*, ***protected salad burnet*** *(off-label)*, ***protected savory*** *(off-label)*, ***protected sorrel*** *(off-label)*, ***protected spike lavender*** *(off-label)*, ***protected st john's wort*** *(off-label)*, ***protected tarragon*** *(off-label)*, ***protected thyme*** *(off-label)*, ***protected violet*** *(off-label)*, ***protected woodruff*** *(off-label)*, ***protected yarrow*** *(off-label)*, ***salad brassicas*** *(off-label - for baby leaf production)*, ***swedes*** *(off-label)*, ***turnips*** *(off-label)* [1]. Rhynchosporium in ***spring barley***, ***winter barley*** [1, 2, 4-6]; ***spring rye*** *(off-label)*, ***triticale***, ***winter rye*** [1]. Ring spot in ***agastache spp.*** *(off-label)*, ***angelica*** *(off-label)*, ***balm*** *(off-label)*, ***basil*** *(off-label)*, ***bay*** *(off-label)*, ***borage*** *(off-label)*, ***camomile*** *(off-label)*, ***caraway*** *(off-label)*, ***catnip*** *(off-label)*, ***chervil*** *(off-label)*, ***chives*** *(off-label)*, ***clary*** *(off-label)*, ***coriander*** *(off-label)*, ***cress*** *(off-label)*, ***curry plant*** *(off-label)*, ***dill*** *(off-label)*, ***dragonhead*** *(off-label)*, ***fennel*** *(off-label)*, ***fenugreek*** *(off-label)*, ***feverfew*** *(off-label)*, ***french lavender*** *(off-label)*, ***frise*** *(off-label)*, ***hyssop*** *(off-label)*, ***korean mint*** *(off-label)*, ***lamb's lettuce*** *(off-label)*, ***land cress*** *(off-label)*, ***lavandin*** *(off-label)*, ***lavender*** *(off-label)*, ***lemon verbena*** *(off-label)*, ***lovage*** *(off-label)*, ***marigold*** *(off-label)*, ***marjoram*** *(off-label)*, ***mint*** *(off-label)*, ***nasturtium*** *(off-label)*, ***nettle*** *(off-label)*, ***oregano*** *(off-label)*, ***origanum heracleoticum*** *(off-label)*, ***parsley*** *(off-label)*, ***radicchio*** *(off-label)*, ***rocket*** *(off-label)*, ***rosemary*** *(off-label)*, ***rue*** *(off-label)*, ***sage*** *(off-label)*, ***salad burnet*** *(off-label)*, ***savory*** *(off-label)*, ***scarole*** *(off-label)*, ***sorrel*** *(off-label)*, ***spike lavender*** *(off-label)*, ***tarragon*** *(off-label)*, ***thyme*** *(off-label)*, ***violet*** *(off-label)*, ***woodruff*** *(off-label)* [1]. Root malformation disorder in ***red beet*** *(off-label)* [1]. Rust in ***agastache spp.*** *(off-label)*, ***angelica*** *(off-label)*, ***balm*** *(off-label)*, ***basil*** *(off-label)*, ***bay*** *(off-label)*, ***borage*** *(off-label)*, ***camomile*** *(off-label)*, ***caraway*** *(off-label)*, ***catnip*** *(off-label)*, ***chervil*** *(off-label)*, ***chives*** *(off-label)*, ***clary*** *(off-label)*, ***coriander*** *(off-label)*, ***cress*** *(off-label)*, ***curry plant*** *(off-label)*, ***dill*** *(off-label)*, ***dragonhead*** *(off-label)*, ***fennel*** *(off-label)*, ***fenugreek*** *(off-label)*, ***feverfew*** *(off-label)*, ***french lavender*** *(off-label)*, ***frise*** *(off-label)*, ***hyssop*** *(off-label)*, ***korean mint*** *(off-label)*, ***lamb's lettuce*** *(off-label)*, ***land cress*** *(off-label)*, ***lavandin*** *(off-label)*, ***lavender*** *(off-label)*, ***leeks***, ***lemon verbena*** *(off-label)*, ***lovage*** *(off-label)*, ***marigold*** *(off-label)*, ***marjoram*** *(off-label)*, ***mint*** *(off-label)*, ***nasturtium*** *(off-label)*, ***nettle*** *(off-label)*, ***oregano*** *(off-label)*, ***origanum heracleoticum*** *(off-label)*, ***parsley*** *(off-label)*, ***radicchio*** *(off-label)*, ***rocket*** *(off-label)*, ***rosemary*** *(off-label)*, ***rue*** *(off-label)*, ***sage*** *(off-label)*, ***salad burnet*** *(off-label)*, ***savory*** *(off-label)*, ***scarole*** *(off-label)*, ***sorrel*** *(off-label)*, ***spike lavender*** *(off-label)*, ***tarragon*** *(off-label)*, ***thyme*** *(off-label)*, ***violet*** *(off-label)*, ***woodruff*** *(off-label)* [1]. Sclerotinia in ***celeriac*** *(off-label)*, ***celery*** *(off-label)*, ***lettuce*** *(off-label)*, ***protected celery*** *(off-label)* [1]. Septoria leaf spot in ***spring wheat***, ***winter wheat*** [1, 2, 4-6]. White rust in ***chrysanthemums*** *(off-label - in pots)*, ***protected chrysanthemums*** *(off-label)* [1]. Yellow rust in ***spring wheat***, ***winter wheat*** [1, 2, 4-6].

Specific Off-Label Approvals (SOLAs)

- ***agastache spp.***, ***angelica***, ***balm***, ***basil***, ***bay***, ***borage***, ***camomile***, ***caraway***, ***catnip***, ***chervil***, ***chives***, ***clary***, ***coriander***, ***cress***, ***curry plant***, ***dill***, ***dragonhead***, ***fennel***, ***fenugreek***, ***feverfew***, ***french lavender***, ***frise***, ***hyssop***, ***korean mint***, ***lamb's lettuce***, ***land cress***, ***lavandin***, ***lavender***, ***lemon verbena***, ***lovage***, ***marigold***, ***marjoram***, ***mint***, ***nasturtium***, ***nettle***, ***oregano***, ***origanum heracleoticum***, ***parsley***, ***radicchio***, ***rocket***, ***rosemary***, ***rue***, ***sage***, ***salad burnet***, ***savory***, ***scarole***, ***sorrel***, ***spike lavender***, ***tarragon***, ***thyme***, ***violet***, ***woodruff*** *(OLA 021293) Jul 2008* [1]

SECTION 4

- ***aubergines*, *protected courgettes*, *protected cucumbers*, *protected gherkins*, *protected tomatoes*** *(OLA 021533) Jul 2008* [1]
- ***broccoli*, *cabbages*, *calabrese*, *cauliflowers*, *red beet*, *swedes*, *turnips*** *(OLA 031126) Jul 2008* [1]
- ***brussels sprouts*** *(OLA 031995) Jul 2008* [1]
- ***celeriac*** *(OLA 031862) Jul 2008* [1]
- ***celery*, *protected celery*** *(OLA 001041) Jul 2008* [1]
- ***chrysanthemums*** *(in pots) (OLA 011684) Jul 2008* [1]
- ***dwarf beans*, *navy beans*, *runner beans*, *spring field beans*, *winter field beans*** *(OLA 022066) Jul 2008* [1]
- ***inert substrate aubergines*, *inert substrate courgettes*, *inert substrate cucumbers*, *inert substrate gherkins*, *inert substrate tomatoes*** *(OLA 011685) Jul 2008* [1]
- ***lettuce*** *(OLA 011465) Jul 2008* [1]
- ***poppies*** *(for morphine production) (OLA 031137) Jul 2008* [1]
- ***potatoes*** *(for baby tuber production) (OLA 030068) Dec 2008* [1]
- ***protected agastache spp.*, *protected angelica*, *protected balm*, *protected basil*, *protected bay*, *protected borage*, *protected camomile*, *protected caraway*, *protected catnip*, *protected chervil*, *protected chives*, *protected clary*, *protected coriander*, *protected curry plant*, *protected dill*, *protected dragonhead*, *protected fennel*, *protected fenugreek*, *protected feverfew*, *protected horehound*, *protected hyssop*, *protected lavandin*, *protected lavender*, *protected lemon verbena*, *protected lettuce*, *protected lovage*, *protected marigold*, *protected marjoram*, *protected mint*, *protected nasturtium*, *protected nettle*, *protected oregano*, *protected origanum heracleoticum*, *protected parsley*, *protected plantain*, *protected rocket*, *protected rosemary*, *protected rue*, *protected sage*, *protected salad burnet*, *protected savory*, *protected sorrel*, *protected spike lavender*, *protected st john's wort*, *protected tarragon*, *protected thyme*, *protected violet*, *protected woodruff*, *protected yarrow*** *(OLA 023630) Jul 2008* [1]
- ***protected agastache spp.*, *protected angelica*, *protected balm*, *protected basil*, *protected bay*, *protected borage*, *protected camomile*, *protected caraway*, *protected catnip*, *protected chervil*, *protected chives*, *protected clary*, *protected coriander*, *protected curry plant*, *protected dill*, *protected dragonhead*, *protected fennel*, *protected fenugreek*, *protected feverfew*, *protected horehound*, *protected hyssop*, *protected land cress*, *protected lavandin*, *protected lavender*, *protected lemon verbena*, *protected lettuce*, *protected lovage*, *protected marigold*, *protected marjoram*, *protected mint*, *protected nasturtium*, *protected nettle*, *protected oregano*, *protected origanum heracleoticum*, *protected parsley*, *protected plantain*, *protected rocket*, *protected rosemary*, *protected rue*, *protected sage*, *protected salad burnet*, *protected savory*, *protected sorrel*, *protected spike lavender*, *protected st john's wort*, *protected tarragon*, *protected thyme*, *protected violet*, *protected woodruff*, *protected yarrow*** *(OLA 030659) Jul 2008* [1]
- ***protected chrysanthemums*** *(OLA 001536) May 2008* [1]
- ***protected chrysanthemums*** *(OLA 011684) Jul 2008* [1]
- ***protected peppers*** *(OLA 021295) Jul 2008* [1]
- ***protected salad brassicas*** *(for baby leaf production) (OLA 030659) Jul 2008* [1]
- ***protected strawberries*, *strawberries*** *(OLA 021294) Jul 2008* [1]
- ***raspberries*** *(OLA 030365) Jul 2008* [1]
- ***salad brassicas*** *(for baby leaf production) (OLA 011465) Jul 2008* [1]
- ***salad brassicas*** *(for baby leaf production) (OLA 023630) Jul 2008* [1]
- ***salad onions*** *(OLA 021687) Jul 2008* [1]

Approval information

- Azoxystrobin included in Annex I under EC Directive 91/414

Efficacy guidance

- Best results obtained from use as a protectant or during early stages of disease establishment or when a predictive assessment indicates a risk of disease development
- In cereals control of established infections can be improved by appropriate tank mixtures or application as part of a programme
- Treatment under poor growing conditions may give less reliable results
- Azoxystrobin is a member of the QoI cross resistance group. Product should be used preventatively and not relied on for its curative potential

- Use product in cereals as part of an Integrated Crop Management strategy incorporating other methods of control, including where appropriate other fungicides with a different mode of action. Do not apply more than two foliar applications of QoI containing products to any cereal crop
- In turf use product at full dose rate in a disease control programme, alternating with fungicides of different modes of action
- There is a significant risk of widespread resistance occurring in *Septoria tritici* populations in UK. Failure to follow resistance management action may result in reduced levels of disease control
- On cereal crops product must always be used in mixture with another product, recommended for control of the same target disease, that contains a fungicide from a different cross resistance group and is applied at a dose that will give robust control
- Strains of barley powdery mildew resistant to QoI's are common in the UK
- For good control of Fusarium patch in amenity turf and grass repeat treamtment at minimum intervals of 2 wk

Restrictions

- Maximum total dose ranges from 2-4 times the single full dose depending on crop and product. See labels for details
- On turf the maximum number of treatments is 4 per yr but they must not exceed one third of the total number of fungicide treatments applied
- Avoid sequential use alone against powdery mildew
- Do not apply to turf when ground is frozen or during drought

Crop-specific information

- In cereals control generally lasts 4-6 wk during period of stem elongation
- Treat pea crops when not under stress and check leaf wax level if necessary. Consult processor before treating pea crops for processing
- Latest use: grain watery ripe (GS 71) for cereals
- HI 3 d for inert substrate vegetables; 10 d for carrots, 14 d for onions, celery, lettuce, salad brassicas, vining peas; 21 d for leeks; 36 d for combining peas

Environmental safety

- Dangerous for the environment [1]
- Dangerous (extremely dangerous [3]) to fish or other aquatic life. Do not contaminate surface waters or ditches with chemical or used container [2, 4-6]
- Avoid spray drift onto surrounding areas or crops, especially apples, plums or privet

Hazard classification and safety precautions

Hazard R14 [1]
Operator protection U02 [2]; U05a [1]; U09a, U19, U20b [1, 2, 4-6]; U20c [3]
Environmental protection E01, E15, E38 [1]; E13a [3]; E13b [2, 4-6]
Consumer protection C03 [1]
Storage and disposal E24 [3, 5]; E26, E30a [1-6]; E31b [3-6]; E31c [1, 2]; E32d [1]
Medical advice M05a [3]

20 azoxystrobin + cyproconazole

A contact and systemic broad spectrum fungicide mixture for cereals

Products

Priori Xtra	Syngenta	200:80 g/l	SC	11518

Uses

Brown rust in ***spring barley***, ***spring rye***, ***spring wheat***, ***winter barley***, ***winter rye***, ***winter wheat***. Crown rust in ***spring oats***, ***winter oats***. Eyespot in ***spring barley*** *(reduction)*, ***spring wheat*** *(reduction)*, ***winter barley*** *(reduction)*, ***winter wheat*** *(reduction)*. Net blotch in ***spring barley***, ***winter barley***. Powdery mildew in ***spring barley***, ***spring oats***, ***spring rye***, ***spring wheat***, ***winter barley***, ***winter oats***, ***winter rye***, ***winter wheat***. Rhynchosporium in ***spring barley*** *(moderate control)*, ***spring rye*** *(moderate control)*, ***winter barley*** *(moderate control)*, ***winter rye*** *(moderate control)*. Septoria diseases in ***spring wheat***, ***winter wheat***. Yellow rust in ***spring wheat***, ***winter wheat***.

Approval information

- Azoxystrobin included in Annex I under EC Directive 91/414

SECTION 4

Efficacy guidance

- Best results obtained from treatment during the early stages of disease development
- A second application may be needed if disease attack is prolonged
- Azoxystrobin is a member of the QoI cross resistance group. Product should be used preventatively and not relied on for its curative potential
- Use product as part of an Integrated Crop Management strategy incorporating other methods of control, including where appropriate other fungicides with a different mode of action. Do not apply more than two foliar applications of QoI containing products to any cereal crop
- There is a significant risk of widespread resistance occurring in *Septoria tritici* populations in UK. Failure to follow resistance management action may result in reduced levels of disease control
- Strains of wheat and barley powdery mildew resistant to QoI's are common in the UK. Control of wheat powdery mildew can only be relied upon from the triazole component
- On cereal crops product must always be used in mixture with another product, recommended for control of the same target disease, that contains a fungicide from a different cross resistance group and is applied at a dose that will give robust control

Restrictions

- Maximum number of treatments 2 per crop

Crop-specific information

- Latest use: up to and including anthesis complete (GS 69) for rye and wheat; up to and including emergence of ear complete (GS 59) for barley and oats

Environmental safety

- Dangerous for the environment
- Very toxic to aquatic organisms. May cause long-term adverse effects in the aquatic environment

Hazard classification and safety precautions

Hazard R03, R14
Risk phrases R03c, R10a, R14c
Operator protection A; U05a, U09a, U19, U20b
Environmental protection E01, E15, E38
Consumer protection C03
Storage and disposal E26, E30a, E31c, E32d, E34
Medical advice M03

21 azoxystrobin + fenpropimorph

A protectant and eradicant fungicide mixture

Products

Amistar Pro	Syngenta	100:280 g/l	SE	10513

Uses

Brown rust in ***spring barley***, ***spring wheat***, ***winter barley***, ***winter wheat***. Late ear diseases in ***spring wheat***, ***winter wheat***. Net blotch in ***spring barley***, ***winter barley***. Powdery mildew in ***spring barley***, ***winter barley***. Rhynchosporium in ***spring barley***, ***winter barley***. Septoria diseases in ***spring wheat***, ***winter wheat***. Yellow rust in ***spring wheat***, ***winter wheat***.

Approval information

- Azoxystrobin included in Annex I under EC Directive 91/414

Efficacy guidance

- Best results obtained from application before infection following a disease risk assessment, or when disease first seen in crop
- Results may be less reliable when used on crops under stress
- Treatments for protection against ear disease should be made at ear emergence
- Azoxystrobin is a member of the QoI cross resistance group. Product should be used preventatively and not relied on for its curative potential
- Use product as part of an Integrated Crop Management strategy incorporating other methods of control, including where appropriate other fungicides with a different mode of action. Do not apply more than two foliar applications of QoI containing products to any cereal crop
- There is a significant risk of widespread resistance occurring in *Septoria tritici* populations in UK. Failure to follow resistance management action may result in reduced levels of disease control

- Strains of barley powdery mildew resistant to QoI's are common in the UK
- On cereal crops product must always be used in mixture with another product, recommended for control of the same target disease, that contains a fungicide from a different cross resistance group and is applied at a dose that will give robust control

Restrictions

- Maximum total dose equivalent to two full dose treatments

Crop-specific information

- Latest use: before early milk stage (GS 73)
- HI 5 wk

Environmental safety

- Dangerous for the environment
- Very toxic to aquatic organisms. May cause long-term adverse effects in the aquatic environment

Hazard classification and safety precautions

Hazard R03, R04, R14
Risk phrases R03b, R04b, R04e, R10a, R14c
Operator protection A, H; U02, U05a, U09a, U14, U15, U19, U20b
Environmental protection E01, E15, E38
Consumer protection C03
Storage and disposal E26, E30a, E31c, E32d
Medical advice M03

22 Bacillus thuringiensis

A bacterial insecticide for control of caterpillars

Products

Dipel DF	Fargro	32000 IU/mg	WG	11184

Uses

Caterpillars in ***agastache spp.*** *(off-label)*, ***amenity vegetation***, ***angelica*** *(off-label)*, ***apples*** *(off-label)*, ***balm*** *(off-label)*, ***basil*** *(off-label)*, ***bay*** *(off-label)*, ***borage*** *(off-label)*, ***broccoli***, ***brussels sprouts***, ***cabbages***, ***camomile*** *(off-label)*, ***caraway*** *(off-label)*, ***cauliflowers***, ***cherries*** *(off-label)*, ***chervil*** *(off-label)*, ***chives*** *(off-label)*, ***clary*** *(off-label)*, ***coriander*** *(off-label)*, ***curry plant*** *(off-label)*, ***dill*** *(off-label)*, ***dragonhead*** *(off-label)*, ***fennel*** *(off-label)*, ***fenugreek*** *(off-label)*, ***feverfew*** *(off-label)*, ***horehound*** *(off-label)*, ***hyssop*** *(off-label)*, ***land cress*** *(off-label)*, ***lavandin*** *(off-label)*, ***lavender*** *(off-label)*, ***lemon verbena*** *(off-label)*, ***lettuce*** *(off-label)*, ***lovage*** *(off-label)*, ***marigold*** *(off-label)*, ***marjoram*** *(off-label)*, ***mint*** *(off-label)*, ***nasturtium*** *(off-label)*, ***nettle*** *(off-label)*, ***oregano*** *(off-label)*, ***origanum heracleoticum*** *(off-label)*, ***ornamental specimens***, ***parsley*** *(off-label)*, ***pears*** *(off-label)*, ***plantain*** *(off-label)*, ***protected agastache spp.*** *(off-label)*, ***protected angelica*** *(off-label)*, ***protected balm*** *(off-label)*, ***protected basil*** *(off-label)*, ***protected bay*** *(off-label)*, ***protected borage*** *(off-label)*, ***protected camomile*** *(off-label)*, ***protected caraway*** *(off-label)*, ***protected chervil*** *(off-label)*, ***protected chives*** *(off-label)*, ***protected clary*** *(off-label)*, ***protected coriander*** *(off-label)*, ***protected cucumbers***, ***protected curry plant*** *(off-label)*, ***protected dill*** *(off-label)*, ***protected dragonhead*** *(off-label)*, ***protected fennel*** *(off-label)*, ***protected fenugreek*** *(off-label)*, ***protected feverfew*** *(off-label)*, ***protected horehound*** *(off-label)*, ***protected hyssop*** *(off-label)*, ***protected land cress*** *(off-label)*, ***protected lavandin*** *(off-label)*, ***protected lavender*** *(off-label)*, ***protected lemon verbena*** *(off-label)*, ***protected lettuce*** *(off-label)*, ***protected lovage*** *(off-label)*, ***protected marigold*** *(off-label)*, ***protected marjoram*** *(off-label)*, ***protected mint*** *(off-label)*, ***protected nasturtium*** *(off-label)*, ***protected nettle*** *(off-label)*, ***protected oregano*** *(off-label)*, ***protected origanum heracleoticum*** *(off-label)*, ***protected ornamentals***, ***protected parsley*** *(off-label)*, ***protected peppers***, ***protected plantain*** *(off-label)*, ***protected rocket*** *(off-label)*, ***protected rosemary*** *(off-label)*, ***protected rue*** *(off-label)*, ***protected sage*** *(off-label)*, ***protected salad burnet*** *(off-label)*, ***protected savory*** *(off-label)*, ***protected sorrel*** *(off-label)*, ***protected spike lavender*** *(off-label)*, ***protected st john's wort*** *(off-label)*, ***protected tarragon*** *(off-label)*, ***protected thyme*** *(off-label)*, ***protected tomatoes***, ***protected violet*** *(off-label)*, ***protected watercress*** *(off-label)*, ***protected woodruff*** *(off-label)*, ***protected yarrow*** *(off-label)*, ***raspberries***, ***rocket*** *(off-label)*, ***rosemary*** *(off-label)*, ***rue*** *(off-label)*, ***sage*** *(off-label)*, ***salad burnet*** *(off-label)*, ***savory*** *(off-label)*, ***sorrel*** *(off-label)*, ***spike lavender*** *(off-label)*, ***st john's wort*** *(off-label)*,

strawberries, ***tarragon*** *(off-label)*, ***thyme*** *(off-label)*, ***violet*** *(off-label)*, ***watercress*** *(off-label)*, ***woodruff*** *(off-label)*, ***yarrow*** *(off-label)*.

Specific Off-Label Approvals (SOLAs)

- ***agastache spp., angelica, apples, balm, basil, bay, borage, camomile, caraway, cherries, chervil, chives, clary, coriander, curry plant, dill, dragonhead, fennel, fenugreek, feverfew, horehound, hyssop, land cress, lavandin, lavender, lemon verbena, lettuce, lovage, marigold, marjoram, mint, nasturtium, nettle, oregano, origanum heracleoticum, parsley, pears, plantain, protected agastache spp., protected angelica, protected balm, protected basil, protected bay, protected borage, protected camomile, protected caraway, protected chervil, protected chives, protected clary, protected coriander, protected curry plant, protected dill, protected dragonhead, protected fennel, protected fenugreek, protected feverfew, protected horehound, protected hyssop, protected land cress, protected lavandin, protected lavender, protected lemon verbena, protected lettuce, protected lovage, protected marigold, protected marjoram, protected mint, protected nasturtium, protected nettle, protected oregano, protected origanum heracleoticum, protected parsley, protected plantain, protected rocket, protected rosemary, protected rue, protected sage, protected salad burnet, protected savory, protected sorrel, protected spike lavender, protected st john's wort, protected tarragon, protected thyme, protected violet, protected woodruff, protected yarrow, rocket, rosemary, rue, sage, salad burnet, savory, sorrel, spike lavender, st john's wort, tarragon, thyme, violet, woodruff, yarrow*** *(OLA 022577) Dec 2008* [1]
- ***protected watercress, watercress*** *(OLA 031071) Dec 2008* [1]

Efficacy guidance

- Pest control achieved by ingestion by caterpillars of the treated plant vegetation. Caterpillars cease feeding and die in 2-3 d
- Apply as soon as larvae appear on crop and repeat every 7-10 d until the end of the hatching period
- Good coverage is essential, especially of undersides of leaves. Spray onto dry foliage and do not apply if rain expected within 6 h

Restrictions

- No restriction on number of treatments on edible crops
- Apply spray mixture as soon as possible after preparation

Crop-specific information

- HI zero

Environmental safety

- Store out of direct sunlight

Hazard classification and safety precautions

Operator protection A, C; U05a, U20c
Environmental protection E01, E15
Consumer protection C03
Storage and disposal E26, E30a, E32a

23 benalaxyl

A phenylamide (acylalanine) fungicide available only in mixtures

24 benalaxyl + mancozeb

A systemic and protectant fungicide mixture

Products

1	Galben M	Sipcam	8:65% w/w	WP	07220
2	Intro Plus	Interfarm	8:65% w/w	WP	11630
3	Tairel	Sipcam	8:65% w/w	WB	07767

Uses

Blight in ***early potatoes***, ***maincrop potatoes***, ***protected potatoes***.

Approval information

- Approved for aerial application on potatoes [1, 3]. See notes in Section 1

Efficacy guidance

- Apply to potatoes at blight warning prior to crop becoming infected and repeat at 10-21 d intervals depending on risk of infection
- Spray irrigated potatoes after irrigation and at 14 d intervals, crops in polythene tunnels at 10 d intervals
- When active potato growth ceases use a non-systemic fungicide to end of season, starting not more than 10 d after last application
- For reduction of downy mildew in oilseed rape apply at seedling to 7-leaf stage, before end Nov as soon as infection seen

Restrictions

- Maximum number of treatments 5 per crop (including other phenylamide-based fungicides) for potatoes, 1 per crop for winter oilseed rape
- Do not treat potatoes showing active blight infection

Crop-specific information

- HI 7 d

Environmental safety

- Dangerous to fish or other aquatic life. Do not contaminate surface waters or ditches with chemical or used container

Hazard classification and safety precautions

Hazard R04
Risk phrases R04a
Operator protection A; U05a, U08, U20b
Environmental protection E01, E13b
Consumer protection C02 [3] (7 d); C03 [1-3]
Storage and disposal E29, E30a [1-3]; E32a [1, 2]

25 benfuracarb

A soil-applied carbamate insecticide and nematicide for beet crops

Products

Oncol 10G	Nufarm UK	10% w/w	GR	08249

Uses

Docking disorder vectors in ***fodder beet***, ***mangels***, ***sugar beet***. Leaf miner in ***fodder beet***, ***mangels***, ***sugar beet***. Millipedes in ***fodder beet***, ***mangels***, ***sugar beet***. Pygmy beetle in ***fodder beet***, ***mangels***, ***sugar beet***. Springtails in ***fodder beet***, ***mangels***, ***sugar beet***. Symphylids in ***fodder beet***, ***mangels***, ***sugar beet***.

Efficacy guidance

- Apply at sowing with suitable applicator so that the granules are mixed with the moving soil closing the seed furrow. See label for recommended applicators and calibration

Restrictions

- Maximum number of treatments 1 per crop

Crop-specific information

- Latest use: at planting

Environmental safety

- Dangerous to game, wild birds and animals
- Extremely dangerous to fish or other aquatic life. Do not contaminate surface waters or ditches with chemical or used container

Hazard classification and safety precautions

Hazard R04
Risk phrases R04a, R04b
Operator protection A, B, C, D, E, H, K, M; U02, U05a, U09a, U13, U19, U20a
Environmental protection E01, E10a, E13a
Consumer protection C03
Storage and disposal E30a, E32a
Medical advice M02

SECTION 4

26 benomyl

A systemic, MBC fungicide with protectant and eradicant activity

Products

Benlate Fungicide	DuPont	50% w/w	WP	00229

Uses

Black spot in ***courgettes*** *(off-label)*. Botrytis in ***protected rhubarb*** *(off-label)*. Chocolate spot in ***spring field beans***, ***winter field beans***. Fungus diseases in ***protected rhubarb*** *(off-label)*. Gummosis in ***courgettes*** *(off-label)*. Leaf and pod spot in ***spring field beans***, ***winter field beans***. Light leaf spot in ***brussels sprouts***. Rhizoctonia in ***watercress*** *(off-label)*. Ring spot in ***brussels sprouts***.

Specific Off-Label Approvals (SOLAs)

- ***courgettes*** *(OLA 961866) Dec 2008* [1]
- ***protected rhubarb*** *(OLA 961806) Dec 2008* [1]
- ***watercress*** *(OLA 970875) Dec 2008* [1]

Approval information

- Following implementation of Directive 98/82/EC, approval for use of benomyl on numerous crops was revoked in 1999
- Approval expiry 25 May 2004 [1]

Efficacy guidance

- Apply as overall spray. One spray effective for some diseases, repeat sprays at 1-4 wk intervals needed for others. Recommended spray programmes vary with disease and crop. See label for details of timing and recommended tank mixes
- Addition of non-ionic wetter recommended for certain uses. See label for details
- To delay appearance of resistant strains in diseases needing more than 2 applications per season, use in programme with fungicide of different mode of action
- Product may lose effectiveness if allowed to become damp during storage

Restrictions

- Maximum number of treatments (including applications of any product containing benomyl, carbendazim or thiophanate-methyl) 2 per crop

Crop-specific information

- Latest use: before drilling for field beans; before emergence or at sowing for watercress
- HI 3 wk for Brussels sprouts, field beans; 3 d for courgettes; 2 d for protected rhubarb

Hazard classification and safety precautions

Operator protection A, C, D, H, M; U20b
Environmental protection E15
Storage and disposal E30a, E32a

27 bentazone

A post-emergence contact diazinone herbicide

Products

1	Agriguard Bentazone 480	AgriGuard	480 g/l	SL	11569
2	Basagran SG	BASF	87% w/w	SG	08360
3	Landgold Bentazone SL	Landgold	480 g/l	SL	10841
4	Standon Bentazone S	Standon	87% w/w	SG	10124

Uses

Annual dicotyledons in ***broad beans***, ***linseed***, ***navy beans***, ***peas***, ***potatoes***, ***runner beans***, ***spring field beans***, ***winter field beans*** [1-4]; ***bulb onions*** *(off-label)*, ***evening primrose*** *(off-label)*, ***leeks*** *(off-label)*, ***salad onions*** *(off-label)* [2]; ***dwarf beans*** [1, 3]; ***french beans*** [2, 4]; ***narcissi*** [1, 2, 4]. Common storksbill in ***leeks*** *(off-label)* [2]. Fool's parsley in ***leeks*** *(off-label)* [2].

Specific Off-Label Approvals (SOLAs)

- ***bulb onions***, ***salad onions*** *(OLA 971234) Dec 2008* [2]
- ***evening primrose*** *(OLA 971232) Dec 2008* [2]
- ***leeks*** *(OLA 021551) Dec 2008* [2]

Approval information

- Bentazone included in Annex I under EC Directive 91/414

Efficacy guidance

- Most effective control obtained when weeds are growing actively and less than 5 cm high or across. Good spray cover is essential
- Various recommendations are made for use in spray programmes with other herbicides or in tank mixes. See label for details
- The addition of specified ajuvant oils is recommended for use on some crops to improve fat hen control. Do not use under hot or humid conditions. See label for details
- Split dose application may be made in all recommended crops except peas and generally gives better weed control. See label for details

Restrictions

- Maximum number of treatments varies with crop and product - see labels
- Crops must be treated at correct stage of growth to avoid danger of scorch. See labels for details
- Not all varieties of recommended crops are fully tolerant. Use only on tolerant varieties named in labels. Do not use on forage or mange-tout varieties of peas
- Do not use on crops which have been affected by drought, waterlogging, frost or other stress conditions
- Do not apply insecticides within 7 d of treatment [1, 3]
- Leave 7 d after using a post-emergence grass herbicide or 14 d where treatment precedes the grass herbicide and carry out a leaf wax test where relevant [1, 3]
- Do not spray at temperatures above 21°C. Delay spraying until evening if necessary
- Do not apply if rain or frost expected, if unseasonably cold, if foliage wet or in drought
- A minimum of 6 h (preferably 12 h) free from rain is required after application
- May be used on selected varieties of maincrop and second early potatoes (see label for details), not on seed crops or first earlies

Crop-specific information

- Best results in narcissi obtained by using a suitable pre-emergence herbicide first
- Consult processor before using on crops for processing
- A satisfactory wax test must be carried out before use on peas
- Do not treat narcissi during flower bud initiation
- Latest use: before shoots exceed 15 cm high for potatoes and spring field beans (or 6-7 leaf pairs), 4 leaf pairs (6 pairs or 15 cm high with split dose) for broad beans, before flower buds visible for French, navy, runner, soya and winter field beans and linseed, before flower buds can be found enclosed in terminal shoot for peas
- HI onions 3 wk

Hazard classification and safety precautions

Hazard R03 [1, 3]; R04 [1-4]
Risk phrases R03c [1, 3]; R04d [2, 4]; R04e [1-4]
Operator protection A [1-4]; C [1, 2, 4]; U05a, U08, U19, U20b
Environmental protection E01, E15
Consumer protection C03
Storage and disposal E26 [1, 3, 4]; E30a [1-4]; E31b, E34 [1, 3]; E32a [2, 4]
Medical advice M03 [3]

28 bentazone + MCPA + MCPB

A post-emergence herbicide for undersown spring cereals and grass

Products

Acumen	BASF	200:80:200 g/l	SL	00028

Uses

Annual dicotyledons in ***newly sown grass***, ***undersown spring cereals***.

Approval information

- Bentazone included in Annex I under EC Directive 91/414

Efficacy guidance

- Best results when weeds small and actively growing provided crop is at correct growth stage
- A minimum of 6 h free from rain is required after treatment

Restrictions

- Maximum number of treatments 1 per crop on undersown cereals; 1 per yr on grassland
- Do not treat red clover after 3-trifoliate leaf stage
- Do not use on crops suffering from herbicide damage or physical stress
- Do not use on seed crops or on cereals undersown with lucerne
- Do not roll or harrow for 7 d before or after spraying
- Clovers may be scorched and undersown crop checked but effects likely to be outgrown
- Do not apply if frost expected, if crop wet or when temperatures at or above 21°C

Crop-specific information

- Apply to cereals from 2-fully expanded leaf stage but before first node detectable (GS 12-30) provided clover has reached 1-trifoliate leaf stage
- On first year grass leys use alone or in mixture with cyanazine on seedling stage weeds before end of Sep
- Apply to newly sown leys after grass has reached 2-leaf stage provided clovers have at least 1 trifoliate leaf and red clover has not passed 3-trifoliate leaf stage
- Latest use: before 1st node detectable for undersown cereals (GS 31), 2 wk before grazing for grassland

Environmental safety

- Harmful to fish or other aquatic life. Do not contaminate surface waters or ditches with chemical or used container
- Keep livestock out of treated areas for at least 2 wk and until foliage of poisonous weeds such as ragwort has died and become unpalatable

Hazard classification and safety precautions

Hazard R03, R04
Risk phrases R03c, R04a, R04b, R04e
Operator protection A, C; U05a, U08, U19, U20b
Environmental protection E01, E13c; E07 (2 wk)
Consumer protection C03
Storage and disposal E30a, E31c, E34
Medical advice M03

29 bentazone + MCPB

A post-emergence herbicide mixture for use in peas

Products

Pulsar	BASF	200:200 g/l	SL	04002

Uses

Annual dicotyledons in ***peas***.

Approval information

- Bentazone included in Annex I under EC Directive 91/414

Efficacy guidance

- To be used in tank mix with cyanazine
- Best results achieved when weeds small and actively growing provided crop at correct stage. Good spray cover is essential
- May be applied as single treatment or as split dose treatment applying first spray when susceptible weeds are not beyond the 2 leaf stage
- Do not apply if rain or frost expected or if foliage wet. A minimum of 24 h free from rain required after treatment
- Do not apply during drought or unseasonably cold weather

Restrictions

- Maximum number of treatments 1 per crop or 2 per crop with split dose
- Apply only to listed pea cultivars (see label) from 3 fully expanded leaf (for full dose) or from 2 node stage (for split dose) to before flower buds can be found enclosed in terminal shoot (GS 102 or 103 to before GS 201)
- Do not treat forage pea cultivars or mange-tout peas

- Apply after a satisfactory wax test. Early drilled crops or crops affected by frost or abrasion may not have sufficiently waxy cuticle
- Allow 7 d after spraying before using a grass weed herbicide, or wait 14 d afterwards and test leaf wax before treating
- Do not use as tank mix with any other product than cyanazine, nor after use of TCA
- Do not apply to any crop that may have been subjected to stress conditions, where foliage damaged or under hot, sunny conditions when temperature exceeds 21°C
- Do not apply insecticides within 7 d of treatment

Crop-specific information

- Latest use: before enclosed bud stage of crop (GS 10x)

Environmental safety

- Harmful to fish or other aquatic life. Do not contaminate surface waters or ditches with chemical or used container
- Keep livestock out of treated areas for 14 d and until foliage of any poisonous weeds such as ragwort has died and become unpalatable

Hazard classification and safety precautions

Hazard R04
Risk phrases R04a, R04b, R04e
Operator protection A, C; U05a, U08, U19, U20a
Environmental protection E01, E07, E13c
Consumer protection C03
Storage and disposal E30a, E31c, E34

30 bentazone + pendimethalin

A contact and residual herbicide mixture for combining peas

Products

Impuls	BASF	480:400 g/l	KL	09720

Uses

Annual dicotyledons in ***combining peas***. Chickweed in ***combining peas***. Cleavers in ***combining peas***. Knotgrass in ***combining peas***. Mayweeds in ***combining peas***.

Approval information

- Bentazone and pendimethalin included in Annex I under EC Directive 91/414

Efficacy guidance

- Best results achieved by treating weeds when they are small and ensuring maximum spray coverage
- Moist soil required for maximum efficacy. Loose or cloddy seedbeds must be consolidated before spraying
- Residual control reduced on soils with more than 6% OM
- Minimum 6 hr free from rain required for optimum foliar activity. Do not spray when foliage wet

Restrictions

- Maximum total dose equivalent to one full dose treatment
- Severe crop damage may occur if mixed with any other product. Leave minimum 7 d before, or 14 d after, using a post-emergence grass herbicide. Do not apply insecticides within 7 d of use. Do not apply if TCA has been used
- Do not use on soils prone to waterlogging
- If day time temperatures likely to exceed 21°C spray in the evening
- Do not treat crops under stress from any cause
- Equipment must be washed out thoroughly after use. Traces of product may damage susceptible crops sprayed later

Crop-specific information

- Crystal violet leaf wax test must be carried out before use. Transient leaf margin scorch may occur after treatment
- Crop damage may occur if heavy rain follows application on stony or gravelly soils
- Latest use: before 3rd node stage (GS 103)

Following crops guidance

- Land must be ploughed to 150 mm before any succeeding crop except cereals. In the event of crop failure spring barley may be sown after ploughing

Environmental safety

- Dangerous to fish or other aquatic life. Do not contaminate surface waters or ditches with chemical or used container

Hazard classification and safety precautions

Hazard R03, R04
Risk phrases R03c, R04e
Operator protection A; U05a, U08, U13, U19, U20b
Environmental protection E01, E13b
Consumer protection C03
Storage and disposal E30a, E31c, E34
Medical advice M03

31 beta-cyfluthrin

A non-systemic pyrethroid insecticide available only in mixtures

32 beta-cyfluthrin + imidacloprid

An insecticide mixture for seed treatment of oilseed rape

Products

1	Chinook Blue	Bayer CropScience	100:100 g/l	LS	11262
2	Chinook Colourless	Bayer CropScience	100:100 g/l	LS	11206

Uses

Cabbage stem flea beetle in ***winter oilseed rape*** *(seed treatment).*

Approval information

- Beta-cyfluthrin included in Annex I under EC Directive 91/414

Efficacy guidance

- Drill treated seed as soon as possible after treatment
- Product will reduce damage by early attacks of flea beetle and may also reduce subsequent larval damage. However a follow-up spray treatment may be required if pest activity is heavy and prolonged

Restrictions

- Maximum number of treatments 1 per batch of seed
- Do not use on seed with more than 9% moisture content, on sprouted seed or on cracked, split or otherwise damaged seed
- Product must be co-applied with a colouring agent [2]
- Must be applied using seed treatment machine recommended by manufacturer
- Allow treated seed to dry before packaging. Drying can be assisted by subsequent application of talc
- Product can cause a transient tingling or numbing sensation to exposed skin. Avoid skin contact with the product, treated seed or dust throughout all operations in the treatment plant and during drilling

Crop-specific information

- Latest use: before drilling

Environmental safety

- Extremely dangerous to fish or other aquatic life. Do not contaminate surface waters or ditches with chemical or used container
- Do not use treated seed as food or feed
- Dangerous to birds, game and other wildlife. Treated seed should not be left on the soil surface. Bury spillages
- Do not broadcast treated seed

Hazard classification and safety precautions

Hazard R03
Risk phrases R03c
Operator protection A, H; U04a, U05a, U07, U20b
Environmental protection E01, E13a
Consumer protection C03
Storage and disposal E26, E30a, E33, E34, E36
Treated seed S01, S02, S03, S04c, S05, S06, S06a, S07, S08
Medical advice M03

33 bifenthrin

A contact and residual pyrethroid acaricide/insecticide for use in agricultural and horticultural crops

Products

1	Starion	FMC	100 g/l	EC	09795
2	Starion	Belchim	100 g/l	EC	11641
3	Talstar	Certis	100 g/l	EC	10835
4	Talstar	Certis	100 g/l	EC	11325

Uses

Aphids in ***broccoli*, *brussels sprouts*, *cabbages*, *calabrese*, *cauliflowers*** [1-4]; ***winter barley*, *winter oats*, *winter wheat*** [1, 2]. Cabbage stem flea beetle in ***winter oilseed rape*** [1, 2]. Caterpillars in ***broccoli*, *brussels sprouts*, *cabbages*, *calabrese*, *cauliflowers*** [1-4]. Clay-coloured weevil in ***protected raspberries*** *(off-label)*, ***raspberries*** *(off-label)* [4]. Damson-hop aphid in ***hops*** [3, 4]. Fruit tree red spider mite in ***apples*, *pears*** [3, 4]. Rape winter stem weevil in ***winter oilseed rape*** [1, 2]. Two-spotted spider mite in ***hops*, *ornamental specimens*, *strawberries*** [3, 4]; ***protected raspberries*** *(off-label)*, ***raspberries*** *(off-label)* [4]. Virus vectors in ***winter barley*, *winter oats*, *winter wheat*** [1, 2]. Whitefly in ***broccoli*, *brussels sprouts*, *cabbages*, *calabrese*, *cauliflowers*** [1-4].

Specific Off-Label Approvals (SOLAs)

- ***protected raspberries*, *raspberries*** *(OLA 030872) Jun 2006* [4]

Efficacy guidance

- Timing of application varies with crop and pest. See label for details
- Good spray cover of upper and lower plant surfaces essential to achieve effective pest control

Restrictions

- Maximum number of treatments 5 per yr for hops; 2 per yr for all other crops

Crop-specific information

- Latest use: before 31 Mar in yr of harvest for cereals
- HI zero for all other crops

Environmental safety

- Flammable
- Extremely dangerous to bees. Do not apply to crops in flower or to those in which bees are actively foraging. Do not apply when flowering weeds are present
- Extremely dangerous to fish or other aquatic life. Do not contaminate surface waters or ditches with chemical or used container
- LERAP Category A
- Keep in original container, tightly closed, in a safe place under lock and key

Hazard classification and safety precautions

Hazard R03, R04, R07d
Risk phrases R03b, R03c, R04a, R04b
Operator protection A, C, H; U02, U04a, U05a, U08, U20b [1-4]; U19 [3, 4]
Environmental protection E01, E12b, E13a, E16c, E16d; E17a (18 m)
Consumer protection C03
Storage and disposal E26, E30b, E31b [1-4]; E34 [3, 4]
Medical advice M03

SECTION 4

34 bitertanol

A conazole fungicide available only in mixtures

35 bitertanol + fuberidazole

A broad spectrum fungicide mixture for seed treatment in cereals

Products

1	Sibutol	Bayer	375:23 g/l	FS	07238
2	Sibutol	Bayer CropScience	375:23 g/l	FS	11305

Uses

Bunt in ***spring wheat*** *(seed treatment)*, ***winter wheat*** *(seed treatment)*. Fusarium root rot in ***spring oats*** *(seed treatment)*, ***spring rye*** *(seed treatment)*, ***spring wheat*** *(seed treatment)*, ***triticale*** *(seed treatment)*, ***winter oats*** *(seed treatment)*, ***winter rye*** *(seed treatment)*, ***winter wheat*** *(seed treatment)*. Loose smut in ***spring wheat*** *(seed treatment - reduction)*, ***winter wheat*** *(seed treatment - reduction)*. Septoria seedling blight in ***spring wheat*** *(seed treatment - reduction)*, ***winter wheat*** *(seed treatment - reduction)*.

Efficacy guidance

- Treated seed should preferably be drilled in the same season
- Control of loose smut may be inadequate for use on seed for multiplication

Restrictions

- Maximum number of treatments 1 per batch of seed
- Do not use on seed with more than 16% moisture content, or on sprouted, cracked or skinned seed
- Must be applied simultaneously with water in the ratio 1 part product to 2 parts water in a recommended seed treatment machine

Crop-specific information

- Before drilling

Environmental safety

- Dangerous to fish or other aquatic life. Do not contaminate surface waters or ditches with chemical or used container
- Do not use treated seed as food or feed
- Treated seed harmful to game and wildlife
- Product also supplied in returnable containers. See label for guidance on handling, storage, protective clothing and precautions [1]

Hazard classification and safety precautions

Operator protection A, H; U20a
Environmental protection E03, E13b
Storage and disposal E30a, E32a, E34
Treated seed S01, S02, S03, S04b, S05, S06, S07

36 bitertanol + fuberidazole + imidacloprid

A broad spectrum fungicide and insecticide seed treatment for winter wheat and winter oats

Products

1	Sibutol Secur	Bayer	140:8.6:87.5 g/l	LS	09131
2	Sibutol Secur	Bayer CropScience	140:8.6:87.5 g/l	LS	11308

Uses

Bunt in ***winter wheat*** *(seed treatment)*. Fusarium root rot in ***winter oats*** *(seed treatment)*, ***winter wheat*** *(seed treatment)*. Loose smut in ***winter wheat*** *(seed treatment - reduction)*. Seedling blight and foot rot in ***winter wheat*** *(seed treatment - reduction)*. Virus vectors in ***winter oats*** *(seed treatment)*, ***winter wheat*** *(seed treatment)*. Wireworm in ***winter oats*** *(reduction of damage)*, ***winter wheat*** *(reduction of damage)*.

Efficacy guidance

- Best applied through recommended seed treatment machines

- Evenness of seed cover improved by simultaneous application of equal volumes of product and water or dilution of product with an equal volume of water
- Drill treated seed in the same season. Use minimum 125 kg treated seed per ha. Lower drilling rates and/or early drilling affect duration of BYDV protection needed and may require follow-up aphicide treatment
- In high risk areas where aphid activity is heavy and prolonged a follow-up aphicide treatment may be required
- Protection against foliar air-borne and splash-borne diseases later in the season will require appropriate fungicide follow-up sprays

Restrictions

- Maximum number of treatments 1 per batch of seed
- Do not use on seed with more than 16% moisture content, or on sprouted, cracked or skinned seed

Crop-specific information

- Latest use: before drilling
- Slightly delayed and reduced emergence may occur but this is normally outgrown
- Field emergence which is delayed for any reason may be accentuated by treatment

Environmental safety

- Dangerous to fish or other aquatic life. Do not contaminate surface waters or ditches with chemical or used container
- Do not use treated seed as food or feed
- Dangerous to birds, game and other wildlife. Treated seed should not be left on the soil surface. Bury spillages
- Treated seed should not be broadcast, but drilled to a depth of 4 cm in a well prepared seedbed
- If seed is left on the soil surface the field should be harrowed and rolled to ensure good incorporation

Hazard classification and safety precautions

Hazard R03
Risk phrases R03c
Operator protection A, H; U04a, U05a, U13, U20b
Environmental protection E01, E03, E13b
Consumer protection C03
Storage and disposal E26, E30a, E32a, E34
Treated seed S01, S02, S03, S04c, S05, S06, S07
Medical advice M03

37 bone oil

A ready-to-use animal repellent

Products

Renardine 72-2	Roebuck Eyot	30% w/w	AL	06769

Uses

Badgers in ***amenity areas***, ***farmland***. Cats in ***agricultural premises***, ***amenity areas***, ***farmland***. Dogs in ***agricultural premises***, ***amenity areas***, ***farmland***. Foxes in ***amenity areas***, ***farmland***. Moles in ***amenity areas***, ***farmland***. Rabbits in ***amenity areas***, ***farmland***

Efficacy guidance

- Soak pieces of stick, rags or sand in product and distribute around area to be protected as directed
- Repeat treatment weekly or after heavy rainfall

Environmental safety

- Do not apply directly to animals or crops
- Used as directed product is purely deterrent and does not harm animals

SECTION 4

38 Bordeaux mixture

A protectant copper sulphate/lime complex fungicide

Products

Wetcol 3	Ford Smith	30 g/l (copper)	SC	02360

Uses

Bacterial canker in ***cherries***. Blight in ***outdoor tomatoes***, ***potatoes***. Cane spot in ***loganberries***, ***raspberries***. Canker in ***apples***, ***pears***. Celery leaf spot in ***celery***. Currant leaf spot in ***blackcurrants***. Downy mildew in ***hops***. Leaf curl in ***apricots***, ***nectarines***, ***peaches***. Scab in ***apples***, ***pears***. Spur blight in ***raspberries***.

Efficacy guidance

- Spray interval normally 7-14 d but varies with disease and crop. See label for details
- Commence spraying potatoes before crop meets in row or immediately first blight period occurs
- For canker control spray monthly from Aug to Oct
- For peach leaf curl control spray at leaf fall in autumn and again in Feb
- Spray when crop foliage dry. Do not spray if rain imminent

Restrictions

- Do not use on copper sensitive cultivars, including Doyenne du Comice pears

Crop-specific information

- HI 7 d

Environmental safety

- Harmful to fish or other aquatic life. Do not contaminate surface waters or ditches with chemical or used container
- Harmful to livestock. Keep all livestock out of treated areas for at least 3 wk

Hazard classification and safety precautions

Hazard R03, R04
Risk phrases R03c, R04a, R04b, R04c
Operator protection U13, U15, U20a
Environmental protection E06b, E13c
Storage and disposal E30a, E32a

39 boscalid

A translocated and translaminar anilide fungicide for oilseed rape

Products

Filan	BASF	50% w/w	WG	11449

Uses

Sclerotinia stem rot in ***spring oilseed rape***, ***winter oilseed rape***.

Efficacy guidance

- For best results apply in high disease risk situations at flowering stage of crop but before symptoms are visible
- Ensure adequate spray penetration and good coverage

Restrictions

- Maximum total dose equivalent to two full dose treatments
- Avoid spray drift onto neighbouring crops

Crop-specific information

- Latest use: up to and including full flower

Environmental safety

- Harmful to fish or other aquatic life. Do not contaminate surface waters or ditches with chemical or used container
- Product represents minimal hazard to bees when used as directed. However local bee-keepers should be notified if crops are to be sprayed when in flower

Hazard classification and safety precautions

Operator protection U05a, U20c

Environmental protection E01, E13c
Consumer protection C03
Storage and disposal E26, E30a, E31c

40 boscalid + pyraclostrobin

A protectant and systemic fungicide mixture

Products

Signum	BASF	26.7:6.7% w/w	WG	11450

Uses

Dark leaf spot in ***cabbages***, ***cauliflowers***. Grey mould in ***strawberries***. Ring spot in ***brussels sprouts***, ***cabbages***, ***cauliflowers***. White blister in ***brussels sprouts***, ***cabbages*** *(qualified minor use)*, ***chinese cabbage*** *(off-label)*, ***collards*** *(off-label)*, ***kale*** *(off-label)*, ***komatsuna*** *(off-label)*, ***pak choi*** *(off-label)*, ***salad brassicas*** *(off-label - baby leaf production)*.

Specific Off-Label Approvals (SOLAs)

- ***chinese cabbage***, ***collards***, ***kale***, ***komatsuna***, ***pak choi*** *(OLA 031595) Nov 2005* [1]
- ***salad brassicas*** *(baby leaf production) (OLA 031595) Nov 2005* [1]

Efficacy guidance

- On brassicas apply as a protectant spray or at the first sign of disease and repeat at 3-4 wk intervals depending on disease pressure
- Ensure adequate spray penetration and coverage by increasing water volume in dense crops
- For best results on strawberries apply as a protectant spray at the white bud stage. Applications should be made in sequence with other products as part of a fungicide spray programme during flowering at 7-10 day intervals
- Pyraclostrobin is a member of the QoI cross-resistance group of fungicides and should be used in programmes with fungicides with a different mode of action

Restrictions

- Maximum total dose equivalent to two full dose treatments on all crops
- On brassicas use a maximum of three applications per yr
- On strawberries do not use consecutive treatments and use a maximum of one QoI containing product in three fungicide applications
- Consult processor before use on crops for processing

Crop-specific information

- HI 14 for brassica crops; 3 d for strawberries

Environmental safety

- Dangerous to fish or other aquatic life. Do not contaminate surface waters or ditches with chemical or used container

Hazard classification and safety precautions

Hazard R03
Risk phrases R03c
Operator protection A; U05a, U20c
Environmental protection E01, E13b, E16a
Consumer protection C03
Storage and disposal E30a, E31c
Medical advice M05a

41 brodifacoum

An anticoagulant coumarin rodenticide

Products

1	Klerat	Sorex	0.005% w/w	RB	H6701
2	Klerat Wax Blocks	Sorex	0.005% w/w	BB	H6703
3	Sorexa Checkatube	Sorex	0.2% w/w	XX	H6702

SECTION 4

Products – continued

4	Talon Rat & Mouse Bait (Cut Wheat)	Killgerm	0.005% w/w	RB	H6709
5	Talon Rat & Mouse Bait (Whole Wheat)	Killgerm	0.005% w/w	RB	H6710

Uses

Mice in ***agricultural premises*** [4, 5]; ***farm buildings*** [1, 2]; ***farm buildings*** *(indoor use only)* [3]. Rats in ***agricultural premises*** [4, 5]; ***farm buildings*** [1, 2].

Efficacy guidance

- Effective against rodents resistant to other commonly used anticoagulants
- A single feed representing a fraction of the pest's normal daily food requirement can be lethal
- Ready-to-use in a baiting programme
- Cover baits by placing in bait boxes, drain pipes or under boards, or use bait tubes [3]
- Inspect bait sites frequently and top up as long as there is evidence of feeding

Restrictions

- For use only by professional pest contractors
- Do not use outdoors. Products must be used in situations where baits are placed within a building or other enclosed structure, and the target is living or feeding predominantly within that building or structure

Environmental safety

- Prevent access to baits by children, birds and other animals
- Do not prepare or lay baits where food, feed or water could become contaminated
- Remove all remains of bait or bait containers after use and burn or bury
- Search for and burn or bury all rodent bodies. Do not place in refuse bins or on rubbish tips
- Contains human taste deterrent denatonium benzoate [1, 2]

Hazard classification and safety precautions

Operator protection U13 [1-5]; U20a [4, 5]; U20b [1-3]
Consumer protection C03 [1, 2, 4, 5]
Storage and disposal E30b [1-5]; E32a [1, 2, 4, 5]
Vertebrate/rodent control products V01a [1, 2]; V01b, V02 [3-5]; V03a, V04a [1-3]; V03b, V04b [4, 5]
Medical advice M03 [4, 5]; M05 [1-3]

42 bromadiolone

An anti-coagulant coumarin-derivative rodenticide

Products

1	Endorats Premium Rat Killer	Irish Drugs	0.005% w/w	GB	H6725
2	Slaymor	Novartis A H	0.005% w/w	AB	H6729
3	Slaymor Bait Bags	Novartis A H	0.005% w/w	AB	H6730
4	Tomcat 2	Antec	0.005% w/w	PT	H6736
5	Tomcat 2 Blox	Antec	0.005% w/w	BB	H6731

Uses

Mice in ***farm buildings*** [2-5]; ***farmyards*** [2, 3]. Rats in ***farm buildings*** [1-5]; ***farmyards*** [2, 3].

Efficacy guidance

- Ready-to-use baits are formulated on a mould-resistant, whole-wheat base
- Use in baiting programme. Place baits in protected situations, sufficient for continuous feeding between treatments
- Chemical is effective against warfarin- and coumatetralyl-resistant rats and mice and does not induce bait shyness
- Use bait bags where loose baiting inconvenient (eg behind ricks, silage clamps etc)

Environmental safety

- Access to baits by children, birds and animals, particularly cats, dogs, pigs and poultry, must be prevented
- Baits must not be placed where food, feed or water could become contaminated

- Remains of bait and bait containers must be removed after treatment and burned or buried
- Rodent bodies must be searched for and burned or buried. They must not be placed in refuse bins or on rubbish tips

Hazard classification and safety precautions

Operator protection U13 [1-5]; U20a [2, 3]; U20b [4, 5]
Environmental protection E15 [4, 5]
Storage and disposal E30a [1-5]; E31a [2, 3]; E32a [4, 5]
Treated seed S06 [1]
Vertebrate/rodent control products V01a, V02, V04a [1-5]; V03a [1, 4, 5]
Medical advice M03 [4, 5]; M04 [2, 3]

43 bromoxynil

A contact acting HBN herbicide

Products

1	Alpha Bromolin 225 EC	Makhteshim	225 g/l	EC	08255
2	Alpha Bromotril P	Makhteshim	250 g/l	LI	07099
3	Barclay Mutiny	Barclay	250 g/l	SC	11379
4	Bravado	DAPT	225 g/l	LI	10487
5	Emblem	Nufarm UK	20% w/w	WB	11559
6	Flagon 400 EC	Makhteshim	400 g/l	LI	08875
7	Greencrop Tassle	Greencrop	240 g/l	SC	09659

Uses

Annual dicotyledons in ***forage maize*** [3, 5, 7]; ***linseed*** [1, 4]; ***maize***, ***sweetcorn*** *(off-label)* [2], ***spring barley***, ***spring oats***, ***spring wheat***, ***winter barley***, ***winter oats***, ***winter wheat*** [4, 6].

Specific Off-Label Approvals (SOLAs)

- ***sweetcorn*** *(OLA 972230) Dec 2008* [2]

Approval information

- Bromoxynil was reviewed in 1995 and approvals for home garden use, and most hand held applications revoked

Efficacy guidance

- Spray when main weed flush has germinated and the largest are at the 4 leaf stage
- Weed control can be enhanced by using a split treatment spraying each application when the weeds are seedling to 2 true leaves. Apply the second treatment before the crop canopy covers the ground [2, 3]

Restrictions

- Maximum number of treatments 1 per crop or yr [1, 4, 6]
- Maximum total dose equivalent to one full dose treatment [2, 3, 7]
- Do not apply with oils or other adjuvants
- Do not apply using hand-held equipment or at concentrations higher than those recommended
- Do not apply during frosty weather, drought, when soil is waterlogged, when rain expected within 4 h or to crops under any stress
- Take particular care to avoid drift onto neighbouring susceptible crops or open water surfaces

Crop-specific information

- Latest use: before 10 fully expanded leaf stage of crop for maize, sweetcorn; before crop 20 cm tall and before flower buds visible for linseed; before 2nd node detectable (GS 32) for cereals
- Apply to spring sown maize from 2 fully expanded leaves up to 9 fully expanded leaves [2, 3]
- Apply to spring sown linseed from 1 fully expanded leaf to 20 cm tall [1, 4]
- Foliar scorch, which rapidly disappears without affecting growth, will occur if treatment made in hot weather or during rapid growth

Environmental safety

- Flammable [4, 6]
- Dangerous to fish or other aquatic life. Do not contaminate surface waters or ditches with chemical or used container
- Harmful (high risk [5]) to bees. Do not apply to crops in flower or to those in which bees are actively foraging. Do not apply when flowering weeds are present

SECTION 4

- LERAP Category B [5]
- Keep livestock out of treated areas for at least 6 wk

Hazard classification and safety precautions

Hazard R03, R04 [1-4, 6, 7]; R07d [4, 6]
Risk phrases R03c, R04a [1-4, 6, 7]; R04e [4, 6]; R04g [4]
Operator protection A, C [1-7]; H [1, 4, 6]; U02 [1-4, 7]; U05a [1-7]; U08, U13, U20a [1-4, 6, 7]; U14, U15 [1, 4-6]; U19 [1-6]; U22a, U23a [5]
Environmental protection E01 [2-7]; E07 [1, 4, 6] (6 wk); E12a, E16a [5]; E12e [1-4, 6, 7]; E13b [1-7]
Consumer protection C02 [4, 6] (6 wk); C03 [1-7]
Storage and disposal E26 [1, 3, 4, 6, 7]; E30a [1-7]; E31b, E34 [1-4, 6, 7]
Medical advice M03 [2-4, 6, 7]

44 bromoxynil + clopyralid

A post-emergence contact and translocated herbicide mixture

Products

Vindex	Dow	240:50 g/l	EC	05470

Uses

Annual dicotyledons in ***linseed***.

Efficacy guidance

- Best results achieved when weeds small and growing actively in warm, moist weather
- Vigorous crop competition enhances control of more resistant weeds and prevents those that germinate after treatment becoming a problem
- Weed spectrum can be widened by mixture with bentazone. See label for details

Restrictions

- Maximum number of treatments 1 per crop
- Avoid spraying during drought, waterlogging, frost, extremes of temperature or if rain is imminent or falling
- Spraying in frosty weather or when hard frost occurs within 3-4 wk may result in leaf scorch
- Do not roll or harrow within 7 d before or after spraying
- Do not apply any other product within 7 d of application (10 d if crop is stressed)
- Do not apply by hand-held equipment or at concentrations higher than those recommended
- Wash equipment thoroughly with water and an authorised non-ionic wetting agent according to manufacturer's instructions immediately after use. Traces of product could harm susceptible crops sprayed later

Crop-specific information

- Before first flower and before crop exceeds 30 cm tall

Environmental safety

- Dangerous to fish or other aquatic life. Do not contaminate surface waters or ditches with chemical or used container
- Harmful to bees. Do not apply to crops in flower or to those in which bees are actively foraging. Do not apply when flowering weeds are present
- Do not harvest crops for animal or human consumption for at least 6 wk after last application
- Keep livestock out of treated areas for at least 6 wk after treatment

Hazard classification and safety precautions

Hazard R03, R04, R07d
Risk phrases R03a, R03c, R04a, R04b, R04g
Operator protection A, C; U05a, U08, U13, U19, U20b
Environmental protection E01, E12e, E13b; E07 (6 wk)
Consumer protection C02 (6 wk); C03
Storage and disposal E26, E30a, E31b, E34
Medical advice M03, M05

45 bromoxynil + diflufenican + ioxynil

A selective contact and translocated herbicide for cereals

Products

Capture	Aventis	300:50:200 g/l	SC	09982

Uses

Annual dicotyledons in ***spring barley***, ***spring wheat***, ***triticale***, ***winter barley***, ***winter rye***, ***winter wheat***. Chickweed in ***spring barley***, ***spring wheat***, ***triticale***, ***winter barley***, ***winter rye***, ***winter wheat***. Knotgrass in ***spring barley***, ***spring wheat***, ***triticale***, ***winter barley***, ***winter rye***, ***winter wheat***. Mayweeds in ***spring barley***, ***spring wheat***, ***triticale***, ***winter barley***, ***winter rye***, ***winter wheat***. Speedwells in ***spring barley***, ***spring wheat***, ***triticale***, ***winter barley***, ***winter rye***, ***winter wheat***.

Efficacy guidance

- Best results obtained on small weeds and when competition removed early
- Good spray coverage essential for good activity

Restrictions

- Maximum number of treatments 1 per crop
- Do not treat crops undersown or to be undersown
- Do not treat frosted crops or those that are under stress from any cause
- Do not apply by hand-held equipment or at concentrations higher than those recommended

Crop-specific information

- Latest use: before 2nd node detectable (GS 32)

Following crops guidance

- In the event of crop failure winter wheat may be drilled immediately after normal cultivations. Winter barley may be re-drilled after ploughing
- Land must be ploughed and an interval of 12 wk elapse after treatment before planting spring crops of wheat, barley, oilseed rape, peas, field beans, sugar beet, potatoes, carrots, edible brassicas or onions
- Successive treatments of any products containing diflufenican can lead to soil build-up and inversion ploughing must precede sowing any following non-cereal crop. Even where ploughing occurs some crops may be damaged

Environmental safety

- Harmful to bees. Do not apply to crops in flower or to those in which bees are actively foraging. Do not apply when flowering weeds are present
- Dangerous to fish or other aquatic life. Do not contaminate surface waters or ditches with chemical or used container
- LERAP Category B
- Keep livestock out of treated areas for at least 2 wk and until foliage of any poisonous weeds such as ragwort has died and become unpalatable

Hazard classification and safety precautions

Hazard R03
Risk phrases R03c
Operator protection A, C, H; U05a, U08, U13, U19, U20b
Environmental protection E01, E12e, E13b, E16a; E07 (2 wk)
Consumer protection C02 (2 wk); C03
Storage and disposal E26, E30a, E31b, E34
Medical advice M03

46 bromoxynil + ethofumesate + ioxynil

A post-emergence herbicide for new grass leys

Products

Leyclene	Aventis	50:200:25 g/l	EC	07263

SECTION 4

Uses

Annual dicotyledons in ***grass seed crops, seedling leys***. Annual meadow grass in ***grass seed crops, seedling leys***.

Approval information

- Ethofumesate included in Annex I under EC Directive 91/414

Efficacy guidance

- Best results when weeds small and growing actively in a vigorous crop, soil moist and further rain within 10 d. Mid Oct to end Dec normally suitable
- Annual meadow grass controlled during early crop establishment. Spray weed grasses before fully tillered
- Ash or trash should be burned, buried or removed before spraying

Restrictions

- Maximum number of treatments 2 per yr
- Do not apply by hand-held equipment or at concentrations higher than those recommended
- Do not use on crops under stress, during periods of very dry weather, prolonged frost or waterlogging
- Do not use where clovers or other legumes are valued components of ley
- Do not roll for 7 d before or after spraying
- Do not spray in cold conditions or when heavy rain or frost imminent
- Do not use on soils with more than 10% organic matter
- Do not cut for 14 d after spraying or graze in Jan-Feb after spraying in Oct-Dec

Crop-specific information

- Latest use: 6 wk before cutting or grazing
- Apply to healthy ryegrasses or tall fescue after 2-3 leaf stage, to cocksfoot, timothy and meadow fescue at least 60 d after emergence and after 2-3 leaf stage

Following crops guidance

- Any crop may be sown 5 mth after application following ploughing to at least 15 cm

Environmental safety

- Dangerous to fish or other aquatic life. Do not contaminate surface waters or ditches with chemical or used container
- Harmful to bees. Do not apply to crops in flower or to those in which bees are actively foraging. Do not apply when flowering weeds are present
- Keep livestock out of treated areas for at least 6 wk after treatment

Hazard classification and safety precautions

Hazard R03, R04, R07d
Risk phrases R03c, R04a, R04b
Operator protection A, C; U05a, U08, U13, U19, U20b
Environmental protection E01, E12e, E13b; E07 (6 wk)
Consumer protection C02 (6 wk); C03
Storage and disposal E30a, E31b, E34
Medical advice M03

47 bromoxynil + fluroxypyr + ioxynil

A post-emergence contact and translocated herbicide for cereals

Products

Treble	Barclay	100:90:100 g/l	SL	10577

Uses

Annual dicotyledons in ***spring barley, spring wheat, winter barley, winter wheat***. Chickweed in ***spring barley, spring wheat, winter barley, winter wheat***. Cleavers in ***spring barley, spring wheat, winter barley, winter wheat***. Hemp-nettle in ***spring barley, spring wheat, winter barley, winter wheat***. Speedwells in ***spring barley, spring wheat, winter barley, winter wheat***.

Approval information

- Fluroxypyr included in Annex I under EC Directive 91/414

Efficacy guidance

- Best results from good spray coverage when weeds small and growing actively in warm humid weather
- Do not spray when weed growth is hard from cold or drought as efficacy will be impaired

Restrictions

- Maximum number of treatments 1 per crop
- Do not spray undersown crops or crops to be undersown with clover or legume mixtures
- Do not spray crops stressed by frost, drought, mineral deficiency, pest or disease attack
- Do not roll or harrow for 7 d before or after spraying
- Avoid drift onto non-target crops as damage may occur

Crop-specific information

- Latest use: before second node detectable (GS 32)

Environmental safety

- Flammable
- Harmful to bees. Do not apply to crops in flower or to those in which bees are actively foraging. Do not apply when flowering weeds are present
- Dangerous to fish or other aquatic life. Do not contaminate surface waters or ditches with chemical or used container
- Do not apply by hand-held equipment or at concentrations higher than those recommended
- Do not harvest crops for animal or human consumption for at least 6 wk after last application
- Keep livestock out of treated areas for at least 6 wk following treatment

Hazard classification and safety precautions

Hazard R03, R04, R07d
Risk phrases R03c, R04a, R04c, R04g
Operator protection A, C; U05a, U08, U13, U19, U20a
Environmental protection E01, E12e, E13b; E07 (6 wk)
Consumer protection C02 (6 wk); C03
Storage and disposal E26, E30a, E31b, E34
Medical advice M03, M05

48 bromoxynil + ioxynil

A contact acting post-emergence HBN herbicide for cereals

Products

1	Alpha Briotril 24/16	Makhteshim	240:160 g/l	EC	04870
2	Alpha Briotril Plus 19/19	Makhteshim	190:190 g/l	EC	04740
3	Biotite 380	DAPT	190:190 g/l	EC	10647
4	Mextrol Biox	Nufarm UK	200:200 g/l	EC	09470
5	Oxytril CM	Aventis	200:200 g/l	EC	10005

Uses

Annual dicotyledons in ***grass re-seeds, undersown barley, undersown oats, undersown rye, undersown wheat*** [1, 2]; ***spring barley, spring oats, spring wheat, winter barley, winter oats, winter wheat*** [1-5]; ***spring rye, triticale, winter rye*** [1, 2, 4, 5]; ***undersown spring cereals, undersown winter cereals*** [4, 5].

Efficacy guidance

- Best results achieved on young weeds growing actively in a highly competitive crop
- Do not apply during periods of drought or when rain imminent (some labels say 'if likely within 4 or 6 h')
- Recommended for tank mixture with hormone herbicides to extend weed spectrum. See labels for details

Restrictions

- Maximum number of treatments 1 per crop
- Do not spray crops stressed by drought, waterlogging or other factors
- Do not roll or harrow for several days before or after spraying. Number of days specified varies with product. See label for details
- Do not apply by hand-held equipment or at concentrations higher than those recommended

Crop-specific information

- Latest use: before 2nd node detectable stage (GS 31).
- HI (animal consumption) 14 d
- Apply to winter or spring cereals from 1-2 fully expanded leaf stage, but before second node detectable (GS 32)
- Spray oats in spring when danger of frost past. Do not spray winter oats in autumn
- Apply to undersown cereals pre-sowing or pre-emergence of legume provided cover crop is at correct stage. Only spray trefoil pre-sowing
- On crops undersown with grasses alone or on direct re-seeds apply from 2-leaf stage of grass

Environmental safety

- Flammable [1-3]
- LERAP Category B [1-3]
- Dangerous to fish or other aquatic life. Do not contaminate surface waters or ditches with chemical or used container
- Harmful to bees. Do not apply to crops in flower or to those in which bees are actively foraging. Do not apply when flowering weeds are present
- Keep livestock out of treated areas for at least 6 wk [1-3, 5], 14 d [4]

Hazard classification and safety precautions

Hazard R03, R04 [1-5]; R07d [1-3]
Risk phrases R03c, R04a [1-5]; R04g [1-3]
Operator protection A, C; U02, U14, U15, U20a [1-3]; U05a, U08, U13, U19 [1-5]; U20b [4, 5]
Environmental protection E01, E12e, E13b [1-5]; E07 [1-3] (6 wk); E07 [4] (2 wk); E07 [5]; E16a [1-3]; E16b [3]
Consumer protection C02 [1-3] (6 wk); C02 [4, 5] (14 d); C03 [1-5]
Storage and disposal E26, E31b [1-3]; E30a, E34 [1-5]; E32a [4, 5]
Medical advice M03 [1-5]; M05 [1-3]

49 bromoxynil + prosulfuron

A contact and residual herbicide mixture for maize and sweetcorn

Products

Jester	Syngenta	60:3 % w/w	WG	08681

Uses

Annual dicotyledons in ***maize***. Black bindweed in ***maize***. Chickweed in ***maize***. Hemp-nettle in ***maize***. Knotgrass in ***maize***. Mayweeds in ***maize***.

Approval information

- Prosulfuron included in Annex I under EC Directive 91/414

Efficacy guidance

- Prosulfuron is a member of the ALS-inhibitor group of herbicides and products should be used in a planned Resistance Management strategy. See Section 1 for more information

Restrictions

- Maximum number of treatments 1 per crop
- Do not treat crops grown for seed production
- Consult before use on crops intended for processing
- Do not use in frosty weather or on crops under stress
- Do not tank mix with organophosphate insecticides or apply in tank mix or sequence with any other sulfonylurea
- Do not apply by knapsack sprayer or in volumes less than those recommended
- Take care to wash out sprayers thoroughly. See label for details

Crop-specific information

- Latest use: before 5 crop leaves unfolded for maize, sweetcorn
- Apply post-emergence up to when the crop has four unfolded leaves
- Product must be used with Agral
- Apply in cool conditions, or during the evening, to avoid scorch

Environmental safety

- Extremely dangerous to fish or other aquatic life. Do not contaminate surface waters or ditches with chemical or used container
- LERAP Category B
- Take special care to avoid drift outside the target area

Hazard classification and safety precautions

Hazard R03, R04
Risk phrases R03c, R04a
Operator protection A, C; U05a, U11, U15, U20c
Environmental protection E01, E13a, E16a
Consumer protection C03
Storage and disposal E29, E30a, E31a

50 bromuconazole

A systemic conazole fungicide for cereals

Products

Granit	Aventis	200 g/l	SC	09995

Uses

Brown rust in ***spring barley***, ***spring wheat***, ***winter barley***, ***winter wheat***. Eyespot in ***spring barley*** *(reduction)*, ***spring wheat*** *(reduction)*, ***winter barley*** *(reduction)*, ***winter wheat*** *(reduction)*. Fusarium ear blight in ***spring wheat***, ***winter wheat***. Glume blotch in ***spring wheat***, ***winter wheat***. Net blotch in ***spring barley***, ***winter barley***. Powdery mildew in ***spring barley***, ***spring wheat***, ***winter barley***, ***winter wheat***. Rhynchosporium in ***spring barley***, ***winter barley***. Septoria leaf spot in ***spring wheat***, ***winter wheat***. Sooty moulds in ***spring wheat***, ***winter wheat***. Yellow rust in ***spring barley***, ***spring wheat***, ***winter barley***, ***winter wheat***.

Efficacy guidance

- Best results achieved from applications before disease becomes established
- If disease pressure persists a second treatment may be required to prevent late season attacks
- Best control of ear diseases in wheat obtained from application at or after the ear completely emerged stage
- More consistent control of powdery mildew obtained from tank mixture with fenpropimorph. See label

Restrictions

- Maximum number of treatments 2 per crop

Crop-specific information

- Latest use varies with dose used and crop - see label
- Apply at any time from the tillering stage up to the latest times indicated for each crop

Following crops guidance

- Only cereals, oilseed rape, field beans, peas, potatoes, linseed or Italian ryegrass may be sown as following crops

Environmental safety

- Dangerous to fish or other aquatic life. Do not contaminate surface waters or ditches with chemical or used container
- LERAP Category B

Hazard classification and safety precautions

Operator protection A, C, H; U08, U19, U20b
Environmental protection E13b, E16a, E16b
Storage and disposal E26, E30a, E31b

51 bupirimate

A systemic pyrimidine fungicide active against powdery mildew

Products

Nimrod	Makhteshim	250 g/l	EC	10563

SECTION 4

Uses

Powdery mildew in ***apples***, ***begonias***, ***blackcurrants***, ***chrysanthemums***, ***courgettes***, ***cucumbers***, ***gooseberries***, ***hops***, ***marrows***, ***pears***, ***protected tomatoes*** *(off-label)*, ***pumpkins*** *(off-label)*, ***raspberries***, ***roses***, ***squashes*** *(off-label)*, ***strawberries***.

Specific Off-Label Approvals (SOLAs)

- ***protected tomatoes*** *(OLA 010910) Dec 2008* [1]
- ***pumpkins***, ***squashes*** *(OLA 010909) Dec 2008* [1]

Efficacy guidance

- On apples during periods that favour disease development lower doses applied weekly give better results than higher rates fortnightly
- Not effective in protected crops against strains of mildew resistant to bupirimate

Restrictions

- Maximum number of treatments or maximum total dose depends on crop and variety (see label for details)

Crop-specific information

- HI depends on crop and variety (see label for details)
- Apply before or at first signs of disease and repeat at 7-14 d intervals. Timing and maximum dose vary with crop. See label for details
- With apples, hops and ornamentals cultivars may vary in sensitivity to spray. See label for details
- If necessary to spray cucurbits in winter or early spring spray a few plants 10-14 d before spraying whole crop to test for likelihood of leaf spotting problem
- On roses some leaf puckering may occur on young soft growth in early spring or under low light intensity. Avoid use of high rates or wetter on such growth
- Never spray flowering begonias (or buds showing colour) as this can scorch petals
- Do not mix with other chemicals for application to begonias, cucumbers or gerberas

Environmental safety

- Flammable
- Harmful to fish or other aquatic life. Do not contaminate surface waters or ditches with chemical or used container
- Product has negligible effect on *Phytoseiulus* and *Encarsia* and may be used in conjunction with biological control of red spider mite

Hazard classification and safety precautions

Hazard R03, R04, R07d
Risk phrases R04a, R04b, R04g
Operator protection A, C; U05a, U20b
Environmental protection E01, E13c
Consumer protection C03
Storage and disposal E30a, E31c
Medical advice M05

52 buprofezin

A moulting inhibitor, thiadiazine insecticide for whitefly control

Products

Applaud	Certis	250 g/l	SC	10930

Uses

Glasshouse whitefly in ***aubergines***, ***ornamental plant production***, ***protected cucumbers***, ***protected ornamentals***, ***protected peppers***, ***protected tomatoes***. Tobacco whitefly in ***aubergines***, ***ornamental plant production***, ***protected cucumbers***, ***protected ornamentals***, ***protected peppers***, ***protected tomatoes***.

Efficacy guidance

- Product has contact, residual and some vapour activity
- Whitefly most susceptible at larval stages but residual effect can also kill nymphs emerging from treated eggs and application to pupae reduces emergence
- Adult whitefly not directly affected. Resistant strains of tobacco whitefly are known and where present control likely to be reduced or ineffective

Restrictions

- Maximum number of treatments up to 8 per crop for tomatoes and cucumbers; up to 2 per crop for aubergines and peppers
- Do not apply more than 2 sprays within a 65 d period on tomatoes, or within a 45 d period on cucumbers
- Do not treat *Dieffenbachia* or *Closmoplictrum*
- Do not apply to crops under stress
- Do not leave spray liquid in sprayer for long periods
- Do not apply as fog or mist

Crop-specific information

- HI edible crops 3 d
- In IPM programme apply as single application and allow at least 60 d before re-applying
- In All Chemical programme apply twice at 7-14 d interval and allow at least 60 d before re-applying
- See label for list of ornamentals successfully treated but small scale test advised to check varietal tolerance. This is especially important if spraying flowering ornamentals with buds showing colour

Environmental safety

- Product may be used either in IPM programme in association with *Encarsia formosa* or in All Chemical programme

Hazard classification and safety precautions

Operator protection U20c
Environmental protection E15
Storage and disposal E30a, E32a, E34

53 captan

A protectant dicarboximide fungicide with horticultural uses

Products

1	Alpha Captan 80 WDG	Makhteshim	80% w/w	WG	07096
2	PP Captan 80-WG	Tomen	80% w/w	WG	11006

Uses

Black spot in ***roses*** [1, 2]. Botrytis in ***strawberries*** [1]. Gloeosporium in ***apples*** [2]. Gloeosporium rot in ***apples*** [1]. Scab in ***apples, pears*** [1, 2].

Restrictions

- Maximum number of treatments 12 per yr on apples and pears as pre-harvest sprays
- Product must not be used as dip or drench on apples or pears [2]
- Do not use on apple cultivars Bramley, Monarch, Winston, King Edward, Spartan, Kidd's Orange or Red Delicious or on pear cultivar D'Anjou
- Do not mix with alkaline materials or oils
- Do not use on fruit for processing
- Powered visor respirator with hood and neck cape must be used when handling concentrate

Crop-specific information

- HI apples, pears 14 d
- For control of scab apply at bud burst and repeat at 10-14 d intervals until danger of scab infection ceased
- For suppression of fruit storage rots apply from late Jul and repeat at 2-3 wk intervals
- For black spot control in roses apply after pruning with 3 further applications at 14 d intervals or spray when spots appear and repeat at 7-10 d intervals
- For grey mould in strawberries spray at first open flower and repeat every 7-10 d [1]
- Do not leave diluted material for more than 2 h. Agitate well before and during spraying

Environmental safety

- Harmful to fish or other aquatic life. Do not contaminate surface waters or ditches with chemical or used container

Hazard classification and safety precautions

Hazard R03 [1]; R04 [1, 2]
Risk phrases R04a [2]; R04c, R04d [1]; R04e [1, 2]
Operator protection A, D, E, H [1, 2]; C, K, M [2]; U05a [1, 2]; U09a, U19, U20b [1]; U20c [2]

Environmental protection E01, E13c
Consumer protection C03
Storage and disposal E30a [1, 2]; E31b, E34 [2]; E32a [1]

54 captan + penconazole

A protectant fungicide for use on apple trees

Products

Topas C 50 WP	Novartis	47.5:2.5% w/w	WB	08459

Uses

Powdery mildew in ***apples***. Scab in ***apples***.

Efficacy guidance

- Use as a protective spray every 7-14 d from bud burst until extension growth ceases
- High antisporulant activity reduces development of primary mildew and controls spread of secondary mildew
- Penconazole is recommended alone for mildew control if scab is not a problem

Restrictions

- Maximum number of treatments 10 per yr
- When making air-assisted applications spray operator must work within a closed cab on a tractor or self-propelled sprayer
- The water soluble bags must be dropped straight into the sprayer tank and not opened beforehand

Crop-specific information

- HI apples 14 d

Environmental safety

- Dangerous to fish or other aquatic life. Do not contaminate surface waters or ditches with chemical or used container
- Does not affect beneficial insects so can be used in integrated control programmes

Hazard classification and safety precautions

Hazard R04
Risk phrases R04a, R04b, R04c
Operator protection A, C; U02, U05a, U08, U19, U20b
Environmental protection E01, E13b
Consumer protection C02 (14 d); C03
Storage and disposal E30a, E32a

55 carbendazim

A systemic benzimidazole fungicide with curative and protectant activity

Products

1	Bavistin DF	BASF	50% w/w	WG	03848
2	Bavistin FL	BASF	500 g/l	SC	00218
3	Delsene 50 Flo	Nufarm UK	500 g/l	SC	09469
4	Mascot Systemic	Rigby Taylor	500 g/l	SC	08776
5	Nuturf Carbendazim	Nufarm UK	500 g/l	SC	11469
6	Ringer	SumiAgro Amenity	500 g/l	SC	10692
7	Tripart Defensor FL	Tripart	500 g/l	SC	02752
8	Turf Systemic Fungicide	SumiAgro Amenity	500 g/l	SC	09349

Uses

Anthracnose in ***dwarf beans*** [1, 2]. Big vein in ***lettuce*** *(off-label)*, ***protected lettuce*** *(off-label)* [1]. Black root rot in ***bedding plants*** *(off-label)*, ***chrysanthemums*** *(off-label)*, ***container-grown ornamentals*** *(off-label)*, ***corms*** *(off-label)*, ***flower bulbs*** *(off-label)*, ***freesias*** *(off-label)*, ***ornamental plant production*** *(off-label - soil grown)*, ***pot plants*** *(off-label)*, ***protected ornamentals*** *(off-label - container grown)*, ***protected ornamentals*** *(off-label - NFT)*, ***protected ornamentals*** *(off-label - soil grown)* [1]. Botrytis in ***dwarf beans*** [1, 2]; ***protected celery*** *(off-label)* [1]. Cane blight in ***blackberries*** *(off-label)*, ***loganberries*** *(off-label)*, ***raspberries*** *(off-label)*, ***rubus hybrids*** *(off-label)* [1]. Canker in ***apples*** [7]. Didymella in ***protected tomatoes*** *(off-label)* [1]. Dollar spot in

managed amenity turf [4-6, 8]. Eyespot in ***spring barley*** [3]; ***winter barley*** [1-3]; ***winter rye*** [1, 2]; ***winter wheat*** [1-3, 7]. Fusarium patch in ***managed amenity turf*** [4-6, 8]. Gloeosporium in ***apples*** [1]. Late ear diseases in ***winter wheat*** [1, 2]. Light leaf spot in ***spring oilseed rape***, ***winter oilseed rape*** [3, 7]; ***spring oilseed rape*** *(reduction)*, ***winter oilseed rape*** *(reduction)* [1, 2]. Mycogone in ***mushrooms*** [1, 2]. Powdery mildew in ***apples*** [7]; ***roses*** [1, 2]. Rhynchosporium in ***spring barley***, ***winter barley*** [1-3]. Root diseases in ***inert substrate cucumbers*** *(off-label)* [1]. Scab in ***apples*** [1, 7]; ***pears*** *(off-label)* [1]. Sooty moulds in ***winter wheat*** [1, 2]. Storage rots in ***apples*** [1, 7]; ***pears*** *(off-label)* [1]. Trichoderma in ***mushrooms*** *(off-label)* [1]. Wormcast formation in ***managed amenity turf*** [4-6].

Specific Off-Label Approvals (SOLAs)

- ***bedding plants***, ***chrysanthemums***, ***container-grown ornamentals***, ***corms***, ***flower bulbs***, ***freesias***, ***pot plants*** *(OLA 990009) Dec 2008* [1]
- ***blackberries***, ***loganberries***, ***raspberries***, ***rubus hybrids*** *(OLA 031590) Jul 2008* [1]
- ***inert substrate cucumbers*** *(OLA 951476) Dec 2008* [1]
- ***lettuce***, ***protected lettuce*** *(OLA 021689) Dec 2008* [1]
- ***mushrooms*** *(OLA 951144) Dec 2008* [1]
- ***ornamental plant production***, ***protected ornamentals*** *(soil grown) (OLA 990009) Dec 2008* [1]
- ***pears*** *(OLA 030045) Dec 2008* [1]
- ***pears*** *(OLA 992256) Dec 2008* [1]
- ***protected celery*** *(OLA 992078) Dec 2008* [1]
- ***protected ornamentals*** *(container grown) (OLA 990009) Dec 2008* [1]
- ***protected ornamentals*** *(NFT) (OLA 990009) Dec 2008* [1]
- ***protected tomatoes*** *(OLA 000965) Dec 2008* [1]

Approval information

- Approved for aerial application on winter wheat and oilseed rape [7]. See notes in Section 1
- Following implementation of Directive 98/82/EC, approval for use of carbendazim on numerous crops was revoked in 1999. UK approvals for use of carbendazim on strawberries revoked in 2001 as a result of implementation of the MRL Directives

Efficacy guidance

- Products vary in the diseases listed as controlled for several crops. Labels must be consulted for full details and for rates and timings
- Mostly applied as spray or drench. Spray treatments normally applied at first sign of disease and repeated after 1 mth if required
- Apply as drench rather than spray where red spider mite predators are being used
- Do not apply during drought conditions. Rain or irrigation after treatment may improve control
- Leave 4 mth intervals where used only for worm cast control [4-6]
- To delay appearance of resistant strains alternate treatment with non-MBC fungicide. Eyespot in cereals and *Botrytis cinerea* in many crops are now widely resistant

Restrictions

- Maximum number of treatments (including applications of any product containing benomyl, carbendazim or thiophanate-methyl) varies with crop treated and product used - see labels for details
- Do not treat crops or turf suffering from drought or other physical or chemical stress
- Consult processors before using on crops for processing
- Not compatible with alkaline products such as lime sulphur

Crop-specific information

- Latest use varies with crop and product used. See labels for details
- HI 2 d for inert substrate cucumbers, protected tomatoes; 7-14 d (depends on dose) for apples, pears, 14 d for mushrooms, protected celery; 21 d for oilseed rape, dwarf beans
- On turf apply as preventative treatment in spring or autumn during periods of high disease risk
- Apply as a drench to control soil-borne diseases in cucumbers and tomatoes and as a pre-planting dip treatment for bulbs
- Apply by incorporation into casing to control mushroom diseases [1]

Environmental safety

- Harmful to fish or other aquatic life. Do not contaminate surface waters or ditches with chemical or used container [1-6, 8]
- After use dipping suspension must not be discharged directly into ditches or drains

Hazard classification and safety precautions

Operator protection A, C, H [1-8]; B, K [1, 2, 7]; D, P [1]; M [1-7]; U05a [1, 2]; U20b [3, 5]; U20c [1, 2, 4, 6, 8]

Environmental protection E01 [1-3, 5, 7]; E13c [1-6, 8]; E15 [3]

Consumer protection C03 [3, 5]

Storage and disposal E29, E31c [2]; E30a [1-6, 8]; E31a [4, 7]; E31b [6, 8]; E32a [1, 3, 5]; E34 [7]

56 carbendazim + epoxiconazole

A systemic fungicide mixture for managed amenity turf

Products

Capricorn	Bayer Environ.	125:125 g/l	SC	10035

Uses

Fusarium patch in ***managed amenity turf***.

Efficacy guidance

- Treat at the first sign of disease and repeat as necessary
- Apply after cutting and delay further mowing for at least 48 h to allow systemic distribution within the plant
- Only moderate control of Fusarium patch claimed

Restrictions

- Maximum number of treatments 2 per yr
- Do not apply during drought conditions or to frozen turf
- Use only on established turf

Environmental safety

- Harmful to fish or other aquatic life. Do not contaminate surface waters or ditches with chemical or used container
- LERAP Category B

Hazard classification and safety precautions

Hazard R04

Risk phrases R04e

Operator protection A, C, H, M; U04a, U05a, U09b, U20c

Environmental protection E01, E13c, E16a, E16b

Consumer protection C03

Storage and disposal E26, E30a, E31b, E34

57 carbendazim + flusilazole

A broad-spectrum systemic and protectant fungicide for cereals

Products

1	Contrast	DuPont	125:250 g/l	SC	06150
2	Landgold Flusilazole MBC	Landgold	125:250 g/l	SC	08528
3	Punch C	DuPont	125:250 g/l	SC	06801
4	Standon Flusilazole Plus	Standon	125:250 g/l	SC	07403

Uses

Brown rust in ***spring barley***, ***winter barley*** [1, 3]; ***spring wheat***, ***winter wheat*** [1-4]. Canker in ***spring oilseed rape*** *(reduction)*, ***winter oilseed rape*** *(reduction)* [1, 3]. Eyespot in ***spring barley***, ***winter barley*** [1, 3]; ***spring wheat***, ***winter wheat*** [1-4]. Light leaf spot in ***spring oilseed rape***, ***winter oilseed rape*** [1-4]. Phoma leaf spot in ***spring oilseed rape***, ***winter oilseed rape*** [1, 3]. Powdery mildew in ***spring barley***, ***sugar beet***, ***winter barley*** [1, 3]; ***spring wheat***, ***winter wheat*** [1-4]. Rust in ***sugar beet*** [1, 3]. Septoria in ***spring wheat***, ***winter wheat*** [1-4]. Yellow rust in ***spring barley***, ***winter barley*** [1, 3]; ***spring wheat***, ***winter wheat*** [1-4].

Efficacy guidance

- Apply at early stage of disease development or in routine preventive programme
- Most effective timing of treatment on cereals varies with disease. See label for details
- Higher rate active against both MBC-sensitive and MBC-resistant eyespot

- Rain occurring within 2-3 h of spraying may reduce effectiveness
- To prevent build-up of resistant strains of cereal mildew tank mix with approved morpholine fungicide [1, 3]

Restrictions

- Maximum total dose (including applications of any product containing flusilazole) equivalent to 3 full dose treatments for winter wheat, 2 full dose treatments for spring wheat, oilseed rape
- Do not apply to crops under stress or during frosty weather

Crop-specific information

- Latest use normally up to and including grain watery ripe stage (GS 71) for wheat, ear emergence complete stage (GS 59) for barley, before first flower opened stage for oilseed rape. However, timings vary with crop and dose - see labels for details
- HI 7 wk for sugar beet
- Treat oilseed rape in autumn when leaf lesions first appear and spring from the start of stem extension when disease appears

Environmental safety

- Dangerous to fish or other aquatic life. Do not contaminate surface waters or ditches with chemical or used container

Hazard classification and safety precautions

Hazard R03, R04
Risk phrases R03c, R04a
Operator protection A, C [1-4]; H, M [1, 2, 4]; U05a, U11, U19, U20b
Environmental protection E01, E13b
Consumer protection C03
Storage and disposal E26 [1, 3]; E30a, E31b, E34 [1-4]
Medical advice M03

58 carbendazim + iprodione

A systemic and contact fungicide mixture

Products

Vitesse	Bayer Environ.	87.5:175 g/l	SC	10042

Uses

Anthracnose in ***managed amenity turf***. Fusarium patch in ***managed amenity turf***. Pink patch in ***managed amenity turf***. Red thread in ***managed amenity turf***. Timothy leaf spot in ***managed amenity turf***.

Approval information

- Iprodione included in Annex I under EC Directive 91/414

Efficacy guidance

- Maximum efficacy against Anthracnose in turf achieved by treatment at early disease development stage. Curative treatments for well-established Anthracnose are not recommended

Crop-specific information

- May be used on turf all year round, but is best suited for spring or late summer/early autumn application
- Where grass is being mown, apply after mowing. Delay further mowing for at least 48 h after treatment

Environmental safety

- Harmful to fish or other aquatic life. Do not contaminate surface waters or ditches with chemical or used container

Hazard classification and safety precautions

Operator protection A, C, H, M; U20c
Environmental protection E13c
Storage and disposal E26, E30a, E31b

59 carbendazim + prochloraz

A broad-spectrum systemic and contact fungicide for cereals and oilseed rape

Products

1	Novak	Aventis	100:267 g/l	SE	08020
2	Novak	BASF	100:267 g/l	SE	11693

Uses

Botrytis in ***spring oilseed rape***, ***winter oilseed rape***. Canker in ***spring oilseed rape***, ***winter oilseed rape***. Dark leaf spot in ***spring oilseed rape***, ***winter oilseed rape***. Eyespot in ***winter wheat***. Glume blotch in ***winter wheat***. Light leaf spot in ***spring oilseed rape***, ***winter oilseed rape***. Powdery mildew in ***winter wheat***. Sclerotinia stem rot in ***spring oilseed rape***, ***winter oilseed rape***. Septoria leaf spot in ***winter wheat***. White leaf spot in ***spring oilseed rape***, ***winter oilseed rape***.

Efficacy guidance

- To protect against eyespot in high risk situations apply to winter barley from when leaf sheaths begin to become erect to first node detectable (GS 30-31). May also be used to control eyespot already in crop (up to 10% of tillers affected)
- Applied for eyespot control, product will also control Rhynchosporium and mildew and protect against new infections of net blotch, mildew and Septoria. Where MBC resistance is not a problem product can be used against eyespot and leaf spot but prochloraz alone is preferable where these diseases are resistant
- Tank-mix with tridemorph, fenpropimorph or fenpropidin to control established mildew

Restrictions

- Maximum number of treatments (including applications of any product containing benomyl, carbendazim or thiophanate-methyl) 2 per crop for cereals; 2 at normal rate or 2 at split plus 1 at normal rate for oilseed rape
- A period of at least 3 h without rain should follow spraying

Crop-specific information

- Latest use: before flowering (GS 60) for wheat
- HI 6 wk for oilseed rape
- May be applied to winter barley up to full ear emergence (GS 59), but if used earlier in season, prochloraz alone or plus a non-MBC fungicide is preferred treatment
- Apply in oilseed rape at first signs of disease and repeat if necessary. Timing varies with disease - see label for details.

Environmental safety

- Dangerous to fish or other aquatic life. Do not contaminate surface waters or ditches with chemical or used container

Hazard classification and safety precautions

Hazard R04
Risk phrases R04a, R04e
Operator protection A, C, H, M; U02, U05a, U08, U20b
Environmental protection E01, E13b
Consumer protection C03
Storage and disposal E30a, E31b

60 carbendazim + vinclozolin

A protectant and systemic fungicide for oilseed rape

Products

Konker	BASF	165:250 g/l	SC	03988

Uses

Alternaria in ***spring oilseed rape*** *(reduction)*, ***winter oilseed rape*** *(reduction)*. Botrytis in ***spring oilseed rape*** *(reduction)*, ***winter oilseed rape*** *(reduction)*. Light leaf spot in ***spring oilseed rape*** *(reduction)*, ***winter oilseed rape*** *(reduction)*. Sclerotinia in ***soya beans*** *(off-label)*. Sclerotinia stem rot in ***spring oilseed rape***, ***winter oilseed rape***.

Specific Off-Label Approvals (SOLAs)
- ***soya beans*** *(OLA 010246) Feb 2006* [1]

Efficacy guidance
- Best results achieved if applied before any disease becomes well established

Restrictions
- Maximum total dose equivalent to two full dose treatments
- Do not apply if rain or frost is expected or when the crop is wet
- Must be applied only by vehicle mounted or trailed hydraulic sprayers
- Vehicles must be fitted with a cab and a forced air filtration unit with a pesticide filter complying with HSE Guidance Note PM74, or equivalent

Crop-specific information
- HI 7 wk for oilseed rape
- Apply from start of rapid spring growth up to and including full flower
- Timing depends on diseases present and weather conditions. See label for details

Environmental safety
- Harmful to fish or other aquatic life. Do not contaminate surface waters or ditches with chemical or used container

Hazard classification and safety precautions

Hazard R04
Risk phrases R04e
Operator protection A, C, H, K, M; U05a, U08, U19, U20b
Environmental protection E01, E13c
Consumer protection C03
Storage and disposal E30a, E31c

61 carbetamide

A residual pre- and post-emergence carbamate herbicide for a range of field crops

Products

Carbetamex	Makhteshim	70% w/w	WP	11150

Uses

Annual grasses in ***cabbage seed crops, collards, fodder rape seed crops, kale seed crops, lucerne, red clover, sainfoin, spring cabbage, sugar beet seed crops, swede seed crops, turnip seed crops, white clover, winter field beans, winter oilseed rape***. Some annual dicotyledons in ***cabbage seed crops, collards, fodder rape seed crops, kale seed crops, lucerne, red clover, sainfoin, spring cabbage, sugar beet seed crops, swede seed crops, turnip seed crops, white clover, winter field beans, winter oilseed rape***. Volunteer cereals in ***cabbage seed crops, collards, fodder rape seed crops, kale seed crops, lucerne, red clover, sainfoin, spring cabbage, sugar beet seed crops, swede seed crops, turnip seed crops, white clover, winter field beans, winter oilseed rape***.

Efficacy guidance
- Best results pre- or early post-emergence of weeds under cool, moist conditions. Adequate soil moisture is essential
- Dicotyledons controlled include chickweed, cleavers and speedwell
- Weed growth stops rapidly after treatment but full effects may take 6-8 wk to develop
- Various tank mixes are recommended to broaden the weed spectrum. See label for details

Restrictions
- Maximum number of treatments 1 per crop for all crops
- Do not treat any crop on waterlogged soil
- Do not use on soils with more than 10% organic matter as residual activity is impaired
- Do not apply during prolonged periods of cold weather when weeds are dormant

Crop-specific information
- HI 6 wk for all crops
- Apply to brassicas from mid-Oct to end-Feb provided crop has at least 4 true leaves (spring cabbage, spring greens), 3-4 true leaves (seed crops, oilseed rape)

- Apply to established lucerne or sainfoin from Nov to end-Feb
- Apply to established red or white clover from Feb to mid-Mar

Following crops guidance

- Succeeding crops may be sown 2 wk after treatment for brassicas, field beans, 8 wk after treatment for peas, runner beans, 16 wk after treatment for cereals, maize

Hazard classification and safety precautions

Operator protection U20c
Environmental protection E15
Storage and disposal E30a, E32a

62 carbetamide + diflufenican + oxadiazon

A residual herbicide mixture for amenity vegetation

Products

Helmsman	Bayer Environ.	1:0.1:2% w/w	GR	09934

Uses

Annual dicotyledons in ***amenity trees and shrubs***, ***amenity vegetation***. Annual meadow grass in ***amenity trees and shrubs***, ***amenity vegetation***. Perennial dicotyledons in ***amenity trees and shrubs***, ***amenity vegetation***.

Efficacy guidance

- Apply to soil surface in late winter or early spring using hand shaker or gravity fed applicator. Uniform application is essential for satisfactory weed control
- Chemical absorbed by emerging shoots of susceptible weeds as they grow through the treated soil
- Adequate moisture is essential. In dry conditions irrigate after application. Effectiveness may be reduced if drought conditions follow application
- Do not disturb treated soil by hoeing otherwise weed control may be impaired
- In some cases heavy rain shortly after application may splash granules onto the lower leaves of shrubs which may result in transient damage

Restrictions

- Maximum number of treatments 1 per yr
- Do not apply to plants grown in containers
- Do not apply to areas under-planted with bulbs or annual plants

Environmental safety

- Not to be used on food crops
- Extremely dangerous to fish or other aquatic life. Do not contaminate surface waters or ditches with chemical or used container
- Do not apply on slopes where heavy rain soon after application could cause surface run-off

Hazard classification and safety precautions

Hazard R04
Risk phrases R04a
Operator protection A; U05a, U19, U20b
Environmental protection E01, E13a
Consumer protection C01, C03
Storage and disposal E30a, E32a

63 carbon dioxide (commodity substance)

A gas for the control of trapped rodents and other vertebrates

Products

carbon dioxide	various	99.9%	GA

Uses

Birds in ***traps***. Mice in ***traps***. Rats in ***traps***.

Approval information

- Approval for the use of carbon dioxide as a commodity substance was granted on 8 October 1993 by Ministers under regulation 5 of the Control of Pesticides Regulations 1986
- Only to be used where a licence has been issued in accordance with Section 16(1) of the Wildlife and Countryside Act 1981

Efficacy guidance

- Use to destroy trapped rodent pests
- Use to control birds covered by general licences issued by the Agriculture and Environment Departments under Section 16(1) of the Wildlife and Countryside Act (1981) for the control of opportunistic bird species, where birds have been trapped or stupefied with alphachloralose/seconal

Restrictions

- Operators must wear self-contained breathing apparatus when carbon dioxide levels are greater than 0.5% v/v
- Operators must be suitably trained and competent

Environmental safety

- Unprotected persons and non-target animals must be excluded from the treatment enclosures and surrounding areas unless the carbon dioxide levels are below 0.5% v/v

Hazard classification and safety precautions

Operator protection G

64 carbosulfan

A systemic carbamate insecticide for control of soil and stem pests

Products

1	Marshal 10G	Belchim	10% w/w	GR	11682
2	Marshal Soil Insecticide suSCon CR granules	Fargro	10% w/w	GR	06978
3	Marshal Soil Insecticide suSCon CR Sachets	Fargro	10% w/w	WB	11754
4	Posse 10G	Bayer	10% w/w	GR	10725
5	Posse 10G	Belchim	10% w/w	GR	11640

Uses

Aphids in ***broccoli*, *brussels sprouts*, *cabbages*, *calabrese*, *cauliflowers*, *collards*, *kale*** [1, 5]; ***carrots*, *fodder beet*, *mangels*, *parsnips*, *sugar beet*** [1, 4, 5]. Cabbage root fly in ***broccoli*, *brussels sprouts*, *cabbages*, *calabrese*, *cauliflowers*, *collards*, *kale*** [1, 5]. Cabbage stem weevil in ***broccoli*, *brussels sprouts*, *cabbages*, *calabrese*, *cauliflowers*, *collards*, *kale*** [1, 5]. Flea beetle in ***broccoli*, *brussels sprouts*, *cabbages*, *calabrese*, *cauliflowers*, *collards*, *kale*** [1, 5]; ***fodder beet*, *mangels*, *sugar beet*** [1, 4, 5]. Free-living nematodes in ***carrots*, *fodder beet*, *mangels*, *parsnips*, *sugar beet*** [1, 4, 5]. Large pine weevil in ***forest nurseries*** *(off-label)* [2]; ***forestry*** [2, 3]. Mangold fly in ***fodder beet*, *mangels*, *sugar beet*** [1, 4, 5]. Millipedes in ***fodder beet*, *mangels*, *sugar beet*** [1, 4, 5]. Pygmy mangold beetle in ***fodder beet*, *mangels*, *sugar beet*** [1, 4, 5]. Springtails in ***fodder beet*, *mangels*, *sugar beet*** [1, 4, 5]. Symphylids in ***fodder beet*, *mangels*, *sugar beet*** [1, 4, 5]. Wireworm in ***fodder beet*, *mangels*, *sugar beet*** [1, 4, 5]

Specific Off-Label Approvals (SOLAs)

- ***forest nurseries*** *(OLA 992383) Dec 2008* [2]

Approval information

- Approval expiry 30 Nov 2004 [4]

Efficacy guidance

- At recommended rates seed of drilled crops is not damaged by contact with product [1, 4, 5]
- Products formulated as controlled release granules to give pine weevil control throughout the 2-year establishment phase [2, 3]

Restrictions

- This product contains an anticholinesterase carbamate compound. Do not use if under medical advice not to work with such compounds
- Maximum number of treatments 1 per crop or tree

Crop-specific information

- Latest use: before or at planting for forestry uses
- HI 100 d for beet crops, carrots, parsnips; 12 wk for broccoli, cabbages, calabrese, cauliflowers; 14 wk for Brussels sprouts; 23 wk for collards
- Apply to drilled crops with suitable granule applicator feeding directly into seed furrow or immediately behind drill coulter (behind seed drill boot for brassicas) or use bow-wave technique [1, 4, 5]
- See label for details of suitable applicators and settings. Correct calibration is essential [1, 4, 5]
- Where used in forestry, granules should be placed in the planting hole by hand or metered applicator before or after placing the tree [2]. Sachets should be placed in the planting hole by hand before placement of the tree [3]
- Safe to Douglas Fir and Sitka Spruce. Other conifers may vary in their sensitivity [2, 3]
- Forest trees may take 10-15 d to achieve full protection. An additional pre-planting insecticide dip or spray is advised, but post-planting sprays during the 2-yr establishment should not be necessary [2, 3]

Environmental safety

- Dangerous to game, wild birds and animals. Bury or remove spillages
- Dangerous to fish or other aquatic life. Do not contaminate surface waters or ditches with chemical or used container
- Keep in original container, tightly closed, in a safe place, under lock and key

Hazard classification and safety precautions

Hazard R03 [1, 4, 5]
Risk phrases R03c [1, 4, 5]
Operator protection A [2, 3]; B, K, M [1, 2, 4, 5]; C [1] (or D+E); C, D, E [2]; C [4, 5] (or D+E); H; U02, U04a, U05a, U13, U20a [1-5]; U09a, U19 [1, 4, 5]; U10 [2, 3]; U11 [2]; U12, U22b [3]
Environmental protection E01, E10a, E13b
Consumer protection C03
Storage and disposal E30a [2]; E30b, E32a, E34 [1, 4, 5]
Medical advice M02 [1-5]; M03 [1, 4, 5]

65 carboxin

A carboxamide fungicide available only in mixtures

66 carboxin + thiram

A fungicide seed dressing for cereals

Products

Anchor	Crompton	200:200 g/l	FS	08684

Uses

Bunt in ***spring wheat*** *(seed treatment)*, ***winter wheat*** *(seed treatment)*. Covered smut in ***spring barley*** *(seed treatment)*, ***spring oats*** *(seed treatment)*, ***winter barley*** *(seed treatment)*, ***winter oats*** *(seed treatment)*. Fusarium foot rot and seedling blight in ***spring barley*** *(seed treatment)*, ***spring oats*** *(seed treatment)*, ***spring rye*** *(seed treatment)*, ***spring wheat*** *(seed treatment)*, ***triticale*** *(seed treatment)*, ***winter barley*** *(seed treatment)*, ***winter oats*** *(seed treatment)*, ***winter rye*** *(seed treatment)*, ***winter wheat*** *(seed treatment)*. Leaf stripe in ***spring barley*** *(seed treatment)*, ***winter barley*** *(seed treatment - reduction)*. Loose smut in ***spring barley*** *(seed treatment - reduction)*, ***spring oats*** *(seed treatment - reduction)*, ***winter barley*** *(seed treatment - reduction)*, ***winter oats*** *(seed treatment - reduction)*. Net blotch in ***spring barley*** *(seed treatment)*. Septoria seedling blight in ***spring wheat*** *(seed treatment)*, ***winter wheat*** *(seed treatment)*.

Efficacy guidance

- Apply through suitable liquid flowable seed treating equipment of the batch treatment or continuous flow type where a secondary mixing auger is fitted
- Drill flow may be affected by treatment. Always re-calibrate seed drill before use

Restrictions
- Maximum number of treatments 1 per batch of seed
- Do not treat seed with moisture content above 16%
- Do not apply to cracked, split or sprouted seed
- Do not store treated seed from one season to the next

Crop-specific information
- Latest use: pre-drilling

Environmental safety
- Harmful to fish or other aquatic life. Do not contaminate surface waters or ditches with chemical or used container
- Do not use treated seed as food or feed
- Treated seed harmful to game and wildlife

Hazard classification and safety precautions

Operator protection A, D, H; U09a, U20c
Environmental protection E13c
Storage and disposal E26, E27, E30a, E31a, E34
Treated seed S01, S02, S04b, S05, S06, S07

67 carfentrazone-ethyl

A triazolinone contact herbicide for cereals and potato haulm destruction

Products

1	Aurora 50 WG	FMC	50% w/w	WG	09807
2	Aurora 50 WG	Belchim	50% w/w	WG	11613
3	Shark	FMC	60 g/l	ME	11054
4	Shark	Belchim	60 g/l	ME	11616
5	Spotlight 24 EC	FMC	240 g/l	EC	10702
6	Spotlight 24 EC	Belchim	240 g/l	EC	11617

Uses

Annual dicotyledons in ***potatoes*** [3, 4]. Black bindweed in ***potatoes*** [3, 4]. Cleavers in ***durum wheat, spring barley, spring oats, spring wheat, triticale, winter barley, winter oats, winter wheat*** [1, 2]; ***potatoes*** [3, 4]. Fat hen in ***potatoes*** [3, 4]. Haulm destruction in ***seed potatoes*** [6]; ***ware potatoes*** [5, 6]. Ivy-leaved speedwell in ***durum wheat, spring barley, spring oats, spring wheat, triticale, winter barley, winter oats, winter wheat*** [1, 2]; ***potatoes*** [3, 4]. Knotgrass in ***potatoes*** [3, 4]. Redshank in ***potatoes*** [3, 4]. Volunteer oilseed rape in ***potatoes*** [3, 4].

Approval information
- Carfentrazone-ethyl included in Annex I under EC Directive 91/414

Efficacy guidance
- Best results achieved from good spray cover applied to small actively growing weeds [1-4] or to potato foliage [5, 6]
- Carfentrazone-ethyl acts by contact only. Weeds emerging after application will not be controlled
- Recommended for use as two spray programme with one application in autumn and one in spring [1, 2]
- For potato crops with very dense vigorous haulm or where regrowth occurs following a single application a second application may be necessary to achieve satisfactory desiccation. A minimum interval between applications of 7 days should be observed to achieve optimum performance [5, 6]
- Efficacy will be reduced where flailed haulm covers the stems at application [6]

Restrictions
- Maximum number of treatments 2 per crop for cereals [1, 2]; 2 per crop for ware potato haulm destruction [6]; 1 per crop for seed potato haulm destruction [5, 6], or for potato weed control [3, 4]
- Do not treat crops under stress from drought, waterlogging, cold, pests, diseases, nutrient or lime deficiency or any factors reducing plant growth [1, 2]
- Do not treat cereals undersown with clover or other legumes [1, 2]
- Allow at least 2-3 wk between application and lifting potatoes to allow skins to set if potatoes are to be stored [5, 6]
- Follow label instructions for sprayer cleaning

Crop-specific information

- Latest use: before 3rd node detectable (GS 33) on cereals [1]; 5% emergence for early potatoes; 10% emergence for potatoes [3]
- HI 14 d for ware potatoes; 7 d for seed potatoes
- Emerged potato shoots may be burned off within 7-10 d, but subsequent crop growth and yield are unaffected [3, 4]
- For potato haulm destruction of ware crops apply at the onset of crop senescence
- Seed potato crops should be flailed when tubers have reached the desired size and sprayed once the flailed haulm is clear of the stems that remain
- Treat from 2 leaf stage of cereal crops [1, 2]; up to 10% crop emergence (maincrop potatoes) or 5% emergence (earlies) [3, 4]

Following crops guidance

- In the event of crop failure cereals, ryegrass, maize, oilseed rape, peas, sunflowers, *Phacelia*, vetches, carrots or onions may be planted within 1 mth of treatment. Any crop may be sown 3 mth after treatment [1, 2]
- No restrictions apply on the planting of succeeding crops 1 mth after application to potatoes for weed control or haulm destruction [3-6]

Environmental safety

- Extremely dangerous to fish or other aquatic life. Do not contaminate surface waters or ditches with chemical or used container [1, 2]
- Dangerous to fish or other aquatic life. Do not contaminate surface waters or ditches with chemical or used container [3-6]

Hazard classification and safety precautions

Hazard R04
Risk phrases R04e [3, 4]; R04g [5, 6]
Operator protection A, H [1-4]; U05a, U13 [1-4]; U08, U14 [1, 2]; U20b [3, 4]
Environmental protection E01 [1-6]; E13a [1, 2]; E13b [3-6]
Consumer protection C03
Storage and disposal E26, E31b [3, 4]; E30a [1-6]; E31c [5, 6]; E32a, E34 [1, 2]
Medical advice M05 [5, 6]

68 carfentrazone-ethyl + flupyrsulfuron-methyl

A foliar and residual acting herbicide for cereals

Products

Lexus Class WSB	DuPont	33.3:16.7% w/w	WB	08637

Uses

Annual dicotyledons in ***triticale***, ***winter oats***, ***winter rye***, ***winter wheat***. Blackgrass in ***triticale***, ***winter oats***, ***winter rye***, ***winter wheat***.

Approval information

- Carfentrazone-ethyl and flupyrsulfuron-methyl included in Annex I under EC Directive 91/414

Efficacy guidance

- Best results obtained when applied to small actively growing weeds
- Good spray cover of weeds must be obtained
- Increased degradation of active ingredient in high soil temperatures reduces residual activity
- Weed control may be reduced in dry soil conditions but susceptible weeds germinating soon after treatment will be controlled if adequate soil moisture present
- Blackgrass should be treated from 1 leaf but before first node. Strains of blackgrass resistant to other herbicides may not be controlled
- Flupyrsulfuron-methyl is a member of the ALS-inhibitor group of herbicides and products should be used in a planned Resistance Management strategy. See Section 1 for more information

Restrictions

- Maximum number of treatments 1 per crop
- Do not use on barley, on crops undersown with grasses or legumes, or any other broad-leaved crop

- Do not apply within 7 d of rolling
- Do not treat any crop suffering from drought, waterlogging, pest or disease attack, nutrient deficiency, or any other stress factors

Crop-specific information

- Latest use: before 31 Dec in yr of sowing for winter oats, winter rye, triticale; before 1st node detectable (GS 31) for winter wheat
- Apply to recommended crops from 2 leaf stage (GS 12) up to 31 Dec on oats, rye, triticale, or before first node detectable (GS 31) on wheat
- Slight chlorosis and stunting may occur in certain conditions. Recovery is rapid and yield not affected

Following crops guidance

- Only cereals, oilseed rape, field beans, clover or grass may be sown in the yr of harvest of a treated crop
- In the event of crop failure only winter wheat may be sown 1-3 mth after treatment. Land should be ploughed and cultivated to 15 cm minimum before resowing

Environmental safety

- Dangerous to fish or other aquatic life. Do not contaminate surface waters or ditches with chemical or used container
- Take extreme care to avoid damage by drift outside the target area or onto surface waters or ditches
- Spraying equipment should not be drained or flushed onto land planted, or to be planted, with trees or crops other than cereals and should be thoroughly cleansed after use - see label for instructions

Hazard classification and safety precautions

Hazard R04
Risk phrases R04e
Operator protection A; U05a, U08, U14, U19, U20b, U22a
Environmental protection E01, E13b, E15
Consumer protection C03
Storage and disposal E30a

69 carfentrazone-ethyl + mecoprop-P

SECTION 4

A foliar applied herbicide for cereals

Products

1	Platform S	FMC	1.5:60% w/w	WG	09706
2	Platform S	Belchim	1.5:60% w/w	WG	11611

Uses

Charlock in ***spring barley***, ***spring oats***, ***spring wheat***, ***winter barley***, ***winter oats***, ***winter wheat***. Chickweed in ***spring barley***, ***spring oats***, ***spring wheat***, ***winter barley***, ***winter oats***, ***winter wheat***. Cleavers in ***spring barley***, ***spring oats***, ***spring wheat***, ***winter barley***, ***winter oats***, ***winter wheat***. Red dead-nettle in ***spring barley***, ***spring oats***, ***spring wheat***, ***winter barley***, ***winter oats***, ***winter wheat***. Speedwells in ***spring barley***, ***spring oats***, ***spring wheat***, ***winter barley***, ***winter oats***, ***winter wheat***.

Approval information

- Carfentrazone-ethyl and mecoprop-P included in Annex I under EC Directive 91/414
- Approval expiry 31 Oct 2004 [1]

Efficacy guidance

- Best results obtained when weeds have germinated and growing vigorously in warm moist conditions
- Treatment of large weeds and poor spray coverage may result in reduced weed control

Restrictions

- Maximum number of treatments 2 per crop. The total amount of mecoprop-P applied in a single yr must not exceed the maximum total dose approved for any single product for the crop/situation
- Do not treat crops suffering from stress from any cause
- Do not treat crops undersown or to be undersown

Crop-specific information

- Latest use: before 3rd node detectable (GS 33)
- Can be used on all varieties of wheat and barley in autumn or spring from the beginning of tillering
- Early sown crops may be prone to damage if treated after period of rapid growth in autumn

Following crops guidance

- In the event of crop failure, any cereal, maize, oilseed rape, peas, vetches or sunflowers may be sown 1 mth after a spring treatment. Any crop may be planted 3 mth after treatment

Environmental safety

- Dangerous to fish or other aquatic life. Do not contaminate surface waters or ditches with chemical or used container

Hazard classification and safety precautions

Hazard R04
Risk phrases R04d, R04e
Operator protection A, C, H, M; U05a, U08, U11, U13, U14, U20b
Environmental protection E01, E13b; E07 (2 wk)
Consumer protection C03
Storage and disposal E30a, E31b

70 carfentrazone-ethyl + metsulfuron-methyl

A foliar applied herbicide mixture for cereals

Products

Ally Express	DuPont	40:10% w/w	WG	08640

Uses

Annual dicotyledons in ***durum wheat***, ***spring barley***, ***spring oats***, ***spring wheat***, ***triticale***, ***winter barley***, ***winter oats***, ***winter wheat***.

Approval information

- Carfentrazone-ethyl and metsulfuron-methyl included in Annex I under EC Directive 91/414

Efficacy guidance

- Best results achieved from applications made in good growing conditions
- Good spray cover of weeds must be obtained
- Growth of weeds is inhibited within hours of treatment but the time taken for visible colour changes to appear will vary according to species and weather
- Product has short residual life in soil. Under normal moisture conditions susceptible weeds germinating soon after treatment will be controlled
- Metsulfuron-methyl is a member of the ALS-inhibitor group of herbicides and products should be used in a planned Resistance Management strategy. See Section 1 for more information

Restrictions

- Maximum number of treatments 1 per crop
- Product must only be used after 1 Feb
- Do not use on any crop suffering stress from drought, waterlogging, cold, pest or disease attack, or nutrient deficiency
- Do not use on crops undersown with grass or legumes or on any broad-leaved crop
- Do not apply within 7 d of rolling
- Do not apply to a crop already treated with any sulfonylurea product except those containing flupyrsulfuron alone or in mixture with carfentrazone-ethyl or thifensulfuron-methyl

Crop-specific information

- Latest use: before 3rd node detectable (GS 33)
- Can be used on all soil types
- Apply in the spring from the 3-leaf stage on all crops
- Slight necrotic spotting of crops can occur under certain crop and soil conditions. Recovery is rapid and there is no effect on grain yield or quality

Following crops guidance

- Only cereals, oilseed rape, field beans or grass may be sown as a following crop in the same calendar yr. Any crop may follow in the next spring. In the event of failure of a treated crop, only winter wheat may be sown 1-3 mth later after ploughing and cultivating to at least 15 cm

Environmental safety

- Extremely dangerous to fish or other aquatic life. Do not contaminate surface waters or ditches with chemical or used container
- LERAP Category B
- Take extreme care to avoid drift onto broad-leaved plants outside the target area or onto ponds, waterways or ditches, or onto land intended for cropping
- Spraying equipment should not be drained or flushed onto land planted, or to be planted, with trees or crops other than cereals and should be thoroughly cleansed after use - see label for instructions

Hazard classification and safety precautions

Hazard R04
Risk phrases R04e
Operator protection A, H; U05a, U08, U14, U19, U20b
Environmental protection E01, E13a, E16a, E16b
Consumer protection C03
Storage and disposal E30a, E32a

71 carfentrazone-ethyl + thifensulfuron-methyl

A foliar applied herbicide mixture for cereals

Products

Harmony Express	DuPont	25:25% w/w	WG	09167

Uses

Annual dicotyledons in ***spring barley***, ***spring oats***, ***spring wheat***, ***winter barley***, ***winter oats***, ***winter wheat***. Cleavers in ***spring barley***, ***spring oats***, ***spring wheat***, ***winter barley***, ***winter oats***, ***winter wheat***. Knotgrass in ***spring barley***, ***spring oats***, ***spring wheat***, ***winter barley***, ***winter oats***, ***winter wheat***. Mayweeds in ***spring barley***, ***spring oats***, ***spring wheat***, ***winter barley***, ***winter oats***, ***winter wheat***. Speedwells in ***spring barley***, ***spring oats***, ***spring wheat***, ***winter barley***, ***winter oats***, ***winter wheat***.

Efficacy guidance

- Best results achieved from applications made in good growing conditions
- Good spray cover of weeds must be obtained
- Growth of weeds is inhibited within hours of treatment but the time taken for visible colour changes to appear will vary according to species and weather
- Thifensulfuron-methyl is a member of the ALS-inhibitor group of herbicides and products should be used in a planned Resistance Management strategy. See Section 1 for more information

Restrictions

- Maximum number of treatments 1 per crop
- Do not use on any crop suffering stress from drought, waterlogging, cold, pest or disease attack or nutrient deficiency
- Do not use on crops undersown with grass or legumes or on any broad-leaved crop
- Do not apply within 7 d of rolling
- Do not apply in tank mix with any sulfonylurea or in sequence to a crop already treated with any sulfonylurea product except those containing flupyrsulfuron or metsulfuron-methyl alone or in mixture with carfentrazone-ethyl

Crop-specific information

- Latest use: before 3rd node detectable (GS 33)
- Product must only be used on winter barley after 1 Feb. Treat other crops in the autumn or spring from the 3-leaf stage
- Slight necrotic spotting of crops can occur under certain crop and soil conditions. Recovery is rapid and there is no effect on grain yield or quality

Following crops guidance

- Only cereals, oilseed rape, field beans or grass may be sown as a following crop in the harvest yr
- Any crop may follow in the next spring. In the event of failure of a treated crop, only winter wheat may be sown before the normal harvest date, after ploughing and cultivating to at least 15 cm

Environmental safety

- Extremely dangerous to fish or other aquatic life. Do not contaminate surface waters or ditches with chemical or used container
- LERAP Category B

Hazard classification and safety precautions

Operator protection U05a, U20b
Environmental protection E01, E13a, E16a, E16b
Consumer protection C03
Storage and disposal E30a

72 chloridazon

A residual pyridazinone herbicide for beet crops

Products

1	Better DF	Sipcam	65% w/w	SG	06250
2	Better Flowable	Sipcam	430 g/l	SC	04924
3	Burex 430 SC	Agricola	430 g/l	SC	09494
4	Gladiator DF	Tripart	65% w/w	SG	06342
5	Pyramin DF	BASF	65% w/w	WG	03438
6	Sculptor	Sipcam	430 g/l	SC	08836
7	Takron	BASF	430 g/l	SC	06237
8	Tripart Gladiator 2	AgriGuard	430 g/l	SC	06618

Uses

Annual dicotyledons in ***bulb onions*** *(off-label)* [5]; ***fodder beet*, *mangels*, *sugar beet*** [1-8]; ***leeks*** *(off-label)*, ***salad onions*** *(off-label)* [1, 5]; ***onions*** *(off-label)* [1]. Annual meadow grass in ***bulb onions*** *(off-label)* [5]; ***fodder beet*, *mangels*, *sugar beet*** [1, 2, 4-8]; ***leeks*** *(off-label)*, ***salad onions*** *(off-label)* [1, 5]; ***onions*** *(off-label)* [1].

Specific Off-Label Approvals (SOLAs)

- ***bulb onions*, *leeks*, *salad onions*** *(OLA 970732) Dec 2008* [5]
- ***leeks*, *onions*, *salad onions*** *(OLA 012570) Dec 2008* [1]

Approval information

- Approval expiry 31 Oct 2004 [7]

Efficacy guidance

- Absorbed by roots of germinating weeds and best results achieved pre-emergence of weeds or crop when soil moist and adequate rain falls after application
- Application rate depends on soil type. See label for details

Restrictions

- Maximum number of treatments generally 1 per crop (pre-emergence) for fodder beet and mangels; 1 (pre-emergence) + 3 (post-emergence) per crop for sugar beet, but labels vary slightly. Check labels for maximum total dose for the crop to be treated
- Maximum total dose for onions and leeks equivalent to one full dose recommended for these crops [5]
- Do not use on Coarse Sands, Sands or Fine Sands or where organic matter exceeds 5%

Crop-specific information

- Latest use: pre-emergence for fodder beet and mangels; normally before leaves of crop meet in row for sugar beet, but labels vary slightly; up to and including second true leaf stage for onions and leeks [5]
- Where used pre-emergence spray as soon as possible after drilling in mid-Mar to mid-Apr on fine, firm, clod-free seedbed
- Where crop drilled after mid-Apr or soil dry apply pre-drilling and incorporate to 2.5 cm immediately afterwards
- Various tank mixes recommended on sugar beet for pre- and post-emergence use and as repeated low dose treatments. See label for details
- Crop vigour may be reduced by treatment of crops growing under unfavourable conditions including poor tilth, drilling at incorrect depth, soil capping, physical damage, pest or disease damage, excess seed dressing, trace-element deficiency or a sudden rise in temperature after a cold spell

Following crops guidance

- In the event of crop failure only sugar or fodder beet, mangels or maize may be re-drilled on treated land after cultivation
- Winter cereals may be sown in autumn after ploughing. Any spring crop may follow treated beet crops harvested normally

Environmental safety

- Dangerous for the environment [7]
- Very toxic to aquatic organisms. May cause long-term adverse effects in the aquatic environment [7]
- Harmful to fish or other aquatic life. Do not contaminate surface waters or ditches with chemical or used container [1-6]

Hazard classification and safety precautions

Hazard R04 [3, 6-8]; R14 [7]
Risk phrases R04a, R04b [6, 8]; R04c [8]; R04e [3, 7]; R14c [7]
Operator protection A [3, 6-8]; U05a [3, 6-8]; U08 [1-7]; U14 [6, 7]; U15 [6]; U19 [6-8]; U20a [2-4]; U20b [1, 5-8]
Environmental protection E01 [3, 6-8]; E13c [1-6]; E15, E38 [7]
Consumer protection C03 [3, 6-8]
Storage and disposal E26 [3, 6, 8]; E30a [1-8]; E31b [3, 6-8]; E32a [1, 2, 4, 5]; E32d [7]
Medical advice M05a [7]

73 chloridazon + chlorpropham + metamitron

A residual herbicide mixture for beet crops

Products

Newtron	Nufarm UK	83:42:333 g/l	SE	11208

Uses

Annual dicotyledons in ***fodder beet***, ***mangels***, ***sugar beet***. Annual meadow grass in ***fodder beet***, ***mangels***, ***sugar beet***.

Efficacy guidance

- Apply as part of a spray programme. For best results a full programme of pre- and post-emergence treatments is required
- Pre-emergence weed control will be reduced under very dry conditions but subsequent rain may activate the chemical
- Best post-emergence performance requires tank mixture with phenmedipham. See label for details
- Correct post-emergence timing important for good control. First treatment should be made when weeds at early cotyledon stage and subsequent applications made when new weed flushes appear

Restrictions

- Subject to the maximum total dose, maximum number of treatments 1 pre-emergence followed by 3 post-emergence, or 5 post-emergence treatments unless another product has been used pre-emergence, in which case 4 post-emergence sprays may be applied
- Must be tank mixed when used post-emergence. See label
- Do not roll or harrow treated soil. Spray after rolling

Crop-specific information

- Latest use: before crop leaves meet between rows

Following crops guidance

- Only sugar beet, fodder beet or mangels should be sown within 4 mth of last application
- After harvesting treated crops land should be mould-board ploughed to 15 cm and any spring crop may then be drilled

Environmental safety

- Harmful to fish or other aquatic life. Do not contaminate surface waters or ditches with chemical or used container

Hazard classification and safety precautions

Hazard R04

Risk phrases R04a, R04b, R04e
Operator protection A, C; U05a, U08, U11, U14, U19, U20a
Environmental protection E01, E13c
Consumer protection C03
Storage and disposal E26, E30a, E31c

74 chloridazon + ethofumesate

A contact and residual herbicide for beet crops

Products

1	Gremlin	Sipcam	285:176 g/l	SC	09468
2	Magnum	BASF	275:170 g/l	SC	11727

Uses

Annual dicotyledons in ***fodder beet*** [2]; ***sugar beet*** [1, 2]. Annual meadow grass in ***fodder beet*** [2]; ***sugar beet*** [1, 2].

Approval information

- Ethofumesate included in Annex I under EC Directive 91/414

Efficacy guidance

- Best results achieved on a fine firm clod-free seedbed when soil moist and adequate rain falls after spraying. Efficacy may be reduced if heavy rain falls just after incorporation
- Effectiveness may be reduced under conditions of low pH
- May be applied by conventional or repeat low dose method. See label for details

Restrictions

- Maximum number of treatments 1 pre-emergence for fodder beet, mangels; 1 pre-emergence plus 3 post-emergence for sugar beet [2]; 1 pre-emergence or 3 post-emergence on sugar beet [1]
- May be used on soil classes Loamy Sand - Silty Clay Loam [2]. Additional restrictions apply for some tank mixtures

Crop-specific information

- Latest use: pre-crop emergence for fodder beet, mangels; before crop leaves meet between rows for sugar beet
- Apply up to cotyledon stage of weeds
- Crop vigour may be reduced by treatment of crops growing under unfavourable conditions including poor tilth, drilling at incorrect depth, soil capping, physical damage, excess nitrogen, excess seed dressing, trace element deficiency or a sudden rise in temperature after a cold spell. Frost after pre-emergence treatment may check crop growth

Following crops guidance

- In the event of crop failure only sugar beet, fodder beet or mangels may be re-drilled
- Any crop may be sown 3 mth after spraying following ploughing to 15 cm

Environmental safety

- Dangerous for the environment [2]
- Very toxic to aquatic organisms. May cause long-term adverse effects in the aquatic environment [2]
- Harmful to fish or other aquatic life. Do not contaminate surface waters or ditches with chemical or used container [1]

Hazard classification and safety precautions

Hazard R03 [1, 2]; R14 [2]
Risk phrases R03c [1, 2]; R14c [2]
Operator protection A, H [1, 2]; C [1]; U05a [2]; U08, U19, U20a [1, 2]
Environmental protection E01 [1, 2]; E13c [1]; E15, E38 [2]
Consumer protection C03
Storage and disposal E26, E32d [2]; E30a, E31b [1, 2]; E34 [1]
Medical advice M05a [2]

75 chloridazon + metamitron

A contact and residual herbicide mixture for sugar beet

Products

Volcan Combi	Sipcam	300:280 g/l	SC	10256

Uses

Annual dicotyledons in ***sugar beet***. Annual meadow grass in ***sugar beet***.

Efficacy guidance

- Best results achieved from a sequential programmme of sprays
- Pre-emergence use improves efficacy of post-emergence programme. Best results obtained from application to a moist seed bed
- First post-emergence treatment should be made when weeds at early cotyledon stage and subsequent applications made when new weed flushes reach this stage
- Weeds surviving an earlier treatment should be treated again after 7-10 d even if no new weeds have appeared

Restrictions

- Maximum number of treaments 1 pre-emergence followed by 3 post-emergence
- Take advice if light soils have a high proportion of stones

Crop-specific information

- Latest use: when leaves of crop meet in rows
- Tolerance of crops growing under stress from any cause may be reduced
- Crops treated pre-emergence and subsequently subjected to frost may be checked and recovery may not be complete

Following crops guidance

- After the last application only sugar beet or mangels may be sown within 4 mth; cereals may be sown after 16 wk. Land should be mouldboard ploughed to 15 cm and thoroughly cultivated before any succeeding crop

Environmental safety

- Harmful to fish or other aquatic life. Do not contaminate surface waters or ditches with chemical or used container

Hazard classification and safety precautions

Hazard R04
Risk phrases R04a, R04b
Operator protection U08, U14, U15, U20a
Environmental protection E13c
Storage and disposal E30a, E31b, E34

76 chloridazon + propachlor

A residual pre-emergence herbicide for use in onions and leeks

Products

Ashlade CP	Nufarm UK	80:400 g/l	SC	06481

Uses

Annual dicotyledons in ***bulb onions***, ***bulb onions*** *(off-label - post-emergence)*, ***leeks***, ***leeks*** *(off-label - post-emergence)*, ***salad onions***. Annual grasses in ***bulb onions***, ***leeks***, ***salad onions***.

Specific Off-Label Approvals (SOLAs)

- ***bulb onions***, ***leeks*** *(post-emergence) (OLA 981362) Dec 2008* [1]

Efficacy guidance

- Best results achieved from application to firm, moist, weed-free seedbed when adequate rain falls afterwards

Restrictions

- Maximum number of treatments 1 per crop
- Do not use on soils with more than 10% organic matter
- Ensure crops are drilled to 20 mm depth

Crop-specific information

- Latest use: pre-emergence of crop; before 2 true leaf stage for onions and leeks (off-label).
- HI 12 wk for chives
- Apply pre-emergence of sown crops, preferably soon after drilling, before weeds emerge. Loose or fluffy seedbeds must be consolidated before application
- Apply to transplanted crops when soil has settled after planting
- Crops stressed by nutrient deficiency, pests or diseases, poor growing conditions or pesticide damage may be checked by treatment, especially on sandy or gravelly soils

Following crops guidance

- In the event of crop failure only onions, leeks or maize should be planted
- Any crop can follow a treated onion or leek crop harvested normally as long as the ground is cultivated thoroughly before drilling

Environmental safety

- Harmful to fish or other aquatic life. Do not contaminate surface waters or ditches with chemical or used container

Hazard classification and safety precautions

Hazard R04
Risk phrases R04a, R04b
Operator protection A, C; U02, U04a, U05a, U08, U13, U19, U20a
Environmental protection E01, E13c
Consumer protection C03
Storage and disposal E30a, E31b
Medical advice M03

77 chloridazon + quinmerac

A herbicide mixture for use in beet crops

Products

Fiesta T	BASF	360:60 g/l	SC	10260

Uses

Annual dicotyledons in ***fodder beet, mangels, sugar beet***. Annual meadow grass in ***fodder beet, mangels, sugar beet***. Chickweed in ***fodder beet, mangels, sugar beet***. Cleavers in ***fodder beet, mangels, sugar beet***. Mayweeds in ***fodder beet, mangels, sugar beet***. Poppies in ***fodder beet, mangels, sugar beet***. Speedwells in ***fodder beet, mangels, sugar beet***.

Efficacy guidance

- Best results obtained when adequate soil moisture is present at application and afterwards to form an active herbicidal layer in the soil
- A programme of pre-emergence treatment followed by post-emergence application(s) in mixture with a contact herbicide optimises weed control and is essential for some difficult weed species such as cleavers
- Treatment pre-emergence only may not provide sufficient residual activity to give season-long weed control
- Effectiveness may be reduced under conditions of low pH

Restrictions

- Maximum total dose on sugar beet equivalent to one treatment pre-emergence plus two treatments post-emergence, all at maximum individual dose; maximum total dose on fodder beet and mangels equivalent to one full dose pre-emergence treatment
- Heavy rain shortly after treatment may check crop growth particularly when it leaves water standing in surface depressions
- Treatment of stressed crops or those growing in unfavourable conditions may depress crop vigour and possibly reduce stand

Crop-specific information

- Latest use: before plants meet between the rows for sugar beet; pre-emergence for fodder beet, mangels

Environmental safety

- Harmful to fish or other aquatic life. Do not contaminate surface waters or ditches with chemical or used container
- To reduce movement to groundwater do not apply to dry soil or when heavy rain is forecast

Hazard classification and safety precautions

Hazard R04
Risk phrases R04e
Operator protection A; U05a, U08, U19, U20b
Environmental protection E01, E13c
Consumer protection C03
Storage and disposal E26, E30a, E31c

78 chlormequat

A plant-growth regulator for reducing stem growth and lodging

Products

1	Adjust	Mandops	620 g/l	SL	05589
2	Agriguard Chlormequat 700	AgriGuard	700 g/l	SL	09782
3	Agriguard Chlormequat 720	AgriGuard	720 g/l	SL	09919
4	Agriguard Chlormequat 760	AgriGuard	760 g/l	SL	10290
5	Barclay Holdup	Barclay	700 g/l	SL	11365
6	Barclay Holdup 600	Barclay	600 g/l	SL	11373
7	Barclay Holdup 640	Barclay	640 g/l	SL	11374
8	Barleyquat B	Mandops	620 g/l	SL	07051
9	BASF 3C Chlormequat 720	BASF	720 g/l	SL	06514
10	Bettaquat B	Mandops	620 g/l	SL	07050
11	Fargro Chlormequat	Fargro	460 g/l	SL	02600
12	Greencrop Carna	Greencrop	600 g/l	SL	09403
13	Greencrop Coolfin	Greencrop	700 g/l	SL	09449
14	Hive	Nufarm UK	730 g/l	SL	10953
15	K2	Mandops	620 g/l	SL	10370
16	Landgold CCC 720	Landgold	720 g/l	SL	08527
17	Mandops Chlormequat 700	Mandops	700 g/l	SL	06002
18	Manipulator	Mandops	620 g/l	SL	05871
19	Mirquat	Nufarm UK	730 g/l	SL	10604
20	Sigma PCT	Nufarm UK	460 g/l	SL	11209
21	Stabilan 700	Nufarm UK	700 g/l	SL	11393
22	Terbine	Nufarm UK	730 g/l	SL	11407
23	Tripart Brevis	Tripart	700 g/l	SL	03754
24	Uplift	United Phosphorus	700 g/l	SL	07527

Uses

Increasing yield in ***spring barley*** [1, 8, 15, 17, 18]; ***winter barley*** [1, 2, 4-9, 12, 13, 15, 17, 18, 23]. Lodging control in ***spring barley*** [1, 8, 15, 18]; ***spring oats***, ***winter oats*** [2-7, 9, 12-14, 16, 17, 19, 21-24]; ***spring rye***, ***triticale*** [9]; ***spring wheat***, ***winter wheat*** [1-7, 9, 10, 12-24]; ***winter barley*** [1-5, 7-9, 13-16, 18-24]; ***winter rye*** [2-4, 9, 13, 23]. Stem shortening in ***bedding plants***, ***camellias***, ***hibiscus trionum***, ***lilies*** [11, 14, 19, 21, 22]; ***geraniums***, ***poinsettias*** [9, 11, 14, 19, 21, 22].

Approval information

- Approved for aerial application on wheat and oats [13, 23]; on wheat, oats and barley [5-7, 12, 17, 24]; wheat and barley [20]; on wheat, oats, barley, rye and triticale [9]. See notes in Section 1
- Approval expiry 30 Jun 2004 [13]
- Approval expiry 30 Nov 2004 [16]

Efficacy guidance

- Most effective results on cereals normally achieved from application from Apr onwards, on wheat and rye from leaf sheath erect to first node detectable (GS 30-31), on oats at second node detectable (GS 32), on winter barley from mid-tillering to leaf sheath erect (GS 25-30). However, recommendations vary with product. See label for details
- Results on barley can be variable
- In tank mixes with other pesticides optimum timing for herbicide action may differ from that for growth reduction. See label for details of tank mix recommendations
- Addition of approved non-ionic wetter recommended for products approved on oats

FOR FULL CONDITIONS OF USE ALWAYS READ THE PRODUCT LABEL

Restrictions

- Maximum number of treatments 1 per crop for cereals (2 per crop for split applications on winter wheat or winter barley); varies with species for ornamentals [9]
- Maximum total dose equivalent to one full dose treatment [1, 10, 15]
- Do not use on very late sown spring wheat or oats or on crops under stress
- Do not use straw from treated cereals as horticultural growth medium or mulch
- Mixtures with liquid nitrogen fertilizers may cause scorch and are specifically excluded on some labels
- At least 6 h, preferably 24 h, required before rain for maximum effectiveness. Do not apply to wet crops
- Check labels for tank mixtures known to be incompatible

Crop-specific information

- Latest use varies with product. See label for details
- May be used on cereals undersown with grass or clovers
- Ornamentals to be treated must be well established and growing vigorously. Do not treat in strong sunlight or when temperatures are likely to fall below 10°C
- Temporary yellow spotting may occur on poinsettias. It can be minimised by use of a non-ionic wetting agent - see label

Environmental safety

- Wash equipment thoroughly with water and wetting agent immediately after use and spray out. Spray out again before storing or using for another product. Traces can cause harm to susceptible crops sprayed later

Hazard classification and safety precautions

Hazard R03 [1-24]; R14 [21]

Risk phrases R03a [2-4, 11, 13, 23, 24]; R03c [1-23]; R14a [21]

Operator protection A; U02, U04a [16]; U05a [1-13, 15-18, 20-24]; U05b [14, 19]; U08 [1-8, 10-24]; U13 [5, 11, 17]; U19 [1-24]; U20a [5, 17]; U20b [1-4, 6-16, 18-24]

Environmental protection E01 [1-24]; E08 [5, 17, 23]; E15 [1, 3-24]

Consumer protection C01 [5, 11, 17]; C03 [1-24]

Storage and disposal E26 [2-7, 11-13, 16, 17, 21, 23, 24]; E30a [1-24]; E31a [5, 17, 21, 23]; E31b [1-4, 6-8, 10-16, 18-20, 22, 24]; E31c [9]; E34 [1-5, 7-24]

Medical advice M03 [1-24]; M05a [21]

79 chlormequat + 2-chloroethylphosphonic acid

A plant growth regulator for use in cereals

Products

1	Barclay Banshee XL	Barclay	305:155 g/l	SL	11339
2	Greencrop Tycoon	Greencrop	305:155 g/l	SL	09571
3	Strate	Aventis	360:180 g/l	SL	10020
4	Sypex	BASF	305:155 g/l	SL	04650
5	Upgrade	Aventis	360:180 g/l	SL	10029

Uses

Lodging control in ***spring barley***, ***winter barley***, ***winter wheat***.

Efficacy guidance

- Best results obtained when crops growing vigorously
- Recommended dose varies with growth stage. See labels for details and recommendations for use of sequential treatments

Restrictions

- 2-chloroethylphosphonic acid is an anticholinesterase organophosphorus compound. Do not use if under medical advice not to work with such compounds
- Maximum number of treatments 1 per crop; maximum total dose equivalent to one full dose treatment
- Product must always be used with specified non-ionic wetter - see labels
- Do not use on any crop in sequence with any other product containing 2-chloroethylphosphonic acid
- Do not spray when crop wet or rain imminent

- Do not spray during cold weather or periods of night frost, when soil is very dry, when crop diseased or suffering pest damage, nutrient deficiency or herbicide stress
- If used on seed crops grown for certification inform seed merchant beforehand
- Do not use on wheat variety Moulin or on any winter varieties sown in spring [1, 2, 4]
- Do not use on spring barley Triumph [2]
- Do not treat barley on soils with more than 10% organic matter [1, 2, 4]
- Do not use straw from treated cereals as a horticultural growth medium or as a mulch [1, 2, 4]
- Do not use in programme with any other product containing 2-chloroethylphosphonic acid [1, 2]
- Only crops growing under conditions of high fertility should be treated

Crop-specific information

- Latest use: before flag leaf ligule/collar just visible (GS 39) or 1st spikelet visible (GS 51) for wheat or barley at top dose; or before flag leaf sheath opening (GS 47) for winter wheat at reduced dose
- Apply before lodging has started

Environmental safety

- Harmful to aquatic organisms. May cause long-term adverse effects to the aquatic environment [4]
- Harmful to fish or other aquatic life. Do not contaminate surface waters or ditches with chemical or used container [3, 5]

Hazard classification and safety precautions

Hazard R03 [1-5]; R04 [3-5]; R14 [4]
Risk phrases R03a, R04a [3, 5]; R03c [1-5]; R04c, R14a [4]
Operator protection A [1-5]; C [3, 5]; U05a, U08, U19, U20b [1-5]; U13 [3, 5]
Environmental protection E01 [1-5]; E13c [3, 5]; E15 [1, 2, 4]
Consumer protection C03
Storage and disposal E26, E30a, E31b, E34
Medical advice M01, M03

80 chlormequat + 2-chloroethylphosphonic acid + mepiquat chloride

A plant growth regulator for reducing lodging in cereals

Products

Cyclade	BASF	230:155:75 g/l	SL	08958

Uses

Lodging control in ***spring barley***, ***winter barley***, ***winter wheat***.

Efficacy guidance

- Best results achieved in a vigorous, actively growing crop with adequate fertility and moisture
- Optimum timing on all crops is from second node detectable stage (GS 32)
- Recommended for use as part of an intensive growing system which includes provision for optimum fertilizer treatment and disease control

Restrictions

- 2-chloroethylphosphonic acid is an anticholinesterase organophosphorus compound. Do not use if under medical advice not to work with such compounds
- Maximum number of treatments 1 per crop
- Maximum total dose depends on spraying regime adopted. See label
- Must be used with a non-ionic wetting agent
- Do not apply to stressed crops or those on soils of low fertility unless receiving adequate dressings of fertilizer
- Do not apply in temperatures above 21°C or if crop is wet or if rain expected
- Do not treat variety Moulin nor any winter varieties sown in spring
- Do not use in a programme with any other product containing 2-chloroethylphosphonic acid
- Do not apply to barley on soils with more than 10% organic matter (winter wheat may be treated)
- Notify seed merchant in advance if use on a seed crop is proposed

Crop-specific information

- Latest use: before first spikelet of ear visible (GS 51) using reduced dose on winter barley; before flag leaf sheath opening (GS 47) using reduced dose on winter wheat; before flag leaf just visible on spring barley
- May be applied to crops undersown with grasses or clovers
- Treatment may cause some delay in ear emergence

Environmental safety

- Do not use straw from treated crops as a horticultural growth medium

Hazard classification and safety precautions

Hazard R03, R04
Risk phrases R03c, R04c
Operator protection A; U05a, U08, U19, U20b
Environmental protection E01, E15
Consumer protection C03
Storage and disposal E30a, E31c, E34
Medical advice M01, M03

81 chlormequat + choline chloride + imazaquin

A plant growth regulator mixture for winter wheat

Products

1	Meteor	BASF	368:28:0.8 g/l	SL	10403
2	Standon Imazaquin 5C	Standon	368:28:0.8 g/l	SL	08813

Uses

Increasing yield in ***winter wheat***. Lodging control in ***winter wheat***.

Efficacy guidance

- Apply to crops during good growing conditions or to those at risk from lodging
- On soils of low fertility, best results obtained where adequate nitrogen fertilizer used

Restrictions

- Maximum number of applications 1 per crop (2 per crop at split dose)
- Do not treat durum wheat
- Do not apply to undersown crops
- Do not use straw from treated cereals as horticultural growth medium or mulch
- Do not apply when crop wet or rain imminent

Crop-specific information

- Latest use: before second node detectable (GS 31)
- Apply as single dose from leaf sheath lengthening up to and including 1st node detectable or as split dose, the first from tillers formed to leaf sheath lengthening, the second from leaf sheath erect up to and including 1st node detectable

Hazard classification and safety precautions

Hazard R03, R04
Risk phrases R03c, R04a
Operator protection A, C; U05a, U08, U13, U19 [1, 2]; U20a [2]; U20b [1]
Environmental protection E01, E15
Consumer protection C03
Storage and disposal E26 [1]; E30a, E31b, E34 [1, 2]
Medical advice M03

82 chlormequat + mepiquat chloride

A plant growth regulator for reducing lodging in cereals

Products

Stronghold	BASF	345:115 g/l	SL	09134

Uses

Lodging control in ***winter wheat***.

Efficacy guidance

- Optimum timing is when leaf sheaths erect (GS 30)

Restrictions

- Maximum total dose equivalent to one full dose treatment
- Do not apply to stressed crops or those on soils of low fertility unless receiving adequate dressings of fertilizer
- Do not apply in temperatures above 21°C or if crop is wet or if rain expected
- Do not treat any winter varieties sown in spring
- Notify seed merchant in advance if use on a seed crop is proposed

Crop-specific information

- Latest use: before 3rd node detectable (GS 33)
- Apply during good growing conditions at the correct timings - see label
- May be applied to crops undersown with grasses or clovers
- Treatment may cause some delay in ear emergence
- Mixtures with liquid fertilizers may cause scorching in some circumstances

Environmental safety

- Do not use straw from treated crops as a horticultural growth medium

Hazard classification and safety precautions

Hazard R03
Risk phrases R03c
Operator protection A; U05a, U08, U19, U20b
Environmental protection E01, E15
Consumer protection C03
Storage and disposal E30a, E31c, E34
Medical advice M03, M05a

83 chlormequat with choline chloride

A plant growth regulator for use in cereals and certain ornamentals

Products

1	Agriguard 5C Chlormequat 460	AgriGuard	460 g/l	SL	09851
2	Atlas 5C Quintacel	Nufarm UK	640 g/l	SL	11130
3	Barclay Take 5	Barclay	645 g/l	SL	11368
4	Cropsafe 5C Chlormequat	Certis	645 g/l	SL	11179
5	Fernpath Tangent	AgriGuard	460 g/l	SL	10341
6	New 5C Cycocel	BASF	645 g/l	SL	01482

Uses

Increasing yield in ***winter barley*** [1, 5, 6]. Lodging control in ***spring oats***, ***spring wheat***, ***winter oats***, ***winter wheat*** [1-3, 5, 6]; ***spring rye***, ***spring wheat*** *(autumn sown)* [6]; ***triticale*** [1, 3, 5, 6]; ***winter barley*** [2, 3]; ***winter rye*** [3, 6]. Stem shortening in ***bedding plants***, ***camellias***, ***hibiscus trionum***, ***lilies*** [2]; ***geraniums***, ***poinsettias*** [2, 4, 6].

Approval information

- Approved for aerial application on wheat, oats [2]; wheat, oats, barley, rye, triticale [3, 6]. See notes in Section 1

Efficacy guidance

- Influence on growth varies with crop and growth stage. Risk of lodging reduced by application at early stem extension. Root development and yield can be improved by earlier treatment
- Most effective results normally achieved from spring application. On winter barley an autumn treatment may also be useful. Timing of spray is critical and recommendations vary with product. See label for details
- Often used in tank-mixes with pesticides. Recommendations for mixtures and sequential treatments vary with product. See label for details

Restrictions

- Maximum number of treatments 1 per crop for spring wheat, oats, rye, triticale; 1 or 2 per crop for winter wheat and winter barley (depending on dose); 1-3 per yr for ornamentals. See label for details

- Do not spray very late sown spring crops, crops on soils of low fertility, crops under stress from any cause or if frost expected
- Do not use on spring barley
- Add authorised non-ionic wetter when spraying oats
- At least 6 h required before rain for maximum effectiveness. Do not apply to wet crops
- Do not use straw from treated cereals as horticultural growth medium or mulch

Crop-specific information

- Latest use: before 3rd node detectable (GS 33) for oats, winter wheat, autumn drilled spring wheat; before 2nd node detectable (GS 32) for rye; before 1st node detectable (GS 31) for spring wheat, winter barley, triticale; after potting for Poinsettias
- Mixtures with liquid nitrogen fertilizers may cause scorch
- May be used on cereals undersown with grass or clovers

Environmental safety

- Wash equipment thoroughly with water and wetting agent immediately after use and spray out. Traces can cause harm to susceptible crops sprayed later [2]

Hazard classification and safety precautions

Hazard R03 [1-3, 5, 6]
Risk phrases R03c [1-3, 5, 6]
Operator protection A; U05a, U08, U19 [1-3, 5, 6]; U20a [2]; U20b [1, 3, 5, 6]
Environmental protection E01, E15 [1-3, 5, 6]
Consumer protection C03 [1-3, 5, 6]
Storage and disposal E30a, E34 [1-3, 5, 6]; E31b [1-3, 5]; E31c [6]
Medical advice M03 [1-3, 5, 6]

84 chlormequat with di-1-p-menthene

A plant-growth regulator for reducing stem growth and lodging

See also di-1-p-menthene

Products

Podquat	Mandops	470 g/l	SL	03003

Uses

Increasing yield in ***broad beans***, ***combining peas***, ***spring field beans***, ***winter field beans***.

Restrictions

- Do not apply to plants covered by frost
- Do not use in tank mixture with herbicides on pea crops
- Do not use in tank mixture with liquid nitorgen fertilizer

Crop-specific information

- On beans either apply in autumn as soon as possible after 3-leaf stage (GS 1,3) but before growth ceases followed by spring treatment in mid-Mar to early Apr, or use a single spring spray during stem elongation. See label for details
- On peas treat when standing crop reaches 7-10 cm above ground level
- May be used at temperatures down to 1°C provided spray dries on leaves before rain, frost or snow occurs

Hazard classification and safety precautions

Hazard R03
Risk phrases R03c
Operator protection A; U05a, U08, U19, U20a
Environmental protection E01, E15
Consumer protection C03
Storage and disposal E30a, E31a, E34
Medical advice M03

85 2-chloroethylphosphonic acid

A plant growth regulator for cereals and various horticultural crops

See also chlormequat + 2-chloroethylphosphonic acid
chlormequat + 2-chloroethylphosphonic acid + mepiquat chloride

Products

1	Agriguard Cerusite	AgriGuard	480 g/l	SL	11494
2	Barclay Coolmore	Barclay	480 g/l	SL	11349
3	Cerone	Aventis	480 g/l	SL	09985
4	Ethrel C	Certis	480 g/l	SL	11387

Uses

Basal bud stimulation in ***protected roses*** [4]. Fruit ripening in ***apples*** *(off-label - for cider making)*, ***outdoor tomatoes***, ***protected tomatoes*** [4]. Increasing branching in ***geraniums*** [4]. Inducing flowering in ***bromeliads*** [4]. Lodging control in ***spring barley***, ***triticale***, ***winter barley***, ***winter rye***, ***winter wheat*** [1-3]. Stem shortening in ***narcissi*** [4].

Specific Off-Label Approvals (SOLAs)

- ***apples*** *(for cider making) (OLA 030875) Dec 2008* [4]

Approval information

- Approved for aerial application on winter barley [2, 3]. See notes in Section 1

Efficacy guidance

- Best results achieved on crops growing vigorously under conditions of high fertility
- Optimum timing varies between crops and products. See labels for details
- Do not spray crops when wet or if rain imminent
- Best results on horticultural crops when temperature does not fall below 10°C [4]
- Use on tomatoes 17 d before planned pulling date [4]
- Apply as drench to daffodils when stems average 15 cm [4]
- Apply to glasshouse roses when new growth started after pruning [4]

Restrictions

- 2-chloroethylphosphonic acid is an anticholinesterase organophosphorus compound. Do not use if under medical advice not to work with such compounds
- Maximum number of treatments 1 per crop or yr
- Do not spray crops suffering from stress caused by any factor, during cold weather or period of night frost nor when soil very dry
- Do not apply to cereals within 10 d of herbicide or liquid fertilizer application
- Do not spray wheat or triticale where the leaf sheaths have split and the ear is visible

Crop-specific information

- Latest use: before 1st spikelet visible (GS 51) for spring barley, winter barley, winter rye; before flag leaf sheath opening (GS 47) for triticale, winter wheat
- HI cider apples, tomatoes 5 d

Environmental safety

- Harmful to fish or other aquatic life. Do not contaminate surface waters or ditches with chemical or used container
- Avoid accidental deposits on painted objects such as cars, trucks, aircraft

Hazard classification and safety precautions

Hazard R04
Risk phrases R04a, R04b
Operator protection A, C; U05a, U08, U13 [1-4]; U20a [4]; U20b [3]
Environmental protection E01, E13c
Consumer protection C03
Storage and disposal E30a, E31b [1-4]; E34 [1, 4]
Treated seed S06 [1, 2]
Medical advice M01 [2-4]

86 2-chloroethylphosphonic acid + mepiquat chloride

A plant growth regulator for reducing lodging in cereals

Products

1	Barclay Banshee	Barclay	155:305 g/l	SL	11343
2	Standon Mepiquat Plus	Standon	155:305 g/l	SL	09373
3	Terpal	BASF	155:305 g/l	SL	02103
4	Terpitz	Me2	155:305 g/l	SL	09634

Uses

Increasing yield in ***winter barley*** *(low lodging situations)* [1, 3, 4]. Lodging control in ***spring barley*** [1, 3, 4]; ***triticale***, ***winter barley***, ***winter rye***, ***winter wheat*** [1-4].

Efficacy guidance

- Best results achieved on crops growing vigorously under conditions of high fertility
- Recommended dose and timing vary with crop, cultivar, growing conditions, previous treatment and desired degree of lodging control. See label for details
- May be applied to crops undersown with grass or clovers
- Do not apply to crops if wet or rain expected as efficacy will be impaired

Restrictions

- 2-chloroethylphosphonic acid is an anticholinesterase organophosphorus compound. Do not use if under medical advice not to work with such compounds
- Maximum number of treatments varies with dose and timing - see labels
- Add an authorised non-ionic wetter to spray solution
- Do not treat crops damaged by herbicides or stressed by drought, waterlogging etc
- Do not treat crops on soils of low fertility unless adequately fertilized
- Do not use in a programme with any other product containing 2-chloroethylphosphonic acid
- Do not apply to winter cultivars sown in spring or treat winter barley, triticale or winter rye on soils with more than 10% organic matter (winter wheat may be treated)
- Do not apply at temperatures above 21°C

Crop-specific information

- Latest use: before ear visible (GS 49) for winter barley, spring barley, winter wheat and triticale; flag leaf just visible (GS 37) for winter rye
- Late tillering may be increased with crops subject to moisture stress and may reduce quality of malting barley

Environmental safety

- Do not contaminate surface waters or ditches with chemical or used container
- Do not use straw from treated cereals as a mulch or growing medium

Hazard classification and safety precautions

Hazard R04
Risk phrases R04a, R04c
Operator protection U20b
Environmental protection E15
Storage and disposal E26, E30a, E31c
Medical advice M01

87 chlorophacinone

An anticoagulant rodenticide

Products

1	Drat Rat Bait	B H & B	0.005% w/w	RB	H6743
2	Endorats	Irish Drugs	0.005% w/w	RB	H6744
3	Karate Ready to Use Rat Bait	JohnsonDiversey	0.006% w/w	RB	H6745
4	Karate Ready to Use Rodenticide Sachets	JohnsonDiversey	0.006%w/w	RB	H6746

Uses

Rats in ***farm buildings*** [1-4]. Voles in ***farm buildings*** [1].

Efficacy guidance

- Chemical formulated with oil, thus improving weather resistance of bait
- Use in baiting programme
- Bait stations should be sited where rats active, by rat holes, along runs or in harbourages. Place bait in suitable containers
- Replenish baits every few days and remove unused bait when take ceases or after 7-10 d

Restrictions

- For use only by professional operators
- Resistance status of target population should be taken into account when considering choice of rodenticide
- Bait stations may be sited conveniently but bait should be inaccessible to non-target animals and protected from prevailing weather

Environmental safety

- Prevent access to baits by children, domestic animals and birds; see label for other precautions required
- Harmful to game, wild birds and animals
- Store unused sachets in a safe place. Do not store half-used sachets [4]

Hazard classification and safety precautions

Operator protection U13 [1-4]; U20a [2]; U20b [1, 3]; U20c [4]
Storage and disposal E30a [1-3]; E31a [1]; E32a [2-4]
Vertebrate/rodent control products V01a, V03a, V04a [1, 2]; V01b, V03b, V04b [3, 4]; V02 [1-4]
Medical advice M03 [3, 4]

88 chloropicrin

A highly toxic horticultural soil fumigant

Products

1	Chloropicrin Fumigant	Dewco-Lloyd	99.5% w/w	LI	04216
2	K & S Chlorofume	K & S Fumigation	99.3% w/w	VP	08722

Uses

Crown rot in ***strawberries*** *(soil fumigation)*. Nematodes in ***strawberries*** *(soil fumigation)*. Red core in ***strawberries*** *(soil fumigation)*. Replant disease in ***all top fruit*** *(soil fumigation)*, ***hardy ornamentals*** *(soil fumigation)*. Verticillium wilt in ***strawberries*** *(soil fumigation)*.

Efficacy guidance

- Treat pre-planting
- Apply with specialised injection equipment
- For treating small areas or re-planting a single tree, a hand-operated injector may be used. Mark the area to be treated and inject to 22 cm at intervals of 22 cm
- Polythene sheeting (150 gauge) should be progressively laid over soil as treatment proceeds. The margin of the sheeting around the treated area must be embedded or covered with treated soil. Remove progressively after at least 4 d provided good air movement conditions prevail. Aerate soil for 15 d before planting

Restrictions

- Chloropicrin is subject to the Poisons Rules 1982 and the Poisons Act 1972. See notes in Section 1
- Before use, consult the code of practice for the fumigation of soil with chloropicrin. 2 fumigators must be present
- Remove contaminated gloves, boots or other clothing immediately and ventilate them in the open air until all odour is eliminated

Crop-specific information

- Latest use: at least 20 d before planting

Following crops guidance

- Carry out a cress test before replanting treated soil

Environmental safety

- Avoid treatment or vapour release when persistent still air conditions prevail
- Dangerous to game, wild birds and animals [1]
- Dangerous to bees [1]

- Dangerous to fish. Do not contaminate surface waters or ditches with chemical or used container [1]
- Dangerous to livestock. Keep all livestock out of treated areas until advised otherwise by fumigator

Hazard classification and safety precautions

Hazard R01, R04 [1, 2]; R05 [2]

Risk phrases R01a, R01b, R01c, R04c [1, 2]; R04a, R04b [1]; R05b [2]

Operator protection A, G, H, K, M; U04a, U05a, U10, U19, U20a [1, 2]; U13, U14, U15 [1]

Environmental protection E01, E10a, E12c, E13b [1]; E02 [1, 2] (until advised); E06a [1] (until advised); E06a [2]; E15 [1, 2]

Consumer protection C03

Storage and disposal E30b, E34 [1, 2]; E32a [1]; E33, E36 [2]

Medical advice M04 [1]

89 chlorothalonil

A protectant chlorophenyl fungicide for use in many crops and turf

See also carbendazim + chlorothalonil

Products

1	Agriguard Chlorothalonil	AgriGuard	500 g/l	SC	09390
2	Atlas Cropguard	Nufarm UK	500 g/l	SC	09123
3	Barclay Corrib 500	Barclay	500 g/l	SC	11350
4	Bombardier FL	Unicrop	500 g/l	SC	07910
5	Bravo 500	Syngenta	500 g/l	SC	10518
6	Clortosip 500	Sipcam	500 g/l	SC	09320
7	Daconil Turf	Scotts	500 g/l	SC	09265
8	Flute	Sipcam	720 g/l	SC	08953
9	Fusonil Turf	Rigby Taylor	500 g/l	SC	09695
10	Greencrop Orchid	Greencrop	500 g/l	SC	09566
11	Joules	Nufarm UK	500 g/l	SC	11047
12	Repulse	Certis	500 g/l	SC	07641
13	Repulse	Certis	500 g/l	SC	11328
14	Rover	Sipcam	500 g/l	SC	09848
15	Scorpio 500 SC	DAPT	500 g/l	SC	10927
16	Sipcam Echo 75	Sipcam	75% w/w	WG	08302
17	Standon Chlorothalonil 500	Standon	500 g/l	SC	08597
18	Tripart Faber	Tripart	500 g/l	SC	04549
19	Ultrafaber	Tripart	720 g/l	SC	05627

Uses

Anthracnose in ***managed amenity turf*** [7, 9]. Ascochyta in ***combining peas*** [1, 3-6, 8, 10, 12-14, 16, 17, 19]. Blight in ***potatoes*** [1-6, 8, 10-19]; ***protected tomatoes*** [4, 6, 12-14]. Botrytis in ***blackberries, raspberries*** [4-6, 14, 16, 19]; ***blackcurrants*** [6, 14, 17]; ***broccoli*** [1, 3-6, 8, 14, 16-19]; ***brussels sprouts, cabbages, cauliflowers*** [1, 3-6, 8, 12-14, 16-19]; ***calabrese*** [3, 8, 16-18]; ***cane fruit*** [1]; ***chinese cabbage, collards, kale*** [5, 19]; ***combining peas*** [1, 6, 12-14, 16]; ***gooseberries*** *(qualified minor use)*, ***redcurrants*** *(qualified minor use)*, ***strawberries*** *(outdoor crops only)* [6, 14]; ***onions*** [1, 4, 6, 12-14]; ***protected cucumbers, protected ornamentals*** [1, 4, 6, 12-14, 16]; ***protected tomatoes*** [1, 4, 12, 13, 16]; ***spring oilseed rape*** [1, 14, 16]; ***strawberries*** [5, 12, 13, 16, 19]; ***strawberries*** *(off-label)* [5, 6]; ***winter oilseed rape*** [1, 5, 14, 16, 19]. Cane spot in ***blackberries, raspberries*** [5, 12, 13, 19]. Celery leaf spot in ***celery*** *(qualified minor use)* [4-6, 12-14, 16, 18]. Chocolate spot in ***spring field beans, winter field beans*** [1-6, 8, 10, 11, 14-19]. Currant leaf spot in ***blackcurrants*** [3-6, 8, 12-14, 16-19]; ***gooseberries*** *(qualified minor use)*, ***redcurrants*** *(qualified minor use)* [4-6, 12-14, 16, 19]. Dark leaf spot in ***broccoli, brussels sprouts, cabbages, calabrese, cauliflowers*** [10, 16]. Dollar spot in ***managed amenity turf*** [7, 9]. Downy mildew in ***brassica seed beds*** [5]; ***broccoli*** [1, 3-6, 8, 14, 16-19]; ***brussels sprouts, cabbages, cauliflowers*** [1, 3-6, 8, 12-14, 16-19]; ***calabrese*** [8, 16-18]; ***chinese cabbage, collards, kale*** [5, 19]; ***edible brassicas*** [12, 13]; ***hops*** [1, 4-6, 12-14, 16, 19]; ***spring oilseed rape*** [1, 6, 14, 16]; ***winter oilseed rape*** [1, 5, 6, 14, 16, 19]. Dry bubble in ***mushrooms*** [5, 19]. Fusarium patch in ***managed amenity turf*** [7, 9]. Glume blotch in ***spring wheat*** [2, 3, 6, 8, 10, 11, 14-18]; ***winter wheat*** [1-6, 8, 10, 11, 14-19]. Ink disease in ***irises*** *(qualified minor use)* [4, 6, 14, 16]. Late ear diseases in ***winter wheat*** [1]. Leaf blotch in ***onions*** [18]. Leaf mould in ***protected tomatoes***

[1, 4, 6, 12-14, 16]. Leaf rot in ***onions*** [1, 4-6, 12-14, 16, 19]. Mycosphaerella in ***combining peas*** [1, 2, 4-6, 11-16, 19]. Neck rot in ***onions*** [1, 4-6, 12-14, 16, 19]. Powdery mildew in ***blackberries*** [5, 19]; ***blackcurrants*** [6, 14, 17]; ***gooseberries*** *(qualified minor use)*, ***redcurrants*** *(qualified minor use)* [6, 14]; ***protected cucumbers*** [16]; ***raspberries*** [5, 16, 19]. Red thread in ***managed amenity turf*** [7, 9]. Rhizoctonia in ***broccoli***, ***brussels sprouts***, ***cabbages***, ***calabrese***, ***cauliflowers*** [18]. Rhynchosporium in ***spring barley***, ***winter barley*** [2, 11, 15, 18]. Ring spot in ***broccoli*** [3, 5, 8, 16-19]; ***brussels sprouts***, ***cabbages***, ***cauliflowers*** [3, 5, 8, 12, 13, 16-19]; ***calabrese*** [3, 8, 16-18]; ***chinese cabbage***, ***collards***, ***kale*** [5, 19]; ***edible brassicas*** [12, 13]. Septoria leaf spot in ***celeriac*** *(off-label)* [5]; ***celery*** *(qualified minor use)* [1]; ***spring wheat*** [2, 3, 6, 8, 10, 11, 14-18]; ***winter wheat*** [1-6, 8, 10, 11, 14-19]. Wet bubble in ***mushrooms*** [5, 19].

Specific Off-Label Approvals (SOLAs)

- ***celeriac*** *(OLA 032083) Dec 2008* [5]
- ***strawberries*** *(OLA 011635) Dec 2008* [5]
- ***strawberries*** *(OLA 021253) Dec 2008* [6]
- ***strawberries*** *(OLA 992452) Dec 2008* [5]

Approval information

- Approval for aerial spraying on potatoes [3-6, 8, 10, 12, 13, 16, 18, 19]. See notes in Section 1
- Following implementation of Directive 98/82/EC, approval for use of chlorothalonil on numerous crops was revoked in 1999, although uses on blackcurrants, redcurrants and gooseberries were subsequently reinstated for some products. Uses on loganberries revoked in 2001 following implementation of MRL Directives

Efficacy guidance

- For some crops products differ in diseases listed as controlled. See label for details and for application rates, timing and number of sprays
- Apply as protective spray or as soon as disease appears and repeat as directed
- Activity against Septoria may be reduced where serious mildew or rust present. In such conditions mix with suitable mildew or rust fungicide
- May be used for preventive and curative treatment of turf [7, 9]

Restrictions

- Maximum number of treatments 2 per crop for brassicas; 3 per crop for celery, peas, cereals; 2 per spawning for mushrooms
- Maximum total dose for other crops varies with crop and product - see labels for details
- Operators must use a vehicle fitted with a cab and a forced air filtration unit with a pesticide filter complying with HSE Guidance Note PM74 or to an equivalent or higher standard when making broadcast or air-assisted applications
- Must only be applied by a pedestrian controlled sprayer or vehicle mounted/drawn equipment [7, 9]
- Do not mow or water turf for 24 h after treatment. Do not add surfactant or mix with liquid fertilizer [7, 9]

Crop-specific information

- Latest time of application to cereals varies with product. Consult label
- Latest use for winter oilseed rape before flowering; end Aug in yr of harvest for blackcurrants, gooseberries, redcurrants, blueberries, loganberries, raspberries
- HI 8 wk for field beans; 6 wk for combining peas; 28 d or before 31 Aug for post-harvest treatment for blackcurrants, gooseberries, redcurrants; 14 d for onions, strawberries; 10 d for hops; 7-14 d for broccoli, Brussels sprouts, cabbages, cauliflowers, celery, onions, potatoes; 3 d for hops, cane fruit; 12-48 h for protected cucumbers, protected tomatoes; 24 h for mushrooms
- For Botrytis control in strawberries important to start spraying early in flowering period and repeat at least 3 times at 10 d intervals
- On strawberries some scorching of calyx may occur with protected crops

Environmental safety

- Dangerous for the environment [5]
- Very toxic to aquatic organisms. May cause long-term adverse effects in the aquatic environment [5]
- Dangerous to fish or other aquatic life. Do not contaminate surface waters or ditches with chemical or used container [1-4, 6-19]
- Do not spray from the air within 250 m horizontal distance of surface waters or ditches

- LERAP Category B
- Broadcast air-assisted LERAP (18 m) [1, 3-6, 10, 14, 16, 17]

Hazard classification and safety precautions

Hazard R03, R13, R14 [5]; R04 [1-19]

Risk phrases R03b, R14c [5]; R04a [2, 6, 7, 10, 11, 14, 15, 18, 19]; R04b [1-7, 10-15, 17-19]; R04c [1, 3-7, 10, 12-14, 16-19]; R04d [1-4, 6-17]; R04e [5, 6, 8, 9, 14]

Operator protection A, C [1-19]; D [16]; H [1-17]; J [1, 3-6, 8, 10, 14, 17]; M [1, 3-10, 12-14, 16, 17]; U02, U13, U14, U15 [2-4, 6, 8-11, 14-18]; U05a, U09a, U19, U20a [1-19]

Environmental protection E01, E13b [1-4, 6-19]; E15, E38 [5]; E16a [1-19]; E16b [1-6, 8-19]; E17a [1, 3-6, 10, 14, 16, 17] (18 m); E18 [1, 3-6, 8, 10, 14, 16]

Consumer protection C02 [2, 11, 15] (field beans 14 d; potatoes 7 d); C02 [18] (field beans 7d; celery 7d; onions 7d; blackcurrants 3d); C03 [1-4, 6-19]

Storage and disposal E26 [1-5, 8-11, 15, 17, 18]; E30a [1-19]; E31a [7, 12, 13, 19]; E31b [1-3, 6, 8, 10, 11, 14-18]; E31c [4, 5, 9]; E32d [5]; E34 [1, 2, 7, 11, 15]

90 chlorothalonil + cyproconazole

A systemic protectant and curative fungicide mixture for cereals

Products

Alto Elite	Syngenta	375:40 g/l	SC	08467

Uses

Brown rust in ***spring barley***, ***winter barley***, ***winter wheat***. Chocolate spot in ***spring field beans***, ***winter field beans***. Eyespot in ***winter barley***, ***winter wheat***. Grey mould in ***combining peas***. Net blotch in ***spring barley***, ***winter barley***. Powdery mildew in ***spring barley***, ***winter barley***, ***winter wheat***. Rhynchosporium in ***spring barley***, ***winter barley***. Rust in ***spring field beans***, ***winter field beans***. Septoria diseases in ***winter wheat***. Yellow rust in ***spring barley***, ***winter barley***, ***winter wheat***.

Efficacy guidance

- Apply at first signs of infection or as soon as disease becomes active
- A repeat application may be made if re-infection occurs
- For established mildew tank-mix with an approved mildewicide
- When applied prior to third node detectable (GS 33) a useful reduction of eyespot will be obtained

Restrictions

- Maximum total dose equivalent to 2 full dose treatments
- Do not apply at concentrations higher than recommended

Crop-specific information

- Completion of ear emergence in wheat and barley
- HI peas, field beans 6 wk
- If applied to winter wheat in spring at GS 30-33 straw shortening may occur but yield is not reduced

Environmental safety

- Dangerous to fish or other aquatic life. Do not contaminate surface waters or ditches with chemical or used container
- LERAP Category B

Hazard classification and safety precautions

Hazard R04

Risk phrases R04d

Operator protection A, C, H, M; U05a, U11, U19, U20a

Environmental protection E01, E13b, E16a, E16b

Consumer protection C03

Storage and disposal E26, E30a, E31b, E34

91 chlorothalonil + flutriafol

A systemic eradicant and protectant fungicide for winter wheat

Products

1	Halo	Headland	375:47 g/l	SC	11546
2	Impact XL	Headland	300:47 g/l	SC	11547

Uses

Brown rust in ***winter wheat***. Late ear diseases in ***winter wheat***. Powdery mildew in ***winter wheat***. Septoria in ***winter wheat***. Yellow rust in ***winter wheat***.

Efficacy guidance

- Generally disease control and yield benefit will be optimised when application is made at an early stage of disease development
- Apply as soon as disease is seen establishing in the crop

Restrictions

- Maximum number of treatments 2 per crop (including other products containing flutriafol)

Crop-specific information

- Latest use: before early milk stage (GS 73)
- On certain cultivars with erect leaves high transpiration can result in flag leaf tip scorch. This may be increased by treatment but does not affect yield

Environmental safety

- Dangerous to fish or other aquatic life. Do not contaminate surface waters or ditches with chemical or used container
- LERAP Category B

Hazard classification and safety precautions

Hazard R04
Risk phrases R04b, R04d, R04e
Operator protection A, C, H; U02, U05a, U09a, U19, U20b
Environmental protection E01, E13b, E16a
Consumer protection C03
Storage and disposal E30a, E31b

SECTION 4

92 chlorothalonil + mancozeb

A multi-site protectant fungicide mixture

Products

1	Adagio	Interfarm	201:274 g/l	SC	10796
2	Guru	Interfarm	286: 194 g/l	SC	10801

Uses

Blight in ***potatoes*** [1]. Septoria diseases in ***winter wheat*** [2].

Efficacy guidance

- Start spray treatments on potatoes immediately after a blight warning or just before the haulm meets in the row [1]
- It is essential to start the blight spray programme before the disease appears in the crop [1]
- Repeat treatments for blight at 7, 10 or 14 d intervals depending on blight risk and continue until haulm is to be destroyed [1]
- Best results on winter wheat achieved from protective applications to the flag leaf. If disease already present on lower leaves treat as soon as flag leaf is just visible (GS 37) [2]
- Activity on Septoria may be reduced in presence of severe mildew infection [2]

Restrictions

- Maximum number of treatments 5 per crop for potatoes [1]
- Maximum total dose on winter wheat equivalent to 1.5 full dose treatments [2]
- Do not treat crops under stress from frost, drought, waterlogging, trace element deficiency or pest attack [2]

- Broadcast air assisted applications must only be made by equipment fitted with a cab with a forced air filtration unit plus a pesticide filter complying with HSE Guidance Note PM 74 or an equivalent or higher standard

Crop-specific information

- Before grain watery ripe (GS 71) for winter wheat
- HI potatoes 7 d
- Irrigated potato crops should be sprayed immediately after irrigation [1]

Environmental safety

- Dangerous to fish or other aquatic life. Do not contaminate surface waters or ditches with chemical or used container
- LERAP Category B

Hazard classification and safety precautions

Hazard R04
Risk phrases R04d, R04e
Operator protection A, C, H, M; U02, U05a, U08, U13, U14, U19 [1, 2]; U11, U20b [2]; U15, U20a [1]
Environmental protection E01, E13b, E16a, E16b
Consumer protection C03
Storage and disposal E26, E30a, E31b [1, 2]; E29 [2]

93 chlorothalonil + metalaxyl-M

A systemic and protectant fungicide for various crops

Products

Folio Gold	Syngenta	500:37.5 g/l	SC	10704

Uses

Alternaria in ***brussels sprouts*** *(moderate control)*, ***calabrese*** *(moderate control)*, ***cauliflowers*** *(moderate control)*. Downy mildew in ***broad beans***, ***brussels sprouts***, ***bulb onions*** *(qualified minor use)*, ***calabrese***, ***cauliflowers***, ***shallots*** *(qualified minor use)*, ***spring field beans***, ***winter field beans***. Ring spot in ***brussels sprouts*** *(reduction)*, ***calabrese*** *(reduction)*, ***cauliflowers*** *(reduction)*. White blister in ***brussels sprouts***, ***calabrese***, ***cauliflowers***. White tip in ***leeks*** *(qualified minor use)*.

Approval information

- Metalaxyl-M included in Annex I under EC Directive 91/414

Efficacy guidance

- Apply at first signs of disease or when weather conditions favourable for disease pressure
- Best results obtained when used in a full and well-timed programme. Repeat treatment at 14-21 d intervals if necessary
- Treatment of established disease will be less effective
- Evidence of effectiveness in bulb onions, shallots and leeks is limited

Restrictions

- Maximum total dose equivalent to 2 full doses on broad beans, field beans, cauliflowers, calabrese; 3 full doses on Brussels sprouts, leeks, onions and shallots

Crop-specific information

- HI 14 d for all crops

Environmental safety

- Dangerous for the environment
- Very toxic to aquatic organisms. May cause long-term adverse effects in the aquatic environment
- LERAP Category B

Hazard classification and safety precautions

Hazard R03, R04, R13, R14
Risk phrases R03b, R04a, R04b, R04c, R04e, R14c
Operator protection A, C, H; U05a, U09a, U15, U20a
Environmental protection E01, E15, E16a, E38

Consumer protection C02 (14 d); C03
Storage and disposal E26, E30a, E31c, E32d

94 chlorothalonil + propamocarb hydrochloride

A contact and systemic fungicide mixture for blight control in potatoes

Products

Merlin	Aventis	375:375 g/l	SC	07943

Uses

Blight in ***potatoes***.

Efficacy guidance

- Commence treatment early in the season as soon as there is risk of infection
- In the absence of a blight warning treatment should start just before potatoes meet along the row
- Use only as a protectant. Stop use when blight readily visible (1% leaf area destroyed)
- Repeat sprays at 10-14 d intervals depending on blight infection risk. See label for details
- Complete blight spray programme after end Aug up to haulm destruction with protectant fungicides preferably fentin based

Restrictions

- Maximum number of treatments 5 per crop
- Apply to dry foliage. Do not apply if rainfall or irrigation imminent

Crop-specific information

- HI 7 d

Environmental safety

- Dangerous to fish or other aquatic life. Do not contaminate surface waters or ditches with chemical or used container
- LERAP Category B

Hazard classification and safety precautions

Hazard R04
Risk phrases R04a, R04d, R04e
Operator protection A, C, H; U05a, U08, U11, U19, U20a
Environmental protection E01, E13b, E16a, E16b
Consumer protection C03
Storage and disposal E30a, E31b, E34

SECTION 4

95 chlorothalonil + tetraconazole

A systemic protectant and curative fungicide mixture for cereals

Products

Voodoo	Sipcam	250:62.5 g/l	SE	09414

Uses

Brown rust in ***spring barley***, ***spring barley*** *(autumn sown)*, ***spring wheat***, ***spring wheat*** *(autumn sown)*, ***winter barley***, ***winter wheat***. Crown rust in ***spring oats***, ***winter oats***. Net blotch in ***spring barley***, ***spring barley*** *(autumn sown)*, ***winter barley***. Powdery mildew in ***spring barley***, ***spring barley*** *(autumn sown)*, ***spring wheat***, ***spring wheat*** *(autumn sown)*, ***winter barley***, ***winter wheat***. Rhynchosporium in ***spring barley***, ***spring barley*** *(autumn sown)*, ***winter barley***. Septoria diseases in ***spring wheat***, ***spring wheat*** *(autumn sown)*, ***winter wheat***. Sooty moulds in ***spring wheat***, ***winter wheat***. Yellow rust in ***spring barley***, ***spring barley*** *(autumn sown)*, ***spring wheat***, ***spring wheat*** *(autumn sown)*, ***winter barley***, ***winter wheat***.

Efficacy guidance

- For best results apply before onset of disease attack, or at an early stage of disease development. Further treatments may be necessary if disease pressure remains high
- May be used in a programme in combination with a number of other fungicides to improve overall control and reduce potential for development of resistance

Restrictions

- Maximum total dose equivalent to three full dose treatments on wheat; two full dose treatments on barley

Crop-specific information

- Latest use: before early flowering (GS 63) in wheat; before end of ear emergence (GS 59) in barley

Environmental safety

- Harmful to fish or other aquatic life. Do not contaminate surface waters or ditches with chemical or used container
- LERAP Category B

Hazard classification and safety precautions

Hazard R04
Risk phrases R04a, R04b
Operator protection A, C; U05a, U19, U20a
Environmental protection E01, E13c, E16a, E16b
Consumer protection C03
Storage and disposal E30a, E31b, E34
Medical advice M03

96 chlorotoluron

A contact and residual urea herbicide for cereals

Products

1	Alpha Chlortoluron 500	Makhteshim	500 g/l	SC	04848
2	Atol	Nufarm UK	700 g/l	SC	11138
3	Copal 500	DAPT	500 g/l	SC	10633
4	Headland Tolerate	Headland	500 g/l	SC	10774
5	Lentipur CL 500	Nufarm UK	500 g/l	SC	08743
6	Luxan Chlorotoluron 500 Flowable	Luxan	500 g/l	SC	09165
7	Tolugan 700	Makhteshim	700 g/l	SC	07874

Uses

Annual dicotyledons in ***durum wheat*** [5, 6]; ***triticale, winter barley, winter wheat*** [1-7]. Annual grasses in ***durum wheat*** [5, 6]; ***triticale, winter barley, winter wheat*** [1-7]. Blackgrass in ***durum wheat*** [5, 6]; ***triticale, winter barley, winter wheat*** [1-7]. Rough meadow grass in ***durum wheat*** [5, 6]; ***triticale, winter barley, winter wheat*** [1-7]. Wild oats in ***durum wheat*** [5, 6]; ***triticale, winter barley, winter wheat*** [1-7].

Approval information

- May be applied through CDA equipment. See labels for details [2, 3, 5]
- Approved for aerial application on winter wheat, winter barley, durum wheat, triticale [5, 6]. See notes in Section 1

Efficacy guidance

- Best results achieved by application soon after drilling. Application in autumn controls most weeds germinating in early spring
- For wild oat control apply within 1 wk of drilling, not after 2-leaf stage. Blackgrass and meadow grasses controlled to 5 leaf, ryegrasses to 3 leaf stage
- Any trash or burnt straw should be buried and dispersed during seedbed preparation
- Control may be reduced if prolonged dry conditions follow application
- Harrowing after treatment may reduce weed control

Restrictions

- Maximum number of treatments 1 per crop
- Use only on listed crop varieties. See label. Ensure seed well covered at drilling
- Apply only as pre-emergence spray in durum wheat, pre- or post-emergence in wheat, barley or triticale
- Do not apply pre-emergence to crops sown after 30 Nov
- Do not apply to crops severely checked by waterlogging, pests, frost or other factors
- Do not use on undersown crops or those due to be undersown
- Do not apply post-emergence in mixture with liquid fertilizers

- Do not roll for 7 d before or after application to an emerged crop
- Do not use on soils with more than 10% organic matter
- Crops on stony or gravelly soils may be damaged, especially after heavy rain

Crop-specific information

- Latest use: pre-emergence for durum wheat. Post emergence timings on other crops vary - see labels

Environmental safety

- Dangerous for the environment [5]
- Very toxic to aquatic organisms. May cause long-term adverse effects in the aquatic environment [5]
- Harmful to fish or other aquatic life. Do not contaminate surface waters or ditches with chemical or used container [1-4, 6, 7]

Hazard classification and safety precautions

Hazard R03 [1, 3-5, 7]; R04 [1-4, 7]; R14 [5]
Risk phrases R03c [1, 3, 4, 7]; R04a, R04b [1-4, 7]; R10a, R10e, R14c [5]
Operator protection A, C [1-5, 7]; U05a [1-4, 7]; U08, U19 [1-7]; U20a [2, 7]; U20b [1, 3-6]
Environmental protection E01 [1-5, 7]; E13c [1-4, 6, 7]
Consumer protection C03 [1-5, 7]
Storage and disposal E26 [4, 6]; E30a, E31b [1-7]; E32d [5]; E34 [1-4, 6, 7]
Medical advice M03 [1-4, 7]

97 chlorotoluron + isoproturon

A urea herbicide mixture for cereals

Products

Tolugan Extra	Makhteshim	300:300 g/l	SC	09393

Uses

Annual dicotyledons in ***winter barley***, ***winter wheat***. Annual grasses in ***winter barley***, ***winter wheat***. Blackgrass in ***winter barley***, ***winter wheat***. Rough meadow grass in ***winter barley***, ***winter wheat***. Wild oats in ***winter barley***, ***winter wheat***.

Approval information

- Isoproturon included in Annex I under EC Directive 91/414

Efficacy guidance

- Best results achieved by application soon after drilling. Application in autumn controls most weeds germinating in early spring
- For wild oat control apply within 1 wk of drilling, not after 2-leaf stage. Blackgrass and meadow grasses controlled to 5 leaf, ryegrasses to 3 leaf stage
- Any trash or burnt straw should be buried and dispersed during seedbed preparation
- Control may be reduced if prolonged dry conditions follow application
- Harrowing after treatment may reduce weed control
- Where strains of herbicide-resistant blackgrass occur control may be unsatisfactory

Restrictions

- Maximum number of treatments 1 per crop
- Use only on listed crop varieties. See label. Ensure seed well covered at drilling
- Apply only as post-emergence treatment
- Do not apply to crops severely checked by waterlogging, pests, frost or other factors
- Do not use on undersown crops or those due to be undersown
- Do not apply post-emergence in mixture with liquid fertilizers
- Do not roll for 7 d before or after application to an emerged crop
- Do not use on soils with more than 10% organic matter
- Crops on stony or gravelly soils may be damaged, especially after heavy rain

Crop-specific information

- Latest use: before end of tillering (GS 29)

SECTION 4

Environmental safety

- Dangerous to fish or other aquatic life. Do not contaminate surface waters or ditches with chemical or used container
- Do not apply to dry, cracked or waterlogged soils where heavy rain may lead to contamination of drains by isoproturon

Hazard classification and safety precautions

Hazard R04
Risk phrases R04e
Operator protection A, C, H; U05a, U08, U19, U20b
Environmental protection E01, E13b
Consumer protection C03
Storage and disposal E26, E30a, E31b, E34

98 chlorpropham

A residual carbamate herbicide and potato sprout suppressant

See also chloridazon + chlorpropham + metamitron

Products

1	BL 500	Whyte Agrochemicals	500 g/l	HN	00279
2	Comrade	United Phosphorus	400 g/l	EC	10181
3	Luxan Gro-Stop Basis	Luxan	300 g/l	EC	08601
4	Luxan Gro-Stop Fog	Luxan	300 g/l	HN	09388
5	Luxan Gro-Stop HN	Luxan	300 g/l	HN	07689
6	MSS CIPC 40 EC	Whyte Agrochemicals	400 g/l	EC	01403
7	MSS CIPC 5 G	Whyte Agrochemicals	5% w/w	GR	01402
8	MSS CIPC 50 LF	Whyte Agrochemicals	500 g/l	EC	03285
9	MSS CIPC 50 M	Whyte Agrochemicals	500 g/l	EC	01404
10	Standon CIPC 300 HN	Standon	300 g/l	HN	09187
11	Warefog 25	Whyte Agrochemicals	600 g/l	HN	06776

Uses

Annual dicotyledons in ***acroclinium, african marigolds, blackcurrants, calendula, celery, china asters, chrysanthemums, coreopsis, french marigolds, gooseberries, lucerne, straw flower, strawberries, sweet sultan*** [2]; ***carrots, flower bulbs, lettuce, onions, parsley*** [2, 6]; ***leeks*** [6]. Annual grasses in ***acroclinium, african marigolds, blackcurrants, calendula, celery, china asters, chrysanthemums, coreopsis, french marigolds, gooseberries, lucerne, straw flower, strawberries, sweet sultan*** [2]; ***carrots, flower bulbs, lettuce, onions, parsley*** [2, 6]; ***leeks*** [6]. Chickweed in ***acroclinium, african marigolds, blackcurrants, calendula, celery, china asters, chrysanthemums, coreopsis, french marigolds, gooseberries, lucerne, straw flower, strawberries, sweet sultan*** [2]; ***carrots, lettuce, parsley*** [2, 6]; ***flower bulbs, leeks, onions*** [6]. Polygonums in ***acroclinium, african marigolds, blackcurrants, calendula, celery, china asters, chrysanthemums, coreopsis, french marigolds, gooseberries, lucerne, straw flower, strawberries, sweet sultan*** [2]; ***carrots, lettuce, parsley*** [2, 6]; ***flower bulbs, leeks, onions*** [6]. Sprout suppression in ***potatoes*** [1, 7-9, 11]; ***ware potatoes*** [3, 5, 10]; ***ware potatoes*** *(fog)* [4].

Approval information

- Some products are formulated for application by thermal fogging. See labels for details [1, 5, 10, 11]

Efficacy guidance

- Apply weed control sprays to freshly cultivated soil. Adequate rainfall must occur after spraying. Activity is greater in cold, wet than warm, dry conditions
- For sprout suppression apply with suitable fogging or rotary atomiser equipment or sprinkle granules over dry tubers before sprouting commences. Repeat applications may be needed. See labels for details
- Effectiveness of fogging reduced if air spaces between tubers are blocked. Best results obtained at 5-10°C and 75-80% humidity

Restrictions

- Maximum number of treatments 1 per batch (3 per batch [8]) for fogging potatoes; 2 per yr for flower bulbs; 1 per yr for grass seed crops [2]; 1 per crop for other named crops
- Maximum total dose equivalent to the above multiples of full dose treatments

- Not to be used on grass seed crops if grass to be grazed or cut for fodder before 31 May following treatment [2]
- Excess rainfall after application may result in crop damage
- Do not use on Sands, Very Light soils or soils low in organic matter
- Poor conditions at drilling or planting, soil compaction, surface capping, waterlogging or attack by pests may result in crop damage
- On crops under glass high temperatures and poor ventilation may cause crop damage
- Only clean, mature, disease-free potatoes should be treated for sprout suppression. Use of chlorpropham can inhibit tuber wound healing and the severity of skin spot infection in store may be increased if damaged tubers are treated
- Do not fog potatoes with a high level of skin spot
- Do not use on potatoes for seed. Do not handle, dry or store seed potatoes or any other seed or bulbs in boxes or buildings in which potatoes are being or have been treated
- Do not remove potatoes for sale or processing for at least 21 d (4 wk [5, 10]) after application
- A minimum interval of 45 d (4 wk [5]) must elapse between applications [8, 9]

Crop-specific information

- Latest use: 2 wk before planting or at least 10 d after planting for chrysanthemums [2]; 21 d before removal for sale or processing for potatoes [1, 7-9, 11]
- HI 2 mth for grass seed crops [6]; 8 wk for ware potatoes [3]
- Cure potatoes according to label instructions before treatment and allow 3 wk between completion of loading into store and first treatment
- Apply weed control sprays to seeded crops pre-emergence of crop or weeds, to onions as soon as first crop seedlings visible, to planted crops a few days before planting, to bulbs immediately after planting, to fruit crops in late autumn-early winter. See label for further details

Environmental safety

- Harmful in contact with skin [2, 6]
- Harmful by inhalation [1, 4, 5, 9, 10]
- Harmful if swallowed [1, 2, 4-6, 8-10]
- Irritating to eyes [1, 4-6, 8-11]
- Irritating to skin [1, 4-6, 8-10]
- Irritating to respiratory system [1, 4-6, 9-11]
- May cause sensitization by skin contact [3-5, 10]
- May cause sensitization by inhalation [4, 5, 10]
- May cause lung damage if swallowed [6]
- Highly flammable [1, 9]
- Flammable [8]
- Harmful to fish or other aquatic life. Do not contaminate surface waters or ditches with chemical or used container [2-5, 10]
- Keep unprotected persons out of treated stores for at least 24 h after application

Hazard classification and safety precautions

Hazard R03 [1, 2, 4-6, 8-10]; R04 [1, 3-6, 8-11]; R07c [1, 9]; R07d [8]

Risk phrases R03a [2, 6]; R03b [1, 4, 5, 9, 10]; R03c [1, 2, 4-6, 8-10]; R04a [1, 4-6, 8-11]; R04b [1, 4-6, 8-10]; R04c [1, 4-6, 9-11]; R04e [3-5, 10]; R04f [4, 5, 10]; R04g [6]

Operator protection A [1-6, 8-11]; C [1-3, 6, 8, 9, 11]; D [1, 4, 5, 8-11]; E [1, 4, 5, 9-11]; G [8]; H [1, 3-5, 8-10]; J [1, 8, 9]; M [1, 4, 5, 8-10]; U02, U04a, U13, U14, U17 [4, 5, 10]; U05a [1-11]; U08 [2, 3, 5, 6, 8-11]; U09a [1, 2, 4]; U19 [1, 4-6, 8-11]; U20a [3-5, 10]; U20b [1, 2, 6-9]; U20c [11]

Environmental protection E01 [1-11]; E02 [1, 5, 9, 10] (24 h); E13c [2-5, 10]; E15 [1, 6-9, 11]

Consumer protection C03

Storage and disposal E26 [2-5, 7]; E30a [1-3, 6-9, 11]; E31a [1, 8, 11]; E31b [2, 6, 9]; E32a [3-5, 7, 10]; E34 [1-10]

Medical advice M03 [1, 2, 4-10]; M05 [6]

99 chlorpropham with cetrimide

A soil-acting herbicide for lettuce under cold glass

Products

Croptex Pewter	Certis	80:80 g/l	SC	11181

SECTION 4

Uses

Annual dicotyledons in ***protected lettuce***. Annual grasses in ***protected lettuce***. Chickweed in ***protected lettuce***. Polygonums in ***protected lettuce***.

Efficacy guidance

- Adequate irrigation must be applied before or after treatment
- Best results achieved on firm soil of fine tilth, free from clods and weeds

Restrictions

- Do not apply to crop foliage or use where seed has chitted
- Excess irrigation may cause temporary check to crop under certain circumstances
- Do not apply where tomatoes or brassicas are growing in the same house

Crop-specific information

- Latest use: before crop emergence or pre-planting
- Apply to drilled lettuce under cold glass within 24 h post-drilling, to transplanted crops pre-planting

Following crops guidance

- In the event of crop failure only lettuce should be grown within 2 mth

Environmental safety

- Highly flammable
- Dangerous to fish or other aquatic life. Do not contaminate surface waters or ditches with chemical or used container

Hazard classification and safety precautions

Hazard R03, R04, R07c
Risk phrases R03a, R03c, R04a, R04b, R04c
Operator protection A, C; U05a, U08, U19, U20a
Environmental protection E01, E13b
Consumer protection C03
Storage and disposal E26, E30a, E31a

100 chlorpropham + fenuron

A residual herbicide for runner beans and spinach

Products

Croptex Chrome	Certis	80:15 g/l	EC	11180

Uses

Annual dicotyledons in ***leaf spinach***, ***runner beans*** *(off-label)*. Annual grasses in ***leaf spinach***. Chickweed in ***leaf spinach***.

Specific Off-Label Approvals (SOLAs)

- ***runner beans*** *(OLA 030871) Dec 2008* [1]

Approval information

- Products containing fenuron have been granted derogations for specified 'Essential Uses' for use until 31 December 2007. Sale and supply must cease by 30 June 2007 but growers have no guarantee that the products will continue to be available until then.
  For more information see 'The Review Programme' under 'Pesticide Legislation' in Section 1

Efficacy guidance

- Apply to soil freshly cultivated and free of established weeds. Adequate rainfall must occur after spraying. Activity is greater in cold, wet than warm, dry conditions

Restrictions

- Maximum number of treatments 1 per crop
- Do not use on Sands, Very Light soils or soils low in organic matter
- Poor conditions at drilling or planting, soil compaction, surface capping, waterlogging or attack by pests may result in crop damage
- This product must not be used on any crops other than those listed, including any extrapolations that would normally be permissible under the Long Term Arrangements for Extension of Use (see Section 1)

Crop-specific information

- Latest use: pre crop emergence

Following crops guidance

- In the event of crop failure only recommended crops should be replanted in treated soil
- Plough or cultivate to 15 cm after harvest to dissipate any residues

Environmental safety

- Flammable
- Dangerous for the environment
- Very toxic to aquatic organisms. May cause long-term adverse effects in the aquatic environment

Hazard classification and safety precautions

Hazard R03, R04, R07d, R14
Risk phrases R03a, R03c, R04a, R04b, R04c, R14c
Operator protection A, C; U05a, U08, U13, U19, U20b
Environmental protection E01, E15
Consumer protection C03
Storage and disposal E30a, E31a, E32e, E34
Medical advice M03

101 chlorpropham + metamitron

A contact and residual herbicide mixture for beet crops

Products

Golmet	Nufarm UK	24.5:280 g/l	SC	10684

Uses

Annual dicotylodons in ***fodder beet***, ***mangels***, ***sugar beet***. Annual meadow grass in ***fodder beet***, ***mangels***, ***sugar beet***.

Efficacy guidance

- May be used pre-emergence alone or post-emergence in tank mixture or with an authorised adjuvant oil
- Use as part of a spray programme. For optimum efficacy a full programme of pre- and post-emergence sprays is recommended
- Ideally one pre-emergence application should be followed by a repeat overall low dose programme on mineral soils. On organic soils a programme of up to five post-emergence sprays is likely to be most effective
- Product combines residual and contact activity. Adequate soil moisture and a fine firm seedbed are essential for optimum residual activity

Restrictions

- Maximum number of treatments one pre-emergence plus three post-emergence or, where no pre-emergence treatment has been applied, up to five post-emergence treatments with an adjuvant oil

Crop-specific information

- Latest use: before crop foliage meets across the rows

Following crops guidance

- Only sugar beet, fodder beet or mangels should be sown as following crops within 4 mth of treatment. After normal harvest of a treated crop land should be mouldboard ploughed to 15 cm after which any spring crops may be sown or planted

Environmental safety

- Harmful to fish or other aquatic life. Do not contaminate surface waters or ditches with chemical or used container

Hazard classification and safety precautions

Hazard H04
Risk phrases R04a, R04b
Operator protection A, C; U05a, U08, U14, U19, U20a
Environmental protection E01, E13c
Consumer protection C03
Storage and disposal E26, E30a, E32a

SECTION 4

102 chlorpropham + pentanochlor

A contact and residual herbicide for horticultural crops

Products

Atlas Brown	Nufarm UK	150:300 g/l	EC	07703

Uses

Annual dicotyledons in ***carrots, celeriac, celery, chrysanthemums, narcissi, parsley, parsnips, tulips***. Annual meadow grass in ***carrots, celeriac, celery, chrysanthemums, narcissi, parsley, parsnips, tulips***.

Approval information

- Products containing pentanochlor have been granted derogations for specified 'Essential Uses' for use until 31 December 2007. Sale and supply must cease by 30 June 2007 but growers have no guarantee that the products will continue to be available until then.
  For more information see 'The Review Programme' under 'Pesticide Legislation' in Section 1

Efficacy guidance

- Apply as pre- or post-weed emergence spray
- Best results by application to weeds up to 2-leaf stage on fine, firm, moist seedbed
- Greatest contact action achieved under warm, moist conditions, the short residual action greatest in earlier part of year

Restrictions

- Maximum number of treatments 1 per yr for narcissi, tulips; 2 per yr for carrots, celeriac, celery, fennel, parsnips, chrysanthemums, outdoor leaf herbs
- This product must not be used on any crops other than those listed, including any extrapolations that would normally be permissible under the Long Term Arrangements for Extension of Use (see Section 1), except herbs and ornamentals

Crop-specific information

- Latest use: pre-emergence for narcissi, tulips
- HI 28 d for carrots, parsnips
- Apply to carrots and related crops pre- or post-emergence after fully expanded cotyledon stage, to narcissi and tulips at any time before emergence
- Apply to chrysanthemums either pre-planting or after planting as carefully directed spray, avoiding foliage

Following crops guidance

- Any crop may be sown or planted after 4 wk following ploughing and cultivation

Hazard classification and safety precautions

Hazard R04
Risk phrases R04a
Operator protection C; U05a, U19, U20a
Environmental protection E01, E15
Consumer protection C03
Storage and disposal E30a, E31b

103 chlorpyrifos

A contact and ingested organophosphorus insecticide and acaricide

Products

1	Agriguard Chlorpyrifos	AgriGuard	480 g/l	EC	10626
2	Alpha Chlorpyrifos 48 EC	Makhteshim	480 g/l	EC	04821
3	Ballad	Headland	480 g/l	EC	11659
4	Barclay Clinch II	Barclay	480 g/l	EC	08596
5	Chevron 48	DAPT	480 g/l	EC	10645
6	Crossfire 480	Bayer Environ.	480 g/l	EC	09929
7	Cyren	Cheminova	480 g/l	EC	08358
8	Cyren	Headland	480 g/l	EC	11028
9	Dursban WG	Dow	75% w/w	WG	09153
10	Equity	Dow	480 g/l	EC	11520
11	Greencrop Pontoon	Greencrop	480 g/l	EC	09667

Products – continued

12	Lorsban T	Rigby Taylor	480 g/l	EC	07813
13	Lorsban WG	Landseer	75% w/w	WG	10139
14	Maraud	Scotts	480 g/l	EC	09274
15	Spannit	SumiAgro Amenity	480 g/l	EC	08744
16	Spannit Granules	SumiAgro	6% w/w	GR	10935
17	Standon Chlorpyrifos 48	Standon	480 g/l	EC	08286
18	suSCon Green Soil Insecticide	Fargro	10% w/w	CG	06312
19	suSCon Indigo	Fargro	10% w/w	GR	09902

Uses

Ambrosia beetle in ***cut logs*** [1, 2, 4, 5, 7-9, 11, 12]; ***cut timber*** [4]. Aphids in ***apples***, ***gooseberries***, ***pears***, ***plums***, ***raspberries***, ***strawberries*** [1, 2, 4, 5, 7, 8, 10, 11, 13, 15, 17]; ***broccoli***, ***cabbages***, ***calabrese***, ***cauliflowers*** [1, 2, 4, 5, 7-11, 17]; ***brussels sprouts*** [2, 17]; ***chinese cabbage*** [1, 4, 5, 7-11, 17]; ***currants*** [1, 2, 4, 5, 7, 8, 10, 11, 15, 17]; ***edible brassicas*** [15]; ***redcurrants***, ***whitecurrants*** [11]; ***spring barley***, ***winter barley*** [2, 5, 9, 10, 15]; ***spring oats***, ***spring wheat***, ***winter oats***, ***winter wheat*** [1, 2, 4, 5, 7-11, 15, 17]. Apple blossom weevil in ***apples*** [1, 2, 4, 5, 7, 8, 10, 11, 13, 15, 17]. Apple sucker in ***apples*** [17]. Bean seed fly in ***green beans*** *(off-label - harvested as a dry pulse)* [9]. Black pine beetle in ***forestry transplants*** [2, 4, 11, 12]. Cabbage root fly in ***broccoli***, ***cabbages***, ***cauliflowers*** [1, 2, 4, 5, 7-11, 16, 17]; ***brussels sprouts*** [2, 9, 10, 17]; ***calabrese*** [1, 2, 4, 5, 7-11, 17]; ***chinese cabbage*** [1, 4, 5, 7-11, 17]; ***collards*** *(off-label)*, ***kale*** *(off-label)* [2, 9]; ***edible brassicas*** [15]; ***kohlrabi*** *(off-label)*, ***mooli*** *(off-label)*, ***radishes*** *(off-label)* [9]. Capsids in ***apples***, ***gooseberries***, ***pears*** [1, 2, 4, 5, 7, 8, 10, 11, 13, 15, 17]; ***currants*** [1, 2, 4, 5, 7, 8, 10, 11, 15, 17], ***redcurrants***, ***whitecurrants*** [11]. Caterpillars in ***apples*** [1, 2, 4, 5, 7, 8, 10, 11, 15]; ***broccoli***, ***cabbages***, ***calabrese***, ***cauliflowers*** [1, 2, 4, 5, 7-11, 17]; ***brussels sprouts*** [2, 17]; ***chinese cabbage*** [1, 4, 5, 7-11, 17]; ***currants*** [1, 2, 4, 5, 7, 8, 10, 11, 15, 17]; ***edible brassicas*** [15]; ***gooseberries*** [1, 2, 4, 5, 7, 8, 10, 11, 13, 15, 17]; ***pears*** [1, 2, 4, 5, 7, 8, 10, 11, 13, 15]; ***redcurrants***, ***whitecurrants*** [11]. Codling moth in ***apples***, ***pears*** [1, 2, 4, 5, 7, 8, 10, 11, 13, 15, 17]. Cutworms in ***broccoli***, ***cabbages***, ***calabrese***, ***cauliflowers*** [1, 2, 4, 5, 7-11, 17]; ***brussels sprouts*** [2, 17]; ***carrots***, ***potatoes*** [1, 2, 4, 7-11, 15, 17]; ***chinese cabbage*** [1, 4, 5, 7-11, 17]; ***edible brassicas*** [15]; ***onions*** [1, 2, 4, 7-11, 17]. Damson-hop aphid in ***plums*** [2, 4, 11, 13, 17]. Elm bark beetle in ***cut logs*** [1, 5, 7, 8, 12]. Frit fly in ***amenity grass*** [1, 3, 4, 6-8, 14, 17]; ***amenity turf*** [11, 17]; ***golf courses*** [7, 8, 15]; ***grassland*** [2, 4, 5, 7-11, 15, 17]; ***maize***, ***spring oats***, ***spring wheat***, ***winter oats***, ***winter wheat*** [1, 2, 4, 5, 7-11, 15, 17]; ***managed amenity turf*** [1-8, 12, 14]; ***permanent pasture***, ***rotational grassland*** [1]; ***spring barley*** [10, 15]; ***winter barley*** [10]. Larch shoot beetle in ***cut logs*** [1, 2, 4, 5, 7-12]; ***cut timber*** [4]. Leatherjackets in ***amenity grass*** [1, 3, 4, 6-8, 14, 17]; ***amenity turf*** [11, 17]; ***broccoli***, ***cabbages***, ***calabrese***, ***cauliflowers***, ***chinese cabbage*** [1, 4, 5, 7-10, 17]; ***brussels sprouts*** [17]; ***edible brassicas*** [15], ***golf courses*** [7, 8, 15]; ***grassland*** [2, 4, 5, 7-11, 15, 17]; ***managed amenity turf*** [1-8, 12, 14]; ***permanent pasture***, ***rotational grassland*** [1]; ***spring barley*** [2, 5, 9, 10]; ***spring oats***, ***spring wheat***, ***sugar beet***, ***winter oats***, ***winter wheat*** [1, 2, 4, 5, 7-11, 15, 17]; ***winter barley*** [2, 5, 9, 10, 15]. Mealy aphid in ***plums*** [2, 4, 11, 13]. Narcissus fly in ***flower bulbs*** *(off-label)* [2]. Pear sucker in ***pears*** [17]. Pine shoot beetle in ***cut logs*** [1, 2, 4, 5, 7-12]; ***cut timber*** [4]. Pine weevil in ***forest nurseries*** [2, 4]; ***forestry transplants*** [2, 4, 11, 12]. Pygmy mangold beetle in ***sugar beet*** [1, 5, 7-10, 15]. Raspberry beetle in ***raspberries*** [1, 2, 4, 5, 7, 8, 10, 11, 13, 15, 17]. Raspberry cane midge in ***raspberries*** [1, 2, 4, 5, 7, 8, 10, 11, 13, 15, 17]. Red spider mites in ***apples***, ***gooseberries***, ***pears***, ***plums***, ***raspberries***, ***strawberries*** [1, 2, 4, 5, 7, 8, 10, 11, 13, 15, 17]; ***currants*** [1, 2, 4, 5, 7, 8, 10, 11, 15, 17]; ***redcurrants***, ***whitecurrants*** [11]. Sawflies in ***apples*** [1, 2, 4, 5, 7, 8, 10, 11, 13, 15, 17]; ***pears*** [4, 10, 15]. Sciarid flies in ***ornamental plant production*** [19]. Strawberry blossom weevil in ***strawberries*** [10, 13]. Suckers in ***apples***, ***pears*** [1, 2, 4, 5, 7, 8, 10, 11, 13, 15]. Summer-fruit tortrix moth in ***apples*** [17]. Thrips in ***spring oats***, ***spring wheat***, ***winter oats***, ***winter wheat*** [1, 5, 7, 8, 15]; ***winter barley*** [15]. Tortrix moths in ***apples***, ***pears***, ***plums***, ***strawberries*** [1, 2, 4, 5, 7, 8, 10, 11, 13, 15, 17]. Vine weevil in ***conifers*** [1, 7-12]; ***container-grown ornamentals***, ***ornamental plant production*** [18]; ***strawberries*** [1, 2, 4, 5, 7, 8, 10, 11, 13, 17]. Wheat bulb fly in ***spring barley***, ***winter barley*** [10]; ***spring oats***, ***spring wheat***, ***winter oats***, ***winter wheat*** [1, 2, 4, 5, 7-11, 15, 17]. Wheat-blossom midge in ***spring oats***, ***spring wheat***, ***winter oats***, ***winter wheat*** [1, 2, 4, 5, 7-11, 15, 17]. Whitefly in ***broccoli***, ***cabbages***, ***calabrese***, ***cauliflowers***, ***chinese cabbage*** [1, 5, 7-10]. Winter moth in ***apples***, ***pears***, ***plums*** [1, 2, 4, 5, 7, 8, 10, 11, 13, 15, 17]. Woolly aphid in ***apples*** [1, 2, 4, 5, 7, 8, 10, 11, 13, 15, 17]; ***pears*** [1, 5, 7, 8, 10, 15]. Yellow cereal fly in ***spring wheat***, ***winter wheat*** [9, 10].

Specific Off-Label Approvals (SOLAs)

- ***collards, kale*** *(OLA 021688) Dec 2008* [2]
- ***collards, kale*** *(OLA 031390) Dec 2008* [9]
- ***flower bulbs*** *(OLA 021458) Dec 2008* [2]
- ***green beans*** *(harvested as a dry pulse) (OLA 031391) Dec 2008* [9]
- ***kohlrabi, mooli, radishes*** *(OLA 031391) Dec 2008* [9]

Approval information

- Following implementation of Directive 98/82/EC, approval for use of chlorpyrifos on numerous crops was revoked in 1999

Efficacy guidance

- Brassicas raised in plant-raising beds may require retreatment at transplanting
- Activity may be reduced when soil temperature below 5°C or on organic soils
- In dry conditions the effect of granules applied as a surface band may be reduced [16]
- Control of sciarids in ornamental plant production not guaranteed if cuttings struck into untreated media are potted into treated media [19]
- Where pear suckers resistant to chlorpyrifos occur control is unlikely to be satisfactory
- Efficacy against frit fly and leatherjackets in grass may be reduced if applied during periods of frost

Restrictions

- Maximum number of treatments and timing vary with crop, product and pest. See label for details
- On carrots the maximum total dose applied per crop must not exceed the equivalent of 3 (on mineral soils) or 4 (on organic soils) full dose applications
- Do not treat potatoes under severe drought stress. The variety Desiree is particularly susceptible
- Do not apply to sugar beet under stress or within 4 d of applying a herbicide
- Do not use product in growing media used for aquatic or semi-aquatic plants, and not for propagation of any edible crop [19]
- Do not graze lactating cows on treated pasture for 14 d after treatment

Crop-specific information

- Latest use 4d after transplanting edible brassica crops; end Jul for sugar beet; varies for other crops. See individual labels
- HI range from 7 d to 6 wk depending on crop. See labels
- For vine weevil control in hardy ornamental nursery stock incorporate in growing medium when plants first potted from rooted cutting stage. Treat the fresh growing medium when plants are potted on into larger containers [18]
- For turf pests apply from Nov where high larval populations detected or damage seen [3, 6-8, 12, 14]
- For sciarid control in ornamental plant production it is essential to incorporate thoroughly into growing media by hand or machine shortly before potting. Product controls larvae for several months from a single application [19]
- On lettuce apply only to strong well developed plants when damage first seen, or on professional advice
- In apples use pre-blossom up to pink/white bud and post-blossom after petal fall
- Test tolerance of glasshouse ornamentals before widescale use for propagating unrooted cuttings, or potting unusual plants and new species, and when using any media with a high content of non-peat ingredients [19]

Environmental safety

- Flammable [1, 2, 4-6, 10-12, 14, 15, 17]
- Dangerous for the environment [10]
- Very toxic to aquatic organisms. May cause long-term adverse effects in the aquatic environment [10]
- High risk to bees. Do not apply to crops in flower or to those in which bees are actively foraging. Do not apply when flowering weeds are present [3, 7, 8]
- Dangerous to bees. Do not apply to crops in flower or to those in which bees are actively foraging. Do not apply when flowering weeds are present [1, 2, 4-6, 9-15, 17]
- Extremely dangerous to fish or other aquatic life. Do not contaminate surface waters or ditches with chemical or used container [1-3, 5-9, 11, 13, 14, 18, 19]
- Dangerous to fish or other aquatic life. Do not contaminate surface waters or ditches with chemical or used container [4, 12, 15-17]
- LERAP Category A (except [16, 18, 19])

- Broadcast air-assisted LERAP (18 m) [1, 2, 5-10, 13]
- Keep livestock out of treated areas for at least 14 d after treatment and until foliage of any poisonous weeds such as ragwort has died and become unpalatable [2-5, 7-9, 11, 13, 17]

Hazard classification and safety precautions

Hazard R03 [1-8, 10-12, 14, 15, 17]; R04 [1-8, 10-12, 14, 15, 17, 19]; R07d [1, 2, 4-6, 10-12, 14, 15, 17]; R14 [10]

Risk phrases R03a [1, 4, 6, 11, 12, 14, 15, 17]; R03c, R04b [1-8, 10-12, 14, 15, 17]; R04a [3, 7, 8, 15, 19]; R04c [6, 10, 14]; R04d, R04e [2, 5]; R04g [1, 2, 5-8, 10, 14]; R12a, R14c [10]

Operator protection A [1-19]; C [2, 3, 5, 7, 8, 15]; H [1, 2, 4, 5, 7, 8, 11, 12, 15, 17]; P [10]; U02 [2, 5, 15, 16, 18, 19]; U05a [1, 2, 4-8, 10-12, 14, 15, 17]; U08 [1-15, 17]; U19 [1, 2, 4-9, 11-15, 17]; U20a [15, 16]; U20b [1-14, 17-19]

Environmental protection E01 [1-12, 14-19]; E07, E17a [1]; E07 [2-5, 7-9] (14 d); E07, E17a, E38 [10]; E07 [11, 13, 17] (14 d); E12a [3, 7, 8]; E12c [1, 2, 4-6, 9-15, 17]; E13a [1-3, 5-9, 11, 13, 14, 18, 19]; E13b [4, 12, 15-17]; E15 [10, 14]; E16c, E16d [1-15, 17]; E17a [2, 5-9] (18 m); E17a [11, 13, 14] (18 m)

Consumer protection C01 [18, 19]; C02 [15, 16]; C03 [1-12, 14, 15, 17]

Storage and disposal E26 [5, 17-19]; E30a [1-19]; E31b [1, 2, 4-15, 17]; E32a [15, 16, 18, 19]; E32d [10]; E34 [1-15, 17]

Medical advice M01 [2, 3, 5-9, 11-19]; M03 [1-15, 17]; M05 [1-3, 5-8, 10]

104 chlorpyrifos-methyl

An organophosphorus insecticide and acaricide for grain store use

Products

1	Greencrop Storeclean 225	Greencrop	225 g/l	EC	11098
2	Reldan 22	Dow	225 g/l	EC	08191

Uses

Grain storage pests in ***stored grain***. Pre-harvest hygiene in ***grain stores***.

Approval information

- Approvals for chlorpyrifos-methyl allowed to continue following the UK Review programme of cholinesterase compounds

Efficacy guidance

- Apply to grain after drying to moisture content below 14%, cooling and cleaning
- Insecticide may become depleted at grain surface if grain is being cooled by continuous extraction of air from the base leading to reduced control of grain store pests especially mites
- Resistance to organophosphorus compounds sometimes occurs in insect and mite pests of stored products

Restrictions

- This product contains an anticholinesterase organophosphorus compound. Do not use if under medical advice not to work with such compounds
- Maximum number of treatments 1 per batch or 1 per store, prior to storage
- Only treat grain in good condition
- Do not treat grain intended for sowing
- Minimum of 90 d must elapse between treatment of grain and removal from store for consumption or processing

Crop-specific information

- May be applied pre-harvest to surfaces of empty store and grain handling machinery and as admixture with grain
- May be used in wheat, barley, oats, rye or triticale stores

Environmental safety

- Extremely dangerous to fish or other aquatic life. Do not contaminate surface waters or ditches with chemical or used container

Hazard classification and safety precautions

Hazard R03 [1]; R04 [1, 2]

Risk phrases R04d, R04g

Operator protection A, C, H, M; U05a, U08, U19, U20b

SECTION 4

Environmental protection E01, E13a
Consumer protection C03
Storage and disposal E26, E30a, E31b, E34
Medical advice M01, M05 [1, 2]; M03 [1]

105 chlorthal-dimethyl

A residual benzoic acid herbicide for use in horticulture

Products

1	Dacthal W-75	Certis	75% w/w	WP	10623
2	Dacthal W-75	Certis	75% w/w	WP	11323

Uses

Annual dicotyledons in ***angelica*** *(off-label)*, ***balm*** *(off-label)*, ***basil*** *(off-label)*, ***bay*** *(off-label)*, ***bedding plants***, ***blackcurrants***, ***borage*** *(off-label)*, ***broccoli***, ***brussels sprouts***, ***cabbages***, ***calabrese***, ***cauliflowers***, ***celery leaves*** *(off-label)*, ***chervil*** *(off-label)*, ***chives*** *(off-label)*, ***clary*** *(off-label)*, ***coriander*** *(off-label)*, ***dandelions*** *(off-label)*, ***dill*** *(off-label)*, ***fennel*** *(off-label)*, ***fenugreek*** *(off-label)*, ***feverfew*** *(off-label)*, ***fodder rape***, ***gooseberries***, ***hyssop*** *(off-label)*, ***kale***, ***land cress*** *(off-label)*, ***leeks***, ***lemon verbena*** *(off-label)*, ***lovage*** *(off-label)*, ***marjoram*** *(off-label)*, ***mint*** *(off-label)*, ***mustard***, ***nasturtium*** *(off-label)*, ***onions***, ***oregano*** *(off-label)*, ***parsley*** *(off-label)*, ***raspberries***, ***rocket*** *(off-label)*, ***rosemary*** *(off-label)*, ***roses***, ***rue*** *(off-label)*, ***runner beans***, ***sage***, ***salad burnet*** *(off-label)*, ***savory*** *(off-label)*, ***shrubs***, ***sorrel*** *(off-label)*, ***spring oilseed rape***, ***strawberries***, ***swedes***, ***sweet ciceley*** *(off-label)*, ***tarragon*** *(off-label)*, ***thyme*** *(off-label)*, ***trees***, ***turnips***, ***winter oilseed rape***, ***woodruff*** *(off-label)*. Black nightshade in ***dwarf beans*** *(off-label)*, ***navy beans*** *(off-label)*. Slender speedwell in ***managed amenity turf***. Volunteer potatoes in ***dwarf beans*** *(off-label)*, ***navy beans*** *(off-label)*.

Specific Off-Label Approvals (SOLAs)

- ***angelica, balm, basil, bay, borage, celery leaves, chervil, chives, clary, coriander, dandelions, dill, fennel, fenugreek, feverfew, hyssop, land cress, lemon verbena, lovage, marjoram, mint, nasturtium, oregano, parsley, rocket, rosemary, rue, salad burnet, savory, sorrel, sweet ciceley, tarragon, thyme, woodruff*** *(OLA 011264) Dec 2008* [1]
- ***angelica, balm, basil, bay, borage, celery leaves, chervil, chives, clary, coriander, dandelions, dill, fennel, fenugreek, feverfew, hyssop, land cress, lemon verbena, lovage, marjoram, mint, nasturtium, oregano, parsley, rocket, rosemary, rue, salad burnet, savory, sorrel, sweet ciceley, tarragon, thyme, woodruff*** *(OLA 030873) Dec 2008* [2]
- ***dwarf beans, navy beans*** *(OLA 021190) Dec 2008* [1]
- ***dwarf beans, navy beans*** *(OLA 030874) Dec 2008* [2]

Efficacy guidance

- Best results on fine firm weed-free soil when adequate rain or irrigation follows
- Apply after drilling or planting prior to weed emergence. Rates and timing vary with crop and soil type. See label for details
- For control of slender speedwell in turf apply when weeds growing actively. Do not mow for at least 3 d after treatment

Restrictions

- Maximum number of treatments 1 per crop for brassicas, oilseed rape, mustard, leeks, onions, sage, newly planted bush and cane fruit; 1 per season for established strawberries
- Do not apply mixture with propachlor to newly planted strawberries after rolling or application of other herbicides
- Do not use on strawberries between flowering and harvest
- Many types of ornamental have been treated successfully. See label for details. For species of unknown susceptibility treat a small number of plants first
- Do not use on turf where bent grasses form a major constituent of sward
- Do not plant lettuce within 6 mth of application, seeded turf within 2 mth, other crops within 3 mth. In the event of crop failure deep plough before re-drilling or planting
- Do not use on dwarf French beans
- Do not use on organic soils

Crop-specific information

- Latest use: before crop emergence for sown crops

- Recommended alone on roses, runner beans, various ornamentals, strawberries and turf, in tank-mix with propachlor on brassicas, onions, leeks, sage, ornamentals, established strawberries and newly planted soft fruit
- Apply to brassicas pre-emergence or after 3-4 true leaf stage

Hazard classification and safety precautions

Operator protection U20a
Environmental protection E15
Storage and disposal E30a, E32a

106 chlorthal-dimethyl + propachlor

A residual herbicide mixture for a range of vegetable crops

Products

Decimate	Certis	225:216 g/l	SC	11008

Uses

Annual dicotyledons in ***broccoli***, ***brussels sprouts***, ***bulb onions***, ***cabbages***, ***calabrese***, ***cauliflowers***, ***fodder rape***, ***kale***, ***mustard***, ***swedes***, ***turnips***. Annual meadow grass in ***broccoli***, ***brussels sprouts***, ***bulb onions***, ***cabbages***, ***calabrese***, ***cauliflowers***, ***fodder rape***, ***kale***, ***mustard***, ***swedes***, ***turnips***.

Efficacy guidance

- Product may be applied pre- or post-emergence of the crop but pre-emergence of the weeds. Emerged weeds are not controlled
- Soils should be fine, firm and free of weed growth
- For best results apply to moist soil and before rain. In dry weather irrigate lightly after application and before seedling weeds appear
- Weed control may be reduced under very wet, dry or windy conditions
- Do not use on soils with more than 10% organic matter as weed control will be impaired

Restrictions

- Maximum number of treatments 1 per crop for all crops

Crop-specific information

- Use on brassica crops pre-crop emergence may result in a check to growth but this is usually outgrown within 8 wk without effect on yield or quality
- Post-emergence treatment on brassica crops should not be made until they have 3-4 true leaves. Post-emergence treatment on onions should be made post-crook stage
- Young transplanted plants, particularly cabbages and cauliflowers, raised under glass should be hardened off before treatment
- Care should be taken with modular transplants to ensure that that roots are not exposed to spray

Following crops guidance

- The interval between treatment and planting succeeding crops should be 2 mth for seeded turf, 6 mth for lettuce, 3 mth for all other crops. In the event of failure of a treated crop the land must be deep mouldboard ploughed before redrilling or replanting

Environmental safety

- Dangerous to fish or other aquatic life. Do not contaminate surface waters or ditches with chemical or used container

Hazard classification and safety precautions

Hazard R04
Risk phrases R04a, R04b
Operator protection A, C; U02, U04a, U05a, U09b, U19, U20a
Environmental protection E01, E13b
Consumer protection C03
Storage and disposal E26, E30a, E31a

SECTION 4

107 cholecalciferol

A hypercalcaemic rodenticide

Products

Sorex Fatal	Sorex	0.1% w/w	RB	H7087

Uses

Rats in ***farm buildings/yards***.

Efficacy guidance

- Carry out pre-baiting with blank pellets at each proposed baiting point in runs, burrows and where droppings seen, for at least 3 d
- Inspect baiting points daily and replenish for up to 5 d
- Remove all remaining bait and containers after 5 d to prevent bait shyness
- If rat activity continues switch to a rodenticide with a different active ingredient

Environmental safety

- Prevent access to bait by children, birds and non-target animals particularly dogs, cats, pigs and poultry

Hazard classification and safety precautions

Operator protection A; U13, U20b
Storage and disposal E30a, E32a
Vertebrate/rodent control products V01a, V02, V03b, V04a
Medical advice M03

108 cholecalciferol + difenacoum

A mixture of rodenticides with different modes of action

Products

Sorexa CD Mouse Bait	Sorex	0.075:0.0025% w/w	RB	H6751

Uses

Mice in ***farm buildings/yards***.

Efficacy guidance

- Lay baits in mouse runs in many locations throughout infested area
- It is important to lay many small baits as mice are sporadic feeders
- Cover baits to protect from moisture in outdoor situations
- Inspect baits frequently and replace or top up with fresh material as necessary

Environmental safety

- Protect baits from access by children, birds and non-target animals, particularly dogs, cats, pigs and poultry
- Product contains Bitrex to help prevent accidental human consumption

Hazard classification and safety precautions

Operator protection A; U13, U20b
Storage and disposal E30a, E32a
Vertebrate/rodent control products V01b, V02, V03b, V04b
Medical advice M03

109 choline chloride

A plant growth regulator available only in mixtures

See also chlormequat + choline chloride + imazaquin

110 cinidon-ethyl

A novel phthalimide contact herbicide for cereals

Products

Lotus	BASF	200 g/l	EC	09231

Uses

Annual dicotyledons in ***durum wheat, spring barley, winter barley, winter rye, winter wheat***. Chickweed in ***durum wheat, spring barley, winter barley, winter rye, winter wheat***. Cleavers in ***durum wheat, spring barley, winter barley, winter rye, winter wheat***. Speedwells in ***durum wheat, spring barley, winter barley, winter rye, winter wheat***.

Approval information

- Cinidon-ethyl included in Annex I under EC Directive 91/414

Efficacy guidance

- Controls sensitive species that have emerged at time of treatment
- Ensure good even spray cover for best results
- Weed spectrum can be widened by mixture with mecoprop-P (see label)

Restrictions

- Maximum total dose equivalent to one full dose application
- Do not treat crops suffering from herbicide damage or physical stress
- Do not roll or harrow for 7d before or after treatment

Crop-specific information

- HI 35 d for cutting for stock feed or grazing
- Latest use: before first node detectable (GS 31) for spring barley and winter rye; before third node detectable (GS 33) for winter wheat, durum wheat and winter barley
- Some necrotic spotting may occur after treatment but has no effect on final yield. Effect can be greater when mixture with mecoprop-P used, especially in spring, but yield not normally affected

Following crops guidance

- In the event of crop failure only winter and spring wheat and barley, oilseed rape, maize or sunflowers may be planted

Environmental safety

- Dangerous to fish or other aquatic life. Do not contaminate surface waters or ditches with chemical or used container
- Treated cereals must not be used as animal feed or bedding for at least 35 d after last application
- Do not allow animals to graze for 35 d after treatment

Hazard classification and safety precautions

Hazard R03, R04
Risk phrases R04b, R04d, R04e, R04f, R04g
Operator protection A, C, H; U05a, U08, U20b
Environmental protection E01, E13b
Consumer protection C03
Storage and disposal E30a, E31c, E34
Medical advice M03, M05

111 citronella oil

A natural plant extract herbicide

Products

Barrier H	Barrier	22.9% w/w	EW	10136

Uses

Ragwort in ***grassland, land temporarily removed from production***.

Efficacy guidance

- Best results obtained from spot treatment of ragwort in the rosette stage, during dry still conditions

SECTION 4

- Aerial growth of ragwort is rapidly destroyed. Longer term control depends on overall management strategy
- Check for regrowth after 28 d and re-apply as necessary

Crop-specific information

- Contact with grasses will result in transient scorch which is outgrown in good growing conditions

Environmental safety

- Apply away from bees
- Harmful to fish or other aquatic life. Do not contaminate surface waters or ditches with chemical or used container
- Keep livestock out of treated areas for at least 2 wk and until foliage of any poisonous weeds such as ragwort has died and become unpalatable

Hazard classification and safety precautions

Hazard R04
Risk phrases R04a, R04b
Operator protection A, C; U05a, U20c
Environmental protection E01, E12f, E13c; E07 (2 wk)
Consumer protection C03
Storage and disposal E26, E30a, E32a

112 clodinafop-propargyl

A contact acting herbicide for annual grass weed control in wheat, triticale and rye

Products

1	Greencrop Boulevard	Greencrop	240 g/l	EC	09960
2	Landgold Clodinafop	Landgold	240 g/l	EC	10172
3	Marathon	Me2	240 g/l	EC	09959
4	Standon Clodinafop 240	Standon	240 g/l	EC	10174
5	Topik	Syngenta	240 g/l	EC	08461

Uses

Blackgrass in ***durum wheat***, ***spring rye***, ***spring wheat***, ***triticale***, ***winter rye***, ***winter wheat***. Rough meadow grass in ***durum wheat***, ***spring rye***, ***spring wheat***, ***triticale***, ***winter rye***, ***winter wheat***. Wild oats in ***durum wheat***, ***spring rye***, ***spring wheat***, ***triticale***, ***winter rye***, ***winter wheat***.

Efficacy guidance

- Spray when majority of weeds have germinated but before competition reduces yield
- Products contain a herbicide safener (cloquintocet-mexyl) that improves crop tolerance to clodinafop-propargyl
- Optimum control achieved when all grass weeds emerged. Wait for delayed germination on dry or cloddy seedbed
- A mineral oil additive is recommended to give more consistent control of very high blackgrass populations or for late season treatments. See label for details
- Weed control not affected by soil type, organic matter or straw residues
- Control may be reduced if rain falls within 1 h of treatment

Restrictions

- Maximum total dose equivalent to one full dose treatment
- Do not use on barley or oats
- Do not treat crops under stress or suffering from waterlogging, pest attack, disease or frost
- Do not treat crops undersown with grass mixtures
- Do not mix with products containing MCPA, mecoprop-P, 2,4-D or 2,4-DB
- MCPA, mecoprop, 2,4-D or 2,4-DB should not be applied within 21 d before, or 7 d after, treatment

Crop-specific information

- Latest use: before second node detectable stage (GS 32) for durum wheat, triticale, rye; before flag leaf sheath extending (GS 41) for wheat
- Spray in autumn, winter or spring from 1 true leaf stage (GS 11) to before second node detectable (GS 32) on durum, rye, triticale; before flag leaf sheath extends (GS 41) on wheat

Following crops guidance

- Only a broad leaved crop may be sown after failure of a treated crop. After normal harvest any broad leaved crop or wheat, durum wheat, rye, triticale or barley should be sown

Environmental safety

- Flammable [2, 5]
- Harmful (dangerous [1, 3, 4]) to fish or other aquatic life. Do not contaminate surface waters or ditches with chemical or used container

Hazard classification and safety precautions

Hazard R04
Risk phrases R04a, R04b [1-5]; R04e [1, 3, 4]
Operator protection A, C, H, K; U02, U05a, U09a [1-5]; U20a [2, 5]; U20b [1, 3, 4]
Environmental protection E01, E13b [1, 3, 4]; E13c [2, 5]
Consumer protection C03
Storage and disposal E26 [1, 3, 4]; E30a, E32a [1-5]

113 clodinafop-propargyl + diflufenican

A contact and residual herbicide mixture for broad spectrum weed control in wheat, triticale and rye

Products

Amazon	Aventis	30:50 g/l	EC	10266

Uses

Annual dicotyledons in ***durum wheat, triticale, winter rye, winter wheat***. Blackgrass in ***durum wheat, triticale, winter rye, winter wheat***. Rough meadow grass in ***durum wheat, triticale, winter rye, winter wheat***. Wild oats in ***durum wheat, triticale, winter rye, winter wheat***.

Efficacy guidance

- Optimum weed control obtained when all grass weeds emerged and broad leaved weeds at susceptible stage before they compete with the crop
- Products contain a herbicide safener (cloquintocet-mexyl) that improves crop tolerance to clodinafop-propargyl
- Delay treatment if dry or cloddy seedbeds favour late weed germination
- Grass weed control unaffected by seedbed condition but control of broad leaved weeds requires a fine, firm seedbed cleared of trash and straw
- Always use a recommended adjuvant. See label
- Speed of action depends on temperature and growing conditions and may appear slow in dry or cold weather

Restrictions

- Maximum number of treatments 1 per crop
- Do not treat crops under stress or suffering from waterlogging, pest attack, disease or frost
- Do not treat broadcast crops or those undersown with grass mixtures
- Do not mix with products containing MCPA, mecoprop-P, 2,4-D or 2,4-DB
- MCPA, mecoprop, 2,4-D or 2,4-DB should not be applied within 21 d before, or 7 d after, treatment
- Do not use on Sands, stony or gravelly soils or soil with over 10% organic matter
- Do not roll autumn treated crops until spring. Do not harrow at any time after treatment
- Only listed crops may be sown after failure of a treated crop - see label

Crop-specific information

- Latest use: before second node detectable (GS 32) for winter wheat and durum wheat; before end Feb in yr of harvest for rye and triticale
- Spray in autumn, winter or spring from first leaf unfolded stage (GS 11)

Following crops guidance

- After normal harvest cereals, field beans, oilseed rape, onions, leaf brassicas or sugar beet seed crops may be sown in the autumn. Land must be ploughed or cultivated to 15 cm except for cereals
- Successive treatments of any products containing diflufenican can lead to soil build up and inversion ploughing must precede sowing any non-cereal crop. Even where ploughing occurs some crops may be damaged - see label

Environmental safety
- Extremely dangerous to fish or other aquatic life. Do not contaminate surface waters or ditches with chemical or used container
- LERAP Category B

Hazard classification and safety precautions

Hazard R04
Risk phrases R04a, R04b
Operator protection A, C, H, K; U02, U05a, U09a, U20b
Environmental protection E13a, E16a, E16b
Consumer protection C03
Storage and disposal E26, E30a, E32a

114 clodinafop-propargyl + trifluralin

A contact and residual herbicide for winter wheat

Products

Hawk	Syngenta	12:383 g/l	EC	08417

Uses

Annual dicotyledons in ***winter wheat***. Blackgrass in ***winter wheat***. Rough meadow grass in ***winter wheat***. Wild oats in ***winter wheat***.

Efficacy guidance
- Product contains a herbicide safener (cloquintocet-mexyl) that improves crop tolerance to clodinafop-propargyl
- Optimum weed control obtained when most grass weeds emerged and broad leaved weeds at susceptible stage before they compete with the crop. Pre-emergence control of annual meadow-grass and broad leaved weeds may not be satisfactory on medium and heavy soils
- Delay treatment if dry or cloddy seedbeds favour late weed germination
- Optimum weed control obtained in crops growing in a fine, firm seedbed cleared of trash and straw
- Wild oats and other weeds germinating from beneath treated zone not controlled
- Speed of action depends on temperature and growing conditions and may appear slow in dry or cold weather
- Rain within 1 h of application may reduce grass weed control

Restrictions
- Maximum number of treatments 1 per crop
- Do not treat crops under stress or suffering from waterlogging, pest attack, disease or frost
- Do not treat crops undersown with grass mixtures
- Do not mix with products containing MCPA, mecoprop-P, 2,4-D or 2,4-DB
- MCPA, mecoprop, 2,4-D or 2,4-DB should not be applied within 21 d before, or 7 d after, treatment
- Do not use on Sands, stony or gravelly soils or soils with over 10% organic matter
- Product must be applied with specified adjuvant - see label

Crop-specific information
- Latest use: before ear at 1 cm stage (GS 30)
- Spray in autumn, winter or spring from first to third leaf unfolded stage (GS 11-13) but before ear at 1 cm stage (GS 30)

Following crops guidance
- After normal harvest any broad leaved crop (except sugar beet), barley, wheat, durum wheat, rye or triticale may be sown. Before drilling or planting subsequent crops soil must be mouldboard ploughed to 15 cm
- See label for details of crops that may be safely drilled or planted within 5 mth of treatment. If a treated crop should fail after 5 mth but before normal harvest, sow only a broad-leaved crop (not sugar beet)
- 12 mth must elapse after treatment before sugar beet is drilled

Environmental safety

- Harmful to fish or other aquatic life. Do not contaminate surface waters or ditches with chemical or used container
- Crops must not be treated if there is risk of run-off (eg from frozen foliage)

Hazard classification and safety precautions

Hazard R04
Risk phrases R04a, R04b, R04e
Operator protection A, C, H, K; U02, U05a, U09a, U20a
Environmental protection E13c
Consumer protection C03
Storage and disposal E26, E30a, E32a

115 clofentezine

A selective ovicidal tetrazine acaricide for use in top fruit

Products

Apollo 50 SC	Makhteshim	500 g/l	SC	10590

Uses

Red spider mites in ***apples***, ***cherries***, ***pears***, ***plums***. Spider mites in ***blackcurrants*** *(off-label)*, ***protected blackberries*** *(off-label)*, ***protected peppers*** *(off-label)*, ***protected raspberries*** *(off-label)*, ***protected strawberries*** *(off-label)*, ***raspberries*** *(off-label)*, ***strawberries*** *(off-label)*.

Specific Off-Label Approvals (SOLAs)

- ***blackcurrants***, ***raspberries***, ***strawberries*** *(OLA 010912) Dec 2008* [1]
- ***blackcurrants***, ***raspberries***, ***strawberries*** *(OLA 012269) Dec 2008* [1]
- ***protected blackberries***, ***protected raspberries*** *(OLA 010911) Dec 2008* [1]
- ***protected blackberries***, ***protected raspberries*** *(OLA 012268) Dec 2008* [1]
- ***protected peppers*** *(OLA 011431) Jul 2007* [1]
- ***protected strawberries*** *(OLA 012271) Dec 2008* [1]

Efficacy guidance

- Acts on eggs and early motile stages of mites. For effective control total cover of plants is essential, particular care being needed to cover undersides of leaves

Restrictions

- Maximum number of treatments 1 per yr for top fruit; 3 per yr for soft and cane fruit

Crop-specific information

- HI protected cane fruit 7 d; blackcurrants, raspberries, strawberries 14 d; apples, pears 28 d, plums, cherries 8 wk
- For red spider mite control spray apples and pears between bud burst and pink bud, plums and cherries between white bud and first flower. Rust mite is also suppressed
- On established infestations apply in conjunction with an adult acaricide

Environmental safety

- Product safe on predatory mites, bees and other predatory insects

Hazard classification and safety precautions

Operator protection U08, U20b
Environmental protection E15
Storage and disposal E26, E30a, E32a

116 clomazone

An isoxazolidinone residual herbicide for oilseed rape and field beans

Products

1	Centium 360 CS	FMC	360 g/l	CS	10720
2	Centium 360 CS	Belchim	360 g/l	CS	11607

Uses

Chickweed in ***carrots***, ***combining peas***, ***potatoes***, ***spring field beans***, ***spring oilseed rape***, ***vining peas***, ***winter field beans***, ***winter oilseed rape***. Cleavers in ***carrots***, ***combining peas***,

potatoes*, *spring field beans*, *spring oilseed rape*, *vining peas*, *winter field beans*, *winter oilseed rape. Fool's parsley in ***carrots*, *combining peas*, *potatoes*, *spring field beans*, *spring oilseed rape*, *vining peas***. Red dead-nettle in ***carrots*, *combining peas*, *potatoes*, *spring field beans*, *spring oilseed rape*, *vining peas*, *winter field beans*, *winter oilseed rape***. Shepherd's purse in ***carrots*, *combining peas*, *potatoes*, *spring field beans*, *spring oilseed rape*, *vining peas*, *winter field beans*, *winter oilseed rape***.

Efficacy guidance

- Best results obtained from application as soon as possible after sowing crop and before emergence of crop or weeds
- Uptake is via roots and shoots. Seedbeds should be firm, level and free from clods. Loose puffy seedbeds should be consolidated before spraying
- Efficacy is reduced on organic soils and on dry cloddy seedbeds
- Clomazone acts by inhibiting synthesis of chlorophyll pigments. Susceptible weeds emerge but are chlorotic and die shortly afterwards
- Season-long control of weeds may not be achieved

Restrictions

- Maximum number of treatments one per crop
- Crops must be covered by a minimum of 20 mm settled soil. Do not apply to broadcast crops. Direct-drilled crops should be harrowed across the slits to cover seed before spraying
- Do not treat two consecutive crops of carrots in one calendar yr
- Do not use on Sands or Very Light soils

Crop-specific information

- Latest use: pre-emergence of crop
- Crop plants emerged at time of treatment may be severely damaged
- Some transient crop bleaching may occur under certain climatic conditions and can be severe where heavy rain follows application. This is normally rapidly outgrown and has no effect on final crop yield. Overlapping spray swaths may cause severe damage to field beans

Following crops guidance

- Following normal harvest of a spring or autumn treated crop, cereals, oilseed rape, field beans, combining peas, potatoes, maize, turnips, linseed or sugar beet may be sown
- In the event of failure of an autumn treated crop, winter cereals or winter beans may be sown in the autumn if 6 wk have elapsed since treatment. In the spring following crop failure combining peas, field beans or potatoes may be sown if 6 wk have elapsed since treatment, and spring cereals, maize, turnips, onions, carrots or linseed may be sown if 7 mth have elapsed since treatment
- In the event of a failure of a spring treated crop a wide range of crops may be sown provided intervals of 6-9 wk have elapsed since treatment. See label for details
- Prior to resowing any listed replacement crop the soil should be ploughed and cultivated to 15 cm

Environmental safety

- Take extreme care to avoid drift outside the target area, or on to ponds, waterways or ditches as considerable damage may occur

Hazard classification and safety precautions

Hazard R04
Risk phrases R04e
Operator protection A, H; U05a, U14, U20b
Environmental protection E01, E15
Consumer protection C03
Storage and disposal E30a, E31b

117 clopyralid

A foliar translocated picolinic herbicide for a wide range of crops

See also bromoxynil + clopyralid

Products

1	Agriguard Clopyralid	AgriGuard	200 g/l	SL	10625
2	Barclay Karaoke	Barclay	200 g/l	SL	11357
3	Dow Shield	Dow	200 g/l	SL	10988

Products – continued

4	Fernpath Torate	AgriGuard	200 g/l	SL	11033
5	Greencrop Champion	Greencrop	200 g/l	SL	11755
6	Loncid	I T Agro	200 g/l	SL	10832

Uses

Annual dicotyledons in ***broccoli, brussels sprouts, cabbages, calabrese, fodder beet, fodder rape, kale, linseed, maize, mangels, onions, ornamental plant production, permanent pasture, red beet, spring oilseed rape, strawberries, sugar beet, swedes, sweetcorn, turnips, winter oilseed rape*** [1-6]; ***cauliflowers, conifers and broadleaved trees*** *(off-label)* [3]; ***rotational grassland*** [1, 5]; ***spring barley, spring oats, spring wheat, winter wheat*** [1, 3, 5, 6]; ***winter barley, winter oats*** [1, 3-6]. Corn marigold in ***broccoli, brussels sprouts, cabbages, calabrese, cauliflowers, fodder beet, fodder rape, kale, linseed, maize, mangels, onions, ornamental plant production, permanent pasture, red beet, spring oilseed rape, strawberries, sugar beet, swedes, sweetcorn, turnips, winter oilseed rape*** [1-6]; ***rotational grassland*** [1, 5]; ***spring barley, spring oats, spring wheat, winter wheat*** [1, 3, 5, 6]; ***winter barley, winter oats*** [1, 3-6]. Creeping thistle in ***apples*** *(off-label)*, ***pears*** *(off-label)* [3]; ***broccoli, brussels sprouts, cabbages, calabrese, cauliflowers, fodder beet, fodder rape, kale, linseed, maize, mangels, onions, ornamental plant production, permanent pasture, red beet, spring oilseed rape, strawberries, sugar beet, swedes, sweetcorn, turnips, winter oilseed rape*** [1-6]; ***rotational grassland*** [1, 5]; ***spring barley, spring oats, spring wheat, winter wheat*** [1, 3, 5, 6]; ***winter barley, winter oats*** [1, 3-6]. Groundsel in ***angelica*** *(off-label)*, ***balm*** *(off-label)*, ***basil*** *(off-label)*, ***bay*** *(off-label)*, ***borage*** *(off-label)*, ***camomile*** *(off-label)*, ***caraway*** *(off-label)*, ***chervil*** *(off-label)*, ***chives*** *(off-label)*, ***clary*** *(off-label)*, ***coriander*** *(off-label)*, ***dill*** *(off-label)*, ***fennel*** *(off-label)*, ***fenugreek*** *(off-label)*, ***feverfew*** *(off-label)*, ***hyssop*** *(off-label)*, ***land cress*** *(off-label)*, ***leaf spinach*** *(off-label)*, ***lemon verbena*** *(off-label)*, ***lovage*** *(off-label)*, ***marjoram*** *(off-label)*, ***mint*** *(off-label)*, ***nettle*** *(off-label)*, ***oregano*** *(off-label)*, ***parsley*** *(off-label)*, ***rocket*** *(off-label)*, ***rosemary*** *(off-label)*, ***rue*** *(off-label)*, ***sage*** *(off-label)*, ***salad burnet*** *(off-label)*, ***savory*** *(off-label)*, ***sorrel*** *(off-label)*, ***spinach beet*** *(off-label)*, ***tarragon*** *(off-label)*, ***thyme*** *(off-label)*, ***woodruff*** *(off-label)* [3]. Mayweeds in ***angelica*** *(off-label)*, ***balm*** *(off-label)*, ***basil*** *(off-label)*, ***bay*** *(off-label)*, ***borage*** *(off-label)*, ***camomile*** *(off-label)*, ***caraway*** *(off-label)*, ***chervil*** *(off-label)*, ***chives*** *(off-label)*, ***clary*** *(off-label)*, ***coriander*** *(off-label)*, ***dill*** *(off-label)*, ***fennel*** *(off-label)*, ***fenugreek*** *(off-label)*, ***feverfew*** *(off-label)*, ***honesty*** *(off-label)*, ***hyssop*** *(off-label)*, ***land cress*** *(off-label)*, ***leaf spinach*** *(off-label)*, ***lemon verbena*** *(off-label)*, ***lovage*** *(off-label)*, ***marjoram*** *(off-label)*, ***mint*** *(off-label)*, ***nettle*** *(off-label)*, ***oregano*** *(off-label)*, ***parsley*** *(off-label)*, ***rocket*** *(off-label)*, ***rosemary*** *(off-label)*, ***rue*** *(off-label)*, ***sage*** *(off-label)*, ***salad burnet*** *(off-label)*, ***savory*** *(off-label)*, ***sorrel*** *(off-label)*, ***spinach beet*** *(off-label)*, ***tarragon*** *(off-label)*, ***thyme*** *(off-label)*, ***woodruff*** *(off-label)* [3]; ***broccoli, brussels sprouts, cabbages, calabrese, cauliflowers, fodder beet, fodder rape, kale, linseed, maize, mangels, onions, ornamental plant production, permanent pasture, red beet, spring oilseed rape, strawberries, sugar beet, swedes, sweetcorn, turnips, winter oilseed rape*** [1-6]; ***rotational grassland*** [1, 5]; ***spring barley, spring oats, spring wheat, winter wheat*** [1, 3, 5, 6]; ***winter barley, winter oats*** [1, 3-6]. Thistles in ***asparagus*** *(off-label)*, ***grass seed crops*** *(off-label)*, ***honesty*** *(off-label)* [3].

Specific Off-Label Approvals (SOLAs)

- ***angelica, balm, basil, bay, borage, camomile, caraway, chervil, chives, clary, coriander, dill, fennel, fenugreek, feverfew, hyssop, land cress, lemon verbena, lovage, marjoram, mint, nettle, oregano, parsley, rocket, rosemary, rue, sage, salad burnet, savory, sorrel, tarragon, thyme, woodruff*** *(OLA 010377) Dec 2008* [3]
- ***apples, pears*** *(OLA 002082) Dec 2008* [3]
- ***asparagus*** *(OLA 000520) Dec 2008* [3]
- ***conifers and broadleaved trees*** *(OLA 920757) Dec 2008* [3]
- ***grass seed crops*** *(OLA 920662) Dec 2008* [3]
- ***honesty*** *(OLA 930480) Dec 2008* [3]
- ***leaf spinach, spinach beet*** *(OLA 010094) Dec 2008* [3]

Efficacy guidance

- Best results achieved by application to young actively growing weed seedlings. Treat creeping thistle at rosette stage and repeat 3-4 wk later as directed
- High activity on weeds of Compositae family. For most crops recommended for use in tank mixes. See label for details

Restrictions

- Maximum total dose varies between the equivalent of one and two full dose treatments, depending on the crop treated. See labels for details
- Do not apply to cereals later than the second node detectable stage (GS 32)
- Do not apply when crop damp or when rain expected within 6 h
- Do not use straw from treated cereals in compost or any other form for glasshouse crops. Straw may be used for strawing down strawberries
- Straw from treated grass seed crops or linseed should be baled and carted away. If incorporated do not plant winter beans in same year
- Do not use on onions at temperatures above 20°C or when under stress
- Do not treat maiden strawberries or runner beds or apply to early leaf growth during blossom period or within 4 wk of picking. Aug or early Sep sprays may reduce yield

Crop-specific information

- Latest use: before 3rd node detectable (GS 33) for cereals; before flower buds visible from above for oilseed rape and linseed
- HI grassland 7 d; apples, pears, strawberries 4 wk; maize, sweetcorn, onions, Brussels sprouts, broccoli, cabbage, cauliflowers, calabrese, kale, fodder rape, oilseed rape, swedes, turnips, sugar beet, red beet, fodder beet, mangels, sage, honesty 6 wk
- Timing of application varies with weed problem, crop and other ingredients of tank mixes. See labels for details
- Apply as directed spray in woody ornamentals, avoiding leaves, buds and green stems. Do not apply in root zone of families Compositae or Leguminosae

Following crops guidance

- Do not plant susceptible autumn-sown crops in same year as treatment. Do not apply later than Jul where susceptible crops are to be planted in spring. See label for details

Environmental safety

- Wash spray equipment thoroughly with water and detergent immediately after use. Traces of product can damage susceptible plants sprayed later
- Keep livestock out of treated areas for at least 7 d and until foliage of any poisonous weeds such as ragwort has died and become unpalatable

Hazard classification and safety precautions

Operator protection A, C [1-4, 6]; U08, U19, U20b [1-4, 6]
Environmental protection E07 [1-4, 6] (7 d); E15 [1-4, 6]
Storage and disposal E26, E30a, E31b [1-4, 6]

118 clopyralid + 2,4-D + MCPA

A translocated herbicide mixture for grassland

Products

Lonpar	Dow	35:150:175 g/l	SL	08686

Uses

Creeping thistle in ***established grassland***.

Approval information

- 2,4-D included in Annex I under EC Directive 91/414

Efficacy guidance

- Treatment must be made when weeds and grass are actively growing
- Important to ensure sufficient leaf area for uptake especially on established thistles with extensive root system. Treat at rosette stage
- On large well established thistles and where there is a large soil seed reservoir further treatment in the following year may be needed
- To allow maximum translocation do not cut grass for 28 d after treatment

Restrictions

- Maximum number of treatments 1 or 2 per yr depending on dose used - see label
- Do not spray in drought, very hot or very cold weather or if rain expected within 6 h
- Do not treat grass less than 1 yr old or sports or amenity turf

- Product kills or severely checks clover and should not be used where clover is an important constituent of the sward
- Do not drill grass, grass mixtures, kale, swedes or turnips into the sward within 6 wk of spraying

Crop-specific information

- Very occasionally some yellowing of sward may occur after treatment which is quickly outgrown

Following crops guidance

- Susceptible crops (see label) must not be planted in the same calendar yr as treatment. Ensure that remains of treated crop have completely decayed before planting a subsequent susceptible crop. Where a susceptible crop is to be planted in the spring, product must not be sprayed later than end Jul in previous yr

Environmental safety

- Harmful to fish or other aquatic life. Do not contaminate surface waters or ditches with chemical or used container
- Wash spray equipment thoroughly with water and detergent immediately after use. Traces of product can damage susceptible plants sprayed later

Hazard classification and safety precautions

Hazard R03, R04
Risk phrases R03c, R04d
Operator protection A, C, H, M; U05a, U15
Environmental protection E01, E13c; E07 (2 wk)
Consumer protection C03
Storage and disposal E26, E34

119 clopyralid + diflufenican + MCPA

A selective herbicide for use in established turf

Products

Spearhead	Bayer Environ.	20:15:300 g/l	SL	09941

Uses

Annual dicotyledons in ***managed amenity turf***. Perennial dicotyledons in ***managed amenity turf***.

Efficacy guidance

- Best results achieved by application when grass and weeds are actively growing
- Treatment during early part of the season is recommended, but not during drought

Restrictions

- Maximum number of treatments 1 per yr
- Only use when sward is satisfactorily established and regular mowing has begun
- Turf sown in spring or early summer may be ready for treatment after 2 mth. Later sown turf should not be sprayed until growth is resumed in the following spring
- Avoid mowing within 3-4 d before or after treatment
- Do not use cuttings from treated area as a mulch for any crop

Environmental safety

- Harmful to fish or other aquatic life. Do not contaminate surface waters or ditches with chemical or used container
- LERAP Category B
- Avoid drift. Small amounts of spray can cause serious injury to herbaceous plants, vegetables, fruit and glasshouse crops
- Keep livestock out of treated areas

Hazard classification and safety precautions

Hazard R04
Risk phrases R04a, R04b
Operator protection A, C; U02, U05a, U08, U19, U20b
Environmental protection E01, E07, E13c, E16a, E16b
Consumer protection C03
Storage and disposal E26, E30a, E31b

SECTION 4

120 clopyralid + fluroxypyr + MCPA

A translocated herbicide mixture for use in sports and amenity turf

Products

Greenor	Rigby Taylor	20:40:200 g/l	ME	07848

Uses

Annual dicotyledons in ***managed amenity turf***.

Approval information

- Fluroxypyr included in Annex I under EC Directive 91/414

Efficacy guidance

- Best results achieved when weeds actively growing and turf grass competitive
- Treatment should normally be between Apr-Sep when the soil is moist
- Do not apply during drought unless irrigation is applied
- Allow 3 d before or after mowing established turf to ensure sufficient weed leaf surface present to allow uptake and movement

Restrictions

- Maximum number of treatments 2 per yr
- Do not treat grass under stress from frost, drought, waterlogging, trace element deficiency, disease or pest attack
- Do not treat if night temperatures are low, when frost is imminent or during prolonged cold weather

Crop-specific information

- Treat young turf only in spring when at least 2 mth have elapsed since sowing
- Allow 5 d after mowing young turf before treatment
- Product selective on a number of turf grass species (see label) but consultation or testing recommended before treatment of any cultivar

Environmental safety

- Harmful to fish or other aquatic life. Do not contaminate surface waters or ditches with chemical or used container
- Wash spray equipment thoroughly with water and detergent immediately after use. Traces of product can damage susceptible plants sprayed later

Hazard classification and safety precautions

Hazard R04
Risk phrases R04a, R04b
Operator protection A, C; U05a, U08, U19, U20a
Environmental protection E01, E13c
Consumer protection C03
Storage and disposal E26, E30a, E31b

121 clopyralid + fluroxypyr + triclopyr

A foliar acting herbicide mixture for grassland

Products

Pastor	Dow	50:75:100 g/l	EC	07440

Uses

Annual dicotyledons in ***newly sown grass***. Docks in ***established grassland***, ***newly sown grass***. Stinging nettle in ***established grassland***. Thistles in ***established grassland***.

Approval information

- Fluroxypyr included in Annex I under EC Directive 91/414

Efficacy guidance

- Apply in spring or autumn depending on weeds present or as a split treatment in spring followed by autumn
- Treatment must be made when weeds and grass are actively growing
- It is important to ensure sufficient leaf area is present for uptake especially on established docks and thistles

- On large well established docks and where there is a large soil seed reservoir further treatment in the following year may be needed
- To allow maximum translocation in the weeds do not cut grass for 4 wk after treatment

Restrictions
- Maximum number of treatments 1 per yr at full dose or 2 per yr at half dose
- Do not spray in drought, very hot or very cold weather
- Do not treat sports or amenity turf
- Product kills or severely checks clover and should not be used where clover is an important constituent of the sward
- Do not roll or harrow 10 d before or 7 d after treatment
- Do not allow spray or drift to reach other crops, amenity plantings, gardens, ponds, lakes or water courses

Crop-specific information
- HI 7 d
- Application during active growth ensures minimal check to grass. Newly sown grass may be treated from the third leaf visible stage
- Product may be used in established grassland which is under non-rotational setaside arrangements
- Ver occasionally some yellowing of the sward may occur after treatment but is quickly outgrown

Following crops guidance
- Residues in incompletely decayed plant tissue may affect succeeding susceptible crops such as peas, beans and other legumes, carrots and related crops, potatoes, tomatoes, lettuce
- Do not plant susceptible autumn-sown crops in the same yr as treatment with product. Spring sown crops may follow if treatment was before end Jul in the previous yr

Environmental safety
- Harmful to fish or other aquatic life. Do not contaminate surface waters or ditches with chemical or used container
- Wash spray equipment thoroughly with water and detergent immediately after use. Traces of product can damage susceptible plants sprayed later
- Keep livestock out of treated areas for at least 7 d following treatment and until foliage of poisonous weeds such as ragwort has died and become unpalatable

Hazard classification and safety precautions

Hazard R04
Risk phrases R04a, R04b, R04e, R04g
Operator protection A, C; U02, U05a, U08, U19, U20b
Environmental protection E01, E13c; E07 (7 d)
Consumer protection C01, C03
Storage and disposal E26, E30a, E31b, E34

122 clopyralid + propyzamide

A post-emergence herbicide for winter oilseed rape

Products

1	Matrikerb	Rohm & Haas	4.3:43% w/w	WP	02443
2	Matrikerb	Dow	4.3:43% w/w	WP	09604

Uses

Annual dicotyledons in ***winter oilseed rape***. Annual grasses in ***winter oilseed rape***. Barren brome in ***winter oilseed rape***. Mayweeds in ***winter oilseed rape***.

Approval information
- Propyzamide included in Annex I under EC Directive 91/414

Efficacy guidance
- Apply from Oct to end Jan. Mayweed and groundsel may be controlled after emergence but before crop large enough to shield seedlings
- Best results achieved on fine, firm, moist soils when weeds germinating or small
- Effectiveness reduced by surface organic debris, burnt straw or ash
- Do not apply if rainfall imminent or frost present on foliage

Restrictions

- Maximum number of treatments 1 per crop
- Do not use on soils with more than 10% organic matter

Crop-specific information

- HI oilseed rape 6 wk; evening primrose 14 wk
- Apply to crop as soon as possible after 3-true leaf stage (GS 1,3)

Following crops guidance

- Minimum period between spraying and drilling a following crop varies from 10 to 40 wk. See label for details

Hazard classification and safety precautions

Operator protection A, C; U09a, U19, U20a
Environmental protection E01, E15
Consumer protection C02
Storage and disposal E30a, E32a, E34

123 clopyralid + triclopyr

A perennial and woody weed herbicide for use in grassland

Products

1	Grazon 90	Dow	60:240 g/l	EC	05456
2	Thistlex	Dow	200:200 g/l	EC	11533

Uses

Brambles in ***amenity grass, established grassland*** [1]. Broom in ***amenity grass, established grassland*** [1]. Creeping thistle in ***permanent pasture, rotational grassland*** [2]. Docks in ***amenity grass, established grassland*** [1]. Gorse in ***amenity grass, established grassland*** [1]. Perennial dicotyledons in ***amenity grass, established grassland, established grassland*** *(off-label)* [1]. Stinging nettle in ***amenity grass, established grassland*** [1]. Thistles in ***amenity grass, established grassland*** [1].

Specific Off-Label Approvals (SOLAs)

- ***established grassland*** *(OLA 950692) Dec 2008* [1]

Efficacy guidance

- Must be applied to actively growing weeds
- Correct timing crucial for good control. Spray stinging nettle before flowering, docks in rosette stage in spring, creeping thistle before flower stems 15-20 cm high, brambles, broom and gorse in Jun-Aug
- Allow 2-3 wk regrowth after grazing or mowing before spraying perennial weeds
- Do not cut grass for 21 d before or 28 d after spraying
- Where there is a large reservoir of weed seed in the soil further treatment in the following yr may be needed

Restrictions

- Maximum number of treatments 1 per yr
- Only use on permanent pasture or rotational grassland established for at least 1 yr
- Do not apply where clover is an important constituent of sward
- Do not roll or harrow within 10 d before or 7 d after spraying
- Do not direct drill kale, swedes, turnips, grass or grass mixtures within 6 wk of spraying
- Do not use any treated plant material for composting or mulching, and do not use manure for composting from animals fed on treated crops [2]
- Do not apply by hand-held rotary atomiser equipment
- Do not allow drift onto other crops, amenity plantings or gardens, ponds, lakes or water courses. All conifers, especially pine and larch, are very sensitive

Crop-specific information

- 7 d before grazing or cutting
- Some transient yellowing of treated swards may occur but is quickly outgrown

Following crops guidance

- Residues in plant tissues which have not completely decayed may affect succeeding susceptible crops such as peas, beans, other legumes, carrots, parsnips, potatoes, tomatoes, lettuce, glasshouse and protected crops
- Do not plant susceptible autumn-sown crops (eg winter beans) in same year as treatment
- Do not spray after end Jul where susceptible crops to be planted next spring

Environmental safety

- Dangerous to fish or other aquatic life. Do not contaminate surface waters or ditches with chemical or used container [1]
- Harmful to fish or other aquatic life. Do not contaminate surface waters or ditches with chemical or used container [2]
- Keep livestock out of treated areas for at least 7 d after spraying and until foliage of any poisonous weeds such as ragwort or buttercup has died down and become unpalatable

Hazard classification and safety precautions

Hazard R03 [1]; R04 [1, 2]
Risk phrases R03c, R04b, R04e [1]; R04d [1, 2]
Operator protection A, C [1, 2]; H, M [1]; U02, U05a, U20b [1, 2]; U08, U19 [1]; U15, U23a [2]
Environmental protection E01 [1, 2]; E07 [1, 2] (7 d); E13b, E23 [1]; E13c [2]
Consumer protection C01 [1]; C03 [1, 2]
Storage and disposal E26, E30a, E31b [1, 2]; E34 [1]
Medical advice M03 [2]

124 copper ammonium carbonate

A protectant copper fungicide

Products

Croptex Fungex	Certis	8% w/w (copper)	SL	11049

Uses

Blight in ***outdoor tomatoes***. Cane spot in ***loganberries***, ***raspberries***. Celery leaf spot in ***celery***. Currant leaf spot in ***blackcurrants***. Damping off in ***seedlings of ornamentals***. Leaf curl in ***peaches***. Leaf mould in ***protected tomatoes***. Powdery mildew in ***chrysanthemums***, ***cucumbers***.

Efficacy guidance

- Apply spray to both sides of foliage
- With protected crops keep foliage dry before and after spraying
- Ventilate glasshouse immediately after spraying

Restrictions

- Maximum number of treatments 5 per crop for celery; 3 per yr for blackcurrants and loganberries; 2 per yr for peaches and raspberries
- Do not spray plants which are dry at the roots

Environmental safety

- Harmful to fish or other aquatic life. Do not contaminate surface waters or ditches with chemical or used container
- Keep all livestock out of treated areas for at least 3 wk. Bury or remove spillages

Hazard classification and safety precautions

Hazard R03, R04
Risk phrases R03c, R04d
Operator protection U05a, U20a
Environmental protection E01, E06b, E13c
Consumer protection C03
Storage and disposal E30a, E31a, E34
Medical advice M03

SECTION 4

125 copper hydroxide

An inorganic root pruning and root development agent

Products

Spin Out	Fargro	71 g/l	LI	07610

Uses

Root control in ***container-grown stock***.

Efficacy guidance

- Product should be painted on inner surfaces of containers prior to planting
- Used containers should be free from any loose soil or dirt
- Apply using conventional painting techniques such as brush, sponge, pad or airless paint sprayer
- Thorough coverage with a single coating to a minimum thickness of 0.075 mm is sufficient for root control

Restrictions

- Tested on a wide range of ornamental species (see label). Efficacy and tolerance should be tested on a small scale for species not listed
- Performance characteristics may vary under some conditions such as when substrates with a pH of less than 5.0 are used
- Product must not be diluted
- Knapsack sprayer or similar equipment is not suitable

Crop-specific information

- Latest use: before planting
- Can be used in production of container grown forestry seedlings, hardy ornamental and herbaceous plant species
- Use at any stage of plant development from seedlings to large container grown trees

Environmental safety

- Harmful to fish or other aquatic life. Do not contaminate surface waters or ditches with chemical or used container

Hazard classification and safety precautions

Operator protection A, D; U11, U12, U20b
Environmental protection E13c
Storage and disposal E26, E30a, E32a

126 copper oxychloride

A protectant copper fungicide and bactericide

Products

1	Cuprokylt	Unicrop	50% w/w (copper)	WP	00604
2	Cuprokylt FL	Unicrop	270 g/l (copper)	SC	08299
3	Headland Inorganic Liquid Copper	Headland	256 g/l (copper)	SC	07799

Uses

Bacterial canker in ***cherries*** [1-3]; ***cob nuts*** *(off-label)*, ***hazel nuts*** *(off-label)*, ***walnuts*** *(off-label)* [2, 3]; ***plums*** [1, 2]. Bacterial rot in ***bulb onions*** *(off-label)*, ***garlic*** *(off-label)*, ***leeks*** *(off-label)*, ***salad onions*** *(off-label)*, ***shallots*** *(off-label)* [1, 3]. Black rot in ***broccoli*** *(off-label)*, ***brussels sprouts*** *(off-label)*, ***cabbages*** *(off-label)*, ***calabrese*** *(off-label)*, ***cauliflowers*** *(off-label)*, ***chinese cabbage*** *(off-label)*, ***collards*** *(off-label)*, ***kale*** *(off-label)* [1, 3]; ***protected brassica seedlings*** *(off-label)* [2]; ***protected broccoli*** *(off-label)*, ***protected brussels sprouts*** *(off-label)*, ***protected cabbages*** *(off-label)*, ***protected calabrese*** *(off-label)*, ***protected cauliflowers*** *(off-label)*, ***protected chinese cabbage*** *(off-label)*, ***protected collards*** *(off-label)*, ***protected kale*** *(off-label)* [3]. Blight in ***cob nuts*** *(off-label)*, ***hazel nuts*** *(off-label)*, ***walnuts*** *(off-label)* [2, 3]; ***outdoor tomatoes***, ***potatoes*** [1-3]; ***protected tomatoes*** [1, 2]. Buck-eye rot in ***outdoor tomatoes***, ***protected tomatoes*** [1-3]. Cane spot in ***loganberries***, ***raspberries*** [1, 2]. Canker in ***apples***, ***pears*** [1-3]. Celery leaf spot in ***celery*** [1-3]. Collar rot in ***apples*** *(off-label)* [2]. Damping off in ***outdoor tomatoes***, ***protected tomatoes*** [1, 2]. Downy mildew in ***grapevines*** [3]; ***hops*** [1-3]. Foot rot in ***outdoor tomatoes***, ***protected tomatoes*** [1, 2]. Leaf curl in ***peaches*** [1, 2]. Pseudomonas storage rots in ***bulb onions*** *(off-label)*,

garlic *(off-label)*, ***leeks*** *(off-label)*, ***salad onions*** *(off-label)*, ***shallots*** *(off-label)* [1, 3]. Purple blotch in ***blackberries*** [3]. Pythium in ***watercress*** *(off-label)* [3]; ***watercress*** *(off-label - during propagation)* [2]. Rhizoctonia in ***watercress*** *(off-label)* [3]; ***watercress*** *(off-label - during propagation)* [2]. Rust in ***blackcurrants*** [1-3]. Spear rot in ***broccoli*** *(off-label)*, ***brussels sprouts*** *(off-label)*, ***cabbages*** *(off-label)*, ***calabrese*** *(off-label)*, ***cauliflowers*** *(off-label)*, ***chinese cabbage*** *(off-label)*, ***collards*** *(off-label)*, ***kale*** *(off-label)* [1, 3]; ***protected brassica seedlings*** *(off-label)* [2]; ***protected broccoli*** *(off-label)*, ***protected brussels sprouts*** *(off-label)*, ***protected cabbages*** *(off-label)*, ***protected calabrese*** *(off-label)*, ***protected cauliflowers*** *(off-label)*, ***protected chinese cabbage*** *(off-label)*, ***protected collards*** *(off-label)*, ***protected kale*** *(off-label)* [3].

Specific Off-Label Approvals (SOLAs)

- ***apples*** *(OLA 982336) Dec 2008* [2]
- ***broccoli, brussels sprouts, cabbages, calabrese, cauliflowers, chinese cabbage, collards, kale*** *(OLA 010115) Dec 2008* [1]
- ***broccoli, brussels sprouts, bulb onions, cabbages, calabrese, cauliflowers, chinese cabbage, cob nuts, collards, garlic, hazel nuts, kale, leeks, protected broccoli, protected brussels sprouts, protected cabbages, protected calabrese, protected cauliflowers, protected chinese cabbage, protected collards, protected kale, salad onions, shallots, walnuts, watercress*** *(OLA 020415) Dec 2008* [3]
- ***bulb onions, garlic, leeks, salad onions, shallots*** *(OLA 991127) Dec 2008* [1]
- ***cob nuts, hazel nuts, walnuts*** *(OLA 990385) Dec 2008* [2]
- ***protected brassica seedlings*** *(OLA 010117) Dec 2008* [2]
- ***watercress*** *(during propagation) (OLA 001538) Dec 2008* [2]

Approval information

- Approved for aerial application on potatoes [1]. See notes in Section 1

Efficacy guidance

- Spray crops at high volume when foliage dry but avoid run off. Do not spray if rain expected soon
- Spray interval commonly 10-14 d but varies with crop, see label for details
- If buck-eye rot occurs, spray soil surface and lower parts of tomato plants to protect unaffected fruit [1, 2]
- A follow-up spray in the following spring should be made to top fruit severely infected with bacterial canker

Restrictions

- Maximum number of treatments 3 per crop for apples, blackberries, grapevines, pears; 2 per crop for watercress. Not specified for other crops
- Some peach cultivars are sensitive to copper. Treat non-sensitive varieties only [1, 2]

Crop specific information

- Latest use: before bud burst for apples and pears
- HI calabrese 3 d; hops 7 d; onions, garlic, leeks, shallots 14 d; nuts 3 mth
- Slight damage may occur to leaves of cherries and plums [1, 2]

Environmental safety

- Harmful to fish or other aquatic life. Do not contaminate surface waters or ditches with chemical or used container
- Keep all livestock out of treated areas for at least 3 wk

Hazard classification and safety precautions

Operator protection A [3]; U20a [1, 3]; U20c [2]
Environmental protection E06b (3 wk); E13c
Storage and disposal E26, E31c [2]; E30a [1-3]; E31a [1]

127 copper sulphate

See Bordeaux Mixture

SECTION 4

128 coumatetralyl

An anticoagulant coumarin rodenticide

Products

Racumin Contact Powder	Bayer Environ.	0.75% w/w	CP	H6755

Uses

Rats in ***farm buildings***.

Efficacy guidance

- Chemical is available as a bait and tracking powder and it is recommended that both formulations are used together
- Place bait in runs in sheltered positions near rat holes. Use at least 250 g per baiting point and examine at least every 2 d. Replenish as long as bait being eaten
- When no signs of activity seen for about 10 d remove unused bait
- Apply tracking powder in a 2.5-5 cm wide layer inside and across entrance to holes or blow into holes with dusting machine. Lay a 3 mm thick layer along runs in 30 cm long patches. Only use on exposed runs if well away from buildings

Environmental safety

- Prevent access to bait or dust by children, birds and non-target animals, particularly dogs, cats, pigs
- Do not lay baits or dust where food, feed or water could become contaminated
- Remove all remains of bait and bait containers or exposed dust after treatment (except where used in sewers) and dispose of safely (eg burn/bury). Do not dispose of in refuse sacks or on open rubbish tips
- Search for rodent bodies (except where used in sewers) and dispose of safely (eg burn/bury). Do not dispose of in refuse sacks or on open rubbish tips

Hazard classification and safety precautions

Operator protection U13, U20c

Storage and disposal E30a, E32a

Vertebrate/rodent control products V01b, V02, V03b, V04b

129 cyanazine

A contact and residual triazine herbicide

Products

1	Fortrol	Makhteshim	500 g/l	SC	11174
2	I T Cyanazine	I T Agro	500 g/l	SC	10179

Uses

Annual dicotyledons in ***broccoli*** *(off-label)*, ***cabbages*** *(off-label)*, ***calabrese*** *(off-label)*, ***cauliflowers*** *(off-label)*, ***collards*** *(off-label)*, ***combining peas***, ***farm forestry*** *(off-label)*, ***forest nurseries*** *(off-label)*, ***kale*** *(off-label)*, ***ornamental plant production***, ***salad onions*** *(fen soils only)*, ***vining peas*** [1]; ***bulb onions***, ***narcissi***, ***winter oilseed rape*** [1, 2]; ***peas***, ***salad onions*** [2]. Annual grasses in ***bulb onions***, ***narcissi***, ***winter oilseed rape*** [1, 2]; ***combining peas***, ***farm forestry*** *(off-label)*, ***forest nurseries*** *(off-label)*, ***ornamental plant production***, ***salad onions*** *(fen soils only)*, ***vining peas*** [1]; ***peas***, ***salad onions*** [2].

Specific Off-Label Approvals (SOLAs)

- ***broccoli***, ***cabbages***, ***calabrese***, ***cauliflowers***, ***collards***, ***kale*** *(OLA 031074) Dec 2007* [1]
- ***farm forestry*** *(OLA 031454) Dec 2007* [1]
- ***forest nurseries*** *(OLA 031455) Dec 2007* [1]

Efficacy guidance

- Weeds controlled before emergence or at young seedling stage. Soil should be moist and some rain should follow application
- Best results achieved when applied during mild, bright weather. Avoid applications in dull, cold or wet conditions

- Numerous tank mixtures recommended to broaden weed spectrum. See label for details of tank mix partners and timings
- Some strains of blackgrass have developed resistance to many blackgrass herbicides which may lead to poor control

Restrictions

- Maximum number of treatments 1 per crop for edible crops
- Maximum total dose equivalent to one full dose treatment on onions
- Variety restrictions apply in peas. See label for details of tolerant varieties
- Do not treat crops suffering from physical damage or stress from pest attack
- Treatment directly after conditions that reduce wax formation may lead to damage
- Do not use as pre-emergence treatment on soils with more than 10% organic matter
- Do not use on Sands, Very Light or stony soils
- These products must not be used on any crops other than those listed, including any extrapolations that would normally be permissible under the Long Term Arrangements for Extension of Use (see Section 1)

Crop-specific information

- Latest use: pre-emergence for broad beans; before 31 Jan for winter oilseed rape; pre-emergence (high dose) or before flower buds visible for peas; before flower buds visible for field beans
- HI 8 wk for onions, salad onions; 11 wk for brassica crops
- In early drillings of peas in late Feb or Mar application may be delayed until immediately prior to emergence to increase effective duration of control
- Peas should be drilled to give 20-25 mm of settled soil above the seed. On light soils crops may be damaged if heavy rain falls soon after spraying
- Apply to oilseed rape after 1 Nov from 5-leaf stage when winter hardened. Do not apply after 31 Jan
- Apply post-emergence to onions after 2 true leaf stage only on fen soils with more than 10% organic matter

Environmental safety

- Harmful to fish or other aquatic life. Do not contaminate surface waters or ditches with chemical or used container

Hazard classification and safety precautions

Hazard R03
Risk phrases R03a, R03c
Operator protection A, C; U05a, U08, U20b [1, 2]; U19 [2]
Environmental protection E01, E13c
Consumer protection C03
Storage and disposal E26, E31b [2]; E30a, E34 [1, 2]; E31a [1]
Medical advice M03

130 cyanazine + pendimethalin

A contact and residual herbicide mixture

Products

Bullet	Makhteshim	150.204 g/l	SC	11204

Uses

Annual dicotyledons in ***combining peas***, ***spring field beans***. Annual meadow grass in ***combining peas***, ***spring field beans***. Rough meadow grass in ***combining peas***, ***spring field beans***.

Approval information

- Pendimethalin included in Annex I under EC Directive 91/414
- Certain products containing cyanazine have been granted derogations for specified 'Essential Uses' for use until 31 December 2007. Sale and supply must cease by 30 June 2007 but growers have no guarantee that the products will continue to be available until then.
  For more information see 'The Review Programme' under 'Pesticide Legislation' in Section 1

Efficacy guidance

- Best results achieved in crops growing on firm, fine seedbeds, where light rain falls soon after treatment. Do not disturb soil after application. Weed control may be reduced if prolonged dry conditions follow application
- Weeds present before crop emergence should be controlled with a contact herbicide as a tank-mix or in sequence
- Weed control may be reduced on soils with a high Kd factor, where OM exceeds 6% or ash content is high. Surface organic crop residues should be dispersed
- Weeds germinating more than 2 mth after spraying may not be controlled

Restrictions

- Maximum number of treatments 1 per crop
- Do not use on very light, very stony or gravelly soils or those with more than 10% organic matter
- Do not apply where crop root growth is likely to be restricted
- Seed should be covered by a minimum of 25 mm of settled soil after seedbed consolidation
- Do not apply later than when the growing point of the crop is within 13 mm of the soil surface
- Do not treat pea variety Vedette

Crop-specific information

- Latest use: pre-emergence of crop

Following crops guidance

- After a dry season land must be ploughed to 150 mm before drilling ryegrass
- Before drilling a winter crop plough or cultivate to 150 mm
- In the event of crop failure land must be ploughed or cultivated to 150 mm and then an interval of 8 wk must elapse before drilling any crop

Environmental safety

- Dangerous to fish or other aquatic life. Do not contaminate surface waters or ditches with chemical or used container

Hazard classification and safety precautions

Hazard R03
Risk phrases R03c
Operator protection A; U05a, U08, U13, U19, U20b
Environmental protection E01, E13b
Consumer protection C03
Storage and disposal E26, E30a, E31b, E34
Medical advice M03

131 cyazofamid

A cyanoimidazolesulfonamide protectant fungicide for potatoes

Products

1	Me2 Notin	Me2	400 g/l	SC	11655
2	Ranman Twinpack	Belchim	400 g/l	KL	11851

Uses

Blight in ***potatoes***.

Approval information

- Cyazofamid included in Annex I under EC Directive 91/414

Efficacy guidance

- Apply as a protectant treatment before blight enters the crop and repeat every 7-10 d depending on severity of disease pressure
- Commence spray programme immediately the risk of blight in the locality occurs, usually when the crop meets along the rows
- Product must always be used with organosilicone adjuvant provided in the twin pack
- To minimise the chance of development of resistance no more than three applications should be made consecutively (out of a permissible total of six) in the blight control programme. For more information on Resistance Management see Section 1

Restrictions

- Maximum number of treatments 6 per crop (no more than three of which should be consecutive)
- Consult processor before using on crops intended for processing

Crop-specific information

- HI 7 d

Environmental safety

- Dangerous to fish or other aquatic life. Do not contaminate surface waters or ditches with chemical or used container

Hazard classification and safety precautions

Hazard R04
Risk phrases R04d
Operator protection A, C; U05a, U11, U15, U20a
Environmental protection E01, E13b
Consumer protection C03
Storage and disposal E30a, E31c, E34
Medical advice M03

132 cycloxydim

A translocated post-emergence oxime herbicide for grass weed control

Products

1	Greencrop Pomeroy	Greencrop	200 g/l	EC	11769
2	Greencrop Valentia	Greencrop	200 g/l	EC	10197
3	Landgold Cycloxydim	Landgold	200 g/l	EC	06269
4	Laser	BASF	200 g/l	EC	05251
5	Marnoch Clodim	Me2	200 g/l	EC	11460
6	Standon Cycloxydim	Standon	200 g/l	EC	08830

SECTION 4

Uses

Annual grasses in ***amenity vegetation*** *(off-label)*, ***forestry*** *(off-label)* [4]; ***brussels sprouts, cabbages, cauliflowers, fodder beet, mangels, spring field beans, sugar beet, swedes, winter field beans, winter oilseed rape*** [1-6]; ***bulb onions, carrots, dwarf beans, farm forestry, forestry, leeks, linseed, parsnips, salad onions, spring oilseed rape, strawberries*** [1, 2, 4-6]; ***calabrese, peas*** [2]; ***combining peas, vining peas*** [3-5]; ***early potatoes*** [1, 3, 4]; ***flower bulbs*** [2, 4-6]; ***forest nurseries, ornamental plant production*** [1]; ***maincrop potatoes*** [2-5]. Annual meadow grass in ***combining peas, maincrop potatoes, vining peas*** [1, 6]. Black bent in ***brussels sprouts, cabbages, cauliflowers, fodder beet, maincrop potatoes, mangels, spring field beans, sugar beet, swedes, winter field beans, winter oilseed rape*** [1-6]; ***bulb onions, carrots, dwarf beans, flower bulbs, leeks, linseed, parsnips, salad onions, spring oilseed rape, strawberries*** [1, 2, 4-6]; ***calabrese, peas*** [2]; ***combining peas, vining peas*** [1, 3-6]; ***early potatoes*** [1, 3, 4]; ***ornamental plant production*** [1]. Blackgrass in ***brussels sprouts, cabbages, cauliflowers, fodder beet, maincrop potatoes, mangels, spring field beans, sugar beet, swedes, winter field beans, winter oilseed rape*** [1-6]; ***bulb onions, carrots, dwarf beans, flower bulbs, leeks, linseed, parsnips, salad onions, spring oilseed rape, strawberries*** [1, 2, 4-6]; ***calabrese, peas*** [2]; ***combining peas, vining peas*** [1, 3 6]; ***early potatoes*** [1, 3, 4]; ***ornamental plant production*** [1]. Couch in ***brussels sprouts, cabbages, cauliflowers, fodder beet, maincrop potatoes, mangels, spring field beans, sugar beet, swedes, winter field beans, winter oilseed rape*** [1-6]; ***bulb onions, carrots, dwarf beans, flower bulbs, leeks, linseed, parsnips, salad onions, spring oilseed rape, strawberries*** [1, 2, 4-6]; ***calabrese, peas*** [2]; ***combining peas, vining peas*** [1, 3-6]; ***early potatoes*** [1, 3, 4]; ***ornamental plant production*** [1]. Creeping bent in ***brussels sprouts, cabbages, cauliflowers, fodder beet, maincrop potatoes, mangels, spring field beans, sugar beet, swedes, winter field beans, winter oilseed rape*** [1-6]; ***bulb onions, carrots, dwarf beans, flower bulbs, leeks, linseed, parsnips, salad onions, spring oilseed rape, strawberries*** [1, 2, 4-6]; ***calabrese, peas*** [2]; ***combining peas, vining peas*** [1, 3-6]; ***early potatoes*** [1, 3, 4]; ***ornamental plant production*** [1]. Green cover in ***land temporarily removed from production*** [1, 2, 4, 5]. Onion couch in ***brussels sprouts, bulb onions, cabbages, carrots, cauliflowers, dwarf beans, flower bulbs, fodder beet, leeks, linseed, maincrop potatoes, mangels, parsnips, salad onions, spring field beans, spring oilseed rape, strawberries, sugar beet, swedes, winter field beans, winter***

oilseed rape [1, 2, 4-6]; ***calabrese, peas*** [2]; ***combining peas, vining peas*** [1, 4-6]; ***early potatoes*** [1, 4]; ***ornamental plant production*** [1]. Perennial grasses in ***amenity vegetation*** *(off-label)*, ***forestry*** *(off-label)* [4]; ***farm forestry, forestry*** [1, 2, 4-6]; ***forest nurseries*** [1]. Volunteer cereals in ***brussels sprouts, cabbages, cauliflowers, fodder beet, maincrop potatoes, mangels, spring field beans, sugar beet, swedes, winter field beans, winter oilseed rape*** [1-6]; ***bulb onions, carrots, dwarf beans, flower bulbs, leeks, linseed, parsnips, salad onions, spring oilseed rape, strawberries*** [1, 2, 4-6]; ***calabrese, peas*** [2]; ***combining peas, vining peas*** [1, 3-6]; ***early potatoes*** [1, 3, 4]; ***ornamental plant production*** [1]. Wild oats in ***brussels sprouts, cabbages, cauliflowers, fodder beet, maincrop potatoes, mangels, spring field beans, sugar beet, swedes, winter field beans, winter oilseed rape*** [1-6]; ***bulb onions, carrots, dwarf beans, flower bulbs, leeks, linseed, parsnips, salad onions, spring oilseed rape, strawberries*** [1, 2, 4-6]; ***calabrese, peas*** [2]; ***combining peas, vining peas*** [1, 3-6]; ***early potatoes*** [1, 3, 4]; ***ornamental plant production*** [1].

Specific Off-Label Approvals (SOLAs)

- ***amenity vegetation, forestry*** *(OLA 962585) Dec 2008* [4]

Efficacy guidance

- Best results achieved when weeds small and have not begun to compete with crop. Effectiveness reduced by drought, cool conditions or stress. Weeds emerging after application are not controlled
- Foliage death usually complete after 3-4 wk but longer under cool conditions, especially late treatments to winter oilseed rape
- Perennial grasses should have sufficient foliage to absorb spray and should not be cultivated for at least 14 d after treatment
- On established couch pre-planting cultivation recommended to fragment rhizomes and encourage uniform emergence
- Split applications to volunteer wheat and barley at GS 12-14 will often give adequate control in winter oilseed rape. See label for details
- Apply to dry foliage when rain not expected for at least 2 h

Restrictions

- Maximum number of treatments 1 per crop for spring oilseed rape, early potatoes; 2 per crop (the second at reduced dose) for other crops. See label for details
- Must be used with Actipron (see label)
- Do not apply to crops damaged or stressed by adverse weather, pest or disease attack or other pesticide treatment
- Prevent drift onto other crops, especially cereals and grass

Crop-specific information

- HI cabbage, cauliflower, calabrese, salad onions 4 wk; peas, dwarf beans 5 wk; bulb onions, carrots, parsnips, strawberries, early potatoes 6 wk; sugar and fodder beet, leeks, mangels, maincrop potatoes, field beans, swedes, Brussels sprouts, winter field beans 8 wk; oilseed rape, soya beans, linseed 12 wk
- Recommended time of application varies with crop. See label for details
- On peas a crystal violet wax test should be done if leaf wax likely to have been affected by weather conditions or other chemical treatment. The wax test is essential if other products are to be sprayed before or after treatment
- May be used on ornamental bulbs when crop 5-10 cm tall. Product has been used on tulips, narcissi, hyacinths and irises but some subjects may be more sensitive and growers advised to check tolerance on small number of plants before treating the rest of the crop
- May be applied to land temporarily removed from production where the green cover is made up predominantly of tolerant crops listed on label. Use on industrial crops of linseed and oilseed rape on land temporarily removed from production also permitted.

Following crops guidance

- Guideline intervals for sowing succeeding crops after failed treated crop: field beans, peas, sugar beet, rape, kale, swedes, radish, white clover, lucerne 1 wk; dwarf French beans 4 wk; wheat, barley, maize 8 wk
- Oats should not be sown after failure of a treated crop

Environmental safety

- Dangerous for the environment [4]
- Toxic to aquatic organisms. May cause long-term adverse effects to the aquatic environment [4]
- Harmful to fish or other aquatic life. Do not contaminate surface waters or ditches with chemical or used container [1, 2, 5, 6]]

Hazard classification and safety precautions

Hazard R03, R14 [4]; R04 [1-6]
Risk phrases R04a [1-3, 5, 6]; R04b [1-6]; R04g, R14b [4]
Operator protection A, C; U05a, U08 [1-6]; U20a [3]; U20b [1, 2, 4-6]
Environmental protection E01 [1-6]; E13c [1, 2, 5, 6]; E15 [3, 4]; E38 [4]
Consumer protection C03
Storage and disposal E26 [1, 2]; E30a [1-6]; E31b [1-3]; E31c [4-6]; E32d [4]
Medical advice M05, M05a [4]

133 cymoxanil

A urea fungicide for potatoes

Products

Sipcam C 50	Sipcam	50% w/w	WP	10610

Uses

Blight in ***potatoes***.

Efficacy guidance

- Product to be used in mixture with recommended partner containing mancozeb or fluazinam in order to combine systemic and protective activity
- Commence spray programme before infection appears as soon as weather conditions favourable for disease development occur. At latest the first treatment should be made as the foliage meets along the rows
- Repeat treatments at 7-10 day intervals according to disease incidence and weather conditions

Restrictions

- Maximum number of treatments 4 per crop

Crop-specific information

- HI 7 d

Environmental safety

- Harmful to fish or other aquatic life. Do not contaminate surface waters or ditches with chemical or used container

Hazard classification and safety precautions

Hazard R03
Risk phrases R03c
Operator protection A; U05a, U08, U20b
Environmental protection E01, E13c
Consumer protection C03
Storage and disposal E30a, E32a
Medical advice M03

134 cymoxanil + famoxadone

A preventative and curative fungicide mixture for potatoes

Products

Tanos	DuPont	25:25% w/w	WG	10677

Uses

Blight in ***potatoes***.

Approval information

- Famoxadone included in Annex I under EC Directive 91/414

Efficacy guidance

- Commence spray programme before infection appears as soon as weather conditions favourable for disease development occur. At latest the first treatment should be made as the foliage meets along the rows
- Repeat treatments at 7-10 day intervals according to disease incidence and weather conditions
- Reduce spray interval if conditions are conducive to the spread of blight
- Spray as soon as possible after irrigation
- Increase water volume in dense crops
- Product combines active substances with different modes of action and is effective against strains of potato blight that are insensitive to phenylamide fungicides
- To minimise the likelihood of development of resistance to strobilurin fungicides these products should be used in a planned Resistance Management strategy. See Section 1 for more information

Restrictions

- Maximum number of treatments 12 per crop
- Consult processors before using on crops for processing

Crop-specific information

- HI 14 d for potatoes

Environmental safety

- Dangerous to fish or other aquatic life. Do not contaminate surface waters or ditches with chemical or used container
- LERAP Category B

Hazard classification and safety precautions

Operator protection A, D; U05a, U20b
Environmental protection E01, E13b, E16a, E16b
Consumer protection C03
Storage and disposal E30a, E31b, E34
Medical advice M03

135 cymoxanil + fludioxonil + metalaxyl-M

A fungicide seed dressing for peas

Products

Wakil XL	Syngenta	10:5:17.5% w/w	WS	10562

Uses

Ascochyta in ***combining peas***, ***vining peas***. Damping off in ***combining peas***, ***vining peas***. Downy mildew in ***broad beans*** *(off-label - seed treatment)*, ***combining peas***, ***spring field beans*** *(off-label - seed treatment)*, ***vining peas***, ***winter field beans*** *(off-label - seed treatment)*. Pythium in ***carrots*** *(off-label)*, ***parsnips*** *(off-label)*.

Specific Off-Label Approvals (SOLAs)

- ***broad beans***, ***spring field beans***, ***winter field beans*** *(seed treatment) (OLA 023932) Dec 2008* [1]
- ***carrots***, ***parsnips*** *(OLA 021191) Dec 2008* [1]

Approval information

- Metalaxyl-M included in Annex I under EC Directive 91/414

Efficacy guidance

- Apply through continuous flow seed treaters which should be calibrated before use

Restrictions

- Max no of treatments 1 per seed batch
- Ensure moisture content of treated seed satisfactory and store in a dry place
- Check calibration of seed drill with treated seed before drilling and sow as soon as possible after treatment
- Consult before using on crops for processing

Crop-specific information

- Latest use: pre-drilling

Environmental safety

- Harmful to fish or other aquatic life. Do not contaminate surface waters or ditches with chemical or used container
- Do not use treated seed as food or feed

Hazard classification and safety precautions

Operator protection A, D, H; U02, U20b
Environmental protection E03, E13c
Storage and disposal E26, E30a, E32a
Treated seed S02, S04b, S05, S07

136 cymoxanil + mancozeb

A protectant and systemic fungicide for potato blight control

Products

1	Agriguard Cymoxanil Plus	AgriGuard	4.5:68% w/w	WP	10893
2	Besiege WSB	DuPont	4.5:68% w/w	WB	08075
3	Curzate M68	DuPont	4.5:68% w/w	WP	08072
4	Curzate M68 WSB	DuPont	4.5:68% w/w	WB	08073
5	Globe	Sipcam	6:70 % w/w	WP	10339
6	Me2 Cymoxeb	Me2	4.5:68% w/w	WP	09486
7	Rhythm	Interfarm	4.5:68% w/w	WP	09636
8	Rhythm	Interfarm	4.5:68% w/w	WP	11669
9	Standon Cymoxanil Extra	Standon	4.5:68% w/w	WP	09442
10	Systol M	DuPont	4.5:68% w/w	WP	08085

Uses

Blight in ***potatoes***.

Approval information

- Approved for aerial application on potatoes [9]. See notes in Section 1
- Approval expiry 31 May 2004 [2]

Efficacy guidance

- Apply immediately after blight warning or as soon as local conditions dictate and repeat at 7-14 d intervals until haulm dies down or is burnt off
- Spray interval should not be more than 7-10 d in irrigated crops (see product label). Apply treatment after irrigation
- To minimise the likelihood of development of resistance these products should be used in a planned Resistance Management strategy. See Section 1 for more information

Restrictions

- Maximum number of treatments 6 per crop
- Do not allow packs to become wet during storage [2-4, 6-10]

Crop-specific information

- HI 7 d [5]
- Destroy and remove any haulm that remains after harvest of early varieties to reduce blight pressure on neighbouring maincrop potatoes

Environmental safety

- Harmful (Dangerous [5]) to fish or other aquatic life. Do not contaminate surface waters or ditches with chemical or used container [2-4, 6-10]
- Keep product away from fire or sparks [2]

Hazard classification and safety precautions

Hazard R04
Risk phrases R04a [1-4, 6-10]; R04e [1-10]
Operator protection A [1-10]; C [1, 3, 5-10]; H [5]; U02 [5]; U05a [1-10]; U08 [1-6, 9, 10]; U19 [1-4, 6-10]; U20a [1]; U20b [2-10]; U22a [2, 4]
Environmental protection E01 [1-10]; E13b, E15 [5]; E13c [1-4, 6-10]
Consumer protection C03
Storage and disposal E26 [7, 8]; E30a [1-10]; E31a [6]; E32a [1-5, 7-10]

SECTION 4

137 cypermethrin

A contact and stomach acting pyrethroid insecticide

Products

1	Agriguard Cypermethrin	AgriGuard	100 g/l	EC	10093
2	Cyperguard 100 EC	Nufarm UK	100 g/l	EC	10779
3	Cyperkill 5	Mitchell Cotts	50 g/l	EC	00625
4	Fernpath Banjo	AgriGuard	100 g/l	EC	10182
5	I T Cyper	I T Agro	100 g/l	EC	10557
6	Jundi 100 EC	DAPT	100 g/l	EC	10848
7	Permasect C	Nufarm UK	100 g/l	EC	11121
8	Toppel 10	United Phosphorus	100 g/l	EC	08772

Uses

Aphids in ***apples***, ***vining peas*** [1-8]; ***broccoli***, ***brussels sprouts***, ***cabbages***, ***calabrese***, ***cauliflowers***, ***durum wheat***, ***hops***, ***lettuce***, ***pears***, ***triticale***, ***winter oats***, ***winter rye*** [1, 2, 4-6, 8]; ***cherries***, ***ornamental plant production***, ***plums*** [1, 4-6, 8]; ***chinese cabbage*** *(off-label)*, ***leaf spinach*** *(off-label)*, ***protected chinese cabbage*** *(off-label)*, ***protected spinach*** *(off-label)* [8]; ***lettuce*** *(outdoor crops)* [3]; ***ornamental specimens*** [2]; ***winter barley***, ***winter wheat*** [1-6, 8]. Asparagus beetle in ***asparagus*** *(off-label)* [8]. Barley yellow dwarf vectors in ***triticale***, ***winter oats***, ***winter rye*** [7]. Barley yellow dwarf virus vectors in ***durum wheat***, ***triticale***, ***winter oats***, ***winter rye*** [1, 4-6, 8]; ***spring barley*** *(autumn sown)*, ***spring wheat*** *(autumn sown)* [7]; ***winter barley***, ***winter wheat*** [1-8]. Bladder pod midge in ***spring oilseed rape***, ***winter oilseed rape*** [1, 2, 4-6, 8]. Cabbage stem flea beetle in ***winter oilseed rape*** [1-8]. Capsids in ***apples*** [1, 2, 4-6, 8]; ***ornamental plant production*** [1, 4-6, 8]; ***ornamental specimens*** [2]; ***pears*** [1-8]. Caterpillars in ***apples*** [1-6, 8]; ***broccoli***, ***brussels sprouts***, ***cabbages***, ***calabrese***, ***cauliflowers*** [1-8]; ***cherries***, ***lettuce***, ***pears***, ***plums*** [1, 2, 4-6, 8]; ***chinese cabbage*** *(off-label)*, ***leaf spinach*** *(off-label)*, ***protected chinese cabbage*** *(off-label)*, ***protected spinach*** *(off-label)* [8]; ***fodder beet***, ***kale***, ***mangels***, ***potatoes***, ***red beet***, ***sugar beet*** [3, 7]; ***lettuce*** *(outdoor crops)* [3]; ***ornamental plant production*** [1, 4-6, 8]; ***ornamental specimens*** [2]. Codling moth in ***apples*** [1-8]. Cutworms in ***carrots*** *(off-label)*, ***parsnips*** *(off-label)* [8]; ***fodder beet***, ***mangels***, ***red beet*** [3, 7]; ***lettuce*** [1, 2, 4-6, 8]; ***lettuce*** *(outdoor crops)* [3]; ***ornamental plant production*** [1, 4-6, 8]; ***ornamental specimens*** [2]; ***potatoes***, ***sugar beet*** [1-8]. Frit fly in ***grass re-seeds*** [1-8]. Leaf miner in ***chicory*** *(off-label)*, ***endives*** *(off-label)*, ***lettuce*** *(off-label)*, ***ornamental specimens*** *(off-label)*, ***protected courgettes*** *(off-label)*, ***protected cucumbers*** *(off-label)*, ***protected gherkins*** *(off-label)* [3]. Pea and bean weevil in ***spring field beans***, ***vining peas***, ***winter field beans*** [1-8]. Pea moth in ***vining peas*** [1-8]. Pod midge in ***spring oilseed rape***, ***winter oilseed rape*** [3, 7]. Pollen beetle in ***spring oilseed rape***, ***winter oilseed rape*** [1-8]. Rape winter stem weevil in ***winter oilseed rape*** [1, 2, 4-6, 8]. Sawflies in ***apples*** [1, 2, 4-6, 8]. Seed weevil in ***spring oilseed rape***, ***winter oilseed rape*** [1-8]. Suckers in ***apples*** [7]; ***pears*** [1, 2, 4-6, 8]. Thrips in ***ornamental plant production*** [1, 4-6, 8]; ***ornamental specimens*** [2]. Tortrix moths in ***apples*** [1-8]; ***pears*** [3, 7]. Whitefly in ***broccoli***, ***brussels sprouts***, ***cabbages***, ***calabrese***, ***cauliflowers***, ***protected cucumbers***, ***protected lettuce***, ***protected ornamentals*** [1, 2, 4-6, 8]; ***ornamental plant production*** [1, 4-6, 8]; ***ornamental specimens***, ***protected celery*** [2]. Winter moth in ***apples***, ***pears*** [7]; ***cherries***, ***plums*** [2]. Yellow cereal fly in ***durum wheat*** [1, 2, 4-6, 8]; ***spring barley***, ***spring wheat*** [3]; ***spring barley*** *(autumn sown)*, ***spring wheat*** *(autumn sown)* [7]; ***triticale***, ***winter oats***, ***winter rye*** [1, 2, 4-8]; ***winter barley***, ***winter wheat*** [1-8].

Specific Off-Label Approvals (SOLAs)

- ***asparagus*** *(OLA 023085) Dec 2008* [8]
- ***asparagus*** *(OLA 030445) Dec 2008* [8]
- ***carrots***, ***parsnips*** *(OLA 982184) Dec 2008* [8]
- ***chicory***, ***endives***, ***lettuce***, ***ornamental specimens***, ***protected courgettes***, ***protected cucumbers***, ***protected gherkins*** *(OLA 992260) Dec 2008* [3]
- ***chinese cabbage***, ***leaf spinach***, ***protected chinese cabbage***, ***protected spinach*** *(OLA 983133) Dec 2008* [8]

Approval information

- Following implementation of Directive 98/82/EC, approval for use of cypermethrin on numerous crops was revoked in 1999
- Approval expiry 30 Jun 2004 [1, 4]

Efficacy guidance

- Products combine rapid action, good persistence, and high activity on Lepidoptera.
- As effect is mainly via contact good coverage is essential for effective action. Spray volume should be increased on dense crops
- A repeat spray after 10-14 d is needed for some pests of outdoor crops, several sprays at shorter intervals for whitefly and other glasshouse pests
- Rates and timing of sprays vary with crop and pest. See label for details
- Add a non-ionic wetter to improve results on leaf brassicas. In Brussels sprouts use of a drop-leg sprayer may be beneficial
- Where aphids in hops, pear suckers or glasshouse whitefly resistant to cypermethrin occur control is unlikely to be satisfactory

Restrictions

- Maximum number of treatments varies with crop and product. See label or approval notice for details

Crop-specific information

- Latest use: normally before end Mar in yr of harvest, but varies with product and crop. See labels for details
- HI asparagus 6 mth; apples, pears 14 d; vining peas 7 d; chicory, lettuce, courgettes, cucumbers, gherkins, endives 1 d; other crops 0 d
- Test spray sample of new or unusual ornamentals before committing whole batches

Environmental safety

- Flammable
- Dangerous to bees. Do not apply to crops in flower or to those in which bees are actively foraging except as directed on peas and oilseed rape. Do not apply when flowering weeds are present [5, 8]
- Extremely dangerous to fish or other aquatic life. Do not contaminate surface waters or ditches with chemical or used container
- LERAP Category A
- Broadcast air-assisted LERAP (18 m) [6, 7]
- Do not spray cereals after 31 Mar within 6 m of the edge of the growing crop

Hazard classification and safety precautions

Hazard R03 [1, 3, 4, 6, 7]; R04, R07d [1-8]
Risk phrases R03c, R04c, R04e [3, 7]; R04a, R04b [1-8]; R04g [2, 3, 6, 7]
Operator protection A, C; U02, U04a, U11 [3, 7]; U05a, U10, U19, U20b [1-8]
Environmental protection E01, E13a, E16c, E16d [1-8]; E12c [5, 8]; E12d [1, 4] (vining peas, oilseed rape); E17a [6, 7] (18 m)
Consumer protection C03
Storage and disposal E26, E30a, E31b [1-8]; E34 [3, 7]
Medical advice M03 [3, 7]; M05 [1, 2, 4, 6]

138 cyproconazole

A contact and systemic conazole fungicide for cereals and other field crops

See also *azoxystrobin + cyproconazole*
chlorothalonil + cyproconazole

Products

1	Caddy 240 EC	Bayer	240 g/l	EC	10631
2	Caddy 240 EC	Bayer CropScience	240 g/l	EC	11256
3	Fort	Bayer CropScience	240 g/l	EC	11606

Uses

Brown rust in ***red beet*** *(off-label)* [1, 2]; ***spring barley***, ***spring rye***, ***winter barley***, ***winter rye***, ***winter wheat*** [1-3]. Chocolate spot in ***spring field beans*** *(with chlorothalonil)*, ***winter field beans*** *(with chlorothalonil)* [1-3]. Crown rust in ***spring oats***, ***winter oats*** [1-3]. Eyespot in ***winter barley***, ***winter wheat*** [1-3]. Late ear diseases in ***winter wheat*** [1-3]. Light leaf spot in ***winter oilseed rape*** [1-3]. Net blotch in ***spring barley***, ***winter barley*** [1-3]. Phoma leaf spot in ***winter oilseed rape*** [1-3]. Powdery mildew in ***spring barley***, ***spring oats***, ***spring rye***, ***sugar beet***, ***winter***

barley, ***winter oats***, ***winter rye***, ***winter wheat*** [1-3]. Ramularia leaf spots in ***sugar beet*** [1-3]. Rhynchosporium in ***spring barley***, ***winter barley*** [1-3]. Rust in ***leeks***, ***spring field beans***, ***sugar beet***, ***winter field beans*** [1-3]. Septoria diseases in ***winter wheat*** [1-3]. Yellow rust in ***spring barley***, ***winter barley***, ***winter wheat*** [1-3].

Specific Off-Label Approvals (SOLAs)

- ***red beet*** *(OLA 020224) Dec 2008* [1]
- ***red beet*** *(OLA 031883) Dec 2008* [2]

Efficacy guidance

- Apply at start of disease development or as preventive treatment and repeat as necessary
- Most effective time of treatment varies with disease and use of tank mixes may be desirable. See label for details
- Product alone gives useful reduction of cereal eyespot. Where high infections probable a tank mix with prochloraz recommended
- On oilseed rape a two spray autumn/spring programme recommended for high risk situations and on susceptible varieties

Restrictions

- Maximum total dose equivalent to 1-3 full dose treatments depending on crop treated. See labels for details

Crop-specific information

- HI 6 wk for field beans; 14 d for sugar beet, red beet, leeks
- Latest use: completion of flowering for rye, wheat; emergence of ear complete (GS 59) for barley, oats; before lowest pods more than 2 cm long (GS 5,1) for winter oilseed rape
- Application to winter wheat in spring between the start of stem elongation and the third node detectable stage (GS 30-33) may cause straw shortening, but does not cause loss of yield

Environmental safety

- Dangerous to fish or other aquatic life. Do not contaminate surface waters or ditches with chemical or used container

Hazard classification and safety precautions

Hazard R04
Risk phrases R04a, R04b
Operator protection A, C, H; U05a, U08, U20b
Environmental protection E01, E13b
Consumer protection C03
Storage and disposal E26, E30a, E31b

139 cyproconazole + cyprodinil

A broad spectrum fungicide mixture for cereals

Products

Radius	Syngenta	5.33:40 % w/w	WG	09387

Uses

Brown rust in ***winter wheat***. Eyespot in ***spring barley***, ***winter barley***, ***winter wheat***. Net blotch in ***spring barley***, ***winter barley***. Powdery mildew in ***spring barley***, ***winter barley***, ***winter wheat***. Rhynchosporium in ***spring barley***, ***winter barley***. Septoria diseases in ***winter wheat***. Yellow rust in ***winter wheat***.

Efficacy guidance

- Best results obtained from treatment at early stages of development. Later treatments may be required if disease pressure remains high
- Eyespot is fully controlled by treatment in spring during stem extension. Control may be reduced when very dry conditions follow application

Restrictions

- Maximum total dose equivalent to two full dose treatments
- Cyprodinil may produce an allergic reaction

Crop-specific information

- Latest use: before first spikelet of inflorescence visible (GS 51) for barley; before caryopsis watery ripe (GS 71) for winter wheat

Environmental safety

- Dangerous to fish or other aquatic life. Do not contaminate surface waters or ditches with chemical or used container
- LERAP Category B

Hazard classification and safety precautions

Hazard R04
Risk phrases R04b, R04e
Operator protection A, H; U05a, U20a
Environmental protection E01, E13b, E16a, E16b
Consumer protection C03
Storage and disposal E26, E29, E30a

140 cyproconazole + prochloraz

A broad spectrum protective and curative fungicide for cereals

Products

1	Profile	Aventis	48:320 g/l	EC	08134
2	Sportak Delta 460 HF	Aventis	48:320 g/l	EC	07431

Uses

Brown rust in ***spring barley***, ***spring wheat***, ***winter barley***, ***winter wheat***. Eyespot in ***spring barley***, ***spring wheat***, ***winter barley***, ***winter wheat***. Glume blotch in ***spring wheat***, ***winter wheat***. Net blotch in ***spring barley***, ***winter barley***. Powdery mildew in ***spring barley***, ***spring wheat***, ***winter barley***, ***winter wheat***. Rhynchosporium in ***spring barley***, ***winter barley***. Septoria leaf spot in ***spring wheat***, ***winter wheat***. Yellow rust in ***spring barley***, ***spring wheat***, ***winter barley***, ***winter wheat***.

Efficacy guidance

- To control foliar diseases apply before infection starts spreading to younger leaves and repeat if necessary
- For established mildew use in tank mixture with an approved morpholine fungicide
- For eyespot control apply in spring from when leaf sheaths start to erect up to and including 3rd node detectable stage (GS 30-33). Effective against eyespot and *Septoria tritici* resistant to MBC fungicides
- Eyespot already in crop on up to 10% tillers also controlled
- When applied to control eyespot may give some control of sharp eyespot and Fusarium if present
- See label for details of timing for foliar disease control. Where infection from Septoria may occur following a rain-splash event treat as soon as possible afterwards to protect second leaf and flag leaf
- Good spray cover of stem bases and leaves is essential
- A period of at least 3 h without rain should follow spraying for full effectiveness

Restrictions

- Maximum number of treatments 2 per crop

Crop-specific information

- Latest use: up to and including emergence of ear complete (GS 59) on barley; before beginning of anthesis (GS 60) on wheat
- Application to winter wheat in spring at GS 30-33 may cause straw shortening but does not cause loss of yield

Environmental safety

- Harmful to fish or other aquatic life. Do not contaminate surface waters or ditches with chemical or used container

Hazard classification and safety precautions

Hazard R04
Risk phrases R04b
Operator protection A; U05a, U09a, U20b
Environmental protection E01, E13c

Consumer protection C03
Storage and disposal E26 [2]; E30a, E31b [1, 2]

141 cyproconazole + propiconazole

A broad-spectrum mixture of conazole fungicides for wheat and barley

Products

Menara	Syngenta	160:250 g/l	EC	09321

Uses

Brown rust in ***spring barley***, ***winter barley***, ***winter wheat***. Late ear diseases in ***winter wheat***. Net blotch in ***spring barley***, ***winter barley***. Powdery mildew in ***spring barley***, ***winter barley***. Rhynchosporium in ***spring barley***, ***winter barley***. Septoria diseases in ***winter wheat***. Yellow rust in ***winter wheat***.

Approval information

- Propiconazole included in Annex I under EC Directive 91/414

Efficacy guidance

- Treatment should be made at first sign of disease
- Useful reduction of eyespot when applied in spring during stem extension (GS 30-32)

Restrictions

- Maximum total dose equivalent to one full dose treatment
- Propiconazole may produce an allergic reaction

Crop-specific information

- Latest use: before beginning of anthesis (GS 60) for barley; before grain watery ripe (GS 71) for wheat
- Application to winter wheat in spring may cause straw shortening but does not cause loss of yield

Environmental safety

- Dangerous for the environment
- Very toxic to aquatic organisms. May cause long-term adverse effects in the aquatic environment

Hazard classification and safety precautions

Hazard R03, R04, R14
Risk phrases R04a, R04b, R10a, R12a, R14c
Operator protection A, C, H; U05a, U08, U20b
Environmental protection E01, E15, E38
Consumer protection C03
Storage and disposal E26, E30a, E31c, E32d

142 cyproconazole + trifloxystrobin

A conazole and strobilurin fungicide mixture for wheat and barley

Products

Sphere	Bayer CropScience	80:187.5 g/l	EC	11429

Uses

Brown rust in ***spring barley***, ***winter barley***, ***winter wheat***. Net blotch in ***spring barley***, ***winter barley***. Powdery mildew in ***spring barley***, ***winter barley***, ***winter wheat***. Rhynchosporium in ***spring barley***, ***winter barley***. Septoria diseases in ***winter wheat***. Yellow rust in ***spring barley***, ***winter barley***, ***winter wheat***.

Approval information

- Trifloxystrobin included in Annex I under EC Directive 91/414

Efficacy guidance

- Best results obtained from treatment at early stages of disease development. Further treatment may be needed if disease attack is prolonged
- Kresoxim-methyl is a member of the QoI cross resistance group. Product should be used preventatively and not relied on for its curative potential

- Use product as part of an Integrated Crop Management strategy incorporating other methods of control, including where appropriate other fungicides with a different mode of action. Do not apply more than two foliar applications of QoI containing products to any cereal crop
- There is a significant risk of widespread resistance occurring in *Septoria tritici* populations in UK. Failure to follow resistance management action may result in reduced levels of disease control
- Strains of barley powdery mildew resistant to QoI's are common in the UK

Restrictions

- Maximum total dose per crop equivalent to two full dose treatments

Crop-specific information

- HI 35 d

Environmental safety

- Dangerous to fish or other aquatic life. Do not contaminate surface waters or ditches with chemical or used container
- LERAP Category B

Hazard classification and safety precautions

Hazard R04
Risk phrases R04a
Operator protection A, C, H; U02, U05a, U09a, U19, U20b
Environmental protection E01, E13b, E16a, E16b
Consumer protection C02 (35 d); C03
Storage and disposal E26, E29, E30a, E31b

143 cyprodinil

An anilinopyrimidine systemic broad spectrum fungicide for cereals

See also cyproconazole + cyprodinil

Products

1	Barclay Amtrak	Barclay	75% w/w	WG	11338
2	Standon Cyprodinil	Standon	75% w/w	WG	09345
3	Unix	Syngenta	75% w/w	WG	08764

Uses

Eyespot in ***winter barley***, ***winter wheat***. Net blotch in ***spring barley***, ***winter barley***. Powdery mildew in ***spring barley***, ***spring wheat***, ***winter barley***, ***winter wheat***. Rhynchosporium in ***spring barley***, ***winter barley***

Efficacy guidance

- Best results obtained from treatment at early stages of disease development
- For best control of eyespot spray before or during the period of stem extension in spring. Control may be reduced if very dry conditions follow treatment

Restrictions

- Maximum number of treatments equal to one and two thirds full dose on wheat and twice full dose on barley
- Cyprodinil may produce an allergic reaction

Crop-specific information

- Latest use: up to and including first awns visible (GS 49) for spring barley, winter barley; up to and including grain watery ripe (GS 71) for spring wheat, winter wheat

Environmental safety

- Dangerous for the environment [3]
- Very toxic to aquatic organisms. May cause long-term adverse effects in the aquatic environment [3]
- Dangerous to fish or other aquatic life. Do not contaminate surface waters or ditches with chemical or used container [1, 2]
- LERAP Category B

Hazard classification and safety precautions

Hazard R14 [3]
Risk phrases R14c [3]

Operator protection A, H; U05a [1-3]; U14 [2]; U20a [2, 3]; U20b [1]
Environmental protection E01, E16a, E16b [1-3]; E13b [1, 2]
Consumer protection C03
Storage and disposal E26, E32d [3]; E30a [1-3]; E32a [1, 2]

144 cyromazine

A triazine insect larvicide

Products

Neporex 2SG	Novartis A H	2% w/w	SG	H7222

Uses

Flies in ***livestock houses***, ***manure heaps***.

Efficacy guidance

- Commence treatment when houseflies begin to breed
- Ensure that all breeding sites are treated
- Repeat after 2-3 wk for heavy populations
- Product affects housefly larvae moulting and has no effect on adults. Use an adulticide if large numbers of adult flies present

Restrictions

- Do not use continuously in intensive or controlled environment animal units. If treatment in subsequent stocking cycles is required use different a.i. and different control method
- Do not feed treated poultry manure to stock
- Spread treated manure only onto grassland or arable land prior to cultivation
- Allow at least 4 wk between spreading and grazing or cropping
- Do not use on manure intended for production of mushroom compost

Crop-specific information

- Latest use: 21 d before spreading manure from treated livestock houses on grass or arable land
- Apply to surface of manure
- Granules can be scattered dry or applied in water

Environmental safety

- Harmful to fish. Do not contaminate surface waters or ditches with chemical or used container

Hazard classification and safety precautions

Operator protection A, B, C, H, M; U05a, U20a
Environmental protection E01, E13c
Consumer protection C03
Storage and disposal E30a, E31a

145 2,4-D

A translocated phenoxy herbicide for cereals, grass and amenity use

See also *amitrole + 2,4-D + diuron*
clopyralid + 2,4-D + MCPA

Products

1	Agricorn D II	FCC	500 g/l	SL	09415
2	Depitox	Nufarm UK	490 g/l	SL	11149
3	Dicotox Extra	Bayer Environ.	400 g/l	EC	09930
4	Dioweed 50	United Phosphorus	500 g/l	SL	08050
5	Dormone	Bayer Environ.	465 g/l	SL	09932
6	Easel	Nufarm UK	500 g/l	EC	11146
7	Headland Staff	Headland	470 g/l	SL	07189
8	Herboxone	Headland	500 g/l	SL	10032
9	HY-D	Agrichem	470 g/l	SL	06278
10	HY-D Super	Agrichem	500 g/l	SL	11618
11	Luxan 2,4-D	Luxan	490 g/l	SL	09379
12	Ragox	Nufarm UK	490 g/l	SL	11145
13	Syford	Vitax	500 g/l	SL	02062

Uses

Annual dicotyledons in ***amenity grass*** [2, 12]; ***amenity turf*** [1, 3-6, 13]; ***apple orchards, pear orchards*** [2, 11]; ***conifer plantations, forestry*** [3, 6]; ***established grassland*** [1, 2, 4, 6-11, 13]; ***grass seed crops*** [13]; ***managed amenity turf*** [1, 2, 4, 5, 12]; ***spring barley, spring wheat, winter barley, winter wheat*** [1, 2, 4, 6-11]; ***spring oats*** [9]; ***spring rye*** [1, 2, 4, 6, 11]; ***winter oats*** [2, 6-10]; ***winter rye*** [1, 2, 4, 7-11]. Aquatic weeds in ***aquatic situations*** [2]; ***water or waterside areas*** [5, 12]. Heather in ***conifer plantations, forestry*** [3, 6]. Perennial dicotyledons in ***amenity grass*** [2, 12]; ***amenity turf*** [1, 3-6, 13]; ***apple orchards, pear orchards*** [2, 11]; ***conifer plantations, forestry*** [3, 6]; ***established grassland*** [1, 2, 4, 6-11, 13]; ***grass seed crops*** [13]; ***managed amenity turf*** [1, 2, 4, 5, 12]; ***spring barley, spring wheat, winter barley, winter wheat*** [1, 2, 4, 6-11]; ***spring oats*** [9]; ***spring rye*** [1, 2, 4, 6, 11]; ***water or waterside areas*** [5, 12]; ***winter oats*** [2, 6-10]; ***winter rye*** [1, 2, 4, 7-11]. Willows in ***conifer plantations, forestry*** [3, 6]. Woody weeds in ***conifer plantations, forestry*** [3, 6].

Approval information

- 2,4-D included in Annex I under EC Directive 91/414
- May be applied through CDA equipment (see label for details) [13]. See notes in Section 1 on ULV application
- Approved for aquatic weed control [2, 5, 12]. See notes in Section 1 on use of herbicides in or near water

Efficacy guidance

- Best results achieved by spraying weeds in seedling to young plant stage when growing actively in a strongly competing crop
- Most effective stage for spraying perennials varies with species. See label for details
- Spray aquatic weeds when in active growth between May and Sep

Restrictions

- Maximum number of treatments normally 1 per crop and in forestry, 2 per yr in grassland and 2 or 3 per yr in amenity turf. Check individual labels
- Do not use on newly sown leys containing clover
- Do not spray grass seed crops after ear emergence
- Do not spray within 6 mth of laying turf or sowing fine grass
- Do not dump surplus herbicide in water or ditch bottoms
- Do not plant conifers until at least 1 mth after treatment
- Do not spray crops stressed by cold weather or drought or if frost expected
- Do not roll or harrow within 7 d before or after spraying
- Do not spray if rain falling or imminent
- Do not mow amenity grass 2 d before or 1 d after spraying. Do not mow grassland or graze for at least 10 d after spraying

Crop-specific information

- Latest use: before 1st node detectable in cereals; end Aug for conifer plantations
- Spray winter cereals in spring when leaf-sheath erect but before first node detectable (GS 31), spring cereals from 5-leaf stage to before first node detectable (GS 15-31)
- Cereals undersown with grass and/or clover, but not with lucerne, may be treated
- Selective treatment of resistant conifers can be made in Aug when growth ceased and plants hardened off, spray must be directed if applied earlier. See label for details

Following crops guidance

- Do not use shortly before or after sowing any crop
- Do not direct drill brassicas or grass/clover mixtures within 3 wk of application

Environmental safety

- 2,4-D is active at low concentrations. Take extreme care to avoid drift onto neighbouring crops, especially beet crops, brassicas, most market garden crops including lettuce and tomatoes under glass, pears and vines
- May be used to control aquatic weeds in presence of fish if used in strict accordance with directions for waterweed control and precautions needed for aquatic use [2, 5, 12]
- Dangerous to aquatic higher plants. Do not contaminate surface waters or ditches with chemical or used container [8]

- Water containing the herbicide must not be used for irrigation purposes within 3 wk of treatment or until the concentration in water is below 0.05 ppm
- Keep livestock out of treated areas for at least 2 wk and until foliage of any poisonous weeds such as ragwort has died and become unpalatable

Hazard classification and safety precautions

Hazard R03 [1-13]; R04 [2-4, 8, 10, 11]

Risk phrases R03a [1, 3-7, 9, 12, 13]; R03c [1-13]; R04a [3, 4]; R04b [2, 3, 11]; R04d [2, 8, 10, 11]; R04e [3]

Operator protection A, C, H, M [1-13]; D [1, 4, 9, 11]; U05a, U08 [1-13]; U20a [1, 4, 9]; U20b [2, 3, 5-8, 10-13]

Environmental protection E01 [1-13]; E07 [1-9] (2 wk); E07 [10]; E07 [11-13] (2 wk); E13b [2]; E13c [1, 3-13]; E14b [8]; E19 [5, 9]; E21 [5, 9] (3 wk)

Consumer protection C03

Storage and disposal E26 [2, 6-8, 10-12]; E30a, E34 [1-13]; E31a [1, 4, 7-10]; E31b [1-4, 6, 9, 11-13]

Medical advice M03

146 2,4-D + dicamba

A translocated herbicide for use on turf

Products

New Estermone	Vitax	200:35 g/l	EC	06336

Uses

Annual dicotyledons in ***amenity turf***, ***lawns***. Perennial dicotyledons in ***amenity turf***, ***lawns***.

Approval information

- 2,4-D included in Annex I under EC Directive 91/414

Efficacy guidance

- Best results achieved by application when weeds growing actively in spring or early summer (later with irrigation and feeding)
- More resistant weeds may need repeat treatment after 3 wk
- Do not use during drought conditions or mow for 3 d before or after treatment

Restrictions

- Maximum number of treatments 3 per yr
- Do not treat newly sown or turfed areas
- Avoid spray drift onto cultivated crops or ornamentals
- Do not re-seed for 6 wk after application

Environmental safety

- Harmful to fish or other aquatic life. Do not contaminate surface waters or ditches with chemical or used container
- Keep livestock out of treated areas for at least 2 wk and until foliage of any poisonous weeds such as ragwort has died and become unpalatable

Hazard classification and safety precautions

Operator protection A, C, H, M; U08, U20b

Environmental protection E07 (2 wk); E13c

Storage and disposal E30a, E31a

147 2,4-D + dicamba + fluroxypyr

A translocated and contact herbicide mixture for amenity turf

Products

Holster	SumiAgro Amenity	285:52.2:105 g/l	SL	10593

Uses

Annual dicotyledons in ***managed amenity turf***. Buttercups in ***managed amenity turf***. Chickweed in ***managed amenity turf***. Clover in ***managed amenity turf***. Dandelions in ***managed amenity turf***.

Approval information

- 2,4-D and fluroxypyr included in Annex I under EC Directive 91/414

Efficacy guidance

- Apply when weeds actively growing (normally between Apr and Sep) and when soil is moist
- Best results obtained from treatment in spring or early summer before weeds begin to flower
- Do not apply if turf is wet or if rainfall expected within 4 h of treatment. Both circumstances will reduce weed control

Restrictions

- Maximum number of treatments 2 per yr
- Do not mow for 3 d before or after treatment
- Avoid overlapping or overdosing, especially on newly sown turf
- Do not spray in drought conditions or if turf under stress from frost, waterlogging, trace element deficiency, pest or disease attack

Crop-specific information

- New turf may be treated in spring provided at least 2 mth have elapsed since sowing

Environmental safety

- Harmful to fish or other aquatic life. Do not contaminate surface waters or ditches with chemical or used container
- Avoid drift onto non-target plants or outside the target area
- Keep livestock out of treated areas for at least 14 d and until foliage of any poisonous weeds such as ragwort has died and become unpalatable

Hazard classification and safety precautions

Hazard R03, R04
Risk phrases R03c, R04a, R04b, R04g
Operator protection A, C, H, M; U05a, U08, U19, U20b
Environmental protection E01, E13c; E07 (14 d)
Consumer protection C03
Storage and disposal E26, E30a, E31b
Medical advice M03

148 2,4-D + dicamba + triclopyr

A translocated herbicide for perennial and woody weed control

Products

1	Broadsword	United Phosphorus	200:85:65 g/l	EC	09140
2	Greengard	SumiAgro Amenity	200:85:65 g/l	EC	11715
3	Nu-Shot	Nufarm UK	200:85:65 g/l	EC	11148

Uses

Annual dicotyledons in ***amenity grass*** [1-3]; ***permanent pasture, rotational grassland*** [1, 3]. Brambles in ***farm forestry, forestry, natural surfaces not intended to bear vegetation*** [1-3]. Docks in ***amenity grass*** [1-3]; ***permanent pasture, rotational grassland*** [1, 3]. Gorse in ***farm forestry, forestry, natural surfaces not intended to bear vegetation*** [1-3]. Japanese knotweed in ***farm forestry, forestry, natural surfaces not intended to bear vegetation*** [1-3]. Perennial dicotyledons in ***amenity grass, farm forestry, forestry, natural surfaces not intended to bear vegetation*** [1-3]; ***permanent pasture, rotational grassland*** [1, 3]. Plantains in ***amenity grass*** [1-3]; ***permanent pasture, rotational grassland*** [1, 3]. Rhododendrons in ***farm forestry, forestry, natural surfaces not intended to bear vegetation*** [1-3]. Stinging nettle in ***amenity grass*** [1-3]; ***permanent pasture, rotational grassland*** [1, 3]. Thistles in ***amenity grass*** [1-3]; ***permanent pasture, rotational grassland*** [1, 3]. Woody weeds in ***farm forestry, forestry, natural surfaces not intended to bear vegetation*** [1-3].

Approval information

- 2,4-D included in Annex I under EC Directive 91/414

Efficacy guidance

- Apply as foliar spray to herbaceous or woody weeds. Timing and growth stage for best results vary with species. See label for details

- Dilute with water for stump treatment and apply after felling up to the start of regrowth. Treat any regrowth with a spray to the growing foliage
- May be applied at 1/3 dilution in weed wipers or 1/8 dilution with ropewick applicators

Restrictions

- Do not use on pasture established less than 1 yr or on grass grown for seed
- Where clover a valued constituent of sward only use as a spot treatment
- Do not graze for 7 d or mow for 14 d after treatment
- Do not direct drill grass, clover or brassicas for at least 6 wk after grassland treatment
- Do not plant trees for 1-3 mth after spraying depending on dose applied. See label
- Avoid spray drift into greenhouses or onto crops or ornamentals. Vapour drift may occur in hot conditions

Crop-specific information

- Sprays may be applied in pines, spruce and fir providing drift is avoided. Optimum time is mid-autumn when tree growth ceased but weeds not yet senescent

Environmental safety

- Take extreme care to avoid drift onto neighbouring crops, especially beet crops, brassicas, most market garden crops including lettuce and tomatoes under glass, pears and vines
- Dangerous to fish or other aquatic life. Do not contaminate surface waters or ditches with chemical or used container
- Docks and other weeds may become increasingly palatable after treatment and may be preferentially grazed. Where ragwort or other poisonous weeds are present keep livestock out until the weeds have died and become unpalatable
- Keep livestock out of treated area for at least 2 wk or until foliage of any poisonous weeds such as ragwort or buttercup has died and become unpalatable

Hazard classification and safety precautions

Hazard R03, R04, R07d [1-3]; R14 [2]

Risk phrases R03c, R04a, R04b [1-3]; R04g, R14b [2]

Operator protection A, C, H, M [1-3]; D [1, 3]; U02, U04a, U11, U13, U20a [2]; U05a, U08, U19 [1-3]; U20b [1, 3]

Environmental protection E01 [1-3]; E07 [1] (7 d); E07 [2]; E07 [3] (7 d); E13b [1, 3]

Consumer protection C01 [2]; C03 [1-3]

Storage and disposal E26, E30a, E31b, E34

Medical advice M03 [1-3]; M05 [2]

149 2,4-D + dichlorprop-P + MCPA + mecoprop-P

A translocated herbicide for use in apple and pear orchards

Products

UPL Camppex	United Phosphorus	34.5:66.6:52.8:82.1 g/l	SL	11661

Uses

Annual dicotyledons in ***apple orchards***, ***pear orchards***. Chickweed in ***apple orchards***, ***pear orchards***. Cleavers in ***apple orchards***, ***pear orchards***. Perennial dicotyledons in ***apple orchards***, ***pear orchards***.

Approval information

- 2,4-D and mecoprop-P included in Annex I under EC Directive 91/414

Efficacy guidance

- Use on emerged weeds in established apple and pear orchards (from 1 yr after planting) as directed application
- Spray when weeds in active growth and at growth stage recommended on label
- Effectiveness may be reduced by rain within 12 h

Restrictions

- Applications must be made around, and not directly to, trees. Do not allow drift onto trees
- Do not spray during blossom period. Do not spray to run-off
- Do not roll, harrow or cut grass crops on orchard floor within at least 3 d before or after spraying

Crop-specific information

- Latest use: before blossom period

Environmental safety

- Take extreme care to avoid drift onto neighbouring sensitive crops
- Harmful to fish or other aquatic life. Do not contaminate surface waters or ditches with chemical or used container
- Avoid spray drift onto neighbouring crops
- Keep livestock out of treated areas for at least 2 wk and until foliage of any poisonous weeds such as ragwort has died and become unpalatable

Hazard classification and safety precautions

Hazard R04
Risk phrases R04a
Operator protection A, C, H, M; U05a, U08, U20b
Environmental protection E01, E13c; E07 (2 wk)
Consumer protection C03
Storage and disposal E30a, E31b, E34
Medical advice M03

150 2,4-D + MCPA

A translocated herbicide mixture for cereals and grass

Products

Headland Polo	Headland	360:315 g/l	SL	10283

Uses

Annual dicotyledons in ***amenity turf***, ***established grassland***, ***spring barley***, ***spring wheat***, ***winter barley***, ***winter oats***, ***winter wheat***. Perennial dicotyledons in ***amenity turf***, ***established grassland***, ***spring barley***, ***spring wheat***, ***winter barley***, ***winter oats***, ***winter wheat***.

Approval information

- 2,4-D included in Annex I under EC Directive 91/414

Efficacy guidance

- Best results achieved by spraying weeds in seedling to young plant stage when growing actively in a strongly competing crop
- Most effective stage for spraying perennials varies with species. See label for details

Restrictions

- Maximum number of treatments 1 per crop in cereals and established grassland; 3 per yr in amenity turf
- Do not spray if rain falling or imminent
- Do not cut grass or graze for at least 10 d after spraying
- Do not use on newly sown leys containing clover or other legumes
- Do not spray crops stressed by cold weather or drought or if frost expected
- Do not use shortly before or after sowing any crop
- Do not roll or harrow within 7 d before or after spraying

Crop-specific information

- Spray winter cereals in spring when leaf-sheath erect but before first node detectable (GS 31), spring cereals from 5-leaf stage to before first node detectable (GS 15-31)
- Latest use: before first node detectable (GS 31) in cereals

Environmental safety

- 2,4-D and MCPA are active at low concentrations. Take extreme care to avoid drift onto neighbouring crops, especially beet crops, brassicas, maost market garden crops including lettuce and tomatoes under glass, pears and vines
- Harmful to fish or other aquatic life. Do not contaminate surface waters or ditches with chemical or used container
- Keep livestock out of treated areas for at least 2 weeks and until foliage of any poisonous weeds such as ragwort has died and become unpalatable

Hazard classification and safety precautions

Hazard R03, R04
Risk phrases R03c, R04d
Operator protection A, C, H, M; U05a, U08, U20b

SECTION 4

Environmental protection E01, E07, E13c
Consumer protection C03
Storage and disposal E26, E30a, E31a, E34
Medical advice M03

151 2,4-D + mecoprop-P

A translocated herbicide for use in amenity turf

Products

1	Supertox 30	Bayer Environ.	93.5:95 g/l	SL	09946
2	Sydex	Vitax	125:125 g/l	SL	06412

Uses

Annual dicotyledons in ***amenity turf***, ***established grassland***, ***grass seed crops*** [2]; ***managed amenity turf*** [1, 2]. Perennial dicotyledons in ***amenity turf***, ***established grassland***, ***grass seed crops*** [2]; ***managed amenity turf*** [1, 2].

Approval information

- 2,4-D and mecoprop-P included in Annex I under EC Directive 91/414

Efficacy guidance

- May be applied from Apr to Sep, best results in May-Jun when weeds in active growth. Less susceptible weeds may require second treatment not less than 6 wk later
- For best results apply fertilizer 1-2 wk before treatment
- Do not spray during drought conditions or when rain imminent as efficacy will be impaired
- Do not close mow infrequently mown turf for 3-4 d before or after treatment. On fine turf do not mow for 24 h before or after treatment

Restrictions

- Maximum number of treatments 2 per yr. The total amount of mecoprop-P applied in a single yr must not exceed the maximum total dose approved for any single product for the crop/situation
- Do not use first 4 mowings for mulching unless composted for at least 6 mth (do not use first 2 mowings for composting on some labels)
- Grass cuttings should not be used for mulching but may be composted and used 6 mth later
- Do not apply during frosty weather
- Do not roll or harrow within 7 d before or after treatment

Crop-specific information

- Newly laid turf or newly sown grass should not be treated for at least 6 mth

Environmental safety

- Take extreme care to avoid drift onto neighbouring crops, especially beet crops, brassicas, most market garden crops including lettuce and tomatoes under glass, pears and vines
- Harmful (dangerous [2]) to fish or other aquatic life. Do not contaminate surface waters or ditches with chemical or used container [1]
- Keep livestock out of treated areas for at least 2 wk and until foliage of any poisonous weeds such as ragwort has died and become unpalatable

Hazard classification and safety precautions

Hazard R03 [2]; R04 [1, 2]
Risk phrases R03a, R03c [2]; R04a [1, 2]
Operator protection A, C, H, M; U05a, U08, U20b [1, 2]; U11, U15 [1]
Environmental protection E01 [1, 2]; E07 [1, 2] (2 wk); E13b [2]; E13c, E23 [1]
Consumer protection C03
Storage and disposal E26, E30a, E31b [1, 2]; E34 [2]
Medical advice M03 [2]

152 2,4-D + picloram

A persistent translocated herbicide for non-crop land

Products

Atladox HI	n2n	240:65 g/l	SL	05559

Uses

Annual dicotyledons in ***amenity grass, non-crop grass, road verges***. Brambles in ***amenity grass, non-crop grass, road verges***. Creeping thistle in ***amenity grass, non-crop grass, road verges***. Docks in ***amenity grass, non-crop grass, road verges***. Japanese knotweed in ***amenity grass, non-crop grass, road verges***. Perennial dicotyledons in ***amenity grass, non-crop grass, road verges***. Ragwort in ***amenity grass, non-crop grass, road verges***. Scrub clearance in ***amenity grass, non-crop grass, road verges***. Woody weeds in ***amenity grass, non-crop grass, road verges***.

Approval information

- 2,4-D included in Annex I under EC Directive 91/414

Efficacy guidance

- Apply as overall foliar spray during period of active growth when foliage well developed

Restrictions

- Maximum number of treatments 1 per yr
- Do not apply around desirable trees or shrubs where roots may absorb chemical
- Prevent leaching into areas where desirable plants are present
- Avoid drift of spray onto desirable plants
- Do not use cuttings from treated grass for mulching or composting

Environmental safety

- Harmful to fish or other aquatic life. Do not contaminate surface waters or ditches with chemical or used container
- Keep livestock out of treated areas for at least 2 wk and until foliage of any poisonous weeds such as ragwort has died and become unpalatable

Hazard classification and safety precautions

Hazard R04, R07d
Risk phrases R04a
Operator protection A, C; U05a, U08, U20b
Environmental protection E01, E07, E13c
Consumer protection C03
Storage and disposal E26, E30a, E31a

SECTION 4

153 daminozide

A hydrazido plant growth regulator for use in certain ornamentals

Products

1	B-Nine	Certis	85% w/w	SP	11471
2	Dazide	Fine	85% w/w	SP	02691

Uses

Internode reduction in ***azaleas, bedding plants, chrysanthemums, hydrangeas*** [1, 2]; ***ornamental specimens, poinsettias*** [1]; ***pot plants*** [2].

Approval information

- Sales of daminozide for use on food crops were halted worldwide by the manufacturer in Oct 1989. Sales for use on flower crops were not affected

Efficacy guidance

- Best results obtained by application in late afternoon when glasshouse has cooled down
- Spray when foliage dry
- A reduced rate tank mix with chlormequat + choline chloride is recommended for use on poinsettias. See label for details [1]

Restrictions

- Maximum number of treatments 2 per crop
- Apply only to turgid, well watered plants. Do not water for 24 h after spraying
- Do not use on chrysanthemum Fandango
- Do not mix with other spray chemicals except as recommended above

Crop-specific information

- Latest use: end Sep for Poinsettias [1]

Environmental safety

- Harmful to fish or other aquatic life. Do not contaminate surface waters or ditches with chemical or used container [2]

Hazard classification and safety precautions

Hazard R04
Risk phrases R04a
Operator protection A, C [1, 2]; H [2]; U05a, U19 [1, 2]; U09a, U20a [2]
Environmental protection E01 [1, 2]; E13c [2]; E15 [1]
Consumer protection C03
Storage and disposal E30a [1, 2]; E31a [2]

154 dazomet

A methyl isothiocyanate releasing soil fumigant

Products

Basamid	Certis	97% w/w	GR	07204

Uses

Nematodes in ***field crops*** *(soil fumigation)*, ***protected crops*** *(soil fumigation)*, ***vegetables*** *(soil fumigation)*. Soil insects in ***field crops*** *(soil fumigation)*, ***protected crops*** *(soil fumigation)*, ***vegetables*** *(soil fumigation)*. Soil-borne diseases in ***field crops*** *(soil fumigation)*, ***protected crops*** *(soil fumigation)*, ***vegetables*** *(soil fumigation)*. Weed seeds in ***field crops*** *(soil fumigation)*, ***protected crops*** *(soil fumigation)*, ***vegetables*** *(soil fumigation)*.

Efficacy guidance

- Dazomet acts by releasing methyl isothiocyanate in contact with moist soil
- Soil sterilization is carried out after harvesting one crop and before planting the next
- The soil must be of fine tilth, free of clods and evenly moist to the depth of sterilization
- Soil moisture must not be less than 50% of water-holding capacity or oversaturated. If too dry, water at least 7-14 d before treatment
- In order to obtain short treatment times it is recommended to treat soils when soil temperature is above 7°C. Treatment should be used outdoors before winter rains make soil too wet to cultivate - usually early Nov
- For club root control treat only in summer when soil temperature above 10°C
- Where onion white rot a problem unlikely to give effective control where inoculum level high or crop under stress
- Apply granules with suitable applicators, mix into soil immediately to desired depth and seal surface with polythene sheeting, by flooding or heavy rolling. See label for suitable application and incorporation machinery
- With 'planting through' technique polythene seal is left in place to form mulch into which new crop can be planted

Restrictions

- Maximum number of treatments 1 per crop or batch of soil
- Do not treat ground where water table may rise into treated layer

Crop-specific information

- Latest use: pre-planting

Following crops guidance

- With 'planting through' technique no gas release cultivations are made and safety test with cress is particularly important. Conduct cress test on soil samples from centre as well as edges of bed. Observe minimum of 30 d from application to cress test in soils at 10°C or above, at least 50 d in soils below 10°C. See label for details
- In all other situations 14-28 d after treatment cultivate lightly to allow gas to disperse and conduct cress test after a further 14-28 d (timing depends on soil type and temperature). Do not treat structures containing live plants or any ground within 1 m of live plants

Environmental safety

- Do not contaminate surface waters or ditches with chemical or used container

Hazard classification and safety precautions

Hazard R03

Risk phrases R03c
Operator protection A, M; U04a, U05a, U09a, U19, U20a
Environmental protection E01, E15
Consumer protection C03
Storage and disposal E30a, E32a, E34
Medical advice M03

155 2,4-DB

A translocated phenoxy herbicide for use in lucerne

Products

DB Straight	United Phosphorus	300 g/l	SL	07523

Uses

Annual dicotyledons in ***lucerne***, ***undersown spring cereals*** *(undersown with lucerne)*, ***undersown winter cereals*** *(undersown with lucerne)*. Thistles in ***lucerne***, ***undersown spring cereals*** *(undersown with lucerne)*, ***undersown winter cereals*** *(undersown with lucerne)*.

Approval information

- 2,4-DB included in Annex I under EC Directive 91/414

Efficacy guidance

- Best results achieved on young seedling weeds under good growing conditions. Treatment less effective in cold weather and dry soil conditions
- Rain within 12 h may reduce effectiveness

Restrictions

- Do not allow spray drift onto neighbouring crops

Crop-specific information

- In direct sown lucerne spray when seedlings have reached first trifoliate leaf stage. Optimum time 3-4 trifoliate leaves
- Do not treat any lucerne after fourth trifoliate leaf
- In spring barley and spring oats undersown with lucerne spray from when cereal has 1 leaf unfolded and lucerne has first trifoliate leaf
- In spring wheat undersown with lucerne spray from when cereal has 3 leaves unfolded and lucerne has first trifoliate leaf
- Latest use: before first node detectable (GS 31) for undersown cereals; fourth trifoliate leaf for lucerne

Environmental safety

- Harmful to fish or other aquatic life. Do not contaminate surface waters or ditches with chemical or used container
- Keep livestock out of treated areas until foliage of any poisonous weeds such as ragwort has died and become unpalatable

Hazard classification and safety precautions

Operator protection U08, U20a
Environmental protection E07 (2 wk); E13c
Storage and disposal E26, E30a, E31b

156 2,4-DB + MCPA

A translocated herbicide for cereals, clovers and leys

Products

1	Agrichem DB Plus	Agrichem	243:40 g/l	SL	00044
2	Headland Cedar	Headland	240:40 g/l	SL	10489
3	Redlegor	United Phosphorus	244:44 g/l	SL	07519

Uses

Annual dicotyledons in ***all cereals***, ***rotational grassland*** [1]; ***grass re-seeds***, ***spring barley***, ***spring oats***, ***spring wheat***, ***undersown barley*** *(red or white clover)*, ***undersown oats*** *(red or white clover)*, ***undersown wheat*** *(red or white clover)*, ***winter barley***, ***winter oats***, ***winter***

wheat [2]; ***seedling leys*** [3]; ***undersown spring cereals*, *undersown winter cereals*** [1, 3]. Perennial dicotyledons in ***all cereals*, *rotational grassland*** [1]; ***grass re-seeds*, *spring barley*, *spring oats*, *spring wheat*, *undersown barley*** *(red or white clover)*, ***undersown oats*** *(red or white clover)*, ***undersown wheat*** *(red or white clover)*, ***winter barley*, *winter oats*, *winter wheat*** [2]; ***seedling leys*** [3]; ***undersown spring cereals*, *undersown winter cereals*** [1, 3]. Polygonums in ***seedling leys*, *undersown spring cereals*, *undersown winter cereals*** [3].

Approval information

- 2,4-DB included in Annex I under EC Directive 91/414

Efficacy guidance

- Best results achieved on young seedling weeds under good growing conditions
- Spray thistles and other perennials when 10-20 cm high provided clover at correct stage
- Effectiveness may be reduced by rain within 12 h, by very cold conditions or drought

Restrictions

- Do not spray established clover crops or lucerne
- Do not roll or harrow within 7 d before or after spraying
- Do not spray immediately before or after sowing any crop
- Avoid drift onto neighbouring sensitive crops

Crop-specific information

- Apply in spring to winter cereals from leaf sheath erect stage, to spring barley or oats from 2-leaf stage (GS 12), to spring wheat from 5-leaf stage (GS 15)
- Spray clovers as soon as possible after first trifoliate leaf, grasses after 2-3 leaf stage
- Red clover may suffer temporary distortion after treatment
- Latest use: before first node detectable (GS 31) for cereals; 4th trifoliate leaf stage of clover for grass re-seeds

Environmental safety

- Harmful to fish or other aquatic life. Do not contaminate surface waters or ditches with chemical or used container
- Keep livestock out of treated areas until foliage of any poisonous weeds such as ragwort has died and become unpalatable

Hazard classification and safety precautions

Hazard R03, R04 [1, 2]
Risk phrases R03c, R04b, R04d [1, 2]
Operator protection A, C [1, 2]; U05a [2]; U08, U20b [1-3]
Environmental protection E01, E07 [2]; E07 [1] (2 wk); E07 [3] (2 wk); E13c [1-3]
Storage and disposal E26 [1-3]; E30a [1, 2]; E31a, E34 [2]; E31b [3]; E31c [1]

157 deltamethrin

A pyrethroid insecticide with contact and residual activity

Products

1	Agriguard Deltamethrin	AgriGuard	25 g/l	EC	10770
2	Bandu	Cheminova	25 g/l	EC	10629
3	Bandu	Headland	25 g/l	EC	10994
4	Decis	Aventis	25 g/l	EC	07172
5	Decis Protech	Bayer CropScience	15 g/l	EW	11502
6	Landgold Deltaland	Landgold	25 g/l	EC	09906
7	Pearl Micro	Aventis	6.25% w/w	GR	08620
8	Thripstick	Aquaspersions	0.125 g/l	AL	02134

Uses

American serpentine leaf miner in ***aubergines*** *(off-label)*, ***non-edible ornamentals*** *(off-label)*, ***rooting beds*** *(off-label)* [2, 4]; ***non-edible glasshouse crops*** *(off-label)*, ***protected aubergines*** *(off-label)* [5]; ***protected courgettes*** *(off-label)*, ***protected cucumbers*** *(off-label)*, ***protected gherkins*** *(off-label)*, ***protected peppers*** *(off-label)*, ***protected tomatoes*** *(off-label)* [2, 4, 5]. Aphids in ***amenity vegetation*, *ornamental plant production*, *protected cucumbers*, *protected peppers*, *protected tomatoes*** [1-5]; ***apples*** [1-6]; ***bulb onions*** *(off-label)*, ***carrots*** *(off-label)*, ***celery*** *(off-label)*, ***garlic*** *(off-label)*, ***leeks*** *(off-label)*, ***parsnips*** *(off-label)*, ***peas*, *protected celery*** *(off-label)*, ***protected chinese cabbage*** *(off-label)*, ***protected salad onions*** *(off-label)*, ***salad onions*** *(off-label)* [7]; ***chinese cabbage*** *(off-label)*, ***lettuce*** *(off-label)* [4, 5, 7]; ***evening primrose*** *(off-label)* [5]; ***grass seed crops*** *(off-label)* [2, 4, 5, 7]; ***protected ornamentals*** [2, 5]; ***protected pot plants*** [1, 3]; ***radicchio*** *(off-label)* [2, 4, 7]; ***spring barley***, ***spring oats***, ***spring wheat***, ***winter barley***, ***winter oats***, ***winter wheat*** [1-7]. Apple sucker in ***apples*** [1-6]. Barley yellow dwarf vectors in ***winter barley***, ***winter wheat*** [1-5]. Barley yellow dwarf virus vectors in ***spring barley***, ***spring oats***, ***spring wheat***, ***winter oats*** [7]; ***winter barley***, ***winter wheat*** [6, 7]. Beet virus yellows vectors in ***winter oilseed rape*** [1-7]. Brassica pod midge in ***mustard***, ***spring oilseed rape***, ***winter oilseed rape*** [7]. Cabbage seed weevil in ***mustard***, ***spring oilseed rape***, ***winter oilseed rape*** [1-7]. Cabbage stem flea beetle in ***winter oilseed rape*** [1-7]. Cabbage stem weevil in ***mustard***, ***spring oilseed rape***, ***winter oilseed rape*** [1-6]. Capsids in ***amenity vegetation***, ***ornamental plant production*** [1-5]; ***apples*** [1-6]. Caterpillars in ***amenity vegetation***, ***ornamental plant production***, ***protected cucumbers***, ***protected peppers***, ***protected tomatoes*** [1-5]; ***apples***, ***plums*** [1-6]; ***broccoli***, ***brussels sprouts***, ***cabbages***, ***cauliflowers***, ***kale***, ***swedes***, ***turnips*** [1-7]; ***bulb onions*** *(off-label)*, ***carrots*** *(off-label)*, ***celery*** *(off-label)*, ***garlic*** *(off-label)*, ***leeks*** *(off-label)*, ***parsnips*** *(off-label)*, ***protected celery*** *(off-label)*, ***protected chinese cabbage*** *(off-label)*, ***protected salad onions*** *(off-label)*, ***salad onions*** *(off-label)* [7]; ***chinese cabbage*** *(off-label)*, ***lettuce*** *(off-label)*, ***protected salad brassicas*** *(off-label - for baby leaf production)* [4, 5, 7]; ***marjoram*** *(off-label)*, ***sorrel*** *(off-label)* [5]; ***protected ornamentals*** [2, 5]; ***protected pot plants*** [1, 3]; ***radicchio*** *(off-label)* [2, 4, 5, 7]; ***salad brassicas*** *(off-label)* [4]; ***salad brassicas*** *(off-label - for baby leaf production)* [5, 7]; ***tomato houses*** *(off-label)* [8]. Codling moth in ***apples*** [1-6]. Colorado beetle in ***potatoes*** *(off-label)* [4]. Cutworms in ***lettuce*** [1-5]. Damson-hop aphid in ***hops***, ***plums*** [1-6]. Flea beetle in ***broccoli***, ***brussels sprouts***, ***cabbages***, ***cauliflowers***, ***kale***, ***swedes***, ***turnips*** [4, 7]; ***cress*** *(off-label)*, ***protected herbs***, ***salad brassicas*** *(off-label)* [4]; ***evening primrose*** *(off-label)*, ***sorrel*** *(off-label)* [2, 4, 5, 7]; ***marjoram*** *(off-label)*, ***protected agastache spp.*** *(off-label)*, ***protected angelica*** *(off-label)*, ***protected balm*** *(off-label)*, ***protected basil*** *(off-label)*, ***protected bay*** *(off-label)*, ***protected borage*** *(off-label)*, ***protected camomile*** *(off-label)*, ***protected caraway*** *(off-label)*, ***protected catnip*** *(off-label)*, ***protected chervil*** *(off-label)*, ***protected chives*** *(off-label)*, ***protected clary*** *(off-label)*, ***protected coriander*** *(off-label)*, ***protected curry plant*** *(off-label)*, ***protected dill*** *(off-label)*, ***protected dragonhead*** *(off-label)*, ***protected fennel*** *(off-label)*, ***protected fenugreek*** *(off-label)*, ***protected feverfew*** *(off-label)*, ***protected frise*** *(off-label)*, ***protected horehound*** *(off-label)*, ***protected hyssop*** *(off-label)*, ***protected lamb's lettuce*** *(off-label)*, ***protected land cress*** *(off-label)*, ***protected lavandin*** *(off-label)*, ***protected lavender*** *(off-label)*, ***protected lemon verbena*** *(off-label)*, ***protected lovage*** *(off-label)*, ***protected marigold*** *(off-label)*, ***protected marjoram*** *(off-label)*, ***protected mint*** *(off-label)*, ***protected nasturtium*** *(off-label)*, ***protected nettle*** *(off-label)*, ***protected oregano*** *(off-label)*, ***protected origanum heracleoticum*** *(off-label)*, ***protected parsley*** *(off-label)*, ***protected plantain*** *(off-label)*, ***protected rocket*** *(off-label)*, ***protected rosemary*** *(off-label)*, ***protected rue*** *(off-label)*, ***protected sage*** *(off-label)*, ***protected salad burnet*** *(off-label)*, ***protected savory*** *(off-label)*, ***protected scarole*** *(off-label)*, ***protected sorrel*** *(off-label)*, ***protected spike lavender*** *(off-label)*, ***protected st john's wort*** *(off-label)*, ***protected tarragon*** *(off-label)*, ***protected thyme*** *(off-label)*, ***protected violet*** *(off-label)*, ***protected watercress*** *(off-label)*, ***protected woodruff*** *(off-label)*, ***protected yarrow*** *(off-label)*, ***radicchio*** *(off-label)* [5]; ***protected herbs*** *(off-label)* [2, 7]; ***protected lettuce*** *(off-label)*, ***protected radicchio*** *(off-label)*, ***salad brassicas*** *(off-label - for baby leaf production)* [5, 7]; ***protected salad brassicas*** *(off-label - for baby leaf production)* [4, 5, 7]; ***sugar beet*** [1-7]. Fruit tree tortrix moth in ***plums*** [1-6]. Insect pests in ***bulb onions*** *(off-label)*, ***carrots*** *(off-label)*, ***garlic*** *(off-label)*, ***leeks*** *(off-label)* [4, 5]; ***celery*** *(off-label)*, ***protected celery*** *(off-label)*, ***protected spring onions*** *(off-label)*, ***salad onions*** *(off-label)* [2, 4, 5]; ***chinese cabbage*** *(off-label)* [4]; ***protected chinese cabbage*** *(off-label)* [2, 5]. Leaf miner in ***tomato houses*** *(off-label)* [8]. Leafhoppers in ***cress*** *(off-label)*, ***protected herbs*** [4]; ***marjoram*** *(off-label)* [2, 4, 5, 7]; ***protected agastache spp.*** *(off-label)*, ***protected angelica*** *(off-label)*, ***protected balm*** *(off-label)*, ***protected basil*** *(off-label)*, ***protected bay*** *(off-label)*, ***protected borage*** *(off-label)*, ***protected camomile*** *(off-label)*, ***protected caraway*** *(off-label)*, ***protected catnip*** *(off-label)*, ***protected chervil*** *(off-label)*, ***protected chives*** *(off-label)*, ***protected clary*** *(off-label)*, ***protected coriander*** *(off-label)*, ***protected curry plant*** *(off-label)*, ***protected dill*** *(off-label)*, ***protected dragonhead*** *(off-label)*, ***protected fennel*** *(off-label)*, ***protected fenugreek*** *(off-label)*, ***protected feverfew*** *(off-label)*, ***protected frise*** *(off-label)*, ***protected horehound*** *(off-label)*, ***protected hyssop*** *(off-label)*, ***protected lamb's lettuce*** *(off-label)*, ***protected land cress*** *(off-label)*, ***protected lavandin*** *(off-label)*, ***protected lavender*** *(off-label)*, ***protected lemon verbena*** *(off-label)*, ***protected lovage*** *(off-label)*, ***protected marigold*** *(off-label)*, ***protected marjoram*** *(off-label)*, ***protected mint*** *(off-*

SECTION 4

label), ***protected nasturtium*** *(off-label)*, ***protected nettle*** *(off-label)*, ***protected oregano*** *(off-label)*, ***protected origanum heracleoticum*** *(off-label)*, ***protected parsley*** *(off-label)*, ***protected plantain*** *(off-label)*, ***protected rocket*** *(off-label)*, ***protected rosemary*** *(off-label)*, ***protected rue*** *(off-label)*, ***protected sage*** *(off-label)*, ***protected salad burnet*** *(off-label)*, ***protected savory*** *(off-label)*, ***protected scarole*** *(off-label)*, ***protected sorrel*** *(off-label)*, ***protected spike lavender*** *(off-label)*, ***protected st john's wort*** *(off-label)*, ***protected tarragon*** *(off-label)*, ***protected thyme*** *(off-label)*, ***protected violet*** *(off-label)*, ***protected watercress*** *(off-label)*, ***protected woodruff*** *(off-label)*, ***protected yarrow*** *(off-label)*, ***radicchio*** *(off-label)*, ***sorrel*** *(off-label)* [5]; ***protected herbs*** *(off-label)* [2, 7]; ***protected lettuce*** *(off-label)*, ***protected radicchio*** *(off-label)* [5, 7]. Mealybugs in ***amenity vegetation***, ***ornamental plant production***, ***protected cucumbers***, ***protected peppers***, ***protected tomatoes*** [1-5]; ***protected ornamentals*** [2, 5]; ***protected pot plants*** [1, 3]. Pea and bean weevil in ***broad beans***, ***peas***, ***spring field beans***, ***winter field beans*** [1-7]. Pea moth in ***peas*** [1-7]. Pear sucker in ***pears*** [1-6]. Phorid flies in ***mushrooms*** *(off-label)* [4, 5]. Pollen beetle in ***evening primrose*** *(off-label)* [2, 4, 5, 7]; ***mustard***, ***spring oilseed rape***, ***winter oilseed rape*** [1-7]. Rape winter stem weevil in ***winter oilseed rape*** [7]. Raspberry beetle in ***raspberries*** [1-6]. Sawflies in ***apples***, ***plums*** [1-6]. Scale insects in ***amenity vegetation***, ***ornamental plant production***, ***protected cucumbers***, ***protected peppers***, ***protected tomatoes*** [1-5]; ***protected ornamentals*** [2, 5]; ***protected pot plants*** [1, 3]. Sciarid flies in ***mushrooms*** *(off-label)* [4, 5]. Thrips in ***amenity vegetation***, ***ornamental plant production*** [1-5]; ***protected cucumbers*** [8]. Tortrix moths in ***apples*** [1-6]. Western flower thrips in ***aubergines*** *(off-label)*, ***non-edible ornamentals*** *(off-label)*, ***rooting beds*** *(off-label)* [2, 4]; ***non-edible glasshouse crops*** *(off-label)*, ***protected aubergines*** *(off-label)* [5]; ***protected courgettes*** *(off-label)*, ***protected cucumbers*** *(off-label)*, ***protected gherkins*** *(off-label)*, ***protected peppers*** *(off-label)*, ***protected tomatoes*** *(off-label)* [2, 4, 5]. Whitefly in ***amenity vegetation***, ***ornamental plant production***, ***protected cucumbers***, ***protected peppers***, ***protected tomatoes*** [1-5]; ***protected ornamentals*** [2, 5]; ***protected pot plants*** [1, 3]. Yellow cereal fly in ***spring barley***, ***spring oats***, ***spring wheat***, ***winter barley***, ***winter oats***, ***winter wheat*** [7].

Specific Off-Label Approvals (SOLAs)

- ***aubergines***, ***non-edible ornamentals***, ***protected courgettes***, ***protected cucumbers***, ***protected gherkins***, ***protected peppers***, ***protected tomatoes***, ***rooting beds*** *(OLA 001839) Dec 2008* [4]
- ***aubergines***, ***celery***, ***evening primrose***, ***grass seed crops***, ***marjoram***, ***non-edible ornamentals***, ***protected celery***, ***protected chinese cabbage***, ***protected courgettes***, ***protected cucumbers***, ***protected gherkins***, ***protected peppers***, ***protected spring onions***, ***protected tomatoes***, ***radicchio***, ***rooting beds***, ***salad onions***, ***sorrel*** *(OLA 011755) Dec 2008* [2]
- ***bulb onions***, ***garlic*** *(OLA 001847) Dec 2008* [4]
- ***bulb onions***, ***carrots***, ***celery***, ***chinese cabbage***, ***evening primrose***, ***garlic***, ***grass seed crops***, ***leeks***, ***lettuce***, ***marjoram***, ***parsnips***, ***radicchio***, ***salad onions***, ***sorrel*** *(OLA 011795) Dec 2008* [7]
- ***bulb onions***, ***carrots***, ***celery***, ***chinese cabbage***, ***evening primrose***, ***garlic***, ***grass seed crops***, ***leeks***, ***lettuce***, ***marjoram***, ***protected agastache spp.***, ***protected angelica***, ***protected balm***, ***protected basil***, ***protected bay***, ***protected borage***, ***protected camomile***, ***protected caraway***, ***protected catnip***, ***protected celery***, ***protected chervil***, ***protected chinese cabbage***, ***protected chives***, ***protected clary***, ***protected coriander***, ***protected curry plant***, ***protected dill***, ***protected dragonhead***, ***protected fennel***, ***protected fenugreek***, ***protected feverfew***, ***protected frise***, ***protected horehound***, ***protected hyssop***, ***protected lamb's lettuce***, ***protected land cress***, ***protected lavandin***, ***protected lavender***, ***protected lemon verbena***, ***protected lettuce***, ***protected lovage***, ***protected marigold***, ***protected marjoram***, ***protected mint***, ***protected nasturtium***, ***protected nettle***, ***protected oregano***, ***protected origanum heracleoticum***, ***protected parsley***, ***protected plantain***, ***protected radicchio***, ***protected rocket***, ***protected rosemary***, ***protected rue***, ***protected sage***, ***protected salad burnet***, ***protected savory***, ***protected scarole***, ***protected sorrel***, ***protected spike lavender***, ***protected spring onions***, ***protected st john's wort***, ***protected tarragon***, ***protected thyme***, ***protected violet***, ***protected watercress***, ***protected woodruff***, ***protected yarrow***, ***radicchio***, ***salad onions***, ***sorrel*** *(OLA 031140) Dec 2008* [5]
- ***carrots*** *(OLA 001845) Dec 2008* [4]
- ***celery***, ***chinese cabbage***, ***protected celery***, ***protected spring onions***, ***salad onions*** *(OLA 001849) Dec 2008* [4]
- ***chinese cabbage***, ***lettuce*** *(OLA 001843) Dec 2008* [4]
- ***cress*** *(OLA 022908) Dec 2008* [4]

- ***evening primrose, grass seed crops, marjoram, radicchio, sorrel*** *(OLA 001841) Dec 2008* [4]
- ***leeks*** *(OLA 001851) Dec 2008* [4]
- ***mushrooms*** *(OLA 030039) Dec 2008* [4]
- ***mushrooms*** *(OLA 030040) Dec 2008* [4]
- ***mushrooms*** *(OLA 031138) Dec 2008* [5]
- ***non-edible glasshouse crops, protected aubergines, protected courgettes, protected cucumbers, protected gherkins, protected peppers, protected tomatoes*** *(OLA 031139) Dec 2008* [5]
- ***potatoes*** *(OLA 001855) Dec 2008* [4]
- ***protected celery, protected chinese cabbage, protected salad onions*** *(OLA 011796) Dec 2008* [7]
- ***protected herbs*** *(OLA 011756) Apr 2004* [2]
- ***protected herbs, protected lettuce, protected radicchio*** *(OLA 011797) Apr 2004* [7]
- ***protected salad brassicas*** *(for baby leaf production) (OLA 001853) Dec 2008* [4]
- ***protected salad brassicas*** *(for baby leaf production) (OLA 011796) Dec 2008* [7]
- ***protected salad brassicas, salad brassicas*** *(for baby leaf production) (OLA 031140) Dec 2008* [5]
- ***salad brassicas*** *(for baby leaf production) (OLA 011795) Dec 2008* [7]
- ***salad brassicas*** *(OLA 001853) Dec 2008* [4]
- ***tomato houses*** *(OLA 920180) Dec 2008* [8]

Approval information

- Deltamethrin included in Annex I under EC Directive 91/414

Efficacy guidance

- A contact and stomach poison with 3-4 wk persistence, particularly effective on caterpillars and sucking insects
- Normally applied at first signs of damage with follow-up treatments where necessary at 10-14 d intervals. Rates, timing and recommended combinations with other pesticides vary with crop and pest. See label for details
- Spray is rainfast within 1 h [4]
- May be applied in frosty weather provided foliage not covered in ice
- Temperatures above 35°C may reduce effectiveness or persistence
- Spray Thripstick without dilution onto polythene covering on glasshouse floor. Spray before planting cucumbers where thrips are a known problem or when pest problem seen. Repeat every 8 wk to ensure complete control [8]

Restrictions

- Maximum number of treatments varies with crop and pest, 4 per crop for wheat and barley, only 1 application between 1 Apr and 31 Aug. See label or off-label approval notice for other crops
- Do not apply more than 1 aphicide treatment to cereals in summer
- Do not spray crops suffering from drought or other physical stress
- Consult processer before treating crops for processing
- Do not apply to a cereal crop if any product containing a pyrethroid insecticide or dimethoate has been applied to that crop after the start of ear emergence (GS 51)
- Do not spray cereals after 31 Mar in the year of harvest within 6 m of the outside edge of the crop
- Reduced volume spraying must not be used on cereals after 31 Mar in yr of harvest

Crop-specific information

- Latest use: early dough (GS 83) for barley, oats, wheat; before flowering for mustard, oilseed rape, evening primrose; before 31 Mar for grass seed crops, marjoram, radicchio, sorrel
- HI leeks, protected spring onions 3 d; chinese cabbage, most herbs, protected celery, protected lettuce, protected salad brassicas, radicchio, sorrel 7 d; marjoram, radicchio, sorrel 14 d; carrots, parsnips 21 d

Environmental safety

- Flammable [1-4, 6]
- Dangerous for the environment [4]
- Toxic to aquatic organisms. May cause long-term adverse effects to the aquatic environment [4]
- High risk to bees. Do not apply to crops in flower or to those in which bees are actively foraging. Do not apply when flowering weeds are present [5]
- Dangerous to bees. Do not apply to crops in flower or to those in which bees are actively foraging except as directed on cereals, oilseed rape, peas, beans. Do not apply when flowering weeds are present [1-4, 6, 7]

- High risk to non-target insects or other arthropods. Do not spray within 6m of the field boundary [5]
- Extremely dangerous to fish or other aquatic life. Do not contaminate surface waters or ditches with chemical or used container
- LERAP Category A [1-7]
- Broadcast air-assisted LERAP (18 m) [4, 5]

Hazard classification and safety precautions

Hazard R03, R04 [1-4, 6, 7]; R07d [1-4, 6]; R14 [4]
Risk phrases R03a [1-4, 6]; R03c, R04a, R04b [1-4, 6, 7]; R04c, R04d, R04g, R14b [4]
Operator protection A, C, H; U02, U05b, U06, U09a [8]; U04a, U08 [1-4, 6, 7]; U05a [1-4, 6-8]; U11 [4]; U14, U15 [5]; U19 [1-4, 7, 8]; U20b [1-8]
Environmental protection E01, E13a [1-8]; E12a, E22a [5]; E12d [1-4, 6, 7] (cereals, oilseed rape, peas, beans); E16c, E16d [1-7]; E17a [4, 5] (18 m)
Consumer protection C03
Storage and disposal E26, E30a [1-8]; E27, E31a [8]; E31b [6]; E31c [1-5]; E32a [7]; E32d [4]; E34 [1-4, 6, 7]
Medical advice M03 [1-4, 6, 7]; M05 [4]

158 deltamethrin + pirimicarb

An insecticide mixture combining systemic, contact and stomach activity

Products

1	Evidence	Aventis	7.5:100 g/l	EC	06934
2	Patriot EC	Aventis	5:100 g/l	EC	08990

Uses

Aphids in ***spring barley*** *(on ears)*, ***spring oats*** *(on ears)*, ***spring wheat*** *(on ears)*, ***winter barley*** *(on ears)*, ***winter oats*** *(on ears)*, ***winter wheat*** *(on ears)* [2]. Blackfly in ***sugar beet*** [1, 2]. Caterpillars in ***broccoli***, ***brussels sprouts***, ***cabbages***, ***calabrese***, ***cauliflowers***, ***kale***, ***swedes***, ***turnips*** [1, 2]. Green aphid in ***broccoli***, ***brussels sprouts***, ***cabbages***, ***calabrese***, ***cauliflowers***, ***kale***, ***potatoes***, ***sugar beet***, ***swedes***, ***turnips*** [1, 2]; ***spring oilseed rape***, ***winter oilseed rape*** [2]. Mealy aphid in ***broccoli***, ***brussels sprouts***, ***cabbages***, ***calabrese***, ***cauliflowers***, ***kale***, ***swedes***, ***turnips*** [1, 2]; ***spring oilseed rape***, ***winter oilseed rape*** [2]. Pea aphid in ***peas*** [1, 2]. Virus vectors in ***potatoes*** [1]. Virus yellows vectors in ***sugar beet*** [1, 2].

Approval information

- Deltamethrin included in Annex I under EC Directive 91/414

Efficacy guidance

- Best results obtained from treatment in warm, calm conditions, ideally in morning or evening when uptake by the plant will be at its highest
- Peas should be inspected from beginning of flowering and treated according to current thresholds
- Treat potatoes when first aphids found to control virus spread and preserve processing quality. Other ware crops should be treated from late Jun onwards before population build-up occurs and seed crops must be treated at 80% emergence
- Higher water volume recommended where crop canopy dense
- Treat other crops when pests first seen or when warnings are issued
- Treat cereals when at least two-thirds of heads infested and numbers increasing [2]
- Where aphids resistant to deltamethrin + pirimicarb occur repeat treatments are likely to result in lower levels of control

Restrictions

- Maximum number of treatments 4 per crop on potatoes, edible brassicas; 2 per crop on peas and sugar beet; 1 per crop on cereals. See label for restrictions on oilseed rape
- Do not spray any crop when wilting or in temperatures above 25°C
- Do not treat wet crops liable to run-off
- Must not be applied to cereals if a pyrethroid or dimethoate product has been used at ear emergence (GS 51)
- Product formulation can release traces of other pesticides left in the sprayer. Sprayer should be thoroughly washed out before spraying this product

Crop-specific information

- Latest use: before grain watery ripe (GS 71) for barley, oats, wheat; end of flowering for oilseed rape
- HI 3 d for brassicas, peas, potatoes, sugar beet, swedes, turnips
- Sugar beet may be treated as a band spray provided the pressure and water volume requirements are met

Environmental safety

- Flammable
- Dangerous to bees. Do not apply to crops in flower or to those in which bees are actively foraging except as directed on peas. Do not apply when flowering weeds are present
- Extremely dangerous to fish or other aquatic life. Do not contaminate surface waters or ditches with chemical or used container.
- LERAP Category A
- Harmful to livestock. Livestock should be kept out of treated areas for at least 7 d

Hazard classification and safety precautions

Hazard R03, R04, R07d
Risk phrases R03b [1]; R03c, R04c, R04d, R04g [1, 2]
Operator protection A, D, E, H [1, 2]; C [2]; U04a, U05a, U19, U20b [1, 2]; U08 [1]; U11 [2]
Environmental protection E01, E12d, E13a, E16c, E16d [1, 2]; E06b [1]; E06b [2] (7 d)
Consumer protection C03
Storage and disposal E26, E30a, E31b, E34
Medical advice M02 [1]; M03, M05 [1, 2]

159 desmedipham

A contact carbamate herbicide available only in mixtures

160 desmedipham + ethofumesate + phenmedipham

A selective contact and residual herbicide for beet

Products

Betanal Expert	Aventis	25:151:75 g/l	EC	10592

Uses

Annual dicotyledons in ***sugar beet***. Annual meadow grass in ***sugar beet***.

Approval information

- Ethofumesate included in Annex I under EC Directive 91/414

Efficacy guidance

- Product recommended for low-volume overall application in a planned spray programme
- Product acts mainly by contact action. A full programme also gives some residual control but this may be reduced on soils with more than 5% organic matter
- Best results achieved from treatments applied at fully expanded cotyledon stage of largest weeds present. Occasional larger weeds will usually be controlled from a full programme of sprays
- Where a pre-emergence band spray has been applied treatment must be timed according to size of the untreated weeds between the rows
- Susceptible weeds may not all be killed by the first spray. Repeat applications as each flush of weeds reaches cotyledon size normally necessary for season long control
- Sequential treatments should be applied when the previous one is still showing an effect on the weeds
- Various mixtures with other beet herbicides are recommended. See label for details

Restrictions

- Maximum total dose equivalent to three full dose treatments
- Do not spray crops stressed by nutrient deficiency, wind damage, pest or disease attack, or previous herbicide treatments. Stressed crops treated under conditions of high light intensity may be checked and not recover fully
- If temperature likely to exceed 21°C spray after 5 pm

- Crystallisation may occur if spray volume exceeds that recommended or spray mixture not used within 2 h, especially if the water temperature is below 5°C
- Before use, wash out sprayer to remove all traces of previous products, especially hormone and sulfonyl urea weedkillers

Crop-specific information
- Latest use: before crop meets between rows
- Apply first treatment when majority of crop plants have reached the fully expanded cotyledon stage
- Frost within 7 d of treatment may cause check from which the crop may not recover

Environmental safety
- Harmful to fish or aquatic life. Do not contaminate surface waters or ditches with chemical or used container

Hazard classification and safety precautions

Operator protection A, H; U08, U20a
Environmental protection E13c
Storage and disposal E26, E30a, E31b
Medical advice M03

161 desmedipham + phenmedipham

A mixture of contact herbicides for use in sugar beet

Products

Betanal Carrera	Aventis	40:125 g/l	SE	11029

Uses

Annual dicotyledons in ***sugar beet***.

Efficacy guidance
- Product is recommended for low volume overall spraying in a planned programme involving pre and/or post-emergence treatments at doses recommended for low volume programmes
- Best results obtained from treatment when earliest germinating weeds have reached cotyledon stage
- Further treatments must be applied as each flush of weeds reaches cotyledon stage but allowing a minimum of 7 d between each spray
- Where a pre-emergence band spray has been applied, the first treatment should be timed according to the size of the weeds in the untreated area between the rows
- Product is absorbed by leaves of emerged weeds which are killed by scorching action in 2-10 d. Apply overall as a fine spray to optimise weed cover and spray retention
- Various tank mixtures and sequences recommended to widen weed spectrum and add residual activity - see label for details

Restrictions
- Maximum total dose equivalent to three full dose treatments
- Do not spray crops stressed by nutrient deficiency, frost, wind damage, pest or disease attack, or previous herbicide treatments. Stressed crops may be checked and not recover fully
- If temperature likely to exceed 21°C spray after 5 pm
- Before use, wash out sprayer to remove all traces of previous products, especially hormone and sulfonyl urea weedkillers
- Crystallisation may occur if spray volume exceeds that recommended or spray mixture not used within 2 h, especially if the water temperature is below 5°C
- Product may cause non-reinforced PVC pipes and hoses to soften and swell. Wherever possible, use reinforced PVC or synthetic rubber hoses

Crop-specific information
- Latest use: before crop leaves meet between rows
- Product safe to use on all soil types

Environmental safety
- Harmful to fish or other aquatic life. Do not contaminate surface waters or ditches with chemical or used container
- Risk to certain non-target insects or other arthropods. Avoid spraying within 6 m of field boundary

Hazard classification and safety precautions

Operator protection A, H; U05a, U08, U19, U20b

Environmental protection E01, E13c, E22b

Consumer protection C03

Storage and disposal E26, E30a, E31b

162 dicamba

A translocated benzoic herbicide for control of bracken and perennial weeds

See also *2,4-D + dicamba*
2,4-D + dicamba + fluroxypyr
2,4-D + dicamba + triclopyr

Products

I T Dicamba	I T Agro	480 g/l	SL	10976

Uses

Bracken in ***amenity grass***, ***forestry***, ***grassland***, ***land not intended to bear vegetation***, ***permanent pasture***. Docks in ***amenity grass*** *(wiper application)*, ***grassland*** *(wiper application)*, ***permanent pasture*** *(wiper application)*.

Efficacy guidance

- For weed control in established leys and pasture apply when weeds actively growing before flowering shoots appear. Large well-established docks may require a second treatment in the following yr
- For dock control in grass using a ropewick applicator treat vigorously growing weeds before flower stems senesce
- Bracken is controlled by application to the ground litter from Mar to early May before any new frond growth appears. Subsequent rainfall is necessary to wash the treatment down to the rhizomes. Any bracken with rhizomes in the area to which the treatment is carried are controlled or severely repressed
- Bracken control may be reduced where rhizomes are broken up by cultivation or ploughing

Restrictions

- Maximum total dose equivalent to one full dose per yr
- Avoid drift onto susceptible crops. Tomatoes may be affected by vapour drift at a considerable distance
- Do not roll crops within 1 wk of spraying. Do not treat crops where clover forms an important part of the sward
- Do not apply in the root zone of new trees or mature trees standing on site and do not allow any spray to contact tree foliage

Crop-specific information

- Trees may be treated in a band between the rows before planting or in the yr of planting. Delay treatment until the yr after planting on trees planted on ploughed lines

Environmental safety

- Keep livestock out of treated areas for at least 2-3 wk and until foliage of any poisonous weeds such as ragwort and buttercups has died and become unpalatable

Hazard classification and safety precautions

Hazard H04

Risk phrases R04a

Operator protection A, C; U05a, U08, U11, U20b

Environmental protection E01, E15; E07 (14 d min)

Consumer protection C03

Storage and disposal E26, E30a, E31b

163 dicamba + dichlorprop-P + ferrous sulphate + MCPA

A herbicide/fertilizer combination for moss and weed control in turf

Products

Renovator 3	Scotts	0.03:0.22:23.08: 0.22% w/w	GR	11411

Uses

Annual dicotyledons in ***amenity turf, golf courses***. Mosses in ***amenity turf, golf courses***. Perennial dicotyledons in ***amenity turf, golf courses***.

Efficacy guidance

- Apply from mid-Apr to mid-Aug when weeds are growing
- For best control of moss scarify vigorously after 2 wk to remove dead moss
- Where regrowth of moss or weeds occurs a repeat treatment may be made after 6 wk
- Avoid treatment of wet grass or during drought. If no rain falls within 48 h water in thoroughly

Restrictions

- Do not treat new turf until established for 6-9 mth
- The first 4 mowings after treatment should not be used to mulch cultivated plants unless composted at least 6 mth
- Avoid walking on treated areas until it has rained or they have been watered
- Do not re-seed or turf within 8 wk of last treatment
- Do not cut grass for at least 3 d before and at least 4 d after treatment
- Do not apply during freezing conditions or when rain imminent

Crop-specific information

- Apply with a suitable calibrated fertilizer distributor

Hazard classification and safety precautions

Operator protection U09a, U20a
Environmental protection E15
Storage and disposal E30a, E32a

164 dicamba + dichlorprop-P + MCPA

A translocated herbicide mixture for turf

Products

1	Cleanrun 3	Scotts	0.03:0.22:0.22% w/w	MG	11410
2	Intrepid 2	Scotts	20.8:167:167 g/l	SL	11594

Uses

Annual dicotyledons in ***managed amenity turf***. Perennial dicotyledons in ***managed amenity turf***.

Efficacy guidance

- Apply as directed on label between Apr and Sep when weeds growing actively
- If no rain falls within 48 h of treatment irrigate thoroughly [1]
- Do not mow for at least 3 d before and 3-4 d after treatment so that there is sufficient leaf growth for spray uptake and sufficient time for translocation. See label for details
- If re-growth occurs or new weeds germinate re-treatment recommended

Restrictions

- Do not use during drought unless irrigation is carried out before and after treatment
- Do not apply during freezing conditions or when heavy rain is imminent
- Do not treat new turf until established for about 6 mth after seeding or turfing
- The first 4 mowings after treatment should not be used to mulch cultivated plants unless composted for at least 6 mth
- Avoid walking where possible on treated areas until after rain or irrigation [1]
- Do not re-seed turf within 8 wk of last treatment

Crop-specific information

- Latest use: end Sep for managed amenity turf

Environmental safety

- Keep livestock out of treated areas for at least 2 weeks and until foliage of any poisonous weeds such as ragwort has died and become unpalatable

Hazard classification and safety precautions

Hazard R03, R04 [2]
Risk phrases R03c, R04b, R04d, R04e [2]
Operator protection A, C [2]; U05a, U08, U19, U20a [2]; U09a, U20b [1]
Environmental protection E01 [2]; E07 [2] (2 wk); E15 [1, 2]
Consumer protection C03 [2]
Storage and disposal E30a [1, 2]; E31b, E34 [2]; E32a [1]
Medical advice M03 [2]

165 dicamba + maleic hydrazide + MCPA

A herbicide/plant growth regulator mixture for amenity grass

Products

Mazide Selective	Vitax	6:200:75 g/l	SL	05753

Uses

Annual dicotyledons in ***amenity grass***, ***road verges***. Growth retardation in ***amenity grass***, ***road verges***. Perennial dicotyledons in ***amenity grass***, ***road verges***.

Approval information

- Maleic hydrazide included in Annex I under EC Directive 91/414

Efficacy guidance

- Best results achieved by application in Apr-May when grass and weeds growing actively but before weeds have started to flower
- May be used either as one annual spray or as spring spray repeated after 8-10 wk

Restrictions

- Maximum number of treatments 2 per yr
- Do not use on fine turf

Environmental safety

- Harmful to fish or other aquatic life. Do not contaminate surface waters or ditches with chemical or used container
- Keep livestock out of treated areas until foliage of any poisonous weeds such as ragwort has died and become unpalatable

Hazard classification and safety precautions

Operator protection U08, U20b
Environmental protection E07, E13c
Storage and disposal E26, E30a, E31a

166 dicamba + MCPA + mecoprop-P

A translocated herbicide for cereals, grassland, amenity grass and orchards

Products

1	Banlene Super	Aventis	18:252:42 g/l	SL	10053
2	Field Marshal	United Phosphorus	18:360:80	SL	08956
3	Greencrop Triathlon	Greencrop	25:200:200 g/l	SL	10956
4	Headland Relay P	Headland	25:200:200 g/l	SL	08580
5	Headland Transfer	Headland	18:315:50 g/l	SL	11010
6	Headland Trinity	Headland	18:315:50 g/l	SL	10842
7	Hycamba Plus	Agrichem	20.4:125.9:240.2 g/l	SL	10180
8	Hyprone-P	Agrichem	16:101:92 g/l	SL	09125
9	Hysward-P	Agrichem	16:101:92 g/l	SL	09052
10	Mircam Plus	Nufarm UK	19.5:245:43.3 g/l	SL	11004
11	Nocweed	Bayer Environ.	18:252:42 g/l	SL	10609
12	Outrun	SumiAgro Amenity	20.4:125.9:240.2 g/l	SL	10581
13	Super Selective Plus	Rigby Taylor	19.5:245:43.3 g/l	SL	10427
14	Tribute	n2n	18:252:42 g/l	SL	06921
15	Tribute Plus	n2n	18:252:42 g/l	SL	09493

Products – continued

16	Trireme	Nufarm UK	19.5:245:43.3 g/l	SL	11524
17	Tritox	Scotts	15:178:54 g/l	SL	07764
18	UPL Grassland Herbicide	United Phosphorus	25:200:200	SL	08934

Uses

Annual dicotyledons in ***amenity grass*** [3, 4, 6, 7, 9, 12, 13, 16]; ***amenity turf*** [9, 10, 14, 15, 17]; ***apple orchards*** [1, 6, 10, 11]; ***established grassland*** [1, 9, 11]; ***grass seed crops*** [1, 5, 6, 8, 10, 11]; ***grassland*** [10]; ***managed amenity turf*** [3, 4, 6, 7, 12-17]; ***newly sown grass*** [8]; ***pear orchards***, ***spring rye***, ***winter rye*** [1, 6, 11]; ***permanent pasture*** [1, 3-6, 11, 18]; ***rotational grassland*** [1-6, 10, 11, 18]; ***spring barley***, ***spring oats***, ***spring wheat***, ***winter barley***, ***winter oats***, ***winter wheat*** [1, 2, 5, 6, 8, 10, 11]; ***undersown barley***, ***undersown oats***, ***undersown wheat*** [8, 11]; ***undersown rye*** [11]; ***undersown spring cereals***, ***undersown winter cereals*** [1, 6]; ***undersown spring cereals*** *(grass only)*, ***undersown winter cereals*** *(grass only)* [2, 5]. Chickweed in ***amenity turf***, ***apple orchards***, ***grassland*** [10]; ***established grassland*** [1, 11]; ***grass seed crops*** [1, 5, 6, 8, 10, 11]; ***newly sown grass*** [8]; ***permanent pasture*** [1, 5, 6, 11, 18]; ***rotational grassland*** [1, 2, 5, 6, 10, 11, 18]; ***spring barley***, ***spring oats***, ***spring wheat***, ***winter barley***, ***winter oats***, ***winter wheat*** [1, 2, 5, 6, 8, 10, 11]; ***spring rye***, ***winter rye*** [1, 6, 11]; ***undersown barley***, ***undersown oats***, ***undersown wheat*** [8, 11]; ***undersown rye*** [11]; ***undersown spring cereals***, ***undersown winter cereals*** [1, 6]; ***undersown spring cereals*** *(grass only)*, ***undersown winter cereals*** *(grass only)* [2, 5]. Cleavers in ***amenity turf***, ***apple orchards***, ***grassland*** [10]; ***grass seed crops*** [5, 8, 10]; ***newly sown grass*** [8]; ***permanent pasture*** [5, 18]; ***rotational grassland*** [10, 18]; ***spring barley***, ***spring oats***, ***spring wheat***, ***winter barley***, ***winter oats***, ***winter wheat*** [1, 2, 5, 6, 8, 10, 11]; ***spring rye***, ***winter rye*** [1, 6, 11]; ***undersown barley***, ***undersown oats***, ***undersown wheat*** [8, 11]; ***undersown rye*** [11]; ***undersown spring cereals***, ***undersown winter cereals*** [1, 6]. Docks in ***amenity grass***, ***amenity turf*** [9]; ***apple orchards***, ***pear orchards*** [1, 6, 11]; ***established grassland*** [1, 9, 11]; ***grass seed crops*** [1, 6, 8, 11]; ***permanent pasture***, ***rotational grassland*** [1, 6, 11, 18]. Mayweeds in ***amenity turf***, ***apple orchards***, ***grassland*** [10]; ***established grassland*** [1, 11]; ***grass seed crops*** [1, 5, 6, 8, 10, 11]; ***newly sown grass*** [8]; ***permanent pasture*** [1, 5, 6, 11, 18]; ***rotational grassland*** [1, 2, 5, 6, 10, 11, 18]; ***spring barley***, ***spring oats***, ***spring wheat***, ***winter barley***, ***winter oats***, ***winter wheat*** [1, 2, 5, 6, 8, 10, 11]; ***spring rye***, ***winter rye*** [1, 6, 11]; ***undersown barley***, ***undersown oats***, ***undersown wheat*** [8, 11]; ***undersown rye*** [11]; ***undersown spring cereals***, ***undersown winter cereals*** [1, 6]; ***undersown spring cereals*** *(grass only)*, ***undersown winter cereals*** *(grass only)* [2, 5]. Perennial dicotyledons in ***amenity grass*** [3, 4, 6, 7, 9, 12, 13, 16]; ***amenity turf*** [9, 10, 14, 15, 17]; ***apple orchards*** [1, 6, 10, 11]; ***established grassland*** [1, 9, 11]; ***grass seed crops*** [1, 5, 6, 8, 10, 11]; ***grassland*** [10]; ***managed amenity turf*** [3, 4, 6, 7, 12-17]; ***newly sown grass*** [8]; ***pear orchards***, ***spring rye***, ***winter rye*** [1, 6, 11]; ***permanent pasture*** [1, 3-6, 11, 18]; ***rotational grassland*** [1-6, 10, 11, 18]; ***spring barley***, ***spring oats***, ***spring wheat***, ***winter barley***, ***winter oats***, ***winter wheat*** [1, 2, 5, 6, 8, 10, 11]; ***undersown barley***, ***undersown oats***, ***undersown wheat*** [8, 11]; ***undersown rye*** [11]; ***undersown spring cereals***, ***undersown winter cereals*** [1, 6]; ***undersown spring cereals*** *(grass only)*, ***undersown winter cereals*** *(grass only)* [2, 5]. Polygonums in ***amenity turf***, ***apple orchards***, ***grassland*** [10]; ***grass seed crops*** [5, 8, 10]; ***newly sown grass*** [8]; ***permanent pasture*** [5, 18]; ***rotational grassland*** [1, 2, 5, 6, 10, 11, 18]; ***spring barley***, ***spring oats***, ***spring wheat***, ***winter barley***, ***winter oats***, ***winter wheat*** [1, 2, 5, 6, 8, 10, 11]; ***spring rye***, ***winter rye*** [1, 6, 11]; ***undersown barley***, ***undersown oats***, ***undersown wheat*** [8, 11]; ***undersown rye*** [11]; ***undersown spring cereals***, ***undersown winter cereals*** [1, 6]; ***undersown spring cereals*** *(grass only)*, ***undersown winter cereals*** *(grass only)* [2, 5].

Approval information

- Mecoprop-P included in Annex I under EC Directive 91/414
- Approval expiry 31 Oct 2004 [13]

Efficacy guidance

- Treatment should be made when weeds growing actively. Weeds hardened by winter weather may be less susceptible
- For best results apply in fine warm weather, preferably when soil is moist. Do not spray if rain expected within 6 h or in drought
- Application of fertilizer 1-2 wk before spraying aids weed control in turf
- Where a second treatment later in the season is needed in amenity situations and on grass allow 4-6 wk between applications to permit sufficient foliage regrowth for uptake

Restrictions

- Maximum number of treatments (including other mecoprop-P products) or maximum total dose varies with crop and product. See label for details. The total amount of mecoprop-P applied in a single yr must not exceed the maximum total dose approved for any single product for the crop/situation
- Do not apply to cereals after the first node is detectable (GS 31), or to grass under stress from drought or cold weather
- Do not spray cereals undersown with clovers or legumes, to be undersown with grass or legumes or grassland where clovers or other legumes are important
- Do not spray leys established less than 18 mth or orchards established less than 3 yr
- Do not roll or harrow within 7 d before or after treatment, or graze for at least 7 d afterwards (longer if poisonous weeds present)
- Do not use on turf in year of establishment. Allow 6-8 wk after treatment before seeding bare patches [4, 7, 9, 12, 13, 15-17]
- The first mowings after use should not be used for mulching unless composted for 6 mth [4, 7, 9, 12, 13, 15-17]
- Turf should not be mown for 24 h before or after treatment (3-4 d for closely mown turf) [4, 7, 9, 12, 13, 15-17]
- Avoid drift onto all broad-leaved plants outside the target area

Crop-specific information

- Latest use: before first node detectable (GS 31) for cereals; 5-6 wk before head emergence for grass seed crops; mid-Oct for established grass
- HI 7-14 d before cutting or grazing for leys, permanent pasture
- Apply to winter cereals from the leaf sheath erect stage (GS 30), and to spring cereals from the 5 expanded leaf stage (GS 15)
- Spray grass seed crops 4-6 wk before flower heads begin to emerge (timothy 6 wk)
- Turf containing bulbs may be treated once the foliage has died down completely [4, 9, 15, 17]

Environmental safety

- Harmful to fish or other aquatic life. Do not contaminate surface waters or ditches with chemical or used container

Hazard classification and safety precautions

Hazard R03 [2-6, 10, 13-16, 18]; R04 [1-18]

Risk phrases R03a [2-4, 10, 13-16, 18]; R03c [2-6, 10, 13, 16, 18]; R04a [1-4, 8-11, 13-18]; R04b [5-10, 12, 13, 16]; R04d [5-7, 12]

Operator protection A, C [1-18]; H, M [1-10, 12-18]; U05a [1, 3-18]; U08 [1-10, 12-18]; U09a [11]; U20b [1, 2, 5 18]

Environmental protection E01, E13c [1-18]; E07 [1-7] (2 wk); E07 [8]; E07 [9-18] (2 wk)

Consumer protection C03

Storage and disposal E26 [1-16, 18]; E30a [1-18]; E31a [5, 6, 17]; E31b [1-4, 10, 11, 13-16, 18]; E31c [7-9, 12]; E34 [1-6, 10, 11, 13-18]

Treated seed S06 [3, 4]

Medical advice M03 [2-6, 10, 13, 16, 18]

167 dicamba + mecoprop-P

A translocated post-emergence herbicide for cereals and grassland

Products

1	Camber	Headland	42:319 g/l	SL	09901
2	Di-Farmon R	Headland	42:319 g/l	SL	08472
3	Dockmaster	Nufarm UK	18.7:150 g/l	SL	11651
4	Foundation	Syngenta	84:600 g/l	SL	08475
5	Foundation	Headland	84:600 g/l	SL	11708
6	Headland Saxon	Headland	84:600 g/l	SL	10430
7	Hyban-P	Agrichem	18.7:150 g/l	SL	09129
8	Hygrass-P	Agrichem	18.7:150 g/l	SL	09130
9	Mircam	Nufarm UK	18.7:150 g/l	SL	11707
10	Prompt	Headland	84:600 g/l	SL	11535

Uses

Annual dicotyledons in ***established grassland*** [1, 2, 4-6, 10]; ***permanent pasture, rotational grassland*** [8, 9]; ***spring barley, spring oats, spring wheat, winter barley, winter oats, winter wheat*** [1-7, 9, 10]; ***spring rye, triticale, undersown barley, undersown oats, undersown rye, undersown wheat, winter rye*** [7]. Buttercups in ***permanent pasture, rotational grassland*** [7]. Chickweed in ***established grassland*** [1, 2]; ***permanent pasture, rotational grassland, spring rye, triticale, undersown barley, undersown oats, undersown rye, undersown wheat, winter rye*** [7]; ***spring barley, spring oats, spring wheat, winter barley, winter oats, winter wheat*** [1-7, 9, 10]. Cleavers in ***established grassland*** [1, 2]; ***spring barley, spring oats, spring wheat, winter barley, winter oats, winter wheat*** [1-7, 9, 10]; ***spring rye, triticale, undersown barley, undersown oats, undersown rye, undersown wheat, winter rye*** [7]. Creeping thistle in ***permanent pasture, rotational grassland*** [7]. Docks in ***permanent pasture, rotational grassland*** [7-9]. Mayweeds in ***established grassland*** [1, 2]; ***spring barley, spring oats, spring wheat, winter barley, winter oats, winter wheat*** [1-7, 9, 10]; ***spring rye, triticale, undersown barley, undersown oats, undersown rye, undersown wheat, winter rye*** [7]. Perennial dicotyledons in ***established grassland*** [4-6, 10]; ***permanent pasture, rotational grassland*** [8, 9]; ***spring barley, spring oats, spring wheat, winter barley, winter oats, winter wheat*** [3, 7, 9]; ***spring rye, triticale, undersown barley, undersown oats, undersown rye, undersown wheat, winter rye*** [7]. Polygonums in ***established grassland*** [1, 2]; ***spring barley, spring oats, spring wheat, winter barley, winter oats, winter wheat*** [1-7, 9, 10]; ***spring rye, triticale, undersown barley, undersown oats, undersown rye, undersown wheat, winter rye*** [7]. Scentless mayweed in ***permanent pasture, rotational grassland*** [7]. Stinking smut in ***permanent pasture, rotational grassland*** [7]. Thistles in ***permanent pasture, rotational grassland*** [8, 9].

Approval information

- Mecoprop-P included in Annex I under EC Directive 91/414

Efficacy guidance

- Best results by application in warm, moist weather when weeds are actively growing

Restrictions

- Maximum number of treatments 1 per crop for cereals and 1 or 2 per yr on grass depending on label. The total amount of mecoprop-P applied in a single yr must not exceed the maximum total dose approved for any single product for the crop/situation
- Do not spray in cold or frosty conditions
- Do not spray if rain expected within 6 h
- Do not treat undersown grass until tillering begins
- Do not spray cereals undersown with clover or legume mixtures
- Do not roll, harrow within 7 d before or after spraying
- Do not treat crops suffering from stress from any cause
- Use product immediately following dilution; do not allow diluted product to stand before use [6]
- Avoid treatment when drift may damage neighbouring susceptible crops

Crop-specific information

- Latest use: before 1st node detectable for cereals; 7 d before cutting or 14 d before grazing grass
- Apply to winter sown crops from 5 expanded leaf stage (GS 15)
- Apply to spring sown cereals from 5 expanded leaf stage but before first node is detectable (GS 15-31)
- Treat grassland just before perennial weeds flower
- Transient crop prostration may occur after spraying but recovery is rapid

Environmental safety

- Harmful to fish or other aquatic life. Do not contaminate surface waters or ditches with chemical or used container
- Keep livestock out of treated areas for at least 2 wk and until foliage of poisonous weeds such as ragwort has died and become unpalatable

Hazard classification and safety precautions

Hazard R03 [1-6, 9, 10]; R04 [1-10]

Risk phrases R03a [1-4, 9]; R03c [1-6, 9, 10]; R04a [1-4, 7-9]; R04b [1-3, 5, 6, 9, 10]; R04d [5, 6, 10]

Operator protection A, C [1-10]; H [1-3, 5-10]; M [1-3, 5, 6, 8-10]; U05a [1-10]; U08 [1-6, 9, 10]; U09a [7, 8]; U20a [2]; U20b [1, 3-10]

Environmental protection E01, E13c [1-10]; E07 [1]; E07 [2-4] (2 wk); E07 [5]; E07 [6-10] (2 wk)

Consumer protection C03
Storage and disposal E26, E30a [1-10]; E31b [1-6, 9, 10]; E31c [7, 8]; E34 [1-7, 9, 10]
Medical advice M03

168 dichlobenil

A residual benzonitrile herbicide for woody crops and non-crop uses

Products

1	Casoron G	n2n	6.75% w/w	GR	09023
2	Casoron G	Scotts	6.75% w/w	GR	10709
3	Casoron G	Certis	6.75% w/w	GR	11632
4	Casoron G4	Crompton	4% w/w	GR	09215
5	Casoron G4	Scotts	4% w/w	GR	10708
6	Embargo G	SumiAgro Amenity	6.75% w/w	GR	10367
7	Luxan Dichlobenil Granules	Luxan	6.75% w/w	GR	09250
8	Midstream GSR	Scotts	20% w/w	GR	11674
9	Sierraron G	Scotts	6.75% w/w	GR	09675
10	Standon Dichlobenil 6G	Standon	6.75% w/w	GR	08874

Uses

Annual dicotyledons in ***grapevines*** *(off-label)*, ***rhubarb*** *(off-label - established)* [3]. Annual grasses in ***rhubarb*** *(off-label - established)* [3]. Annual weeds in ***apple orchards***, ***blackberries***, ***blackcurrants***, ***gooseberries***, ***loganberries***, ***pear orchards***, ***raspberries***, ***redcurrants*** [3, 6, 7, 10]; ***established woody ornamentals***, ***roses*** [1, 2, 4-7, 9, 10]; ***ornamental trees*** [4-6, 10]. Aquatic weeds in ***aquatic situations*** [1-3, 7-9]. Perennial dicotyledons in ***apple orchards***, ***blackberries***, ***blackcurrants***, ***gooseberries***, ***loganberries***, ***pear orchards***, ***raspberries***, ***redcurrants*** [3, 6, 7, 10]; ***established woody ornamentals***, ***roses*** [1, 2, 4-7, 9, 10]; ***grapevines*** *(off-label)*, ***rhubarb*** *(off-label - established)* [3]; ***ornamental trees*** [4-6, 10]. Perennial grasses in ***apple orchards***, ***blackberries***, ***blackcurrants***, ***gooseberries***, ***loganberries***, ***pear orchards***, ***raspberries***, ***redcurrants*** [3, 6, 7, 10]; ***established woody ornamentals***, ***roses*** [1, 2, 4-7, 9, 10]; ***ornamental trees*** [4-6, 10]; ***rhubarb*** *(off-label - established)* [3]. Total vegetation control in ***land not intended to bear vegetation*** [1-3, 6, 7, 9, 10]; ***non-crop areas*** [1, 2, 9]. Volunteer potatoes in ***potato dumps*** *(blight prevention)* [3, 6, 7, 10].

Specific Off-Label Approvals (SOLAs)

- ***grapevines*** *(OLA 030216) Jan 2008* [3]
- ***rhubarb*** *(established) (OLA 010913) Dec 2008* [3]

Approval information

- Approved for aquatic weed control [1, 3, 7-9]. See notes in Section 1 on use of herbicides in or near water

Efficacy guidance

- Best results achieved by application in winter to moist soil during cool weather, particularly if rain follows soon after. Do not disturb treated soil by hoeing
- In aquatic situations apply as soon as active growth commences to reduce the likelihood of subsequent water deoxygenation caused by die-back of heavy weed stands. Later partial treatment for rooted species may be effective [1, 3, 7-9]
- Where mulching is practised, apply beforehand
- Lower rates control annuals, higher rates perennials, see label for details
- Residual activity lasts 3-6 mth with selective, up to 12 mth with non-selective rates
- Do not use on fen peat or moss soils as efficacy will be severely impaired
- For control of emergent, floating and submerged aquatics apply to water surface in early spring. Intended for use in still or sluggish flowing water [1, 3, 7-9]
- Control in water is effective where flow does not exceed 90 m/hr (2.5 cm/second). Treatment of faster flowing water will only be effective if the flow can be checked for a minimum of 7 d after application

Restrictions

- Maximum number of treatments 1 per yr
- Maximum total dose equivalent to one full dose treatment
- Do not treat crops established for less than 2 yr. See label for lists of resistant and sensitive species. Do not treat stone fruit or Norway spruce (Christmas trees)

SECTION 4

- Do not apply within 300 m of glasshouses or hops or near areas underplanted with bulbs, annuals or herbaceous stock
- Do not apply to frozen, snow-covered or waterlogged ground or when crop foliage wet
- Do not apply to sites less than 18 mth before replanting or sowing
- Store well away from corms, bulbs, tubers and seed

Crop-specific information

- Latest use: before end of spring for aquatic situations; before onset of spring growth for potato dumps and all other crops. See labels for precise details
- Apply to crops in dormant period. See label for details of timing and rates
- Apply evenly by hand applicator or suitable mechanical spreader
- Apply to potato dump sites before emergence of potatoes for blight prevention [3-5, 7, 10]

Environmental safety

- Harmful to aquatic organisms. May cause long-term adverse effects to the aquatic environment [8]
- Harmful to fish or other aquatic life. Do not contaminate surface waters or ditches with chemical or used container [1-10]
- If fish are known to be present in a water body intended for treatment for aquatic weed control, partial treatment should be adopted. See label for details [8]
- Do not dump surplus herbicide in water or ditch bottoms [1-3, 7, 9]
- Use not permitted on reservoirs that form part of a water supply system [1, 3, 7-9]
- Do not use treated water for irrigation purposes within 2 wk of treatment or until concentration of dichlobenil falls below 0.3 ppm [1, 3, 7-9]
- Use as an aquatic herbicide subject to requirements set out in *Guidelines for the use of herbicides in or near watercourses or lakes* available from Defra

Hazard classification and safety precautions

Hazard R04 [6, 7]; R14 [8]
Risk phrases R04e [6, 7]; R14a [8]
Operator protection A, H [6, 7]; U05a [6, 7]; U20b [2, 9]; U20c [1, 3-8, 10]
Environmental protection E01 [6, 7]; E13c [1-10]; E19 [1-3, 7-9]; E21 [2, 8]; E21 [9] (2 wk)
Consumer protection C03 [6, 7]
Storage and disposal E25 [2, 4, 5, 10]; E29 [2, 4, 5, 8, 10]; E30a, E32a [1-10]; E34 [7]
Medical advice M03 [7]

169 dichlorophen

A chlorophenol moss-killer, fungicide, bactericide and algicide

Products

1	50/50 Liquid Mosskiller	Vitax	360 g/l	SL	07191
2	Enforcer	Scotts	360 g/l	SL	09288
3	Fungo	Dax	170 g/l	SL	H4768
4	Mascot Mosskiller	Rigby Taylor	360 g/l	SL	02439
5	Mossicide	SumiAgro Amenity	360 g/l	SL	09606
6	n2n Mosskiller	n2n	360 g/l	SL	11823
7	Panacide M	Coalite	360 g/l	SL	05611
8	Panacide TS	Coalite	480 g/l	SL	05612
9	Super Mosstox	Bayer Environ.	360 g/l	SL	09942

Uses

Algae in ***hard surfaces, paths and drives*** [3]. Fungus diseases in ***hard surfaces*** [1, 2, 6]; ***managed amenity turf, paths and drives*** [1]. Mosses in ***hard surfaces*** [1-6, 9]; ***lawns*** [5]; ***managed amenity turf*** [1, 2, 4, 6-9]; ***paths and drives*** [1, 3, 4, 9]. Red thread in ***managed amenity turf*** [9].

Efficacy guidance

- For moss control in turf spray at any time when moss is growing. Supplement spraying with other measures to improve fertility and drainage
- Rake out dead moss 2-3 wk after treatment
- Treat hard surfaces when rain not expected and brush away dead material when treatment fully dried
- Efficacy will be impaired if used during freezing, frosty or drought conditions

Restrictions
- Dose must not exceed 1 litre in 20 litres water on turf, or 1 litre in 60 litres on hard surfaces [2]
- Keep unprotected persons out of treated areas for 48 h or until surfaces are dry

Environmental safety
- Prevent any surface run-off from entering storm drains
- Dangerous (harmful [9]) to fish or other aquatic life. Do not contaminate surface waters or ditches with chemical or used container [1-4, 6-8]

Hazard classification and safety precautions

Hazard R04
Risk phrases R04a, R04b [1-9]; R04d [3]
Operator protection A, C, H [1-4, 6-9]; M [1, 2, 4, 6-9]; U02 [1, 2, 4-8]; U04a [1, 2, 4, 5, 7, 8]; U05a [1-9]; U08, U15 [1, 4, 6]; U09a [2, 3, 5, 7-9]; U14 [1, 2, 4, 6]; U17 [2, 4, 6]; U20a [5, 7, 8]; U20b [1-4, 6]; U20c [9]
Environmental protection E01 [1-9]; E02 [1, 2, 4-6] (48 h); E02 [7, 8]; E13b [1-4, 6-8]; E13c [9]; E15 [3]; E20 [1-7]
Consumer protection C03 [1-9]; C09 [1-8]
Storage and disposal E26 [1, 4, 9]; E30a [1-9]; E31a [1, 2, 4, 6-9]
Medical advice M03 [6-8]; M05 [3]

170 dichlorophen + ferrous sulphate

A mosskiller/fertilizer mixture for use on turf

Products

SHL Lawn Sand Plus	Sinclair	0.3:10% w/w	GR	04439

Uses

Mosses in ***amenity turf***.

Efficacy guidance
- Apply to established turf from late spring to early autumn when soil moist but not when grass wet or damp with dew
- Heavy infestations may need a repeat treatment

Restrictions
- Maximum number of treatments 2 per yr
- Do not treat newly sown grass or freshly laid turf for the first year
- Do not apply during drought or freezing conditions or when rain imminent
- Avoid walking on treated areas until it has rained or turf has been watered
- Do not mow for 3-4 d before or after application

Environmental safety
- Harmful to fish or other aquatic life. Do not contaminate surface waters or ditches with chemical or used container

Hazard classification and safety precautions

Operator protection U20c
Environmental protection E01, E13d
Storage and disposal E30a, E32a

171 1,3-dichloropropene

A halogenated hydrocarbon soil nematicide

Products

Telone II	Dow	94% w/w	VP	05749

Uses

Free-living nematodes in ***potatoes***. Nematodes in ***hops***, ***raspberries***, ***strawberries***. Potato cyst nematode in ***potatoes***. Stem and bulb nematodes in ***narcissi***. Stem nematodes in ***strawberries***. Virus vectors in ***hops***, ***raspberries***, ***strawberries***.

Efficacy guidance

- Before treatment soil should be in friable seed bed condition above 5°C, with adequate moisture and all crop remains decomposed. Tilling to 30 cm improves results
- Apply with sub-surface A-blade injector combined with soil-sealing roller or, in hops, with a hollow tined injector. See label for details of suitable machinery
- Leave soil undisturbed for 21 d after treatment. Wet or cold soils need longer exposure
- For control of potato-cyst nematode contact firm's representative, check nematode level and use in integrated control programme
- For use in hops apply in May or Jun following the yr of grubbing and leave as long as possible before replanting
- Do not use water to clean out apparatus nor use aluminium or magnesium alloy containers which may corrode

Restrictions

- Maximum number of treatments 2 per yr for hops, 1 per yr for other crops
- Do not drill or plant until odour of fumigant is eliminated
- Do not use fertilizer containing ammonium salts after treatment. Use only nitrate nitrogen fertilizers until crop well established and soil temperature above 18°C
- Do not use on heavy clays or soils with many large stones

Crop-specific information

- Latest use: 6 wk before planting for raspberries, strawberries; pre-planting of crop for hops, narcissi, potatoes

Following crops guidance

- Allow at least 6 wk after treatment before planting raspberries or strawberries

Environmental safety

- Harmful to fish or other aquatic life. Do not contaminate surface waters or ditches with chemical or used container

Hazard classification and safety precautions

Hazard R02, R03, R04, R07d
Risk phrases R02c, R03a, R03b, R04a, R04b, R04c, R04e, R04g
Operator protection A, C, H, M; U02, U04a, U05a, U09a, U14, U15, U19, U20a
Environmental protection E01, E13c
Consumer protection C03
Storage and disposal E24, E28, E30b, E33, E34
Medical advice M04, M05

172 dichlorprop-P

A translocated phenoxy herbicide for use in cereals

See also *2,4-D + dichlorprop-P + MCPA + mecoprop-P*
dicamba + dichlorprop-P + ferrous sulphate + MCPA
dicamba + dichlorprop-P + MCPA

Products

Headland Link	Headland	500 g/l	SL	11091

Uses

Annual dicotyledons in ***spring barley, spring oats, spring wheat, winter barley, winter oats, winter wheat***. Charlock in ***spring barley, spring oats, spring wheat, winter barley, winter oats, winter wheat***. Chickweed in ***spring barley, spring oats, spring wheat, winter barley, winter oats, winter wheat***. Fat hen in ***spring barley, spring oats, spring wheat, winter barley, winter oats, winter wheat***. Perennial dicotyledons in ***spring barley, spring oats, spring wheat, winter barley, winter oats, winter wheat***. Redshank in ***spring barley, spring oats, spring wheat, winter barley, winter oats, winter wheat***.

Efficacy guidance

- Best results achieved by application to young seedling weeds in good growing conditions in a strongly competing crop

- Where cleavers are to be controlled increase water volume if crop or weed is dense
- Control of redshank is best achieved in warm temperatures even if this means allowing the weed to grow beyond the seedling stage

Restrictions

- Maximum number of treatments 1 per crop
- Do not apply during cold weather, drought, rain or when rain is expected
- Do not roll or harrow within 7 d before or after spraying
- Do not apply immediately before sowing any crop

Crop-specific information

- Latest use: before second node detectable (GS 32)
- Spray winter cereals in the spring from the leaf sheath erect stage; spray spring cereals from the one leaf stage

Following crops guidance

- Keep livestock out of treated areas until foliage of any poisonous weeds such as ragwort has died and become unpalatable

Environmental safety

- Harmful to fish or other aquatic life. Do not contaminate surface waters or ditches with chemical or used container
- Do not spray in windy conditions when drift may cause damage to neighbouring sensitive crops

Hazard classification and safety precautions

Hazard R03, R04
Risk phrases R03c, R04d, R04e
Operator protection A, C, H; U05a, U08, U20b
Environmental protection E01, E07, E13c
Consumer protection C03
Storage and disposal E26, E30a, E31a, E34
Medical advice M03

173 dichlorprop-P + ferrous sulphate + MCPA

A herbicide/fertilizer combination for moss and weed control in turf

Products

1	SHL Granular Feed, Weed & Mosskiller	Sinclair	0.2:10.9:0.3% w/w	GR	10972
2	SHL Turf Feed, Weed & Mosskiller	Sinclair	0.2:10.9:0.3% w/w	DP	10973

Uses

Annual dicotyledons in ***managed amenity turf***. Buttercups in ***managed amenity turf***. Mosses in ***managed amenity turf***. Perennial dicotyledons in ***managed amenity turf***.

Efficacy guidance

- Apply between Mar and Sep when grass in active growth and soil moist
- A repeat treatment may be needed after 4-6 wk to control perennial weeds or if moss regrows
- Water in if rainfall does not occur within 48 h [2]

Restrictions

- Do not treat newly sown grass for 6 mth after establishment [1]
- Do not apply during drought or freezing conditions or when rain imminent
- Avoid walking on treated areas until it has rained or turf has been watered
- Do not mow for 3-4 d before or after application
- Do not use first 4 mowings after treatment for mulching. Mowings should be composted for 6 mth before use
- Do not treat areas of fine turf such as golf or bowling greens [1]
- Avoid contact with tarmac surfaces as staining may occur

Environmental safety

- Avoid drift onto nearby plants and borders [1]

Hazard classification and safety precautions

Operator protection U20c

Environmental protection E01
Storage and disposal E30a, E32a

174 dichlorprop-P + MCPA

A translocated herbicide for use in cereals and turf

Products

1	SHL Granular Feed and Weed	Sinclair	0.2:0.3% w/w	GR	10970
2	SHL Turf Feed and Weed	Sinclair	0.2:0.31% w/w	DP	10963

Uses

Annual dicotyledons in ***managed amenity turf*** [1, 2]. Buttercups in ***managed amenity turf*** [1, 2]. Clovers in ***managed amenity turf*** [2]. Daisies in ***managed amenity turf*** [2]. Dandelions in ***managed amenity turf*** [2]. Perennial dicotyledons in ***managed amenity turf*** [1, 2].

Efficacy guidance

- Best results achieved by application to young seedling weeds in active growth between Mar and Sep
- Water in if rainfall does not occur within 48 h [2]

Restrictions

- Do not use on newly sown turf for 6 mth after establishment
- Do not cut or mow for 2-3 d before and after treatment
- Do not spray during cold weather, if rain or frost expected, if crop wet or in drought
- The first four mowings should not be used for mulching and should be composted for 6 mth before use
- Avoid close mowing for 3-4 d before or after treatment
- Do not treat areas of fine turf such as golf or bowling greens [1]

Environmental safety

- Avoid drift onto nearby plants and borders [2]
- Avoid contact with tarmac surfaces as staining may occur

Hazard classification and safety precautions

Operator protection U20c
Environmental protection E01
Storage and disposal E30a, E32a

175 dichlorprop-P + MCPA + mecoprop-P

A translocated herbicide mixture for winter and spring cereals

Products

Hymec Triple	Agrichem	310:160:130 g/l	SL	09949

Uses

Annual dicotyledons in ***durum wheat***, ***spring barley***, ***spring wheat***, ***winter barley***, ***winter wheat***. Chickweed in ***durum wheat***, ***spring barley***, ***spring wheat***, ***winter barley***, ***winter wheat***. Cleavers in ***durum wheat***, ***spring barley***, ***spring wheat***, ***winter barley***, ***winter wheat***. Field pansy in ***durum wheat***, ***spring barley***, ***spring wheat***, ***winter barley***, ***winter wheat***. Mayweeds in ***durum wheat***, ***spring barley***, ***spring wheat***, ***winter barley***, ***winter wheat***. Poppies in ***durum wheat***, ***spring barley***, ***spring wheat***, ***winter barley***, ***winter wheat***.

Efficacy guidance

- Best results obtained if application is made while majority of weeds are at seedling stage but not if temperatures are too low
- Optimum results achieved by spraying when temperature is above 10°C. If temperatures are lower delay spraying until growth becomes more active

Restrictions

- Maximum number of treatments 1 per crop
- Do not spray in windy conditions where spray drift may cause damage to neighbouring crops, especially sugar beet, oilseed rape, peas, turnips and most horticultural crops including lettuce and tomatoes under glass

Crop-specific information

- Latest use: before second node detectable (GS 32) for all crops

Environmental safety

- Harmful to fish or other aquatic life. Do not contaminate surface waters or ditches with chemical or used container

Hazard classification and safety precautions

Hazard R03, R04
Risk phrases R03c, R04d
Operator protection A, C, H, M; U05a, U08, U20b
Environmental protection E01, E13c
Consumer protection C03
Storage and disposal E26, E30a, E31a, E34
Medical advice M03

176 dichlorvos

A contact and fumigant organophosphorus insecticide

Products

1	Luxan Dichlorvos 600	Luxan	600 g/l	EC	08297
2	Luxan Dichlorvos Aerosol 15	Luxan	120 g/l	AE	08298
3	Nuvan 500 EC	Novartis A H	500 g/l	EC	08590

Uses

Beetles in ***poultry houses*** [3]. Flies in ***poultry houses*** [3]. Leaf miner in ***aubergines*** *(off-label)*, ***basil*** *(off-label)*, ***bay*** *(off-label)*, ***borage*** *(off-label)*, ***camomile*** *(off-label)*, ***caraway*** *(off-label)*, ***chervil*** *(off-label)*, ***chives*** *(off-label)*, ***coriander*** *(off-label)*, ***dill*** *(off-label)*, ***fenugreek*** *(off-label)*, ***feverfew*** *(off-label)*, ***lovage*** *(off-label)*, ***marjoram*** *(off-label)*, ***mint*** *(off-label)*, ***parsley*** *(off-label)*, ***protected beans*** *(off-label)*, ***protected broccoli*** *(off-label)*, ***protected brussels sprouts*** *(off-label)*, ***protected cabbages*** *(off-label)*, ***protected calabrese*** *(off-label)*, ***protected cauliflowers*** *(off-label)*, ***protected courgettes*** *(off-label)*, ***protected endives*** *(off-label)*, ***protected gherkins*** *(off-label)*, ***protected kohlrabi*** *(off-label)*, ***protected lettuce*** *(off-label)*, ***protected ornamentals*** *(off-label)*, ***protected peas*** *(off-label)*, ***protected peppers*** *(off-label)*, ***protected pumpkins*** *(off-label)*, ***protected radicchio*** *(off-label)*, ***protected tomatoes*** *(off-label)*, ***rosemary*** *(off-label)*, ***sage*** *(off-label)*, ***savory*** *(off-label)*, ***sorrel*** *(off-label)*, ***tarragon*** *(off-label)*, ***thyme*** *(off-label)* [1]. Mosquitoes in ***poultry houses*** [3]. Poultry ectoparasites in ***poultry houses*** [3]. Western flower thrips in ***aubergines*** *(off-label)*, ***basil*** *(off-label)*, ***bay*** *(off-label)*, ***borage*** *(off-label)*, ***camomile*** *(off-label)*, ***caraway*** *(off-label)*, ***chervil*** *(off-label)*, ***chives*** *(off-label)*, ***coriander*** *(off-label)*, ***dill*** *(off-label)*, ***fenugreek*** *(off-label)*, ***feverfew*** *(off-label)*, ***lovage*** *(off-label)*, ***marjoram*** *(off-label)*, ***mint*** *(off-label)*, ***parsley*** *(off-label)*, ***protected beans*** *(off-label)*, ***protected broccoli*** *(off-label)*, ***protected brussels sprouts*** *(off-label)*, ***protected cabbages*** *(off-label)*, ***protected calabrese*** *(off-label)*, ***protected cauliflowers*** *(off-label)*, ***protected courgettes*** *(off-label)*, ***protected endives*** *(off-label)*, ***protected garlic*** *(off-label)*, ***protected gherkins*** *(off-label)*, ***protected kohlrabi*** *(off-label)*, ***protected lettuce*** *(off-label)*, ***protected onions*** *(off-label)*, ***protected ornamentals*** *(off-label)*, ***protected peas*** *(off-label)*, ***protected peppers*** *(off-label)*, ***protected pumpkins*** *(off-label)*, ***protected radicchio*** *(off-label)*, ***protected salad onions*** *(off-label)*, ***protected shallots*** *(off-label)*, ***protected spinach*** *(off-label)*, ***protected tomatoes*** *(off-label)*, ***rosemary*** *(off-label)*, ***sage*** *(off-label)*, ***savory*** *(off-label)*, ***sorrel*** *(off-label)*, ***tarragon*** *(off-label)*, ***thyme*** *(off-label)* [1]; ***protected cucumbers*** [1, 2].

Specific Off-Label Approvals (SOLAs)

- ***aubergines, protected beans, protected courgettes, protected endives, protected gherkins, protected kohlrabi, protected lettuce, protected ornamentals, protected peas, protected peppers, protected pumpkins, protected radicchio, protected tomatoes*** *(OLA 990625) Dec 2008* [1]
- ***aubergines, basil, bay, borage, camomile, caraway, chervil, chives, coriander, dill, fenugreek, feverfew, lovage, marjoram, mint, parsley, protected beans, protected broccoli, protected brussels sprouts, protected cabbages, protected calabrese, protected cauliflowers, protected courgettes, protected endives, protected gherkins, protected kohlrabi, protected lettuce, protected ornamentals, protected peas, protected peppers, protected***

SECTION 4

***pumpkins**, **protected radicchio**, **protected tomatoes**, **rosemary**, **sage**, **savory**, **sorrel**, **tarragon**, **thyme** (OLA 990626) Dec 2008* [1]
- ***protected garlic**, **protected onions**, **protected salad onions**, **protected shallots**, **protected spinach** (OLA 990627) Dec 2008* [1]

Efficacy guidance
- For use in poultry houses apply as surface spray, cold fog or mist using suitable applicator. Direct away from poultry. See label for details [3]
- Product mainly active in vapour phase and is inactive on insect eggs. Best results obtained from programme of treatments on dry crops when temperatures 15-25°C [1, 2]

Restrictions
- Dichlorvos is subject to the Poisons Rules 1982 and the Poisons Act 1972. See notes in Section 1
- This product contains an anticholinesterase organophosphorus compound. Do not use if under medical advice not to work with such compounds
- To be used only by operators trained in the use of the chemical and familiar with the precautionary measures to be observed
- Do not enter treated areas within 12 h and thoroughly ventilate before doing so [1, 2]
- Do not use on chrysanthemums in flower or on roses. Phytotoxicity may occur on cucumbers and other cucurbits
- Do not allow temperature to drop after treatment as condensation may lead to phytotoxicity. Inspect for phytotoxicity before retreatment [1, 2]
- Do not apply directly to growing cucumbers; apply between rows [1, 2]
- Do not apply other pesticides with product or one day before or after treatment, especially sulphur based formulations [1, 2]

Crop-specific information
- HI Brussels sprouts, cabbage 14 wk; cauliflower 9 wk; broccoli, calabrese 7 wk; kohlrabi 1 wk; cucurbits 48 h; other edible crops 24 h
- Treatment of cucumbers at flowering may cause flower or fruit abortion [1, 2]

Environmental safety
- Flammable [1, 3]
- Corrosive [1, 2]
- Keep in original container, tightly closed, in a safe place, under lock and key
- Keep unprotected persons out of treated areas for at least 12 h
- Harmful to game, wild birds and animals
- Extremely dangerous to bees. Do not apply to crops in flower or to those in which bees are actively foraging. Do not apply when flowering weeds are present
- Extremely dangerous to fish or other aquatic life. Do not contaminate surface waters or ditches with chemical or used container

Hazard classification and safety precautions

Hazard R01, R04, R05 [1, 2]; R02 [1-3]; R03 [1]; R07d [1, 3]
Risk phrases R01b, R04e [1, 2]; R02a, R02c [1-3]; R02b [3]; R04g [1]
Operator protection A, H [1-3]; C [1, 3]; D, J, K, M [1, 2]; U02, U04a, U05a, U08, U13, U19, U20a [1-3]; U10, U11 [1, 2]
Environmental protection E01, E10b, E12b, E13a [1-3]; E02 [1, 2] (12 h); E02 [3]
Consumer protection C02 [3]; C03 [1-3]
Storage and disposal E30b, E34 [1-3]; E31b [1, 3]
Medical advice M01, M04 [1-3]; M05 [1]

177 diclofop-methyl

A translocated phenoxypropionic herbicide available only in mixtures

178 diclofop-methyl + fenoxaprop-P-ethyl

A foliar acting herbicide mixture for grass control in wheat and barley

Products

1	Corniche	Aventis	250:20 g/l	EW	08947
2	Tigress Ultra	Aventis	250:20 g/l	EW	08946

Uses

Blackgrass in ***spring barley***, ***winter barley***, ***winter wheat***. Ryegrass in ***spring barley***, ***winter barley***, ***winter wheat***. Wild oats in ***spring barley***, ***winter barley***, ***winter wheat***.

Efficacy guidance

- Apply from emergence of crop up to before second node detectable (GS 32)
- Wild oats and blackgrass controlled from 2 fully expanded leaves to end of tillering but before 1st node detectable
- Ryegrass from seed controlled from 2 fully expanded leaves up to 3 tillers
- Spray is rainfast from 1 h after spraying. Do not apply to wet or icy foliage
- Any conditions resulting in moisture stress may reduce effectiveness especially with later applications to spring barley
- Performance not affected by soils with high OM content or high Kd factor

Restrictions

- Maximum number of treatments 1 per crop
- Do not apply to undersown crops or those to be undersown
- Broadcast crops should be sprayed post-emergence after root system well established
- Do not roll or harrow within 1 wk
- Do not spray crops under stress from drought, waterlogging etc
- Do not spray immediately before or after a sudden drop in temperature, a period of warm days/cold nights or when extremely low temperatures forecast
- Interval of 7 d must elapse before or after application of any other product
- Must not be mixed with weedkillers containing hormones or bifenox

Crop-specific information

- Latest use: before 2nd node detectable (GS 32)
- Can be sprayed in frosty weather provided crop hardened off
- Applications may be followed by transient leaf discoloration

Environmental safety

- Dangerous to fish or other aquatic life. Do not contaminate surface waters or ditches with chemical or used containers
- Keep livestock out of treated areas for at least 7 d

Hazard classification and safety precautions

Hazard R04
Risk phrases R04a
Operator protection A, C, H; U05a, U20b
Environmental protection E01, E13b; E07 (7 d)
Consumer protection C02 (6 wk); C03
Storage and disposal E26, E30a, E31b

179 dicloran

A protectant nitroaniline fungicide used as a glasshouse fumigant

Products

Fumite Dicloran Smoke	Certis	40% w/w	FU	09291

Uses

Botrytis in ***protected lettuce***. Rhizoctonia in ***protected lettuce***.

Efficacy guidance

- Treat at first sign of disease and at 14 d intervals if necessary
- For best results fumigate in late afternoon or evening when temperature is at least 16°C. Keep house closed overnight or for at least 4 h. Do not fumigate in bright sunshine, windy conditions or when temperature too high
- Water plants, paths and straw used as mulches several hours before fumigating. Ensure that foliage is dry before treatment

Restrictions

- Do not fumigate young seedlings or plants being hardened off
- Treat only crops showing strong growth. Do not treat crops which are dry at the roots
- Ventilate glasshouse thoroughly before re-entering

SECTION 4

Crop-specific information

- HI lettuce 14 d

Environmental safety

- Harmful to fish or other aquatic life. Do not contaminate surface waters or ditches with chemical or used container

Hazard classification and safety precautions

Hazard R04
Risk phrases R04a, R04c
Operator protection U05a, U19, U20a
Environmental protection E13c
Consumer protection C02, C03
Storage and disposal E30a, E32a

180 difenacoum

An anticoagulant coumarin rodenticide

See also cholecalciferol + difenacoum

Products

1	Deosan Rataway	JohnsonDiversey	0.005% w/w	RB	08803
2	Deosan Rataway Bait Bags	JohnsonDiversey	0.005% w/w	RB	08805
3	Neosorexa	Sorex	0.005% w/w	AB	H6773
4	Neosorexa Ratpacks	Sorex	0.005% w/w	AB	H6782
5	Sakarat D (Whole Wheat)	Killgerm	0.005% w/w	RB	H7109
6	Sakarat D Wax Bait	Killgerm	0.005% w/w	RB	H7489
7	Sorexa Gel	Sorex	0.005% w/w	RB	H6790

Uses

Mice in ***farm buildings*, *farmyards*** [1-7]. Rats in ***farm buildings*, *farmyards*** [1-5].

Efficacy guidance

- Difenacoum is a chronic poison and rodents need to feed several times before accumulating a lethal dose. Effective against rodents resistant to other commonly used anticoagulants
- Ready-to-use in baiting programme for indoor and outdoor use
- Lay small baits about 1 m apart throughout infested areas for mice, larger baits for rats near holes and along runs; place 2-4 blocks 5-10 m apart depending on severity of infestation
- Cover baits by placing in bait boxes, drain pipes or under boards
- Inspect bait sites frequently and top up as long as there is evidence of feeding
- Product formulated for application through a skeleton or caulking gun [6]

Restrictions

- Only for use by farmers, horticulturists and other professional users
- When working in rodent infested areas wear synthetic rubber/PVC gloves to protect against rodent-borne diseases

Environmental safety

- Cover bait to prevent access by children, animals or birds

Hazard classification and safety precautions

Operator protection A [7]; U13 [1-7]; U20a [1, 2, 4]; U20b [3, 5-7]
Storage and disposal E30a [1-7]; E32a [1, 2, 4-7]
Vertebrate/rodent control products V01a, V03a, V04a [1-4, 7]; V01b, V03b, V04b [6]; V02 [1-4, 6, 7]
Medical advice M03 [5, 6]

181 difenoconazole

A diphenyl-ether triazole protectant and curative fungicide

Products

1	Landgold Difenoconazole	Landgold	250 g/l	EC	09964
2	Plover	Syngenta	250 g/l	EC	08429

Uses

Alternaria in ***broccoli*** [2]; ***brussels sprouts, cabbages, calabrese, cauliflowers, spring oilseed rape, winter oilseed rape*** [1, 2]. Brown rust in ***winter wheat*** [1, 2]. Fungus diseases in ***protected hybrid pinks*** *(off-label - ground grown)*, ***protected hybrid pinks*** *(off-label - pot grown)*, ***protected sweet williams*** *(off-label - ground grown)*, ***protected sweet williams*** *(off-label - pot grown)* [2]. Late blight in ***celeriac*** *(off-label)*, ***celery*** *(off-label)* [2]. Light leaf spot in ***spring oilseed rape, winter oilseed rape*** [1, 2]. Ring spot in ***broccoli*** [2]; ***brussels sprouts, cabbages, calabrese, cauliflowers*** [1, 2]. Rust in ***asparagus*** *(off-label)* [2]. Septoria in ***winter wheat*** [1, 2]. Stem canker in ***spring oilseed rape, winter oilseed rape*** [1, 2]. Yellow rust in ***winter wheat*** [1, 2].

Specific Off-Label Approvals (SOLAs)

- ***asparagus*** *(OLA 011666) Dec 2008* [2]
- ***celeriac, celery*** *(OLA 021320) Apr 2006* [2]
- ***protected hybrid pinks, protected sweet williams*** *(ground grown) (OLA 011664) Dec 2008* [2]
- ***protected hybrid pinks, protected sweet williams*** *(pot grown) (OLA 011665) Dec 2008* [2]

Efficacy guidance

- For most effective control of Septoria, apply as part of a programme of sprays which includes a suitable flag leaf treatment
- Adequate control of yellow rust may require an earlier appropriate treatment
- Improved control of established infections on oilseed rape achieved by mixture with carbendazim. See label
- In cabbages a 3-spray programme should be used starting at the first sign of disease and repeated at 14-21 d intervals
- Product is fully rainfast 2 h after application

Restrictions

- Maximum number of treatments 3 per crop for cabbages; 2 per crop for oilseed rape; 1 per crop for wheat
- Apply to wheat any time from ear fully emerged stage but before early milk-ripe stage (GS 59-73)

Crop-specific information

- Latest use: before grain early milk-ripe stage (GS 73) for cereals; end of flowering for oilseed rape; not before harvest in yr of cropping for asparagus
- HI ornamentals 7 d; celery 14 d; brassicas, celeriac 21 d
- Recommended for use on all varieties of winter and spring sown oilseed rape and cabbage
- Treat oilseed rape in autumn from 4 expanded true leaf stage (GS 1,4). A repeat spray may be made in spring at the beginning of stem extension (GS 2,0) if visible symptoms develop

Environmental safety

- Harmful to fish or other aquatic life. Do not contaminate surface waters or ditches with chemical or used container

Hazard classification and safety precautions

Hazard R04
Risk phrases R04a, R04e
Operator protection A, C, H; U05a, U09a, U20b [1, 2]; U07 [2]
Environmental protection E01, E13c
Consumer protection C03
Storage and disposal E26, E29 [1, 2]; E30a, E31b [1]; E31a, E33, E34, E36 [2]

182 diflubenzuron

A selective, persistent, contact and stomach acting insecticide

Products

1	Dimilin Flo	Crompton	480 g/l	SC	08760
2	Dimilin Flo	Certis	480 g/l	SC	11056

Uses

Browntail moth in ***amenity trees and shrubs, hedges, nursery stock, ornamental specimens*** [2]. Bud moth in ***apples, pears*** [2]. Carnation tortrix moth in ***amenity trees and shrubs, hedges, nursery stock, ornamental specimens*** [2]. Caterpillars in ***broccoli, brussels sprouts,***

SECTION 4

cabbages, calabrese, cauliflowers [2]. Clouded drab moth in ***apples, pears*** [2]. Codling moth in ***apples, pears*** [2]. Fruit tree tortrix moth in ***apples, pears*** [2]. Houseflies in ***livestock houses, manure heaps, refuse tips*** [2]. Lackey moth in ***amenity trees and shrubs, hedges, nursery stock, ornamental specimens*** [2]. Oak leaf roller moth in ***forestry*** [2]. Pear sucker in ***pears*** [2]. Pine beauty moth in ***forestry*** [2]. Pine looper in ***forestry*** [2]. Plum fruit moth in ***plums*** [2]. Rust mite in ***apples, pears, plums*** [2]. Sciarid flies in ***mushrooms*** [1]. Small ermine moth in ***amenity trees and shrubs, hedges, nursery stock, ornamental specimens*** [2]. Tortrix moths in ***plums*** [2]. Winter moth in ***amenity trees and shrubs, apples, blackcurrants, forestry, hedges, nursery stock, ornamental specimens, pears, plums*** [2].

Approval information

- Approved for aerial application in forestry when average wind velocity does not exceed 18 knots and gusts do not exceed 20 knots [2]. See notes in Section 1

Efficacy guidance

- Most active on young caterpillars and most effective control achieved by spraying as eggs start to hatch
- Dose and timing of spray treatments vary with pest and crop. See label for details
- Addition of wetter recommended for use on brassicas and for pear sucker control in pears

Restrictions

- Maximum number of treatments 3 per yr for apples, pears; 2 per yr for plums, blackcurrants; 2 per crop for brassicas; 1 per spawning for mushrooms
- Before treating ornamentals check varietal tolerance on a small sample
- Do not use as a compost drench or incorporated treatment on ornamental crops
- Do not spray protected plants in flower or with flower buds showing colour
- For use only on the food crops specified on the label
- Do not apply directly to livestock/poultry

Crop-specific information

- HI apples, pears, plums, blackcurrants, brassicas 14 d
- Apply as casing mixing treatment (use immediately) on mushrooms or as post-casing drench

Environmental safety

- Do not spray from the air within 250 m horizontal distance of surface waters or ditches [2]
- LERAP Category B
- Broadcast air-assisted LERAP (20 m in orchards, 10 m in blackcurrants forestry or ornamentals) [2]

Hazard classification and safety precautions

Operator protection U20c

Environmental protection E05, E15, E16a, E16b [1, 2]; E17a [2] (20 m when used in orchards; 10 m when used in blackcurrants, forestry or ornamentals); E18 [2]

Storage and disposal E26 [2]; E30a, E32a [1, 2]

183 diflufenican

A shoot absorbed pyridinecarboxamide herbicide available only in mixtures

See also *bromoxynil + diflufenican + ioxynil*
carbetamide + diflufenican + oxadiazon
clodinafop-propargyl + diflufenican
clopyralid + diflufenican + MCPA

184 diflufenican + flurtamone

A contact and residual herbicide mixture for cereals

Products

1	Bacara	Aventis	100:250 g/l	SC	09976
2	Bacara	Bayer CropScience	100:250 g/l	SC	10744
3	Graduate	Aventis	80:400 g/l	SC	10776
4	Me2 Snap!	Me2	100:250 g/l	SC	10476

Uses

Annual dicotyledons in ***winter barley, winter wheat*** [1-4]. Annual meadow grass in ***winter barley, winter wheat*** [1-4]. Blackgrass in ***winter barley, winter wheat*** [1, 2, 4]. Loose silky bent in ***winter barley, winter wheat*** [1, 2, 4]. Volunteer oilseed rape in ***winter barley, winter wheat*** [1-4].

Efficacy guidance

- Apply pre-emergence or from when crop has first leaf unfolded before susceptible weeds pass recommended size
- Best results obtained on firm, fine seedbeds with adequate soil moisture present at and after application. Increase water volume where crop or weed foliage is dense
- Good weed control requires ash, trash and burnt straw to be buried during seed bed preparation
- Loose fluffy seedbeds should be rolled before application and the final seed bed should be fine and firm without large clods
- Speed of control depends on weather conditions and activity can be slow under cool conditions

Restrictions

- Maximum number of treatments 1 per crop. Maximum total dose of isoproturon 2.5 kg a.i./ha per crop
- Crops should be drilled to a normal depth of 25 mm and the seed well covered. Do not treat broadcast crops as uncovered seed may be damaged
- Do not use on crops undersown or to be undersown
- Do not treat frosted crops or when frost is imminent. Severe frost after application, or any other stress, may lead to transient discoloration or scorch
- Do not use on Sands or Very Light soils or those that are very stony or gravelly
- Do not use on waterlogged soils, or on crops subject to temporary waterlogging by heavy rainfall, as there is risk of persistent crop damage which may result in yield loss
- Do not use on soils with more than 10% organic matter
- Do not harrow at any time after application and do not roll autumn treated crops until spring

Crop-specific information

- Latest use: before 2nd node detectable (GS32)
- Take particular care to match spray swaths otherwise crop discoloration and biomass reduction may occur which may lead to yield reduction

Following crops guidance

- In the event of crop failure winter wheat may be redrilled immediately after normal cutlivation, and winter barley may be sown after ploughing. Fields must be ploughed and 20 wk must elapse before sowing spring crops of wheat, barley, oilseed rape, peas, field beans or potatoes
- Do not broadcast or direct drill oilseed rape or other brassica crops as a following crop on treated land. See label for detailed advice on preparing land for subsequent autumn cropping in the normal rotation
- Successive treatments of any products containing diflufenican can lead to soil build-up and inversion ploughing must precede sowing any following non-cereal crop. Even where ploughing occurs some crops may be damaged

Environmental safety

- Extremely dangerous to fish or other aquatic life. Do not contaminate surface waters or ditches with chemical or used container
- LERAP Category B

Hazard classification and safety precautions

Operator protection A, H; U20c
Environmental protection E13a, E16a, E16b
Storage and disposal E26, E30a, E31b

185 diflufenican + flurtamone + isoproturon

A contact and residual herbicide for winter cereals

Products

Ingot	Aventis	27:67:400 g/l	SC	09997

Uses

Annual dicotyledons in ***winter barley***, ***winter wheat***. Annual meadow grass in ***winter barley***, ***winter wheat***. Blackgrass in ***winter barley***, ***winter wheat***. Loose silky bent in ***winter barley***, ***winter wheat***. Volunteer oilseed rape in ***winter barley***, ***winter wheat***.

Approval information

- Isoproturon included in Annex I under EC Directive 91/414

Efficacy guidance

- Apply from when crop has first leaf unfolded before susceptible weeds pass recommended size
- Best results obtained on firm, fine seedbeds with adequate soil moisture. Good spray coverage of soil and weeds is essential
- Speed of control depends on weather conditions and activity will be reduced during prolonged dry periods especially in spring. Weed control, especially of grasses, may also be reduced in wet seasons or where heavy rain falls shortly after treatment

Restrictions

- Maximum number of treatments 1 per crop. Maximum total dose of isoproturon 2.5 kg a.i./ha per crop
- Crops should be drilled to a normal depth of 25 mm and the seed well covered. Do not treat broadcast crops
- Do not apply pre-emergence to wheat or barley
- Do not use on crops undersown or to be undersown
- Do not treat frosted crops or when frost is imminent. Severe frost after application may cause transient discoloration or scorch
- Do not use on Sands or Very Light soils or those that are very stony or gravelly as there is risk of crop damage
- Do not treat on soils with more than 10% organic matter
- Do not use on waterlogged soils, or on crops subject to temporary waterlogging by heavy rainfall, as there is risk of persistent crop damage which may result in yield loss
- Do not roll autumn treated crops until the spring and do not harrow at any time after application
- Take particular care to match spray swaths otherwise crop discoloration and biomass reduction may occur which may lead to yield reduction

Crop-specific information

- Latest use: before 2nd node detectable (GS32)

Following crops guidance

- Do not broadcast oilseed rape or other brassica crops as a following crop on treated land. See label for details of time and land preparation before sowing other crops in the event of failure of a treated crop
- Successive treatments of any products containing diflufenican can lead to soil build-up and inversion ploughing must precede sowing any following non-cereal crop. Even where ploughing occurs some crops may be damaged

Environmental safety

- Extremely dangerous to fish or other aquatic life. Do not contaminate surface waters or ditches with chemical or used container
- LERAP Category B
- Do not apply to dry, cracked or waterlogged soils where heavy rain may lead to contamination of drains by isoproturon

Hazard classification and safety precautions

Operator protection A, C, H; U20c
Environmental protection E13a, E16a, E16b
Storage and disposal E26, E30a, E31b

186 diflufenican + isoproturon

A contact and residual herbicide for use in winter cereals

Products

1	Fernpath Ipex	AgriGuard	50:500 g/l	SC	11391
2	Javelin	Aventis	62.5:500 g/l	SC	09998
3	Javelin	Aventis	62.5:500 g/l	SC	10440

Products – continued

4	Javelin Gold	Aventis	20:500 g/l	SC	09999
5	Panther	Aventis	50:500 g/l	SC	10008
6	Panther	Aventis	50:500 g/l	SC	10745
7	Panther WDG	Aventis	9.75:78.5% w/w	WG	10650
8	Standon Diflufenican 625	Standon	62.5:500 g/l	SC	10753
9	Standon Diflufenican-IPU	Standon	50:500 g/l	SC	11109

Uses

Annual dicotyledons in ***triticale, winter rye*** [1-6, 8, 9]; ***winter barley, winter wheat*** [1-9]. Annual grasses in ***triticale, winter rye*** [1-6, 8, 9]; ***winter barley, winter wheat*** [1-9]. Blackgrass in ***triticale, winter rye*** [1-6, 8, 9]; ***winter barley, winter wheat*** [1-9]. Wild oats in ***triticale, winter rye*** [1-6, 8, 9]; ***winter barley, winter wheat*** [1-9].

Approval information

- Isoproturon included in Annex I under EC Directive 91/414

Efficacy guidance

- May be applied in autumn or spring (but only post-emergence on wheat or barley). Best control normally achieved by early post-emergence treatment
- Best results by application to fine, firm seedbed moist at or after application
- Weeds controlled from before emergence to 6 true leaf stage
- Apply to moist, but not waterlogged, soils
- Any trash or ash should be buried during seedbed preparation

Restrictions

- Maximum number of treatments 1 per crop. Maximum total dose of isoproturon 2.5 kg a.i./ha per yr
- On triticale and winter rye only treat named varieties and apply as pre-emergence spray
- Do not use on other cereals, broadcast or undersown crops or crops to be undersown
- Do not use on Sands or Very Light soils, or those that are very stony or gravelly or on soils with more than 10% organic matter
- Do not spray when heavy rain is forecast or on crops suffering from stress, frost, deficiency, pest or disease attack
- Do not harrow after application nor roll autumn-treated crops until spring

Crop-specific information

- Latest use: before second node detectable (GS 32) [4-6, 9], or before end Feb [2, 3, 7], for wheat and barley; pre-emergence of crop for triticale and rye
- Spray winter wheat and barley post-emergence before end Feb [2, 4], before crop reaches second node detectable stage (GS 32) [5]
- Drill crop to normal depth (25 mm) and ensure seed well covered

Following crops guidance

- See label for details of time and land preparation needed before sowing other crops (12 wk for most crops). In the event of crop failure winter wheat can be redrilled without ploughing, other crops only after ploughing - see label for details
- Successive treatments of any products containing diflufenican can lead to soil build up and inversion ploughing must precede sowing any non-cereal crop. Even where ploughing occurs some crops may be damaged

Environmental safety

- Extremely dangerous (dangerous [8]) to fish or other aquatic life. Do not contaminate surface waters or ditches with chemical or used container [1-3, 6, 7, 9]
- Do not apply to dry, cracked or waterlogged soils where heavy rain may lead to contamination of drains by isoproturon
- LERAP Category B

Hazard classification and safety precautions

Operator protection A, C, H [1-3, 6-9]; D [7]; U08, U19 [9]; U20a [1]; U20b [4, 5, 8, 9]; U20c [2, 3, 6, 7]

Environmental protection E13a [1-3, 6, 7, 9]; E13b [8]; E13c [4, 5]; E16a, E16b [1-9]

Storage and disposal E26 [2-9]; E30a [1-9]; E31b [1-6, 8, 9]; E32a [7]

SECTION 4

187 diflufenican + trifluralin

A contact and residual herbicide for use in winter cereals

Products

Ardent	Aventis	40:400 g/l	SC	09968

Uses

Annual dicotyledons in ***triticale***, ***winter barley***, ***winter rye***, ***winter wheat***. Annual meadow grass in ***winter barley***, ***winter wheat***.

Efficacy guidance

- Weeds controlled pre-emergence up to 4 leaves (2 leaves for annual meadow grass)
- Best results achieved on firm, fine moist seedbeds
- Rolling after autumn treatment will reduce weed control. Roll in spring if necessary
- Speed of control depends on temperature and growing conditions; activity can be slow under cool conditions

Restrictions

- Maximum number of treatments 1 per crop
- Ensure that crop is evenly drilled and seed well covered. Do not treat broadcast crops or those that are frosted or stressed
- Do not treat undersown crops or those to be undersown
- Do not harrow after application
- Do not use on oats or any spring sown cereals
- Do not use on Sands or Very Light soils, or those that are very stony or gravelly
- Do not use on soils with more than 10% organic matter; this may include newly ploughed grassland for a time

Crop-specific information

- Latest use: before 2nd tiller stage (GS 22), or end Nov, whichever is sooner
- Apply pre- or early post-emergence of crop up to maximum recommended weed size
- Crops occasionally show transient leaf discoloration after treatment. Symptoms are quickly outgrown and yield not affected

Following crops guidance

- See label for details of time and land preparation needed in the event of crop failure and for normal subsequent cropping
- Successive treatments of any products containing diflufenican can lead to soil build-up and inversion ploughing must precede sowing any following non-cereal crop. Even where ploughing occurs some crops may be damaged

Environmental safety

- Harmful to fish or other aquatic life. Do not contaminate surface waters or ditches with chemical or used container
- LERAP Category B

Hazard classification and safety precautions

Hazard R04
Risk phrases R04a
Operator protection A, C; U05a, U08, U13, U20b
Environmental protection E01, E13c, E16a, E16b
Consumer protection C03
Storage and disposal E26, E30a, E31b

188 di-1-p-menthene

An antitranspirant and coating agent available only in mixtures

See also chlormequat with di-1-p-menthene

189 dimethoate

A contact and systemic organophosphorus insecticide and acaricide

Products

1	BASF Dimethoate 40	BASF	400 g/l	EC	00199
2	Danadim	Headland	400 g/l	EC	11550
3	Rogor L40	Isagro	400 g/l	EC	07611

Uses

Aphids in ***broccoli***, ***brussels sprouts***, ***calabrese***, ***cauliflowers***, ***ornamental specimens*** [2]; ***fodder beet*** *(excluding Myzus persicae)*, ***mangels*** *(excluding Myzus persicae)*, ***sugar beet*** *(excluding Myzus persicae)*, ***winter wheat*** [1-3]; ***grass seed crops***, ***mangel seed crops***, ***ornamental plant production***, ***spring rye*** [1]; ***leaf spinach***, ***mangel seed crops*** *(excluding Myzus persicae)*, ***peas***, ***potatoes*** *(excluding Myzus persicae)*, ***red beet*** [3]; ***lettuce***, ***red beet*** *(excluding Myzus persicae)*, ***triticale***, ***winter rye*** [1, 2]; ***spring wheat***, ***sugar beet seed crops*** *(excluding Myzus persicae)* [1, 3]. Cabbage aphid in ***broccoli***, ***brussels sprouts***, ***calabrese***, ***cauliflowers*** [1]. Insect pests in ***bulb onions*** *(off-label)*, ***bulb onions*** *(off-label - seedlings)*, ***calabrese*** *(off-label)*, ***celeriac*** *(off-label)*, ***chinese cabbage*** *(off-label)*, ***collards*** *(off-label)*, ***garlic*** *(off-label)*, ***garlic*** *(off-label - seedlings)*, ***kale*** *(off-label)*, ***kohlrabi*** *(off-label)*, ***leeks*** *(off-label)*, ***leeks*** *(off-label - seedlings)*, ***protected broccoli*** *(off-label)*, ***protected brussels sprouts*** *(off-label)*, ***protected cabbages*** *(off-label)*, ***protected calabrese*** *(off-label)*, ***protected cauliflowers*** *(off-label)*, ***protected celeriac*** *(off-label)*, ***protected chinese cabbage*** *(off-label)*, ***protected collards*** *(off-label)*, ***protected garlic*** *(off-label - seedlings)*, ***protected kale*** *(off-label)*, ***protected kohlrabi*** *(off-label)*, ***protected leeks*** *(off-label - seedlings)*, ***protected onions*** *(off-label - seedlings)*, ***protected roscoff cauliflowers*** *(off-label)*, ***protected salad onions*** *(off-label - seedlings)*, ***protected savoy cabbage*** *(off-label)*, ***protected shallots*** *(off-label - seedlings)*, ***protected spinach*** *(off-label)*, ***roscoff cauliflowers*** *(off-label)*, ***salad onions*** *(off-label - seedlings)*, ***savoy cabbage*** *(off-label)*, ***shallots*** *(off-label)*, ***shallots*** *(off-label - seedlings)* [1]. Leaf miner in ***fodder beet***, ***mangels***, ***red beet***, ***sugar beet*** [1-3]; ***leaf spinach***, ***mangel seed crops***, ***sugar beet seed crops*** [3]; ***ornamental plant production*** [1]; ***ornamental specimens*** [2]. Mangold fly in ***fodder beet***, ***mangels***, ***red beet***, ***sugar beet*** [2]. Pea midge in ***peas*** [3]. Red spider mites in ***ornamental plant production*** [1]. Thrips in ***peas*** [3]. Wheat bulb fly in ***spring wheat*** [1, 3]; ***winter wheat*** [1-3].

Specific Off-Label Approvals (SOLAs)

- ***bulb onions***, ***garlic***, ***leeks***, ***shallots*** *(OLA 001511) May 2005* [1]
- ***bulb onions***, ***garlic***, ***leeks***, ***protected garlic***, ***protected leeks***, ***protected onions***, ***protected salad onions***, ***protected shallots***, ***salad onions***, ***shallots*** *(seedlings) (OLA 001511) May 2005* [1]
- ***calabrese***, ***celeriac***, ***chinese cabbage***, ***collards***, ***kale***, ***kohlrabi***, ***protected broccoli***, ***protected brussels sprouts***, ***protected cabbages***, ***protected calabrese***, ***protected cauliflowers***, ***protected celeriac***, ***protected chinese cabbage***, ***protected collards***, ***protected kale***, ***protected kohlrabi***, ***protected roscoff cauliflowers***, ***protected savoy cabbage***, ***protected spinach***, ***roscoff cauliflowers***, ***savoy cabbage*** *(OLA 940389) Dec 2008* [1]

Approval information

- Approved for aerial application on cereals, peas, ware potatoes, sugar beet [1, 3]. See notes in Section 1

Efficacy guidance

- Chemical has quick knock-down effect and systemic activity lasts for up to 14 d
- With some crops, products differ in range of pests listed as controlled. Uses section above provides summary. See labels for details
- For most pests apply when pest first seen and repeat 2-3 wk later or as necessary. Timing and number of sprays varies with crop and pest. See labels for details
- Best results achieved when crop growing vigorously. Systemic activity reduced when crops suffering from drought or other stress
- In hot weather apply in early morning or late evening
- Where aphids or spider mites resistant to organophosphorus compounds occur control is unlikely to be satisfactory and repeat treatments may result in lower levels of control

SECTION 4

Restrictions

- Maximum number of treatments 6 per crop for Brussels sprouts, broccoli, cauliflower, calabrese, lettuce; 4 per crop for grass seed crops; 1 per crop for beet crops (see also below), cereals, watercress
- Contains an anticholinesterase organophosphorus compound. Do not use if under medical advice not to work with such compounds
- On beet crops only one treatment per crop may be made for control of leaf miners (max 84 g a.i./ha) and black bean aphid (max 400 g a.i./ha)
- In beet crops resistant strains of peach-potato aphid (*Myzus persicae*) are common and dimethoate products must not be used to control this pest
- Test for varietal susceptibility on all unusual plants or new cultivars
- Do not tank mix with alkaline materials. See label for recommended tank-mixes
- Consult processor before spraying crops grown for processing
- Must not be applied to cereals if any product containing a pyrethroid insecticide or dimethoate has been sprayed after the start of ear emergence (GS 51)

Crop-specific information

- Latest use: before 30 Jun in yr of harvest for beet crops; flowering just complete stage (GS 30) for ground applications to cereals; before 31 Mar in yr of harvest for aerial application to cereals; before 6 true leaves for onions, leeks, shallots, garlic; before transfer to cropping beds for watercress
- HI brassicas 21 d; grass seed crops, lettuce 14 d; onions, garlic, leeks, shallots, watercress 7 d

Environmental safety

- Flammable
- Harmful to game, wild birds and animals [1, 2]
- Harmful to livestock. Keep all livestock out of treated areas for at least 7 d
- Likely to cause adverse effects on beneficial arthropods
- Dangerous (high risk [2]) to bees. Do not apply to crops in flower, except as directed on peas, or to those in which bees are actively foraging. Do not apply when flowering weeds are present [1, 3]
- Surface residues may also cause bee mortality following spraying
- Dangerous to fish or other aquatic life. Do not contaminate surface waters or ditches with chemical or used container
- LERAP Category A
- Broadcast air-assisted LERAP (18 m) [2]
- Do not treat cereals after 1 Apr within 6 m of edge of crop

Hazard classification and safety precautions

Hazard R03, R07d [1-3]; R04 [1, 3]
Risk phrases R03a, R03c [1-3]; R03b [2]; R04e [1, 3]
Operator protection A, H, M [1-3]; C, J [1, 3]; U02, U04a, U19 [1-3]; U05a, U13, U20a [1, 3]; U08 [2, 3]; U10 [1]; U20b [2]
Environmental protection E01, E13b, E16c, E16d [1-3]; E06b [1-3] (7 d); E10b [1, 2]; E12a, E22a [2]; E12c [3]; E12d [1]; E17 [2] (18 m)
Consumer protection C02 [1]; C03 [1-3]
Storage and disposal E26, E31b [1, 3]; E30a [3]; E30b [1, 2]; E34 [1-3]
Treated seed S04b [3]
Medical advice M01 [1-3]; M03 [2, 3]

190 dimethomorph

A cinnamic acid fungicide with translaminar activity available only in mixtures

191 dimethomorph + mancozeb

A systemic and protectant fungicide for potato blight control

Products

Invader	BASF	7.5:66.7% w/w	WG	10390

Uses

Blight in ***potatoes***. Downy mildew in ***bulb onions*** *(off-label)*, ***garlic*** *(off-label)*, ***shallots*** *(off-label)*.

Specific Off-Label Approvals (SOLAs)

- ***bulb onions*, *garlic*, *shallots*** *(OLA 011446) Dec 2008* [1]

Efficacy guidance

- Commence treatment as soon as there is a risk of blight infection
- In the absence of a warning treatment should start before the crop meets along the rows
- Repeat treatments every 7-14 d depending in the degree of infection risk
- Irrigated crops should be regarded as at high risk and treated every 7 d
- For best results good spray coverage of the foliage is essential
- To minimise the likelihood of development of resistance these products should be used in a planned Resistance Management strategy. See Section 1 for more information

Restrictions

- Maximum total dose equivalent to eight full dose treatments

Crop-specific information

- HI 7 d for all crops

Environmental safety

- Dangerous to fish or other aquatic life. Do not contaminate surface waters or ditches with chemical or used container
- LERAP Category B

Hazard classification and safety precautions

Hazard R04
Risk phrases R04e
Operator protection A; U02, U04a, U05a, U08, U13, U19, U20b
Environmental protection E01, E13b, E16a, E16b
Consumer protection C03
Storage and disposal E30a

192 dimoxystrobin

A protectant strobilurin fungicide for cereals available only in mixtures

193 dimoxystrobin + epoxiconazole

A protectant and curative fungicide mixture for cereals

Products

Swing Gold	BASF	133:50 g/l	SC	11658

Uses

Brown rust in ***winter wheat***. Fusarium ear blight in ***winter wheat***. Septoria in ***winter wheat***. Tan spot in ***winter wheat***.

Efficacy guidance

- For best results apply from the start of ear emergence
- Dimoxystrobin is a member of the QoI cross resistance group. Product should be used preventatively and not relied on for its curative potential
- Use product as part of an Integrated Crop Management strategy incorporating other methods of control, including where appropriate other fungicides with a different mode of action. Do not apply more than two foliar applications of QoI containing products to any cereal crop
- There is a significant risk of widespread resistance occurring in *Septoria tritici* populations in UK. Failure to follow resistance management action may result in reduced levels of disease control

Restrictions

- Maximum number of treatments 1 per crop
- Do not apply before the start of ear emergence

Crop-specific information

- Latest use: up to and including flowering (GS 69)

Environmental safety

- Dangerous to fish or other aquatic life. Do not contaminate surface waters or ditches with chemical or used container
- LERAP Category B
- Avoid drift onto neighbouring crops

Hazard classification and safety precautions

Hazard R03
Risk phrases R03b, R03c
Operator protection A; U05a, U20b
Environmental protection E01, E13b, E16c
Consumer protection C03
Storage and disposal E26, E30a, E31c, E34
Medical advice M03, M05a

194 dinocap

A protectant dinitrophenyl fungicide for powdery mildew control

Products

Karathane Liquid	Landseer	350 g/l	EC	09262

Uses

Powdery mildew in ***apples***, ***chrysanthemums***, ***grapevines*** *(off-label)*, ***roses***, ***strawberries***.

Specific Off-Label Approvals (SOLAs)

- ***grapevines*** *(OLA 021410) Dec 2008* [1]
- ***grapevines*** *(OLA 991543) Dec 2008* [1]

Efficacy guidance

- Spray at 7-14 d intervals to maintain protective film
- Product must be applied to roses before disease becomes established
- Regular use on apples suppresses red spider mites and rust mites

Restrictions

- Maximum total dose on apples equivalent to 10 full dose treatments
- Maximum number of treatments on strawberries 5 per yr
- When applied to apples during blossom period may cause spotting but no adverse effect on pollination or fruit set. Do not apply to Golden Delicious during blossom period
- Do not apply when temperature is above 24°C. Do not apply with white oils
- Certain chrysanthemum cultivars may be susceptible
- May cause petal spotting on white roses

Crop-specific information

- HI apples 14 d, strawberries 7 d

Environmental safety

- Dangerous to fish or other aquatic life. Do not contaminate surface waters or ditches with chemical or used container
- LERAP Category B
- Broadcast air-assisted LERAP (18 m)

Hazard classification and safety precautions

Hazard R03, R04
Risk phrases R03c, R04a, R04b, R04e
Operator protection A, C, H, J, M; U04a, U05a, U10, U11, U19, U20a
Environmental protection E01, E13b, E16a; E17a (18 m)
Consumer protection C03
Storage and disposal E26, E30a, E31a, E34
Medical advice M03

195 diquat

A non-residual bipyridyl contact herbicide and crop desiccant

Products

1	Agriguard Diquat	AgriGuard	200 g/l	SL	10836
2	Greencrop Boomerang	Greencrop	200 g/l	SL	10691
3	Landgold Diquat	Landgold	200 g/l	SL	10649
4	Reglone	Syngenta	200 g/l	SL	10534
5	Standon Diquat SL	Standon	200 g/l	SL	10732
6	Waterloo	Interfarm	200 g/l	SL	11196

Uses

Annual dicotyledons in ***asparagus*** *(off-label)*, ***mustard*** *(off-label)*, ***sage*** *(off-label)*, ***sunflowers*** *(off-label)* [4]; ***flower bulbs***, ***sugar beet*** [1, 2, 6]; ***potatoes*** [1, 2, 5]; ***row crops*** *(between row treatment)* [1, 2, 4-6]. Aquatic weeds in ***areas of water*** [1]. Chemical stripping in ***hops*** [1, 2, 4, 6]. General weed control in ***potatoes*** [4]. General weed growth in ***laid barley*** *(stockfeed only)*, ***laid oats*** *(stockfeed only)* [1-3, 5, 6]. Haulm destruction in ***potatoes*** [3-5]. Pre-harvest desiccation in ***asparagus*** *(off-label)*, ***borage*** *(off-label)*, ***echium plantaginium*** *(off-label)*, ***honesty*** *(off-label)*, ***laid barley*** *(stockfeed only)*, ***laid oats*** *(stockfeed only)*, ***mustard*** *(off-label)*, ***navy beans*** *(off-label)*, ***peas***, ***poppies*** *(off-label - for morphine production)*, ***sage*** *(off-label)*, ***soya beans*** *(off-label)*, ***sunflowers*** *(off-label)* [4]; ***clover seed crops*** [1, 2, 4, 6]; ***combining peas*** [1-3, 5, 6]; ***linseed***, ***spring field beans*** *(stock or pigeon feed only)*, ***spring oilseed rape***, ***winter field beans*** *(stock or pigeon feed only)*, ***winter oilseed rape*** [1-6]; ***potatoes*** [1, 2, 6].

Specific Off-Label Approvals (SOLAs)

- ***asparagus***, ***mustard***, ***sage***, ***sunflowers*** *(OLA 011651) Dec 2008* [4]
- ***borage*** *(OLA 011654) Dec 2008* [4]
- ***echium plantaginium*** *(OLA 011650) Dec 2008* [4]
- ***echium plantaginium*** *(OLA 991844) Sep 2004* [4]
- ***honesty*** *(OLA 011655) Dec 2008* [4]
- ***navy beans*** *(OLA 022039) Dec 2008* [4]
- ***poppies*** *(for morphine production) (OLA 031410) Dec 2008* [4]
- ***soya beans*** *(OLA 011653) Dec 2008* [4]

Approval information

- Diquat included in Annex I under EC Directive 91/414
- Approved for aquatic weed control [1]. See notes in Section 1 on use of herbicides in or near water

Efficacy guidance

- Acts rapidly on green parts of plants and rainfast in 15 min
- Best results for potato desiccation achieved by spraying in bright light and low humidity conditions
- For weed control in row crops apply as overall spray before crop emergence or before transplanting or as an inter-row treatment using a spray guard
- Treatments for weed control normally require co-application with paraquat. Use of authorised non-ionic wetter for improved weed control essential for most uses except potato desiccation, treatment of hops and aquatic weed control (except *Lemna*)
- Do not apply in muddy water or activity will be reduced [1]
- Apply undiluted to submerged weeds in fast-flowing streams and rivers and in static or sluggishly flowing waterbodies. Heavy infestations may need successive treatments at 14-21 d intervals. Use special equipment (see label) [1]

Restrictions

- Maximum number of treatments 3 per crop on hops; 1 per crop in most other situations - see labels for details
- Do not apply potato haulm destruction treatment when soil dry. Tubers may be damaged if spray applied during or shortly after dry periods. See label for details of maximum allowable soil-moisture deficit and varietal drought resistance scores
- Do not add wetters to desiccant sprays for potatoes or for water weed control except for *Lemna* control.
- Consult processor before adding Agral or other wetter on peas. Staining may be increased
- Do not use treated straw or haulm as animal feed or bedding within 4 d of spraying
- Do not use on barley, oats and field beans intended for human consumption [2-5]

Crop-specific information

- Latest use varies with crop and product to ensure best results - see labels
- HI zero when used as a crop desiccant but see label for advisory intervals
- For pre-harvest desiccation, apply to potatoes when tubers the desired size and to other crops when mature or approaching maturity. See label for details of timing and of period to be left before harvesting potatoes and for timing of pea and bean desiccation sprays
- Spray linseed when seed matured evenly over whole field; direct combining can normally begin 10-20 d after spraying
- Apply as hop stripping treatment when shoots have reached top wire
- Apply to floating and submerged aquatic weeds in still or slow moving water [1]
- Treated laid barley and oats may be used only for stock feed; treated peas must be harvested dry; treated field beans may be used for pigeon and animal feed only
- If potato tubers are to be stored leave 14 d after treatment before lifting
- Use on thin or stressed crops of oilseed rape may increase seed losses and decrease oil yield to an unacceptable level
- Treatment of hops under stress or when leaves wet with a heavy dew may cause damage

Environmental safety

- Dangerous for the environment [4]
- Very toxic to aquatic organisms. May cause long-term adverse effects in the aquatic environment [4]
- Do not dump surplus herbicide in water or ditch bottoms
- Do not use treated water for human consumption within 24 h or for overhead irrigation within 10 d of treatment [1]
- Do not contaminate surface waters or ditches with chemical or used container
- Keep all livestock out of treated areas and away from treated water for at least 24 h [2-5]

Hazard classification and safety precautions

Hazard R02, R14 [4]; R03, R04 [1-6]

Risk phrases R02d, R04c, R14c [4]; R03c [1-6]; R04a, R04b [1-3, 5, 6]

Operator protection A, C; U02, U05a, U08, U19, U20b [1-6]; U04a [1, 2, 4-6]; U04b [3]

Environmental protection E01, E15 [1-6]; E06b [1, 2] (24 h); E06b, E08 [3]; E06b [4] (24 h); E06b, E09 [5]; E06b [6] (24 h); E08 [4] (4 d); E19 [1]; E38 [4]

Consumer protection C03

Storage and disposal E26 [1-3, 5, 6]; E30a, E34 [1-6]; E31a [1-4, 6]; E31b [5]; E32d [4]

Medical advice M03 [1-6]; M04 [4]

196 diquat + paraquat

A non-selective non-residual bipyridyl contact herbicide

Products

1	ASAP	Me2	80:120 g/l	SC	11485
2	PDQ	Syngenta	80:120 g/l	SL	10532
3	Speedway 2	Scotts	2.5:2.5% w/w	WG	09273

Uses

Annual dicotyledons in ***all edible crops, all non-edible crops, forest nurseries, natural surfaces not intended to bear vegetation, ornamental plant production*** [1]; ***apple orchards, bush fruit, cane fruit, cherries, cultivated land/soil, damsons, flower bulbs, pear orchards, plums, row crops, strawberries, stubbles*** [2]; ***forestry, hops, potatoes*** [1, 2]; ***managed amenity turf*** *(sward destruction/turf renovation)* [3]; ***non-crop areas*** [2, 3]. Annual grasses in ***all edible crops, all non-edible crops, forest nurseries, natural surfaces not intended to bear vegetation, ornamental plant production*** [1]; ***apple orchards, bush fruit, cane fruit, cherries, cultivated land/soil, damsons, flower bulbs, pear orchards, plums, row crops, strawberries, stubbles*** [2]; ***forestry, hops, potatoes*** [1, 2]; ***managed amenity turf*** *(sward destruction/turf renovation)* [3]; ***non-crop areas*** [2, 3]. Chemical stripping in ***hops*** [2]. Grass weeds in ***field margins*** [2]. Green cover in ***land temporarily removed from production*** [1, 2]. Perennial grasses in ***managed amenity turf*** *(sward destruction/turf renovation)*, ***non-crop areas*** [3]. Perennial non-rhizomatous grasses in ***apple orchards, bush fruit, cane fruit, cherries, cultivated land/soil, damsons, flower bulbs, forestry, hops, non-crop areas, pear orchards, plums, potatoes, row crops, strawberries, stubbles*** [2]. Sward destruction in ***grassland*** [2]; ***permanent pasture, rotational***

grassland [1]. Volunteer cereals in ***apple orchards, bush fruit, cane fruit, cherries, cultivated land/soil, damsons, forestry, hops, non-crop areas, pear orchards, plums, potatoes, row crops, strawberries, stubbles*** [2].

Approval information

- Diquat included in Annex I under EC Directive 91/414
- Approval expiry 30 Sep 2004 [3]

Efficacy guidance

- Rapid kill obtained under bright conditions but most effective results on difficult weeds obtained from slower action in winter
- Apply to young emerged weeds less than 15 cm high, annual grasses must have at least 2 leaves when sprayed [2]
- Addition of approved non-ionic wetter recommended for control of certain species and with low dose rates. See label for details [2]
- Interval between spraying and cultivation varies. See label for details [2]
- For chemical stripping apply in Jul or after hops have reached top wire. Do not use on hops under drought conditions [2]
- Chemical rapidly inactivated in moist soil, activity reduced in dirty or muddy water
- Spray is rainfast in 10 min

Restrictions

- Paraquat is subject to the Poisons Rules 1982 and the Poisons Act 1972. See notes in Section 1
- Diquat may cause an allergic reaction
- Maximum number of treatments 2 per yr for grassland destruction
- In non-crop areas do not use around green bark [3]
- Do not put in a food or drinks container

Crop-specific information

- Latest use varies according to situation - see label
- Apply up to just before sown crops emerge or just before planting [2]
- In potatoes spray earlies up to 10% emergence, maincrop to 40% emergence, provided plants are less than 15 cm high. Do not use post-emergence on potatoes from diseased or small tubers or under very hot, dry conditions [2]
- Use guarded no-drift sprayers to kill inter-row weeds and strawberry runners [2]
- Apply to fruit crops as a directed spray, preferably in dormant season [2]
- If spraying bulbs at end of season ensure all crop foliage is detached from bulbs. Do not use on very sandy soils [2]

Following crops guidance

- On sandy or immature peat soils and on forest nursery seedbeds allow 3 d between spraying and planting [2]
- Where trash or dying weeds are left on surface allow at least 3 d before planting

Environmental safety

- Dangerous for the environment [2]
- Very toxic to aquatic organisms. May cause long-term adverse effects in the aquatic environment [2]
- Harmful to livestock. Paraquat may be harmful to hares; stubbles must be sprayed early in the day
- Keep all livestock out of treated areas for at least 24 h
- Keep in original container, tightly closed, in a safe place, under lock and key

Hazard classification and safety precautions

Hazard R02, R14 [2]; R03, R04 [1, 2]
Risk phrases R02d, R04c, R14c [2]; R03a, R03c, R04b, R04d [1, 2]
Operator protection A, C [1-3]; H, M [1, 2]; U04b, U05a, U09a, U19, U20b [1-3]; U04c [2]
Environmental protection E01, E15 [1-3]; E06b [1-3] (24 h); E11 [1, 2]; E38 [2]
Consumer protection C03 [1, 2]
Storage and disposal E30b, E34 [1-3]; E31a [1, 3]; E31c, E32d [2]
Medical advice M03, M05c [2]; M04 [1-3]

SECTION 4

197 dithianon

A protectant and eradicant dicarbonitrile fungicide for scab control

Products

1	Barclay Cluster	Barclay	750 g/l	SC	11347
2	Dithianon Flowable	BASF	750 g/l	SC	10219

Uses

Scab in ***apples, pears***.

Efficacy guidance

- Apply at bud-burst and repeat every 10-14 d until danger of scab infection ceases
- Application at high rate within 48 h of a Mills period prevents new infection
- Spray programme also reduces summer infection with apple canker

Restrictions

- Maximum number of treatments 8 per crop
- Do not use on Golden Delicious apples after green cluster
- Do not mix with lime sulphur or highly alkaline products

Crop-specific information

- HI 4 wk

Environmental safety

- Dangerous for the environment [2]

Hazard classification and safety precautions

Hazard R03 [1, 2]; R14 [2]
Risk phrases R03b [2]; R03c [1, 2]
Operator protection U05a, U08, U19 [1, 2]; U20a [2]; U20b [1]
Environmental protection E01, E15
Consumer protection C03
Storage and disposal E26, E30a, E31b, E34
Medical advice M03

198 diuron

A residual urea herbicide for non-crop areas and woody crops

See also amitrole + 2,4-D + diuron

Products

1	Diurex 50SC	Makhteshim	500 g/l	SC	11472
2	Diuron 80 WP	Bayer Environ.	80% w/w	WP	09931
3	Freeway	Bayer Environ.	500 g/l	SC	09933
4	Freeway	Bayer Environ.	500 g/l	SC	11129
5	MSS Diuron 500 FL	Nufarm UK	500 g/l	SC	11493
6	MSS Diuron 80 WP	Nufarm UK	80% w/w	WP	11503
7	n2n Diuron 80	n2n	80% w/w	WP	11821
8	n2n Diuron Flowable	n2n	500 g/l	SC	11822
9	Unicrop Flowable Diuron	Unicrop	500 g/l	SC	02270
10	UPL Diuron 80	United Phosphorus	80% w/w	WP	07619

Uses

Annual and perennial weeds in ***asparagus*** *(off-label)*, ***blackcurrants*** *(off-label)* [1]. Annual dicotyledons in ***amenity trees and shrubs*** [1, 3, 4, 7]; ***amenity vegetation*** [5, 6]; ***apple orchards, pear orchards*** [1, 3, 4, 9]; ***asparagus*** *(off-label)* [9]; ***established woody ornamentals*** [2]; ***hard surfaces*** [1, 5]; ***land not intended to bear vegetation*** [2-7]; ***natural surfaces not intended to bear vegetation*** [1]; ***permeable surfaces overlying soil*** [1, 5, 6]; ***trees and shrubs*** [3, 4]; ***woody nursery stock*** [3]; ***woody nursery stock*** *(not container grown)* [4]. Annual grasses in ***amenity trees and shrubs*** [1, 3, 4]; ***apple orchards, pear orchards*** [1, 3, 4, 9]; ***asparagus*** *(off-label)* [9]; ***established woody ornamentals*** [2]; ***hard surfaces, natural surfaces not intended to bear vegetation, permeable surfaces overlying soil*** [1]; ***land not intended to bear vegetation*** [2-4]; ***trees and shrubs, woody nursery stock*** [3, 4]. Annual meadow grass in ***amenity trees and shrubs, land not intended to bear vegetation*** [7]. Annual weeds in ***amenity***

trees and shrubs, farm buildings, industrial sites, land not intended for cropping, non-crop areas, paths, railway tracks, road verges [8]. Perennial weeds in ***amenity vegetation, land not intended to bear vegetation, permeable surfaces overlying soil*** [5, 6]; ***farm buildings, industrial sites, land not intended for cropping, non-crop areas, paths, railway tracks, road verges*** [8]; ***hard surfaces*** [5]. Total vegetation control in ***industrial sites, non-crop areas*** [10]. Willowherb in ***blackcurrants*** *(off-label)* [1, 9].

Specific Off-Label Approvals (SOLAs)

- ***asparagus, blackcurrants*** *(OLA 032172) Dec 2008* [1]
- ***asparagus*** *(OLA 982352) Dec 2008* [9]
- ***blackcurrants*** *(OLA 951318) Dec 2008* [9]

Efficacy guidance

- Best results when applied to moist soil and rain falls soon afterwards
- Length of residual activity may be reduced on heavy or highly organic soils or those with ash substrates
- Selective rates must be applied to weed-free soil and activity persists for 2-3 mth
- Application for total vegetation control may be at any time of year, best results obtained in late winter to early spring. See label for list of resistant species
- Application to frozen ground not recommended

Restrictions

- Maximum number of treatments 1 per yr on land not intended for cropping (high rate), 2 per yr around trees and shrubs (low rate); one per year for apples, pears; 2 per yr (low rate) for blackcurrants [9]
- Do not treat trees and shrubs less than 5 cm tall or established less than 12 mth or in areas of container grown trees and shrubs. See label for list of sensitive ornamental species that should not be treated
- Do not use around trees and shrubs on Sands or Very Light soils or those that are gravelly or where less than 1% organic matter
- Do not use on lawns, grass tennis courts or similar areas of turf
- Do not apply non-selective rates on or near desirable plants where chemical may be washed into contact with roots
- Application for amenity use should only take place between the beginning of Feb and end of Apr (May in Scotland)

Crop-specific information

- Latest use: 30 April (31 May in Scotland) on non-porous surfaces; at any time on porous surfaces for land not intended to bear vegetation; before bud-burst for blackcurrants [9]; pre-emergence for asparagus
- Around trees and shrubs spray in late winter or early spring
- Treat hard surfaces between Feb and end May
- Apply to weed-free soil in apple and pear orchards established for at least 1 yr during Feb-Mar

Following crops guidance

- Succeeding crops should not be planted for at least 12 mth after treatment. Do not plant susceptible crops such as vegetables for at least 2 yr

Environmental safety

- Dangerous for the environment [5]
- Very toxic to aquatic organisms. May cause long-term adverse effects in the aquatic environment [5]
- Do not apply over drains or in drainage channels or gullies
- Avoid run-off when using on paved and similar surfaces

Hazard classification and safety precautions

Hazard R03, R06, R14 [5]; R04 [1-3, 7-10]
Risk phrases R04a, R04b [1-3, 7-9]; R04c [7, 9, 10]; R10e, R14c [5]
Operator protection A [1, 3-5, 7, 8]; C [7, 8]; D [10]; H, M [1, 3-5]; N [1, 5]; U05a [1-10]; U08 [1-8, 10]; U09a [9]; U19 [2-4, 6-10]; U20a [1, 5, 6]; U20b [2-4, 7-10]
Environmental protection E01 [1 10]; E13c [1 4, 6 10]; E38 [5]
Consumer protection C03

SECTION 4

Storage and disposal E26 [8-10]; E30a [1-10]; E31b [1, 3-5, 8]; E31c [9]; E32a [2, 6, 7, 10]; E32d [5]; E34 [5, 9, 10]
Medical advice M05a [5]

199 diuron + glyphosate

A non-selective residual herbicide mixture for non-crop and amenity use

Products

1	Touché	n2n	217.6:109 g/l	RH	07913
2	Xanadu	Bayer Environ.	125:100 g/l	EW	09943

Uses

Annual dicotyledons in ***amenity trees and shrubs, land not intended to bear vegetation***. Annual grasses in ***amenity trees and shrubs, land not intended to bear vegetation***. Perennial dicotyledons in ***amenity trees and shrubs, land not intended to bear vegetation***. Perennial grasses in ***amenity trees and shrubs, land not intended to bear vegetation***.

Approval information

- Glyphosate included in Annex I under EC Directive 91/414
- Approved for use through ULV applicators [1]

Efficacy guidance

- Best results achieved from treatment when weeds are green and actively growing. Symptoms may be slow to appear in poor growing conditions
- Perennial grasses are susceptible when tillering and making new rhizome growth, normally when plants have 4-5 new leaves
- Perennial dicotyledons most susceptible if treated at or near flowering but will be severely checked if treated at other times when growing actively
- Most species of germinating weeds are controlled. See label for more resistant species
- Weed control may be reduced on heavy or highly organic soils and when weeds are suffering stress in any situation
- At least 6 and preferably 24 h rain-free must follow spraying. Rain soon after treatment may reduce initial weed control
- Apply with manufacturer's specialist equipment [1]

Restrictions

- Maximum number of treatments 1 per yr
- Ornamental trees and shrubs should be established for at least 12 mth and be at least 50 mm tall
- Among ornamental trees and shrubs take care to avoid foliage or the stems of young plants
- Do not allow spray to contact desired plants or crops
- Do not use on tree and shrub nurseries or before planting
- Do not treat ornamental trees and shrubs on Sands or Very Light soils or those that are gravelly or with less than 1% organic matter [1]
- Do not decant, connect directly to Nomix applicator [1]

Crop-specific information

- Treat hard surfaces between Feb and end May

Environmental safety

- Not to be used on food crops [2]
- Harmful to fish or other aquatic life. Do not contaminate surface waters or ditches with chemical or used container
- Do not apply over drains or in drainage channels, gullies or similar structures

Hazard classification and safety precautions

Operator protection A [2]; H, M; U02 [1]; U09a, U19, U20b [1, 2]
Environmental protection E01 [2]; E13c [1, 2]
Consumer protection C01, C03 [2]
Storage and disposal E26 [1]; E30a, E32a [1, 2]; E34 [2]

200 diuron + paraquat

A total herbicide with contact and residual activity

Products

Dexuron	n2n	300:100 g/l	SC	07169

Uses

Annual dicotyledons in ***established woody ornamentals, woody nursery stock***. Annual grasses in ***established woody ornamentals, woody nursery stock***. Perennial dicotyledons in ***established woody ornamentals, woody nursery stock***. Perennial grasses in ***established woody ornamentals, woody nursery stock***. Total vegetation control in ***non-crop areas, paths***.

Efficacy guidance

- Apply to emerged weeds at any time of year. Best results achieved by application in spring or early summer
- Effectiveness not reduced by rain soon after treatment

Restrictions

- Paraquat is subject to the Poisons Rules 1982 and the Poisons Act 1972. See notes in Section 1
- Maximum number of treatments 1 per yr
- Avoid contact of spray with green bark, buds or foliage of desirable trees or shrubs

Crop-specific information

- In nurseries use low dose rate as inter-row spray, not more than once per year

Environmental safety

- Keep in original container, tightly closed, in a safe place, under lock and key
- Harmful to fish or other aquatic life. Do not contaminate surface waters or ditches with chemical or used container
- Keep livestock out of treated areas for at least 2 wk and until foliage of any poisonous weeds such as ragwort has died and become unpalatable

Hazard classification and safety precautions

Hazard R02, R03, R04
Risk phrases R02c, R03a, R04a, R04b
Operator protection A, C; U02, U04a, U05a, U09a, U19, U20b
Environmental protection E01, E13c; E06b (24 h)
Consumer protection C03
Storage and disposal E26, E30b, E31a, E34
Medical advice M04

201 dodemorph

A systemic morpholine fungicide for powdery mildew control

Products

F238	BASF	385 g/l	EC	00206

Uses

Powdery mildew in ***roses***.

Efficacy guidance

- Spray roses every 10-14 d during mildew period or every 7 d and at increased dose if cleaning up established infection or if disease pressure high
- Add Citowett when treating rose varieties which are difficult to wet
- Product has negligible effect on *Phytoseiulus* spp being used to control red spider mites

Restrictions

- Do not use on seedling roses
- Do not apply to roses under hot, sunny conditions, particularly under glass, but spray early in the morning or during the evening. Increase the humidity some hours before spraying
- Check tolerance of new varieties before treating rest of crop

SECTION 4

FOR FULL CONDITIONS OF USE ALWAYS READ THE PRODUCT LABEL

Environmental safety

- Flammable
- Harmful to fish or other aquatic life. Do not contaminate surface waters or ditches with chemical or used container

Hazard classification and safety precautions

Hazard R04, R07d
Risk phrases R04b, R04d
Operator protection A, C; U04a, U05a, U09a, U14, U15, U19, U20c
Environmental protection E01, E13c
Consumer protection C03
Storage and disposal E26, E30a, E31c

202 dodine

A protectant and eradicant guanidine fungicide

Products

1	Barclay Dodex	Barclay	450 g/l	SC	11351
2	Greencrop Budburst	Greencrop	450 g/l	SC	11042
3	Radspor FL	Truchem	450 g/l	SC	01685

Uses

Currant leaf spot in ***blackcurrants***. Scab in ***apples, pears***.

Efficacy guidance

- Apply protective spray on apples and pears at bud-burst and at 10-14 d intervals until late Jun to early Jul
- Apply post-infection spray within 36 h of rain responsible for initiating infection. Where scab already present spray prevents production of spores
- On blackcurrants commence spraying at early grape stage and repeat at 2-3 wk intervals, and at least once after picking

Restrictions

- Do not apply in very cold weather (under 5°C) or under slow drying conditions to pears or dessert apples during bloom or immediately after petal fall
- Do not mix with lime sulphur or tetradifon
- Consult processors before use on crops grown for processing

Crop-specific information

- Latest use: early Jul for culinary apples; pre-blossom for dessert apples and pears

Environmental safety

- Harmful to fish or other aquatic life. Do not contaminate surface waters or ditches with chemical or used container

Hazard classification and safety precautions

Hazard R03, R04
Risk phrases R03a, R03c, R04a, R04b, R04e [1-3]; R04c [2, 3]
Operator protection A, C; U05a, U08, U19, U20b
Environmental protection E01, E13c
Consumer protection C03
Storage and disposal E26, E30a, E34 [1-3]; E31a [1]; E31b [2, 3]
Medical advice M03

203 epoxiconazole

A systemic, protectant and curative triazole fungicide for use in cereals

See also *carbendazim + epoxiconazole*
dimoxystrobin + epoxiconazole

Products

1	Agriguard Epoxiconazole	AgriGuard	125 g/l	SC	09407
2	Epic	BASF	125 g/l	SC	08320
3	Landgold Epoxiconazole	Landgold	125 g/l	SC	09821

Products – continued

4	Opus	BASF	125 g/l	SC	08319
5	Standon Epoxiconazole	Standon	125 g/l	SC	09517

Uses

Brown rust in ***spring barley***, ***winter barley***, ***winter wheat*** [1-5]; ***spring rye***, ***spring wheat***, ***triticale***, ***winter rye*** [2, 4]. Eyespot in ***winter barley*** *(reduction)*, ***winter wheat*** *(reduction)* [1-5]. Fusarium ear blight in ***spring wheat***, ***winter wheat*** [2, 4]; ***winter wheat*** *(reduction)* [1, 3, 5]. Net blotch in ***spring barley***, ***winter barley*** [1-5]. Powdery mildew in ***spring barley***, ***winter barley***, ***winter wheat*** [1-5]; ***spring oats***, ***spring rye***, ***spring wheat***, ***triticale***, ***winter oats***, ***winter rye*** [2, 4]. Rhynchosporium in ***spring barley***, ***winter barley*** [1-5]; ***spring rye***, ***winter rye*** [2, 4]. Septoria in ***spring wheat***, ***triticale*** [2, 4]; ***winter wheat*** [1-5]. Sooty moulds in ***spring wheat*** *(reduction)* [2, 4]; ***winter wheat*** *(reduction)* [1-5]. Yellow rust in ***spring barley***, ***winter barley***, ***winter wheat*** [1-5]; ***spring rye***, ***spring wheat***, ***triticale***, ***winter rye*** [2, 4].

Efficacy guidance

- Apply at the start of foliar disease attack
- Optimum effect against eyespot achieved by spraying between leaf-sheath erect and second node detectable stages (GS 30-32)
- Best control of ear diseases of wheat obtained by treatment during ear emergence
- For Septoria spray after third node detectable stage (GS 33) when weather favouring disease development has occurred
- Mildew control improved by use of tank mixtures. See label for details

Restrictions

- Maximum total dose equivalent to two full dose treatments
- Product may cause damage to broad-leaved plant species
- Avoid spray drift onto neighbouring crops

Crop-specific information

- Latest use: up to and including flowering just complete (GS 69) in wheat, rye, triticale; up to and including emergence of ear just complete (GS 59) in barley, oats

Environmental safety

- Dangerous to fish or other aquatic life. Do not contaminate surface waters or ditches with chemical or used container
- LERAP Category B

Hazard classification and safety precautions

Operator protection A; U05a [1-5]; U20a [1]; U20b [2-5]
Environmental protection E01, E13b, E16a [1-5]; E16b [1-3, 5]
Consumer protection C03
Storage and disposal E26 [1, 2]; E30a [1-5]; E31b [1]; E31c [2-5]

SECTION 4

204 epoxiconazole + fenpropimorph

A systemic, protectant and curative fungicide mixture for cereals

Products

1	Barclay Riverdance	Barclay	84:250 g/l	SE	11341
2	Eclipse	BASF	84:250 g/l	SE	07361
3	Greencrop Galore	Greencrop	84:250 g/l	SE	09561
4	Landgold Epoxiconazole FM	Landgold	84:250 g/l	SE	08806
5	Opus Team	BASF	84:250 g/l	SE	07362
6	Standon Epoxifen	Standon	84:250 g/l	SE	08972

Uses

Brown rust in ***spring barley***, ***winter barley***, ***winter wheat*** [1-6]; ***spring rye***, ***spring wheat***, ***triticale***, ***winter rye*** [2, 5]. Eyespot in ***winter barley*** *(reduction)*, ***winter wheat*** *(reduction)* [1-6]. Fusarium ear blight in ***spring wheat***, ***winter wheat*** [2, 5]; ***winter wheat*** *(reduction)* [1, 3, 4, 6]. Net blotch in ***spring barley***, ***winter barley*** [1-6]. Powdery mildew in ***spring barley***, ***winter barley***, ***winter wheat*** [1-6]; ***spring oats***, ***spring rye***, ***spring wheat***, ***triticale***, ***winter oats***, ***winter rye*** [2, 5]. Rhynchosporium in ***spring barley***, ***winter barley*** [1-6]; ***spring rye***, ***winter rye*** [2, 5]. Septoria in ***spring wheat***, ***triticale*** [2, 5]; ***winter wheat*** [1-6]. Sooty moulds in ***spring***

wheat *(reduction)* [2, 5]; ***winter wheat*** *(reduction)* [2, 3, 5]. Yellow rust in ***spring barley, winter barley, winter wheat*** [1-6]; ***spring rye, spring wheat, triticale, winter rye*** [2, 5].

Efficacy guidance

- Apply at the start of foliar disease attack
- Optimum effect against eyespot achieved by spraying between leaf-sheath erect and second node detectable stages (GS 30-32)
- Best control of ear diseases obtained by treatment during ear emergence
- For Septoria spray after third node detectable stage (GS 33) when weather favouring disease development has occurred

Restrictions

- Maximum total dose equivalent to two full dose treatments
- Product may cause damage to broad-leaved plant species

Crop-specific information

- Latest use: up to and including flowering just complete (GS 69) in wheat, rye, triticale; up to and including emergence of ear just complete (GS 59) in barley, oats

Environmental safety

- Harmful (dangerous [4, 6]) to fish or other aquatic life. Do not contaminate surface waters or ditches with chemical or used container [1-3, 5]
- LERAP Category B
- Avoid spray drift onto neighbouring crops

Hazard classification and safety precautions

Hazard R03 [2, 5]; R04 [1-6]
Risk phrases R04b
Operator protection A; U05a, U20b
Environmental protection E01, E16a, E16b [1-6]; E13b [4, 6]; E13c [1-3, 5]
Consumer protection C03
Storage and disposal E26, E30a [1-6]; E31a [3]; E31b [1, 4, 6]; E31c [2, 5]

205 epoxiconazole + fenpropimorph + kresoxim-methyl

A protectant, systemic and curative fungicide mixture for cereals

Products

1	Mantra	BASF	125:150:125 g/l	SE	11728
2	Standon Kresoxim Super	Standon	125:150:125 g/l	SE	09794

Uses

Brown rust in ***spring barley, winter barley, winter wheat*** [1, 2]; ***spring rye, spring wheat, triticale, winter rye*** [1]. Eyespot in ***spring rye*** *(reduction)*, ***triticale*** *(reduction)*, ***winter oats*** *(reduction)*, ***winter rye*** *(reduction)* [1]; ***winter barley*** *(reduction)*, ***winter wheat*** *(reduction)* [1, 2]. Fusarium ear blight in ***spring wheat*** *(reduction)* [1]; ***winter wheat*** *(reduction)* [1, 2]. Net blotch in ***spring barley, winter barley*** [1, 2]. Powdery mildew in ***spring barley, winter barley*** [1, 2]; ***spring oats, spring rye, triticale, winter oats, winter rye*** [1]. Rhynchosporium in ***spring barley, winter barley*** [1, 2]; ***spring rye, winter rye*** [1]. Septoria diseases in ***spring wheat, triticale*** [1]; ***winter wheat*** [1, 2]. Sooty moulds in ***winter wheat*** *(reduction)* [1, 2]. Yellow rust in ***spring barley, winter barley, winter wheat*** [1, 2]; ***spring rye, spring wheat, triticale, winter rye*** [1].

Approval information

- Kresoxim-methyl included in Annex I under EC Directive 91/414

Efficacy guidance

- For best results spray at the start of foliar disease attack and repeat if infection conditions persist
- Optimum effect against eyespot obtained by treatment between leaf sheaths erect and first node detectable stages (GS 30-32)
- For protection against ear diseases apply during ear emergence
- Kresoxim-methyl is a member of the QoI cross resistance group. Product should be used preventatively and not relied on for its curative potential
- Use product as part of an Integrated Crop Management strategy incorporating other methods of control, including where appropriate other fungicides with a different mode of action. Do not apply more than two foliar applications of QoI containing products to any cereal crop

- There is a significant risk of widespread resistance occurring in *Septoria tritici* populations in UK. Failure to follow resistance management action may result in reduced levels of disease control
- Strains of barley powdery mildew resistant to QoI's are common in the UK

Restrictions

- Maximum total dose equivalent to two full dose treatments

Crop-specific information

- Latest use: mid flowering (GS 65) for wheat, rye, triticale; completion of ear emergence (GS 59) for barley, oats

Environmental safety

- Dangerous for the environment [1]
- Dangerous to fish or other aquatic life. Do not contaminate surface waters or ditches with chemical or used container [2]
- LERAP Category B
- Avoid spray drift onto neighbouring crops. Product may damage broad-leaved species

Hazard classification and safety precautions

Hazard R04 [2]; R13, R14 [1]
Risk phrases R04e [2]; R10a, R10c [1]
Operator protection A; U05a, U14, U20b
Environmental protection E01, E16a [1, 2]; E13b, E16b [2]
Consumer protection C03
Storage and disposal E26, E30a [1, 2]; E29, E31b [2]; E31c [1]

206 epoxiconazole + fenpropimorph + pyraclostrobin

A protectant and systemic fungicide mixture for cereals

Products

Diamant	BASF	42:214:114 g/l	SE	11557

Uses

Brown rust in ***spring barley***, ***spring wheat***, ***winter barley***, ***winter wheat***. Crown rust in ***spring oats***, ***winter oats***. Fusarium ear blight in ***spring wheat*** *(reduction)*, ***winter wheat*** *(reduction)*. Net blotch in ***spring barley***, ***winter barley***. Powdery mildew in ***spring barley***, ***spring oats***, ***winter barley***, ***winter oats***. Rhynchosporium in ***spring barley***, ***winter barley***. Septoria diseases in ***spring wheat***, ***winter wheat***. Yellow rust in ***spring barley***, ***spring wheat***, ***winter barley***, ***winter wheat***.

Efficacy guidance

- For best results spray at the start of foliar disease attack and repeat if infection conditions persist
- For protection against ear diseases apply during ear emergence
- Pyraclostrobin is a member of the QoI cross resistance group. Product should be used preventatively and not relied on for its curative potential
- Use product as part of an Integrated Crop Management strategy incorporating other methods of control, including where appropriate other fungicides with a different mode of action. Do not apply more than two foliar applications of QoI containing products to any cereal crop
- There is a significant risk of widespread resistance occurring in *Septoria tritici* populations in UK. Failure to follow resistance management action may result in reduced levels of disease control
- Strains of barley powdery mildew resistant to QoI's are common in the UK

Restrictions

- Maximum number of treatments 2 per crop

Crop-specific information

- Latest use: up to and including ear emergence just complete (GS 59) for barley, oats; up to and including flowering just complete (GS 69) for wheat

Environmental safety

- Dangerous to fish or other aquatic life. Do not contaminate surface waters or ditches with chemical or used container
- LERAP Category B
- Avoid spray drift onto neighbouring crops. Product may damage broad-leaved species

Hazard classification and safety precautions

Hazard R03, R04
Risk phrases R03b, R03c, R04b
Operator protection A; U05a, U14, U20b
Environmental protection E01, E13b, E16a, E16b
Consumer protection C03
Storage and disposal E26, E30a, E31c, E34
Medical advice M03

207 epoxiconazole + kresoxim-methyl

A protectant, systemic and curative fungicide mixture for cereals

Products

1	Barclay Avalon	Barclay	125:125 g/l	SC	11337
2	Landgold Strobilurin KE	Landgold	125:125 g/l	SC	09908
3	Landmark	BASF	125:125 g/l	SC	11730
4	Me2 KME	Me2	125:125 g/l	SC	09594
5	Standon Kresoxim-Epoxiconazole	Standon	125:125 g/l	SC	09281

Uses

Brown rust in ***spring barley***, ***winter barley***, ***winter wheat*** [1-5]; ***spring rye***, ***spring wheat***, ***triticale***, ***winter rye*** [3]. Eyespot in ***spring rye*** *(reduction)*, ***triticale*** *(reduction)*, ***winter oats*** *(reduction)*, ***winter rye*** *(reduction)* [3]; ***winter barley*** *(reduction)*, ***winter wheat*** *(reduction)* [1-5]. Fusarium ear blight in ***spring wheat*** *(reduction)* [3]; ***winter wheat*** *(reduction)* [1-5]. Net blotch in ***spring barley***, ***winter barley*** [1-5]. Powdery mildew in ***spring barley***, ***winter barley*** [1-5]; ***spring oats***, ***spring rye***, ***triticale***, ***winter oats***, ***winter rye*** [3]. Rhynchosporium in ***spring barley***, ***winter barley*** [1-5]; ***spring rye***, ***winter rye*** [3]. Septoria diseases in ***spring wheat***, ***triticale*** [3]; ***winter wheat*** [1-5]. Sooty moulds in ***winter wheat*** *(reduction)* [1-5]. Yellow rust in ***spring barley***, ***winter barley***, ***winter wheat*** [1-5]; ***spring rye***, ***spring wheat***, ***triticale***, ***winter rye*** [3].

Approval information

- Kresoxim-methyl included in Annex I under EC Directive 91/414

Efficacy guidance

- For best results spray at the start of foliar disease attack and repeat if infection conditions persist
- Optimum effect against eyespot obtained by treatment between leaf sheaths erect and first node detectable stages (GS 30-32)
- For protection against ear diseases apply during ear emergence
- Kresoxim-methyl is a member of the QoI cross resistance group. Product should be used preventatively and not relied on for its curative potential
- Use product as part of an Integrated Crop Management strategy incorporating other methods of control, including where appropriate other fungicides with a different mode of action. Do not apply more than two foliar applications of QoI containing products to any cereal crop
- There is a significant risk of widespread resistance occurring in *Septoria tritici* populations in UK. Failure to follow resistance management action may result in reduced levels of disease control
- Strains of barley powdery mildew resistant to QoI's are common in the UK

Restrictions

- Maximum total dose equivalent to two full dose treatments

Crop-specific information

- Latest use: mid flowering (GS 65) for wheat, rye, triticale; completion of ear emergence (GS 59) for barley, oats

Environmental safety

- Dangerous for the environment [3]
- Very toxic to aquatic organisms. May cause long-term adverse effects in the aquatic environment [3]
- Dangerous to fish or other aquatic life. Do not contaminate surface waters or ditches with chemical or used container [1, 2, 4, 5]
- LERAP Category B
- Avoid spray drift onto neighbouring crops. Product may damage broad-leaved species

Hazard classification and safety precautions

Hazard R04 [1-5]; R13, R14 [3]
Risk phrases R04e [1, 2, 4, 5]; R10a, R10c, R14c [3]
Operator protection A; U05a, U14 [1-5]; U20a [2, 5]; U20b [1, 3, 4]
Environmental protection E01, E16a [1-5]; E13b, E16b [1, 2, 4, 5]; E15, E38 [3]
Consumer protection C03
Storage and disposal E26, E30a [1-5]; E31a [4]; E31b [2, 5]; E31c [1, 3]; E32d [3]
Medical advice M05a [3]

208 epoxiconazole + kresoxim-methyl + pyraclostrobin

A protectant, systemic and curative fungicide mixture for cereals

Products

Opponent	BASF	50:67:133 g/l	SE	10877

Uses

Brown rust in ***spring barley, spring wheat, winter barley, winter wheat***. Crown rust in ***spring oats, winter oats***. Net blotch in ***spring barley, winter barley***. Rhynchosporium in ***spring barley, winter barley***. Septoria diseases in ***spring wheat, winter wheat***. Yellow rust in ***spring barley, spring wheat, winter barley, winter wheat***.

Approval information

- Kresoxim-methyl included in Annex I under EC Directive 91/414

Efficacy guidance

- For best results apply at the start of foliar disease attack
- Best results on Septoria leaf blotch when treated in the latent phase
- Good reduction of Fusarium ear blight can be obtained from treatment during flowering
- Yield response may be obtained in the absence of visual disease symptoms
- Kresoxim-methyl and pyraclostrobin are members of the QoI cross resistance group. Product should be used preventatively and not relied on for its curative potential
- Use product as part of an Integrated Crop Management strategy incorporating other methods of control, including where appropriate other fungicides with a different mode of action. Do not apply more than two foliar applications of QoI containing products to any cereal crop
- There is a significant risk of widespread resistance occurring in *Septoria tritici* populations in UK. Failure to follow resistance management action may result in reduced levels of disease control

Restrictions

- Maximum total dose equivalent to two full dose treatments

Crop-specific information

- Latest use: before grain watery ripe (GS 71) for wheat; up to and including emergence of ear just complete (GS 59) for barley and oats

Environmental safety

- Dangerous to fish or other aquatic life. Do not contaminate surface waters or ditches with chemical or used container
- LERAP Category B

Hazard classification and safety precautions

Hazard R03
Risk phrases R03b, R03c
Operator protection A; U05a, U14, U20b
Environmental protection E01, E13b, E16a, E16b
Consumer protection C03
Storage and disposal E26, E30a, E31c, E34

SECTION 4

209 epoxiconazole + pyraclostrobin

A protectant, systemic and curative fungicide mixture for cereals

Products

1	Euro	Me2	50:133 g/l	SE	11511
2	Ibex	BASF	50:133 g/l	SE	10901
3	Opera	BASF	50:133 g/l	SE	10876

Uses

Brown rust in ***spring barley***, ***spring wheat***, ***winter barley***, ***winter wheat***. Crown rust in ***spring oats***, ***winter oats***. Net blotch in ***spring barley***, ***winter barley***. Rhynchosporium in ***spring barley***, ***winter barley***. Septoria diseases in ***spring wheat***, ***winter wheat***. Yellow rust in ***spring barley***, ***spring wheat***, ***winter barley***, ***winter wheat***.

Efficacy guidance

- For best results apply at the start of foliar disease attack
- Good reduction of Fusarium ear blight can be obtained from treatment during flowering
- Yield response may be obtained in the absence of visual disease symptoms
- Pyraclostrobin is a member of the QoI cross resistance group. Product should be used preventatively and not relied on for its curative potential
- Use product as part of an Integrated Crop Management strategy incorporating other methods of control, including where appropriate other fungicides with a different mode of action. Do not apply more than two foliar applications of QoI containing products to any cereal crop
- There is a significant risk of widespread resistance occurring in *Septoria tritici* populations in UK. Failure to follow resistance management action may result in reduced levels of disease control

Restrictions

- Maximum total dose equivalent to two full dose treatments

Crop-specific information

- Latest use: before grain watery ripe (GS 71) for wheat; up to and including emergence of ear just complete (GS 59) for barley and oats

Environmental safety

- Dangerous to fish or other aquatic life. Do not contaminate surface waters or ditches with chemical or used container
- LERAP Category B

Hazard classification and safety precautions

Hazard R03, R04
Risk phrases R03b, R03c, R04b
Operator protection A; U05a, U14, U20b
Environmental protection E01, E13b, E16a, E16b
Consumer protection C03
Storage and disposal E26, E30a, E31c, E34
Medical advice M03

210 esfenvalerate

A contact and ingested pyrethroid insecticide

Products

1	Greencrop Cajole	Greencrop	25 g/l	EC	11402
2	Sumi-Alpha	BASF	25 g/l	EC	10401

Uses

Aphids in ***winter barley***, ***winter wheat***.

Approval information

- Esfenvalerate included in Annex I under EC Directive 91/414

Efficacy guidance

- Crops at high risk (e.g. after grass or in areas with history of BYDV) should be treated when aphids first seen or by mid-Oct. Otherwise treat in late Oct-early Nov

- High risk crops will need a second treatment
- Product also recommended between onset of flowering and milky ripe stages (GS 61-73) for control of summer cereal aphids

Restrictions

- Do not use if another pyrethroid or dimethoate has been applied to crop after start of ear emergence (GS 51)

Crop-specific information

- Latest use: 31 Mar in yr of harvest (winter use)
- HI 20 d (summer use)

Environmental safety

- Dangerous for the environment [2]
- Extremely dangerous to bees. Do not apply to crops in flower or to those in which bees are actively foraging. Do not apply when flowering weeds are present
- Extremely dangerous to fish or other aquatic life. Do not contaminate surface waters or ditches with chemical or used container
- High risk to non-target insects or other arthropods. Do not spray within 6 m of the field boundary
- LERAP Category A
- Store product in dark away from direct sunlight

Hazard classification and safety precautions

Hazard R03, R04 [1, 2]; R14 [2]
Risk phrases R03b, R04d, R04e [2]; R03c, R04b [1, 2]; R04a [1]
Operator protection A, C, H; U04a, U05a, U08, U19, U20b
Environmental protection E01, E12b, E13a, E16c, E16d, E22a
Consumer protection C03
Storage and disposal E26, E30a, E34 [1, 2]; E31b [1]; E31c [2]
Medical advice M03

211 ethofumesate

A benzofuran herbicide for grass weed control in various crops

See also *bromoxynil + ethofumesate + ioxynil*
chloridazon + ethofumesate
desmedipham + ethofumesate + phenmedipham

Products

1	Agriguard Ethofumesate 200	AgriGuard	200 g/l	EC	10698
2	Agriguard Ethofumesate Flo	AgriGuard	500 g/l	SC	09478
3	Barclay Keeper	Barclay	200 g/l	EC	11375
4	Barclay Keeper 500 FL	Barclay	500 g/l	SC	11376
5	Ethosat 500	Makhteshim	500 g/l	SC	11501
6	Kubist 2	Nufarm UK	200 g/l	EC	11126
7	Kubist Flo	Nufarm UK	500 g/l	SC	11167
8	Landgold Ethofumesate 200	Landgold	200 g/l	EC	08980
9	Nortron Flo	Aventis	500 g/l	SC	08154
10	Salute	United Phosphorus	200 g/l	EC	07660
11	Standon Ethofumesate 200	Standon	200 g/l	EC	09360

Uses

Annual dicotyledons in ***fodder beet***, ***mangels***, ***red beet***, ***sugar beet*** [1-11]; ***garlic*** *(off-label)*, ***horseradish*** *(off-label)*, ***onions*** *(off-label)*, ***strawberries*** *(off-label)* [9]. Annual grasses in ***amenity turf***, ***established grassland*** [4]; ***garlic*** *(off-label)*, ***horseradish*** *(off-label)*, ***onions*** *(off-label)*, ***strawberries*** *(off-label)* [9]; ***grass seed crops***, ***rotational grassland*** [1, 4]; ***permanent pasture*** [1]. Annual meadow grass in ***fodder beet***, ***mangels***, ***red beet***, ***sugar beet*** [1-11]; ***garlic*** *(off-label)*, ***onions*** *(off-label)* [7]. Blackgrass in ***amenity turf***, ***established grassland*** [4, 6, 10, 11]; ***fodder beet***, ***mangels***, ***red beet***, ***sugar beet*** [1-11]; ***grass seed crops***, ***rotational grassland*** [1, 3, 4, 6, 10, 11]; ***permanent pasture*** [1]. Chickweed in ***amenity turf***, ***established grassland*** [4, 6, 10, 11]; ***garlic*** *(off-label)*, ***onions*** *(off-label)* [7]; ***grass seed crops***, ***rotational grassland*** [1, 3, 4, 6, 10, 11]; ***permanent pasture*** [1]. Cleavers in ***amenity turf***, ***established grassland*** [4, 6, 10, 11]; ***grass seed crops***, ***rotational grassland*** [1, 3, 4, 6, 10, 11]; ***permanent pasture*** [1]. Clover in ***strawberries*** *(off-label)* [7]. General weed control in ***horseradish*** *(off-label)* [7]. Grass weeds in

strawberries *(off-label)* [7]. Volunteer cereals in ***amenity turf***, ***established grassland*** [4, 6, 10, 11]; ***grass seed crops***, ***rotational grassland*** [3, 4, 6, 10, 11].

Specific Off-Label Approvals (SOLAs)

- ***garlic***, ***horseradish***, ***onions***, ***strawberries*** *(OLA 001916) Dec 2008* [9]
- ***garlic***, ***horseradish***, ***onions***, ***strawberries*** *(OLA 032082) Dec 2008* [7]

Approval information

- Ethofumesate included in Annex I under EC Directive 91/414
- Approval expiry 1 Sep 2004 [4, 11]

Efficacy guidance

- Most products may be applied pre- or post-emergence of crop or weeds but some restricted to pre-emergence or post-emergence use only. Check label
- Volunteer cereals not well controlled pre-emergence, weed grasses should be sprayed before fully tillered
- Grass crops may be sprayed during rain or when wet. Not recommended in very dry conditions or prolonged frost

Restrictions

- Maximum number of treatments 1 pre- plus 1 or 2 post-emergence per crop or yr for beet crops and grass seed crops; 1 per crop for horseradish, strawberries; 2 per yr for amenity turf and established grassland
- Do not use on Sands or Heavy soils, Very Light soils containing a high percentage of stones, or soils with more than 5-10% organic matter (percentage varies according to label)
- Do not use on swards reseeded without ploughing
- Clovers will be killed or severely checked
- Do not graze or cut grass for 14 d after, or roll less than 7 d before or after spraying

Crop-specific information

- Latest use: before crops meet across rows for beet crops and mangels; 14 d before cutting or grazing for grass leys; pre-emergence for horseradish
- HI strawberries 11 wk; garlic, onions 3 mth [9]
- Apply in beet crops in tank mixes with other pre- or post-emergence herbicides. Recommendations vary for different mixtures. See label for details
- Safe timing on beet crops varies with other ingredient of tank mix. See label for details
- In grass crops apply to moist soil as soon as possible after sowing or post-emergence when crop in active growth, normally mid-Oct to mid-Dec. See label for details
- May be used in Italian, hybrid and perennial ryegrass, timothy, cocksfoot, meadow fescue and tall fescue. Apply pre-emergence to autumn-sown leys, post-emergence after 2-3 leaf stage. See label for details

Following crops guidance

- Any crop may be sown 3 mth after application of mixtures in beet crops following ploughing, 5 mth after application in grass crops

Environmental safety

- Flammable [1, 3, 6, 8, 10, 11]
- Harmful to fish or other aquatic life. Do not contaminate surface waters or ditches with chemical or used container

Hazard classification and safety precautions

Hazard R07d [1, 3, 6, 8, 10, 11]
Operator protection A [1, 3]; C, H [1]; U05a [1, 3, 4]; U08, U19 [1-11]; U20a [1-3, 5-11]; U20b [4]
Environmental protection E01 [1, 3, 4]; E13c [1-11]
Consumer protection C03 [1, 3, 4]
Storage and disposal E26 [2-11]; E30a, E31b [1-11]; E34 [1, 5, 8]
Medical advice M03 [1, 3]; M05 [1]

212 ethofumesate + metamitron

A contact and residual herbicide mixture for beet crops

Products

1	Galahad	Bayer CropScience	100:400 g/l	SC	10727
2	Torero	Makhteshim	150:350 g/l	SC	11158

Uses

Annual dicotyledons in ***fodder beet***, ***sugar beet*** [1, 2]; ***mangels*** [1]. Annual meadow grass in ***fodder beet***, ***sugar beet*** [1, 2]; ***mangels*** [1].

Approval information

- Ethofumesate included in Annex I under EC Directive 91/414

Efficacy guidance

- Best results obtained from a series of treatments applied as an overall fine spray commencing when earliest germinating weeds are no larger than fully expanded cotyledon and the majority of the crop at fully expanded cotyledon
- Apply subsequent sprays as each new flush of weeds reaches early cotyledon and continue until weed emergence ceases (maximum 3 sprays)
- Product may be used on all soil types but residual activity may be reduced on those with more than 5% organic matter

Restrictions

- Maximum number of treatments three per crop [2]
- Maximum total dose equivalent to three full dose treatments [1]

Crop-specific information

- Latest use: before crop leaves meet between rows
- Crop tolerance may be reduced by stress caused by growing conditions, effects of pests, disease or other pesticides, nutrient deficiency etc

Following crops guidance

- Beet crops may be sown at any time after treatment. Any other crop may be sown after mouldboard ploughing to 15 cm and a minimum interval of 3 mth after treatment

Environmental safety

- Harmful if swallowed [1]
- May cause sensitization by skin contact [1]
- Harmful to fish or other aquatic life. Do not contaminate surface waters or ditches with chemical or used container

Hazard classification and safety precautions

Hazard R03, R04 [1]
Risk phrases R03c, R04e [1]
Operator protection A, H [1]; U05a, U08, U19 [1, 2]; U14, U15, U20a [2]; U20b [1]
Environmental protection E01, E13c
Consumer protection C03
Storage and disposal E26, E30a, E31b, E34
Medical advice M03 [1]

213 ethofumesate + phenmedipham

A contact and residual herbicide for use in beet crops

Products

1	Agriguard Duo	AgriGuard	100:80 g/l	EC	10453
2	Barclay Goalpost	Barclay	100:80 g/l	EC	11382
3	Powertwin	Makhteshim	200:200 g/l	SC	11187
4	Twin	Makhteshim	94:97 g/l	EC	11172

Uses

Annual dicotyledons in ***fodder beet***, ***mangels***, ***sugar beet***. Annual meadow grass in ***fodder beet***, ***mangels***, ***sugar beet***. Blackgrass in ***fodder beet***, ***mangels***, ***sugar beet***.

Approval information
- Ethofumesate included in Annex I under EC Directive 91/414
- Approval expiry 1 Sep 2004 [1, 2]

Efficacy guidance
- Best results achieved by repeat applications to cotyledon stage weeds. Larger susceptible weeds not killed by first treatment usually checked and controlled by second application
- Apply on all soil types at 7-10 d intervals
- On soils with more than 5-10% organic matter residual activity may be reduced

Restrictions
- Maximum number of treatments normally 3 per crop or maximum total dose equivalent to three full dose treatments - see labels for details
- Do not spray wet foliage or if rain imminent
- Spray must be applied low volume. See label for details
- Apply reduced dose from when the crop has fully expanded cotyledons or full dose from 2 fully expanded true leaf stage [1, 2]
- Spray in evening if daytime temperatures above 21°C expected
- Avoid or delay treatment if frost expected within 7 d
- Avoid or delay treating crops under stress from wind damage, manganese or lime deficiency, pest or disease attack etc
- Spray volumes must not exceed those recommended [1, 2]
- Interval between mixing spray and completion of spraying should not exceed 2 h to avoid crystallization [1, 2]
- Product may cause non-reinforced PVC pipes and hoses to soften and swell. Wherever possible use reinforced PVC or synthetic rubber [1, 2]

Crop-specific information
- Latest use: before crop foliage meets in the rows
- Check from which recovery may not be complete may occur if treatment made during conditions of sharp diurnal temperature fluctuation

Following crops guidance
- Beet crops may be sown at any time after treatment. Any other crop may be sown after mouldboard ploughing to 15 cm and a minimum interval of 3 mth after treatment

Environmental safety
- Flammable [1, 2]
- Harmful (Dangerous [3]) to fish or aquatic life. Do not contaminate surface waters or ditches with chemical or used container [1, 2, 4]
- Extra care necessary to avoid drift because product is recommended for use as a fine spray

Hazard classification and safety precautions

Hazard R04 [1-4]; R07d [1, 2]
Risk phrases R04a, R04b [1, 2]; R04c [4]; R04e [3, 4]
Operator protection A [1-4]; C [1, 2]; H [3]; U05a, U19 [1-4]; U08, U20a [3, 4]; U10, U11, U12, U20b [1, 2]
Environmental protection E01 [1-4]; E13b [3]; E13c [1, 2, 4]
Consumer protection C03
Storage and disposal E26 [1-3]; E30a, E31b [1-4]; E34 [3, 4]

214 ethoprophos

An organophosphorus nematicide and insecticide

Products

Mocap 10G	Aventis	10% w/w	GR	10003

Uses

Potato cyst nematode in ***potatoes***. Wireworm in ***potatoes***.

Efficacy guidance
- Broadcast shortly before or during final soil preparation with suitable fertilizer spreader and incorporate immediately to 10-15 cm. Deeper incorporation may reduce efficacy. See label for details

- Treatment can be applied on all soil types. Control of pests reduced on organic soils
- Effectiveness dependent on soil moisture. Drought after application may reduce control
- Where pre-planting nematode populations are high, loss of crop vigour and subsequent loss of yield may occur in spite of treatment

Restrictions

- This product contains an anticholinesterase organophosphorous compound. Do not use if under medical advice not to work with such compounds
- Maximum number of treatments 1 per crop
- Do not harvest crops for human or animal consumption for at least 8 wk after application
- Vehicles with a closed cab must be used when applying and incorporating
- Do not apply by hand or hand held equipment

Crop-specific information

- Latest use: pre-planting of crop
- HI 8 wk
- Field scale application must only be through positive displacement type specialist granule applicators or dual purpose fertilizer applicators with microgranule setting

Environmental safety

- Dangerous to game, wild birds and animals
- Dangerous to fish or other aquatic life. Do not contaminate surface waters or ditches with chemical or used container
- Product supplied in returnable/refillable containers. Follow instructions on label

Hazard classification and safety precautions

Hazard R03, R04
Risk phrases R03a, R03c, R04e
Operator protection A, D, H; U02, U04a, U05a, U07, U08, U13, U19, U20a
Environmental protection E01, E10a, E13b
Consumer protection C02 (8 wk); C03
Storage and disposal E30a, E33, E34, E36
Medical advice M03

SECTION 4

215 etridiazole

A protective thiadiazole fungicide for soil or compost incorporation

Products

1	Standon Etridiazole 35	Standon	35% w/w	WP	08778
2	Terrazole 35 WP	Scotts	35% w/w	WP	10408

Uses

Damping off and foot rot in ***cabbages***, ***cauliflowers*** *(seeds and seedlings)*, ***celery*** *(seeds and seedlings)*, ***cucumbers*** *(seeds and seedlings)*, ***mustard and cress***, ***ornamental plant production*** *(seeds and seedlings)*, ***outdoor tomatoes*** *(seeds and seedlings)*, ***protected tomatoes*** *(seeds and seedlings)* [1, 2]. Phytophthora in ***watercress*** *(off-label)* [1]. Phytophthora root rot in ***cabbages***, ***cauliflowers*** *(transplants)*, ***celery*** *(transplants)*, ***cucumbers*** *(transplants)*, ***hardy ornamental nursery stock***, ***ornamental plant production*** *(rooted cuttings and transplants)*, ***outdoor tomatoes*** *(transplants)*, ***protected tomatoes*** *(transplants)*, ***tulips*** [1, 2]. Phytophthora wilt in ***cabbages***, ***cauliflowers*** *(transplants)*, ***celery*** *(transplants)*, ***cucumbers*** *(transplants)*, ***hardy ornamental nursery stock***, ***ornamental plant production*** *(rooted cuttings and transplants)*, ***outdoor tomatoes*** *(transplants)*, ***protected tomatoes*** *(transplants)*, ***tulips*** [1, 2]. Pythium in ***watercress*** *(off-label)* [1]. Root diseases in ***inert substrate tomatoes*** *(off-label)* [1].

Specific Off-Label Approvals (SOLAs)

- ***inert substrate tomatoes***, ***watercress*** *(OLA 020409) Dec 2008* [1]

Efficacy guidance

- Best results obtained when used as prophylactic treatment
- Compost/soil incorporation more effective than when applied as drench
- After drenching wash spray residue from crop foliage
- When treating compost for blocking reduce dose by 50% to allow for compaction
- Uniform distribution and thorough incorporation essential for best results. Dilute powder with dry sand to facilitate handling

- Treat compost as soon as possible before use
- Do not apply to wet soil

Restrictions

- Maximum number of treatments 1 per crop or yr for compost incorporation. No restrictions on drenches
- Do not use on *Escallonia, Pyracantha, Gloxinia* spp., pansies or lettuces
- Do not drench seedlings until well established
- When using compost containing more than 20% inert material or a high proportion of young peat, check crop safety before widespread use
- Test on small numbers of plants in advance when treating subjects of unknown susceptibility or using compost with more than 20% inert material

Crop-specific information

- Latest use: normally before sowing or transplanting. 24 h after seeding watercress
- HI tomatoes, cucumbers, mustard and cress 3 d; inert substrate tomatoes 24 h
- Germination of lettuce in previously treated soil may be impaired

Environmental safety

- Dangerous to fish or other aquatic life. Do not contaminate surface waters or ditches with chemical or used container

Hazard classification and safety precautions

Hazard R04
Risk phrases R04a, R04b
Operator protection A, C; U02, U04a, U05a, U08, U20b [1, 2]; U19 [2]
Environmental protection E01, E15
Consumer protection C03
Storage and disposal E30a, E32a

216 famoxadone

A strobilurin fungicide available only in mixtures

See also *cymoxanil + famoxadone*

217 famoxadone + flusilazole

A contact, preventative and curative fungicide mixture for cereals

Products

Charisma	DuPont	100:106.7 g/l	EC	10415

Uses

Brown rust in ***spring barley***, ***winter barley***, ***winter wheat***. Net blotch in ***spring barley***, ***winter barley***. Rhynchosporium in ***spring barley***, ***winter barley***. Septoria diseases in ***winter wheat***. Yellow rust in ***spring barley***, ***winter barley***, ***winter wheat***.

Approval information

- Famoxadone included in Annex I under EC Directive 91/414

Efficacy guidance

- Best results obtained from application at early stage of disease development before infection spreads to new growth
- Famoxadone is a member of the QoI cross resistance group. Product should be used preventatively and not relied on for its curative potential
- Use product as part of an Integrated Crop Management strategy incorporating other methods of control, including where appropriate other fungicides with a different mode of action. Do not apply more than two foliar applications of QoI containing products to any cereal crop
- There is a significant risk of widespread resistance occurring in *Septoria tritici* populations in UK. Failure to follow resistance management action may result in reduced levels of disease control

Restrictions

- Maximum total dose equivalent to three full dose treatments in wheat; two full dose treatments in barley

- Do not apply to crops under stress
- Do not apply during frosty weather

Crop-specific information

- Latest use: before flowering (GS 60) for winter wheat; before quarter ear emergence (GS 53) for barley

Environmental safety

- Dangerous to fish or other aquatic life. Do not contaminate surface waters or ditches with chemical or used container
- LERAP Category B

Hazard classification and safety precautions

Hazard R03
Risk phrases R03c
Operator protection A, C; U05a, U20b
Environmental protection E01, E13b, E16a, E16b
Consumer protection C03
Storage and disposal E26, E30a, E31b, E34
Medical advice M03

218 fatty acids

A soap concentrate insecticide and acaricide

Products

Savona	Koppert	49% w/w	SL	06057

Uses

Aphids in ***broad beans, brussels sprouts, cabbages, cucumbers, fruit trees, lettuce, outdoor tomatoes, peas, peppers, protected tomatoes, pumpkins, runner beans, woody ornamentals***. Mealybugs in ***broad beans, brussels sprouts, cabbages, cucumbers, fruit trees, lettuce, outdoor tomatoes, peas, peppers, protected tomatoes, pumpkins, runner beans, woody ornamentals***. Scale insects in ***broad beans, brussels sprouts, cabbages, cucumbers, fruit trees, lettuce, outdoor tomatoes, peas, peppers, protected tomatoes, pumpkins, runner beans, woody ornamentals***. Spider mites in ***broad beans, brussels sprouts, cabbages, cucumbers, fruit trees, lettuce, outdoor tomatoes, peas, peppers, protected tomatoes, pumpkins, runner beans, woody ornamentals***. Whitefly in ***broad beans, brussels sprouts, cabbages, cucumbers, fruit trees, lettuce, outdoor tomatoes, peas, peppers, protected tomatoes, pumpkins, runner beans, woody ornamentals***.

Efficacy guidance

- Use only soft or rain water for diluting spray
- Pests must be sprayed directly to achieve any control. Spray all plant parts thoroughly to run off
- For glasshouse use apply when insects first seen and repeat as necessary. For scale insects apply several applications at weekly intervals after egg hatch
- To control whitefly spray when required and use biological control after 12 h

Restrictions

- Do not use on new transplants, newly rooted cuttings or plants under stress
- Do not use on specified susceptible shrubs. See label for details

Crop-specific information

- HI zero

Environmental safety

- Harmful to fish or other aquatic life. Do not contaminate surface waters or ditches with chemical or used container

Hazard classification and safety precautions

Operator protection U20c
Environmental protection E13c
Storage and disposal E26, E30a, E31a

219 fenamidone

A strobilurin fungicide available only in mixtures

220 fenamidone + mancozeb

A fungicide mixture for potatoes

Products

Sonata	Bayer CropScience	10:50% w/w	WG	11570

Uses

Blight in ***potatoes***.

Approval information

- Fenamidone included in Annex I under EC Directive 91/414

Efficacy guidance

- Apply as soon as there is risk of blight infection or immediately after a blight warning
- To minimise the likelihood of development of resistance these products should be used in a planned Resistance Management strategy. See Section 1 for more information

Restrictions

- Maximum number of treatments 6 per yr but no more than three should be applied consecutively. Use in alternation with fungicides from a different cross-resistance group
- Observe a 7 d interval between treatments
- Use only as a protective treatment. Do not use when blight has become readily visible (1% leaf area destroyed)

Crop-specific information

- HI 7 d for potatoes

Environmental safety

- Dangerous for the environment
- Very toxic to aquatic organisms. May cause long-term adverse effects in the aquatic environment
- Risk to certain non-target insects or other arthropods. See directions for use
- LERAP Category B

Hazard classification and safety precautions

Hazard R04, R14
Risk phrases R04a, R04c, R14c
Operator protection D, E, H; U05a, U08, U11, U19, U20a
Environmental protection E01, E16a, E22b
Consumer protection C03
Storage and disposal E30a, E31b, E32d, E34

221 fenarimol

A systemic curative and protective pyrimidine fungicide

Products

Rimidin	Rigby Taylor	120 g/l	SC	05907

Uses

Dollar spot in ***managed amenity turf***. Fusarium patch in ***managed amenity turf***. Red thread in ***managed amenity turf***.

Efficacy guidance

- For turf disease control spray as preventive treatment and repeat as necessary. See label for details

Restrictions

- Maximum number of treatments 4 per yr on managed amenity turf
- Do not mow within 24 h after treatment

Environmental safety

- Dangerous to fish or other aquatic life. Do not contaminate surface waters or ditches with chemical or used container

Hazard classification and safety precautions

Operator protection A, C; U08, U19
Environmental protection E13b
Consumer protection C02 (14 d)
Storage and disposal E26, E30a, E31b

222 fenbuconazole

A systemic protectant and curative triazole fungicide for top fruit

Products

1	Indar 5EW	Landseer	50 g/l	EW	09518
2	Kruga 5EC	Interfarm	50 g/l	EC	09863
3	Reward 5EC	Interfarm	50 g/l	EC	09862
4	Surpass 5EC	Interfarm	50 g/l	EC	09861

Uses

Blossom wilt in ***cherries*** *(off-label)*, ***mirabelles*** *(off-label)*, ***plums*** *(off-label)* [1]. Brown rot in ***cherries*** *(off-label)*, ***mirabelles*** *(off-label)*, ***plums*** *(off-label)* [1]. Brown rust in ***spring barley***, ***winter barley***, ***winter wheat*** [2-4]. Glume blotch in ***winter wheat*** [2-4]. Powdery mildew in ***apples*** *(reduction)*, ***grapevines*** *(off-label)*, ***pears*** *(reduction)* [1]; ***winter wheat*** *(reduction)* [2-4]. Rhynchosporium in ***spring barley***, ***winter barley*** [2-4]. Scab in ***apples***, ***pears*** [1]. Septoria leaf spot in ***winter wheat*** [2-4]. Yellow rust in ***winter wheat*** [2-4].

Specific Off-Label Approvals (SOLAs)

- ***cherries***, ***mirabelles***, ***plums*** *(OLA 031372) Dec 2008* [1]
- ***grapevines*** *(OLA 032081) Dec 2008* [1]

Efficacy guidance

- Most effective when used as part of a routine preventative programme from bud burst to onset of petal fall [1]
- After petal fall, tank mix with other protectant fungicides to enhance scab control [1]
- See label for recommended spray intervals. In periods of rapid growth or high disease pressure, a 7 d interval should be used [1]

Restrictions

- Maximum total dose on top fruit equivalent to ten full doses per yr [1]
- Consult processors before using on pears for processing [1]
- Do not harvest for human or animal consumption for at least 4 wk after last application

Crop-specific information

- HI 28 d for top fruit
- Safe to use on all main commercial varieties of apples and pears in UK [1]

Environmental safety

- Dangerous (harmful [1]) to fish or other aquatic life. Do not contaminate surface waters or ditches with chemical or used container

Hazard classification and safety precautions

Hazard R04
Risk phrases R04b [2-4]; R04d [1-4]
Operator protection A, C; U05a, U08 [1-4]; U19, U20b [2-4]; U20a [1]
Environmental protection E01 [1-4]; E13b [2-4]; E13c [1]
Consumer protection C02 [1] (4 wk); C03 [1-4]
Storage and disposal E26 [1-4]; E30a, E31b [2-4]; E32a [1]

223 fenbutatin oxide

A selective contact and ingested organotin acaricide

Products

Torq	Fargro	50% w/w	WP	08370

Uses

Spider mites in ***protected peppers*** *(off-label)*. Two-spotted spider mite in ***protected cucumbers***, ***protected ornamentals***, ***protected tomatoes***, ***tunnel grown strawberries***.

Specific Off-Label Approvals (SOLAs)

- ***protected peppers*** *(OLA 022124) Dec 2008* [1]

Efficacy guidance

- Active on larvae and adult mites. Spray may take 7-10 d to effect complete kill but mites cease feeding and crop damage stops almost immediately
- Apply as soon as mites first appear and repeat as necessary
- On tunnel-grown strawberries apply when mites first appear, usually before flowering starts, and repeat 10-14 d later. A post-harvest spray is also recommended

Restrictions

- Maximum number of treatments 2 per crop pre-harvest + 1 post harvest for tunnel grown strawberries
- Allow at least 10 d between spray applications and do not apply within 10 d of a previous spray
- Do not add wetters or mix with anything other than water
- Do not use white petroleum oil within 28 d of treatment, or any other pesticide within 7 d
- Do not apply to crops which are under stress for any reason
- On subjects of unknown susceptibility test treat on a small number of plants in advance

Crop-specific information

- HI glasshouse cucumbers, tomatoes 3 d; tunnel-grown strawberries 7 d

Environmental safety

- Extremely dangerous to fish or other aquatic life. Do not contaminate surface waters or ditches with chemical or used container
- Suitable for use in IPM programmes with *Encarsia* or *Phytoseiulus* being used for biological control

Hazard classification and safety precautions

Hazard R01, R04
Risk phrases R01b, R04a, R04b, R04c
Operator protection A, C, D, H, M; U05a, U08, U19, U20a
Environmental protection E01, E13a
Consumer protection C03
Storage and disposal E30a, E32a, E34
Medical advice M04

224 fenhexamid

A protectant fungicide for soft fruit

Products

1	Teldor	Bayer	50% w/w	WG	08955
2	Teldor	Bayer CropScience	50% w/w	WG	11229

Uses

Botrytis in ***blackberries***, ***blackcurrants***, ***cherries*** *(off-label)*, ***gooseberries***, ***grapevines***, ***loganberries***, ***plums*** *(off-label)*, ***raspberries***, ***redcurrants***, ***rubus hybrids***, ***strawberries***, ***whitecurrants***.

Specific Off-Label Approvals (SOLAs)

- ***cherries***, ***plums*** *(OLA 021948) Nov 2005* [1]
- ***cherries***, ***plums*** *(OLA 031866) May 2011* [2]

Approval information

- Fenhexamid included in Annex I under EC Directive 91/414

Efficacy guidance

- Use as part of a programme of sprays throughout the flowering period to achieve effective control of Botrytis
- To minimise possibility of development of resistance, no more than two sprays of the product may be applied consecutively. Other fungicides from a different chemical group should then be used for at least two consecutive sprays. If only two applications are made on grapevines, only one may include fenhexamid

- Complete spray cover of all flowers and fruitlets throughout the blossom period is essential for successful control of Botrytis
- Spray programmes should normally start at the start of flowering

Restrictions

- Maximum number of treatments 2 per yr on grapevines; 4 per yr on other listed crops but no more than 2 sprays may be applied consecutively

Crop-specific information

- HI 1 d for strawberries; 7 d for raspberries, loganberries, blackberries, Rubus hybrids, blackcurrants, redcurrants, whitecurrants, gooseberries; 21d for outdoor grapes

Environmental safety

- Harmful to fish or other aquatic life. Do not contaminate surface waters or ditches with chemical or used container

Hazard classification and safety precautions

Operator protection U08, U19, U20b
Environmental protection E13c
Storage and disposal E30a, E32a

225 fenhexamid + tolylfluanid

A protectant fungicide mixture for soft fruit

Products

Talat	Bayer CropScience	16.7:33.3% w/w	WG	11311

Uses

Grey mould in ***blackberries***, ***blackcurrants***, ***gooseberries***, ***loganberries***, ***raspberries***, ***red-currants***, ***rubus hybrids***, ***strawberries***, ***whitecurrants***.

Approval information

- Fenhexamid included in Annex I under EC Directive 91/414

Efficacy guidance

- Effective control of grey mould in soft fruit requires a programme of treatment throughout the blossom period
- Complete even spray coverage of all flowers and developing fruitlets is essential for good results
- To minimise the likelihood of development of resistance use in a planned Resistance Management strategy. See Section 1 for more details

Restrictions

- Maximum total dose equivalent to four full dose treatments for strawberries, raspberries, loganberries, blackberries and other Rubus hybrids, and to two full dose treatments for currants and gooseberries
- Do not apply more than two treatments consecutively (including other products containing fenhexamid or tolylfluanid)
- Do not treat crops growing under glass or polythene
- Consult processor before using on crops for processing

Crop-specific information

- HI 14 days for blackberries, loganberries, raspberries, Rubus hybrids, strawberries; 21 days for currants, gooseberries

Environmental safety

- Dangerous to fish or other aquatic life. Do not contaminate surface waters or ditches with chemical or used container
- Risk to certain non-target insects or other arthropods. See directions for use
- LERAP Category B
- Broadcast air-assisted LERAP (7.5 m)

Hazard classification and safety precautions

Operator protection A, H, M; U02, U05a, U10, U20b
Environmental protection E01, E13b, E16a, E16b, E22b; E17a (7.5 m)

SECTION 4

Consumer protection C03
Storage and disposal E30a, E32a

226 fenitrothion

A broad spectrum, contact organophosphorus insecticide

Products

1	Killgerm Fenitrothion 40 WP	Killgerm	38.69% w/w	WP	H4858
2	Killgerm Fenitrothion 50 EC	Killgerm	500 g/l	EC	H4722
3	Micromite	Killgerm	500 g/l	EC	H4480
4	Sectacide 50 EC	Killgerm	500 g/l	EC	H4939

Uses

Ants in ***food storage areas, livestock houses*** [3]. Bedbugs in ***food storage areas, livestock houses*** [3]. Beetles in ***food storage areas, livestock houses*** [3]. Cockroaches in ***food storage areas, livestock houses*** [3]. Crickets in ***food storage areas, livestock houses*** [3]. Earwigs in ***food storage areas, livestock houses*** [3]. Fleas in ***food storage areas, livestock houses*** [3]. Flies in ***food storage areas*** [3]; ***livestock houses*** [2-4]; ***poultry houses*** [1]; ***refuse tips*** [1, 2, 4]. Mites in ***food storage areas, livestock houses*** [3]. Moths in ***food storage areas, livestock houses*** [3]. Poultry house pests in ***poultry houses*** [3]. Silverfish in ***food storage areas, livestock houses*** [3].

Efficacy guidance

- For best results apply spray to floors, walls and ceilings paying particular attention to floor-wall interfaces and the bases of structures which may act as insect harbourages
- To spray cracks and crevices use a pin-stream nozzle or special crack and crevice extension

Restrictions

- This product contains an anticholinesterase organophosphorus compound. Do not use if under medical advice not to work with such compounds
- Products for use in livestock houses etc must be used by professional operators
- Do not apply directly to livestock and poultry. Remove exposed milk and collect eggs before application. Protect milk machinery and milk containers from contamination

Environmental safety

- These products must not be used on grain or in grain storage areas
- Flammable [2-4]
- In animal husbandry the need to keep animals away from treated surfaces does not apply provided all other precautions and label recommendations are observed
- Dangerous to bees [1]
- Harmful to fish or other aquatic life. Do not contaminate surface waters or ditches with chemical or used container
- Keep unprotected persons and animals out of treated areas for 48 h or until surfaces are dry

Hazard classification and safety precautions

Hazard R03, R04, R07d [2-4]
Risk phrases R03a, R03c, R04g [2-4]
Operator protection A, H [1-4]; C, D [3]; U02, U09a, U19, U20b [1-4]; U04a, U05a [2-4]
Environmental protection E01 [2-4]; E02, E12c [1]; E02 [2-4] (until surfaces dry); E04, E05, E13c, E15 [1-4]
Consumer protection C03, C09 [1-4]; C07 [1]
Storage and disposal E28 [2-4]; E30a, E32a [1-4]
Medical advice M01, M03 [1-4]; M05 [2-4]

227 fenoxaprop-P-ethyl

A phenoxypropionic acid herbicide for use in wheat

See also diclofop-methyl + fenoxaprop-P-ethyl

Products

1	Cheetah Super	Aventis	55 g/l	EW	08723
2	Triumph	Aventis	120 g/l	EC	08740
3	Triumph	Aventis	120 g/l	EC	10902

Uses

Blackgrass in ***spring wheat, winter wheat***. Canary grass in ***spring wheat, winter wheat***. Rough meadow grass in ***spring wheat, winter wheat***. Wild oats in ***spring wheat, winter wheat***.

Efficacy guidance

- Treat weeds from 2 fully expanded leaves up to flag leaf ligule just visible; for awned canary-grass and rough meadow-grass from 2 leaves to the end of tillering
- A second application may be made in spring where susceptible weeds emerge after an autumn application
- Spray is rainfast 1 h after application
- Dry conditions resulting in moisture stress may reduce effectiveness

Restrictions

- Maximum total dose equivalent to two full dose treatments
- Do not apply to barley, durum wheat, undersown crops or crops to be undersown
- Do not roll or harrow within 1 wk of spraying
- Do not spray crops under stress, suffering from drought, waterlogging or nutrient deficiency or those grazed or if soil compacted
- Avoid spraying immediately before or after a sudden drop in temperature or a period of warm days/cold nights
- Do not mix with hormone weedkillers

Crop-specific information

- Latest use: before flag leaf sheath extending (GS 41)
- Treat from crop emergence to flag leaf fully emerged (GS 41).
- Product may be sprayed in frosty weather provided crop hardened off but do not spray wet foliage or leaves covered with ice
- Broadcast crops should be sprayed post-emergence after plants have developed well-established root system

Environmental safety

- Harmful to fish or other aquatic life. Do not contaminate surface waters or ditches with chemical or used container

Hazard classification and safety precautions

Hazard R04
Risk phrases R04a [2, 3]; R04b [1-3]
Operator protection A, C, H; U05a, U20b [1-3]; U08 [2, 3]; U09a [1]
Environmental protection E01, E13c
Consumer protection C03
Storage and disposal E26 [2, 3]; E30a, E31b [1-3]

228 fenoxycarb

An insect specific growth regulator for top fruit

Products

Insegar WG	Syngenta	25% w/w	WG	09789

Uses

Codling moth in ***apples***. Summer-fruit tortrix moth in ***apples, pears***.

Efficacy guidance

- Best results from application at 5th instar stage before pupation. Product prevents transformation from larva to pupa

- Correct timing best identified from pest warnings
- Because of mode of action rapid knock-down of pest is not achieved and larvae continue to feed for a period after treatment
- Adequate water volume necessary to ensure complete coverage of leaves

Restrictions

- Maximum number of treatments 2 per crop
- Consult processors before use

Crop-specific information

- HI 42 d
- Use on all varieties of apples and pears

Environmental safety

- Dangerous for the environment
- Toxic to aquatic organisms. May cause long-term adverse effects to the aquatic environment
- High risk to bees. Do not apply to crops in flower or to those in which bees are actively foraging. Do not apply when flowering weeds are present
- Risk to certain non-target insects or other arthropods. See directions for use
- Broadcast air-assisted LERAP (8 m)
- Apply to minimise off-target drift to reduce effects on non-target organisms. Some margin of safety to beneficial arthropods is indicated.

Hazard classification and safety precautions

Hazard R14
Risk phrases R14b
Operator protection A, C, H; U05a, U20a, U23a
Environmental protection E01, E12a, E15, E22b, E38; E17a (8 m)
Consumer protection C03
Storage and disposal E26, E30a, E32a, E32d

229 fenpropathrin

A contact and ingested pyrethroid acaricide and insecticide

Products

Meothrin	BASF	100 g/l	EC	10400

Uses

Blackcurrant gall mite in ***blackcurrants***. Blackcurrant leaf midge in ***blackcurrants***. Capsids in ***blackcurrants***. Currant sawfly in ***blackcurrants***. Two-spotted spider mite in ***blackcurrants***.

Approval information

- Products containing this active ingredient have been granted derogations for specified 'Essential Uses' for use until 31 December 2007. Sale and supply must cease by 30 June 2007 but growers have no guarantee that the products will continue to be available until then.
  For more information see 'The Review Programme' under 'Pesticide Legislation' in Section 1

Efficacy guidance

- Acts on motile stages of mites and gives rapid kill

Restrictions

- Maximum number of treatments 3 per yr for blackcurrants
- This product must not be used on any crops other than those listed, including any extrapolations that would normally be permissible under the Long Term Arrangements for Extension of Use (see Section 1)

Crop-specific information

- Latest use: 14 d after end of flowering for blackcurrants

Environmental safety

- Flammable
- High risk to bees. Do not apply to crops in flower or to those in which bees are actively foraging. Do not apply when flowering weeds are present
- Extremely dangerous to fish or other aquatic life. Do not contaminate surface waters or ditches with chemical or used container

- Do not operate air-assisted sprayers within 18 m of surface water or ditches. Direct spray away from water
- LERAP Category A

Hazard classification and safety precautions

Hazard R02, R04, R07d
Risk phrases R02a, R02c, R04a, R04b
Operator protection A, C, H, J; U02, U04a, U05a, U08, U19, U20b
Environmental protection E01, E12a, E13a, E16c, E16d
Consumer protection C02, C03
Storage and disposal E30a, E31a, E34
Medical advice M04

230 fenpropidin

A systemic, curative and protective piperidine (morpholine) fungicide

Products

1	Landgold Fenpropidin 750	Landgold	750 g/l	EC	08973
2	Tern	Syngenta	750 g/l	EC	08660

Uses

Brown rust in ***spring barley***, ***spring wheat***, ***winter barley***, ***winter wheat***. Glume blotch in ***spring wheat***, ***winter wheat***. Leaf blotch in ***spring barley***, ***winter barley***. Powdery mildew in ***spring barley***, ***spring wheat***, ***winter barley***, ***winter wheat***. Septoria leaf spot in ***spring wheat***, ***winter wheat***. Yellow rust in ***spring barley***, ***spring wheat***, ***winter barley***, ***winter wheat***.

Efficacy guidance

- Best results obtained when applied at early stage of disease development. See label for details of recommended timing alone and in mixtures
- Disease control enhanced by vapour-phase activity. Control can persist for 4-6 wk
- Alternate with triazole fungicides to discourage build-up of resistance

Restrictions

- Maximum number of treatments 3 per crop (up to 2 in yr of harvest) for winter crops; 2 per crop for spring crops

Crop-specific information

- Latest use: up to and including ear emergence complete (GS 59).
- HI 5 wk

Environmental safety

- Dangerous for the environment [2]
- Very toxic to aquatic organisms. May cause long-term adverse effects in the aquatic environment [2]
- Dangerous to fish or other aquatic life. Do not contaminate surface waters or ditches with chemical or used container [1]

Hazard classification and safety precautions

Hazard R03, R04 [1, 2]; R14 [2]
Risk phrases R03a, R04b, R04d [1]; R03b, R04a, R14c [2]; R03c, R04e [1, 2]
Operator protection A, C, H; U02, U04a, U05a, U10, U20a [1, 2]; U05b [1]; U14 [2]
Environmental protection E01, E13b [1]; E15, E38 [2]
Consumer protection C02 [1] (5 wk); C03 [1]
Storage and disposal E26, E32d [2]; E29, E31c [1]; E30a, E34 [1, 2]
Medical advice M03

SECTION 4

231 fenpropimorph

A contact and systemic morpholine fungicide

See also *azoxystrobin + fenpropimorph*
epoxiconazole + fenpropimorph
epoxiconazole + fenpropimorph + kresoxim-methyl
epoxiconazole + fenpropimorph + pyraclostrobin

Products

1	Corbel	BASF	750 g/l	EC	00578
2	Landgold Fenpropimorph 750	Landgold	750 g/l	EC	10472
3	Marnoch Phorm	Me2	750 g/l	EC	11087
4	Standon Fenpropimorph 750	Standon	750 g/l	EC	08965

Uses

Alternaria in ***carrots*** *(off-label)*, ***horseradish*** *(off-label)*, ***parsley root*** *(off-label)*, ***parsnips*** *(off-label)*, ***salsify*** *(off-label)* [1]. Brown rust in ***spring barley, spring wheat, triticale, winter barley, winter wheat*** [1-4]. Crown rot in ***carrots*** *(off-label)*, ***horseradish*** *(off-label)*, ***parsley root*** *(off-label)*, ***parsnips*** *(off-label)*, ***salsify*** *(off-label)* [1]. Powdery mildew in ***bilberries*** *(off-label)*, ***blackcurrants*** *(off-label)*, ***carrots*** *(off-label)*, ***cranberries*** *(off-label)*, ***dewberries*** *(off-label)*, ***gooseberries*** *(off-label)*, ***hops*** *(off-label)*, ***horseradish*** *(off-label)*, ***loganberries*** *(off-label)*, ***parsley root*** *(off-label)*, ***parsnips*** *(off-label)*, ***raspberries*** *(off-label)*, ***redcurrants*** *(off-label)*, ***salsify*** *(off-label)*, ***strawberries*** *(off-label)*, ***tayberries*** *(off-label)*, ***whitecurrants*** *(off-label)* [1]; ***spring barley, spring oats, spring rye, spring wheat, winter barley, winter oats, winter rye, winter wheat*** [1-4]. Rhynchosporium in ***spring barley, winter barley*** [1-4]. Rust in ***leeks*** [1, 3, 4]; ***red beet*** *(off-label)*, ***sugar beet seed crops*** *(off-label)* [1]; ***spring field beans, winter field beans*** [1-4]. Yellow rust in ***spring barley, spring wheat, triticale, winter barley, winter wheat*** [1-4].

Specific Off-Label Approvals (SOLAs)

- ***bilberries, blackcurrants, cranberries, dewberries, gooseberries, loganberries, raspberries, redcurrants, strawberries, tayberries, whitecurrants*** *(OLA 023755) Dec 2008* [1]
- ***bilberries, blackcurrants, cranberries, dewberries, gooseberries, loganberries, raspberries, redcurrants, strawberries, tayberries, whitecurrants*** *(OLA 023933) Dec 2008* [1]
- ***bilberries, blackcurrants, cranberries, dewberries, gooseberries, loganberries, raspberries, redcurrants, strawberries, tayberries, whitecurrants*** *(OLA 950787) Dec 2008* [1]
- ***carrots, horseradish, parsley root, parsnips, salsify*** *(OLA 023753) Dec 2008* [1]
- ***carrots, horseradish, parsley root, parsnips, salsify*** *(OLA 962483) Dec 2008* [1]
- ***hops*** *(OLA 010595) Dec 2008* [1]
- ***hops*** *(OLA 023759) Dec 2008* [1]
- ***red beet*** *(OLA 023751) Dec 2008* [1]
- ***red beet*** *(OLA 941246) Dec 2008* [1]
- ***sugar beet seed crops*** *(OLA 023757) Dec 2008* [1]
- ***sugar beet seed crops*** *(OLA 961807) Dec 2008* [1]

Efficacy guidance

- On cereals spray at start of disease attack. See labels for recommended tank mixes. Follow-up treatments may be needed if disease pressure remains high
- On field beans a second application may be needed after 2-3 wk
- Product rainfast after 2 h

Restrictions

- Maximum number of treatments 2 per crop for spring cereals, field beans, red beet; 3 per crop at 1.0 l/ha or 4 at 0.75 l/ha (with 1 treatment applied in autumn) for winter cereals; 3 per crop for bilberries, blackcurrants, carrots, cranberries, dewberries, gooseberries, horseradish, loganberries, parsley root, parsnips, raspberries, redcurrants, salsify, strawberries, tayberries, whitecurrants; 6 per crop for leeks
- Consult processors before using on crops for processing
- An interval of at least 10 d on field beans and 14 d on leeks must elapse between applications

Crop-specific information

- Latest use: before 2 true leaves for sugar beet seed crops

- HI cereals, field beans 5 wk; carrots, horseradish, parsley root, parsnips, salsify 4 wk; leeks, red beet 3 wk; bilberries, currants, cranberries, dewberries, gooseberries, loganberries, raspberries, redcurrants, strawberries, tayberries 2 wk
- Scorch may occur if applied during frosty weather or in high temperatures

Environmental safety

- Dangerous to fish or other aquatic life. Do not contaminate surface waters or ditches with chemical or used container

Hazard classification and safety precautions

Hazard R04
Risk phrases R04a [2-4]; R04b [1-4]
Operator protection A, C; U05a, U08, U14, U15, U19 [1-4]; U20a [3, 4]; U20b [1, 2]
Environmental protection E01, E13b
Consumer protection C03
Storage and disposal E26, E30a [1-4]; E31b [3, 4]; E31c [1, 2]

232 fenpropimorph + flusilazole

A broad-spectrum eradicant and protectant fungicide mixture for cereals

Products

Colstar	DuPont	375:160 g/l	EC	06783

Uses

Brown rust in ***spring barley***, ***winter barley***, ***winter wheat***. Net blotch in ***spring barley***, ***winter barley***. Powdery mildew in ***spring barley***, ***winter barley***, ***winter wheat***. Rhynchosporium in ***spring barley***, ***winter barley***. Septoria diseases in ***winter wheat***. Yellow rust in ***spring barley***, ***winter barley***, ***winter wheat***.

Efficacy guidance

- Disease control is more effective if treatment made at an early stage of disease development
- Treat winter cereals in spring or early summer before diseases spread to new growth
- Spring barley should be treated when diseases are first evident
- Treatment may be repeated after 3-4 wk if necessary

Restrictions

- Maximum number of treatments (including other products containing flusilazole) 3 per crop (winter wheat), 2 per crop (winter or spring barley)
- Do not apply to crops under stress
- Do not apply during frosty weather

Crop-specific information

- Latest use: before beginning of anthesis (GS 60) for winter wheat; up to and including completion of ear emergence (GS 59) for barley

Environmental safety

- Dangerous to fish or other aquatic life. Do not contaminate surface waters or ditches with chemical or used container

Hazard classification and safety precautions

Hazard R04
Risk phrases R04a
Operator protection A, C; U05a, U11, U19, U20b
Environmental protection E01, E13b
Consumer protection C03
Storage and disposal E26, E30a, E31b

233 fenpropimorph + kresoxim-methyl

A protectant and systemic fungicide mixture for cereals

Products

1	Ensign	BASF	300:150 g/l	SE	11729
2	Greencrop Monsoon	Greencrop	300:150 g/l	SE	09573

SECTION 4

Products – continued

3	Landgold Strobilurin KF	Landgold	300:150 g/l	SE	09196
4	Standon Kresoxim FM	Standon	300:150 g/l	SE	08922

Uses

Powdery mildew in ***spring barley***, ***spring oats***, ***spring rye***, ***triticale***, ***winter barley***, ***winter oats***, ***winter rye***. Rhynchosporium in ***spring barley***, ***spring rye***, ***triticale***, ***winter barley***, ***winter rye***. Septoria in ***spring rye*** *(reduction)*, ***triticale*** *(reduction)*, ***winter rye*** *(reduction)*, ***winter wheat*** *(reduction)*.

Approval information

- Kresoxim-methyl included in Annex I under EC Directive 91/414

Efficacy guidance

- For best results spray at the start of foliar disease attack and repeat if infection conditions persist
- To minimise the likelihood of development of resistance to strobilurin fungicides these products should be used in a planned Resistance Management strategy. See Section 1 for more information

Restrictions

- Maximum total dose equivalent to two full dose treatments for all crops

Crop-specific information

- Latest use: completion of ear emergence (GS 59) for barley and oats; completion of flowering (GS 69) for wheat, rye and triticale

Environmental safety

- Dangerous for the environment [1]
- Very toxic to aquatic organisms. May cause long-term adverse effects in the aquatic environment [1]
- Dangerous to fish or other aquatic life. Do not contaminate surface waters or ditches with chemical or used container [2-4]

Hazard classification and safety precautions

Hazard R03, R13, R14 [1]; R04 [2-4]
Risk phrases R04e [2-4]; R10a, R14c [1]
Operator protection A; U05a, U14, U20b
Environmental protection E01 [1-4]; E13b [2-4]; E15, E38 [1]
Consumer protection C03
Storage and disposal E26, E30a [1-4]; E31b [2]; E31c [1, 3, 4]; E32d [1]
Medical advice M05a [1]

234 fenpropimorph + quinoxyfen

A systemic fungicide mixture for cereals

Products

Orka	Dow	250:66.7 g/l	EW	08879

Uses

Powdery mildew in ***durum wheat***, ***spring barley***, ***spring oats***, ***spring rye***, ***spring wheat***, ***triticale***, ***winter barley***, ***winter oats***, ***winter rye***, ***winter wheat***.

Efficacy guidance

- For best results treat at early stage of disease development before infection spreads to new crop growth. Further treatment may be necessary if disease pressure remains high
- For control of established infections and broad spectrum disease control use in tank mixtures. See label
- Product rainfast after 1 h
- Systemic activity may be reduced in severe drought

Restrictions

- Maximum total dose 3.0 l product per ha
- Apply only in the spring from mid-tillering stage (GS 25)

Crop-specific information

- Latest use: when first awns visible (GS 49)
- Crop scorch may occur when treatment made in high temperatures

Environmental safety

- Dangerous to fish or other aquatic life. Do not contaminate surface waters or ditches with chemical or used container
- LERAP Category B

Hazard classification and safety precautions

Hazard R04
Risk phrases R04e
Operator protection A, C, H; U05a, U14
Environmental protection E01, E13b, E16a, E16b
Consumer protection C03
Storage and disposal E34

235 fenpyroximate

A mitochondrial electron transport inhibitor acaricide for apples

Products

Sequel	Certis	51.3 g/l	SC	11408

Uses

Fruit tree red spider mite in ***apples***, ***plums*** *(off-label)*.

Specific Off-Label Approvals (SOLAs)

- ***plums*** *(OLA 030869) Jan 2006* [1]

Efficacy guidance

- Kills motile stages of fruit tree red spider mite. Best results achieved if applied in warm weather
- Total spray cover of trees essential. Use higher volumes for large trees
- Apply when majority of winter eggs have hatched

Restrictions

- Maximum number of treatments 1 per yr
- Other mitochondrial electron transport inhibitor (METI) acaricides should not be applied to the same crop in the same calendar yr either separately or in mixture
- Do not apply when apple crops or pollinator are in flower
- Consult processor before use on crops for processing

Crop-specific information

- HI 2 wk

Environmental safety

- Dangerous to fish or other aquatic life. Do not contaminate surface waters or ditches with chemical or used container
- Broadcast air-assisted LERAP (38 m)
- Risk to certain non-target insects or other arthropods. See directions for use

Hazard classification and safety precautions

Hazard R04
Risk phrases R04d, R04e
Operator protection A, C, H; U05a, U08, U14, U20b
Environmental protection E01, E13b, E22b; E17a (38 m)
Consumer protection C03
Storage and disposal E26, E30a, E31c

236 fenuron

A urea herbicide available only in mixtures

See also *chlorpropham + fenuron*

237 ferrous sulphate

A herbicide/fertilizer combination for moss control in turf

See also *dicamba + dichlorprop-P + ferrous sulphate + MCPA*
dichlorophen + ferrous sulphate

Products

1	Elliott's Lawn Sand	Elliott	17% w/w	SA	04860
2	Elliott's Mosskiller	Elliott	45% w/w	GR	04909
3	Greenmaster Autumn	Scotts	18.1% w/w	GR	07508
4	Greenmaster Mosskiller	Scotts	26.8% w/w	GR	07509
5	SHL Lawn Sand	Sinclair	10% w/w	SA	05254
6	Taylors Lawn Sand	Rigby Taylor	7.35% w/w	SA	04451
7	Vitax Microgran 2	Vitax	20% w/w	MG	04541
8	Vitax Turf Tonic	Vitax	15% w/w	SA	04354

Uses

Mosses in ***managed amenity turf***.

Efficacy guidance

- Apply autumn treatment from Sep onward but not when heavy rain expected or in frosty weather [3]
- Apply from Mar to Sep except during drought or when soil frozen [4]
- For best results apply when light showers expected, mow 3 d before treatment and do not mow for 3-4 d afterwards
- Water after 2 d if no rain
- Rake out dead moss thoroughly 7-14 d after treatment. See label
- Fertilizer component of autumn treatment encourages root growth, that of mosskiller formulations promotes tillering

Restrictions

- Maximum number of treatments - see labels
- Do not walk on treated areas until well watered

Crop-specific information

- If spilt on paving, concrete, clothes etc brush off immediately to avoid discolouration
- Observe label restrictions for interval before cutting after treatment

Environmental safety

- Harmful to fish or other aquatic life. Do not contaminate surface waters or ditches with chemical or used container [3, 4]

Hazard classification and safety precautions

Operator protection U20a [1, 2]; U20b [3, 4, 7, 8]; U20c [3-6]
Environmental protection E01 [5, 8]; E13c [3, 4]; E15 [1, 2, 6-8]
Storage and disposal E30a, E32a

238 fipronil

A phenylpyrazole insecticide for horticulture

Products

Vi-Nil	Certis	0.1% w/w	FG	11077

Uses

Vine weevil in ***container-grown ornamentals***.

Efficacy guidance

- Product may be used at any stage of plant propagation to potting on

- Treated compost will give control of vine weevil larvae into second yr after treatment
- Incorporate into fresh compost each time the subject is re-potted
- Control not guaranteed if untreated liners or plugs are potted into treated compost

Restrictions
- Maximum number of treatments 1 per batch of compost
- Check tolerance on sample plants before large scale treatment
- Do not use in compost for aquatic plants or marginals

Crop-specific information
- Latest use: before planting in treated compost
- Product may be used in all compost types but dose should be reduced in proportion if more than 20% inert material included

Environmental safety
- Harmful to fish or other aquatic life. Do not contaminate surface waters or ditches with chemical or used container

Hazard classification and safety precautions

Operator protection A, D; U13, U20b
Environmental protection E01, E13c
Consumer protection C01, C03
Storage and disposal E32a

239 florasulam

A triazolopyrimidine herbicide for cereals

Products

Boxer	Dow	50 g/l	SC	09819

Uses

Annual dicotyledons in ***spring barley***, ***spring oats***, ***spring wheat***, ***winter barley***, ***winter oats***, ***winter wheat***. Charlock in ***spring barley***, ***spring oats***, ***spring wheat***, ***winter barley***, ***winter oats***, ***winter wheat***. Chickweed in ***spring barley***, ***spring oats***, ***spring wheat***, ***winter barley***, ***winter oats***, ***winter wheat***. Cleavers in ***spring barley***, ***spring oats***, ***spring wheat***, ***winter barley***, ***winter oats***, ***winter wheat***. Volunteer oilseed rape in ***spring barley***, ***spring oats***, ***spring wheat***, ***winter barley***, ***winter oats***, ***winter wheat***.

Approval information
- Florasulam included in Annex I under EC Directive 91/414

Efficacy guidance
- Best results obtained from treatment of small actively growing weeds in good conditions
- Apply in autumn or spring once crop has 3 leaves
- Product is mainly absorbed by leaves of weeds and is effective on all soil types
- Florasulam is a member of the ALS-inhibitor group of herbicides

Restrictions
- Maximum total dose on any crop equivalent to one full dose treatment
- Do not roll or harrow within 7 d before or after application
- Do not spray when crops under stress from cold, drought, pest damage, nutrient deficiency or any other cause
- Do not spray in tank mixture with a product containing any other ALS-inhibitor (eg sulfonylurea herbicides) except those containing metsulfuron-methyl, thifensulfuron-methyl or tribenuron-methyl. Product may be sprayed in sequence with any of the above ALS-inhibitors or flupyrsulfuron-methyl

Crop-specific information
- Latest use: before flag leaf sheath extending stage (GS 41) for all crops

Following crops guidance
- Only cereals, oilseed rape, field beans or grass may be sown as a following crop in the same calendar yr as treatment. Oilseed rape may show some temporary reduction of vigour after a dry summer, but yields are not affected

- Cereals, oilseed rape, field beans, grass, linseed, peas, sugar beet, potatoes, maize or clover (for use in grass/clover mixtures) may be sown in the calendar yr following treatment
- In the event of failure of a treated crop in spring only spring wheat, spring barley, spring oats, maize or ryegrass may be sown

Environmental safety

- Dangerous for the environment
- Very toxic to aquatic organisms. May cause long-term adverse effects in the aquatic environment
- See label for detailed instructions on tank cleaning

Hazard classification and safety precautions

Hazard R14
Risk phrases R14c
Operator protection U05a
Environmental protection E01, E15, E38
Consumer protection C03
Storage and disposal E26, E30a, E31b, E32d, E34

240 florasulam + fluroxypyr

A post-emergence herbicide mixture for cereals

Products

1	Starane Gold	Dow	1.0:100 g/l	SE	10879
2	Starane XL	Dow	2.5:100 g/l	SE	10921

Uses

Annual dicotyledons in ***spring barley***, ***spring oats***, ***spring wheat***, ***winter barley***, ***winter oats***, ***winter wheat*** [2]. Chickweed in ***spring barley***, ***spring oats***, ***spring wheat***, ***winter barley***, ***winter oats***, ***winter wheat*** [1, 2]. Cleavers in ***spring barley***, ***spring oats***, ***spring wheat***, ***winter barley***, ***winter oats***, ***winter wheat*** [1, 2]. Mayweeds in ***spring barley***, ***spring oats***, ***spring wheat***, ***winter barley***, ***winter oats***, ***winter wheat*** [2].

Approval information

- Florasulam and fluroxypyr included in Annex I under EC Directive 91/414

Efficacy guidance

- Best results obtained when weeds are small and growing actively
- Products are mainly absorbed through weed foliage. Cleavers emerging after application will not be controlled
- Florasulam is a member of the ALS-inhibitor group of herbicides

Restrictions

- Maximum total dose equivalent to one full dose treatment
- Product must not be applied before 1 Mar in yr of harvest
- Do not roll or harrow 7 d before or after application
- Do not spray when crops are under stress from cold, drought, pest damage or nutrient deficiency
- Do not apply through CDA applicators
- These products may be applied in sequence with products containing only one of flupyrsulfuron-methyl, metsulfuron-methyl, thifensulfuron-methyl or tribenuron-methyl
- Apart from the above, these products must not be applied in tank mix or sequence with any other ALS-inhibitor (eg sulfonylurea herbicides)
- Only one other ALS-inhibitor product may be applied to the same crop

Crop-specific information

- Latest use: before flag leaf sheath extended for barley and wheat; before second node detectable for oats

Following crops guidance

- In the event of failure of a treated crop in the spring, only spring cereals, maize or ryegrass may be planted
- Cereals, oilseed rape, field beans or grass may follow treated crops in the same yr. Oilseed rape may suffer temporary vigour reduction after a dry summer
- In addition to the above, linseed, peas, sugar beet, potatoes, maize or clover may be sown in the calendar yr following treatment

Environmental safety

- Dangerous to fish or other aquatic life. Do not contaminate surface waters or ditches with chemical or used container
- Take extreme care to avoid drift onto non-target crops or plants

Hazard classification and safety precautions

Hazard R04
Risk phrases R04a, R04b
Operator protection A, C; U05a, U11, U19
Environmental protection E01, E13b
Consumer protection C03
Storage and disposal E30a, E31c, E34

241 fluazifop-P-butyl

A phenoxypropionic acid grass herbicide for broadleaved crops

Products

1	Fusilade 250 EW	Syngenta	250 g/l	EW	10525
2	Fusilade Max	Syngenta	125 g/l	EC	11519

Uses

Annual grasses in ***blackcurrants***, ***broad beans*** *(off-label)*, ***carrots***, ***chinese cabbage*** *(off-label)*, ***collards*** *(off-label)*, ***combining peas***, ***farm forestry***, ***field margins***, ***flax***, ***flax for industrial use***, ***fodder beet***, ***garlic*** *(off-label)*, ***gooseberries***, ***haricot beans*** *(off-label)*, ***hops***, ***kale*** *(stockfeed only)*, ***linseed***, ***linseed for industrial use***, ***navy beans*** *(off-label)*, ***onions***, ***ornamental plant production*** *(off-label)*, ***parsnips*** *(off-label)*, ***protected ornamentals*** *(off-label)*, ***raspberries***, ***red beet*** *(off-label)*, ***spring field beans***, ***spring oilseed rape***, ***spring oilseed rape for industrial use***, ***strawberries***, ***sugar beet***, ***swedes*** *(stockfeed only)*, ***turnips*** *(stockfeed only)*, ***vining peas***, ***winter field beans***, ***winter oilseed rape***, ***winter oilseed rape for industrial use*** [1, 2]; ***kale*** *(off-label)*, ***non-edible flowers*** *(off-label)*, ***spring field beans*** *(off-label)*, ***swedes*** *(off-label)*, ***tic beans*** *(off-label)*, ***turnips*** *(off-label)*, ***winter field beans*** *(off-label)* [1]. Annual meadow grass in ***lucerne*** *(off-label)* [1]. Barley cover crops in ***row crops*** [1]. Barren brome in ***field margins*** [1, 2]. Blackgrass in ***hops***, ***spring field beans***, ***winter field beans*** [1, 2]; ***lucerne*** *(off-label)* [1]. General weed control in ***farm forestry*** *(off-label)* [1]. Green cover in ***land temporarily removed from production*** [1, 2]. Perennial grasses in ***blackcurrants***, ***broad beans*** *(off-label)*, ***carrots***, ***chinese cabbage*** *(off-label)*, ***collards*** *(off-label)*, ***combining peas***, ***farm forestry***, ***flax***, ***flax for industrial use***, ***fodder beet***, ***garlic*** *(off-label)*, ***gooseberries***, ***haricot beans*** *(off-label)*, ***hops***, ***kale*** *(stockfeed only)*, ***linseed***, ***linseed for industrial use***, ***navy beans*** *(off-label)*, ***onions***, ***ornamental plant production*** *(off-label)*, ***parsnips*** *(off-label)*, ***protected ornamentals*** *(off-label)*, ***raspberries***, ***red beet*** *(off-label)*, ***spring field beans***, ***spring oilseed rape***, ***spring oilseed rape for industrial use***, ***strawberries***, ***sugar beet***, ***swedes*** *(stockfeed only)*, ***turnips*** *(stockfeed only)*, ***vining peas***, ***winter field beans***, ***winter oilseed rape***, ***winter oilseed rape for industrial use*** [1, 2]; ***kale*** *(off-label)*, ***non-edible flowers*** *(off-label)*, ***spring field beans*** *(off-label)*, ***swedes*** *(off-label)*, ***tic beans*** *(off-label)*, ***turnips*** *(off-label)*, ***winter field beans*** *(off-label)* [1]. Perennial ryegrass in ***lucerne*** *(off-label)* [1]. Volunteer cereals in ***blackcurrants***, ***carrots***, ***combining peas***, ***farm forestry***, ***field margins***, ***flax***, ***flax for industrial use***, ***fodder beet***, ***gooseberries***, ***hops***, ***kale*** *(stockfeed only)*, ***linseed***, ***linseed for industrial use***, ***onions***, ***raspberries***, ***spring field beans***, ***spring oilseed rape***, ***spring oilseed rape for industrial use***, ***strawberries***, ***sugar beet***, ***swedes*** *(stockfeed only)*, ***turnips*** *(stockfeed only)*, ***vining peas***, ***winter field beans***, ***winter oilseed rape***, ***winter oilseed rape for industrial use*** [1, 2]; ***lucerne*** *(off-label)* [1]. Wild oats in ***blackcurrants***, ***carrots***, ***combining peas***, ***farm forestry***, ***field margins***, ***flax***, ***flax for industrial use***, ***fodder beet***, ***gooseberries***, ***hops***, ***kale*** *(stockfeed only)*, ***linseed***, ***linseed for industrial use***, ***onions***, ***raspberries***, ***spring field beans***, ***spring oilseed rape***, ***spring oilseed rape for industrial use***, ***strawberries***, ***sugar beet***, ***swedes*** *(stockfeed only)*, ***turnips*** *(stockfeed only)*, ***vining peas***, ***winter field beans***, ***winter oilseed rape***, ***winter oilseed rape for industrial use*** [1, 2]; ***lucerne*** *(off-label)* [1].

Specific Off-Label Approvals (SOLAs)

- ***broad beans, chinese cabbage, collards, garlic, haricot beans, kale, navy beans, non-edible flowers, ornamental plant production, parsnips, protected ornamentals, red beet, spring field beans, swedes, tic beans, turnips, winter field beans*** *(OLA 011638) Dec 2008* [1]
- ***broad beans, chinese cabbage, collards, garlic, haricot beans, navy beans, ornamental plant production, parsnips, protected ornamentals, red beet*** *(OLA 032138) Dec 2008* [2]
- ***lucerne*** *(OLA 011637) Dec 2008* [1]
- ***lucerne*** *(OLA 982699) Dec 2008* [1]

Efficacy guidance

- Best results achieved by application when weed growth active under warm conditions with adequate soil moisture. Agral or other specified adjuvant must always be added to spray. See label
- Spray weeds from 2-expanded leaf stage to fully tillered, couch from 4 leaves when majority of shoots have emerged, with a second application if necessary
- Control may be reduced under dry conditions. Do not cultivate for 2 wk after spraying couch
- Annual meadow grass is not controlled
- May also be used to remove grass cover crops

Restrictions

- Maximum number of treatments normally 2 per crop for sugar and fodder beet, winter oilseed rape (including crops for industrial use); 2 per yr for strawberries and farm forestry; 1 per crop or yr for spring oilseed rape (including crops for industrial use) and other crops
- Do not sow cereals for at least 8 wk after application of high rate or 2 wk after low rate
- Do not apply through CDA sprayer, with hand-held equipment or from air
- Avoid treatment before spring growth has hardened or when buds opening
- Do not treat bush and cane fruit or hops between flowering and harvest
- Oilseed rape, linseed and flax for industrial use must not be harvested for human or animal consumption nor grazed
- Do not use for forestry establishment on land not previously under arable cultivation or improved grassland
- Treated vegetation in field margins, land temporarily removed from production etc, must not be grazed or harvested for human or animal consumption and unprotected persons must be kept out of treated areas for at least 24 h

Crop-specific information

- Latest use: before 5 leaf stage for spring oilseed rape; before flowering for blackcurrants, gooseberries, hops, raspberries, strawberries; before flower buds visible for field beans, peas, linseed, flax, oilseed rape; 2 wk before sowing cereals or grass for field margins, land temporarily removed from production, oilseed rape for industrial use, grass crops for farm forestry
- HI beet crops, brassicas, carrots, parsnips, 8 wk; garlic, onions 4 wk
- Apply to sugar and fodder beet from 1-true leaf to 50% ground cover
- Apply to winter oilseed rape from 1-true leaf to established plant stage
- Apply to spring oilseed rape from 1-true leaf but before 5-true leaves
- Apply in fruit crops after harvest. See label for timing details on other crops
- Before using on onions or peas use crystal violet test to check that leaf wax is sufficient

Environmental safety

- Dangerous for the environment [2]
- Toxic to aquatic organisms. May cause long-term adverse effects to the aquatic environment [1]
- Very toxic to aquatic organisms. May cause long-term adverse effects in the aquatic environment [2]

Hazard classification and safety precautions

Hazard R03 [2]; R04, R14 [1, 2]
Risk phrases R04b [1, 2]; R10a, R12a, R14b [1]; R10b, R14c [2]
Operator protection A, C, H [1, 2]; M [2]; U05a [2]; U05b [1]; U08, U20b [1, 2]
Environmental protection E01 [1]; E15 [2]; E38 [1, 2]
Consumer protection C03 [1]
Storage and disposal E26, E30a, E31c, E32d

242 fluazinam

A pyridinamine fungicide for use in potatoes

Products

1	Barclay Cobbler	Barclay	500 g/l	SC	11348
2	Greencrop Solanum	Greencrop	500 g/l	SC	11052
3	Landgold Fluazinam	Landgold	500 g/l	SC	08060
4	Shirlan	Syngenta	500 g/l	SC	10573
5	Standon Fluazinam 500	Standon	500 g/l	SC	08670

Uses

Blight in ***potatoes*** [1-5]. Root rot in ***protected blackberries*** *(off-label)*, ***protected raspberries*** *(off-label)*, ***protected rubus hybrids*** *(off-label)* [4].

Specific Off-Label Approvals (SOLAs)

- ***protected blackberries***, ***protected raspberries***, ***protected rubus hybrids*** *(OLA 032168) Dec 2008* [4]

Efficacy guidance

- Commence treatment at the first blight risk warning (before blight enters the crop). Products are rainfast within 1 h
- In the absence of a warning, treatment should start before foliage of adjacent plants meets in the rows
- Spray at 7-14 d intervals (5-10 d intervals [4]) depending on severity of risk (see label)
- Ensure complete coverage of the foliage and stems, increasing volume as haulm growth progresses, in dense crops and if blight risk increases

Restrictions

- Maximum number of treatments 10 per crop
- Do not use with hand-held sprayers

Crop-specific information

- HI 7 d (0 d [4])

Environmental safety

- Dangerous to fish or other aquatic life. Do not contaminate surface waters or ditches with chemical or used container
- LERAP Category B

Hazard classification and safety precautions

Hazard R04
Risk phrases R04e
Operator protection A, C, H; U02, U04a, U08 [1, 2, 4, 5]; U05a [1-5]; U20a [3, 4]; U20b [1, 2, 5]
Environmental protection E01, E13b, E16a
Consumer protection C03
Storage and disposal E26, E30a [1-5]; E31b [1-3, 5]; E31c, E34 [4]
Medical advice M03 [2, 4]

243 fluazinam + metalaxyl-M

A mixture of contact and systemic fungicides for potatoes

Products

Epok	Belchim	400:200 g/l	EC	10429

Uses

Blight in ***potatoes***.

Efficacy guidance

- Commence treatment at the first blight risk warning (before blight enters the crop). Products are rainfast within 1 h
- In the absence of a warning, treatment should start before foliage of adjacent plants meets in the rows
- Spray at 7-14 d intervals depending on severity of risk (see label)
- Ensure complete coverage of the foliage and stems, increasing volume as haulm growth progresses, in dense crops and if blight risk increases

SECTION 4

- Metalaxyl-M works best on young actively growing foliage and efficacy declines with the onset of senescence. Therefore it is recommended that the product is only used for the first part of the blight control program, which should be completed with a reliable protectant fungicide

Restrictions

- Maximum number of treatments 5 per crop

Crop-specific information

- HI 7d for potatoes

Hazard classification and safety precautions

Hazard R04
Risk phrases R04a, R04b, R04e
Operator protection A, C, H; U05a, U07, U20a
Environmental protection E01, E13b, E16a, E16b
Consumer protection C03
Storage and disposal E30a, E31b, E34

244 fludioxonil

A cyanopyrrole fungicide seed treatment for wheat and barley

See also cymoxanil + fludioxonil + metalaxyl-M

Products

Beret Gold	Syngenta	25 g/l	FS	08390

Uses

Bunt in ***winter wheat*** *(seed treatment)*. Covered smut in ***spring barley*** *(seed treatment)*, ***winter barley*** *(seed treatment)*. Fusarium foot rot and seedling blight in ***spring barley*** *(seed treatment)*, ***spring oats*** *(seed treatment)*, ***spring wheat*** *(seed treatment)*, ***winter barley*** *(seed treatment)*, ***winter oats*** *(seed treatment)*, ***winter wheat*** *(seed treatment)*. Leaf stripe in ***spring barley*** *(seed treatment - reduction)*, ***winter barley*** *(seed treatment - reduction)*. Pyrenophora leaf spot in ***spring oats*** *(seed treatment)*, ***winter oats*** *(seed treatment)*. Snow mould in ***spring barley*** *(seed treatment)*, ***winter barley*** *(seed treatment)*, ***winter wheat*** *(seed treatment)*.

Efficacy guidance

- Apply direct to seed using conventional seed treatment equipment. Continuous flow treaters should be calibrated using product before use
- Effective against benzimidazole-resistant strains of *Fusarium nivale*

Restrictions

- Maximum number of treatments 1 per seed batch
- Do not apply to cracked, split or sprouted seed
- Sow treated seed within 6 mth

Crop-specific information

- Latest use: before drilling
- Product may reduce flow rate of seed through drill. Recalibrate with treated seed before drilling

Environmental safety

- Harmful to fish or other aquatic life. Do not contaminate surface waters or ditches with chemical or used container
- Do not use treated seed as food or feed
- Treated seed harmful to game and wildlife

Hazard classification and safety precautions

Hazard R14
Risk phrases R14a
Operator protection A, H; U05a, U20b
Environmental protection E01, E03, E15
Consumer protection C03
Storage and disposal E26, E30a, E32a
Treated seed S02, S05, S07

245 flufenacet

A broad spectrum oxyacetamide herbicide available only in mixtures

246 flufenacet + metribuzin

A herbicide mixture for potatoes

Products

1	Artist	Bayer	24:17.5% w/w	WP	10658
2	Artist	Bayer CropScience	24:17.5% w/w	WP	11239

Uses

Annual dicotyledons in ***early potatoes***, ***maincrop potatoes***. Annual meadow grass in ***early potatoes***, ***maincrop potatoes***.

Efficacy guidance

- Product acts through root uptake and needs sufficient soil moisture at and shortly after application
- Effectiveness is reduced under dry soil conditions
- Residual activity is reduced on mineral soils with a high organic matter content and on peaty or organic soils
- Ensure application is made evenly to both sides of potato ridges
- Perennial weeds are not controlled

Restrictions

- Maximum total dose equivalent to one full dose treatment
- Potatoes must be sprayed before emergence of crop and weeds
- See label for list of tolerant varieties. Do not treat Maris Piper grown on Sands or Very Light soils
- Do not use on Sands
- On stony or gravelly soils there is risk of crop damage especially if heavy rain falls soon after application

Crop-specific information

- Latest use: before crop emergence

Following crops guidance

- Before drilling or planting any succeeding crop soil must be mouldboard ploughed to at least 15 cm as soon as possible after lifting and no later than end Dec
- In W Cornwall early potatoes may be followed by summer planted brassica crops 14 wk after spraying and after mouldboard ploughing. Elsewhere cereals or winter beans may be grown in the same year if at least 16 wk have elapsed since treatment
- In the yr following treatment any crop may be grown except lettuce or radish, or vegetable brassica crops on silt soils in Lincs

Environmental safety

- Take care to avoid spray drift onto neighbouring crops, especially lettuce or brassicas

Hazard classification and safety precautions

Hazard R03
Risk phrases R03c
Operator protection A, D, H; U05a, U08, U13, U19, U20b
Environmental protection E01, E15
Consumer protection C03
Storage and disposal E30a, E32a
Medical advice M03

247 flufenacet + pendimethalin

A broad spectrum residual and contact herbicide mixture for winter cereals

Products

1	Crystal	BASF	60:300 g/l	EC	10657
2	Ice	BASF	60:300 g/l	EC	10681

Products – continued

3	Standon FFA60 Plus	Standon	60:300 g/l	EC	11649
4	Trooper	BASF	60:300 g/l	EC	10682

Uses

Annual dicotyledons in ***winter barley***, ***winter wheat***. Annual grasses in ***winter barley***, ***winter wheat***. Annual meadow grass in ***winter barley***, ***winter wheat***. Blackgrass in ***winter barley***, ***winter wheat***. Chickweed in ***winter barley***, ***winter wheat***. Corn marigold in ***winter barley***, ***winter wheat***. Speedwells in ***winter barley***, ***winter wheat***.

Approval information

- Pendimethalin included in Annex I under EC Directive 91/414

Efficacy guidance

- Best results achieved when applied from pre-emergence of weeds to 2-leaf stage but post emergence treatment is not recommended on clay soils
- Product requires some soil moisture to be activated ideally from rain within 7 d of application. Prolonged dry conditions may reduce residual control
- Product is slow acting and final level of weed control may take some time to appear
- For effective weed control seed bed preparations should ensure even incorporation of any trash, straw and ash to 15 cm
- Efficacy may be reduced on soils with more than 6% organic matter

Restrictions

- Maximum total dose equivalent to one full dose treatment
- For pre-emergence treatments seed should be covered with min 32 mm settled soil. Shallow drilled crops should be treated post-emergence only
- Do not treat undersown crops
- Avoid spraying during periods of prolonged or severe frosts
- Do not use on stony or gravelly soils or those with more than 10% organic matter
- Pre-emergence treatment may only be used on crops drilled before 30 Nov. All crops must be treated before 31 Dec in yr of planting.
- Concentrated or diluted product may stain clothing or skin

Crop-specific information

- Latest use: before third tiller stage (GS 23) and before 31 Dec in yr of planting
- Very wet weather before and after treatment may result in loss of crop vigour and reduced yield, particularly where soils become waterlogged

Following crops guidance

- Any crop may follow a failed or normally harvested treated crop provided ploughing to at least 15 cm is carried out beforehand

Environmental safety

- Harmful if swallowed [1-4]
- Irritating to skin [1-4]
- May cause lung damage if swallowed [1, 2, 4]
- Limited evidence of a carcinogenic effect [1, 2, 4]
- Dangerous for the environment [1, 2, 4]
- Very toxic to aquatic organisms. May cause long-term adverse effects in the aquatic environment [1, 2, 4]
- Dangerous to fish or other aquatic life. Do not contaminate surface waters or ditches with chemical or used container [3]
- Risk to certain non-target insects or other arthropods. See directions for use [1-4]
- LERAP Category B
- Some products supplied in small volume returnable packs. Follow instructions for use

Hazard classification and safety precautions

Hazard R03, R04 [1-4]; R13, R14 [1, 2, 4]
Risk phrases R03c, R04b [1-4]; R04g, R14c [1, 2, 4]
Operator protection A, C, H; U02, U05a, U20c
Environmental protection E01, E16a, E16b, E22b [1-4]; E13b [3]; E15, E38 [1, 2, 4]
Consumer protection C03
Storage and disposal E26 [3]; E30a, E31a, E34 [1-4]; E32d [1, 2, 4]
Medical advice M03 [1-4]; M05 [1, 2, 4]

248 flupyrsulfuron-methyl

A sulfonylurea herbicide for winter wheat

See also carfentrazone-ethyl + flupyrsulfuron-methyl

Products

Lexus 50 DF	DuPont	50% w/w	WG	09026

Uses

Annual dicotyledons in ***winter wheat***. Blackgrass in ***winter wheat***.

Approval information

- Flupyrsulfuron-methyl included in Annex I under EC Directive 91/414

Efficacy guidance

- Best results achieved from applications made in good growing conditions
- Good spray cover of weeds must be obtained
- Growth of weeds is inhibited within hours of treatment but visible symptoms may not be apparent for up to 4 wk
- Product has moderate residual life in soil. Under normal moisture conditions susceptible weeds germinating soon after treatment will be controlled
- Product may be used on all soil types but residual activity and weed control is reduced on highly alkaline soils
- Flupyrsulfuron-methyl is a member of the ALS-inhibitor group of herbicides and products should be used in a planned Resistance Management strategy. See Section 1 for more information

Restrictions

- Maximum number of treatments 1 per crop
- Do not use on wheat undersown with grasses or legumes, or any other broad-leaved crop
- Do not apply within 7 d of rolling
- Do not treat any crop suffering from drought, waterlogging, pest or disease attack, nutrient deficiency, or any other stress factors
- After treatment other 'ALS-inhibiting' herbicides should not be used on the same crop except metsulfuron-methyl, thifensulfuron-methyl or tribenuron-methyl (alone or in mixtures)

Crop-specific information

- Latest use: before 1st node detectable (GS 31)
- Apply in the spring from the 3-leaf stage on all crops
- Slight chlorosis and stunting may occur in certain conditions. Recovery is rapid and yield not affected

Following crops guidance

- Only cereals, oilseed rape, field beans, clover or grass may be sown in the yr of harvest of a treated crop
- In the event of crop failure only winter or spring wheat may be sown within 3 mth of treatment. Land should be ploughed and cultivated to 15 cm minimum before resowing

Environmental safety

- Extremely dangerous to fish or other aquatic life. Do not contaminate surface waters or ditches with chemical or used container
- Take extreme care to avoid drift onto broad-leaved plants outside the target area or onto ponds, waterways or ditches, or onto land intended for cropping
- Spraying equipment should not be drained or flushed onto land planted, or to be planted, with trees or crops other than cereals and should be thoroughly cleansed after use - see label for instructions

Hazard classification and safety precautions

Operator protection U08, U19, U20b
Environmental protection E13a
Storage and disposal E30a, E32a

249 flupyrsulfuron-methyl + metsulfuron-methyl

A sulfonylurea herbicide mixture for winter wheat

Products

1	Lexus XPE WSB	DuPont	33.3:16.7% w/w	WB	08542
2	Standon Flupyrsulfuron MM	Standon	33.3:16.7% w/w	WG	09098

Uses

Annual dicotyledons in ***winter wheat***. Blackgrass in ***winter wheat***.

Approval information

- Flupyrsulfuron-methyl and metsulfuron-methyl included in Annex I under EC Directive 91/414
- Approval expiry 31 May 2004 [1]

Efficacy guidance

- Best results obtained when applied to small actively growing weeds
- Good spray cover of weeds must be obtained
- Increased degradation of active ingredient in high soil temperatures reduces residual activity
- Product has moderate residual life in soil. Under normal moisture conditions susceptible weeds germinating soon after treatment will be controlled
- Product may be used on all soil types but residual activity and weed control is reduced on highly alkaline soils
- Blackgrass should be treated from 1 leaf up to mid-tillering. Strains of blackgrass resistant to other herbicides may not be controlled
- Flupyrsulfuron-methyl and metsulfuron-methyl are members of the ALS-inhibitor group of herbicides and products should be used in a planned Resistance Management strategy. See Section 1 for more information

Restrictions

- Maximum number of treatments 1 per crop
- Must only be applied in spring after 1 Feb, from 2 leaf stage of crop
- Do not use on wheat undersown with grasses or legumes, or any other broad-leaved crop
- Do not apply within 7 d of rolling
- Do not treat any crop suffering from drought, waterlogging, pest or disease attack, nutrient deficiency, or any other stress factors
- Do not apply in sequence, or in tank mixture, with any product containing any other sulfonyl urea

Crop-specific information

- Latest use: beginning of stem extension (GS 30)
- Slight chlorosis and stunting may occur in certain conditions. Recovery is rapid and yield not affected

Following crops guidance

- Only cereals, oilseed rape, field beans, clover or grass may be sown in the yr of harvest of a treated crop
- In the event of crop failure only winter wheat may be sown within 3 mth of treatment. Land should be ploughed and cultivated to 15 cm minimum before resowing

Environmental safety

- Extremely dangerous to fish or other aquatic life. Do not contaminate surface waters or ditches with chemical or used container
- LERAP Category B
- Take extreme care to avoid drift onto broad-leaved plants outside the target area or onto ponds, waterways or ditches, or onto land intended for cropping
- Spraying equipment should not be drained or flushed onto land planted, or to be planted, with trees or crops other than cereals and should be thoroughly cleansed after use - see label for instructions

Hazard classification and safety precautions

Operator protection U08, U19, U20b
Environmental protection E13a, E16a, E16b
Storage and disposal E30a, E32a

250 flupyrsulfuron-methyl + thifensulfuron-methyl

A sulfonylurea herbicide mixture for winter wheat

Products

Lexus Millenium	DuPont	10:40% w/w	WG	09206

Uses

Annual dicotyledons in ***winter wheat***. Blackgrass in ***winter wheat***.

Approval information

- Flupyrsulfuron-methyl and thifensulfuron-methyl included in Annex I under EC Directive 91/414

Efficacy guidance

- Best results obtained when applied to small actively growing weeds
- Good spray cover of weeds must be obtained
- Increased degradation of active ingredient in high soil temperatures reduces residual activity
- Product has moderate residual life in soil. Under normal moisture conditions susceptible weeds germinating soon after treatment will be controlled
- Product may be used on all soil types but residual activity and weed control is reduced on highly alkaline soils
- Blackgrass should be treated from 1 leaf up to mid-tillering. Strains of blackgrass resistant to other herbicides may not be controlled
- Flupyrsulfuron-methyl and thifensulfuron-methyl are members of the ALS-inhibitor group of herbicides and products should be used in a planned Resistance Management strategy. See Section 1 for more information

Restrictions

- Maximum number of treatments 1 per crop
- Do not treat winter wheat at or after 1st node detectable stage (GS 31)
- Do not use on wheat undersown with grasses or legumes, or any other broad-leaved crop
- Do not apply within 7 d of rolling
- Do not treat any crop suffering from drought, waterlogging, pest or disease attack, nutrient deficiency, or any other stress factors
- Do not apply to a crop already treated with any other product containing thifensulfuron-methyl. Only specified products containing metsulfuron-methyl or tribenuron-methyl may follow treatment. See label

Crop-specific information

- Latest use: before 1st node detectable (GS 31)
- Slight chlorosis, speckling and stunting may occur in certain conditions. Recovery is rapid and yield not affected

Following crops guidance

- Only cereals, oilseed rape, field beans or grass may be sown in the yr of harvest of a treated crop. Any crop may be sown the following spring
- In the event of crop failure only winter wheat may be sown before normal harvest date. Land should be ploughed and cultivated to 15 cm minimum before resowing

Environmental safety

- Extremely dangerous to fish or other aquatic life. Do not contaminate surface waters or ditches with chemical or used container
- LERAP Category B

Hazard classification and safety precautions

Hazard R04
Risk phrases R04e
Operator protection A, H; U05a, U08, U20b
Environmental protection E01, E13a, E16a, E16b
Consumer protection C03
Storage and disposal E26, E30a, E32a

SECTION 4

251 fluquinconazole

A protectant, eradicant and systematic triazole fungicide for winter wheat

Products

1	Flamenco	Aventis	100 g/l	SC	09913
2	Flamenco	BASF	100 g/l	SC	11699
3	Galmano	Bayer CropScience	167 g/l	FS	11650
4	Jockey F	Aventis	167 g/l	FS	10074
5	Jockey F	BASF	167 g/l	FS	11690

Uses

Brown rust in ***winter wheat*** [1, 2]. Bunt in ***winter wheat*** *(seed treatment)* [3-5]. Powdery mildew in ***winter wheat*** [1, 2]. Septoria diseases in ***winter wheat*** [1, 2]. Septoria leaf spot in ***winter wheat*** *(seed treatment)* [3-5]. Take-all in ***winter wheat*** *(seed treatment - reduction)* [3-5]. Yellow rust in ***winter wheat*** [1, 2]; ***winter wheat*** *(seed treatment)* [3-5].

Efficacy guidance

- Best results obtained from treatment when disease first becomes active in crop but before infection spreads to younger leaves [1, 2]
- Adequate disease protection throughout the season will usually require a programme of at least two fungicide treatments [1, 2]
- May also be applied as a protectant at end of ear emergence (but no later) if crop still disease free [1, 2]
- Ensure good even coverage of seed to obtain reliable disease control [3-5]
- Seed treatment provides early control of Septoria leaf spot and yellow rust but may require foliar treatment for later infection [3-5]

Restrictions

- Maximum total spray dose equivalent to two full doses [1, 2]
- Maximum number of seed treatments 1 per batch [3-5]
- Do not apply after the start of anthesis (GS 59) [1, 2]
- Do not treat cracked, split or sprouted seed [3-5]
- Do not use treated seed as food or feed [4, 5]

Crop-specific information

- Latest use: before beginning of anthesis (GS 59) for spray treatment; before drilling for seed treatment
- Ensure good spray coverage and increase volume in dense crops [1, 2]
- Delayed emergence may result when treated seed is drilled into heavy or poorly drained soils which then become wet or waterlogged [3-5]

Environmental safety

- Dangerous to fish or other aquatic life. Do not contaminate surface waters or ditches with chemical or used container
- High risk to non-target insects or other arthropods. Do not spray within 6 m of the field boundary [1, 2]
- Product supplied in returnable packs for which special PPE requirements and precautions apply. Check label [4, 5]

Hazard classification and safety precautions

Hazard R03 [1-5]; R04 [1, 2]

Risk phrases R03c [1-5]; R04a, R04e [1, 2]

Operator protection A, H [1-5]; C, M [1, 2]; D [3-5]; U05a [1-5]; U07 [3]; U07 [4] (returnable pack only); U07 [5]; U11, U14, U15 [1, 2]; U20b [3-5]

Environmental protection E01, E13b [1-5]; E22a [1, 2]

Consumer protection C03

Storage and disposal E26 [1, 2, 4, 5]; E29, E30a [3-5]; E31c [1, 2]; E32a [4] (disposable packs only); E33, E36 [3]; E33, E36 [4] (returnable pack only); E33, E36 [5]; E34 [1-5]

Treated seed S02, S05, S06, S07, S08 [3-5]

Medical advice M03

252 fluquinconazole + prochloraz

A broad-spectrum triazole mixture for use as a spray and seed treatment in winter wheat

Products

1	Foil	Aventis	54:174 g/l	SE	10905
2	Foil	BASF	54:174 g/l	SE	11700
3	Galmano Plus	Bayer CropScience	167:31 g/l	FS	11645
4	Jockey	Aventis	167:31 g/l	FS	10076
5	Jockey	BASF	167:31 g/l	FS	11689

Uses

Brown rust in ***winter barley***, ***winter wheat*** [1, 2]. Bunt in ***winter wheat*** *(seed treatment)* [3-5]. Fusarium root rot in ***winter wheat*** *(seed treatment)* [3-5]. Powdery mildew in ***winter barley***, ***winter wheat*** *(moderate control)* [1, 2]. Rhynchosporium in ***winter barley*** [1, 2]. Septoria leaf spot in ***winter wheat*** [1, 2]; ***winter wheat*** *(seed treatment)* [3-5]. Take-all in ***winter wheat*** *(seed treatment - reduction)* [3-5]. Yellow rust in ***winter wheat*** [1-5].

Efficacy guidance

- Best results obtained from treatment when disease first becomes active in crop but before infection spreads to younger leaves [1, 2]
- Ensure good even coverage of seed to obtain reliable disease control [3-5]
- Seed treatment provides early control of Septoria leaf spot and yellow rust but may require foliar treatment for later infection [3-5]

Restrictions

- Maximum number of treatments 1 per seed batch (seed dressing)
- Maximum total dose equivalent to two full doses (spray)
- Do not treat cracked, split or sprouted seed [3-5]
- Do not use treated seed as food or feed

Crop-specific information

- Latest use: before 1st awns visible (GS 47) for winter barley; before beginning of anthesis (GS 59) for winter wheat; before drilling (seed treatment)
- Ensure good spray coverage and increase volume in dense crops [1, 2]
- Delayed emergence may result when treated seed is drilled into heavy or poorly drained soils which then become wet or waterlogged [3-5]

Environmental safety

- Dangerous to fish or other aquatic life. Do not contaminate surface waters or ditches with chemical or used container
- Product supplied in returnable packs for which special PPE requirements and precautions apply. Check label

Hazard classification and safety precautions

Hazard R03 [1-5]; R04 [1, 2]; R06, R14 [2]

Risk phrases R03c [1-5]; R04a, R04e [1, 2]; R04g [1]; R12b, R14b [2]

Operator protection A, H [1-5]; C [1, 2]; D [3-5]; U05a [1-5]; U07 [3]; U07 [4] (returnable pack only); U07 [5]; U20b [3-5]

Environmental protection E01 [1-5]; E13b [1, 3-5]; E15, E38 [2]

Consumer protection C03

Storage and disposal E26 [1, 2, 4, 5]; E29 [3-5]; E30a, E34 [1-5]; E31c [1, 2]; E32a [4] (disposable packs only); E32d [2]; E33, E36 [3]; E33, E36 [4] (returnable pack only); E33, E36 [5]

Treated seed S02, S05, S06, S07, S08 [3-5]

Medical advice M03 [1-5]; M05 [1, 2]

253 fluroxypyr

A post-emergence aryloxyalkanoic acid herbicide

See also *2,4-D + dicamba + fluroxypyr*
bromoxynil + fluroxypyr + ioxynil
clopyralid + fluroxypyr + MCPA
clopyralid + fluroxypyr + triclopyr
florasulam + fluroxypyr

Products

1	Agriguard Fluroxypyr 200	AgriGuard	200 g/l	EC	09298
2	Barclay Hurler	Barclay	200 g/l	EC	11356
3	Fernpath Hatchet	AgriGuard	200 g/l	EC	11510
4	Greencrop Reaper	Greencrop	200 g/l	EC	09359
5	Greencrop Reaper 2	Greencrop	200 g/l	EC	10060
6	Landgold Fluroxypyr	Landgold	200 g/l	EC	10490
7	Standon Fluroxypyr	Standon	200 g/l	EC	08923
8	Standon Fluroxypyr 200	Standon	200 g/l	EC	11534
9	Starane 2	Dow	200 g/l	EC	05496
10	Tomahawk	Makhteshim	200 g/l	EC	09249
11	Vizzler	Interfarm	200 g/l	EC	11409

Uses

Annual dicotyledons in ***bulb onions*** *(off-label)*, ***leeks*** *(off-label)* [9]; ***durum wheat, spring barley, spring oats, spring rye, spring wheat, triticale, winter barley, winter oats, winter rye, winter wheat*** [1-11]; ***established grassland*** [1-5, 7-11]; ***forage maize*** [2, 4-8, 10]; ***maize*** [9, 11]; ***newly sown grass*** [8]; ***seedling leys*** [1-5, 7, 9-11]. Black bindweed in ***durum wheat, spring barley, spring oats, spring rye, spring wheat, triticale, winter barley, winter oats, winter rye, winter wheat*** [1-11]; ***established grassland*** [1-5, 7-11]; ***forage maize*** [2, 4-8, 10]; ***maize*** [9, 11]; ***newly sown grass*** [8]; ***seedling leys*** [1-5, 7, 9-11]. Chickweed in ***durum wheat, spring barley, spring oats, spring rye, spring wheat, triticale, winter barley, winter oats, winter rye, winter wheat*** [1-11]; ***established grassland*** [1-5, 7-11]; ***forage maize*** [2, 4-8, 10]; ***maize*** [9, 11]; ***newly sown grass*** [8]; ***seedling leys*** [1-5, 7, 9-11]. Cleavers in ***apple orchards*** *(off-label)*, ***pear orchards*** *(off-label)* [9]; ***durum wheat, spring barley, spring oats, spring rye, spring wheat, triticale, winter barley, winter oats, winter rye, winter wheat*** [1-11]; ***established grassland*** [1-5, 7-11]; ***forage maize*** [2, 4-8, 10]; ***maize*** [9, 11]; ***newly sown grass*** [8]; ***seedling leys*** [1-5, 7, 9-11]. Docks in ***apple orchards*** *(off-label)*, ***pear orchards*** *(off-label)* [9]; ***durum wheat, spring barley, spring oats, spring rye, spring wheat, triticale, winter barley, winter oats, winter rye, winter wheat*** [1-11]; ***established grassland*** [1-5, 7-11]; ***forage maize*** [2, 4-8, 10]; ***maize*** [9, 11]; ***newly sown grass*** [8]; ***seedling leys*** [1-5, 7, 9-11]. Forget-me-not in ***durum wheat, spring barley, spring oats, spring rye, spring wheat, triticale, winter barley, winter oats, winter rye, winter wheat*** [1-11]; ***established grassland*** [1-5, 7-11]; ***forage maize*** [2, 4-8, 10]; ***maize*** [9, 11]; ***newly sown grass*** [8]; ***seedling leys*** [1-5, 7, 9-11]. Hemp-nettle in ***durum wheat, spring barley, spring oats, spring rye, spring wheat, triticale, winter barley, winter oats, winter rye, winter wheat*** [1-11]; ***established grassland*** [1-5, 7-11]; ***forage maize*** [2, 4-8, 10]; ***maize*** [9, 11]; ***newly sown grass*** [8]; ***seedling leys*** [1-5, 7, 9-11]. Stinging nettle in ***apple orchards*** *(off-label)*, ***pear orchards*** *(off-label)* [9]. Volunteer potatoes in ***bulb onions*** *(off-label)*, ***leeks*** *(off-label)* [9]; ***durum wheat, spring barley, spring oats, spring rye, spring wheat, triticale, winter barley, winter oats, winter rye, winter wheat*** [1-11]; ***established grassland, seedling leys*** [1-5, 7, 9-11]; ***forage maize*** [2, 4-8, 10]; ***maize*** [9, 11].

Specific Off-Label Approvals (SOLAs)

- ***apple orchards, pear orchards*** *(OLA 002710) Dec 2008* [9]
- ***bulb onions*** *(OLA 012137) Dec 2008* [9]
- ***leeks*** *(OLA 023656) Dec 2008* [9]

Approval information

- Fluroxypyr included in Annex I under EC Directive 91/414

Efficacy guidance

- Best results achieved under good growing conditions in a strongly competing crop

- A number of tank mixtures with HBN and other herbicides are recommended for use in autumn and spring to extend range of species controlled. See label for details
- Spray is rainfast in 1 h

Restrictions

- Maximum number of treatments 1 per crop or yr or maximum total dose equivalent to one full dose treatment
- Do not use on crops undersown with clovers or other legumes
- Do not treat crops suffering stress caused by any factor
- Do not roll or harrow for 7 d before or after treatment
- Do not spray if frost imminent

Crop-specific information

- Latest use: before flag leaf sheath opening (GS 47) for winter wheat and barley; before flag leaf sheath extending (GS 41) for spring wheat and barley; before second node detectable (GS 32) for oats, rye, triticale and durum wheat; before 7 leaves unfolded and before buttress roots appear for maize
- HI apples, pears 4 wk; onions 11 wk
- Apply to new leys from 3 expanded leaf stage
- Timing varies in tank mixtures. See label for details
- Crops undersown with grass may be sprayed provided grasses are tillering

Following crops guidance

- Keep livestock out of treated areas for at least 3 d and until foliage of poisonous weeds such as ragwort has died and become unpalatable

Environmental safety

- Flammable
- Harmful to fish or other aquatic life. Do not contaminate surface waters or ditches with chemical or used container
- Wash spray equipment thoroughly with water and detergent immediately after use. Traces of product can damage susceptible plants sprayed later

Hazard classification and safety precautions

Hazard R03, R04 [3, 5, 6, 8, 9, 11]; R07d [1-11]
Risk phrases R04c [3, 11]; R04g [3, 5, 6, 8, 9, 11]
Operator protection U08, U20b [1-11]; U19 [2-11]
Environmental protection E07 [1, 2] (3 d); E07 [3]; E07 [4] (3 d); E07 [5, 6]; E07 [7-11] (3 d); E13c [1-11]
Storage and disposal E26, E31b [1-11]; E30a [1, 2, 4-11]; E34 [1, 3, 5, 8]
Medical advice M05 [3, 5, 6, 8, 9, 11]

SECTION 4

254 fluroxypyr + mecoprop-P

A post-emergence herbicide for broadleaved weeds in amenity turf

Products

Bastion T	Rigby Taylor	72:300 g/l	EC	06011

Uses

Annual dicotyledons in ***amenity turf, managed amenity turf***. Perennial dicotyledons in ***amenity turf, managed amenity turf***. Slender speedwell in ***amenity turf, managed amenity turf***.

Approval information

- Fluroxypyr included in Annex I under EC Directive 91/414

Efficacy guidance

- Best results achieved when soil moist and weeds in active growth, normally Apr-Sep

Restrictions

- Maximum number of treatments 2 per yr. The total amount of mecoprop-P applied in a single yr must not exceed the maximum total dose approved for any single product for the crop/situation
- Do not treat turf under stress from any cause, if night temperatures are low, if ground frost imminent or during prolonged cold weather
- Do not apply in drought period unless irrigation applied
- Do not spray if turf wet

Crop-specific information

- Avoid mowing for 3 d before or after spraying (5 d before for young turf)
- Young turf must only be treated in spring provided that at least 2 mth have elapsed between sowing and application

Environmental safety

- Harmful to fish or other aquatic life. Do not contaminate surface waters or ditches with chemical or used container
- Wash spray equipment thoroughly with water and detergent immediately after use. Traces of product can damage susceptible plants sprayed later
- Avoid drift onto all broad-leaved plants outside the target area

Hazard classification and safety precautions

Hazard R03, R04
Risk phrases R03a, R03c, R04a
Operator protection A, C; U05a, U08, U19, U20b
Environmental protection E01, E13c
Consumer protection C03
Storage and disposal E26, E30a, E31b, E34
Medical advice M03

255 fluroxypyr + triclopyr

A foliar acting herbicide for docks in grassland

Products

1	Doxstar	Dow	100:100 g/l	EC	06050
2	Evade	SumiAgro Amenity	20:60 g/l	ME	08071

Uses

Annual dicotyledons in ***amenity grass*** [2]. Chickweed in ***newly sown grass*** [1]. Docks in ***established grassland***, ***newly sown grass*** [1]. Perennial dicotyledons in ***amenity grass*** [2]. Woody weeds in ***amenity grass*** [2].

Approval information

- Fluroxypyr included in Annex I under EC Directive 91/414

Efficacy guidance

- Seedling docks in new leys are controlled up to 50 mm diameter. In established grass apply in spring or autumn or, at lower dose, in spring and autumn on docks up to 200 mm. A second application in the subsequent yr may be needed [1]
- Allow 2-3 wk after cutting or grazing to allow sufficient regrowth of docks to occur before spraying [1]
- For woody weed control apply between May and end Aug when weeds are actively growing [2]
- Control may be reduced if rain falls within 2 h of application
- To allow maximum translocation to the roots of docks do not cut grass for 28 d after spraying [1]

Restrictions

- Maximum number of treatments 2 per yr [2]; maximum total dose equivalent to one full dose treatment [1]
- Do not roll or harrow for 10 d before or 7 d after spraying
- Do not spray in drought, very hot or very cold weather
- Avoid contact with foliage, stems or branches of desirable vegetation either directly or by drift. Use of a tree guard recommended when applying near trees and shrubs [2]

Crop-specific information

- Latest use: 7 d before grazing or harvest of grass
- Grass less than one yr old may be treated from the third leaf visible stage [1]
- Clover will be killed or severely checked by treatment

Following crops guidance

- Do not sow kale, turnips, swedes or grass mixtures containing clover by direct drilling or minimum cultivation techniques within 6 wk of application [1]

- Do not plant winter beans or legumes in same yr as treatment. Restrictions apply to many other crops planted in spring following treatment. See label for details of cultivations and intervals required [2]

Environmental safety

- Flammable [1]
- Harmful to fish or other aquatic life. Do not contaminate surface waters or ditches with chemical or used container
- LERAP Category B [2]
- Do not allow drift to come into contact with crops, amenity plantings, gardens, ponds, lakes or watercourses
- Wash spray equipment thoroughly with water and detergent immediately after use. Traces of product can damage susceptible plants sprayed later
- Keep livestock out of treated areas for at least 7 d and until foliage of any poisonous weeds such as ragwort has died and become unpalatable [1]

Hazard classification and safety precautions

Hazard R03, R07d [1]; R04 [1, 2]
Risk phrases R04a [2]; R04b [1, 2]; R04c, R04g [1]
Operator protection A, C [1, 2]; H, M [2]; U02, U19 [1]; U05a, U08, U20b [1, 2]
Environmental protection E01, E13c [1, 2]; E07 [1] (7 d); E16a, E16b [2]
Consumer protection C01 [1]; C03 [1, 2]
Storage and disposal E26, E30a, E31b, E34
Medical advice M05 [1]

256 flurtamone

A carotenoid synthesis inhibitor available only in mixtures

See also *diflufenican + flurtamone*
diflufenican + flurtamone + isoproturon

257 flusilazole

A systemic, protective and curative conazole fungicide for cereals and oilseed rape

See also *carbendazim + flusilazole*
famoxadone + flusilazole
fenpropimorph + flusilazole

Products

1	Genie 25	DuPont	250 g/l	EW	10285
2	Lyric	DuPont	250 g/l	EW	08252
3	Sanction 25	DuPont	250 g/l	EW	10284

Uses

Brown rust in ***spring barley***, ***winter barley***, ***winter wheat***. Eyespot in ***spring barley***, ***winter barley***, ***winter wheat***. Light leaf spot in ***spring oilseed rape***, ***winter oilseed rape***. Net blotch in ***spring barley***, ***winter barley***. Powdery mildew in ***spring barley***, ***sugar beet***, ***winter barley***, ***winter wheat***. Rhynchosporium in ***spring barley***, ***winter barley***. Rust in ***sugar beet***. Septoria in ***spring barley***, ***winter barley***, ***winter wheat***. Yellow rust in ***spring barley***, ***winter barley***, ***winter wheat***.

Efficacy guidance

- On cereals use as a routine preventative spray or when disease first develops
- Best control of eyespot achieved by spraying between leaf-sheath erect and second node detectable stages (GS 30-32)
- Product active against both MBC-sensitive and MBC-resistant strains of eyespot
- Rain occurring within 2 h after application may reduce effectiveness
- See label for recommended tank-mixes to give broader spectrum control

Restrictions

- Maximum number of treatments (including other products containing flusilazole) on cereals and oilseed rape depends on dose and timing - see labels for details. High rate must not be used more than once in any crop
- Maximum total dose on sugar beet equivalent to one full dose
- Do not apply to crops under stress or during frosty weather

Crop-specific information

- Latest use: high dose before 3rd node detectable (GS 33) plus reduced dose before early milk stage (GS 72) on winter wheat and barley; before first flowers open (GS 4,0) on oilseed rape
- HI 7 wk for sugar beet
- Treat oilseed rape in autumn and/or spring at stem extension stage
- On sugar beet apply at an early stage of disease development, usually in early Aug

Environmental safety

- Harmful to fish or other aquatic life. Do not contaminate surface waters or ditches with chemical or used container

Hazard classification and safety precautions

Hazard R03, R04
Risk phrases R03c, R04a
Operator protection A, C; U05a, U11, U19, U20a
Environmental protection E01, E13c
Consumer protection C03
Storage and disposal E26, E30a, E31b, E34
Medical advice M03

258 flutolanil

An oxathiin fungicide for treatment of potato seed tubers

Products

Rhino	Certis	460 g/l	FS	11802

Uses

Black scurf in ***potatoes*** *(tuber treatment)*. Stem canker in ***potatoes*** *(tuber treatment)*.

Efficacy guidance

- Apply to clean tubers before chitting, prior to planting, or at planting
- Apply through canopied, hydraulic or spinning disc equipment (with or without electrostatics) mounted on a rolling conveyor or table
- May be diluted with water up to 2.0 l per tonne to improve tuber coverage. Disease in areas not covered by spray will not be controlled

Restrictions

- Maximum number of treatments 1 per batch of seed tubers

Crop-specific information

- Latest use: at planting
- Seed tubers should be of good quality and free from bacterial rots, physical damage or virus infection, and should not be sprouted to such an extent that mechanical damage to the shoots will occur during treatment or planting

Environmental safety

- Harmful to aquatic organisms. May cause long-term adverse effects to the aquatic environment

Hazard classification and safety precautions

Hazard R04, R14
Risk phrases R04e, R14a
Operator protection A, H; U04a, U05a, U19, U20a
Environmental protection E01, E15
Consumer protection C03
Storage and disposal E30a, E31a, E34
Treated seed S01, S03, S04a, S05
Medical advice M03

259 flutriafol

A broad-spectrum conazole fungicide for cereals

Products

Pointer	Headland	125 g/l	SC	11522

Uses

Brown rust in ***spring barley, winter barley, winter wheat***. Powdery mildew in ***spring barley, winter barley, winter wheat***. Rhynchosporium in ***spring barley, winter barley***. Septoria leaf spot in ***winter wheat***. Yellow rust in ***spring barley, winter barley, winter wheat***.

Efficacy guidance

- Best results obtained from treatment in early stages of disease development. See label for detailed guidance on spray timing for specific diseases and the need for repeat treamtments
- Tank mix options available on the label to broaden activity spectrum
- Good spray coverage essential for optimum performance

Restrictions

- Maximum number of treatments 2 per crop (including other products containing flutriafol)

Crop-specific information

- Latest use: before early grain milky ripe stage (GS 73)
- Flag leaf tip scorch on wheat caused by stress may be increased by fungicide treatment

Environmental safety

- Harmful to fish or other aquatic life. Do not contaminate surface waters or ditches with chemical or used container

Hazard classification and safety precautions

Hazard R04
Risk phrases R04e
Operator protection A, H; U05a, U09a, U19, U20b
Environmental protection E01, E13c
Consumer protection C03
Storage and disposal E30a, E31c

SECTION 4

260 fomesafen

A contact and residual diphenyl ether herbicide for use in leguminous crops

Products

Flex	Syngenta	250 g/l	SL	08885

Uses

Annual dicotyledons in ***dwarf beans***. Black nightshade in ***soya beans*** *(off-label)*. Volunteer oilseed rape in ***dwarf beans, soya beans*** *(off-label)*.

Specific Off-Label Approvals (SOLAs)

- ***soya beans*** *(OLA 023930) Nov 2006* [1]

Approval information

- Products containing this active ingredient have been granted derogations for specified 'Essential Uses' for use until 31 December 2007. Sale and supply must cease by 30 June 2007 but growers have no guarantee that the products will continue to be available until then.
  For more information see 'The Review Programme' under 'Pesticide Legislation' in Section 1

Efficacy guidance

- Fomesafen combines contact activity and residual soil uptake
- Best results obtained when seedling weeds are actively growing in warm moist conditions with adequate soil moisture
- Weeds should be treated at seedling stage. Larger weeds, and those hardened off by adverse conditions, may be less well controlled
- Residual activity gives control of later germinating weed seedlings

- Product may be used on all soil types but residual control of weeds germinating after treatment may be reduced on soils with high organic matter
- Weed spectrum may be widened by use in a programme with other herbicides. See label for details

Restrictions

- Maximum total dose 0.9 l/ha
- Apply to healthy crops only. Crops growing under stress from drought, waterlogging etc should not be treated
- This product must not be used on any crops other than those listed, including any extrapolations that would normally be permissible under the Long Term Arrangements for Extension of Use (see Section 1)

Crop-specific information

- HI green beans 5 wk; soya beans 8 wk
- Some crop damage, such as leaf crinkling, may occur. Effect is transient and should not affect yield

Hazard classification and safety precautions

Hazard R04
Risk phrases R04a, R04e
Operator protection A, C, H; U04a, U05a, U11, U14, U19, U20b
Environmental protection E01, E15
Consumer protection C03
Storage and disposal E26, E30a, E31a
Medical advice M03

261 fomesafen + terbutryn

A residual herbicide for use in leguminous crops

Products

Reflex T	Syngenta	80:400 g/l	SC	08884

Uses

Annual dicotyledons in ***broad beans*** *(spring sown)*, ***combining peas*** *(spring sown)*, ***spring field beans***, ***vining peas*** *(spring sown)*.

Approval information

- Products containing fomesafen and terbutryn have been granted derogations for specified 'Essential Uses' for use until 31 December 2007. Sale and supply must cease by 30 June 2007 but growers have no guarantee that the products will continue to be available until then.
  For more information see 'The Review Programme' under 'Pesticide Legislation' in Section 1

Efficacy guidance

- Best results achieved when applied to moist, fine, firm tilth
- Rain soon after application is essential for optimum weed control
- Results may be unsatisfactory in dry, or excessively wet, soil conditions
- Weed control may be reduced on soils with more than 10% organic matter

Restrictions

- Maximum number of treatments 1 per crop
- Use only once every 5 yr
- All varieties of spring sown peas may be treated but forage varieties may be damaged from which recovery may not be complete. All varieties of spring sown field and broad beans may be treated
- Do not use on Sands
- This product must not be used on any crops other than those listed, including any extrapolations that would normally be permissible under the Long Term Arrangements for Extension of Use (see Section 1)

Crop-specific information

- Latest use: pre-emergence of crop
- Apply after planting, but before the crop emerges
- Emerged crop leaves may be severely scorched or killed

Following crops guidance

- Plough or cultivate to at least 150 mm before drilling or planting another crop. Only cereals should be planted in the calendar year of use with a minimum interval of 4 mth after treatment

Environmental safety

- Dangerous for the environment
- Toxic to aquatic organisms. May cause long-term adverse effects to the aquatic environment

Hazard classification and safety precautions

Hazard R03, R04, R14
Risk phrases R03c, R04e, R14b
Operator protection A, C, H; U04a, U05a, U08, U14, U19, U20b
Environmental protection E01, E15
Consumer protection C03
Storage and disposal E26, E31c

262 formaldehyde (commodity substance)

An agricultural/horticultural and animal husbandry fungicide

Products

1	formaldehyde	various	38-40%	SL
2	paraformaldehyde	various	-	ZZ

Uses

Fungus diseases in ***flower bulbs*** *(dip)*, ***glasshouses*** *(spray, dip or fumigant)*, ***mushroom houses*** *(spray or fumigant)* [1]; ***livestock houses*** [2]. Soil-borne diseases in ***soil and compost*** *(drench)* [1].

Approval information

- Approval for the use of formaldehyde as a commodity substance was granted on 1 March 1991 by Ministers under regulation 5 of the Control of Pesticides Regulations 1986

Efficacy guidance

- Use as a dip to sterilize flower bulbs [1]
- Use as drench to sterilize soil and compost, indoors and outdoors [1]
- Use as spray or fumigant in mushroom houses [1]
- Use as spray, dip or fumigant for glasshouse hygiene [1]
- Use as fumigant to sterilize animal houses [2]

Restrictions

- Formaldehyde is subject to the Poisons Rules 1982 and the Poisons Act 1972. See notes in Section 1
- Operators must observe Occupational Exposure Standard as set out in HSE Guidance Note EH40/90 and ACOP 30 *Control of Substances Hazardous to Health in Fumigation*
- Operators must be supplied with a Section 6 (HSW) Safety Data Sheet before commencing work
- Observe maximum permitted concentrations. See PR 12, 1990, pp 9-10

Hazard classification and safety precautions

Operator protection A, D, P [1]

263 fosetyl-aluminium

A systemic phosphonic acid fungicide for various horticultural crops

Products

1	Aliette 80 WG	Certis	80% w/w	WG	09156
2	Aliette 80 WG	Certis	80% w/w	WG	11213
3	Standon Fosetyl-AL 80 WG	Standon	80% w/w	WG	10667

Uses

Collar rot in ***apples*** [1-3]. Crown rot in ***apples*** [1-3]; ***protected strawberries*** *(off-label)*, ***strawberries*** *(off-label)* [1, 2]. Damping off in ***broccoli*** *(off-label)*, ***brussels sprouts*** *(off-label)*, ***cabbages*** *(off-label)*, ***calabrese*** *(off-label)*, ***cauliflowers*** *(off-label)*, ***chinese cabbage*** *(off-label)*, ***collards*** *(off-label)*, ***kale*** *(off-label)* [1, 2]; ***protected brassica seedlings*** *(off-label)* [2]; ***protected broccoli*** *(off-label)*, ***protected brussels sprouts*** *(off-label)*, ***protected cabbages*** *(off-label)*,

SECTION 4

protected calabrese *(off-label)*, ***protected cauliflowers*** *(off-label)*, ***protected chinese cabbage*** *(off-label)*, ***protected collards*** *(off-label)*, ***protected kale*** *(off-label)* [1]. Downy mildew in ***agastache spp.*** *(off-label)*, ***angelica*** *(off-label)*, ***balm*** *(off-label)*, ***basil*** *(off-label)*, ***bay*** *(off-label)*, ***borage*** *(off-label)*, ***camomile*** *(off-label)*, ***caraway*** *(off-label)*, ***catnip*** *(off-label)*, ***chervil*** *(off-label)*, ***chives*** *(off-label)*, ***clary*** *(off-label)*, ***coriander*** *(off-label)*, ***curry plant*** *(off-label)*, ***dill*** *(off-label)*, ***dragonhead*** *(off-label)*, ***endives*** *(off-label)*, ***fennel*** *(off-label)*, ***fenugreek*** *(off-label)*, ***feverfew*** *(off-label)*, ***horehound*** *(off-label)*, ***hyssop*** *(off-label)*, ***land cress*** *(off-label)*, ***lavandin*** *(off-label)*, ***lavender*** *(off-label)*, ***lemon verbena*** *(off-label)*, ***lettuce*** *(off-label)*, ***lovage*** *(off-label)*, ***marigold*** *(off-label)*, ***marjoram*** *(off-label)*, ***mint*** *(off-label)*, ***nasturtium*** *(off-label)*, ***nettle*** *(off-label)*, ***oregano*** *(off-label)*, ***origanum heracleoticum*** *(off-label)*, ***parsley*** *(off-label)*, ***plantain*** *(off-label)*, ***protected agastache spp.*** *(off-label)*, ***protected angelica*** *(off-label)*, ***protected balm*** *(off-label)*, ***protected basil*** *(off-label)*, ***protected bay*** *(off-label)*, ***protected borage*** *(off-label)*, ***protected camomile*** *(off-label)*, ***protected caraway*** *(off-label)*, ***protected catnip*** *(off-label)*, ***protected chervil*** *(off-label)*, ***protected chives*** *(off-label)*, ***protected clary*** *(off-label)*, ***protected coriander*** *(off-label)*, ***protected curry plant*** *(off-label)*, ***protected dill*** *(off-label)*, ***protected dragonhead*** *(off-label)*, ***protected endives*** *(off-label)*, ***protected fennel*** *(off-label)*, ***protected fenugreek*** *(off-label)*, ***protected feverfew*** *(off-label)*, ***protected horehound*** *(off-label)*, ***protected hyssop*** *(off-label)*, ***protected lamb's lettuce*** *(off-label)*, ***protected land cress*** *(off-label)*, ***protected lavandin*** *(off-label)*, ***protected lavender*** *(off-label)*, ***protected lemon verbena*** *(off-label)*, ***protected lettuce*** *(off-label)*, ***protected lovage*** *(off-label)*, ***protected marigold*** *(off-label)*, ***protected marjoram*** *(off-label)*, ***protected mint*** *(off-label)*, ***protected nasturtium*** *(off-label)*, ***protected nettle*** *(off-label)*, ***protected oregano*** *(off-label)*, ***protected origanum heracleoticum*** *(off-label)*, ***protected parsley*** *(off-label)*, ***protected plantain*** *(off-label)*, ***protected rocket*** *(off-label)*, ***protected rosemary*** *(off-label)*, ***protected rue*** *(off-label)*, ***protected sage*** *(off-label)*, ***protected salad burnet*** *(off-label)*, ***protected savory*** *(off-label)*, ***protected scarole*** *(off-label)*, ***protected sorrel*** *(off-label)*, ***protected spike lavender*** *(off-label)*, ***protected st john's wort*** *(off-label)*, ***protected tarragon*** *(off-label)*, ***protected thyme*** *(off-label)*, ***protected violet*** *(off-label)*, ***protected woodruff*** *(off-label)*, ***protected yarrow*** *(off-label)*, ***rocket*** *(off-label)*, ***rosemary*** *(off-label)*, ***rue*** *(off-label)*, ***sage*** *(off-label)*, ***salad burnet*** *(off-label)*, ***savory*** *(off-label)*, ***scarole*** *(off-label)*, ***sorrel*** *(off-label)*, ***spike lavender*** *(off-label)*, ***st john's wort*** *(off-label)*, ***tarragon*** *(off-label)*, ***thyme*** *(off-label)*, ***violet*** *(off-label)*, ***woodruff*** *(off-label)*, ***yarrow*** *(off-label)* [2, 3]; ***broad beans***, ***broccoli*** *(off-label - seedlings)*, ***brussels sprouts*** *(off-label - seedlings)*, ***cabbages*** *(off-label - seedlings)*, ***calabrese*** *(off-label - seedlings)*, ***cauliflowers*** *(off-label - seedlings)*, ***chinese cabbage*** *(off-label - seedlings)*, ***hops***, ***kale*** *(off-label - seedlings)*, ***leaf spinach*** *(off-label)*, ***protected lettuce***, ***spinach beet*** *(off-label)*, ***watercress*** *(off-label - during propagation)* [1-3]; ***broccoli*** *(off-label)*, ***brussels sprouts*** *(off-label)*, ***cabbages*** *(off-label)*, ***calabrese*** *(off-label)*, ***cauliflowers*** *(off-label)*, ***chinese cabbage*** *(off-label)*, ***collards*** *(off-label)*, ***combining peas*** *(off-label - seed treatment)*, ***grapevines*** *(off-label)*, ***kale*** *(off-label)*, ***salad onions*** *(off-label)*, ***vining peas*** *(off-label - seed treatment)* [1, 2]; ***combining peas*** *(off-label)*, ***vining peas*** *(off-label)* [3]; ***protected brassica seedlings*** *(off-label)* [2]; ***protected broccoli*** *(off-label)*, ***protected brussels sprouts*** *(off-label)*, ***protected cabbages*** *(off-label)*, ***protected calabrese*** *(off-label)*, ***protected cauliflowers*** *(off-label)*, ***protected chinese cabbage*** *(off-label)*, ***protected collards*** *(off-label)*, ***protected kale*** *(off-label)* [1]; ***protected broccoli*** *(off-label - seedlings)*, ***protected brussels sprouts*** *(off-label - seedlings)*, ***protected cabbages*** *(off-label - seedlings)*, ***protected calabrese*** *(off-label - seedlings)*, ***protected cauliflowers*** *(off-label - seedlings)*, ***protected chinese cabbage*** *(off-label - seedlings)*, ***protected kale*** *(off-label - seedlings)* [1, 3]. Phytophthora in ***chicory*** *(off-label - for forcing)* [1, 2]; ***chicory*** *(off-label - in forcing sheds)* [3]; ***watercress*** *(off-label - during propagation)* [1-3]. Phytophthora root rot in ***capillary benches***, ***protected pot plants*** [1-3]. Phytophthora stem rot in ***capillary benches***, ***protected pot plants*** [1-3]. Phytophthora wilt in ***hardy ornamental nursery stock*** [1-3]. Pythium in ***watercress*** *(off-label - during propagation)* [1-3]. Red core in ***protected strawberries*** *(off-label)*, ***strawberries*** *(off-label)* [1, 2]; ***strawberries*** [1-3]. Root rot in ***capillary benches***, ***protected pot plants*** [1-3].

Specific Off-Label Approvals (SOLAs)

- ***agastache spp., angelica, balm, basil, bay, borage, camomile, caraway, catnip, chervil, chives, clary, coriander, curry plant, dill, dragonhead, endives, fennel, fenugreek, feverfew, horehound, hyssop, land cress, lavandin, lavender, leaf spinach, lemon verbena, lettuce, lovage, marigold, marjoram, mint, nasturtium, nettle, oregano, origanum heracleoticum, parsley, plantain, protected agastache spp., protected angelica, protected balm, protected basil, protected bay, protected borage, protected camomile, protected caraway, protected***

catnip, protected chervil, protected chives, protected clary, protected coriander, protected curry plant, protected dill, protected dragonhead, protected endives, protected fennel, protected fenugreek, protected feverfew, protected horehound, protected hyssop, protected lamb's lettuce, protected land cress, protected lavandin, protected lavender, protected lemon verbena, protected lettuce, protected lovage, protected marigold, protected marjoram, protected mint, protected nasturtium, protected nettle, protected oregano, protected origanum heracleoticum, protected parsley, protected plantain, protected rocket, protected rosemary, protected rue, protected sage, protected salad burnet, protected savory, protected scarole, protected sorrel, protected spike lavender, protected st john's wort, protected tarragon, protected thyme, protected violet, protected woodruff, protected yarrow, rocket, rosemary, rue, sage, salad burnet, savory, scarole, sorrel, spike lavender, spinach beet, st john's wort, tarragon, thyme, violet, woodruff, yarrow *(OLA 030366) Dec 2008* [3]

- ***agastache spp., angelica, balm, basil, bay, borage, camomile, caraway, catnip, chervil, chives, clary, coriander, curry plant, dill, dragonhead, endives, fennel, fenugreek, feverfew, horehound, hyssop, land cress, lavandin, lavender, lemon verbena, lettuce, lovage, marigold, marjoram, mint, nasturtium, nettle, oregano, origanum heracleoticum, parsley, plantain, protected agastache spp., protected angelica, protected balm, protected basil, protected bay, protected borage, protected camomile, protected caraway, protected catnip, protected chervil, protected chives, protected clary, protected coriander, protected curry plant, protected dill, protected dragonhead, protected endives, protected fennel, protected fenugreek, protected feverfew, protected horehound, protected hyssop, protected lamb's lettuce, protected land cress, protected lavandin, protected lavender, protected lemon verbena, protected lettuce, protected lovage, protected marigold, protected marjoram, protected mint, protected nasturtium, protected nettle, protected oregano, protected origanum heracleoticum, protected parsley, protected plantain, protected rocket, protected rosemary, protected rue, protected sage, protected salad burnet, protected savory, protected scarole, protected sorrel, protected spike lavender, protected st john's wort, protected tarragon, protected thyme, protected violet, protected woodruff, protected yarrow, rocket, rosemary, rue, sage, salad burnet, savory, scarole, sorrel, spike lavender, st john's wort, tarragon, thyme, violet, woodruff, yarrow*** *(OLA 030868) Dec 2008* [2]
- ***broccoli, brussels sprouts, cabbages, calabrese, cauliflowers, chinese cabbage, collards, kale, protected broccoli, protected brussels sprouts, protected cabbages, protected calabrese, protected cauliflowers, protected chinese cabbage, protected collards, protected kale*** *(OLA 002076) Jul 2005* [1]
- ***broccoli, brussels sprouts, cabbages, calabrese, cauliflowers, chinese cabbage, collards, kale, protected brassica seedlings*** *(OLA 030865) Jan 2005* [2]
- ***broccoli, brussels sprouts, cabbages, calabrese, cauliflowers, chinese cabbage, kale, protected broccoli, protected brussels sprouts, protected cabbages, protected calabrese, protected cauliflowers, protected chinese cabbage, protected kale*** *(seedlings) (OLA 001218) Dec 2008* [1]
- ***broccoli, brussels sprouts, cabbages, calabrese, cauliflowers, chinese cabbage, kale, protected broccoli, protected brussels sprouts, protected cabbages, protected calabrese, protected cauliflowers, protected chinese cabbage, protected kale*** *(seedlings) (OLA 030366) Dec 2008* [3]
- ***broccoli, brussels sprouts, cabbages, calabrese, cauliflowers, chinese cabbage, kale*** *(seedlings) (OLA 030863) Dec 2008* [2]
- ***chicory*** *(for forcing) (OLA 021816) Dec 2008* [1]
- ***chicory*** *(for forcing) (OLA 030866) Dec 2008* [2]
- ***chicory*** *(in forcing sheds) (OLA 030366) Dec 2008* [3]
- ***combining peas, vining peas*** *(OLA 030367) Dec 2008* [3]
- ***combining peas, vining peas*** *(seed treatment) (OLA 030861) Dec 2008* [2]
- ***combining peas, vining peas*** *(seed treatment) (OLA 990058) Dec 2008* [1]
- ***grapevines*** *(OLA 030862) Dec 2008* [2]
- ***grapevines*** *(OLA 992907) Dec 2008* [1]
- ***leaf spinach, spinach beet*** *(OLA 001220) Dec 2008* [1]
- ***leaf spinach, spinach beet*** *(OLA 030864) Dec 2008* [2]
- ***protected brassica seedlings*** *(OLA 030863) Dec 2008* [2]

SECTION 4

- ***protected strawberries*, *strawberries*** *(OLA 030578) Dec 2008* [1]
- ***protected strawberries*, *strawberries*** *(OLA 030579) Dec 2008* [2]
- ***salad onions*** *(OLA 023629) Dec 2008* [1]
- ***salad onions*** *(OLA 023925) Dec 2007* [1]
- ***salad onions*** *(OLA 023926) Dec 2007* [2]
- ***watercress*** *(during propagation) (OLA 021817) Dec 2008* [1]
- ***watercress*** *(during propagation) (OLA 030366) Dec 2008* [3]
- ***watercress*** *(during propagation) (OLA 030867) Dec 2008* [2]

Restrictions

- Maximum number of treatments 1 per seed batch for peas; 1 per batch of compost for lettuce; 1 per yr for strawberries (root dip or foliar spray); 1 per yr for apples (bark paste), 2 per yr for apples (foliar spray); 2 per crop for broad beans, peas; 2 per yr for hops (basal spray); 6 per yr for hops (foliar spray); 1 per crop for brassicas and salad brassicas; 1 per crop during propagation, 2 per crop after planting out for lettuce; 7 per yr for grapevines
- Use on strawberries only between harvest and 31 Dec
- Use in compost for protected lettuce only from Sep to Apr and follow all directions carefully to avoid severe crop injury. Crop maturity may be delayed by a few days
- Check tolerance of ornamental species before large-scale treatment

Crop-specific information

- Latest use: pre-planting for dipping autumn planted strawberry runners; up to 31 Dec for spraying autumn planted strawberry runners; pre-sowing for protected lettuce, peas; before transplanting or pre-emergence for brassicas; 24 h after seeding watercress
- HI apples 5 mth (bark paste) or 4 wk (spray); grapevines 35 d; chicory 21 d; broad beans 17 d; hops, strawberries, herbs and salad crops 14 d
- Spray young orchards for crown rot protection after blossom when first leaves fully open and repeat after 4-6 wk. Apply as paste to bark of apples to control collar rot
- Apply to broad beans when infection appears (usually at flowering) and 14 d later. Consult before treating crops to be processed
- Spray autumn-planted strawberry runners 2-3 wk after planting or use dip treatment at planting. Spray established crops in late summer/early autumn after picking and repeat annually
- Apply to hops as early season basal spray or as foliar spray every 10-14 d from when training is completed
- May be used by wet incorporation in blocking compost for protected lettuce
- Apply as drench to rooted cuttings of hardy nursery stock after first potting and repeat mthly. Up to 6 applications may be needed
- See label for details of application to capillary benches and trays of young plants.

Hazard classification and safety precautions

Operator protection A, C, H; U20c
Environmental protection E15
Storage and disposal E30a, E32a

264 fosthiazate

An organophosphorus contact nematicide for potatoes

Products

Nemathorin 10G	Syngenta	10% w/w	FG	08915

Uses

Potato cyst nematode in ***potatoes***. Wireworm in ***potatoes*** *(reduction)*.

Approval information

- Fosthiazate included in Annex I under EC Directive 91/414

Efficacy guidance

- Apply and incorporate granules in one operation to a uniform depth of 10-15 cm. Deeper incorporation will reduce control
- Application best achieved using equipment such as Horstine Farmery Microband Applicator, Matco or Stocks Micrometer applicators together with a rear mounted powered rotary cultivator
- Granules must not become wet or damp before use

Restrictions

- Contains an anticholinesterase organophosphorus compound. Do not use if under medical advice not to work with such compounds
- Maximum number of treatments 1 per crop
- Product must only be applied using tractor-mounted/drawn direct placement machinery. Do not use air assisted broadcast machinery
- Do not apply more than once every four years on the same area of land
- Consult before using on crops intended for processing

Crop-specific information

- Latest use: at planting

Environmental safety

- Dangerous to game, wild birds and animals
- Dangerous to livestock. Keep all livestock out of treated areas for at least 13 wk
- Risk of explosion if dust in the air
- Harmful to fish or other aquatic life. Do not contaminate surface waters or ditches with chemical or used container
- Incorporation to 10-15 cm and ridging up of treated soil must be carried out immediately after application. Powered rotary cultivators are preferred implements for incorporation but discs, power, spring tine or Dutch harrows may be used provided two passes are made at right angles
- Failure completely to bury granules immediately after application is hazardous to wildlife
- Do not allow granules to become wet or damp before application and do not allow granules to stand overnight in the application hopper

Hazard classification and safety precautions

Hazard R03
Risk phrases R03c
Operator protection A, E, G, H, K, M; U02, U04a, U05a, U07, U09a, U13, U19, U20a
Environmental protection E01, E10a, E13c; E06a (13 wk)
Consumer protection C03
Storage and disposal E26, E29, E30a, E32a, E33, E34, E36
Treated seed S04c
Medical advice M01, M03

265 fuberidazole

A benzimidazole (MBC) fungicide available only in mixtures

See also *bitertanol + fuberidazole*
bitertanol + fuberidazole + imidacloprid

266 fuberidazole + imidacloprid + triadimenol

A broad spectrum systemic fungicide and insecticide seed treatment for winter cereals

Products

1	Baytan Secur	Bayer	15:117:125 g/l	FS	09510
2	Baytan Secur	Bayer CropScience	15:117:125 g/l	FS	11253

Uses

Blue mould in ***winter wheat*** *(seed treatment)*. Brown foot rot in ***winter barley*** *(seed treatment)*. Brown rust in ***winter barley*** *(seed treatment)*, ***winter wheat*** *(seed treatment)*. Bunt in ***winter wheat*** *(seed treatment)*. Covered smut in ***winter barley*** *(seed treatment)*. Fusarium foot rot and seedling blight in ***winter barley*** *(seed treatment - reduction)*, ***winter oats*** *(seed treatment - reduction)*, ***winter wheat*** *(seed treatment - reduction)*. Lodging control in ***winter wheat*** *(seed treatment - reduction)*. Loose smut in ***winter barley*** *(seed treatment)*, ***winter oats*** *(seed treatment)*, ***winter wheat*** *(seed treatment)*. Net blotch in ***winter barley*** *(seed treatment - seed-borne only)*. Powdery mildew in ***winter barley*** *(seed treatment)*, ***winter oats*** *(seed treatment)*, ***winter wheat*** *(seed treatment)*. Pyrenophora leaf spot in ***winter oats*** *(seed treatment)*. Septoria leaf spot in ***winter wheat*** *(seed treatment)*. Septoria seedling blight in ***winter wheat*** *(seed treatment - reduction)*. Snow rot in ***winter barley*** *(seed treatment - reduction)*. Virus vectors in

winter barley *(seed treatment)*, ***winter oats*** *(seed treatment)*, ***winter wheat*** *(seed treatment)*. Wireworm in ***winter barley*** *(reduction of damage)*, ***winter oats*** *(reduction of damage)*, ***winter wheat*** *(reduction of damage)*. Yellow rust in ***winter barley*** *(seed treatment)*, ***winter wheat*** *(seed treatment)*.

Efficacy guidance

- Apply through recommended seed treatment machinery
- Evenness of seed cover improved by simultaneous application of equal volumes of product and water
- Calibrate drill for treated seed and drill at 4 cm into firm, well prepared seedbed
- Use minimum 125 kg treated seed per ha and increase seed rate as drilling season progresses. Lower drilling rates and/or early drilling affect duration of BYDV protection needed and may require follow-up aphicide treatment
- When aphid activity unusually late, or is heavy and prolonged in areas of high risk, and mild weather predominates, follow-up treatment may be required
- In addition to seed-borne diseases early attacks of various foliar, air-borne diseases are controlled or suppressed. See label for details

Restrictions

- Maximum number of treatments 1 per batch of seed
- Do not use on naked oats
- Do not drill treated winter wheat seed after end of Nov
- Do not handle seed unnecessarily
- Do not use treated seed as food or feed
- Treated seed should not be broadcast, but drilled to a depth of 4 cm in a well prepared seedbed
- Germination tests should be done on all batches of seed to be treated to ensure seed viability and suitability for treatment
- Do not use on seed with moisture content above 16%, on sprouted, cracked or skinned seed or on seed already treated with another seed treatment

Crop-specific information

- Latest use: before drilling
- Store treated seed in cool, dry, well-ventilated store and drill as soon as possible, preferably in season of purchase
- Treatment may accentuate effects of adverse seedbed conditions on crop emergence

Environmental safety

- Dangerous to fish or other aquatic life. Do not contaminate surface waters or ditches with chemical or used container
- Dangerous to birds, game and other wildlife. Treated seed should not be left on the soil surface. Bury spillages
- If seed is left on the soil surface the field should be harrowed and rolled to ensure good incorporation

Hazard classification and safety precautions

Hazard R03, R04
Risk phrases R03c, R04e
Operator protection A, H; U04a, U05a, U13, U20b
Environmental protection E01, E03, E13b
Consumer protection C03
Storage and disposal E26, E30a, E32a, E34
Treated seed S01, S02, S03, S04c, S05, S06, S07
Medical advice M03

267 fuberidazole + triadimenol

A broad spectrum systemic fungicide seed treatment for cereals

Products

1	Baytan Flowable	Bayer	22.5:187.5 g/l	FS	02593
2	Baytan Flowable	Makhteshim	22.5:187.5 g/l	FS	11714

Uses

Blue mould in ***spring wheat*** *(seed treatment)*, ***triticale*** *(seed treatment)*, ***winter wheat*** *(seed treatment)*. Brown foot rot in ***spring barley*** *(seed treatment)*, ***spring oats*** *(seed treatment)*, ***spring rye*** *(seed treatment)*, ***spring wheat*** *(seed treatment)*, ***triticale*** *(seed treatment)*, ***winter barley*** *(seed treatment)*, ***winter oats*** *(seed treatment)*, ***winter rye*** *(seed treatment)*, ***winter wheat*** *(seed treatment)*. Brown rust in ***spring barley*** *(seed treatment)*, ***spring rye*** *(seed treatment)*, ***spring wheat*** *(seed treatment)*, ***winter barley*** *(seed treatment)*, ***winter rye*** *(seed treatment)*, ***winter wheat*** *(seed treatment)*. Bunt in ***spring wheat*** *(seed treatment)*, ***winter wheat*** *(seed treatment)*. Covered smut in ***spring barley*** *(seed treatment)*, ***winter barley*** *(seed treatment)*. Crown rust in ***spring oats*** *(seed treatment)*, ***winter oats*** *(seed treatment)*. Leaf stripe in ***spring barley*** *(seed treatment)*. Loose smut in ***spring barley*** *(seed treatment)*, ***spring oats*** *(seed treatment)*, ***spring wheat*** *(seed treatment)*, ***winter barley*** *(seed treatment)*, ***winter oats*** *(seed treatment)*, ***winter wheat*** *(seed treatment)*. Mildew in ***spring barley*** *(seed treatment)*, ***spring wheat*** *(seed treatment)*, ***winter barley*** *(seed treatment)*, ***winter wheat*** *(seed treatment)*. Pyrenophora leaf spot in ***spring oats*** *(seed treatment)*, ***winter oats*** *(seed treatment)*. Septoria in ***spring wheat*** *(seed treatment)*, ***winter wheat*** *(seed treatment)*. Yellow rust in ***spring barley*** *(seed treatment)*, ***spring wheat*** *(seed treatment)*, ***winter barley*** *(seed treatment)*, ***winter wheat*** *(seed treatment)*.

Efficacy guidance

- Apply through recommended seed treatment machinery
- Calibrate drill for treated seed and drill at 2.5-4 cm into firm, well prepared seedbed
- In addition to seed-borne diseases early attacks of various foliar, air-borne diseases are controlled or suppressed. See label for details

Restrictions

- Maximum number of treatments 1 per batch of seed
- Do not use on naked oats
- Do not drill treated winter wheat or rye seed after end of Nov. Seed rate should be increased as drilling season progresses
- Germination tests should be done on all batches of seed to be treated to ensure seed viability and suitability for treatment
- Do not use on seed with moisture content above 16%, on sprouted, cracked or skinned seed or on seed already treated with another seed treatment
- Do not use treated seed as food or feed

Crop-specific information

- Latest use: before drilling
- Treated spring wheat may be drilled in autumn up to end of Nov or from Feb onwards
- Store treated seed in cool, dry, well ventilated store and drill as soon as possible, preferably in season of purchase
- Treatment may accentuate effects of adverse seedbed conditions on crop emergence

Environmental safety

- Dangerous to fish or other aquatic life. Do not contaminate surface waters or ditches with chemical or used container

Hazard classification and safety precautions

Operator protection A; U20b
Environmental protection E03, E13b
Storage and disposal E26, E30a, E32a, E34
Treated seed S01, S02, S03, S04a, S05, S06, S07

SECTION 4

268 gibberellins

A plant growth regulator for use in top fruit

Products

Novagib	Fine	10 g/l	SL	08954

Uses

Growth regulation in ***pears*** *(off-label)*. Reducing fruit russeting in ***apples***.

Specific Off-Label Approvals (SOLAs)

- ***pears*** *(OLA 012756) Dec 2008* [1]

Efficacy guidance

- Apply to apples at completion of petal fall and repeat 3 or 4 times at 7-10 d intervals. Number of sprays and spray interval depend on dose (see labels)

Restrictions

- Maximum total dose on apples equivalent to four full dose treatments
- Good results achieved on Cox's Orange Pippin, Discovery, Golden Delicious and Karmijn apples. For other cultivars test on a small number of trees
- Return bloom may be reduced in yr following treatment
- Consult before treating apple crops grown for processing

Crop-specific information

- HI zero for apples

Hazard classification and safety precautions

Operator protection U20b
Environmental protection E01, E15
Storage and disposal E24, E26, E29, E30a, E31b

269 glufosinate-ammonium

A non-selective, non-residual phosphinic acid contact herbicide

Products

1	Challenge	Aventis	150 g/l	SL	07306
2	Challenge 60	Fargro	60 g/l	SL	08236
3	Finale	Bayer Environ.	120 g/l	SL	10092
4	Harvest	Aventis	150 g/l	SL	07321
5	Kaspar	Certis	150 g/l	SL	10615

Uses

Annual dicotyledons in ***apple orchards, cherries, currants, damsons, fallows, forestry, grapevines, headlands*** *(uncropped)*, ***non-crop farm areas, pear orchards, plums, strawberries, sugar beet, tree nuts, vegetables, ware potatoes*** [1, 5]; ***bush fruit, fruit trees, ornamental trees, shrubs*** [2]; ***cane fruit, cultivated land/soil*** [1, 2, 5]; ***combining peas, linseed, potatoes, spring barley, spring field beans, spring oilseed rape, spring wheat, winter barley, winter field beans, winter oilseed rape, winter wheat*** [4]; ***nursery stock, woody ornamentals*** [3]; ***stubbles*** [1]. Annual grasses in ***apple orchards, cherries, currants, damsons, fallows, forestry, grapevines, headlands*** *(uncropped)*, ***non-crop farm areas, pear orchards, plums, strawberries, sugar beet, tree nuts, vegetables, ware potatoes*** [1, 5]; ***bush fruit, fruit trees, ornamental trees, shrubs*** [2]; ***cane fruit, cultivated land/soil*** [1, 2, 5]; ***combining peas, linseed, potatoes, spring barley, spring field beans, spring oilseed rape, spring wheat, winter barley, winter field beans, winter oilseed rape, winter wheat*** [4]; ***nursery stock, woody ornamentals*** [3]; ***stubbles*** [1]. Green cover in ***land temporarily removed from production*** [1, 5]. Harvest management/desiccation in ***combining peas, linseed, spring field beans, spring oilseed rape, winter field beans, winter oilseed rape*** [1, 4, 5]; ***potatoes*** [4]; ***spring barley, spring wheat, winter barley, winter wheat*** [1, 4]; ***ware potatoes*** [1, 5]. Line marking preparation in ***managed amenity turf*** [3]. Perennial dicotyledons in ***apple orchards, cherries, currants, damsons, fallows, forestry, grapevines, headlands*** *(uncropped)*, ***land temporarily removed from production, non-crop farm areas, pear orchards, plums, strawberries, tree nuts*** [1, 5]; ***bush fruit, cultivated land/soil, fruit trees, ornamental trees, shrubs*** [2]; ***cane fruit*** [1, 2, 5]; ***nursery stock, woody ornamentals*** [3]. Perennial grasses in ***apple orchards, cherries, currants, damsons, fallows, forestry, grapevines, headlands*** *(uncropped)*, ***land temporarily removed from production, non-crop farm areas, pear orchards, plums, strawberries, tree nuts*** [1, 5]; ***bush fruit, cultivated land/soil, fruit trees, ornamental trees, shrubs*** [2]; ***cane fruit*** [1, 2, 5]; ***nursery stock, woody ornamentals*** [3]. Sward destruction in ***grassland*** [1, 5].

Efficacy guidance

- Activity quickest under warm, moist conditions. Light rainfall 3-4 h after application will not affect activity. Do not spray wet foliage or if rain likely within 6 h

- For weed control uses treat when weeds growing actively. Deep rooted weeds may require second treatment [2]
- On uncropped headlands apply in May/Jun to prevent weeds invading field

Restrictions

- Maximum number of treatments 4 per crop for potatoes (including 2 desiccant uses); 2 per crop (including 1 desiccant use) for oilseed rape, dried peas, field beans, linseed, wheat and barley; 3 (2 [2]) per yr for fruit and forestry; 2 per yr for strawberries, non-crop land and land temporarily removed from production; 1 per crop for sugar beet, vegetables and other crops; 1 per yr for grassland destruction, on cultivated land prior to planting edible crops
- Pre-harvest desiccation sprays should not be used on seed crops of wheat, barley, peas or potatoes but may be used on seed crops of oilseed rape, field beans and linseed [1, 4, 5]
- Do not desiccate potatoes in exceptionally wet weather or in saturated soil. See label for details
- Do not spray potatoes after emergence if grown from small or diseased seed or under very dry conditions
- Application for weed control and line-marking uses in sports turf must be between 1 Mar and 30 Sep
- Do not allow spray to contact dormant or green buds, suckers, damaged or green bark and foliage of wanted plants [2]
- Do not use straw from treated crops as animal feed or bedding [1, 4]
- Do not spray hedge bottoms [1, 4]

Crop-specific information

- Latest use: 30 Sep for use on non-crop land, top fruit, soft fruit, cane fruit, bush fruit, forestry; pre-drilling, pre-planting or pre-emergence in sugar beet, vegetables and other crops; before winter dormancy for grassland destruction.
- HI potatoes, oilseed rape, combining peas, field beans, linseed 7 d; wheat, barley 14 d
- Ploughing or other cultivations can follow 4 h after spraying
- Crops can normally be sown/planted immediately after spraying or sprayed post-drilling. On sand, very light or immature peat soils allow at least 3 d before sowing/planting or expected emergence
- For weed control in potatoes apply pre-emergence or up to 10% emergence on earlies and seed crops, up to 40% on maincrop, on plants up to 15 cm high
- In sugar beet and vegetables apply just before crop emergence, using stale seedbed technique
- In top and soft fruit, grapevines and forestry apply between 1 Mar and 30 Sep as directed sprays
- For grass destruction apply before winter dormancy occurs. Heavily grazed fields should show active regrowth. Plough from the day after spraying
- Apply pre-harvest desiccation treatments 10-21 d before harvest (14-21 d for oilseed rape). See label for timing details on individual crops
- For potato haulm desiccation apply to listed varieties (not seed crops) at onset of senescence, 14-21 d before harvest

Environmental safety

- Harmful to fish or other aquatic life. Do not contaminate surface waters or ditches with chemical or used container
- Keep livestock out of treated areas until foliage of any poisonous weeds such as ragwort has died and become unpalatable
- Treated pea haulm may be fed to livestock from 7 d after spraying, treated grain from 14 d [1, 4]
- Product supplied in small volume returnable container - see label for filling and mixing instructions [4]

Hazard classification and safety precautions

Hazard R03 [1, 4, 5]; R04 [1, 3-5]
Risk phrases R03a, R03c [1, 4, 5], R04a [1, 3-5]
Operator protection A [1, 2, 4, 5]; B [3]; C, H, M; U02 [2]; U04a, U13, U14, U15, U19 [1, 4, 5]; U05a [1, 3-5]; U08 [1-5]; U20a [1, 2, 4, 5], U20b [3]
Environmental protection E01 [1, 3-5]; E07, E13c [1-5]; E09 [1, 4]; E22b [5]
Consumer protection C03 [1, 3-5]
Storage and disposal E30a [1-5]; E31a [2]; E31b [3, 5]; E34 [1, 4, 5]
Medical advice M03 [1, 4, 5]

270 glyphosate (agricultural uses)

A translocated non-residual phosphonic acid herbicide

See also diuron + glyphosate

Products

1	Amega Pro TMF	Nufarm UK	480 g/l	SL	09630
2	Asteroid	Headland	360 g/l	SL	11118
3	Azural	Cardel	360 g/l	SL	11668
4	Barclay Barbarian	Barclay	360 g/l	SL	11362
5	Barclay Gallup	Barclay	360 g/l	SL	05161
6	Barclay Gallup 360	Barclay	360 g/l	SL	11369
7	Barclay Gallup Biograde 360	Barclay	360 g/l	SL	09840
8	Barclay Gallup Hi-Aktiv	Barclay	490 g/l	SL	10457
9	Barclay Garryowen	Barclay	360 g/l	SL	09869
10	Buggy SG	Sipcam	36% w/w	WG	08573
11	Clinic	Nufarm UK	360 g/l	SL	09378
12	Envision	Headland	450 g/l	SL	10569
13	Flame	Dow	360 g/l	SL	11583
14	Glydate	Nufarm UK	360 g/l	SL	10999
15	Glyfos	Cheminova	360 g/l	SL	07109
16	Glyfos	Headland	360 g/l	SL	10995
17	Glyper	Interfarm	360 g/l	SL	11095
18	Glyphogan	Makhteshim	360 g/l	SL	05784
19	Glyphosate 360	Dow	360 g/l	SL	09151
20	Glyphosate 360	Cardel	360 g/l	SL	11579
21	Greenaway Gly-490	Greenaway	490 g/l	SL	11064
22	Greencrop Gypsy	Greencrop	360 g/l	SL	09432
23	I T Glyphosate	I T Agro	360 g/l	SL	07212
24	Kernel	Cheminova	480 g/l	SL	10616
25	Kernel	Headland	480 g/l	SL	10993
26	Landgold Glyphosate 360	Landgold	360 g/l	SL	05929
27	Manifest	Headland	360 g/l	SL	11041
28	Reliance	Cardel	360 g/l	SL	11667
29	Romany	Greencrop	360 g/l	SL	11000
30	Roundup Biactive	Monsanto	360 g/l	SL	10320
31	Roundup Gold	Monsanto	450 g/l	SL	10975
32	Roundup Max	Monsanto	68% w/w	SG	10231
33	Roundup Ultra ST	Monsanto	450 g/l	SL	10199
34	Samurai	Monsanto	360 g/l	SL	10334
35	Scorpion	Cardel	360 g/l	SL	11656
36	Stacato	Sipcam	360 g/l	SL	05892
37	Standon Glyphosate 360	Standon	360 g/l	SL	05582
38	Sting ECO	Monsanto	120 g/l	SL	10337
39	Touchdown	Syngenta	330 g/l	SL	10538
40	Touchdown Quattro	Syngenta	360 g/l	SL	10608

Uses

Annual and perennial weeds in ***combining peas, durum wheat, spring barley, spring field beans, spring oilseed rape, spring wheat, winter barley, winter field beans, winter oilseed rape, winter wheat*** [2, 4, 40]; ***cultivated land/soil*** [1-3, 11, 13, 14, 19, 20, 22, 29, 34]; ***field crops*** *(wiper application)* [3, 11, 13, 14, 20, 22, 29, 34]; ***grassland*** [1, 3, 10, 11, 13-15, 17-20, 24, 28-31, 34, 35]; ***land not intended to bear vegetation*** [30]; ***land temporarily removed from production*** [2, 10, 30, 31, 40]; ***linseed*** [2, 40]; ***mustard*** [40]; ***non-crop farm areas*** [3-6, 9, 11, 14, 19, 20, 22, 29, 34, 36]; ***spring oats, winter oats*** [2, 4]; ***stubbles*** [1-4, 10, 11, 13-15, 17-24, 26, 28-32, 34-37, 39, 40]. Annual dicotyledons in ***combining peas, durum wheat, leeks, linseed, mustard, onions, spring field beans, spring oilseed rape, sugar beet, swedes, turnips, vining peas, winter field beans, winter oilseed rape*** [32]; ***cultivated land/soil*** [4-9, 21, 38]; ***non-crop farm areas*** [38]; ***spring barley, spring oats, spring wheat, winter barley, winter oats, winter wheat*** [12, 16, 25, 27, 32]; ***stubbles*** [5-9, 33, 38]. Annual grasses in ***combining peas, durum wheat, leeks, linseed, mustard, onions, spring barley, spring field beans, spring oats, spring oilseed rape, spring wheat, sugar beet, swedes, turnips, vining peas, winter barley, winter field beans, winter oats, winter oilseed rape, winter wheat*** [32]; ***cultivated land/soil*** [4-9, 12, 16, 21, 25, 27, 33, 38]; ***non-crop farm areas*** [38]; ***stubbles*** [5-9, 33, 38]. Annual weeds in ***durum wheat, leeks, linseed, mustard, onions, peas, spring barley, spring field beans, spring oats,***

spring oilseed rape, spring wheat, sugar beet, swedes, turnips, winter barley, winter field beans, winter oats, winter oilseed rape, winter wheat [30, 31]; ***grassland*** [28, 30, 31, 35, 36, 38]. Black bent in ***combining peas, spring field beans, spring oilseed rape, stubbles, winter field beans, winter oilseed rape*** [4-9, 12, 16, 21, 25, 27, 33]; ***durum wheat, spring barley, spring oats, spring wheat, winter barley, winter oats, winter wheat*** [4-9, 21, 33]; ***linseed*** [7, 8, 12, 16, 25, 27, 33]. Bolters in ***sugar beet*** *(wiper application)* [3-9, 11, 13, 14, 20-22, 28-31, 33-35]. Bracken in ***grassland*** [39]. Couch in ***combining peas, spring field beans, winter field beans*** [1-22, 24-31, 33-35, 37, 40]; ***durum wheat*** [1, 2, 4-9, 15, 17, 21, 24, 26, 28, 30, 31, 33, 35, 37, 40]; ***grassland*** [1, 4, 10, 15, 17, 24]; ***linseed*** [1, 2, 7, 8, 10, 12, 15-18, 24, 25, 27, 28, 30, 31, 33, 35, 40]; ***mustard*** [10, 17, 18, 28, 30, 31, 35, 40]; ***oilseed rape for industrial use*** [39]; ***spring barley, spring wheat, winter barley, winter wheat*** [1-11, 13-15, 17-22, 24, 26, 28-31, 33-37, 40]; ***spring oats, winter oats*** [1-11, 13-15, 17-22, 24, 26, 28-31, 33-37]; ***spring oilseed rape, winter oilseed rape*** [1-22, 24-31, 33-37, 39, 40]; ***stubbles*** [1-31, 33-37, 39, 40]. Creeping bent in ***combining peas, spring field beans, spring oilseed rape, stubbles, winter field beans, winter oilseed rape*** [4-9, 12, 16, 21, 25, 27, 33]; ***cultivated land/soil*** [1]; ***durum wheat, spring barley, spring oats, spring wheat, winter barley, winter oats, winter wheat*** [4-9, 21, 33]; ***linseed*** [7, 8, 12, 16, 25, 27, 33]. Green cover in ***land temporarily removed from production*** [2, 6-8, 12, 13, 15, 16, 21, 24, 25, 27, 32, 33, 39]. Harvest management/desiccation in ***combining peas, spring field beans, winter field beans*** [1, 3, 10, 11, 13-15, 17-20, 22, 24, 26, 28-31, 34, 35, 37, 39]; ***durum wheat*** [1, 3, 11, 13-15, 17, 19, 20, 22, 24, 26, 28-31, 34, 35, 37]; ***linseed*** [1, 3, 10, 11, 14, 15, 17-21, 24, 28-31, 33-35, 39]; ***mustard*** [1, 3, 10, 11, 13, 14, 17-20, 23, 28-31, 34, 35, 39, 40]; ***oilseed rape for industrial use*** [39]; ***spring barley, winter barley*** [1, 10, 12, 15-18, 22-28, 30, 31, 35-37, 39]; ***spring oats, winter oats*** [1, 3, 10-20, 22-31, 34 37]; ***spring oilseed rape, winter oilseed rape*** [1-4, 10-31, 33-37, 39, 40]; ***spring wheat, winter wheat*** [1, 3, 10-20, 22-31, 34-37, 39]. Perennial dicotyledons in ***combining peas, spring field beans, spring oilseed rape, winter field beans, winter oilseed rape*** [5-9, 12, 16, 25, 27, 33]; ***durum wheat, spring barley, spring oats, spring wheat, winter barley, winter oats, winter wheat*** [5-9, 33]; ***grassland*** *(wiper application)* [28, 30, 31, 35]; ***linseed*** [7, 8, 12, 16, 25, 27, 33]; ***non-crop farm areas*** [7]. Pre-harvest desiccation in ***linseed*** [7, 8]; ***poppies*** *(off-label - for morphine production)* [30]; ***spring oilseed rape, winter oilseed rape*** [5-9]. Rushes in ***grassland*** [39]. Sward destruction in ***grassland*** [1-31, 33-37, 39, 40]. Volunteer cereals in ***combining peas, vining peas*** [32]; ***cultivated land/soil*** [4-9, 12, 16, 21, 25, 27, 33, 38]; ***durum wheat, leeks, linseed, mustard, onions, spring barley, spring field beans, spring oats, spring oilseed rape, spring wheat, sugar beet, swedes, turnips, winter barley, winter field beans, winter oats, winter oilseed rape, winter wheat*** [30-32]; ***peas*** [30, 31]; ***stubbles*** [1-31, 33-40]. Volunteer oilseed rape in ***stubbles*** [39]. Volunteer potatoes in ***stubbles*** [1-11, 13-15, 17-24, 26, 28-31, 33-37, 39, 40]. Weed beet in ***sugar beet*** *(wiper application)* [3, 11, 13, 14, 20, 22, 29, 34]. Wild oats in ***durum wheat*** [3, 13, 19, 20, 29, 34]; ***spring barley, spring oats, spring wheat, winter barley, winter oats, winter wheat*** [3, 11, 13, 14, 19, 20, 29, 34].

Specific Off-Label Approvals (SOLAs)

- ***poppies*** *(for morphine production) (OLA 031409) Dec 2008* [30]

Approval information

- Glyphosate included in Annex I under EC Directive 91/414
- May be applied through CDA equipment. See label for details

Efficacy guidance

- For best results apply to actively growing weeds with enough leaf to absorb chemical
- Products are formulated as either isopropylamine [1, 2, 5, 8, 9, 11-28, 30 38], ammonium [3, 4, 6, 7, 10], or trimesium [39, 40] salts of glyphosate and may vary in the details of efficacy claims. See individual product labels
- When applying products through rotary atomisers the spray droplet spectra must have a minimum Volume Median Diameter (VMD) of 200 microns
- With wiper application weeds should be at least 10 cm taller than crop
- Annual weed grasses should have at least 5 cm of leaf and annual broad-leaved weeds at least 2 expanded true leaves
- Perennial grass weeds should have 4-5 new leaves and be at least 10 cm long when treated. Perennial broad-leaved weeds should be treated at or near flowering but before onset of senescence
- Volunteer potatoes and polygonums are not controlled by harvest-aid rates

- In order to allow translocation, do not cultivate before treating perennials and do not apply other pesticides, lime, fertilizer or farmyard manure within 5 d of treatment
- Recommended intervals after treatment and before cultivation vary. See labels
- A rainfree period of at least 6 h (preferably 24 h) should follow spraying
- Do not tank-mix with other pesticides or fertilizers as such mixtures may lead to reduced control. Adjuvants are obligatory for some products and recommended for some uses with others. See labels
- Do not spray weeds affected by drought, waterlogging, frost or high temperatures

Restrictions

- Maximum number of applications normally 1 per crop (pre-harvest), 1 per yr (grassland destruction, non-crop land, green cover treatment) and 1 in other situations but check label for details
- Do not treat cereals grown for seed or undersown crops
- Do not use on grassland if the crop is to be used as animal feed or bedding
- Disperse decaying matter by thorough cultivation before sowing or planting a following crop
- Do not mix, store or apply in galvanized or unlined mild steel containers or spray tanks
- Do not leave spray in spray tanks for long period and make sure tanks are well vented
- Consult grain merchant before treating crops grown on contract or intended for malting

Crop-specific information

- Latest use: 24 h-14 d before cultivating or drilling depending on product and dose (see labels for details); pre-emergence for autumn crops and spring cereals; 72 h post-drilling for sugar beet, peas, swedes, turnips, onions and leeks; post-leaf fall but before white bud or green cluster for top fruit.
- HI grass 5 d; wheat, barley, oats, field beans, combining peas 7 d; intervals vary from 7-14 d for other crops. Check label

Environmental safety

- Harmful to aquatic organisms [39]
- Harmful to fish or other aquatic life. Do not contaminate surface waters or ditches with chemical or used container
- Do not dump surplus herbicide in water or ditch bottoms
- Take extreme care to avoid drift
- Treated poisonous plants must be removed before grazing or conserving
- Do not use in covered areas such as greenhouses or under polythene
- For field edge treatment direct spray away from hedge bottoms
- Exclude livestock from treated areas. Livestock may not graze or be fed the treated forage nor may it be used for hay, silage or bedding

Hazard classification and safety precautions

Hazard R03 [39, 40]; R04 [1, 3-6, 9-11, 14, 17-20, 22, 23, 26, 28, 29, 32, 34-37, 40]; R05 [40]; R14 [39]

Risk phrases R03c [39, 40]; R04a, R04b [1, 3, 5, 9, 11, 14, 17-20, 22, 23, 26, 28, 29, 34-37, 40]; R04d [4, 6, 10, 32, 40]; R04e [17]; R05a [40]; R14d [39]

Operator protection A [1-37, 39, 40]; C [1-6, 8-32, 34-37, 39]; D [2, 12, 15, 16, 24, 25, 27]; F [1, 3, 11, 13, 14, 17, 18, 20, 28, 34, 35]; H [1-3, 7, 8, 10-18, 20-22, 24, 25, 27, 28, 30-35, 39]; M [1-3, 8, 10-22, 24, 25, 27-35, 39, 40]; N [3, 11, 13, 14, 17-20, 22, 28, 29, 34, 35, 40]; U02, U08 [1, 3-11, 13-29, 32-37]; U05a [1, 3-6, 9-11, 13, 14, 17-20, 22, 23, 26, 28, 29, 32, 34-37, 39]; U07 [39]; U19 [1, 3, 5, 7-11, 14-29, 32-37]; U20a [5, 7, 15, 17, 23, 26, 30, 37, 39, 40]; U20b [1-4, 6, 8-14, 16, 18-22, 24, 25, 27-29, 31-36]; U20c [38]

Environmental protection E01 [1-7, 9-11, 13-20, 22-32, 34-37, 39]; E07 [40]; E13c [1, 3-7, 9-11, 14-18, 20-29, 31-38, 40]; E15 [2, 12, 30, 39]; E19 [2, 17, 28, 35]; E38 [39]

Consumer protection C03 [1, 3-7, 9-11, 13, 14, 17-20, 22, 23, 26, 28-32, 34-37, 39]

Storage and disposal E26 [1, 5, 7-9, 15, 17, 18, 20-22, 24, 26, 28, 29, 32-35, 37-40]; E29, E32a [32]; E30a [1-18, 20-38, 40]; E31a [9]; E31b [1-4, 6-8, 10-14, 17, 18, 20-24, 28-31, 33-36, 38]; E31c [15, 16, 25, 27, 40]; E32d [39]; E34 [5, 18, 22, 26, 37, 38]

Medical advice M03 [39]

271 glyphosate (fruit, horticulture, forestry, amenity etc.)

A translocated non-residual phosphonic acid herbicide

See also diuron + glyphosate

Products

1	Asteroid	Headland	360 g/l	SL	11118
2	Azural	Cardel	360 g/l	SL	11668
3	Barclay Barbarian	Barclay	360 g/l	SL	11362
4	Barclay Gallup	Barclay	360 g/l	SL	05161
5	Barclay Gallup 360	Barclay	360 g/l	SL	11369
6	Barclay Gallup Amenity	Barclay	360 g/l	SL	06753
7	Barclay Gallup Biograde 360	Barclay	360 g/l	SL	11370
8	Barclay Gallup Biograde Amenity	Barclay	360 g/l	SL	10203
9	Barclay Gallup Hi-Aktiv	Barclay	490 g/l	SL	10457
10	Barclay Garryowen	Barclay	360 g/l	SL	09869
11	Buggy SG	Sipcam	36% w/w	WG	08573
12	CDA Vanquish	Bayer Environ.	120 g/l	RH	09927
13	CDA Vanquish Biactive	Bayer Environ.	120 g/l	RH	10950
14	Clinic	Nufarm UK	360 g/l	SL	09378
15	Envision	Headland	450 g/l	SL	10569
16	Flame	Dow	360 g/l	SL	11583
17	Glydate	Nufarm UK	360 g/l	SL	10999
18	Glyfos	Cheminova	360 g/l	SL	07109
19	Glyfos	Headland	360 g/l	SL	10995
20	Glyfos Gold	n2n	360 g/l	SL	10570
21	Glyfos ProActive	n2n	360 g/l	SL	07800
22	Glyper	Interfarm	360 g/l	SL	11095
23	Glyphogan	Makhteshim	360 g/l	SL	05784
24	Glyphosate 360	Dow	360 g/l	SL	09151
25	Glyphosate 360	SumiAgro Amenity	360 g/l	SL	11060
26	Glyphosate 360	Cardel	360 g/l	SL	11579
27	Greenaway Gly-490	Greenaway	490 g/l	SL	11064
28	Greencrop Gypsy	Greencrop	360 g/l	SL	09432
29	Habitat	SumiAgro Amenity	360 g/l	SL	10498
30	Hilite	n2n	144 g/l	RH	06261
31	Kernel	Cheminova	480 g/l	SL	10616
32	Kernel	Headland	480 g/l	SL	10993
33	Manifest	Headland	360 g/l	SL	11041
34	Nufosate	Nufarm UK	360 g/l	SL	10090
35	Reliance	Cardel	360 g/l	SL	11667
36	Romany	Greencrop	360 g/l	SC	11000
37	Roundup Amenity	Monsanto	360 g/l	SL	10319
38	Roundup Biactive	Monsanto	360 g/l	SL	10320
39	Roundup Gold	Monsanto	450 g/l	SL	10975
40	Roundup Max	Monsanto	68% w/w	SG	10231
41	Roundup Pro Biactive	Monsanto	360 g/l	SL	10330
42	Roundup Ultra ST	Monsanto	450 g/l	SL	10199
43	Samurai	Monsanto	360 g/l	SL	10334
44	Scorpion	Cardel	360 g/l	SL	11056
45	Spasor	Bayer Environ.	360 g/l	SL	09945
46	Spasor Biactive	Bayer Environ.	360 g/l	SL	09940
47	Stirrup	n2n	144 g/l	RH	06132
48	Touchdown	Syngenta	330 g/l	SL	10538
49	Touchdown Quattro	Syngenta	360 g/l	SL	10608
50	Tumbleweed Pro	Scotts	480 g/l	SL	11083

Uses

Annual and perennial weeds in ***all top fruit***, ***hedges*** [30, 47]; ***amenity areas*** [6-9, 27, 41, 46]; ***amenity areas*** *(wiper application)*, ***broad-leaved trees***, ***conifers***, ***farm buildings/yards***, ***land clearance*** [41]; ***amenity grass***, ***amenity grass*** *(wiper application)*, ***amenity trees and shrubs*** *(wiper application)* [41, 45]; ***amenity trees and shrubs*** [30, 41, 45, 47]; ***amenity vegetation*** [12, 13, 41]; ***apple orchards***, ***cherries***, ***damsons***, ***pear orchards***, ***plums*** [1-5, 7, 9-11, 14-19, 22-24, 26-28, 31-33, 35, 36, 38, 39, 42-44]; ***asparagus*** *(off-label)*, ***blackcurrants*** *(off-label)*, ***blueberries*** *(off-label)*, ***chestnuts*** *(off-label)*, ***cob nuts*** *(off-label)*, ***grapevines*** *(off-label)*, ***hazel nuts*** *(off-label)*, ***rhubarb*** *(off-label)*, ***walnuts*** *(off-label)* [3]; ***fencelines***, ***walls*** [41, 46]; ***forestry*** [3-11, 14, 15, 17-25,

SECTION 4

27-34, 36, 37, 41, 47, 50]; ***hard surfaces*** [20, 27, 41, 46, 49, 50]; ***industrial sites*** [6-9, 20, 21, 25, 29, 34, 37, 41, 50]; ***land immediately adjacent to aquatic areas*** [27]; ***land not intended to bear vegetation*** [2-13, 15, 16, 18, 19, 21, 24-27, 29, 31-37, 40-44, 48]; ***natural surfaces not intended to bear vegetation, permeable surfaces overlying soil*** [20, 27, 49, 50]; ***non-crop areas*** [22, 39, 41, 46]; ***paths and drives*** [20, 29, 34, 37, 41, 46, 50]; ***road verges*** [20, 21, 29, 34, 37, 41, 46, 50]; ***woody ornamentals*** [30, 41, 45-47]. Aquatic weeds in ***open waters*** [27]. Bracken in ***amenity trees and shrubs, fencelines, land clearance, non-crop areas, paths and drives*** [41]; ***forestry*** [3-11, 14, 15, 17-25, 27-29, 31-34, 36, 37, 41, 50]; ***tolerant conifers*** [11, 18, 21, 31, 41]. Brambles in ***tolerant conifers*** [11, 18, 21, 31, 41]. Chemical thinning in ***forestry*** [6-9, 14, 15, 17, 19, 20, 22, 23, 25, 27, 29, 32-34, 36, 37, 41, 50]. Couch in ***forestry*** [6, 8, 9, 14, 17, 22, 23, 27, 28, 36, 41]; ***land not intended to bear vegetation*** [48]. Heather in ***forestry*** [3-11, 15, 18-21, 24, 25, 27, 29, 31-34, 36, 37, 50]. Japanese knotweed in ***land not intended to bear vegetation*** [48]. Perennial dicotyledons in ***forestry*** [20, 25, 29, 34, 37, 50]; ***forestry*** *(wiper application)* [41]. Perennial grasses in ***aquatic situations*** [1, 3, 5-11, 14-25, 27, 29, 31-34, 36, 37, 41, 45, 46, 50]; ***enclosed waters, land immediately adjacent to aquatic areas*** [20, 50]; ***tolerant conifers*** [11, 18, 21, 31, 41]. Reeds in ***aquatic situations*** [1, 3, 5-11, 14-25, 27, 29, 31-34, 36-38, 41, 45, 46, 50]; ***enclosed waters, land immediately adjacent to aquatic areas*** [20, 27, 50]. Rhododendrons in ***forestry*** [6, 8, 9, 11, 15, 18-21, 24, 25, 27, 29, 31-34, 36, 37, 41, 50]. Rushes in ***aquatic situations*** [1, 3, 5-11, 14-25, 27, 29, 31-34, 36-38, 41, 45, 46, 50]; ***enclosed waters, land immediately adjacent to aquatic areas*** [20, 27, 50]; ***forestry*** [22, 23]. Sedges in ***aquatic situations*** [1, 3, 5-11, 14-25, 27, 29, 31-34, 36-38, 41, 45, 46, 50]; ***enclosed waters, land immediately adjacent to aquatic areas*** [20, 27, 50]. Sucker control in ***apple orchards, cherries, damsons, pear orchards, plums*** [26, 35, 38, 39, 43, 44]. Sward destruction in ***amenity grass*** [30, 41, 47]. Total vegetation control in ***amenity areas*** [30, 41, 45, 47]; ***amenity vegetation, land not intended to bear vegetation*** [12, 13, 21, 41]; ***fencelines, road verges*** [21, 41, 45]; ***hard surfaces*** [21, 30, 41, 47]; ***industrial sites*** [21, 30, 41, 45-47]. Waterlilies in ***aquatic situations*** [1, 3, 5-11, 14-25, 27, 29, 31-34, 36, 37, 41, 45, 46, 50]; ***enclosed waters*** [27]. Woody weeds in ***forestry*** [3-11, 14, 15, 17-25, 27-34, 36, 37, 41, 47, 50]; ***tolerant conifers*** [11, 18, 21, 31, 41].

Specific Off-Label Approvals (SOLAs)

- ***asparagus, blackcurrants, blueberries, chestnuts, cob nuts, grapevines, hazel nuts, rhubarb, walnuts*** *(OLA 032129) Dec 2008* [3]

Approval information

- Glyphosate included in Annex I under EC Directive 91/414
- Approved for aquatic weed control. See notes in Section 1 on use of herbicides in or near water [1, 3, 5-11, 14-27, 29, 31-34, 36-38, 41, 43, 45, 46, 50]
- May be applied through CDA equipment. See label for details

Efficacy guidance

- For best results apply to actively growing weeds with enough leaf to absorb chemical
- Products are formulated as either isopropylamine [1-10, 12-47], ammonium [11], or trimesium [48-50] salts of glyphosate and may vary in the details of efficacy claims. See individual product labels
- Some products formulated for application through specialist CDA applicators. See label for instructions [12, 13, 30]
- When applying products through rotary atomisers the spray droplet spectra must have a minimum Volume Median Diameter (VMD) of 200 microns
- With wiper application weeds should be at least 10 cm taller than crop
- Annual weed grasses should have at least 5 cm of leaf and annual broad-leaved weeds at least 2 expanded true leaves
- Perennial grass weeds should have 4-5 new leaves and be at least 10 cm long when treated. Perennial broad-leaved weeds should be treated at or near flowering but before onset of senescence
- Bracken must be treated at full frond expansion
- In order to allow translocation, do not cultivate before treating perennials and do not apply other pesticides, lime, fertilizer or farmyard manure within 5 d of treatment
- Recommended intervals after treatment and before cultivation vary. See labels
- A rainfree period of at least 6 h (preferably 24 h) should follow spraying
- Most products should not be tank-mixed with other pesticides or fertilizers as such mixtures may lead to reduced control, but some are compatible with listed diuron-containing products. See labels for details

- Adjuvants are obligatory for some products and recommended for some uses with others. See labels
- Fruit tree suckers best treated in late spring
- Chemical thinning treatment can be applied as stump spray or stem injection
- Product formulated ready for use without dilution. Apply only through specified applicator (see label for details) [12, 30, 47]

Restrictions

- Maximum number of treatments normally 1 per crop or season on field and edible crops and unrestricted for non-crop uses. Check labels for details
- Decaying remains of plants killed by spraying must be dispersed before direct drilling
- Do not use treated straw as a mulch or growing medium for horticultural crops
- Do not use trimesium salt products in aquatic situations or forestry
- For use in nursery stock, shrubberies, orchards, grapevines and tree nuts care must be taken to avoid contact with the trees. Do not use in orchards established less than 2 yr and keep off low-lying branches
- Do not spray root suckers in orchards in late summer or autumn
- Do not use under glass or polythene as damage to crops may result
- Do not use wiper techniques in soft fruit crops
- Certain conifers may be sprayed overall in dormant season. See label for details
- Use a tree guard when spraying in established forestry plantations
- Do not spray weeds affected by drought, waterlogging, frost or high temperatures
- Do not mix, store or apply in galvanised or unlined mild steel containers or spray tanks
- Do not leave diluted chemical in spray tanks for long period and make sure that tanks are well vented

Crop-specific information

- Latest use: after harvest (post-leaf fall) but before bud formation in the following season for most fruit crops. See labels for details
- HI 4 wk for blackcurrants, blueberries

Environmental safety

- Harmful to aquatic organisms [48]
- Harmful to fish or other aquatic life. Do not contaminate surface waters or ditches with chemical or used container
- Do not dump surplus herbicide in water or ditch bottoms
- Maximum permitted concentration in treated water 0.2 ppm
- The Environment Agency or Local River Purification Authority must be consulted before use in or near water
- Take extreme care to avoid drift onto other crops
- For field edge treatment direct spray away from hedge bottoms
- Exclude livestock from treated areas

Hazard classification and safety precautions

Hazard R03 [48, 49]; R04 [2-6, 10, 11, 14, 17, 22-26, 28, 34-37, 40, 43-45, 47, 49]; R05 [49]; R14 [48]

Risk phrases R03c [48, 49]; R04a [2, 4, 6, 10, 14, 17, 22-26, 28, 34-37, 43-45, 47, 49]; R04b [2, 4, 6, 10, 14, 17, 22-26, 28, 34-37, 43-45, 49]; R04d [3, 5, 11, 40, 49]; R04e [22], R05a [49]; R14d [48]

Operator protection A [1-50]; C [1-11, 14-41, 43-45, 47, 48, 50]; D [1, 15, 18-21, 31-33, 50]; F [14, 16, 17, 22, 25, 26, 34, 35, 37, 43, 44]; H [1, 7-9, 11-22, 26-29, 31-33, 35, 37-44, 46, 48, 50]; M [1-3, 5, 7-9, 11-22, 24-50]; N [2, 3, 5, 14, 16, 17, 22, 24-26, 28, 30, 34-37, 43-45, 47, 49]; U02 [2-11, 14, 16-37, 40, 42-45, 47, 50]; U05a [2-6, 10, 11, 14, 16, 17, 22-26, 28, 30, 34-37, 40, 43-45, 47, 48]; U07 [48], U08 [2-11, 14, 16-20, 22-37, 40, 42-45, 47, 50]; U19 [2, 4, 6-14, 17-19, 21-37, 40, 42-45, 47]; U20a [4, 6-8, 18, 21-23, 29, 35, 38, 39, 44, 46, 48, 49]; U20b [1-3, 5, 9-17, 19, 20, 24-28, 30-34, 36, 37, 40-43, 45, 47, 50]

Environmental protection E01 [1-8, 10-14, 16-19, 21-26, 28-41, 43-48]; E07 [49], E13c [2-11, 14, 17-23, 25-37, 40, 42-45, 47, 49, 50]; E15 [1, 12, 13, 15, 38, 39, 41, 46, 48]; E19 [1, 3, 5, 7-9, 11, 14, 15, 17-19, 21, 22, 26, 27, 29, 31-34, 36, 37, 43, 45]; E38 [40]

Consumer protection C03 [2-8, 10-14, 16, 17, 22-26, 28-30, 34-41, 43-48]

SECTION 4

Storage and disposal E26 [4, 6-10, 12, 13, 18, 21-23, 25-31, 34, 36, 37, 40, 42, 43, 47-49]; E29 [40]; E30a [1-23, 25-47, 49, 50]; E31a [10, 12, 13, 50]; E31b [1-3, 5, 7-9, 11, 14-17, 20-22, 25-29, 31, 34-39, 41-44, 46]; E31c [18, 19, 32, 33, 49]; E32a [30, 40, 47]; E32d [48]; E34 [4, 6, 12, 13, 23, 28]
Medical advice M03 [48]

272 glyphosate + oxadiazon

A non-selective herbicide for total vegetation control

Products

Zapper	Bayer Environ.	10.8:30% w/w	WB	09944

Uses

Annual and perennial weeds in ***land not intended to bear vegetation*** *(soil surface treatment).* Total vegetation control in ***land not intended to bear vegetation*** *(soil surface treatment).*

Approval information

- Glyphosate included in Annex I under EC Directive 91/414
- Approval expiry 31 Jul 2004 [1]

Efficacy guidance

- Apply post-emergence of weeds at any time when they are actively growing from Mar to end Sep. Control may be reduced if weeds are under stress when treated
- Best results on perennials including docks, perennial sowthistle and willowherb obtained from treatment just before flowering or seed set
- A rainfree period of at least 6 h (preferably 24 h) should follow spraying
- Perennials such as dandelions or docks emerging after application from established rootstocks will not be controlled
- Pre-emergence activity is reduced on soils with more than 10% organic matter or where leaves have collected or a mat of organic matter has built up
- Use of wetting agents or adjuvants may reduce activity

Restrictions

- When used on sites that are to be cleared or grubbed 2 yr should elapse before sowing or planting anything. Soil should be ploughed or dug after clearance

Following crops guidance

- Take care to prevent heavy rain after application washing product onto sensitive areas such as newly sown grass or areas about to be planted

Environmental safety

- Dangerous to fish or other aquatic life. Do not contaminate surface waters or ditches with chemical or used container
- LERAP Category B

Hazard classification and safety precautions

Operator protection A, C; U20c, U22a
Environmental protection E13b, E16a, E16b
Storage and disposal E30a, E32a

273 guazatine

A guanidine fungicide seed dressing for cereals

Products

1	Panoctine	Makhteshim	300 g/l	LS	10094
2	Ravine	Makhteshim	300 g/l	LS	10095

Uses

Brown foot rot in ***spring barley*** *(seed treatment - reduction),* ***spring oats*** *(seed treatment - reduction),* ***winter barley*** *(seed treatment - reduction),* ***winter oats*** *(seed treatment - reduction)* [1]. Foot rot in ***spring barley*** *(seed treatment),* ***winter barley*** *(seed treatment)* [1]. Fusarium foot rot and seedling blight in ***spring barley*** *(seed treatment - reduction),* ***spring oats*** *(seed treatment - reduction),* ***winter barley*** *(seed treatment - reduction),* ***winter oats*** *(seed treatment - reduction)* [1, 2]; ***spring wheat*** *(seed treatment - reduction),* ***winter wheat*** *(seed treatment - reduction)* [2].

Leaf stripe in ***spring barley*** *(seed treatment)*, ***winter barley*** *(seed treatment)* [1]. Net blotch in ***spring barley*** *(seed treatment)*, ***winter barley*** *(seed treatment)* [1]. Pyrenophora leaf spot in ***spring oats*** *(seed treatment)*, ***winter oats*** *(seed treatment)* [1]. Septoria seedling blight in ***spring wheat*** *(seed treatment)*, ***winter wheat*** *(seed treatment)* [2].

Efficacy guidance

- Apply with conventional seed treatment machinery

Restrictions

- Maximum number of treatments 1 per batch
- Do not treat grain with moisture content above 16% and do not allow moisture content of treated seed to exceed 16%
- Do not apply to cracked, split or sprouted seed
- Do not use treated seed as food or feed

Crop-specific information

- Latest use: pre-drilling
- After treating, bag seed immediately and keep in dry, draught-free store
- Treatment may lower germination capacity, particularly if seed grown, harvested or stored under adverse conditions

Environmental safety

- Dangerous to fish or other aquatic life. Do not contaminate surface waters or ditches with chemical or used container

Hazard classification and safety precautions

Hazard R03, R04
Risk phrases R03a, R03c, R04d
Operator protection A, C, D, H; U05a, U20b [1, 2]; U09a [1]
Environmental protection E01, E13b
Consumer protection C03
Storage and disposal E26, E30a, E31a, E34
Treated seed S01, S02, S04b, S05, S06, S07
Medical advice M03

274 guazatine + triticonazole

A fungicide seed treatment for wheat

Products

1	Premis	Aventis	150:12.5 g/l	FS	10177
2	Premis	BASF	150:12.5 g/l	FS	11742

Uses

Bunt in ***spring wheat***, ***winter wheat***. Fusarium foot rot and seedling blight in ***spring wheat***, ***winter wheat***.

Efficacy guidance

- Apply undiluted using conventional seed treatment machinery
- Ensure seed drill is correctly calibrated before sowing

Restrictions

- Maximum number of treatments 1 per seed batch
- Do not treat grain with moisture content above 16% and do not allow moisture content of treated seed to exceed 16%
- Do not store treated seed from one season to the next
- Do not treat cracked, split or sprouted seed
- Do not use treated seed as food or feed

Crop-specific information

- Latest use: before drilling
- After treating, bag seed immediately and keep in dry, draught-free store

SECTION 4

Environmental safety

- Harmful to fish or other aquatic life. Do not contaminate surface waters or ditches with chemical or used container
- Product available in returnable containers which should not be rinsed and returned to the supplier as instructed. Used containers should be treated as if they contained pesticide

Hazard classification and safety precautions

Hazard R03, R04
Risk phrases R03c, R04a
Operator protection A, C, D, H; U05a, U07, U08, U20b
Environmental protection E01, E13c
Consumer protection C03
Storage and disposal E26, E30a, E32c, E33, E34, E36
Treated seed S01, S02, S03, S04b, S05, S06, S07
Medical advice M03

275 hymexazol

A systemic isoxazole fungicide for pelleting sugar beet seed

Products

Tachigaren 70 WP	SumiAgro	70% w/w	WP	10496

Uses

Black leg in ***sugar beet*** *(seed treatment).*

Efficacy guidance

- Incorporate into pelleted seed using suitable seed pelleting machinery

Restrictions

- Maximum number of treatments 1 per batch of seed
- Do not use treated seed as food or feed

Crop-specific information

- Latest use: before planting sugar beet seed

Environmental safety

- Harmful to fish of aquatic life. Do not contaminate surface waters or ditches with chemical or used container
- Treated seed harmful to game and wildlife

Hazard classification and safety precautions

Hazard R04
Risk phrases R04a, R04b, R04c
Operator protection A, C, F; U05a, U20a
Environmental protection E01, E13c
Consumer protection C03
Storage and disposal E30a, E32a
Treated seed S01, S02, S05, S06

276 imazalil

A systemic and protectant conazole fungicide

See also azaconazole + imazalil

Products

1	Fungaflor	Certis	200 g/l	EC	11326
2	Fungaflor Smoke	Certis	15% w/w	FU	11327
3	Fungazil 100 SL	BASF	100 g/l	LS	11762
4	Magnate 100 SL	Makhteshim	100 g/l	SL	11705

Uses

Dry rot in ***seed potatoes*** [3, 4]. Gangrene in ***seed potatoes*** [3, 4]. Powdery mildew in ***protected courgettes*** *(off-label)*, ***protected gherkins*** *(off-label)* [1]; ***protected cucumbers***, ***protected***

ornamentals, ***protected roses*** [1, 2]. Silver scurf in ***seed potatoes*** [3, 4]. Skin spot in ***seed potatoes*** [3, 4].

Specific Off-Label Approvals (SOLAs)

- ***protected courgettes***, ***protected gherkins*** *(OLA 030870) Dec 2008* [1]

Approval information

- Imazalil included in Annex I under EC Directive 91/414
- Approval expiry 31 Dec 2004 [1]

Efficacy guidance

- For best control of skin and wound diseases of ware potatoes treat as soon as possible after harvest, preferably within 7-10 d, before any wounds have healed [3, 4]
- Treat cucurbits before or as soon as disease appears and repeat every 10-14 d or every 7 d if infection pressure great or with susceptible cultivars [1, 2]

Restrictions

- Maximum number of treatments 2 per batch of seed tubers [3, 4]
- Consult processor before treating potatoes for processing
- Do not treat outdoor crops of cucumbers, roses or ornamentals
- Do not spray cucurbits or ornamentals in full, bright sunshine. When spraying in the evening the spray should dry before nightfall. May cause damage if open flowers are sprayed
- Test for crop tolerance before large scale treatment. Do not use on rose cultivars Dr A.J. Verhage and Jack Frost
- With ornamentals of unknown tolerance test on a few plants in first instance

Crop-specific information

- Latest use: during storage and before chitting for seed potatoes
- HI edible crops 1 d
- Apply to clean soil-free potatoes post-harvest before putting into store, or at first grading. A further treatment may be applied in early spring before planting [3, 4]
- Apply through canopied hydraulic or spinning disc equipment prefereably diluted with up to two litres water per tonne of potatoes to obtain maximum skin cover and penetration [3, 4]

Environmental safety

- Flammable [1]; highly flammable [2]
- Keep unprotected persons out of treated areas for at least 2 h [2]
- Harmful to fish or other aquatic life. Do not contaminate surface waters or ditches with chemical or used container [1, 3, 4]
- Personal protective equipment requirements may vary for each pack size. Check label

Hazard classification and safety precautions

Hazard R03, R05, R07d [1]; R04 [2-4]; R07c [2]
Risk phrases R03c, R05b [1]; R04a [2, 4]; R04d [3]
Operator protection A, C [1, 3, 4]; D [4]; H [1]; U04a, U20a [3, 4]; U05a [1-4]; U10, U20b [1]; U11 [1, 3]; U14 [3]; U19 [1, 3, 4]
Environmental protection E01 [1-4]; E02 [2]; E13c [1, 3, 4]
Consumer protection C02 [1] (1 d); C02 [2]; C03 [1-4]
Storage and disposal E26, E34 [1, 3, 4]; E30a [1-4]; E31a [4]; E31b [1]; E31c [3]; E32a [2]
Treated seed S01, S03, S04a, S05, S06 [3, 4]; S02 [3]
Medical advice M03 [1, 4]

277 imazalil + pencycuron

A fungicide mixture for treatment of seed potatoes

Products

1	Monceren IM	Bayer	0.6:12.5% w/w	DS	06259
2	Monceren IM	Bayer CropScience	0.6:12.5% w/w	DS	11426

Uses

Black scurf in ***potatoes*** *(tuber treatment)*. Silver scurf in ***potatoes*** *(tuber treatment - reduction)*. Stem canker in ***potatoes*** *(tuber treatment - reduction)*.

Approval information

- Imazalil included in Annex I under EC Directive 91/414

Efficacy guidance

- Apply to clean seed tubers during planting (see label for suitable method) or sprinkle over tubers in chitting trays before loading into planter. It is essential to obtain an even distribution over tubers
- If seed tubers become damp from light rain distribution of product should not be affected. Tubers in the hopper should be covered if a shower interrupts planting

Restrictions

- Maximum number of treatments 1 per batch
- Operators must wear suitable respiratory equipment and gloves when handling product and when riding on planter. Wear gloves when handling treated tubers
- Do not use on tubers which have previously been treated with a dry powder seed treatment or hot water
- Do not use treated tubers for human or animal consumption

Crop-specific information

- Latest use: immediately before planting
- May be used on seed tubers previously treated with a liquid fungicide but not before 8 wk have elapsed if this contained imazalil

Environmental safety

- Harmful to fish or other aquatic life. Do not contaminate surface waters or ditches with chemical or used container
- Treated seed harmful to game and wildlife

Hazard classification and safety precautions

Operator protection A, D; U20b
Environmental protection E03, E13c
Storage and disposal E30a, E32a
Treated seed S01, S02, S03, S04a, S05, S06

278 imazalil + thiabendazole

A fungicide mixture for treatment of seed potatoes

Products

Extratect Flowable	Banks Cargill	100:300 g/l	FS	08704

Uses

Dry rot in ***seed potatoes*** *(tuber treatment - reduction)*. Gangrene in ***seed potatoes*** *(tuber treatment - reduction)*. Silver scurf in ***potatoes*** *(tuber treatment - reduction)*, ***seed potatoes*** *(tuber treatment - reduction)*. Skin spot in ***potatoes*** *(tuber treatment - reduction)*, ***seed potatoes*** *(tuber treatment - reduction)*. Stem canker in ***potatoes*** *(tuber treatment - reduction)*.

Approval information

- Imazalil and thiabendazole included in Annex I under EC Directive 91/414
- Approved for use through ULV equipment [1]

Efficacy guidance

- Apply immediately after lifting with suitable spinning disc or low volume hydraulic applicator for disease reduction during storage
- Apply within 2 mth of planting for maximum disease reduction in the progeny crop

Restrictions

- Maximum number of treatments 1 per batch
- Do not mix with any other product
- May not be used on seed tubers previously treated with a product containing imazalil
- Do not use treated seed as food or feed

Crop-specific information

- Latest use: pre-planting
- Delayed emergence noted under certain conditions but with no final effect on yield

Environmental safety

- Harmful to fish or other aquatic life. Do not contaminate surface waters or ditches with chemical or used container

Hazard classification and safety precautions

Hazard R04

Risk phrases R04c, R04d

Operator protection A, C, D, H; U05a, U09a, U11, U19, U20c

Environmental protection E01, E13c

Consumer protection C03

Storage and disposal E30a, E31a

Treated seed S01, S02, S05, S06

279 imazalil + triticonazole

A conazole fungicide mixture for seed treatment of barley

Products

1	Robust	Aventis	12.5:12.5 g/l	LS	10176
2	Robust	BASF	12.5:12.5 g/l	FS	11807

Uses

Covered smut in ***spring barley***, ***winter barley***. Leaf stripe in ***spring barley***, ***winter barley***. Loose smut in ***spring barley***, ***winter barley***. Net blotch in ***spring barley*** *(moderate control)*, ***winter barley*** *(moderate control)*. Seedling blight and foot rot in ***spring barley*** *(moderate control)*, ***winter barley*** *(moderate control)*.

Approval information

- Imazalil included in Annex I under EC Directive 91/414

Efficacy guidance

- Apply undiluted through conventional seed treatment machine
- Treated seed may be stored for up to 18 mth but germination should be checked before use and sowing rate adjusted if necessary
- Keep treated seed in a dry, draught free store
- Calibrate seed drill before sowing

Restrictions

- Maximum number of treatments 1 per batch of seed
- Do not treat seed with moisture content above 16%
- Do not apply to cracked, split or sprouted seed
- Do not handle treated seed unnecessarily
- Do not use treated seed as food or feed

Crop-specific information

- Latest use: pre-drilling

Environmental safety

- Harmful to fish or other aquatic life. Do not contaminate surface waters or ditches with chemical or used container
- Product available in large (1000 l) returnable containers. They should be re-circulated for 30 min before use and operated as directed on the label

Hazard classification and safety precautions

Hazard R04

Risk phrases R04e

Operator protection A, C, H; U05a, U09a, U20b

Environmental protection E01, E13c

Consumer protection C03

Storage and disposal E26, E30a, E34

Treated seed S01, S02, S03, S04b, S05, S06, S07

Medical advice M03

280 imazamethabenz-methyl

A post-emergence imidazolinone grass weed herbicide for use in winter cereals

Products

Dagger	BASF	300 g/l	SC	10218

SECTION 4

Uses

Blackgrass in ***winter barley***, ***winter wheat***. Charlock in ***winter barley***, ***winter wheat***. Loose silky bent in ***winter barley***, ***winter wheat***. Onion couch in ***winter barley***, ***winter wheat***. Volunteer oilseed rape in ***winter barley***, ***winter wheat***. Wild oats in ***winter barley***, ***winter wheat***.

Efficacy guidance

- Activity is contact and residual. Grass and broad leaved weeds controlled from pre-emergence to specified post-emergence sizes
- When used alone add authorised non-ionic wetter for improved contact activity
- Best results achieved when applied to fine, firm, clod-free seedbed when soil moist
- Effects of autumn/winter treatment normally persist to control spring flushes of weeds
- Follow-up treatment with a specific blackgrass herbicide may be necessary for complete control of light infestations. For heavy infestations use a specific blackgrass herbicide
- Imazamethabenz-methyl is a member of the ALS-inhibitor group of herbicides and products should be used in a planned Resistance Management strategy. See Section 1 for more information

Restrictions

- Maximum number of treatments 2 per crop (as split dose treatment)
- Do not use on durum wheat
- Do not use on soils where surface water is likely to accumulate
- Do not use on soils with more than 10% organic matter

Crop-specific information

- Latest use: before 4th node detectable (GS 34)
- Apply from 2-fully expanded leaf stage of crop up to three nodes detectable (GS 12-33)

Following crops guidance

- See label for guidance on following crops after normal harvesting
- In case of crop failure land must be ploughed to 15 cm and re-drilled in spring but only with specified crops - see label for details

Environmental safety

- Dangerous for the environment

Hazard classification and safety precautions

Hazard R03, R14
Risk phrases R03b
Operator protection A, C; U05a, U08, U20b
Environmental protection E01, E15, E38
Consumer protection C03
Storage and disposal E26, E30a, E31c, E32d
Medical advice M05a

281 imazaquin

An imidazolinone herbicide and plant growth regulator available only in mixtures

See also *chlormequat + choline chloride + imazaquin*

282 imidacloprid

A neonicotinoid insecticide for seed, soil, peat or foliar treatment

See also *beta-cyfluthrin + imidacloprid*
bitertanol + fuberidazole + imidacloprid
fuberidazole + imidacloprid + triadimenol

Products

1	Admire	Bayer	70% w/w	WG	07481
2	Admire	Bayer CropScience	70% w/w	WG	11234
3	Gaucho	Bayer	70% w/w	WS	06590
4	Gaucho	Bayer CropScience	70% w/w	WS	11281

Products – continued

5	Intercept 5GR	Scotts	5% w/w	GR	08126
6	Intercept 70WG	Scotts	70% w/w	WG	08585

Uses

Aphids in ***bedding plants***, ***hardy ornamental nursery stock***, ***herbaceous perennials***, ***pot plants*** [5, 6]; ***lettuce*** *(off-label)* [4]; ***lettuce*** *(off-label - seed treatment)* [3]. Beet virus yellows vectors in ***sugar beet*** *(seed treatment)* [3, 4]. Cabbage aphid in ***collards*** *(off-label - seed treatment)*, ***kale*** *(off-label - seed treatment)* [4]. Damson-hop aphid in ***hops*** [1, 2]. Flea beetle in ***sugar beet*** *(seed treatment)* [3, 4]. Glasshouse whitefly in ***bedding plants***, ***hardy ornamental nursery stock***, ***herbaceous perennials***, ***pot plants*** [5, 6]. Mangold fly in ***sugar beet*** *(seed treatment)* [3, 4]. Millipedes in ***sugar beet*** *(seed treatment)* [3, 4]. Peach-potato aphid in ***broccoli*** *(off-label - seed treatment)*, ***brussels sprouts*** *(off-label - seed treatment)*, ***cabbages*** *(off-label - seed treatment)*, ***calabrese*** *(off-label - seed treatment)*, ***cauliflowers*** *(off-label - seed treatment)* [3, 4]; ***collards*** *(off-label - seed treatment)*, ***kale*** *(off-label - seed treatment)* [4]. Pygmy beetle in ***sugar beet*** *(seed treatment)* [3, 4]. Sciarid flies in ***bedding plants***, ***hardy ornamental nursery stock***, ***herbaceous perennials***, ***pot plants*** [5, 6]. Springtails in ***sugar beet*** *(seed treatment)* [3, 4]. Symphylids in ***sugar beet*** *(seed treatment)* [3, 4]. Thrips in ***protected chrysanthemums*** *(off-label)* [1, 2]. Tobacco whitefly in ***bedding plants***, ***hardy ornamental nursery stock***, ***herbaceous perennials***, ***pot plants*** [5, 6]. Vine weevil in ***bedding plants***, ***hardy ornamental nursery stock***, ***herbaceous perennials***, ***pot plants*** [5, 6].

Specific Off-Label Approvals (SOLAs)

- ***broccoli***, ***brussels sprouts***, ***cabbages***, ***calabrese***, ***cauliflowers*** *(seed treatment) (OLA 002571) Dec 2008* [3]
- ***broccoli***, ***brussels sprouts***, ***cabbages***, ***calabrese***, ***cauliflowers***, ***collards***, ***kale*** *(seed treatment) (OLA 023927) Dec 2008* [4]
- ***lettuce*** *(OLA 031886) Dec 2008* [4]
- ***lettuce*** *(seed treatment) (OLA 010174) Dec 2008* [3]
- ***protected chrysanthemums*** *(OLA 010584) Dec 2008* [1, 2]

Efficacy guidance

- Treated seed should be drilled within the season of purchase [3, 4]
- Base of hop plants should be free of weeds and debris at application [1, 2]
- Hop bines emerging away from the main stock or adjacent to poles may require a separate application [1, 2]
- Uptake and movement within hops requires soil moisture and good growing conditions [1, 2]
- Control may be impaired in plantations greater than 3640 plants/ha [1, 2]
- To minimise likelihood of resistance do not treat all the hop crop in any one yr [1, 2] and adopt a planned programme to alternate with pesticides of different types or use other measures when using in compost [5, 6]
- When applied as drench or incorporated as granules in moist compost compound is readily absorbed and translocated to aerial parts of plant [5, 6]

Restrictions

- Maximum number of treatments 1 per batch of seed [3, 4] or growing medium [5, 6]; 1 per yr [1, 2]
- Product must not be used in compost that has already been treated with an imidacloprid-containing product [6]
- Product formulated for use only as a compost incorporation treatment into peat-based growing media using suitable automated equipment [5]
- For use only on container grown ornamentals [5, 6]
- The safety of seeds sown into treated compost should not be assumed. Test treat before full-scale use [5]
- Product must not be used on crops for human or animal consumption and treated compost must not be re-used for this purpose [5, 6]
- Do not use treated seed as food or feed [3, 4]

Crop specific information

- Latest use: before bines reach 2 m or before end 1st wk Jun for hops; before drilling for sugar beet, before sowing or planting for brassicas, bedding plants, hardy ornamental nursery stock, pot plants

- Apply to sugar beet seed as part of the normal commercial pelleting process using special treatment machinery [3, 4]
- Apply to hops as a directed stem base spray before most bines reach a height of 2 m. If necessary treat both sides of the crown at half the normal concentration [1, 2]

Environmental safety

- High risk to bees. Do not apply to crops in flower or to those in which bees are actively foraging. Do not apply when flowering weeds are present [1, 2, 6]
- Avoid spillage or other environmental contamination when incorporating into compost [5]
- Treated seed harmful to game and wildlife [3, 4]

Hazard classification and safety precautions

Hazard R03 [6]; R04 [1-4]
Risk phrases R03c [6]; R04e [1-4]
Operator protection A [1-5]; C, D, H [3, 4]; U05a [1, 2, 6]; U20a [1-4]; U20b [5, 6]
Environmental protection E01, E12a [1, 2, 6]; E15 [1-6]
Consumer protection C03 [1, 2, 6]
Storage and disposal E30a, E32a [1-6]; E34 [1, 2, 6]
Treated seed S02, S04b, S05, S06, S07 [3, 4]

283 imidacloprid + tebuconazole + triazoxide

A broad spectrum fungicide and insecticide seed treatment for winter barley

Products

1	Raxil Secur	Bayer	233:20:20 g/l	LS	08966
2	Raxil Secur	Bayer CropScience	233:20:20 g/l	LS	11298

Uses

Leaf stripe in ***winter barley*** *(seed treatment)*. Loose smut in ***winter barley*** *(seed treatment)*. Net blotch in ***winter barley***. Virus vectors in ***winter barley*** *(seed treatment)*. Wireworm in ***winter barley*** *(reduction of damage)*.

Efficacy guidance

- Best applied through recommended seed treatment machines
- Evenness of seed cover improved by simultaneous application of equal volumes of product and water or dilution of product with an equal volume of water
- Drill treated seed in the same season. Use minimum 125 kg treated seed per ha. Lower drilling rates and/or early drilling affect duration of BYDV protection needed and may require follow-up aphicide treatment
- In high risk areas where aphid activity is heavy and prolonged a follow-up aphicide treatment may be required
- Protection against foliar air-borne and splash-borne diseases later in the season will require appropriate fungicide follow-up sprays

Restrictions

- Maximum number of treatments 1 per batch of seed
- Do not use on seed with more than 16% moisture content, or on sprouted, cracked or skinned seed
- Do not use treated seed as food or feed

Crop-specific information

- Latest use: before drilling
- Slightly delayed and reduced emergence may occur but this is normally outgrown
- Any delay in field emergence, for whatever reason, may be accentuated by treatment

Environmental safety

- Harmful to fish or other aquatic life. Do not contaminate surface waters or ditches with chemical or used container
- Dangerous to birds, game and other wildlife. Treated seed should not be left on the soil surface. Bury spillages
- Treated seed should not be broadcast, but drilled to a depth of 4 cm in a well prepared seedbed
- If seed is left on the soil surface the field should be harrowed and rolled to ensure good incorporation

Hazard classification and safety precautions

Hazard R03, R04
Risk phrases R03c, R04e
Operator protection A, H; U04a, U05a, U13, U20b
Environmental protection E01, E03, E13c
Consumer protection C03
Storage and disposal E26, E30a, E32a, E34
Treated seed S01, S02, S03, S04c, S05, S06, S07
Medical advice M03

284 indol-3-ylacetic acid

A plant growth regulator for promoting rooting of cuttings

Products

1	Rhizopon A Powder	Rhizopon	1% w/w	DP	09087
2	Rhizopon A Tablets	Rhizopon	50 mg a.i.	TB	09088

Uses

Rooting of cuttings in ***ornamental specimens***.

Efficacy guidance

- Apply by dipping end of prepared cuttings into powder or dissolved tablet solution
- Shake off excess powder and make planting holes to prevent powder stripping off [1]
- Consult manufacturer for details of application by spray or total immersion

Restrictions

- Maximum number of treatments 1 per cutting
- Store product in a cool, dark and dry place
- Use solutions once only. Discard after use [2]
- Use plastic, not metallic, container for solutions [2]

Crop-specific information

- Latest use: before cutting insertion

Hazard classification and safety precautions

Operator protection U19 [1]; U20a [1, 2]
Environmental protection E15
Storage and disposal E30a, E32a

285 4-indol-3-yl-butyric acid

A plant growth regulator promoting the rooting of cuttings

Products

1	Chryzoplus Grey 0.8%	Rhizopon	0.8% w/w	DP	09094
2	Chryzopon Rose 0.1%	Rhizopon	0.1% w/w	DP	09092
3	Chryzosan White 0.6%	Rhizopon	0.6% w/w	DP	09093
4	Chryzotek Beige	Rhizopon	0.4% w/w	DP	09081
5	Chryzotop Green	Rhizopon	0.25% w/w	DP	09085
6	Rhizopon AA Powder (0.5%)	Rhizopon	0.5% w/w	DP	09082
7	Rhizopon AA Powder (1%)	Rhizopon	1% w/w	DP	09084
8	Rhizopon AA Powder (2%)	Rhizopon	2% w/w	DP	09083
9	Rhizopon AA Tablets	Rhizopon	50 mg a.i.	TB	09086
10	Seradix 1	Certis	0.1% w/w	DP	10422
11	Seradix 1	Certis	0.1% w/w	DP	11330
12	Seradix 2	Certis	0.3% w/w	DP	10423
13	Seradix 2	Certis	0.3% w/w	DP	11331
14	Seradix 3	Certis	0.8% w/w	DP	10424
15	Seradix 3	Certis	0.8% w/w	DP	11332

Uses

Rooting of cuttings in ***ornamental specimens***.

Efficacy guidance

- Dip base of cuttings into powder immediately before planting

SECTION 4

- Powders or solutions of different concentration are required for different types of cutting. Lowest concentration for softwood, intermediate for semi-ripe, highest for hardwood
- See label for details of concentration and timing recommended for different species
- Use of planting holes recommended for powder formulations to ensure product is not removed on insertion of cutting. Cuttings should be watered in if necessary

Restrictions
- Use of too strong a powder or solution may cause injury to cuttings
- No unused moistened powder should be returned to container

Crop-specific information
- Latest use: before cutting insertion for ornamental specimens

Hazard classification and safety precautions

Operator protection U19, U20a
Environmental protection E15
Storage and disposal E30a, E32a

286 4-indol-3-yl-butyric acid + 2-(1-naphthyl)acetic acid with dichlorophen

A plant growth regulator for promoting rooting of cuttings

Products

Synergol	Certis	5.0:5.0 g/l	SL	07386

Uses

Rooting of cuttings in ***ornamental specimens***.

Efficacy guidance
- Dip base of cuttings into diluted concentrate immediately before planting
- Suitable for hardwood and softwood cuttings
- See label for details of concentration and timing for different species

Hazard classification and safety precautions

Hazard R03
Risk phrases R03c
Operator protection A, C; U05a, U13, U20b
Environmental protection E01, E15
Consumer protection C03
Storage and disposal E30a, E32a, E34
Medical advice M03

287 iodosulfuron-methyl-sodium

A post-emergence sulfonylurea herbicide for winter sown cereals

See also amidosulfuron + iodosulfuron-methyl-sodium

Products

Hussar	Bayer CropScience	5% w/w	WG	11205

Uses

Chickweed in ***triticale***, ***winter rye***, ***winter wheat***. Cleavers in ***triticale***, ***winter rye***, ***winter wheat***. Italian ryegrass in ***triticale*** *(from seed)*, ***winter rye*** *(from seed)*, ***winter wheat*** *(from seed)*. Mayweeds in ***triticale***, ***winter rye***, ***winter wheat***. Perennial ryegrass in ***triticale*** *(from seed)*, ***winter rye*** *(from seed)*, ***winter wheat*** *(from seed)*. Red dead-nettle in ***triticale***, ***winter rye***, ***winter wheat***. Speedwells in ***triticale***, ***winter rye***, ***winter wheat***.

Efficacy guidance
- Best results obtained from treatment in warm weather when soil is moist and the weeds are growing actively
- Weeds must be present at application to be controlled
- Weed control is slow especially under cool dry conditions
- Dry conditions resulting in moisture stress may reduce effectiveness

- Occasionally weeds may only be stunted but they will normally have little or no competitive effect on the crop
- Iodosulfuron is a member of the ALS-inhibitor group of herbicides and products should be used in a planned Resistance Management strategy. See Section 1 for more information

Restrictions

- Maximum number of treatments 1 per crop
- Must only be applied between 1 Feb in yr of harvest and specified latest time of application
- Do not apply to undersown crops, or crops to be undersown
- Do not roll or harrow within 1 wk of spraying
- Do not spray crops under stress from any cause or if the soil is compacted
- Treat broadcast crops after the plants have a well-established root system
- Do not spray if rain imminent or frost expected
- Do not apply in mixture or in sequence with any other ALS inhibitor

Crop-specific information

- Latest use: before third node detectable (GS 33)

Environmental safety

- Dangerous to fish or other aquatic life. Do not contaminate surface waters or ditches with chemical or used container
- LERAP Category B
- Take extreme care to avoid damage by drift onto broad-leaved plants outside the target area or onto ponds, waterways and ditches
- Observe carefully label instructions for sprayer cleaning

Hazard classification and safety precautions

Hazard R04
Risk phrases R04d
Operator protection A, C, H; U05a, U08, U14, U15, U20b
Environmental protection E01, E13b, E16a, E16b
Consumer protection C03
Storage and disposal E31a

288 iodosulfuron-methyl-sodium + mesosulfuron-methyl

A sulfonyl urea herbicide mixture for winter wheat

Products

Atlantis	Bayer CropScience	0.6:3.0% w/w	WG	11679

Uses

Annual meadow grass in ***winter wheat***. Blackgrass in ***winter wheat***. Chickweed in ***winter wheat***. Italian ryegrass in ***winter wheat***. Mayweeds in ***winter wheat***. Perennial ryegrass in ***winter wheat***. Rough meadow grass in ***winter wheat***. Wild oats in ***winter wheat***.

Efficacy guidance

- Optimum grass weed control obtained when all grass weeds are emerged at spraying. Activity is primarily via foliar uptake and good spray coverage of the target weeds is essential
- Translocation occurs readily within the target weeds and growth is inhibited within hours of treatment but symptoms may not be apparent for up to 4 wk, depending on weed species, timing of treatment and weather conditions
- Residual activity is important for best results and is optimised by treatment on fine moist seedbeds. Avoid application under very dry conditions
- Residual efficacy may be reduced by high soil temperatures and cloddy seedbeds
- Iodosulfuron-methyl and mesosulfuron-methyl are both members of the ALS-inhibitor group of herbicides and products should be used in a planned Resistance Management strategy. See Section 1 for more information

Restrictions

- Maximum number of treatments 1 per crop with a maximum total dose equivalent to one full dose treatment
- Do not use on crops undersown with grasses, clover or other legumes or any other broad-leaved crop

- Do not use as a stand-alone treatment for blackgrass, ryegrass or chickweed control
- Do not use as the sole means of weed control in successive crops
- Do not use in mixture or in sequence with any other ALS-inhibitor herbicide
- Do not apply to crops under stress from any cause
- Do not apply when rain is imminent or during periods of frosty weather
- Specified adjuvant must be used. See label

Crop-specific information

- Latest use: flag leaf just visible (GS 39)
- Winter wheat may be treated from the two-leaf stage of the crop

Following crops guidance

- In the event of crop failure sow only winter wheat in the same cropping season
- Only winter wheat or winter barley may be sown in the year of harvest of a treated crop
- Spring wheat, spring barley or sugar beet may be drilled in the following spring

Environmental safety

- Dangerous for the environment
- Very toxic to aquatic organisms. May cause long-term adverse effects in the aquatic environment
- Dangerous to fish or other aquatic life. Do not contaminate surface waters or ditches with chemical or used container
- LERAP Category B
- Take extreme care to avoid drift onto plants outside the target area or on to ponds, waterways or ditches

Hazard classification and safety precautions

Hazard R04, R14
Risk phrases R04d, R14c
Operator protection A, C, H; U05a, U11, U20b
Environmental protection E01, E13b, E16a, E38
Consumer protection C03
Storage and disposal E26, E30a, E31a, E32d

289 ioxynil

A contact acting HBN herbicide for use in turf and onions

See also *bromoxynil + diflufenican + ioxynil*
bromoxynil + ethofumesate + ioxynil
bromoxynil + fluroxypyr + ioxynil
bromoxynil + ioxynil

Products

Totril	Aventis	225 g/l	EC	10026

Uses

Annual dicotyledons in ***bulb onions***, ***carrots*** *(off-label)*, ***chives*** *(off-label)*, ***garlic***, ***leeks***, ***parsnips*** *(off-label)*, ***salad onions***, ***shallots***.

Specific Off-Label Approvals (SOLAs)

- ***carrots***, ***parsnips*** *(OLA 001920) Feb 2005* [1]
- ***chives*** *(OLA 001918) Dec 2008* [1]

Approval information

- Ioxynil was reviewed in 1995 and approvals for home garden use, and most hand held applications revoked.

Efficacy guidance

- Best results on seedling to 4-leaf stage weeds in active growth during mild weather

Restrictions

- Maximum number of treatments up to 4 per crop at split doses on onions; 1 per crop on carrots, garlic, leeks, parsnips, shallots
- Do not apply by hand-held equipment or at concentrations higher than those recommended

Crop-specific information

- Latest use: pre-emergence for carrots, parsnips

- HI onions, chives, shallots, garlic, leeks 14 d
- Apply to sown onion crops as soon as possible after plants have 3 true leaves or to transplanted crops when established

Environmental safety

- Dangerous to fish or other aquatic life. Do not contaminate surface waters or ditches with chemical or used container
- Harmful to bees. Do not apply to crops in flower or to those in which bees are actively foraging. Do not apply when flowering weeds are present
- Keep livestock out of treated areas for at least 6 wk after treatment and until foliage of any poisonous weeds such as ragwort has died and become unpalatable

Hazard classification and safety precautions

Hazard R03, R04
Risk phrases R03a, R03c, R04a, R04b, R04c, R04g
Operator protection A, C; U05a, U08, U13, U19, U20b, U23a
Environmental protection E01, E12e, E13b
Consumer protection C03
Storage and disposal E26, E30a, E31b, E34
Medical advice M03, M05

290 iprodione

A protectant dicarboximide fungicide with some eradicant activity

See also carbendazim + iprodione

Products

1	Amenitywise Iprodione Green	Standon	250 g/l	SC	10883
2	Chipko Green	Bayer Environ.	250 g/l	SC	11211
3	I T Iprodione	I T Agro	50% w/w	WP	08267
4	Rovral Flo	Aventis	255 g/l	SC	10013
5	Rovral Flo	BASF	255 g/l	SC	11702
6	Rovral Green	Bayer Environ.	250 g/l	SC	09938
7	Rovral Liquid FS	Aventis	500 g/l	FS	10551
8	Rovral Liquid FS	BASF	500 g/l	FS	11703
9	Rovral WP	Aventis	50% w/w	WP	10015
10	Rovral WP	BASF	50% w/w	WP	11694
11	Standon Iprodione 50 WP	Standon	50% w/w	WP	10656

Uses

Alternaria in ***brassica seed crops*** [3-5, 9-11]; ***edible brassicas*** *(seed treatment)*, ***flower seeds*** *(seed treatment)* [9, 10]; ***fodder rape*** *(seed treatment)*, ***leaf brassicas*** *(seed treatment)*, ***mustard*** *(seed treatment)*, ***ornamental specimens*** *(seed treatment)*, ***stubble turnips*** *(seed treatment)*, ***swedes*** *(seed treatment)*, ***turnips*** *(seed treatment)* [3, 11]; ***leaf brassicas***, ***mustard***, ***spring oilseed rape***, ***stubble turnips***, ***winter oilseed rape***, ***winter wheat*** [4, 5]; ***linseed*** *(seed treatment)*, ***spring oilseed rape*** *(seed treatment)*, ***winter oilseed rape*** *(seed treatment)* [3, 7-11]; ***poppies*** *(off-label - for morphine production)* [4]; ***stored cabbages*** [3, 9-11]. Aphids in ***poppies*** *(off-label - for morphine production)* [4]. Black scurf in ***seed potatoes*** *(seed treatment)* [3, 7, 8, 11]. Black scurf and stem canker in ***potatoes*** *(seed treatment)* [9, 10]. Botrytis in ***brassica seed crops***, ***mustard***, ***spring oilseed rape***, ***winter oilseed rape***, ***winter wheat*** [4, 5]; ***chicory*** *(off-label)*, ***combining peas*** *(off-label)*, ***dwarf beans*** *(off-label)*, ***french beans*** *(off-label)*, ***mange-tout peas*** *(off-label)*, ***pears*** *(off-label)*, ***protected rhubarb*** *(off-label)*, ***red beet*** *(off-label)*, ***runner beans*** *(off-label)*, ***vining peas*** *(off-label)* [9]; ***cress*** *(off-label)*, ***grapevines*** *(off-label)*, ***lamb's lettuce*** *(off-label)*, ***lettuce*** *(off-label)*, ***poppies*** *(off-label - for morphine production)*, ***radicchio*** *(off-label)*, ***scarole*** *(off-label)* [4]; ***cucumbers*** [3, 9, 10]; ***lettuce*** *(outdoor crops)*, ***strawberries*** [3-5, 9-11]; ***outdoor tomatoes***, ***pot plants***, ***protected lettuce***, ***protected tomatoes***, ***raspberries***, ***stored cabbages*** [3, 9-11]. Botrytis bunch rot in ***grapevines*** *(off-label)* [4]. Brown patch in ***amenity grass***, ***managed amenity turf*** [1, 2, 6]. Canker in ***horseradish*** *(off-label)* [9]. Chocolate spot in ***spring field beans***, ***winter field beans*** [4, 5]. Collar rot in ***onions***, ***salad onions*** [4, 5]. Dollar spot in ***amenity grass***, ***managed amenity turf*** [1, 2, 6]. Fusarium patch in ***amenity grass***, ***managed amenity turf*** [1, 2, 6]. Glume blotch in ***winter wheat*** [4, 5]. Grey mould in ***protected cucumbers*** *(off-label)* [9]. Leaf rot in ***onions***, ***salad onions*** [4, 5]. Melting out in ***amenity grass***, ***managed amenity turf*** [1, 2, 6]. Net blotch in ***spring barley***, ***winter barley*** [4, 5]. Phoma in ***red beet*** *(off-label)* [9]. Poppy fire in ***poppies*** *(off-label - for morphine production)* [4]. Red thread in ***amenity grass***, ***managed amenity turf*** [1, 2, 6]. Sclerotinia in ***chicory*** *(off-label)*, ***combining peas*** *(off-label)*, ***dwarf beans*** *(off-label)*, ***french beans*** *(off-label)*, ***mange-tout peas*** *(off-label)*, ***runner beans*** *(off-label)*, ***vining peas*** *(off-label)* [9]; ***cress*** *(off-label)*, ***lamb's lettuce*** *(off-label)*, ***lettuce*** *(off-label)*, ***poppies*** *(off-label - for morphine production)*, ***radicchio*** *(off-label)*, ***scarole*** *(off-label)* [4]. Sclerotinia stem rot in ***spring oilseed rape***, ***winter oilseed rape*** [4, 5]. Snow mould in ***amenity grass***, ***managed amenity turf*** [1, 2, 6]. Stemphylium in ***asparagus*** *(off-label)* [9].

Specific Off-Label Approvals (SOLAs)

- ***asparagus***, ***combining peas***, ***vining peas*** *(OLA 001902) Dec 2008* [9]
- ***chicory*** *(OLA 001906) Dec 2008* [9]
- ***cress***, ***lamb's lettuce***, ***lettuce***, ***radicchio***, ***scarole*** *(OLA 001869) Dec 2008* [4]
- ***dwarf beans***, ***french beans***, ***mange-tout peas***, ***runner beans*** *(OLA 001904) Dec 2008* [9]
- ***grapevines*** *(OLA 001870) Dec 2008* [4]
- ***horseradish*** *(OLA 021254) Dec 2008* [9]
- ***pears*** *(OLA 012525) Dec 2008* [9]
- ***poppies*** *(for morphine production) (OLA 030765) Dec 2008* [4]
- ***poppies*** *(for morphine production) (OLA 030767) Dec 2008* [4]
- ***protected cucumbers*** *(OLA 003070) Dec 2008* [9]
- ***protected rhubarb*** *(OLA 001912) Dec 2008* [9]
- ***red beet*** *(OLA 001910) Dec 2008* [9]

Approval information

- Iprodione included in Annex I under EC Directive 91/414

Efficacy guidance

- Many diseases require a programme of 2 or more sprays at intervals of 2-4 wk. Recommendations vary with disease and crop - see label for details
- Use as a drench to control cabbage storage diseases. Spray ornamental pot plants and cucumbers to run-off
- Apply turf and amenity grass treatments after mowing to dry grass, free of dew. Do not mow again for at least 24 h
- Seed treatments on oilseed rape and linseed should be applied to dry seed prior to sowing
- Best results on seed potatoes achieved using hydraulic sprayer with solid or hollow cone jets with air or liquid pressure atomisation
- Apply to seed potatoes after harvest or after grading out and before traying out for chitting

Restrictions

- Maximum number of treatments 1 per batch for seed treatments; 1 per crop on cereals, vining peas, cabbage (as drench), chicory, borage; 2 per crop on field beans and stubble turnips; 3 per crop on brassicas (including seed crops), oilseed rape, protected winter lettuce (Oct-Feb); 4 per crop on strawberries, grapevines, salad onions, cucumbers; 5 per crop on raspberries; 6 per crop on bulb onions, tomatoes, curcurbits, peppers, aubergines, turf, nuts; 7 per crop on lettuce (Mar-Sep)
- A minimum of 3 wk must elapse between treatments on leaf brassicas
- See label for pot plants showing good tolerance. Check other species before applying on a large scale
- Do not treat oilseed rape seed that is cracked or broken or of low viability
- Do not excessively wet seed potato skins
- Do not treat oats
- Do not use treated seed as food or feed [7, 9]

Crop-specific information

- Latest use: pre-planting for seed treatments; at planting for potatoes; before grain watery ripe (GS 69) for wheat and barley; 5 wk before forcing for chicory; 6 wk before release of pears for human consumption; 16 wk before processing or sale for red beet
- HI strawberries, protected tomatoes 1 d; outdoor tomatoes, cucurbits, aubergines 2 d; lettuce, raspberries, nuts, cress, dwarf beans, endives, fennel, French beans, runner beans, lamb's lettuce (Mar-Sep), lettuce, mange-tout peas, onions, radicchio, scarole 7 d; grapevines, protected rhubarb 14 d; brassicas, brassica seed crops, oilseed rape, mustard, field beans, peas, stubble turnips 21 d; protected lettuce (Oct-Feb) 28 d; asparagus 5 mth

- Turf and amenity grass may become temporarily yellowed if frost or hot weather follows treatment
- Personal protective equipment requirements may vary for each pack size. Check label

Environmental safety

- Harmful to fish or other aquatic life. Do not contaminate surface waters or ditches with chemical or used container
- Treated brassica seed crops not to be used for human or animal consumption
- See label for guidance on disposal of spent drench liquor
- Treatment harmless to *Encarsia* or *Phytoseiulus* being used for integrated pest control

Hazard classification and safety precautions

Hazard R04 [1, 2, 6-8]
Risk phrases R04a, R04b [1, 2, 6]; R04e [7, 8]
Operator protection A [1, 2, 4-8]; C [1, 2, 6]; H [7, 8]; U04a [1, 2, 6]; U05a [1, 2, 4-8]; U08 [1, 2, 4-6]; U19 [3, 9-11]; U20b [4, 5]; U20c [1-3, 6-11]
Environmental protection E01 [1, 2, 4-8]; E13c [1-11]
Consumer protection C03 [1, 2, 4-8]
Storage and disposal E26 [1, 2, 6-8]; E30a [1-11]; E31b [1, 2, 4-6]; E32a [3, 7-11]; E34 [1-3, 6-11]
Treated seed S01, S05, S06 [7-10]; S02 [9, 10]; S03, S04a, S07 [7, 8]

291 iprodione + thiophanate-methyl

A protectant and systemic fungicide for oilseed rape

Products

1	Compass	Aventis	167:167 g/l	SC	10041
2	Compass	BASF	167:167 g/l	SC	11740
3	Snooker	Aventis	150:200 g/l	SC	10018
4	Snooker	BASF	150:200 g/l	SC	11744

Uses

Alternaria in ***carrots*** *(off-label)*, ***horseradish*** *(off-label)*, ***parsnips*** *(off-label)* [1]; ***spring oilseed rape***, ***winter oilseed rape*** [1-4]. Botrytis in ***limnanthes alba (meadowfoam)*** *(off-label)* [1]. Chocolate spot in ***spring field beans***, ***winter field beans*** [1-4]. Crown rot in ***carrots*** *(off-label)*, ***horseradish*** *(off-label)*, ***parsnips*** *(off-label)* [1]. Grey mould in ***spring oilseed rape***, ***winter oilseed rape*** [1-4]. Light leaf spot in ***spring oilseed rape***, ***winter oilseed rape*** [1-4]. Sclerotinia stem rot in ***spring oilseed rape***, ***winter oilseed rape*** [1-4]. Stem canker in ***spring oilseed rape***, ***winter oilseed rape*** [1, 2].

Specific Off-Label Approvals (SOLAs)

- ***carrots***, ***horseradish***, ***parsnips*** *(OLA 001936) Dec 2008* [1]
- ***limnanthes alba (meadowfoam)*** *(OLA 001934) Dec 2008* [1]

Approval information

- Iprodione included in Annex I under EC Directive 91/414

Efficacy guidance

- Timing of sprays on oilseed rape varies with disease, see label for details
- For season-long control product should be applied as part of a disease control programme

Restrictions

- Maximum number of treatments (refers to total sprays containing benomyl, carbendazim and thiophanate methyl) 2 per crop for oilseed rape, field beans, meadowfoam; 3 per crop for carrots, horseradish, parsnips

Crop-specific information

- Latest use: before end of flowering for oilseed rape, field beans
- HI meadowfoam, field beans, oilseed rape 3 wk; carrots, horseradish, parsnips 28 d
- Treatment may extend duration of green leaf in winter oilseed rape

Environmental safety

- Harmful to fish or other aquatic life. Do not contaminate surface waters or ditches with chemical or used container

Hazard classification and safety precautions

Hazard R04 [3, 4]

Risk phrases R04a, R04b, R04e [3, 4]
Operator protection A, C, H, M; U05a, U09a [3, 4]; U20c [1-4]
Environmental protection E01 [3, 4]; E13c [1-4]
Consumer protection C03 [3, 4]
Storage and disposal E26, E30a, E31b

292 isoproturon

A residual urea herbicide for use in cereals

See also *chlorotoluron + isoproturon*
diflufenican + flurtamone + isoproturon
diflufenican + isoproturon

Products

1	Agriguard Isoproturon	AgriGuard	500 g/l	SC	09769
2	Alpha IPU 500	Makhteshim	500 g/l	SC	10733
3	Alpha Isoproturon 500	Makhteshim	500 g/l	SC	05882
4	Arelon 500	Nufarm UK	500 g/l	SC	11639
5	Bison 83 WG	Nufarm UK	83% w/w	WG	11580
6	Fieldgard	Nufarm UK	500 g/l	SC	11541
7	Isoguard	Nufarm UK	500 g/l	SC	10829
8	Isotron 500	DAPT	500 g/l	SC	10668
9	Landgold Isoproturon	Landgold	500 g/l	SC	06012
10	Luxan Isoproturon 500 Flowable	Luxan	500 g/l	SC	07437
11	Primer	AgriGuard	500 g/l	SC	11782
12	Protugan	Makhteshim	500 g/l	SC	10043
13	Protugan 80 WDG	Makhteshim	80% w/w	WG	10430
14	Tolkan Liquid	Aventis	500 g/l	SC	10023
15	Tolkan Liquid	Aventis	500 g/l	SC	10895

Uses

Annual dicotyledons in ***spring wheat*** [4]; ***spring wheat*** *(autumn sown)* [13-15]; ***triticale, winter rye*** [2, 4, 12-15]; ***winter barley, winter wheat*** [1-15]. Annual grasses in ***spring wheat*** [4]; ***spring wheat*** *(autumn sown)* [13-15]; ***triticale, winter rye*** [2, 4, 12-15]; ***winter barley, winter wheat*** [1-15]. Annual meadow grass in ***spring barley*** [12]; ***spring barley*** *(off-label)* [2, 3, 12-14]. Blackgrass in ***spring wheat*** [4]; ***spring wheat*** *(autumn sown)* [13-15]; ***triticale, winter rye*** [2, 4, 12-15]; ***winter barley, winter wheat*** [1-4, 6-15]. Rough meadow grass in ***spring wheat*** [4]; ***spring wheat*** *(autumn sown)* [13-15]; ***triticale, winter rye*** [2, 4, 12-15]; ***winter barley, winter wheat*** [1-4, 6-15]. Wild oats in ***spring wheat*** [4]; ***spring wheat*** *(autumn sown)* [13-15]; ***triticale, winter rye*** [2, 4, 12-15]; ***winter barley, winter wheat*** [1-4, 6-15].

Specific Off-Label Approvals (SOLAs)

- ***spring barley*** *(OLA 001938) Dec 2008* [14]
- ***spring barley*** *(OLA 011481) Dec 2008* [3]
- ***spring barley*** *(OLA 011488) Dec 2008* [12]
- ***spring barley*** *(OLA 012166) Dec 2008* [2]
- ***spring barley*** *(OLA 023410) Dec 2008* [13]

Approval information

- Isoproturon included in Annex I under EC Directive 91/414
- All approvals for aerial use of isoproturon were revoked in 1995
- Approval expiry 30 Jun 2004 [1, 10]
- Approval expiry 31 Oct 2004 [14]
- Approval expiry 30 Nov 2004 [9]

Efficacy guidance

- May be applied in autumn or spring (but only post-emergence on wheat or barley). Best control normally achieved by early post-emergence treatment
- See label for details of rates, timings and tank mixes for different weed problems
- Apply to moist soils. Effectiveness may be reduced in seasons of above average rainfall or by prolonged dry weather
- Residual activity reduced on soils with more than 10% organic matter. Only use on such soils in spring

Restrictions

- Maximum number of treatments 1 per crop
- Maximum total dose 2.5 kg a.i./ha for any crop. The IPU stewardship programme guidelines recommend that this maximum dose should not be used after the end of Oct, and that where possible it should be reduced to 1.5 kg a.i./ha by using mixtures or sequences with other herbicides
- Do not apply pre-emergence to wheat or barley. On triticale and winter rye use pre-emergence only and on named varieties
- Do not use on durum wheats, oats, undersown cereals or crops to be undersown
- Do not apply to very wet or waterlogged soils, or when heavy rainfall is forecast
- Do not use on very cloddy soils or if frost is imminent or after onset of frosty weather
- Do not roll for 1 wk before or after treatment or harrow for 1 wk before or any time after treatment

Crop-specific information

- Latest use: not later than second node detectable stage (GS 32)
- Recommended timing of treatment varies depending on crop to be treated, method of sowing, season of application, weeds to be controlled and product being used. See label for details
- Crop damage may occur on free draining, stony or gravelly soils if heavy rain falls soon after spraying. Early sown crops may be damaged if spraying precedes or coincides with a period of rapid growth

Environmental safety

- Dangerous for the environment [4]
- Very toxic to aquatic organisms. May cause long-term adverse effects in the aquatic environment [4]
- Dangerous (harmful [3]) to fish or other aquatic life. Do not contaminate surface waters or ditches with chemical or used container [1, 2, 5-15]
- Do not apply to dry, cracked or waterlogged soils where heavy rain may lead to contamination of drains by isoproturon

Hazard classification and safety precautions

Hazard R03, R14 [4]; R04 [6, 7]
Risk phrases R04a, R04b [6, 7]; R10e, R14c [4]
Operator protection A, C, H [1-15]; D [5, 13]; U05a [2, 3, 6-8, 12, 13]; U08 [2-8, 12-15]; U19 [3, 6-8, 13-15]; U20a [3, 9, 11]; U20b [1, 2, 4-7, 10, 12-15]
Environmental protection E01 [2, 3, 6-8, 12, 13]; E13b [1, 2, 5-15]; E13c [3]; E15 [4]
Consumer protection C03 [2, 3, 6-8, 12, 13]
Storage and disposal E26, E30a [1-15]; E31b [1-4, 6-12, 14, 15]; E32a [5, 13]; E32d [4]; E34 [2, 12]
Treated seed S06 [8]

293 isoproturon + pendimethalin

A contact and residual herbicide for use in winter cereals

Products

1	Encore	BASF	125:250 g/l	SC	10375
2	Trump	BASF	236:200 g/l	SC	10376

Uses

Annual dicotyledons in ***spring wheat*** *(autumn sown)* [2]; ***triticale, winter barley, winter rye, winter wheat*** [1, 2]. Annual grasses in ***spring wheat*** *(autumn sown)* [2]; ***triticale, winter barley, winter rye, winter wheat*** [1, 2]. Blackgrass in ***spring wheat*** *(autumn sown)* [2]; ***triticale, winter barley, winter rye, winter wheat*** [1, 2]. Wild oats in ***spring wheat*** *(autumn sown)* [2]; ***triticale, winter barley, winter rye, winter wheat*** [1, 2].

Approval information

- Isoproturon and pendimethalin included in Annex I under EC Directive 91/414

Efficacy guidance

- May be applied in autumn or spring (but only post-emergence on wheat or barley)
- Annual grasses controlled from pre-emergence to 4-leaf stage, extended to 3-tiller stage with blackgrass by tank mixing with additional isoproturon. Best results on wild oats post-emergence in autumn

SECTION 4

- Annual dicotyledons controlled pre-emergence and up to 8-leaf stage [1], up to 12-leaf stage [2]. See label for details
- Apply to moist soils. Best results achieved by application to fine, firm seedbed. Trash, ash or straw should have been incorporated evenly
- Contact activity is reduced by rain within 6 h of application
- Activity may be reduced on soil with more than 6% organic matter or ash. Do not use on soils with more than 10% organic matter

Restrictions

- Maximum number of treatments 1 per crop. Maximum total dose of isoproturon 2.5 kg a.i./ha per crop
- Do not apply pre-emergence to wheat or barley. On winter rye and triticale use pre-emergence only on named cultivars
- Do not use on durum wheat or spring cereals
- Do not use pre-emergence on winter rye or triticale drilled after 30 Nov [2]
- Do not use on crops suffering stress from disease, drought, waterlogging, poor seedbed conditions or other causes or apply post-emergence when frost imminent
- Do not apply to very wet or waterlogged soils, or when heavy rain is forecast
- Do not undersow treated crops
- Do not roll or harrow after application

Crop-specific information

- Latest use: pre-emergence for winter rye and triticale; before leaf sheath erect (GS 30) [2] or before 1st node detectable (GS 31) [1] for winter wheat and barley
- Apply before first node detectable (GS 31) [1], to before leaf sheath erect (GS 30) [2]
- May be used on autumn sown spring wheat [2]

Following crops guidance

- In the event of autumn crop failure spring wheat, spring barley, maize, potatoes, beans or peas may be grown following ploughing to at least 150 mm

Environmental safety

- Dangerous for the environment
- Very toxic to aquatic organisms. May cause long-term adverse effects in the aquatic environment
- Do not apply to dry, cracked or waterlogged soils where heavy rain may lead to contamination of drains by isoproturon

Hazard classification and safety precautions

Hazard R03, R14
Risk phrases R10e [2]; R14c [1, 2]
Operator protection A, H [1, 2]; C [1]; U05a, U08, U13, U19, U20b
Environmental protection E01, E38 [1, 2]; E15 [1]
Consumer protection C03
Storage and disposal E26, E34 [1]; E30a, E31b, E32d [1, 2]
Medical advice M05a [2]

294 isoproturon + simazine

A contact and residual herbicide for winter wheat and barley

Products

1	Alpha Protugan Plus	Makhteshim	375:60 g/l	SC	08799
2	Harlequin 500 SC	Makhteshim	450:50 g/l	SC	09779

Uses

Annual dicotyledons in ***winter barley*** [2]; ***winter wheat*** [1, 2]. Annual grasses in ***winter barley*** [2]; ***winter wheat*** [1, 2]. Blackgrass in ***winter barley*** [2]; ***winter wheat*** [1, 2].

Approval information

- Isoproturon included in Annex I under EC Directive 91/414

Efficacy guidance

- Annual grasses controlled up to early tillering, annual dicotyledons to 50 mm

- Apply to fine, firm, even seedbed. Any trash or burnt straw should be dispersed in preparing seedbed
- Apply to moist soils. Weed control may be reduced in excessively wet autumns or if prolonged dry weather follows application to dry soil

Restrictions

- Maximum number of treatments 1 per crop. Maximum total dose of isoproturon 2.5 kg a.i./ha per crop
- Do not use on durum wheat, undersown crops or those due to be undersown
- Do not use on sand or on soils with more than 10% organic matter
- On stony or gravelly soils there is risk of crop damage, especially with heavy rain soon after application
- Do not apply when frost or heavy rain is forecast or to crops severely checked by frost, waterlogging, pest or disease attack.
- Do not harrow for 7 d before treatment or roll for 7 d before or after treatment in spring
- Do not harrow after application

Crop-specific information

- Latest use: end of tillering [1]; leaf sheath erect (GS 30) [2]
- May be applied in autumn or spring (but only post-emergence of the crop)
- Apply from 2-leaf stage of crop to before leaf sheath erect (GS 12-30)
- With direct drilled crops soil surface should be broken by surface cultivation and seed covered by 12-25 mm soil
- Early sown crops may be prone to damage if spraying precedes or coincides with period of rapid growth in autumn

Following crops guidance

- In the event of crop failure land should be inverted by mouldboard ploughing to at least 150 mm and harrowed before drilling/planting another crop

Environmental safety

- Dangerous to fish or other aquatic life. Do not contaminate surface waters or ditches with chemical or used container
- Do not apply to dry, cracked or waterlogged soils where heavy rain may lead to contamination of drains by isoproturon
- LERAP Category B

Hazard classification and safety precautions

Hazard R04 [2]
Risk phrases R04b [2]
Operator protection A, C, H [1, 2]; D, M [2]; U05a, U08, U20b [1, 2]; U19 [2]
Environmental protection E01, E13b, E16a, E16b
Consumer protection C03
Storage and disposal E26, E30a, E31b [1, 2]; E29 [2]

295 isoxaben

A soil-acting amide herbicide for use in grass and fruit

Products

1	Agriguard Isoxaben	AgriGuard	125 g/l	SC	11652
2	Fluxidor 125	Landseer	125 g/l	SC	05104
3	Knot Out	Vitax	125 g/l	SC	05163

Uses

Annual dicotyledons in ***amenity grass*** [3]; ***amenity vegetation***, ***canary grass*** *(off label)*, ***carrots*** *(off-label)*, ***clovers*** *(off-label)*, ***courgettes*** *(off-label)*, ***forage maize*** *(off-label - grown for game cover)*, ***marrows*** *(off-label)*, ***millet*** *(off-label)*, ***protected courgettes*** *(off-label)*, ***protected marrows*** *(off-label)*, ***protected ornamentals*** *(off-label)*, ***protected pumpkins*** *(off-label)*, ***protected squashes*** *(off-label)*, ***pumpkins*** *(off-label)*, ***quinoa*** *(off-label)*, ***squashes*** *(off-label)*, ***sunflowers*** *(off-label)* [2]; ***apple orchards***, ***blackberries***, ***blackcurrants***, ***cherries***, ***container-grown woody ornamentals***, ***forestry transplants***, ***gooseberries***, ***grapevines***, ***hardy ornamental nursery stock***, ***hops***, ***pear orchards***, ***plums***, ***raspberries***, ***strawberries*** [1, 2]. Cleavers in ***asparagus*** *(off-label)* [2]. Fat hen in ***asparagus*** *(off-label)* [2]. Groundsel in ***asparagus*** *(off-label)* [2]. Knotgrass in ***asparagus*** *(off-label)* [2].

Specific Off-Label Approvals (SOLAs)

- ***asparagus*** *(OLA 972318) Dec 2008* [2]
- ***canary grass, clovers, millet, quinoa, sunflowers*** *(OLA 012631) Dec 2008* [2]
- ***carrots*** *(OLA 962209) Dec 2008* [2]
- ***courgettes, marrows, protected courgettes, protected marrows, protected pumpkins, protected squashes, pumpkins, squashes*** *(OLA 982578) Dec 2008* [2]
- ***forage maize*** *(grown for game cover) (OLA 012631) Dec 2008* [2]
- ***protected ornamentals*** *(OLA 940605) Dec 2008* [2]

Efficacy guidance

- When used alone apply pre-weed emergence
- Effectiveness is reduced in dry conditions. Weed seeds germinating at depth are not controlled
- Activity reduced on soils with more than 10% organic matter. Do not use on peaty soils
- Various tank mixtures are recommended for early post-weed emergence treatment (especially for grass weeds). See label for details
- Best results on turf achieved by applying to firm, moist seedbed within 2 d of sowing. Avoid disturbing soil surface after application [3]

Restrictions

- Maximum number of treatments 1 per crop for all edible crops except asparagus; 1 per sowing or yr for amenity grass, ornamentals, strawberries, 2 per yr for hardy ornamental nursery stock and forestry transplants
- Maximum total dose equivalent to one full dose treatment on edible crops except asparagus (2 full dose treatments)

Crop-specific information

- Latest use: 3 true leaves for cucurbits; before 1 Apr in yr of harvest for other edible crops

Following crops guidance

- See label for details of crops which may be sown in the event of failure of a treated crop

Environmental safety

- Keep all livestock out of treated areas for at least 50 d

Hazard classification and safety precautions

Operator protection A, C; U05a [2]; U20b [1-3]
Environmental protection E07 (50 d); E15
Storage and disposal E26, E31a, E34 [3]; E30a, E32a [1-3]

296 isoxaben + trifluralin

A residual herbicide mixture for amenity and horticultural use

Products

1	Axit GR	Fargro	0.5:2.0% w/w	GR	08892
2	Premiere Granules	Dow	0.5:2.0% w/w	GR	07987

Uses

Annual dicotyledons in ***amenity trees and shrubs, amenity vegetation*** [2]; ***container-grown ornamentals*** *(outdoor stock only)*, ***hardy ornamental nursery stock*** *(outdoor stock only)* [1]. Annual meadow grass in ***amenity trees and shrubs, amenity vegetation*** [2]; ***container-grown ornamentals*** *(outdoor stock only)*, ***hardy ornamental nursery stock*** *(outdoor stock only)* [1]. Fat hen in ***amenity trees and shrubs, amenity vegetation*** [2]; ***container-grown ornamentals*** *(outdoor stock only)*, ***hardy ornamental nursery stock*** *(outdoor stock only)* [1]. Groundsel in ***amenity trees and shrubs, amenity vegetation*** [2]. Sowthistle in ***amenity trees and shrubs, amenity vegetation*** [2]; ***container-grown ornamentals*** *(outdoor stock only)*, ***hardy ornamental nursery stock*** *(outdoor stock only)* [1].

Efficacy guidance

- Best results obtained when granules applied to firm, moist soil, free from clods. 20-30 mm irrigation or rainfall required within 3 d after treatment
- Weed control reduced under dry soil conditions
- Treatment should be made before mulching and before weed emergence

- Uniform cover with granules optimises results. Improved uniformity of cover may be achieved by spreading the granules twice at half dose at right angles
- Control may be reduced on soils with high organic matter or where organic manure has been applied

Restrictions
- Maximum number of treatments 1 per yr
- Do not use on soils with more than 10% organic matter (except in container grown hardy nursery stock)
- Do not use on ornamentals grown under protection
- Do not use in the 12 mth prior to lifting field grown nursery stock if edible crops are to be grown as a following crop

Crop-specific information
- Ensure soil has settled and no cracks are present before application to transplanted trees and shrubs [2]
- May be used on Light, Medium and Heavy soils, but on Light soils early growth may be reduced under adverse conditions
- For newly planted containerised stock first irrigate and allow compost to settle and foliage and stems to dry before application [1]
- Products contain up to 1% crystalline silica. The MEL for respirable silica dust is 0.4 mg/cu m

Following crops guidance
- Land must be mouldboard ploughed to at least 20 cm before drilling or planting succeeding plants except well rooted forestry trees and ornamentals

Environmental safety
- Harmful to fish or other aquatic life. Do not contaminate surface waters or ditches with chemical or used container
- Do not allow direct applications of granules from ground based/vehicle mounted applicators to fall within 6 m, or from hand-held applicators to within 2 m, of surface waters or ditches. Direct applications away from water

Hazard classification and safety precautions

Operator protection A; U20b
Environmental protection E01, E13c, E16
Storage and disposal E30a, E32a, E34

SECTION 4

297 kresoxim-methyl

A protectant strobilurin fungicide for apples

See also *epoxiconazole + fenpropimorph + kresoxim-methyl*
epoxiconazole + kresoxim-methyl
epoxiconazole + kresoxim-methyl + pyraclostrobin
fenpropimorph + kresoxim-methyl

Products

Stroby WG	BASF	50% w/w	WG	08653

Uses

Black spot in ***roses***. Powdery mildew in ***apples*** *(reduction)*, ***blackcurrants***, ***protected strawberries***, ***roses***, ***strawberries***. Scab in ***apples***.

Approval information
- Kresoxim-methyl included in Annex I under EC Directive 91/414
- Approved for use in ULV systems

Efficacy guidance
- Activity is protectant. Best results achieved from treatments prior to disease development. See label for timing details on each crop. Treatments should be repeated at 10-14 d intervals but note limitations below
- To minimise the likelihood of development of resistance to strobilurin fungicides these products should be used in a planned Resistance Management strategy. See Section 1 for more information
- Product may be applied in ultra low volumes (ULV) but disease control may be reduced

Restrictions

- Maximum number of treatments 4 per yr on apples; 3 per yr on other crops. See notes in Efficacy about limitations on consecutive treatments
- Consult before using on crops intended for processing

Crop-specific information

- HI 14 d for blackcurrants, protected strawberries, strawberries; 35 d for apples
- On apples do not spray product more than twice consecutively and separate each block of two consecutive treatments with at least two applications from a different cross-resistance group. For all other crops do not apply consecutively and use a maximum of once in every three fungicide sprays
- Product should not be used as final spray of the season on apples

Environmental safety

- Dangerous for the environment
- Broadcast air-assisted LERAP (5 m)
- Harmless to ladybirds and predatory mites
- Harmless to honey bees and may be applied during flowering. Nevertheless local beekeepers should be notified when treatment of orchards in flower is to occur

Hazard classification and safety precautions

Hazard R03, R14
Operator protection U05a, U20b
Environmental protection E01, E13b; E17a (5 m)
Consumer protection C03
Storage and disposal E26, E30a, E31c

298 lambda-cyhalothrin

A quick-acting contact and ingested pyrethroid insecticide

Products

1	Hallmark with Zeon Technology	Syngenta	100 g/l	CS	10480
2	Stealth	Syngenta	2.5% w/w	WG	11514

Uses

Aphids in ***chestnuts*** *(off-label)*, ***cob nuts*** *(off-label)*, ***durum wheat***, ***hazel nuts*** *(off-label)*, ***potatoes***, ***short rotation coppice willow***, ***spring barley***, ***spring oats***, ***spring wheat***, ***walnuts*** *(off-label)*, ***winter barley***, ***winter oats***, ***winter wheat*** [1]. Barley yellow dwarf virus vectors in ***durum wheat***, ***winter barley***, ***winter oats***, ***winter wheat*** [1, 2]; ***spring barley***, ***spring wheat*** [1]. Beet leaf miner in ***sugar beet*** [1]. Beet virus yellows vectors in ***spring oilseed rape*** [1]; ***winter oilseed rape*** [1, 2]. Beetles in ***short rotation coppice willow*** [1]. Cabbage stem flea beetle in ***spring oilseed rape*** [1]; ***winter oilseed rape*** [1, 2]. Cabbage stem weevil in ***radishes*** *(off-label)* [1]. Carrot fly in ***carrots*** *(off-label)*, ***celeriac*** *(off-label)*, ***celery*** *(off-label)*, ***fennel*** *(off-label)*, ***parsnips*** *(off-label)* [1]. Caterpillars in ***broccoli***, ***brussels sprouts***, ***cabbages***, ***calabrese***, ***cauliflowers***, ***chestnuts*** *(off-label)*, ***cob nuts*** *(off-label)*, ***hazel nuts*** *(off-label)*, ***walnuts*** *(off-label)* [1]. Cutworms in ***carrots***, ***celeriac*** *(off-label)*, ***chicory*** *(off-label - for forcing)*, ***fennel*** *(off-label)*, ***lettuce***, ***red beet*** *(off-label)*, ***sugar beet*** [1]. Flea beetle in ***poppies*** *(off-label - for morphine production)*, ***spring oilseed rape***, ***sugar beet*** [1]; ***winter oilseed rape*** [1, 2]. Frit fly in ***sweetcorn*** *(off-label)* [1]. Pea and bean weevil in ***peas***, ***spring field beans***, ***winter field beans*** [1]. Pea aphid in ***peas*** [1]. Pea moth in ***peas*** [1]. Pear sucker in ***pears*** [1]. Pod midge in ***spring oilseed rape***, ***winter oilseed rape*** [1]. Pollen beetle in ***poppies*** *(off-label - for morphine production)*, ***spring oilseed rape***, ***winter oilseed rape*** [1]. Sawflies in ***short rotation coppice willow*** [1]. Seed weevil in ***spring oilseed rape***, ***winter oilseed rape*** [1]. Silver Y moth in ***celery*** *(off-label)*, ***dwarf beans*** *(off-label)*, ***navy beans*** *(off-label)*, ***red beet*** *(off-label)*, ***runner beans*** *(off-label)* [1]. Whitefly in ***broccoli***, ***brussels sprouts***, ***cabbages***, ***calabrese***, ***cauliflowers*** [1]. Yellow cereal fly in ***winter wheat*** [1].

Specific Off-Label Approvals (SOLAs)

- ***carrots***, ***parsnips*** *(OLA 011642) Dec 2008* [1]
- ***celeriac***, ***radishes*** *(OLA 011430) Dec 2008* [1]
- ***celery*** *(OLA 011644) Dec 2008* [1]
- ***chestnuts***, ***cob nuts***, ***hazel nuts***, ***walnuts*** *(OLA 011646) Dec 2008* [1]

- ***chicory*** *(for forcing) (OLA 011649) Dec 2008* [1]
- ***dwarf beans, navy beans, runner beans*** *(OLA 011645) Dec 2008* [1]
- ***fennel*** *(OLA 011648) Dec 2008* [1]
- ***poppies*** *(for morphine production) (OLA 030768) Dec 2008* [1]
- ***red beet*** *(OLA 011643) Dec 2008* [1]
- ***sweetcorn*** *(OLA 011647) Dec 2008* [1]

Approval information

- Lambda-cyhalothrin included in Annex I under EC Directive 91/414

Efficacy guidance

- Best results normally obtained from treatment when pest attack first seen. See label for detailed recommendations on each crop
- Timing for control of barley yellow dwarf virus vectors depends on specialist assessment of the level of risk in the area
- Repeat applications recommended in some crops where prolonged attack occurs, up to maximum total dose. See label for details
- Where strains of aphids resistant to lambda-cyhalothrin occur control is unlikely to be satisfactory
- Addition of wetter recommended for control of certain pests in brassicas and oilseed rape
- Use of sufficient water volume to ensure thorough crop penetration recommended for optimum results
- Use of drop-legged sprayer gives improved results in crops such as Brussels sprouts

Restrictions

- Maximum number of applications or maximum total dose per crop varies - see labels
- Do not apply to a cereal crop if any product containing a pyrethroid insecticide or dimethoate has been applied to the crop after the start of ear emergence (GS 51)
- Do not spray cereals in the spring/summer (ie after 1 Apr) within 6 m of edge of crop

Crop-specific information

- Latest use on cereals and oilseed rape depends on product used. See labels
- HI 3 d for lettuce, radishes, red beet; 7 d for celery, chicory, dwarf beans, fennel, navy beans, runner beans, sweetcorn; 14 d for carrots, celeriac, nuts, parsnips; 6 wk for spring oilseed rape; 8 wk for sugar beet
- If using mixtures, add lambda-cyhalothrin product to the tank last

Environmental safety

- Dangerous for the environment
- Very toxic to aquatic organisms. May cause long-term adverse effects in the aquatic environment
- LERAP Category A
- Broadcast air-assisted LERAP (38 m in pears) [1]

Hazard classification and safety precautions

Hazard R03, R04, R14
Risk phrases R03b, R03c, R04e, R14c [1, 2]; R04a, R04b [2]
Operator protection A, C, H, J, K, M [1, 2]; L [1]; U02, U05a, U08, U14, U20b [1, 2]; U11 [2]
Environmental protection E01, E16c, E16d, E38 [1, 2]; E15, E22a [2]; E17a [1] (pears 38 m)
Consumer protection C03
Storage and disposal E30a, E32d [1, 2]; E31b [1]; E31c, E34 [2]
Medical advice M03 [1, 2]; M05a [2]

299 lambda-cyhalothrin + pirimicarb

An insecticide mixture combining translaminar, contact, fumigant and stomach activity

Products

Dovetail	Syngenta	5:100 g/l	EC	10479

Uses

Aphids in ***broccoli, brussels sprouts, cabbages, calabrese, carrots, cauliflowers, durum wheat, lettuce, peas, potatoes, spring barley, spring oats, spring rye, spring wheat, sugar beet, triticale, winter barley, winter oats, winter rye, winter wheat***. Beet virus yellows vectors in ***spring oilseed rape, winter oilseed rape***. Cabbage stem flea beetle in ***spring oilseed rape, winter oilseed rape***. Caterpillars in ***broccoli, brussels sprouts, cabbages, calabrese,***

cauliflowers. Cutworms in ***carrots, lettuce, potatoes, sugar beet***. Flea beetle in ***spring oilseed rape, sugar beet, winter oilseed rape***. Leaf miner in ***sugar beet***. Mealy aphid in ***broccoli, brussels sprouts, cabbages, calabrese, cauliflowers, spring oilseed rape, winter oilseed rape***. Pea and bean weevil in ***peas, spring field beans, winter field beans***. Pea midge in ***peas***. Pea moth in ***peas***. Pod midge in ***spring oilseed rape, winter oilseed rape***. Pollen beetle in ***broccoli, brussels sprouts, cabbages, calabrese, cauliflowers, spring oilseed rape, winter oilseed rape***. Seed weevil in ***spring oilseed rape, winter oilseed rape***. Whitefly in ***broccoli, brussels sprouts, cabbages, calabrese, cauliflowers***.

Approval information

- Lambda-cyhalothrin included in Annex I under EC Directive 91/414

Efficacy guidance

- Best results obtained from treatment when pest attack first seen or after warning issued
- Repeat applications recommended in some crops where prolonged attacks occur, up to maximum total dose
- Addition of Agral recommended for certain uses in brassicas and oilseed rape. See label
- Use of drop-leg sprayers improves efficacy in crops such as Brussels sprouts
- Control unlikely to be satisfactory if aphids resistant to lambda-cyhalothrin or pirimicarb present

Restrictions

- Maximum total dose equivalent to two full dose treatments on carrots, lettuce, peas, field beans; three full dose treatments on brassicas, oilseed rape, cereals; four full dose treatments on sugar beet; eight full dose treatments on potatoes
- Must not be applied to cereals if any product containing a pyrethroid insecticide or dimethoate has been sprayed after the start of ear emergence (GS 51)

Crop-specific information

- Latest use: before late milky ripe (GS 77) for cereals; before end of flowering for oilseed rape
- HI carrots 9 wk; sugar beet 8 wk; brassicas, lettuce, potatoes, peas, field beans 3 d

Environmental safety

- Flammable
- Harmful to fish or other aquatic life. Do not contaminate surface waters or ditches with chemical or used container
- LERAP Category A
- Do not spray cereals after 1 Apr within 6 m of the edge of the crop
- Harmful to livestock. Keep all livestock out of treated areas for at least 7 d following treatment

Hazard classification and safety precautions

Hazard R03, R04, R07d, R14
Risk phrases R03b, R03c, R04b, R04c, R04d, R04e, R04g, R14c
Operator protection A, C, E, H, J, K, M; U02, U05a, U08, U11, U19, U20b
Environmental protection E01, E15, E16c, E16d, E38; E06b (7 d)
Consumer protection C03
Storage and disposal E30b, E31c, E32d, E34
Medical advice M03, M05

300 lenacil

A residual, soil-acting uracil herbicide for beet and horticultural crops

Products

1	Agricola Lenacil FL	Agricola	440 g/l	SC	09481
2	Agriguard Lenacil	AgriGuard	80% w/w	WP	10488
3	Clayton Lenacil 80 W	Fargro	80% w/w	WP	09488
4	Venzar Flowable	DuPont	440 g/l	SC	06907

Uses

Annual dicotyledons in ***blackcurrants, established woody ornamentals, gooseberries, raspberries, roses, strawberries*** [3]; ***farm woodland*** *(off-label)* [4]; ***fodder beet, mangels, sugar beet*** [1, 2, 4]; ***red beet*** [2, 4]. Annual meadow grass in ***blackcurrants, established woody ornamentals, gooseberries, raspberries, roses, strawberries*** [3]; ***farm woodland*** *(off-label)* [4]; ***fodder beet, mangels, sugar beet*** [1, 2, 4]; ***red beet*** [2, 4].

Specific Off-Label Approvals (SOLAs)

- ***farm woodland*** *(OLA 971282) Dec 2008* [4]

Efficacy guidance

- On beet crops may be used pre- or post-emergence, alone or in mixture to broaden weed spectrum
- Apply overall or as band spray to beet crops pre-drilling incorporated, pre- or post-emergence
- All labels have limitations on soil types that may be treated. Residual activity reduced on soils with high OM content
- Best results achieved on a firm moist tilth. Continuing presence of moisture from rain or irrigation gives improved residual control of later germinating weeds. Effectiveness may be reduced by dry conditions

Restrictions

- Maximum number of treatments in beet crops 1 pre-emergence + 3 post-emergence per crop; 2 per yr on established woody ornamentals and roses
- See label for specific soil type restrictions on named strawberry varieties. Do not treat varieties Ostara, Pantangruella, Senga Gigana [3]
- See label for soil type restrictions
- Do not apply other residual herbicides within 3 mth before or after spraying
- Do not treat crops under stress from drought, low temperatures, nutrient deficiency, pest or disease attack, or waterlogging

Crop-specific information

- Latest use: pre-emergence for red beet, fodder beet, spinach, spinach beet, mangels; before leaves meet over rows when used on these crops post-emergence; 24 h after planting new strawberry runners; in spring before flowering in established crops of strawberries
- May be used on newly planted or established crops of strawberries, bush and cane fruit
- Roses and woody ornamentals must be established before treatment
- Heavy rain after application may cause damage especially if followed by very hot weather
- Reduction in beet stand may occur where crop emergence or vigour is impaired by soil capping or pest attack

Following crops guidance

- Succeeding crops should not be planted or sown for at least 4 mth (6 mth on organic soils) after treatment following ploughing to at least 150 mm

Environmental safety

- Harmful to fish or other aquatic life. Do not contaminate surface waters or ditches with chemical or used container

Hazard classification and safety precautions

Hazard R04
Risk phrases R04a, R04b [1-4]; R04c [2-4]
Operator protection A, C; U05a, U08 [1-4]; U19 [1, 3, 4]; U20a [1, 2, 4]; U20b [3]
Environmental protection E01, E13c
Consumer protection C03
Storage and disposal E26 [1]; E30a [1-4]; E31a [4]; E31b [1, 2]; E32a [3, 4]; E34 [2]

301 lenacil + phenmedipham

A contact and residual herbicide for use in sugar beet

Products

Agricola Lens	Agricola	66.6:95 g/l	SC	10257

Uses

Annual dicotyledons in ***fodder beet***, ***mangels***, ***sugar beet***.

Efficacy guidance

- Apply at any stage of crop when weeds at cotyledon stage
- Germinating weeds controlled by root uptake for several weeks
- Best results achieved under warm, moist conditions on a fine, firm seedbed
- Treatment may be repeated up to a maximum of 3 applications on later weed flushes

- Residual activity may be reduced on highly organic soils or under very dry conditions
- Rain falling 1 h after application does not reduce activity

Restrictions

- Maximum number of treatments 3 per crop
- Do not apply when temperature above or likely to exceed 21°C on day of spraying or under conditions of high light intensity
- Do not spray any crop under stress from drought, waterlogging, cold, wind damage or any other cause

Crop-specific information

- Latest use: before crop leaves meet across rows
- Heavy rain after application may reduce stand of crop particularly in very hot weather. In severe cases yield may be reduced

Following crops guidance

- Do not sow or plant any crop within 4 mth of treatment. In case of crop failure only sow or plant beet crops, strawberries or other tolerant horticultural crop within 4 mth

Environmental safety

- Harmful to fish or other aquatic life. Do not contaminate surface waters or ditches with chemical or used container

Hazard classification and safety precautions

Hazard R04
Risk phrases R04a, R04c
Operator protection A, C; U05a, U08, U19, U20b
Environmental protection E01, E13c
Consumer protection C03
Storage and disposal E30a, E31b
Medical advice M03

302 lindane

An organochlorine insecticide all approvals for which were revoked in 2001 following non-inclusion of the substance in Annex I of EC Directive 91/414

303 linuron

A contact and residual urea herbicide for various field crops

Products

1	Alpha Linuron 50 SC	Makhteshim	500 g/l	SC	06967
2	Lincon 50	DAPT	500 g/l	SC	10639
3	Linuron 500	Nufarm UK	480 g/l	SC	11590
4	UPL Linuron 45% Flowable	United Phosphorus	450 g/l	SC	07435

Uses

Annual dicotyledons in ***bulb onions*** *(off-label)*, ***celeriac*** *(off-label)*, ***garlic*** *(off-label)*, ***leeks*** *(off-label)*, ***soya beans*** *(off-label)* [1]; ***carrots, parsley, potatoes*** [1-4]; ***parsnips*** [1, 2, 4]; ***spring barley, spring wheat*** [2-4]; ***spring oats*** [1, 2]. Annual grasses in ***soya beans*** *(off-label)* [1]. Annual meadow grass in ***carrots, parsley, potatoes, spring barley, spring wheat*** [1-4]; ***parsnips*** [1, 2, 4]; ***spring oats*** [2]. Black bindweed in ***carrots, parsley, potatoes, spring barley, spring wheat*** [1-4]; ***parsnips*** [1, 2, 4]; ***spring oats*** [1, 2]. Chickweed in ***carrots, parsley, potatoes, spring barley, spring wheat*** [1-4]; ***parsnips*** [1, 2, 4]; ***spring oats*** [1, 2]. Corn marigold in ***carrots, parsley, potatoes, spring barley, spring wheat*** [1-3]; ***parsnips, spring oats*** [1, 2]. Fat hen in ***carrots, parsley, potatoes, spring barley, spring wheat*** [1-4]; ***parsnips*** [1, 2, 4]; ***spring oats*** [1, 2]. Redshank in ***carrots, parsley, potatoes, spring barley, spring wheat*** [1-4]; ***parsnips*** [1, 2, 4]; ***spring oats*** [1, 2].

Specific Off-Label Approvals (SOLAs)

- ***bulb onions, celeriac, garlic, leeks*** *(OLA 021748) Dec 2008* [1]
- ***soya beans*** *(OLA 010245) Apr 2004* [1]

Approval information

- Linuron included in Annex I under EC Directive 91/414

Efficacy guidance

- Many weeds controlled pre-emergence or post-emergence to 2-3 leaf stage, some (annual meadow grass, mayweed) only susceptible pre-emergence. See label for details
- Best results achieved by application to firm, moist soil of fine tilth
- Little residual effect on soil with more than 10% organic matter

Restrictions

- Maximum number of treatments 1 per crop for celery, potatoes, cereals; 2 per crop for carrots, parsley, parsnips
- Maximum total dose equivalent to one full dose treatment
- Apply only pre-crop emergence [4]
- Do not use on undersown cereals or crops grown on Sands or Very Light soils or soils heavier than Sandy Clay Loam or with more than 10% organic matter
- Do not apply to emerged crops of carrots, parsnips or parsley under stress
- Do not apply by hand-held sprayers

Crop-specific information

- Latest use: pre-emergence for cereals and linseed; pre- or up to 40% emergence for potatoes (products differ, see label for details).
- HI onions, garlic 8 wk; celeriac 12 wk; leeks 16 wk
- Drill spring cereals at least 3 cm deep and apply pre-emergence of crop or weeds
- Apply to potatoes well earthed up to a rounded ridge pre-crop emergence and do not cultivate after spraying
- Apply to carrots or parsley at any time after drilling on organic soils, and within 4 d of drilling on other soils. Apply post-emergence as soon as weeds appear but after first rough leaf stage.
- Recommendations for parsnips, parsley and celery vary. See label for details

Following crops guidance

- Potatoes, carrots and parsnips may be planted at any time after application. Lettuce should not be grown within 12 mth of treatment. Transplanted brassicas may be grown from 3 mth after treatment

Environmental safety

- Dangerous for the environment [3]
- Very toxic to aquatic organisms. May cause long-term adverse effects in the aquatic environment [3]
- Dangerous to fish or other aquatic life. Do not contaminate surface waters or ditches with chemical or used container [1, 2, 4]
- LERAP Category B
- Keep livestock out of treated areas for at least 5 mth [3]

Hazard classification and safety precautions

Hazard R03, R06, R14 [3]; R04 [4]
Risk phrases R04a, R04b [4]; R10e, R14c [3]
Operator protection A, C, H; U05a [3, 4]; U08, U19 [1-4]; U09a [4]; U20b [1, 3, 4]
Environmental protection E01 [3, 4]; E07 [3] (5 mth); E13b [1, 2, 4]; E16a [1-4]; E16b [2]; E38 [3]
Consumer protection C03 [4]
Storage and disposal E26, E30a [1-4]; E31b [3, 4]; E32a [1, 2]; E32d [3]; E34 [4]
Treated seed S06 [2]
Medical advice M03 [4]

304 linuron + trifluralin

A residual pre-emergence herbicide for use in winter cereals

Products

1	Arizona	Makhteshim	120:240 g/l	EC	11722
2	Uranus	Makhteshim	120:240 g/l	EC	11637

Uses

Annual dicotyledons in ***triticale***, ***winter barley***, ***winter wheat***. Annual grasses in ***triticale***, ***winter barley***, ***winter wheat***.

Approval information

- Linuron included in Annex I under EC Directive 91/414

Efficacy guidance

- Effective against weeds germinating near soil surface. Best results achieved by application to fine, firm, moist seedbed free of clods, crop residues or established weeds
- Effectiveness reduced by long dry periods after application or on waterlogged soil
- Autumn application residual effects normally last until spring but further herbicide treatment may be needed on thin or backward crops
- Results in loose seedbeds on lighter soils improved by rolling after drilling

Restrictions

- Maximum number of treatments 1 per crop
- Do not treat durum wheat or undersown crops
- Do not use on soils classed as Sands. Do not harrow after treatment
- Do not use on peaty soils or where organic matter exceeds 10%

Crop-specific information

- Latest use: pre-emergence of crop
- Apply without incorporation as soon as possible after drilling and before crop emergence, within 3 d on early drilled crops
- Crop seed must be well covered to a minimum depth of 30 mm

Following crops guidance

- In the event of failure of a treated crop, only peas, carrots or sunflowers may be sown within 5 mth of application. Grass and cereals should not be planted until the following autumn, and sugar beet not for one yr
- Before drilling or planting a following crop mouldboard plough to at least 150 mm

Environmental safety

- Flammable
- Dangerous to fish or other aquatic life. Do not contaminate surface waters or ditches with chemical or used container
- LERAP Category B

Hazard classification and safety precautions

Hazard R03, R04, R07d
Risk phrases R03a, R03b, R03c, R04a, R04b, R04c, R04g
Operator protection A, C, H; U05a, U08, U13, U19, U20b
Environmental protection E01, E13b, E16a, E16b
Consumer protection C03
Storage and disposal E26, E30a, E31b
Medical advice M05

305 magnesium phosphide

A phosphine generating compound used to control insect pests in stored commodities

Products

Degesch Plates	Rentokil	56% w/w	GE	07603

Uses

Insect pests in ***stored grain***.

Efficacy guidance

- Product acts as fumigant by releasing poisonous hydrogen phosphide gas on contact with moisture in the air
- Place plates on the floor or wall of the building or on the surface of the commodity. Exposure time varies depending on temperature and pest. See label

Restrictions

- Only to be used by professional operators trained in the use of magnesium phosphide and familiar with the precautionary measures to be observed. See label for full precautions

Environmental safety

- Highly flammable
- Prevent access to buildings under fumigation by livestock, pets and other non-target mammals and birds
- Dangerous to fish or other aquatic life. Do not contaminate surface waters or ditches with chemical or used container
- Keep in original container, tightly closed, in a safe place, under lock and key
- Do not allow plates or their spent residues to come into contact with food other than raw cereal grains
- Remove used plates after treatment. Do not bulk spent plates and residues: spontaneous ignition could result
- Keep livestock out of treated areas

Hazard classification and safety precautions

Hazard R01, R03, R07c
Risk phrases R01b, R01c, R03a
Operator protection A, D, H; U01, U05b, U07, U13, U19, U20a
Environmental protection E01, E07, E13b; E02 (4 h min)
Consumer protection C03
Storage and disposal E26, E29, E30b, E32b, E34
Medical advice M04

306 malathion

A broad-spectrum contact organophosphorus insecticide and acaricide

Products

Fyfanon 440	Headland	440 g/l	EC	11134

Uses

Aphids in ***carnations***, ***chrysanthemums***, ***dahlias***, ***gladioli***, ***outdoor cyclamen***, ***protected carnations***, ***protected chrysanthemums***, ***protected cyclamen***, ***protected dahlias***, ***protected gladioli***, ***protected roses***, ***protected tulips***, ***roses***, ***tulips***.

Efficacy guidance

- Spray when pest first seen and repeat as necessary, usually at 7-14 d intervals
- Number and timing of sprays vary with crop and pest. See label for details
- Repeat spray routinely for scale insect and whitefly control in glasshouses
- Where aphids resistant to malathion occur control is unlikely to be satisfactory

Restrictions

- Malathion is an anticholinesterase organophosphorus compound. Do not use if under medical advice not to work with such compounds
- Do not use on antirrhinums, crassula, ferns, fuchsias, gerberas, petunias, pileas, sweet peas or zinnias as damage may occur
- Do not apply by tractor mounted/drawn or broadcast air assisted sprayers

Environmental safety

- Dangerous to bees. Do not apply to crops in flower or to those in which bees are actively foraging. Do not apply when flowering weeds are present
- Extremely dangerous to fish or other aquatic life. Do not contaminate surface waters or ditches with chemical or used container

Hazard classification and safety precautions

Operator protection A, H, M, P; U08, U19, U20b
Environmental protection E12c, E13a, E22b
Storage and disposal E26, E30a, E31a, E34
Medical advice M01

SECTION 4

307 maleic hydrazide

A pyridazinone plant growth regulator suppressing sprout and bud growth

See also dicamba + maleic hydrazide + MCPA

Products

1	Fazor	Dow	60% w/w	SG	05558
2	Mazide 25	Vitax	250 g/l	SL	02067
3	Regulox K	Bayer Environ.	250 g/l	SL	09937
4	Rouge	Nufarm UK	60% w/w	WB	11090
5	Royal MH	Certis	180 g/l	SL	11062
6	Royal MH 180	Nufarm UK	180 g/l	SL	10982
7	Source II	Chiltern	60% w/w	WB	08314

Uses

Growth retardation in ***hedges*** [2]. Growth suppression in ***amenity grass*** [2, 3, 6]; ***areas around farm buildings*** [6]; ***carrots*** *(off-label)*, ***parsnips*** *(off-label)* [1]; ***golf courses***, ***road verges*** [2, 6]; ***grass near water*** [3, 6]; ***industrial sites***, ***roadside grass***, ***waste ground*** [3]. Sprout suppression in ***bulb onions*** [1, 2, 4, 5, 7]; ***ware potatoes*** [1, 7]. Sucker inhibition in ***amenity trees and shrubs*** [2, 3]. Volunteer suppression in ***ware potatoes*** [1, 4, 7].

Specific Off-Label Approvals (SOLAs)

- ***carrots***, ***parsnips*** *(OLA 012159) Dec 2008* [1]

Approval information

- Maleic hydrazide included in Annex I under EC Directive 91/414
- Approved for use on grass near water [6] . See notes in Section 1 on use of herbicides in or near water

Efficacy guidance

- Apply to grass at any time of yr when growth active, best when growth starting in Apr-May and repeated when growth recommences
- Uniform coverage and 8 h dry weather necessary for effective results
- Mow 2-3 d before and 5-10 d after spraying for best results. Need for mowing reduced for up to 6 wk
- Rain or irrigation within 24 h may reduce effectiveness on onions
- Accurate timing essential for good results on potatoes but rain or irrigation within 24 h may reduce effectiveness
- When used for suppression of volunteer potatoes treatment may also give some suppression of sprouting in store but separate treatment will be necessary if sprouting occurs
- To control suckers wet trunks thoroughly, especially pruned and basal bud areas [2]

Restrictions

- Maximum number of treatments 2 per yr on amenity grass, land not intended for cropping and land adjacent to aquatic areas; 1 per crop on onions, potatoes and on or around tree trunks
- Do not apply in drought or when crops are suffering from pest, disease or herbicide damage. Do not treat fine turf or grass seeded less than 8 mth previously
- Do not treat potatoes within 3 wk of applying a haulm desiccant or if temperatures above 26°C
- Consult processor before use on potato crops for processing
- Do not apply by knapsack sprayer [1]

Crop-specific information

- Latest use: 3 wk before haulm destruction for potatoes; before 50% necking for onions
- HI onions 4 d (7 d [1]); potatoes, carrots, parsnips 3 wk
- Apply to onions at 10% necking and not later than 50% necking stage when the tops are still green
- Only treat onions in good condition and properly cured, and do not treat more than 2 wk before maturing. Treated onions may be stored until Mar but must then be removed to avoid browning
- Apply to second early or maincrop potatoes at least 3 wk before haulm destruction
- Only treat potatoes of good keeping quality; not on seed, first earlies or crops grown under polythene
- Spray hawthorn hedges in full leaf, privet 7 d after cutting, in Apr-May [2]
- May be applied to grass along water courses but not to water surface [3, 6]

Environmental safety

- Only apply to grass not to be used for grazing

- Do not use treated water for irrigation purposes within 3 wk of treatment or until concentration in water falls below 0.02 ppm [3, 5, 6]
- Maximum permitted concentration in water 2 ppm [3, 6]
- Do not dump surplus product in water or ditch bottoms [3, 5, 6]
- Avoid drift onto nearby vegetables, flowers or other garden plants

Hazard classification and safety precautions

Hazard R04 [4, 7]

Risk phrases R04d [4, 7]

Operator protection A [1, 4, 7]; C [4, 7]; H [1]; U02, U05a, U20a, U22a [4, 7]; U08 [1]; U13, U14, U15, U19 [7]; U20b [1, 6]; U20c [2, 3, 5]

Environmental protection E01 [4, 7]; E15 [1-7]; E19, E21 [3, 5, 6]

Consumer protection C03 [4, 7]

Storage and disposal E26 [3, 4, 7]; E30a [1-7]; E31a [2, 5, 6]; E31b [3]; E32a [1]

308 mancozeb

A protective dithiocarbamate fungicide for potatoes and other crops

See also *benalaxyl + mancozeb*
chlorothalonil + mancozeb
cymoxanil + mancozeb
dimethomorph + mancozeb
fenamidone + mancozeb

Products

1	Barclay Manzeb 455	Barclay	455 g/l	SC	11358
2	Dithane 945	Interfarm	80% w/w	WP	09897
3	Dithane 945	Dow	80% w/w	WP	10985
4	Dithane NT Dry Flowable	Interfarm	75% w/w	WG	09898
5	Dithane NT Dry Flowable	Dow	75% w/w	WG	10986
6	Dithane Superflo	Dow	455 g/l	SC	10799
7	Karamate Dry Flo Newtec	Landseer	75% w/w	WG	09759
8	Laminator DG	Interfarm	75% w/w	WG	11073
9	Laminator FL	Interfarm	455 g/l	SC	11072
10	Laminator WP	Interfarm	80% w/w	WP	11071
11	Micene 80	Sipcam	80% w/w	WP	09112
12	Micene DF	Sipcam	77% w/w	WG	09957
13	Penncozeb WDG	Nufarm UK	77% w/w	WG	11622
14	Quell Flo	Interfarm	455 g/l	SC	09894

Uses

Black spot in ***roses*** [7]. Blight in ***potatoes*** [1-6, 8-14]. Brown rust in ***spring barley***, ***winter barley*** [11-13]; ***spring wheat***, ***winter wheat*** [2-6, 8-10, 12-14]. Currant leaf spot in ***blackcurrants***, ***gooseberries*** [7]. Downy mildew in ***bulb onions*** *(off-label)*, ***garlic*** *(off-label)* [3, 12]; ***lettuce***, ***protected lettuce*** [7]; ***winter oilseed rape*** [2-6, 8-10, 14]. Fire in ***tulips*** [7]. Fungus diseases in ***grapevines*** *(off-label)* [7]. Net blotch in ***spring barley*** [11-13]; ***winter barley*** [6, 9, 11-14]. Ray blight in ***chrysanthemums*** [7]. Rhynchosporium in ***spring barley*** [12, 13]; ***winter barley*** [2-6, 8-10, 12-14]. Rust in ***carnations***, ***geraniums***, ***roses*** [7]. Scab in ***apples*** [7, 11-13], ***pears*** [7]. Septoria diseases in ***spring wheat***, ***winter wheat*** [12, 13]. Septoria leaf spot in ***spring wheat***, ***winter wheat*** [2-6, 8-11, 14]. Sooty moulds in ***spring barley*** [12, 13]; ***spring wheat***, ***winter wheat*** [2-6, 8-10, 12-14]; ***winter barley*** [6, 9, 12-14]. Yellow rust in ***spring barley***, ***spring wheat***, ***winter barley*** [12, 13]; ***winter wheat*** [2-6, 8-10, 12-14].

Specific Off-Label Approvals (SOLAs)

- ***bulb onions***, ***garlic*** *(OLA 020158) Jun 2006* [12]
- ***bulb onions***, ***garlic*** *(OLA 021430) Jun 2006* [3]
- ***grapevines*** *(OLA 021426) Dec 2008* [7]

Approval information

- Approved for aerial application on potatoes [1-6, 11-14]. See notes in Section 1

Efficacy guidance

- Mancozeb is a protectant fungicide and will give moderate control, suppression or reduction of the cereal diseases listed if treated before they are established but in many cases mixture with carbendazim is essential to achieve satisfactory results. See labels for details
- May be recommended for suppression or control of mildew in cereals depending on product and tank mix. See label for details

Restrictions

- Maximum number of treatments varies with crop and product used - check labels for details
- Check labels for minimum interval that must elapse between treatments
- On protected lettuce only 2 post-planting applications of mancozeb or of any combination of products containing EBDC fungicide (mancozeb, maneb, thiram, zineb) either as a spray or a dust are permitted within 2 wk of planting out and none thereafter.
- Avoid treating wet cereal crops or those suffering from drought or other stress
- Keep dry formulations away from fire and sparks
- Do not allow dry formulations to become wet during storage as this may reduce effectiveness and create flammable vapours

Crop-specific information

- Latest use: before early milk stage (GS 73) for cereals; before 6 true leaf stage and before 31 Dec for winter oilseed rape.
- HI potatoes 7 d; outdoor lettuce 14 d; apples, bulb onions, garlic 4 wk
- Apply to potatoes before haulm meets across rows (usually mid-Jun) or at earlier blight warning, and repeat every 7-14 d depending on conditions and product used (see label)
- May be used on potatoes up to desiccation of haulm
- On oilseed rape apply as soon as disease develops between cotyledon and 5-leaf stage (GS 1,0-1,5)
- Apply to cereals from 4-leaf stage to before early milk stage (GS 71). Recommendations vary, see labels for details
- Treat winter oilseed rape before 6 true leaf stage (GS 1,6) and before 31 Dec

Environmental safety

- Harmful to fish or other aquatic life. Do not contaminate surface waters or ditches with chemical or used container

Hazard classification and safety precautions

Hazard R04 [1, 3-14]
Risk phrases R04a, R04b [11-13]; R04c [3-5, 8, 10-13]; R04e [1, 3-10, 14]
Operator protection A [1-14]; C [11, 12]; D [3-5, 7, 8, 11, 12]; U05a [1-14]; U08 [1, 6, 7, 9, 11-14]; U13, U19 [11-13]; U20a [2, 11]; U20b [1, 3-10, 12-14]
Environmental protection E01 [1-5, 7, 8, 10-13]; E13c [1-14]
Consumer protection C02 [1] (7 d); C03 [1-14]
Storage and disposal E26 [1, 4-6, 8, 9, 11, 12, 14]; E29, E34 [6, 9, 14]; E30a, E32a [1-14]

309 mancozeb + metalaxyl-M

A systemic and protectant fungicide mixture

Products

Fubol Gold WG	Syngenta	64:4% w/w	WG	10184

Uses

Blight in ***potatoes***. Downy mildew in ***rhubarb*** *(off-label)*. Fungus diseases in ***agastache spp.*** *(off-label)*, ***angelica*** *(off-label)*, ***balm*** *(off-label)*, ***basil*** *(off-label)*, ***bay*** *(off-label)*, ***borage*** *(off-label)*, ***camomile*** *(off-label)*, ***caraway*** *(off-label)*, ***catnip*** *(off-label)*, ***chervil*** *(off-label)*, ***chives*** *(off-label)*, ***clary*** *(off-label)*, ***coriander*** *(off-label)*, ***curry plant*** *(off-label)*, ***dill*** *(off-label)*, ***dragonhead*** *(off-label)*, ***fennel*** *(off-label)*, ***fenugreek*** *(off-label)*, ***feverfew*** *(off-label)*, ***hyssop*** *(off-label)*, ***land cress*** *(off-label)*, ***lavandin*** *(off-label)*, ***lavender*** *(off-label)*, ***lemon verbena*** *(off-label)*, ***lettuce*** *(off-label)*, ***lovage*** *(off-label)*, ***marigold*** *(off-label)*, ***marjoram*** *(off-label)*, ***mint*** *(off-label)*, ***nasturtium*** *(off-label)*, ***nettle*** *(off-label)*, ***oregano*** *(off-label)*, ***origanum heracleoticum*** *(off-label)*, ***parsley*** *(off-label)*, ***protected agastache spp.*** *(off-label)*, ***protected angelica*** *(off-label)*, ***protected balm*** *(off-label)*, ***protected basil*** *(off-label)*, ***protected bay*** *(off-label)*, ***protected borage*** *(off-label)*, ***protected camomile*** *(off-label)*, ***protected caraway*** *(off-label)*, ***protected catnip*** *(off-label)*,

protected chervil *(off-label)*, ***protected chives*** *(off-label)*, ***protected clary*** *(off-label)*, ***protected coriander*** *(off-label)*, ***protected curry plant*** *(off-label)*, ***protected dill*** *(off-label)*, ***protected dragonhead*** *(off-label)*, ***protected fennel*** *(off-label)*, ***protected fenugreek*** *(off-label)*, ***protected feverfew*** *(off-label)*, ***protected hyssop*** *(off-label)*, ***protected land cress*** *(off-label)*, ***protected lavandin*** *(off-label)*, ***protected lavender*** *(off-label)*, ***protected lemon verbena*** *(off-label)*, ***protected lettuce*** *(off-label)*, ***protected lovage*** *(off-label)*, ***protected marigold*** *(off-label)*, ***protected marjoram*** *(off-label)*, ***protected mint*** *(off-label)*, ***protected nasturtium*** *(off-label)*, ***protected nettle*** *(off-label)*, ***protected oregano*** *(off-label)*, ***protected origanum heracleoticum*** *(off-label)*, ***protected parsley*** *(off-label)*, ***protected rocket*** *(off-label)*, ***protected rosemary*** *(off-label)*, ***protected rue*** *(off-label)*, ***protected sage*** *(off-label)*, ***protected salad burnet*** *(off-label)*, ***protected savory*** *(off-label)*, ***protected sorrel*** *(off-label)*, ***protected spike lavender*** *(off-label)*, ***protected tarragon*** *(off-label)*, ***protected thyme*** *(off-label)*, ***protected violet*** *(off-label)*, ***protected woodruff*** *(off-label)*, ***rocket*** *(off-label)*, ***rosemary*** *(off-label)*, ***rue*** *(off-label)*, ***sage*** *(off-label)*, ***salad burnet*** *(off-label)*, ***savory*** *(off-label)*, ***sorrel*** *(off-label)*, ***spike lavender*** *(off-label)*, ***tarragon*** *(off-label)*, ***thyme*** *(off-label)*, ***violet*** *(off-label)*, ***woodruff*** *(off-label)*. Phytophthora fruit rot in ***apples*** *(off-label - applied to orchard floor)*. White blister in ***cabbages*** *(off-label)*, ***protected radishes*** *(off-label)*, ***radishes*** *(off-label)*.

Specific Off-Label Approvals (SOLAs)

- ***agastache spp., angelica, balm, basil, bay, borage, camomile, caraway, catnip, chervil, chives, clary, coriander, curry plant, dill, dragonhead, fennel, fenugreek, feverfew, hyssop, land cress, lavandin, lavender, lemon verbena, lettuce, lovage, marigold, marjoram, mint, nasturtium, nettle, oregano, origanum heracleoticum, parsley, protected agastache spp., protected angelica, protected balm, protected basil, protected bay, protected borage, protected camomile, protected caraway, protected catnip, protected chervil, protected chives, protected clary, protected coriander, protected curry plant, protected dill, protected dragonhead, protected fennel, protected fenugreek, protected feverfew, protected hyssop, protected land cress, protected lavandin, protected lavender, protected lemon verbena, protected lettuce, protected lovage, protected marigold, protected marjoram, protected mint, protected nasturtium, protected nettle, protected oregano, protected origanum heracleoticum, protected parsley, protected rocket, protected rosemary, protected rue, protected sage, protected salad burnet, protected savory, protected sorrel, protected spike lavender, protected tarragon, protected thyme, protected violet, protected woodruff, rocket, rosemary, rue, sage, salad burnet, savory, sorrel, spike lavender, tarragon, thyme, violet, woodruff*** *(OLA 032142) Dec 2008* [1]
- ***apples*** *(applied to orchard floor) (OLA 011610) Dec 2008* [1]
- ***cabbages, protected radishes, radishes*** *(OLA 011610) Dec 2008* [1]
- ***rhubarb*** *(OLA 022011) Aug 2006* [1]

Approval information

- Metalaxyl-M included in Annex I under EC Directive 91/414

Efficacy guidance

- Commence potato blight programme before risk of infection occurs as crops begin to meet along the rows and repeat every 10-14 d according to blight risk. Do not exceed a 14 d interval between sprays
- If infection risk conditions occur earlier than the above growth stage commence spraying potatoes immediately
- Complete the potato blight programme using a protectant fungicide starting no later than 10 d after the last phenylamide spray. At least 2 such sprays should be applied
- Monitor sprayed potato crops and stop using any phenylamide product if active blight is identified. Switch to a tin-based product programme within 7 d and continue up to haulm destruction or harvest
- To minimise the likelihood of development of resistance these products should be used in a planned Resistance Management strategy. See Section 1 for more information

Restrictions

- Maximum number of treatments or maximum total dose varies with crop. See labels

Crop-specific information

- Latest use: before end of active potato haulm growth or before end Aug, whichever is earlier; before transplanting protected herbs; before dormancy in yr of harvest for rhubarb

SECTION 4

- HI 14 d for outdoor herbs; 21 d for protected herbs; 28 d for apples
- After treating early potatoes destroy and remove any remaining haulm after harvest to minimise blight pressure on neighbouring maincrop potatoes

Environmental safety

- Dangerous to fish or other aquatic life. Do not contaminate surface waters or ditches with chemical or used container
- Do not harvest crops for human consumption for at least 7 d after final application

Hazard classification and safety precautions

Hazard R04
Risk phrases R04e
Operator protection A; U05a, U08, U20b
Environmental protection E13b
Consumer protection C02 (7 d)
Storage and disposal E26, E29, E30a, E32a, E34

310 mancozeb + propamocarb hydrochloride

A systemic and contact protectant fungicide for potato blight control

Products

Tattoo	Aventis	301.6:248 g/l	SC	07293

Uses

Blight in ***potatoes***.

Efficacy guidance

- Commence treatment as soon as there is risk of blight infection
- In the absence of a warning treatment should start before the crop meets within the row
- Repeat treatments every 10-14 d depending on degree of infection risk
- Irrigated crops should be regarded as at high risk and treated every 10 d
- Do not spray when rainfall is imminent and apply only to dry foliage
- To complete the spray programme after the end of Aug, make subsequent treatments up to haulm destruction with a protectant fungicide, preferably fentin based
- To minimise the likelihood of development of resistance these products should be used in a planned Resistance Management strategy. See Section 1 for more information

Restrictions

- Maximum number of treatments 5 per crop
- Do not use once blight has become readily visible

Crop-specific information

- HI potatoes 14 d

Environmental safety

- Harmful to fish or other aquatic life. Do not contaminate surface waters or ditches with chemical or used container

Hazard classification and safety precautions

Hazard R04
Risk phrases R04e
Operator protection A, H; U05a, U08, U14, U19, U20a
Environmental protection E01, E13c
Consumer protection C03
Storage and disposal E30a, E31b

311 mancozeb + zoxamide

A protectant fungicide mixture for potatoes

Products

1	Electis 75 WG	Interfarm	66.7:8.3% w/w	WG	10565
2	Roxam 75 WG	Interfarm	66.7:8.3% w/w	WG	10566
3	Unikat 75 WG	Landseer	66.7:8.3% w/w	WG	10567

Uses

Blight in ***potatoes*** [1, 2]. Downy mildew in ***grapevines*** [3].

Efficacy guidance

- Apply as protectant spray on potatoes immediately risk of blight in district or as crops begin to meet along the rows and repeat every 7-14 d according to blight risk [1, 2]
- Do not use if potato blight present in crop. Products are not curative [1, 2]
- Spray irrigated potato crops as soon as possible after irrigation once the crop leaves are dry [1, 2]
- In the absence of an official warning commence treatment of grapevines at the 5-6 leaf stage and repeat at a minimum interval of 10 d [3]
- Increase spray volume with growth of grapevines but avoid excessive application [3]

Restrictions

- Maximum number of treatments 6 per yr for grapevines; 10 per crop for potatoes
- Do not use on grapevines for plant propagation [3]
- Consult processors before use on crops for juice production [3]

Crop-specific information

- HI 7 d for potatoes [1, 2]; 56 d for grapevines [3]

Environmental safety

- Dangerous to fish or other aquatic life. Do not contaminate surface waters or ditches with chemical or used container
- Risk to certain non-target insects or other arthropods. See directions for use
- LERAP Category B [1, 2]
- Broadcast air-assisted LERAP (20 m)
- Keep away from fire and sparks [3]

Hazard classification and safety precautions

Hazard R04
Risk phrases R04e
Operator protection A, H; U05a, U20b
Environmental protection E01, E13b, E16b, E22b; E17a (20 m)
Consumer protection C03
Storage and disposal E26, E29, E30a, E32a

312 maneb

A protectant dithiocarbamate fungicide

Products

1	Maneb 80	Rohm & Haas	80% w/w	WP	01276
2	Trimanzone	Intracrop	80% w/w	WP	09584
3	X-Spor SC	United Phosphorus	480 g/l	SC	08077

Uses

Blight in ***potatoes*** [1-3]. Brown rust in ***spring barley***, ***winter barley*** [1-3]; ***winter wheat*** [1, 3]. Eyespot in ***spring barley***, ***winter barley***, ***winter wheat*** [3]. Net blotch in ***spring barley***, ***winter barley*** [1, 2]. Powdery mildew in ***winter wheat*** [3]. Rhynchosporium in ***spring barley***, ***winter barley*** [3]. Scab in ***apples*** [2]. Septoria in ***spring wheat*** [2]; ***winter wheat*** [1, 2]. Septoria diseases in ***winter wheat*** [3]. Sooty moulds in ***winter wheat*** [3]. White tip in ***leeks*** *(off-label)* [2]. Yellow rust in ***spring barley***, ***winter barley***, ***winter wheat*** [1, 3].

Specific Off-Label Approvals (SOLAs)

- ***leeks*** *(OLA 000849) Dec 2008* [2]

Approval information

- Approved for aerial application on potatoes [1, 3]. See notes in Section 1
- Following implementation of Directive 98/82/EC, approval for use of maneb on several crops was revoked in 1999

Efficacy guidance

- Maneb is a protectant fungicide and will give moderate control, suppression or reduction of the diseases listed if treated before they become established but in many cases mixture with carbendazim is essential to achieve satisfactory results - see labels for details

- Best results on cereals obtained from a programme of a preventative treatment before disease established, an early application at about first node stage (GS 31), a late application after flag leaf emergence (GS 37) and during ear emergence before watery ripe stage (GS 71)

Restrictions

- Maximum number of treatments 2 or 3 per crop (check labels)
- Do not apply if frost or rain expected, if crop wet or suffering from drought or physical or chemical stress
- A minimum of 7 d must elapse between applications to apples, wheat and barley, and 10 d for potatoes
- Keep away from fire and sparks [1, 2]
- Use contents of opened container as soon as possible and do not store opened containers until following season. If product becomes wet, effectiveness may be reduced and inflammable vapour produced [1]

Crop-specific information

- Latest use: before grain milky-ripe (GS 73) for wheat, barley [2]; before flag leaf sheath opening (GS 45) for barley [3]
- HI potatoes 7 d; leeks 14 d; apples 28 d
- Apply to potatoes before blight infection occurs, at blight warning or before haulms meet in row and repeat every 10-14 d
- Apply to apples at bud burst and repeat at 7-10 d intervals to end Jun [2]

Environmental safety

- Harmful to fish or other aquatic life. Do not contaminate surface waters or ditches with chemical or used container [1, 2]

Hazard classification and safety precautions

Hazard R03 [3]; R04 [1-3]
Risk phrases R03c [3]; R04a, R04b [2, 3]; R04c [1-3]; R04e [1]
Operator protection A [1-3]; C [2, 3]; D [2]; U04a [3]; U05a [1, 2]; U08 [1-3]; U13, U19 [2]; U20a [1]; U20b [2, 3]
Environmental protection E01 [1-3]; E13c [1, 2]; E15 [3]
Consumer protection C02 [1] (7 d); C03 [1-3]
Storage and disposal E26, E31b [3]; E29, E34 [1]; E30a [1-3]; E32a [1, 2]
Medical advice M03 [3]

313 MCPA

A translocated phenoxyacetic herbicide for cereals and grassland

See also *2,4-D + dichlorprop-P + MCPA + mecoprop-P*
2,4-D + MCPA
2,4-DB + MCPA
bentazone + MCPA + MCPB
clopyralid + 2,4-D + MCPA
clopyralid + diflufenican + MCPA
clopyralid + fluroxypyr + MCPA
dicamba + dichlorprop-P + ferrous sulphate + MCPA
dicamba + dichlorprop-P + MCPA
dicamba + maleic hydrazide + MCPA
dicamba + MCPA + mecoprop-P
dichlorprop-P + MCPA
dichlorprop-P + MCPA + mecoprop-P

Products

1	Agricorn 500 II	FCC	500 g/l	SL	09155
2	Agritox 50	Nufarm UK	500 g/l	SL	07400
3	Agritox Dry	Nufarm UK	80% w/w	SG	10554
4	Agroxone	Headland	485 g/l	SL	09947
5	BASF MCPA Amine 50	BASF	500 g/l	SL	00209
6	Campbell's MCPA 50	United Phosphorus	500 g/l	SL	00381
7	Circium II	Nufarm UK	485 g/l	SL	11143
8	Headland Spear	Headland	500 g/l	SL	07115

Products – continued

9	HY-MCPA	Agrichem	500 g/l	SL	06293
10	Luxan MCPA 500	Luxan	500 g/l	SL	07470
11	Tasker 75	Headland	750 g/l	SL	10544

Uses

Annual dicotyledons in ***amenity grass*** [2, 3, 7, 8]; ***established grassland, spring barley, spring oats, spring wheat, winter barley, winter oats, winter wheat*** [1-6, 8-11]; ***grass seed crops*** [2-4, 10, 11]; ***land not intended to bear vegetation*** [2]; ***linseed*** [2-4, 9, 11]; ***managed amenity turf*** [2, 3, 8]; ***newly sown grass*** [4, 10]; ***rotational grassland*** [11]; ***spring rye*** [2, 3]; ***undersown barley, undersown oats, undersown wheat*** [5]; ***undersown barley*** *(red clover or grass)*, ***undersown wheat*** *(red clover or grass)* [4, 11]; ***undersown spring cereals, undersown winter cereals*** [9]; ***undersown spring cereals*** *(grass only)*, ***undersown winter cereals*** *(grass only)* [10]; ***winter rye*** [1-3, 5, 6, 8, 10]. Buttercups in ***amenity grass, managed amenity turf*** [3, 7-9]; ***established grassland*** [1, 5, 6, 8, 9]; ***lawns*** [7]. Charlock in ***established grassland*** [2-4, 10, 11]; ***grass seed crops*** [4, 10, 11]; ***linseed*** [4, 9, 11]; ***newly sown grass*** [4, 10]; ***rotational grassland*** [9, 11]; ***spring barley, spring oats, spring wheat, winter barley, winter oats, winter wheat*** [1-6, 8-11]; ***spring rye*** [2, 3]; ***undersown barley, undersown oats, undersown wheat*** [5]; ***undersown barley*** *(red clover or grass)*, ***undersown wheat*** *(red clover or grass)* [4, 11]; ***undersown spring cereals*** *(grass only)*, ***undersown winter cereals*** *(grass only)* [10]; ***winter rye*** [1-3, 5, 6, 8, 10]. Dandelions in ***amenity grass*** [3]; ***established grassland*** [1, 5, 6, 9]; ***lawns, non-crop areas, road verges*** [7]; ***managed amenity turf*** [3, 7]. Docks in ***established grassland*** [1, 5, 6, 8, 9]. Fat hen in ***amenity grass, hard surfaces, managed amenity turf, natural surfaces not intended to bear vegetation, permeable surfaces overlying soil*** [9]; ***established grassland*** [2-4, 10, 11]; ***grass seed crops*** [4, 10, 11]; ***linseed*** [4, 9, 11]; ***newly sown grass*** [4, 10]; ***rotational grassland*** [9, 11]; ***spring barley, spring oats, spring wheat, winter barley, winter oats, winter wheat*** [1-6, 8-11]; ***spring rye*** [2, 3]; ***undersown barley, undersown oats, undersown wheat*** [5]; ***undersown barley*** *(red clover or grass)*, ***undersown wheat*** *(red clover or grass)* [4, 11]; ***undersown spring cereals*** *(grass only)*, ***undersown winter cereals*** *(grass only)* [10]; ***winter rye*** [1-3, 5, 6, 8, 10]. Hemp-nettle in ***established grassland*** [2-4, 10, 11]; ***grass seed crops*** [4, 10, 11]; ***linseed*** [4, 9, 11]; ***newly sown grass*** [4, 10]; ***rotational grassland*** [11]; ***spring barley, spring oats, spring wheat, winter barley, winter oats, winter wheat*** [1-6, 8-11]; ***spring rye*** [2, 3]; ***undersown barley, undersown oats, undersown wheat*** [5]; ***undersown barley*** *(red clover or grass)*, ***undersown wheat*** *(red clover or grass)* [4, 11]; ***undersown spring cereals*** *(grass only)*, ***undersown winter cereals*** *(grass only)* [10]; ***winter rye*** [1-3, 5, 6, 8, 10]. Perennial dicotyledons in ***amenity grass*** [2, 3, 7, 8]; ***established grassland, spring barley, spring wheat, winter barley, winter wheat*** [1-6, 8-11]; ***grass seed crops*** [4, 10, 11]; ***land not intended to bear vegetation*** [2]; ***linseed, undersown barley*** *(red clover or grass)*, ***undersown wheat*** *(red clover or grass)* [4, 11]; ***managed amenity turf*** [2, 3, 8]; ***newly sown grass*** [4, 10]; ***rotational grassland*** [11]; ***spring oats, winter oats*** [2-6, 8-11]; ***spring rye*** [2, 3]; ***undersown spring cereals*** *(grass only)*, ***undersown winter cereals*** *(grass only)* [10]; ***winter rye*** [1-3, 5, 6, 8, 10]. Plantains in ***amenity grass, established grassland, grass seed crops, managed amenity turf*** [9]. Ragwort in ***established grassland*** [8]. Rushes in ***established grassland*** [8]. Stinging nettle in ***hard surfaces, natural surfaces not intended to bear vegetation, permeable surfaces overlying soil*** [9]; ***industrial sites, land not intended to bear vegetation*** [8]; ***non-crop areas, road verges*** [7]. Thistles in ***established grassland*** [5, 6, 8, 9]; ***hard surfaces, natural surfaces not intended to bear vegetation, permeable surfaces overlying soil*** [9]; ***industrial sites, land not intended to bear vegetation*** [8]; ***non-crop areas, road verges*** [7]. Wild radish in ***established grassland*** [2-4, 10, 11]; ***grass seed crops*** [4, 10, 11]; ***linseed*** [4, 9, 11]; ***newly sown grass*** [4, 10]; ***rotational grassland*** [11]; ***spring barley, spring oats, spring wheat, winter barley, winter oats, winter wheat*** [1-6, 8-11]; ***spring rye*** [2, 3]; ***undersown barley, undersown oats, undersown wheat*** [5]; ***undersown barley*** *(red clover or grass)*, ***undersown wheat*** *(red clover or grass)* [4, 11], ***undersown spring cereals*** *(grass only)*, ***undersown winter cereals*** *(grass only)* [10]; ***winter rye*** [1-3, 5, 6, 8, 10].

Efficacy guidance

- Best results achieved by application to weeds in seedling to young plant stage under good growing conditions when crop growing actively
- Spray perennial weeds in grassland before flowering. Most susceptible growth stage varies between species. See label for details
- Do not spray during cold weather, drought, if rain or frost expected or if crop wet

Restrictions

- Maximum number of treatments normally 1 per crop or yr except grass (2 per yr) for some products. See label
- Do not treat grass within 3 mth of germination and preferably not in the first yr of a direct sown ley or after reseeding
- Do not use on cereals before undersowing
- Do not roll, harrow or graze for a few days before or after spraying; see label
- Do not use on grassland where clovers are an important part of the sward
- Do not use on any crop suffering from stress or herbicide damage
- Avoid spray drift onto nearby susceptible crops

Crop-specific information

- Latest use: before 1st node detectable (GS 31) for cereals; 4-6 wk before heading for grass seed crops; before crop 15-25 cm high for linseed
- Apply to winter cereals in spring from fully tillered, leaf sheath erect stage to before first node detectable (GS 31)
- Apply to spring barley and wheat from 5-leaves unfolded (GS 15), to oats from 1-leaf unfolded (GS 11) to before first node detectable (GS 31)
- Apply to cereals undersown with grass after grass has 2-3 leaves unfolded
- Recommendations for crops undersown with legumes vary. Red clover may withstand low doses after 2-trifoliate leaf stage, especially if shielded by taller weeds but white clover is more sensitive. See label for details
- Apply to grass seed crops from 2-3 leaf stage to 5 wk before head emergence
- Temporary wilting may occur on linseed but without long term effects

Following crops guidance

- Do not direct drill brassicas or legumes within 6 wk of spraying grassland

Environmental safety

- MCPA is active at low concentrations. Take extreme care to avoid drift onto neighbouring crops, especially beet crops, brassicas, most market garden crops including lettuce and tomatoes under glass, pears and vines
- Harmful to fish or other aquatic life. Do not contaminate surface waters or ditches with chemical or used container [1, 5]
- Keep livestock out of treated areas until foliage of poisonous weeds such as ragwort has died and become unpalatable

Hazard classification and safety precautions

Hazard R03 [1-10]; R04 [3, 5, 8-10]
Risk phrases R03a [7, 8, 10]; R03c [1-10]; R04b [3, 5, 8, 10]; R04d [3, 5, 8-10]
Operator protection A, C [1-10]; U02 [10]; U05a, U08 [1-10]; U20a [1, 6]; U20b [2-5, 7-10]
Environmental protection E01, E07 [1-10]; E13c [1, 5]; E15 [2-4, 6-10]
Consumer protection C03 [1-10]
Storage and disposal E26, E34 [1, 2, 4-10]; E30a [1-6, 8-10]; E31a [1, 2, 4, 6, 9]; E31b [7, 8, 10]; E31c [5]; E32a [3]
Medical advice M03 [1-10]

314 MCPA + MCPB

A translocated herbicide for undersown cereals, grassland and various legumes

Products

1	Bellmac Plus	United Phosphorus	38:262 g/l	SL	07521
2	Impetus	Headland	25:275 g/l	SL	11021
3	Trifolex-Tra	BASF	34:216 g/l	SL	10396
4	Tropotox Plus	Nufarm UK	37.5:262.5 g/l	SL	11142

Uses

Annual dicotyledons in ***all cereals***, ***clover seed crops***, ***direct-sown seedling clovers***, ***dredge corn containing peas***, ***established leys*** [4]; ***established grassland***, ***peas*** [3]; ***permanent pasture*** [2, 4]; ***red clover***, ***spring barley***, ***spring oats***, ***spring wheat***, ***white clover***, ***winter barley***, ***winter oats***, ***winter wheat*** [2]; ***rotational grassland*** [1-3]; ***sainfoin*** [1, 4]; ***undersown spring cereals***, ***undersown winter cereals*** [1-4]. Perennial dicotyledons in ***all cereals***, ***clover***

seed crops, direct-sown seedling clovers, dredge corn containing peas, established leys [4]; ***established grassland, peas*** [3]; ***permanent pasture*** [2, 4]; ***red clover, spring barley, spring oats, spring wheat, white clover, winter barley, winter oats, winter wheat*** [2]; ***rotational grassland*** [1-3]; ***sainfoin*** [1, 4]; ***undersown spring cereals, undersown winter cereals*** [1-4].

Efficacy guidance

- Best results achieved by application to weeds in seedling to young plant stage under good growing conditions when crop growing actively. Spray perennials when adequate leaf surface before flowering. Retreatment often needed in following year
- Spray leys and sainfoin before crop provides cover for weeds
- Rain, cold or drought may reduce effectiveness

Restrictions

- Maximum number of treatments 1 per crop or yr
- Do not spray clovers for seed
- Do not spray peas [1]
- Do not roll or harrow for a few days before or after spraying

Crop-specific information

- Latest use: before first node detectable (GS 31) for cereals; before flower buds visible for peas
- Apply to cereals from 2-expanded leaf stage to before jointing (GS 12-30) and, where undersown, after 1-trifoliate leaf stage of clover
- Apply to direct-sown seedling clover after 1-trifoliate leaf stage
- Apply to mature white clover for fodder at any stage. Do not spray red clover after flower stalk has begun to form
- Apply to sainfoin after first trifoliate leaf stage

Environmental safety

- Take extreme care to avoid drift onto neighbouring crops, especially beet crops, brassicas, most market garden crops including lettuce and tomatoes under glass, pears and vines
- Harmful to fish or other aquatic life. Do not contaminate surface waters or ditches with chemical or used container
- Keep livestock out of treated areas until foliage of any poisonous weeds such as ragwort has died and become unpalatable

Hazard classification and safety precautions

Hazard R04 [2]
Risk phrases R04b, R04d [2]
Operator protection A, C [2]; U05a [2-4]; U08, U20b [1-4]
Environmental protection E01, E07 [2-4]; E07 [1] (2 wk); E13c [1-4]
Consumer protection C03 [2, 3]
Storage and disposal E26, E30a, E31b [1-4]; E34 [1-3]
Medical advice M03 [2]

315 MCPA + mecoprop-P

A translocated selective herbicide for amenity grass

Products

Greenmaster Extra	Scotts	0.49:0.29 % w/w	GR	11563

Uses

Annual dicotyledons in ***managed amenity turf***. Perennial dicotyledons in ***managed amenity turf***.

Efficacy guidance

- Apply from Apr to Sep, when weeds growing actively and have large leaf area available for chemical absorption

Restrictions

- Maximum total dose 105 g/sq m per yr. The total amount of mecoprop-P applied in a single yr must not exceed the maximum total dose approved for any single product for use on turf
- Avoid contact with cultivated plants
- Do not use first 4 mowings as compost or mulch unless composted for 6 mth
- Do not treat newly sown or turfed areas for at least 6 mth

SECTION 4

- Do not reseed bare patches for 8 wk after treatment
- Do not apply when heavy rain expected or during prolonged drought. Irrigate after 1-2 d unless rain has fallen
- Do not mow within 2-3 d of treatment
- Treat areas planted with bulbs only after the foliage has died down
- Avoid walking on treated areas until it has rained or irrigation has been applied

Crop-specific information

- Granules contain NPK fertilizer to encourage grass growth

Environmental safety

- Take extreme care to avoid drift onto neighbouring crops, especially beet crops, brassicas, most market garden crops including lettuce and tomatoes under glass, pears and vines
- Harmful to fish or other aquatic life. Do not contaminate surface waters or ditches with chemical or used container
- Keep livestock out of treated areas for at least 2 wk and until foliage of any poisonous weeds such as ragwort has died and become unpalatable

Hazard classification and safety precautions

Operator protection A, C, H, M; U20a
Environmental protection E07 (2 wk); E13c
Storage and disposal E30a, E32a

316 MCPB

A translocated phenoxybutyric herbicide

See also *bentazone + MCPA + MCPB*
bentazone + MCPB
MCPA + MCPB

Products

1	Bellmac Straight	United Phosphorus	400 g/l	SL	07522
2	Butoxone	Headland	400 g/l	SL	10501
3	Tropotox	Nufarm UK	400 g/l	SL	11141

Uses

Annual dicotyledons in ***blackberries, combining peas, gooseberries, loganberries, permanent pasture, raspberries, spring barley, spring oats, spring wheat, vining peas, white clover*** *(seed crops)*, ***winter barley, winter oats, winter wheat*** [2]; ***blackcurrants*** [1-3]; ***clover seed crops, peas*** [1, 3]; ***rotational grassland, undersown spring cereals, undersown winter cereals*** [1, 2]. Perennial dicotyledons in ***blackberries, combining peas, gooseberries, loganberries, permanent pasture, raspberries, spring barley, spring oats, spring wheat, vining peas, white clover*** *(seed crops)*, ***winter barley, winter oats, winter wheat*** [2]; ***blackcurrants*** [1-3]; ***clover seed crops, peas*** [1, 3]; ***rotational grassland, undersown spring cereals, undersown winter cereals*** [1, 2].

Efficacy guidance

- Best results achieved by spraying young seedling weeds in good growing conditions
- Best results on perennials by spraying before flowering
- Effectiveness may be reduced by rain within 12 h, by very cold or dry conditions

Restrictions

- Maximum number of treatments 1 per crop or yr. One label allows two treatments on blackcurrants [3]
- Do not roll or harrow for 7-10 d before or after treatment (check label)

Crop-specific information

- Latest use: first node detectable stage (GS 31) for cereals; before flower buds appear in terminal leaf (GS 201) for peas; before flower buds form for clover; before weeds damaged by frost for cane and bush fruit
- Apply to undersown cereals from 2-leaves unfolded to first node detectable (GS 12-31), and after first trifoliate leaf stage of clover
- Red clover seedlings may be temporarily damaged but later growth is normal

- Apply to white clover seed crops in Mar to early Apr, not after mid-May, and allow 3 wk before cutting and closing up for seed
- Apply to peas from 3-6 leaf stage but before flower bud detectable (GS 103-201). Consult PGRO (see Appendix 2) or label for information on susceptibility of cultivars. Do not treat peas grown for seed [2]
- Do not use on leguminous crops not mentioned on the label
- Apply to cane and bush fruit after harvest and after shoot growth ceased but before weeds are damaged by frost, usually in late Aug or Sep; direct spray onto weeds as far as possible

Environmental safety

- Take extreme care to avoid drift onto neighbouring sensitive crops
- Harmful to fish or other aquatic life. Do not contaminate surface waters or ditches with chemical or used container
- Keep livestock out of treated areas until foliage of any poisonous weeds such as ragwort has died and become unpalatable

Hazard classification and safety precautions

Hazard R03 [1-3]; R04 [2]
Risk phrases R03a [1, 3]; R03c [1-3]; R04b, R04d [2]
Operator protection A, C; U05a, U08, U20b [1-3]; U19 [3]
Environmental protection E01, E13c [1-3]; E07 [1]; E07 [2] (2 wk); E07 [3]
Consumer protection C03
Storage and disposal E26 [1, 2]; E30a, E31b, E34 [1-3]
Medical advice M03

317 mecoprop-P

A translocated phenoxypropionic herbicide for cereals and grassland

See also *2,4-D + dichlorprop-P + MCPA + mecoprop-P*
2,4-D + mecoprop-P
carfentrazone-ethyl + mecoprop-P
dicamba + MCPA + mecoprop-P
dicamba + mecoprop-P
dichlorprop-P + MCPA + mecoprop-P
fluroxypyr + mecoprop-P
MCPA + mecoprop-P

Products

1	Clenecorn Super	FCC	600 g/l	SL	09818
2	Clovotox	Bayer Environ.	142.5 g/l	SL	09928
3	Compitox Plus	Nufarm UK	600 g/l	SL	10077
4	Duplosan KV	BASF	600 g/l	SL	09431
5	Isomec	Nufarm UK	600 g/l	SL	09881
6	Landgold Mecoprop-P 600	Landgold	600 g/l	SL	09014
7	Optica	Headland	600 g/l	SL	09963

Uses

Annual dicotyledons in ***established grassland*** [6, 7]; ***grass seed crops, young leys*** [1, 3-5, 7]; ***grassland*** [1, 3-5]; ***managed amenity turf*** [2, 3, 5, 7]; ***rotational grassland*** [6]; ***spring barley, spring oats, spring wheat, winter barley, winter oats, winter wheat*** [1, 3-7]. Chickweed in ***established grassland*** [6, 7]; ***grass seed crops, young leys*** [1, 3-5, 7]; ***grassland*** [1, 3-5]; ***managed amenity turf*** [2, 3, 5, 7]; ***rotational grassland*** [6]; ***spring barley, spring oats, spring wheat, winter barley, winter oats, winter wheat*** [1, 3-7]. Cleavers in ***established grassland*** [6, 7]; ***grass seed crops, young leys*** [1, 3-5, 7]; ***grassland*** [1, 3-5]; ***managed amenity turf*** [3, 5, 7]; ***rotational grassland*** [6]; ***spring barley, spring oats, spring wheat, winter barley, winter oats, winter wheat*** [1, 3-7]. Clovers in ***managed amenity turf*** [2]. Perennial dicotyledons in ***established grassland, rotational grassland*** [6]; ***grass seed crops, grassland, young leys*** [1, 3-5]; ***managed amenity turf*** [2, 3, 5]; ***spring barley, spring oats, spring wheat, winter barley, winter oats, winter wheat*** [1, 3-6].

Approval information

- Mecoprop-P included in Annex I under EC Directive 91/414

SECTION 4

Efficacy guidance

- Best results achieved by application to seedling weeds which have not been frost hardened, when soil warm and moist and expected to remain so for several days

Restrictions

- Maximum number of treatments normally 1 per crop for spring cereals and newly sown grass; 2 per crop for winter cereals and established grass. Check labels for details
- The total amount of mecoprop-P applied in a single yr must not exceed the maximum total dose approved for any single product for the crop/situation
- Do not spray cereals undersown with clovers or legumes or to be undersown with legumes or grasses
- Do not spray grass seed crops within 5 wk of seed head emergence
- Do not spray crops suffering from herbicide damage or physical stress
- Do not spray during cold weather, periods of drought, if rain or frost expected or if crop wet
- Do not roll or harrow for 7 d before or after treatment

Crop-specific information

- Latest use: generally before 1st node detectable (GS 31) for spring cereals and before 3rd node detectable (GS 33) for winter cereals, but individual labels vary; 5 wk before emergence of seed head for grass seed crops
- Spray winter cereals from 1 leaf stage in autumn up to and including first node detectable in spring (GS 10-31) or up to second node detectable (GS 32) if necessary. Apply to spring cereals from first fully expanded leaf stage (GS 11) but before first node detectable (GS 31)
- Spray cereals undersown with grass after grass starts to tiller
- Spray newly sown grass leys when grasses have at least 3 fully expanded leaves and have begun to tiller. Any clovers will be damaged

Environmental safety

- Take extreme care to avoid drift onto neighbouring crops, especially beet crops, brassicas, most market garden crops including lettuce and tomatoes under glass, pears and vines
- Harmful to fish or other aquatic life. Do not contaminate surface waters or ditches with chemical or used container [1-5, 7]
- Do not contaminate surface waters or ditches with chemical or used container [6]
- Keep livestock out of treated areas for at least 2 wk and until foliage of any poisonous weed, such as ragwort, has died and become unpalatable

Hazard classification and safety precautions

Hazard R03 [1, 3-7]; R04 [1-7]
Risk phrases R03a [1, 3, 5, 6]; R03c [1, 3-7]; R04a [1-3, 6, 7]; R04b [1, 3, 4, 6, 7]; R04d [3-5]
Operator protection A, C [1-7]; H [2, 4, 7]; M [4, 7]; U05a, U08 [1-7]; U13 [3]; U20a [6]; U20b [1-5, 7]
Environmental protection E01 [1-7]; E07 [1-4]; E07 [5] (2 wk); E07 [6, 7]; E13c [1-5, 7]; E15 [3, 6]
Consumer protection C03
Storage and disposal E26 [1-7]; E30a [1-6]; E31a [3]; E31b [1, 2, 5-7]; E31c [4]; E34 [1, 3-7]
Medical advice M03 [1, 3-7]

318 mecoprop-P + metsulfuron-methyl

A translocated, contact and residual herbicide mixture for cereals

Products

Headland Neptune	Headland	600 g/l:20% w/w	KK	10230

Uses

Annual dicotyledons in ***spring barley***, ***spring wheat***, ***winter barley***, ***winter wheat***. Black bindweed in ***spring barley***, ***spring wheat***, ***winter barley***, ***winter wheat***. Chickweed in ***spring barley***, ***spring wheat***, ***winter barley***, ***winter wheat***. Cleavers in ***spring barley***, ***spring wheat***, ***winter barley***, ***winter wheat***. Knotgrass in ***spring barley***, ***spring wheat***, ***winter barley***, ***winter wheat***. Mayweeds in ***spring barley***, ***spring wheat***, ***winter barley***, ***winter wheat***. Speedwells in ***spring barley***, ***spring wheat***, ***winter barley***, ***winter wheat***.

Approval information

- Mecoprop-P and metsulfuron-methyl included in Annex I under EC Directive 91/414

Efficacy guidance

- Best results obtained from applications to small actively growing weeds
- Ensure good spray cover for optimum activity
- Weed control may be reduced under dry soil conditions
- Metsulfuron-methyl is a member of the ALS-inhibitor group of herbicides and products should be used in a planned Resistance Management strategy. See Section 1 for more information

Restrictions

- Maximum number of treatments 1 per crop. The total amount of mecoprop-P applied in a single yr must not exceed the maximum total dose approved for that crop for any single product
- Product must only be used after 1 Feb and after crop has three leaves
- Do not apply to any crop under stress from drought, waterlogging, low temperatures, pest or disease attack, or nutrient deficiency
- Do not use on undersown cereals
- Do not apply within 7 d of rolling or if rain is expected within 6 h
- Do not spray in tank mixture, or in sequence, with a product containing any other sulfonylurea herbicide except as directed on the label

Crop-specific information

- Latest use: first node detectable (GS 31) for spring sown cereals; second node detectable (GS 32) for winter sown cereals
- Winter barley variety Igri may be damaged if treated during periods of rapid growth and must not be treated before the leaf sheath erect stage (GS 31)

Following crops guidance

- In the event of failure of a treated crop sow only wheat within 3 mth of treatment. See label for restrictions that apply when imazamethabenz-methyl or amidosulfuron have been used on the same crop
- Only cereals, oilseed rape or grass may be sown to follow a treated crop in the same calendar yr. Any crop may be planted the following spring

Environmental safety

- Extremely dangerous to fish or other aquatic life. Do not contaminate surface waters or ditches with chemical or used container
- LERAP Category B
- Take extreme care to avoid damage by drift onto broad leaved plants outside the target area, or onto adjacent cropping land or water courses

Hazard classification and safety precautions

Hazard R03, R04
Risk phrases R03a, R03c, R04a, R04b, R04d
Operator protection A, C, H, M; U05a, U08, U19, U20b
Environmental protection E01, E13a, E16a, E16b
Consumer protection C03
Storage and disposal E30a, E31b, E34
Medical advice M03

319 mefluidide

A sulfonanilide plant growth regulator for suppression of grass

Products

Embark Lite	Intracrop	24 g/l	SL	08749

Uses

Growth suppression in ***amenity grass, managed amenity turf***.

Efficacy guidance

- Apply when grass growth active in Apr/May and repeat when growth re-commences
- Grass growth suppressed for up to 8 wk
- Mow (not below 2 cm) 2-3 d before or 5-10 d after spraying
- Only apply to dry foliage but do not use in drought
- A rainfree period of 8 h is needed after spraying
- May be tank-mixed with suitable 2,4-D or mecoprop product for control of dicotyledons

Restrictions

- Maximum number of treatments 2 per yr
- Do not use on fine turf areas or on turf established for less than 4 mth
- Do not use on grass suffering from herbicide damage
- Do not reseed within 2 wk after application

Crop-specific information

- Delay application for 3-4 wk after application of high-nitrogen fertilizer
- Treated turf may appear less dense and be temporarily discoloured but normal colour returns in 3-4 wk

Environmental safety

- Treated grass must not be used for animal consumption

Hazard classification and safety precautions

Operator protection U05a, U20c
Environmental protection E01, E15
Storage and disposal E26, E30a, E31b

320 mepanipyrim

An anilinopyrimidine fungicide for use on strawberries

Products

1	Frupica	Certis	50% w/w	WP	10944
2	Frupica	Certis	50% w/w	WP	11178

Uses

Botrytis in ***protected strawberries***, ***strawberries***.

Efficacy guidance

- Product is protectant and should be applied as a preventative spray when conditions favourable for Botrytis development occur
- To maintain Botrytis control use as part of a programme with other fungicides that control the disease
- To minimise the possibility of development of resistance adopt resistance management procedures by using products from different chemical groups as part of a mixed spray programme

Restrictions

- Maximum number of treatments 2 per crop (including other anilinopyrimidine products)
- Consult processor before use on crops for processing
- Use spray mixture immediately after preparation

Crop-specific information

- HI 3 d

Environmental safety

- Dangerous to fish or other aquatic life. Do not contaminate surface waters or ditches with chemical or used container
- LERAP Category B

Hazard classification and safety precautions

Operator protection A, H; U20c
Environmental protection E13b, E16a, E16b
Storage and disposal E26, E32a

321 mepiquat chloride

A quaternary ammonium plant growth regulator available only in mixtures

See also *2-chloroethylphosphonic acid + mepiquat chloride*
chlormequat + 2-chloroethylphosphonic acid + mepiquat chloride
chlormequat + mepiquat chloride

322 mesosulfuron-methyl

A sulfonyl urea herbicide for cereals available in mixtures

See also *iodosulfuron-methyl-sodium + mesosulfuron-methyl*

323 metalaxyl-M

A phenylamide systemic fungicide

See also *chlorothalonil + metalaxyl-M*
cymoxanil + fludioxonil + metalaxyl-M
mancozeb + metalaxyl-M

Products

SL 567A	Syngenta	480 g/l	EC	08811

Uses

Cavity spot in ***carrots***. Crown rot in ***protected water lilies*** *(off-label)*, ***water lilies*** *(off-label)*. Downy mildew in ***grapevines*** *(off-label)*, ***leaf spinach*** *(off-label)*, ***protected leaf spinach*** *(off-label)*, ***protected spinach beet*** *(off-label)*, ***spinach beet*** *(off-label)*.

Specific Off-Label Approvals (SOLAs)

- ***grapevines*** *(OLA 021187) Feb 2004* [1]
- ***leaf spinach***, ***protected leaf spinach***, ***protected spinach beet***, ***spinach beet*** *(OLA 021183) Feb 2004* [1]
- ***protected water lilies***, ***water lilies*** *(OLA 021185) Feb 2004* [1]

Approval information

- Metalaxyl-M included in Annex I under EC Directive 91/414

Efficacy guidance

- Best results achieved when applied to damp soil
- Efficacy may be reduced in prolonged dry weather. Irrigation immediately before or after application may be necessary
- Results may not be satisfactory on soils with high organic matter content
- Control of cavity spot on carrots overwintered in the ground or lifted in winter may be lower than expected

Restrictions

- Maximum number of treatments or maximum total dose varies with crop and nature of treatment. See label for details
- Do not use where carrots have been grown on the same site for either of the previous two yrs
- Consult before use on crops intended for processing

Crop-specific information

- Latest use: 4 wk before removal of apples or pears from store; 6 wk after drilling for carrots; 10 wk before sale or supply (after rinsing) of water lilies
- HI horseradish 8 wk, raspberries 3 mth

Environmental safety

- Harmful to fish or other aquatic life. Do not contaminate surface waters or ditches with chemical or used container

Hazard classification and safety precautions

Hazard R03, R04
Risk phrases R03c, R04a, R04e
Operator protection A, C, H; U02, U04a, U05a, U07, U10, U20b
Environmental protection E01, E13c
Consumer protection C03
Storage and disposal E26, E29, E30a, E31b, E33, E34, E36
Medical advice M03

SECTION 4

324 metaldehyde

A molluscicide bait for controlling slugs and snails

Products

1	Allure	Chiltern	1.5% w/w	PT	11089
2	Brits	Doff Portland	6% w/w	PT	10116
3	Chiltern Blues	Chiltern	6% w/w	PT	10071
4	Chiltern Hundreds	Chiltern	3% w/w	PT	10072
5	Dixie 6	Greencrop	6% w/w	PT	10052
6	Doff Horticultural Slug Killer Blue Mini Pellets	Doff Portland	3% w/w	PT	09666
7	Escar-Go 6	Chiltern	6% w/w	PT	06076
8	Hardy	Chiltern	6% w/w	PT	06948
9	Luxan 9363	Luxan	6% w/w	PT	07359
10	Luxan Deal	Luxan	6% w/w	PT	11101
11	Luxan Metaldehyde	Luxan	6% w/w	PT	06564
12	Luxan Trigger	Luxan	6% w/w	PT	10419
13	Metarex Amba	De Sangosse	5% w/w	PT	11157
14	Metarex Green	De Sangosse	5% w/w	PT	10113
15	Metarex RG	De Sangosse	5% w/w	PT	10115
16	Mifaslug	FCC	6% w/w	PT	10292
17	Molotov	Chiltern	3% w/w	PT	08295
18	Regel	De Sangosse	5% w/w	PT	10114
19	Superflor 6% Metaldehyde Slug Killer Mini Pellets	CMI	6% w/w	PT	09773

Uses

Slugs in ***all edible crops*** [1-19]; ***all non-edible crops*** [1-10, 13-15, 17-19]; ***amenity grass, managed amenity turf*** [9, 11, 12, 16]; ***cultivated land/soil*** [1]; ***land not intended to bear vegetation*** [11]; ***ornamental plant production*** [13-15, 18]; ***protected crops*** [1, 13-15, 18]. Snails in ***all edible crops*** [1-19]; ***all non-edible crops*** [1-10, 13-15, 17-19]; ***amenity grass, managed amenity turf*** [9, 11, 12, 16]; ***cultivated land/soil*** [1]; ***land not intended to bear vegetation*** [11]; ***ornamental plant production*** [13-15, 18]; ***protected crops*** [1, 13-15, 18].

Approval information

- Approved for aerial application on all crops [2, 5, 6, 19]. See notes in Section 1
- Approval expiry 30 Nov 2004 [2, 6]
- Approval expiry 31 Dec 2004 [5, 19]

Efficacy guidance

- Apply pellets by hand, fiddle drill, fertilizer distributor, by air (check label) or in admixture with seed. See labels for rates and timing.
- Best results achieved from an even spread of granules applied during mild, damp weather when slugs and snails most active. May be applied in standing crops
- Varieties of oilseed rape low in glucosinolates can be more acceptable to slugs than "single low" varieties and control may not be as good
- To prevent slug build up apply at end of season to brassicas and other leafy crops
- To reduce tuber damage in potatoes apply twice in Jul and Aug

Restrictions

- Do not apply when rain imminent or water glasshouse crops within 4 d of application
- Take care to avoid lodging of pellets in the foliage when making late applications to edible crops

Environmental safety

- Dangerous to game, wild birds and animals
- Harmful to fish or other aquatic life. Do not contaminate surface waters or ditches with chemical or used container [19]
- Product contains proprietary cat and dog deterrent [5, 6, 9]
- Keep poultry out of treated areas for at least 7 d

Hazard classification and safety precautions

Operator protection A, H [1, 2, 4, 5, 9-18]; J [2, 5, 10]; U20b [3, 7, 8]; U20c [1, 2, 4-6, 9-19]
Environmental protection E01 [1, 3, 4, 7-12, 16, 17, 19]; E05a [1-4] (7 d); E05a [5]; E05a [6-9] (7 d); E05a [10]; E05a [11-19] (7 d); E10a [1-19]; E13c [19]; E15 [1-18]
Storage and disposal E29 [1-8, 13-15, 17, 18]; E30a [1, 3-5, 7-19]; E32a [1-19]
Treated seed S04a [13-15, 18]

325 metamitron

A contact and residual triazinone herbicide for use in beet crops

See also chloridazon + chlorpropham + metamitron
chloridazon + metamitron
chlorpropham + metamitron

Products

1	Agriguard Metamitron	AgriGuard	70% w/w	WG	09859
2	Barclay Seismic	Barclay	70% w/w	WG	11378
3	Bettix 70 WG	United Phosphorus	70% w/w	WG	11019
4	Fernpath Haptol	AgriGuard	70% w/w	WG	10446
5	Fernpath Haptol Flo	AgriGuard	700 g/l	SC	11481
6	Goltix 90	Bayer	90% w/w	WG	08654
7	Goltix 90	Makhteshim	90% w/w	WG	11578
8	Goltix Flowable	Bayer	700 g/l	SC	08986
9	Goltix Flowable	Makhteshim	700 g/l	SC	11540
10	Goltix WG	Bayer	70% w/w	WG	02430
11	Goltix WG	Makhteshim	70% w/w	WG	11539
12	Landgold Metamitron	Landgold	70% w/w	WG	06287
13	Marquise	Makhteshim	70% w/w	WG	08738
14	MM 70	Nufarm UK	70% w/w	WG	11582
15	MM 70 Flo	Nufarm UK	700 g/l	SC	11197
16	Skater	Makhteshim	700 g/l	SC	11159
17	Standon Metamitron	Standon	70% w/w	WG	07885
18	Target SC	Unicrop	700 g/l	SC	10861
19	Target SC	Unicrop	700 g/l	SC	11070
20	Tripart Accendo	Tripart	70% w/w	WG	06110
21	Volcan	Sipcam	70% w/w	WG	09295

Uses

Annual dicotyledons in ***asparagus*** *(off-label)* [10, 13, 16]; ***fodder beet*** [1-13, 15-21]; ***forestry*** *(off-label)* [13, 16]; ***mangels*** [1-3, 5-13, 15-21]; ***red beet*** [1-3, 5-13, 15, 17-21]; ***sugar beet*** [1-21]. Annual grasses in ***asparagus*** *(off-label)* [10, 13, 16]; ***fodder beet*** [1-13, 15-21]; ***forestry*** *(off-label)* [13, 16]; ***mangels*** [1-3, 5-13, 15-21]; ***red beet*** [1-3, 5-13, 15, 17-21]; ***sugar beet*** [1-21]. Annual meadow grass in ***fodder beet, mangels*** [1-3, 5-13, 15-21]; ***red beet*** [1-3, 5-13, 15, 17-21]; ***sugar beet*** [1-3, 5-21]. Fat hen in ***fodder beet, mangels*** [1-3, 5-13, 15-21]; ***red beet*** [1-3, 5-13, 15, 17-21]; ***sugar beet*** [1-3, 5-21]. Groundsel in ***agastache spp.*** *(off-label)*, ***basil*** *(off-label)*, ***borage*** *(off-label)*, ***caraway*** *(off-label)*, ***catnip*** *(off-label)*, ***chervil*** *(off-label)*, ***clary*** *(off-label)*, ***coriander*** *(off-label)*, ***curry plant*** *(off-label)*, ***dill*** *(off-label)*, ***dragonhead*** *(off-label)*, ***fennel*** *(off-label)*, ***fenugreek*** *(off-label)*, ***feverfew*** *(off-label)*, ***lavandin*** *(off-label)*, ***lavender*** *(off-label)*, ***marigold*** *(off-label)*, ***nasturtium*** *(off-label)*, ***origanum heracleoticum*** *(off-label)*, ***parsley*** *(off-label)*, ***rocket*** *(off-label)*, ***savory*** *(off-label)*, ***spike lavender*** *(off-label)*, ***violet*** *(off-label)* [13, 16]; ***angelica*** *(off-label)*, ***balm*** *(off-label)*, ***bay*** *(off-label)*, ***camomile*** *(off-label)*, ***chives*** *(off-label)*, ***hyssop*** *(off-label)*, ***land cress*** *(off-label)*, ***lemon verbena*** *(off-label)*, ***lovage*** *(off-label)*, ***marjoram*** *(off-label)*, ***mint*** *(off-label)*, ***nettle*** *(off-label)*, ***oregano*** *(off-label)*, ***rosemary*** *(off-label)*, ***rue*** *(off-label)*, ***sage*** *(off-label)*, ***salad burnet*** *(off-label)*, ***sorrel*** *(off-label)*, ***tarragon*** *(off-label)*, ***thyme*** *(off-label)*, ***woodruff*** *(off-label)* [10, 13, 16]

Specific Off-Label Approvals (SOLAs)

- ***agastache spp., angelica, asparagus, balm, basil, bay, borage, camomile, caraway, catnip, chervil, chives, clary, coriander, curry plant, dill, dragonhead, fennel, fenugreek, feverfew, forestry, hyssop, land cress, lavandin, lavender, lemon verbena, lovage, marigold, marjoram, mint, nasturtium, nettle, oregano, origanum heracleoticum, parsley, rocket, rosemary, rue, sage, salad burnet, savory, sorrel, spike lavender, tarragon, thyme, violet, woodruff*** *(OLA 030737) Dec 2008* [13]
- ***agastache spp., angelica, asparagus, balm, basil, bay, borage, camomile, caraway, catnip, chervil, chives, clary, coriander, curry plant, dill, dragonhead, fennel, fenugreek, feverfew, forestry, hyssop, land cress, lavandin, lavender, lemon verbena, lovage, marigold, marjoram, mint, nasturtium, nettle, oregano, origanum heracleoticum, parsley, rocket, rosemary, rue, sage, salad burnet, savory, sorrel, spike lavender, tarragon, thyme, violet, woodruff*** *(OLA 030738) Dec 2008* [16]

- ***angelica, balm, bay, camomile, chives, hyssop, land cress, lemon verbena, lovage, marjoram, mint, nettle, oregano, rosemary, rue, sage, salad burnet, sorrel, tarragon, thyme, woodruff*** *(OLA 991208) Dec 2008* [10]
- ***asparagus*** *(OLA 962415) Dec 2008* [10]

Approval information

- Approval expiry 30 Jun 2004 [20]
- Approval expiry 31 Aug 2004 [4]

Efficacy guidance

- May be used pre-emergence alone or post-emergence in tank mixture or with an authorised adjuvant oil
- Low dose programme (LDP). Apply a series of low-dose post-weed emergence sprays, including adjuvant oil, timing each treatment according to weed emergence and size. See label for details and for recommended tank mixes and sequential treatments. On mineral soils the LDP should be preceeded by pre-drilling or pre-emergence treatment
- Traditional application. Apply either pre-drilling before final cultivation with incorporation to 8-10 cm, or pre-crop emergence at or soon after drilling into firm, moist seedbed to emerged weeds from cotyledon to first true leaf stage
- On emerged weeds at or beyond 2-leaf stage addition of adjuvant oil advised
- Up to 3 post-emergence sprays may be used on soils with over 10% organic matter
- For control of wild oats and certain other weeds, tank mixes with other herbicides or sequential treatments are recommended. See label for details

Restrictions

- Maximum total dose equivalent to three full dose treatments for most products; 4 lower maximum dose treatments [13]
- Using traditional method post-crop emergence on mineral soils do not apply before first true leaves have reached 1 cm long

Crop-specific information

- Latest use: before crop foliage meets across rows for beet crops; 12 wk after transplanting asparagus
- HI herbs 6 wk
- Crop tolerance may be reduced by stress caused by growing conditions, effects of pests, disease or other pesticides, nutrient deficiency etc

Following crops guidance

- Only sugar beet, fodder beet or mangels may be drilled within 4 mth after treatment. Winter cereals may be sown in same season after ploughing, provided 16 wk passed since last treatment

Environmental safety

- Flammable [4]
- Harmful to fish or other aquatic life. Do not contaminate surface waters or ditches with chemical or used container [15, 16]

Hazard classification and safety precautions

Hazard R03 [4, 5, 8, 9, 15, 18, 19]; R04 [1, 2, 4, 13]; R07d [4]

Risk phrases R03c [5, 8, 9, 15, 18, 19]; R04a, R04b [4]; R04d [1, 2]; R04e [1, 2, 13]

Operator protection A [1, 2, 5, 8, 9, 11, 15, 16, 18, 19]; C, H [1, 2]; U05a [1, 2, 5, 13, 15, 18, 19]; U08 [5, 13-15, 18, 19]; U14, U19 [5, 13, 15, 18, 19]; U20a [5-7, 10-15, 18-20]; U20b [1, 2, 8, 9, 17]; U20c [3, 4, 16, 21]

Environmental protection E01 [1, 2, 5, 8, 9, 13, 15, 18, 19]; E13c [15, 16]; E15 [1-14, 16-21]

Consumer protection C03 [1, 2, 5, 8, 9, 13, 15, 18, 19]

Storage and disposal E26 [5, 8, 9, 15, 18, 19]; E30a [1-21]; E31c [16]; E32a [1-5, 8, 9, 12-14, 17-19, 21]; E34 [5-11, 20]

Medical advice M03 [5, 8, 9]

326 metam-sodium

A methyl isothiocyanate producing sterilant for glasshouse, nursery and outdoor soils

Products

1	Discovery	United Phosphorus	510 g/l	SL	10416
2	Metham Sodium 400	United Phosphorus	400 g/l	SL	08051
3	Sistan 51	Unicrop	510 g/l	SL	10046

Uses

Nematodes in ***chrysanthemums, ornamental specimens, outdoor tomatoes*** *(Jersey only)*, ***protected carnations, protected crops, protected tomatoes*** [2]; ***glasshouse soils, nursery soils, outdoor soils, potting soils*** [1, 3]. Soil insects in ***chrysanthemums, ornamental specimens, outdoor tomatoes*** *(Jersey only)*, ***protected carnations, protected crops, protected tomatoes*** [2]. Soil pests in ***glasshouse soils, nursery soils, outdoor soils, potting soils*** [1, 3]. Soil-borne diseases in ***chrysanthemums, ornamental specimens, outdoor tomatoes*** *(Jersey only)*, ***protected carnations, protected crops, protected tomatoes*** [2]; ***glasshouse soils, nursery soils, outdoor soils, potting soils*** [1, 3]. Weed seeds in ***chrysanthemums, ornamental specimens, outdoor tomatoes*** *(Jersey only)*, ***protected carnations, protected crops, protected tomatoes*** [2]; ***glasshouse soils, nursery soils, outdoor soils, potting soils*** [1, 3].

Efficacy guidance

- Metam-sodium is a partial soil sterilant and acts by breaking down in contact with soil to release methyl isothiocyanate (MIT)
- Apply to glasshouse soils as a drench, or inject undiluted to 20 cm at 30 cm intervals and seal immediately, or apply to surface and rotavate
- May also be used by mixing into potting soils
- Apply when soil temperatures exceed 7°C, preferably above 10°C, between 1 Apr and 31 Oct. Soil must be of fine tilth, free from debris and with 'potting moisture' content. If soil is too dry postpone treatment and water soil

Restrictions

- No plants must be present during treatment
- Crops must not be planted until a cress germination test has been completed satisfactorily
- Do not treat glasshouses within 2 m of growing crops. Fumes are damaging to all plants
- Container must be handled my mechanical means only and returned to supplier as directed [1]
- Avoid using in equipment incorporating natural rubber parts

Crop-specific information

- Latest use: pre-planting of crop

Following crops guidance

- Do not plant until soil is entirely free of fumes

Environmental safety

- Keep unprotected persons, livestock and pets out of treated areas for at least 24 h following treatment
- When diluted breakdown commences almost immediately. Only quantities for immediate use should be made up
- Divert or block drains which could carry solution under untreated glasshouses
- After treatment allow sufficient time (several weeks) for residues to dissipate and aerate soil by forking. Time varies with soil and season. Soils with high clay or organic matter content will retain gas longer than lighter soils

Hazard classification and safety precautions

Hazard R03 [1, 3]; R04 [1-3]

Risk phrases R03c, R04e [1, 3]; R04a [2]; R04b, R04c [1-3]

Operator protection A, C, H, M [1-3]; D [2]; U02, U05a, U08, U19 [1-3]; U04a [2, 3]; U07 [1]; U20a [2]; U20b [1, 3]

Environmental protection E01, E15 [1-3]; E02 [1, 3] (24 h)

Consumer protection C03

Storage and disposal E30a, E31a [2, 3]; E33, E36 [1]; E34 [1, 3]

Medical advice M03 [1, 3]

327 metazachlor

A residual anilide herbicide for use in brassicas, nurseries and forestry

Products

1	Agriguard Metazachlor	AgriGuard	500 g/l	SC	10417
2	Alpha Metazachlor 50 SC	Makhteshim	500 g/l	SC	10669
3	Barclay Metaza	Barclay	500 g/l	SC	11359
4	Butisan S	BASF	500 g/l	SC	00357
5	Gharda Bonanza	Nufarm UK	500 g/l	SC	11571
6	Landgold Metazachlor 50	Landgold	500 g/l	SC	09726
7	Me2 Booty	Me2	500 g/l	SC	10659
8	Standon Metazachlor 50	Standon	500 g/l	SC	05581
9	Sultan 50 SC	Makhteshim	500 g/l	SC	10418

Uses

Annual dicotyledons in ***broccoli*, *brussels sprouts*, *cabbages*, *calabrese*, *cauliflowers*, *spring oilseed rape*, *swedes*, *turnips*, *winter oilseed rape*** [1-9]; ***farm forestry*** *(off-label)*, ***honesty*** *(off-label)* [4]; ***farm woodland*, *forestry*, *nursery fruit trees and bushes*, *ornamental trees*, *shrubs*** [1, 2, 4, 7, 9]. Annual grasses in ***farm forestry*** *(off-label)*, ***honesty*** *(off-label)* [4]; ***farm woodland*, *forestry*, *ornamental trees*, *shrubs*** [1, 2, 4, 7, 9]. Annual meadow grass in ***broccoli*, *brussels sprouts*, *cabbages*, *calabrese*, *cauliflowers*, *spring oilseed rape*, *swedes*, *turnips*, *winter oilseed rape*** [1-9]; ***nursery fruit trees and bushes*, *ornamental trees*, *shrubs*** [1, 2, 4, 7, 9]. Blackgrass in ***broccoli*, *brussels sprouts*, *cabbages*, *calabrese*, *cauliflowers*, *spring oilseed rape*, *swedes*, *turnips*, *winter oilseed rape*** [1-9]; ***nursery fruit trees and bushes*, *ornamental trees*, *shrubs*** [1, 2, 4, 7, 9].

Specific Off-Label Approvals (SOLAs)

- ***farm forestry*** *(OLA 940227) Dec 2008* [4]
- ***honesty*** *(OLA 930551) Dec 2008* [4]

Efficacy guidance

- Activity is dependent on root uptake. For pre-emergence use apply to firm, moist, clod-free seedbed
- Some weeds (chickweed, mayweed, blackgrass etc) susceptible up to 2- or 4-leaf stage. Moderate control of cleavers achieved provided weeds not emerged and adequate soil moisture present
- Split pre- and post-emergence treatments recommended for certain weeds in winter oilseed rape on light and/or stony soils
- Effectiveness is reduced on soils with more than 10% organic matter
- Various tank-mixtures recommended to broaden spectrum. See label for details

Restrictions

- Maximum number of treatments 1 per crop for spring oilseed rape, swedes, turnips and brassicas; 2 per crop for winter oilseed rape and honesty (split dose treatment); 3 per yr for ornamentals, nursery stock, nursery fruit trees, forestry and farm forestry
- Do not use on sand, very light or poorly drained soils
- Do not treat protected crops or spray overall on ornamentals with soft foliage
- Do not spray crops suffering from wilting, pest or disease
- Do not spray broadcast crops or if a period of heavy rain forecast
- Do not use on nursery fruit trees within 1 yr of fruit production
- When used on nursery fruit trees any fruit harvested within 1 yr of treatment must be destroyed

Crop-specific information

- Latest use: pre-emergence for swedes and turnips; before 10 leaf stage for spring oilseed rape; before end of Jan for winter oilseed rape; before 6 pairs of true leaves visible for honesty
- HI brassicas 6 wk
- On winter oilseed rape may be applied pre-emergence from drilling until seed chits, post-emergence after fully expanded cotyledon stage (GS 1,0) or by split dose technique depending on soil and weeds. See label for details
- On spring oilseed rape, swedes, turnips metazachlor recommended as a pre-emergence sequential treatment following trifluralin
- On spring oilseed rape may also be used pre-weed-emergence from cotyledon to 10-leaf stage of crop (GS 1,0-1,10)

- With pre-emergence treatment ensure seed covered by 15 mm of well consolidated soil. Harrow across slits of direct-drilled crops
- Ensure brassica transplants have roots well covered and are well established. Direct drilled brassicas should not be treated before 3 leaf stage
- In ornamentals and hardy nursery stock apply after plants established and hardened off as a directed spray or, on some subjects, as an overall spray. See label for list of tolerant subjects. Do not treat plants in containers

Following crops guidance

- Any crop can follow normally harvested treated winter oilseed rape. See label for details of crops which may be planted after spring treatment and in event of crop failure

Environmental safety

- Dangerous for the environment [4]
- Very toxic to aquatic organisms. May cause long-term adverse effects in the aquatic environment [4]
- Harmful to fish or other aquatic life. Do not contaminate surface waters or ditches with chemical or used container [1-6, 8, 9]
- Keep livestock out of treated areas until foliage of any poisonous weeds such as ragwort has died and become unpalatable
- Keep livestock out of treated areas of swede and turnip for at least 5 wk following treatment

Hazard classification and safety precautions

Hazard R03 [1-4, 6-9]; R04 [1-9]; R14 [4]

Risk phrases R03c [1-4, 6-9]; R04b [1-3, 6-9]; R04d [7]; R04e [1-6, 9]; R14c [4]

Operator protection A, C [1-9]; H, M [1-4, 6-9]; U05a, U08, U19 [1-9]; U14 [4]; U20a [7]; U20b [1-6, 8, 9]

Environmental protection E01 [1-9]; E07 [1, 2] (2 wk or 5 wk for swedes and turnips); E07 [3-5] (5 wk for swedes, turnips); E07 [6] (swedes/turnips 5 wk); E07 [7] (5 wk for swedes, turnips); E07 [8, 9] (2 wk or 5 wk for swedes and turnips); E13c [1-6, 8, 9]; E38 [4]

Consumer protection C02 [7]; C03 [1-9]

Storage and disposal E26 [1-3, 5, 6, 8, 9]; E30a [1-9]; E31b [1, 3, 5, 6, 8]; E31c [2, 4, 7, 9]; E32d [4]; E34 [1-4, 6-9]

Medical advice M03 [1-4, 6-9]; M05a [4]

328 metazachlor + quinmerac

A residual herbicide mixture for oilseed rape

Products

1	Katamaran	BASF	375:125 g/l	SC	09049
2	Standon Metazachlor-Q	Standon	375:125 g/l	SC	09676

Uses

Annual dicotyledons in ***winter oilseed rape***. Annual meadow grass in ***winter oilseed rape***. Blackgrass in ***winter oilseed rape***. Cleavers in ***winter oilseed rape***. Poppies in ***winter oilseed rape***.

Efficacy guidance

- Activity is dependent on root uptake. Pre-emergence treatments should be applied to firm moist seedbeds. Applications to dry soil do not become effective until after rain has fallen
- Maximum activity achieved from treatment before weed emergence for some species
- Weed control may be reduced if excessive rain falls shortly after application especially on light soils
- May be used on all soil types except Sands, Very Light Soils, and soils containing more than 10% organic matter. Crop vigour and/or plant stand may be reduced on brashy and stony soils

Restrictions

- Maximum total dose equivalent to one full dose treatment
- Damage may occur in waterlogged conditions. Do not use on poorly drained soils
- Do not treat stressed crops. In frosty conditions transient scorch may occur

Crop-specific information

- Latest use: end Jan in yr of harvest

- To ensure crop safety it is essential that crop seed is well covered with soil to 15 mm. Loose or puffy seedbeds must be consolidated before treatment. Do not use on broadcast crops
- Crop vigour and possibly plant stand may be reduced if excessive rain falls shortly after treatment especially on light soils

Following crops guidance

- In the event of crop failure after use, wheat or barley may be sown in the autumn after ploughing to 15 cm. Spring cereals or brassicas may be planted after ploughing in the spring

Environmental safety

- Dangerous to fish or other aquatic life. Do not contaminate surface waters or ditches with chemical or used container
- LERAP Category B
- To reduce risk of movement to water do not apply to dry soil or if heavy rain is forecast. On clay soils create a fine consolidated seedbed
- Keep livestock out of treated areas until foliage of any poisonous weeds such as ragwort has died and become unpalatable

Hazard classification and safety precautions

Hazard R04
Risk phrases R04e
Operator protection A; U05a, U08, U14, U20b [1, 2]; U19 [1]
Environmental protection E01, E07, E13b, E16a, E16b
Consumer protection C03
Storage and disposal E26, E31b [2]; E30a [1, 2]; E31c [1]

329 metconazole

A conazole fungicide

Products

1	Caramba	BASF	60 g/l	SL	10213
2	Sunorg Pro	BASF	90 g/l	SL	11112

Uses

Alternaria in ***spring oilseed rape, winter oilseed rape*** [1, 2]. Brown rust in ***spring barley, winter barley, winter wheat*** [1]. Fusarium ear blight in ***winter wheat*** *(reduction)* [1]. Net blotch in ***spring barley*** *(reduction)*, ***winter barley*** *(reduction)* [1]. Phoma in ***spring oilseed rape*** *(foliar disease - reduction)*, ***winter oilseed rape*** *(foliar disease - reduction)* [1]; ***spring oilseed rape*** *(reduction)*, ***winter oilseed rape*** *(reduction)* [2]. Powdery mildew in ***spring barley*** *(moderate control)*, ***winter barley, winter wheat*** *(moderate control)* [1]. Rhynchosporium in ***spring barley, winter barley*** [1]. Sclerotinia stem rot in ***spring oilseed rape*** *(reduction)*, ***winter oilseed rape*** *(reduction)* [1, 2]. Septoria leaf spot in ***winter wheat*** [1]. Yellow rust in ***winter wheat*** [1].

Efficacy guidance

- Best results from application to healthy, vigorous crops when disease starts to develop
- Treatment for leaf blotch should be made after GS 33 and when weather favouring development of the disease has occurred
- Treat mildew infections before 3% infection on any green leaf. A specific mildewicide will improve control of established infections
- Treat yellow rust before 1% infection on any leaf or as preventive treatment after GS 39
- For brown rust spray susceptible varieties before any of top 3 leaves has more than 2% infection
- Sclerotinia in oilseed rape should be treated at petal fall. A tank mixture with carbendazim may be needed for fully effective control

Restrictions

- Maximum total dose equivalent to two full dose treatments
- Do not apply to oilseed rape crops that are damaged or stressed from previous treatments, adverse weather, nutrient deficiency or pest attack. Spring application may lead to reduction of crop height
- The addition of adjuvants is neither advised nor necessary and can lead to enhanced growth regulatory effects on stressed crops of oilseed rape

- Ensure sprayer is free from residues of previous treatments that may harm the crop, especially oilseed rape. Use of a detergent cleaner is advised before and after use
- Do not apply with pyrethroid insecticides on oilseed rape at flowering [2]

Crop-specific information

- Latest use: up to and including milky ripe stage (GS 71) for cereals; 10% pods at final size for oilseed rape

Following crops guidance

- Only cereals, oilseed rape, sugar beet, linseed, maize, clover, beans, peas, carrots, potatoes or onions may be sown as following crops after treatment

Environmental safety

- Flammable [1]
- Dangerous (harmful [2]) to fish or other aquatic life. Do not contaminate surface waters or ditches with chemical or used container [1]
- Risk to certain non-target insects or other arthropods. See directions for use [1]
- LERAP Category B but avoid treatment close to field boundary, even if permitted by LERAP assessment, to reduce effects on non-target insects or other arthropods [1]

Hazard classification and safety precautions

Hazard R03, R07d [1]; R04 [1, 2]
Risk phrases R04a [2]; R04b, R04d, R04e, R04g [1]
Operator protection A, C [1, 2]; H [1]; U02, U05a, U20c [1, 2]; U11, U14 [1]
Environmental protection E01 [1, 2]; E13b, E16a, E16b [1]; E13c [2]
Consumer protection C03
Storage and disposal E26, E30a, E31a, E34
Medical advice M05 [1]

330 methiocarb

A stomach acting carbamate molluscicide and insecticide

Products

1	Decoy Wetex	Bayer	2% w/w	PT	09707
2	Decoy Wetex	Bayer CropScience	2% w/w	PT	11266
3	Exit Wetex	Bayer	3% w/w	PT	10149
4	Exit Wetex	Bayer CropScience	3% w/w	PT	11276
5	Huron	Bayer	3% w/w	PT	10148
6	Huron	Bayer CropScience	3% w/w	PT	11288
7	Karan	Bayer	3% w/w	PT	00037
8	Karan	Bayer CropScience	3% w/w	PT	11289
9	Lupus	Bayer	3% w/w	PT	09638
10	Lupus	Bayer CropScience	3% w/w	PT	11291
11	New Draza	Bayer	3% w/w	PT	11082
12	New Draza	Bayer CropScience	3% w/w	PT	11294
13	Rivet	Bayer	3% w/w	PT	09512
14	Rivet	Bayer CropScience	3% w/w	PT	11300

Uses

Cutworms in ***sugar beet*** *(reduction)*. Leatherjackets in ***all cereals*** *(reduction)*, ***potatoes*** *(reduction)*, ***ryegrass*** *(seed admixture)*, ***sugar beet*** *(reduction)*. Millipedes in ***sugar beet*** *(reduction)*. Slugs in ***all cereals***, ***all non edible crops*** *(outdoor only)*, ***brussels sprouts***, ***cabbages***, ***cauliflowers***, ***leaf spinach***, ***lettuce***, ***maize***, ***potatoes***, ***ryegrass*** *(seed admixture)*, ***spring oilseed rape***, ***strawberries***, ***sugar beet***, ***sunflowers***, ***winter oilseed rape***. Strawberry seed beetle in ***strawberries***.

Efficacy guidance

- Use as a surface, overall application when pests active (normally mild, damp weather), pre-drilling or post-emergence. May also be used on cereals or ryegrass in admixture with seed at time of drilling
- Also reduces populations of cutworms and millipedes
- Best on potatoes in late Jul to Aug
- Apply to strawberries before strawing down to prevent seed beetles contaminating crop
- See label for details of suitable application equipment

SECTION 4

Restrictions

- This product contains an anticholinesterase carbamate compound. Do not use if under medical advice not to work with such compounds
- Maximum number of treatments 3 per crop for potatoes; 2 per crop for cereals, Brussels sprouts, cauliflowers, maize, oilseed rape, sunflowers; 1 per crop for cabbages, lettuce, leaf spinach, sugar beet; 1 per yr for strawberries
- Do not treat any protected crops
- Do not allow pellets to lodge in edible crops

Crop-specific information

- Latest use: before first node detectable for cereals, maize; before three visibly extended internodes for oilseed rape, sunflowers
- HI 7 d for spinach, strawberrries; 14 d for Brussels sprouts, cabbages, cauliflowers, lettuce; 18 d for potatoes; 6 mth for sugar beet

Environmental safety

- Dangerous to game, wild birds and animals
- Dangerous to fish or other aquatic life. Do not contaminate surface waters or ditches with chemical or used container
- Risk to certain non-target insects or other arthropods
- Avoid surface broadcasting application within 6 m of field boundary to reduce effects on non-target species
- Admixed seed should be drilled and not broadcast, and may not be applied from the air

Hazard classification and safety precautions

Hazard R03
Risk phrases R03c
Operator protection A, H, J; U05a, U20b
Environmental protection E01, E10a, E13b, E22b; E05a (7 d)
Consumer protection C03
Storage and disposal E30a, E32a, E34
Medical advice M02, M03

331 methoprene

A pheromone analogue insect growth regulator

Products

Apex 5E	Novartis A H	600 g/l	EC	08739

Uses

Sciarid flies in ***mushroom houses***.

Efficacy guidance

- Use as surface spray either on the compost or the casing
- Use pre-casing treatment only when early pest infestation occurs
- Product controls sciarids at pupal stage. Treatment will not control the generation of sciarids present at application
- Product prevents development of adults from eggs laid up to 14 d before or after treatment. Alternative treatments may be necessary if adult numbers increase subsequently
- Pre-casing treatment may not prevent development from larvae able to move into untreated casing prior to pupation

Restrictions

- Maximum number of treatments 1 per spawning on compost or casing

Crop-specific information

- Latest use: before casing on the compost or immediately after casing

Environmental safety

- Dangerous to fish or other aquatic life. Do not contaminate surface waters or ditches with chemical or used container

Hazard classification and safety precautions

Operator protection A, C, H; U05a, U19, U20c
Environmental protection E01, E13b

Consumer protection C03
Storage and disposal E26, E30a, E32a

332 methoxyfenozide

A moulting accelerating diacylhydrazine insecticide

Products

Runner	Bayer CropScience	240 g/l	SC	11470

Uses

Codling moth in ***apples, pears***. Tortrix moths in ***apples, pears***. Winter moth in ***apples, pears***.

Efficacy guidance

- To achieve best results uniform coverage of the foliage and full spray penetration of the leaf canopy is important, particularly when spraying post-blossom
- For maximum effectiveness on winter moth and tortrix spray pre-blossom when first signs of active larvae are seen, followed by a further spray in June if larvae of the summer generation are present
- For codling moth spray post-blossom to coincide with early to peak egg deposition. Follow-up treatments will normally be needed
- Methoxyfenozide is a moulting accelerating compound (MAC) and may be used in an anti-resistance strategy with other top fruit insecticides (including chitin biosynthesis inhibitors and juvenile hormones) which have a different mode of action
- To reduce further the likelihood of resistance development use at full recommended dose in sufficient water volume to achieve required spray penetration

Restrictions

- Maximum number of treatments 3 per yr but no more than two should be sprayed consecutively

Crop-specific information

- HI 14 d for apples, pears

Environmental safety

- Broadcast air-assisted LERAP (5 m)
- Do not allow direct spray from hand-held sprayers to fall within 1 m of the top of the bank of a static or flowing waterbody

Hazard classification and safety precautions

Operator protection U20b
Environmental protection E16b, E17a
Storage and disposal E26, E30a, E32a

SECTION 4

333 methyl bromide

A highly toxic alkyl halide fumigant for food and non-food commodities

Products

Sobrom BM 100	Brian Jones	100% w/w	GA	04381

Uses

Food storage pests in ***food storage areas***. Insect pests in ***stored plant material***.

Approval information

- Methyl bromide is currently restricted under the Montreal Protocol, which envisages a phased withdrawal by 2005 followed by a world-wide ban

Efficacy guidance

- Chemical is released from containers as a highly toxic, colourless gas with only a slight odour
- Dose depends on temperature. At lower temperatures higher doses and longer exposures should be used
- See label for dosage and exposure required in different situations

Restrictions

- Methyl bromide is subject to the Poisons Rules 1982 and the Poisons Act 1972. See notes in Section 1
- Not for retail sale

- Special equipment is needed for application and the chemical may only be used by professional operators trained in its use and familiar with the precautionary measures to be observed
- All operations must be in accordance with the HSE guidance note CS 12. A minimum of 2 operators are required
- Maximum number of treatments 1 per infestation
- Air animal feed for at least 5 d following treatment
- Some types of produce, cut flowers and plants are susceptible to damage by methyl bromide fumigation. Check before treating
- Methyl bromide can cause taint to flours and foodstuffs containing reactive sulphur compounds. See label for guidance
- Wear suitable protective clothing (See HSE Guidance Note CS 12)
- Wear suitable respiratory equipment during fumigation and while removing sheets. Do *not* wear gloves or rubber boots
- Do not use application equipment incorporating natural rubber, PVC or aluminium, zinc, magnesium or their alloys
- Extinguish all naked flames, including pilot lights when applying fumigant
- Keep in original container, tightly closed in a safe place under lock and key. Do not re-use container and return to supplier when empty
- Keep unprotected persons out of treated area until it is shown by test to be clear of methyl bromide

Crop-specific information

- Latest use: 3 d before loading grains; 7 d before use of all other commodities and spaces

Environmental safety

- Harmful to fish or other aquatic life. Do not contaminate surface waters or ditches with chemical or used container
- Keep animals and livestock out of treated area until it is shown by test to be clear of methyl bromide

Hazard classification and safety precautions

Hazard R01
Risk phrases R01a, R01b
Operator protection U04a, U05a
Environmental protection E01, E13c; E02 (until clear)
Consumer protection C03
Storage and disposal E27, E30b, E34
Medical advice M04

334 methyl bromide with amyl acetate

A highly toxic alkyl halide fumigant for soil and food storage areas

Products

Fumyl-O-Gas	Brian Jones	99.7% w/w	GA	04833

Uses

Food storage pests in ***food storage areas***. Nematodes in ***glasshouse soils***, ***soils***. Soil pests in ***glasshouse soils***, ***soils***. Soil-borne diseases in ***glasshouse soils***, ***soils***. Weed seeds in ***glasshouse soils***, ***soils***.

Approval information

- Methyl bromide is currently restricted under the Montreal Protocol, which envisages a phased withdrawal by 2005 followed by a world-wide ban

Efficacy guidance

- Chemical is released from containers as a highly volatile, highly toxic gas and must be released under gas-proof sheeting
- Amyl acetate is added as a warning odourant
- Soil must be clear of trash and cultivated to at least 40 cm before treatment
- Gas must be confined in soil for 48-96 h

Restrictions
- Methyl bromide is subject to the Poisons Rules 1982 and the Poisons Act 1972. See notes in Section 1
- Product not for retail sale
- Maximum number of treatments 1 per yr
- Special equipment is needed for application and the chemical may only be used by professional operators trained in its use and familiar with the precautionary measures to be observed. Refer to HSE Guidance Note CS 12 before applying
- Keep in original container, tightly closed, in a safe place, under lock and key
- Wear suitable respiratory equipment during fumigation and while removing sheeting
- Do *not* wear gloves or rubber boots
- Ventilate treated areas thoroughly when gas has cleared
- Keep unprotected persons out of treated areas for at least 24 h Maximum number of treatments 1 per yr

Crop-specific information
- Latest use: before sowing or planting
- Crops may be planted 3-21 d after removal of gas-proof sheeting. For some crops a leaching irrigation is essential. See label for details

Environmental safety
- Dangerous to game, wild birds, animals and bees
- Harmful to fish or other aquatic life. Do not contaminate surface waters or ditches with chemical or used container
- Keep livestock out of treated areas for at least 24 h

Hazard classification and safety precautions

Hazard R01
Risk phrases R01a, R01b
Operator protection U04a, U05a
Environmental protection E01; E02 (until clear)
Consumer protection C03
Storage and disposal E30b, E34
Medical advice M04

SECTION 4

335 methyl bromide with chloropicrin

A highly toxic alkyl halide fumigant for treatment of soil and stored products

Products

Sobrom BM 98	Brian Jones	98% w/w	GA	04189

Uses

Nematodes in ***glasshouse soils***, ***orchards***, ***soils***. Soil insects in ***glasshouse soils***, ***orchards***, ***soils***. Soil-borne diseases in ***glasshouse soils***, ***orchards***, ***soils***. Weed seeds in ***glasshouse soils***, ***orchards***, ***soils***.

Approval information
- Methyl bromide is currently restricted under the Montreal Protocol, which envisages a phased withdrawal by 2005 followed by a world-wide ban

Efficacy guidance
- Chemical is released from containers as a highly volatile, highly toxic gas and for soil fumigation must be released under a plastic tarpaulin
- Chloropicrin is added as a warning odourant tear gas
- Soil must be clear of trash and cultivated to at least 40 cm before treatment
- Gas must be confined in soil for 48-96 h

Restrictions
- Methyl bromide and chloropicrin are subject to the Poisons Rules 1982 and the Poisons Act 1972. See notes in Section 1
- Product not for retail sale
- Maximum number of treatments 1 per yr

- Special equipment is needed for application and the chemical may only be used by professional operators trained in its use and familiar with the precautionary measures to be observed. Refer to HSE Guidance Note CS 12 before applying
- Keep in original container, tightly closed, in a safe place, under lock and key
- Wear suitable respiratory equipment during fumigation and while removing sheeting
- Do *not* wear gloves or rubber boots
- Ventilate treated areas thoroughly when gas has cleared
- Keep unprotected persons out of treated areas for at least 24 h

Crop-specific information

- Latest use: before sowing or planting
- Crops may be planted 3-21 d after removal of tarpaulin. For some crops a leaching irrigation is essential. See labels for details

Environmental safety

- Harmful to fish or other aquatic life. Do not contaminate surface waters or ditches with chemical or used container [1]
- Keep livestock out of treated areas for at least 24 h

Hazard classification and safety precautions

Hazard R01, R04
Risk phrases R01a, R01b, R01c, R04a, R04b, R04c
Operator protection U04a, U05a
Environmental protection E01, E13c; E02, E07 (24 h)
Consumer protection C03
Storage and disposal E30b, E34
Medical advice M04

336 metoxuron

A contact and residual urea herbicide for carrots

Products

Dosaflo	Syngenta	500 g/l	SC	09531

Uses

Annual dicotyledons in ***carrots***. Mayweeds in ***carrots***.

Approval information

- Products containing this active ingredient have been granted derogations for specified 'Essential Uses' for use until 31 December 2007. Sale and supply must cease by 30 June 2007 but growers have no guarantee that the products will continue to be available until then.
  For more information see 'The Review Programme' under 'Pesticide Legislation' in Section 1

Efficacy guidance

- Best results achieved by application to weeds, especially mayweeds, in seedling to young plant stage
- Product will suppress growth of annual meadow-grass and wild oats between emergence and the 3 leaf stage of growth

Restrictions

- Maximum number of treatments not specified
- Carrots should be drilled to an even depth of not less than 1.5 cms otherwise some plants may be treated at cotyledon stage and their growth checked
- Do not spray when soil very dry or wet, when heavy rain is imminent, or when shade temperature exceeds 25°C
- Do not spray carrots on soils with more than 80% sand or less than 1% organic matter
- Do not apply during prolonged frosty weather, when temperatures are below freezing, or when frost is imminent
- Do not spray crops checked by pests, wind, frost or waterlogging until recovered
- This product must not be used on any crops other than those listed, including any extrapolations that would normally be permissible under the Long Term Arrangements for Extension of Use (see Section 1)
- No crop should be sown within 6 wk of treatment

Crop-specific information

- Latest use: not specified for carrots

Hazard classification and safety precautions

Operator protection U08, U19, U20b
Environmental protection E15
Storage and disposal E26, E30a, E31b

337 metribuzin

A contact and residual triazinone herbicide for use in potatoes

See also flufenacet + metribuzin

Products

1	Citation 70	United Phosphorus	70% w/w	WG	09370
2	Inter-Metribuzin WG	I T Agro	70% w/w	WG	09801
3	Lexone 70DF	DuPont	70% w/w	WG	04991
4	Python	Makhteshim	70% w/w	WG	11166
5	Sencorex WG	Bayer	70% w/w	WG	03755
6	Sencorex WG	Bayer CropScience	70% w/w	WG	11304
7	Tuberon	Unicrop	480 g/l	SC	10613

Uses

Annual and perennial weeds in ***asparagus*** *(off-label)* [5, 6]. Annual dicotyledons in ***asparagus*** *(off-label)*, ***carrots*** *(off-label)*, ***parsnips*** *(off-label)* [4], ***early potatoes***, ***maincrop potatoes*** [1-7]. Annual grasses in ***asparagus*** *(off-label)*, ***carrots*** *(off-label)*, ***parsnips*** *(off-label)* [4]; ***early potatoes***, ***maincrop potatoes*** [1-7]. Perennial dicotyledons in ***asparagus*** *(off-label)*, ***carrots*** *(off-label)*, ***parsnips*** *(off-label)* [4]. Volunteer oilseed rape in ***early potatoes***, ***maincrop potatoes*** [1-6].

Specific Off-Label Approvals (SOLAs)

- ***asparagus*** *(OLA 021815) Dec 2008* [5]
- ***asparagus***, ***carrots***, ***parsnips*** *(OLA 030368) Dec 2008* [4]
- ***asparagus*** *(OLA 031888) Dec 2008* [6]

Efficacy guidance

- Water dispersible granule formulations may be applied pre- or post-emergence of crop [1-6]; suspension concentrate product may be applied pre-emergence only [7]. Best results achieved on weeds at cotyledon to 1-leaf stage
- Apply to moist soil with well-rounded ridges and few clods
- Activity reduced by dry conditions and on soils with high organic matter content
- On fen and moss soils pre-planting incorporation to 10-15 cm gives increased activity. Incorporate thoroughly and evenly
- With named maincrop and second early varieties on soils with more than 10% organic matter shallow pre- or post-planting incorporation may be used. See label for details
- Effective control using a programme of reduced doses is made possible by using a spray of smaller droplets, thus improving retention. See label for details [1-6]

Restrictions

- Maximum number of treatments 3 per crop subject to maximum permitted dose for water dispersible granule formulations [1-6]; 2 per crop for suspension concentrate product [7]. See labels for details
- Apply pre-emergence only on named first earlies, pre- or post-emergence on named second earlies. On named maincrop varieties apply pre-emergence (except for certain varieties on Sands or Very Light soils) or post-emergence before longest shoots reach 15 cm. See labels for details [1-6]
- Do not treat any variety on Light soils [7]
- Do not cultivate after treatment
- Some varieties may be sensitive to post-emergence treatment if crop under stress [1-6]

Crop-specific information

- Latest use: pre-emergence for earlies; before most advanced shoots reach 15 cm for maincrop [1-6]; pre-emergence for all crops [7]; whilst crop still dormant in yr of harvest for asparagus [5]
- HI carrots, parsnips 4 wk [5]

- On stony or gravelly soils there is risk of crop damage, especially if heavy rain falls soon after application
- When days are hot and sunny delay spraying until evening

Following crops guidance

- Ryegrass, cereals or winter beans may be sown in same season provided at least 16 wk elapsed after treatment and ground ploughed to 15 cm and thoroughly cultivated soon after harvest
- Do not grow any vegetable brassicas on silt soils in Lincs, and lettuces or radishes anywhere in UK on land treated the previous yr. Other crops may be sown normally in spring of next yr

Environmental safety

- Dangerous to fish or other aquatic life. Do not contaminate surface waters or ditches with chemical or used container [4-7]
- LERAP Category B

Hazard classification and safety precautions

Hazard R03 [7]; R04 [4]
Risk phrases R03c [7]; R04e [4]
Operator protection A [4, 7]; H [4]; U05a [4, 7]; U08, U13, U19 [1-7]; U20a [3, 5, 6]; U20b [1, 2, 4, 7]
Environmental protection E01 [4, 7]; E13b [4-7]; E15 [1-3, 5, 6]; E16a, E16b [1-3, 5-7]
Consumer protection C03 [4, 7]
Storage and disposal E26, E31c, E34 [7]; E29 [2]; E30a [1-7]; E32a [1-6]

338 metsulfuron-methyl

A contact and residual sulfonylurea herbicide for use in cereals and land temporarily removed from production

See also *carfentrazone-ethyl + metsulfuron-methyl*
flupyrsulfuron-methyl + metsulfuron-methyl
mecoprop-P + metsulfuron-methyl

Products

1	Agriguard Metsulfuron	AgriGuard	20% w/w	WG	09569
2	Ally	DuPont	20% w/w	WG	02977
3	Ally WSB	DuPont	20% w/w	WB	06588
4	Barclay Flumen	Barclay	20% w/w	WG	11355
5	Jubilee (WSB)	DuPont	20% w/w	WB	06082
6	Jubilee 20 DF	DuPont	20% w/w	WG	06136
7	Landgold Metsulfuron	Landgold	20% w/w	WG	06280
8	Lorate 20 DF	DuPont	20% w/w	WG	06135
9	Standon Metsulfuron	Standon	20% w/w	WG	05670

Uses

Annual dicotyledons in ***linseed*** [2-6, 8]; ***spring barley, spring oats, spring wheat, winter barley, winter oats, winter wheat*** [1-9]; ***triticale*** [1-6, 8]. Chickweed in ***linseed*** [2-6, 8]; ***spring barley, spring oats, spring wheat, winter barley, winter oats, winter wheat*** [1-9]; ***triticale*** [1-6, 8]. Green cover in ***land temporarily removed from production*** [2, 4, 6, 8]. Mayweeds in ***linseed*** [2-6, 8]; ***spring barley, spring oats, spring wheat, winter barley, winter oats, winter wheat*** [1-9]; ***triticale*** [1-6, 8].

Approval information

- Metsulfuron-methyl included in Annex I under EC Directive 91/414

Efficacy guidance

- Best results achieved on small, actively growing weeds up to 6-true leaf stage. Good spray cover is important
- Commonly used in tank-mixture on wheat and barley with other cereal herbicides to improve control of resistant dicotyledons (cleavers, fumitory, ivy-leaved speedwell), larger weeds and grasses. See label for recommended mixtures
- When tank mixing, always add metsulfuron-methyl product to tank first

- May be used on land temporarily removed from production where the green cover is made up predominantly of grass, wheat barley, oats or triticale [2, 4, 6, 8]
- Metsulfuron-methyl is a member of the ALS-inhibitor group of herbicides and products should be used in a planned Resistance Management strategy. See Section 1 for more information

Restrictions

- Maximum number of treatments 1 per crop or per yr on land temporarily removed from production
- Product must only be used after 1 Feb
- Do not use on cereal crops undersown with grass or legumes
- Do not use on any crop suffering stress from drought, waterlogging, frost, deficiency, pest or disease attack or apply within 7 d of rolling
- Do not spray a cereal crop in tank mixture, or in sequence, with a product containing any other sulfonylurea except as directed on the label
- Do not open the water soluble bag or touch it with wet hands or gloves [3, 5]
- Spraying equipment should not be drained or flushed onto land planted, or to be planted, with trees or crops other than cereals and should be thoroughly cleansed after use - see label for instructions

Crop-specific information

- Latest use: before flag leaf sheath extending stage for cereals (GS 41); before flower bud visible or up to 30 cm tall, whichever is earlier, for linseed; before 1 Aug in yr of treatment for land temporarily removed from production
- Apply to wheat and oats from 2-leaf (GS 12), to barley and triticale from 3-leaf stage (GS 13) until flag-leaf fully emerged (GS 39). Do not spray Igri barley before leaf sheath erect stage (GS 30)
- Apply to linseed from first pair of true leaves unfolded up to 30 cm high or before flower bud visible, whichever is the sooner
- Recommendations for oats, triticale and linseed apply to product alone

Following crops guidance

- Only cereals, oilseed rape, field beans or grass may be sown in same calendar year after treating cereals with the product alone. Other restrictions apply to tank mixtures. See label for details
- In the event of crop failure sow only wheat within 3 mth after treatment
- Only cereals should be planted within 16 mth of applying to a linseed crop

Environmental safety

- Extremely dangerous to fish or other aquatic life. Do not contaminate surface waters or ditches with chemical or used container [1, 4]
- Extremely dangerous to aquatic higher plants. Do not contaminate surface waters or ditches with chemical or used container
- LERAP Category B
- Take extreme care to avoid damage by drift onto broad-leaved plants outside the target area, onto surface waters or ditches or onto land intended for cropping
- A range of broad leaved species will be fully or partially controlled when used in land temporarily removed from production, hence product may be suitable where wild flower borders or other forms of conservation headland are being developed [2, 4, 6, 8]
- Before use on land temporarily removed from production as part of grant-aided scheme, ensure compliance with the management rules [2, 4, 6, 8]
- Green cover on land temporarily removed from production must not be grazed by livestock or harvested for human or animal consumption or used for animal bedding [2, 4, 6, 8]

Hazard classification and safety precautions

Operator protection U05a, U20a [2, 5-9]; U08, U19 [1-9]; U20b [1, 4]; U22a [3, 5]
Environmental protection E01, E14a [2, 3, 5-9]; E13a [1, 4]; E16a, E16b [1-9]
Consumer protection C03 [2, 3, 5-9]
Storage and disposal E30a, E32a [1-9]; E34 [1]

339 metsulfuron-methyl + thifensulfuron-methyl

A contact residual and translocated sulfonylurea herbicide mixture for use in cereals

Products

Harmony M	DuPont	7:68% w/w	WG	03990

Uses

Annual dicotyledons in ***spring barley, spring wheat, winter wheat***. Cleavers in ***spring barley, spring wheat, winter wheat***. Field pansy in ***spring barley, spring wheat, winter wheat***. Polygonums in ***spring barley, spring wheat, winter wheat***. Speedwells in ***spring barley, spring wheat, winter wheat***.

Approval information

- Metsulfuron-methyl and thifensulfuron-methyl included in Annex I under EC Directive 91/414

Efficacy guidance

- Best results by application to small, actively growing weeds up to 6-true leaf stage
- Effectiveness may be reduced if heavy rain occurs within 4 h of application or if soil conditions very dry
- Tank-mixture with mecoprop, mecoprop-P or reduced rate of fluroxypyr improves control of cleavers and other problem weeds
- Metsulfuron-methyl and thifensulfuron-methyl are members of the ALS-inhibitor group of herbicides and products should be used in a planned Resistance Management strategy. See Section 1 for more information

Restrictions

- Maximum number of treatments 1 per crop
- Product must only be used after 1 Feb and after crop has three leaves
- Do not use on durum wheat or winter barley or on any crop suffering stress from drought, waterlogging, frost, deficiency, pest or disease attack
- Do not spray in tank mixture, or in sequence, with a product containing any other sulfonylurea herbicide except as directed on the label
- Do not apply within 7 d of rolling

Crop-specific information

- Latest use: before flag leaf ligule/collar just visible (GS 39)
- When tank mixing, always add metsulfuron-methyl/thifensulfuron-methyl product to tank first and fully disperse before adding second product
- Apply in spring to crops from 3-leaf stage to flag-leaf fully emerged (GS 13-39)

Following crops guidance

- Only cereals, oilseed rape, field beans or grass may be sown in same calendar year after treatment. In the event of crop failure sow only wheat within 3 mth after treatment

Environmental safety

- Extremely dangerous to aquatic higher plants. Do not contaminate surface waters or ditches with chemical or used container
- LERAP Category B
- Take extreme care to avoid drift onto broad-leaved crops or contamination of land planted or to be planted with any crops other than cereals
- Spraying equipment should not be drained or flushed onto land planted, or to be planted, with trees or crops other than cereals and should be thoroughly cleansed after use - see label for instructions

Hazard classification and safety precautions

Operator protection U08, U19, U20a
Environmental protection E14a, E16a, E16b
Storage and disposal E30a, E32a

340 myclobutanil

A systemic, protectant and curative conazole fungicide

Products

1	Systhane 20EW	Landseer	200 g/l	EW	09396
2	Systhane 6 Flo	Landseer	60 g/l	EW	07334
3	Systhane 6 W	Dow	6% w/w	WP	10808

Uses

American gooseberry mildew in ***blackcurrants, gooseberries*** [1, 2]. Black spot in ***roses*** [3]. Blossom wilt in ***cherries*** *(off-label)*, ***mirabelles*** *(off-label)* [1]. Mildew in ***roses*** [3]. Plum rust in

plums *(off-label)* [1]. Powdery mildew in ***apples***, ***strawberries*** [1, 2]; ***grapevines*** *(off-label)*, ***hops*** *(off-label)*, ***pears***, ***protected blackberries*** *(off-label)*, ***protected raspberries*** *(off-label)*, ***protected rubus hybrids*** *(off-label)* [1]. Rust in ***roses*** [3]. Scab in ***apples***, ***pears*** [1, 2].

Specific Off-Label Approvals (SOLAs)

- ***cherries***, ***mirabelles*** *(OLA 991535) Dec 2008* [1]
- ***grapevines***, ***protected blackberries***, ***protected raspberries***, ***protected rubus hybrids*** *(OLA 023195) Dec 2008* [1]
- ***hops*** *(OLA 021412) Dec 2008* [1]
- ***plums*** *(OLA 012459) Dec 2008* [1]

Efficacy guidance

- Best results achieved when used as part of routine preventive spray programme from bud burst to end of flowering in apples and pears and from just before the signs of mildew infection in blackcurrants and gooseberries [1]
- In strawberries commence spraying at, or just prior to, first flower. Post-harvest sprays may be required on mildew-susceptible varieties where mildew is present and likely to be damaging [1]
- Spray at 7-14 d intervals depending on disease pressure and dose applied [1]
- For improved scab control in post-blossom period tank-mix with mancozeb or captan [1]
- Apply alone from mid-Jun for control of secondary mildew on apples and pears [1]
- On roses spray at first signs of disease and repeat every 2 wk. In high risk areas spray when leaves emerge in spring, repeat 1 wk later and then continue normal programme [3]

Restrictions

- Maximum total dose equivalent to ten full dose treatments in apples and pears; six full dose treatments in blackcurrants, gooseberries, strawberries; 5 full dose treatments in plums; 3.5 full dose treatments in cherries, mirabelles

Crop-specific information

- HI 21 d (cherries, mirabelles); 14 d (apples, pears, blackcurrants, gooseberries, hops); 3 d (plums, protected blackberries, protected raspberries, protected Rubus hybrids, strawberries)

Environmental safety

- Harmful to fish or other aquatic life. Do not contaminate surface waters or ditches with chemical or used container

Hazard classification and safety precautions

Operator protection A, H, J [1, 2]; U08 [1, 2]; U20a [1-3]
Environmental protection E01 [1, 2]; E13c [1-3]
Consumer protection C03 [1, 2]
Storage and disposal E26 [1]; E30a [3]; E32a [1-3]

341 2-(1-naphthyl)acetic acid

A plant growth regulator to promote rooting of cuttings

See also 4-indol-3-yl-butyric acid + 2-(1-naphthyl)acetic acid with dichlorophen

Products

1	Rhizopon B Powder (0.1%)	Rhizopon	0.1% w/w	SP	09089
2	Rhizopon B Powder (0.2%)	Rhizopon	0.2% w/w	SP	09090
3	Rhizopon B Tablets	Rhizopon	25 mg a.i.	TB	09091

Uses

Rooting of cuttings in ***ornamental specimens***.

Efficacy guidance

- Dip moistened base of cuttings into powder immediately before planting [1, 2]
- See label for details of concentrations recommended for promotion of rooting in cuttings of different species [3]
- Dip prepared cuttings in solution for 4-24 h depending on species [3]

Restrictions

- Maximum number of treatments 1 per cutting

Crop-specific information

- Latest use: before cutting insertion for ornamental specimens [1, 2]

SECTION 4

Hazard classification and safety precautions

Operator protection U08, U19, U20a
Environmental protection E15
Storage and disposal E30a, E32a, E34

342 (2-naphthyloxy)acetic acid

A plant growth regulator for setting tomato fruit

Products

Betapal Concentrate	Vitax	16 g/l	SL	00234

Uses

Increasing fruit set in ***outdoor tomatoes***, ***protected tomatoes***.

Efficacy guidance

- Apply with any fine sprayer or syringe when first half dozen flowers are open

Restrictions

- Do not spray growing head of tomato plants

Crop-specific information

- Spray actual trusses in flower and repeat every 2 wk

343 napropamide

A soil applied amide herbicide for oilseed rape, fruit and woody ornamentals

Products

Devrinol	United Phosphorus	450 g/l	SC	09374

Uses

Annual dicotyledons in ***apple orchards***, ***blackcurrants***, ***forest nurseries***, ***gooseberries***, ***ornamental plant production***, ***pear orchards***, ***plums***, ***raspberries***, ***strawberries***, ***winter oilseed rape***. Annual grasses in ***apple orchards***, ***blackcurrants***, ***forest nurseries***, ***gooseberries***, ***ornamental plant production***, ***pear orchards***, ***plums***, ***raspberries***, ***strawberries***, ***winter oilseed rape***. Cleavers in ***apple orchards***, ***blackcurrants***, ***forest nurseries***, ***gooseberries***, ***ornamental plant production***, ***pear orchards***, ***plums***, ***raspberries***, ***strawberries***, ***winter oilseed rape***. Groundsel in ***apple orchards***, ***blackcurrants***, ***forest nurseries***, ***gooseberries***, ***ornamental plant production***, ***pear orchards***, ***plums***, ***raspberries***, ***strawberries***, ***winter oilseed rape***.

Efficacy guidance

- Best results obtained from treatment pre-emergence of weeds but product may be used in conjunction with contact herbicide such as paraquat for control of emerged weeds. Otherwise remove existing weeds before application
- Product broken down by sunlight, so application during summer not recommended. Most crops recommended for treatment between Nov and end-Feb
- Apply to winter oilseed rape as pre-drilling treatment in tank-mixture with trifluralin and incorporate within 30 min. Post-emergence use of specific grass weedkilller recommended where volunteer cereals are a serious problem
- Increase water volume to ensure adequate dampening of compost when treating containerised nursery stock with dense leaf canopy

Restrictions

- Maximum number of treatments 1 per crop or yr
- Do not use on Sands
- Do not use on soils with more than 10% organic matter
- Do not treat ornamentals in containers of less than 1 litre
- Some phytotoxicity seen on yellow and golden varieties of conifers and container grown alpines. On any ornamental variety treat only a small number of plants in first season

Crop-specific information

- Latest use: at drilling for winter oilseed rape; before end Feb for apples, pears, strawberries, bush and cane fruit; before end of Apr for field and container grown ornamental trees and shrubs

- Where minimal cultivation used to establish oilseed rape, tank-mixture may be applied directly to stubble and mixed into top 25 mm as part of surface cultivations
- Apply up to 14 d prior to drilling winter oilseed rape
- Apply to strawberries established for at least one season or to maiden crops as long as planted carefully and no roots exposed, between 1 Nov and end Feb. Do not treat runners of poor vigour or with shallow roots, or runner beds
- Newly planted ornamentals should have no roots exposed. Do not treat stock of poor vigour or with shallow roots. Treatments made in Mar and Apr must be followed by 25 mm irrigation within 24 h
- Bush and cane fruit must be established for at least 10 mth before spraying and treated between 1 Nov and end Feb

Following crops guidance

- After use in fruit or ornamentals no crop can be drilled within 7 mth of treatment. Leaf, flowerhead, root and fodder brassica crops may be drilled after 7 mth; potatoes, maize, peas or dwarf beans after 9 mth; autumn sown wheat or grass after 18 mth; any crop after 2 yr
- After use in oilseed rape only oilseed rape, swedes, fodder turnips, brassicas or potatoes should be sown within 12 mth of application
- Soil should be mould-board ploughed to a depth of at least 200 mm before drilling or planting any following crop

Environmental safety

- Harmful to fish or other aquatic life. Do not contaminate surface waters or ditches with chemical or used container

Hazard classification and safety precautions

Operator protection A, C; U09a, U20b
Environmental protection E13c
Storage and disposal E26, E30a, E31a, E34

344 natural plant extracts

A contact insecticide that works by physical action

Products

Majestik	Certis	SL	–

Uses

Aphids in ***fruit crops***, ***ornamental plant production***, ***protected fruit***, ***protected ornamentals***, ***protected vegetables***, ***vegetables***. Leafhoppers in ***fruit crops***, ***ornamental plant production***, ***protected fruit***, ***protected ornamentals***, ***protected vegetables***, ***vegetables***. Mites in ***fruit crops***, ***ornamental plant production***, ***protected fruit***, ***protected ornamentals***, ***protected vegetables***, ***vegetables***. Thrips in ***fruit crops***, ***ornamental plant production***, ***protected fruit***, ***protected ornamentals***, ***protected vegetables***, ***vegetables***. Whitefly in ***fruit crops***, ***ornamental plant production***, ***protected fruit***, ***protected ornamentals***, ***protected vegetables***, ***vegetables***.

Approval information

- Product not controlled by Control of Pesticides Regulations/Plant Protection Products Regulations because it acts by physical means only

Efficacy guidance

- Product acts by coating target pests and inhibiting respiration or movement by physical action
- Effective on all growth stages of mites, aphids, thrips, whitefly and leafhoppers. Treat as soon as insects are seen and repeat as necessary
- Ensure thorough spray coverage of plant, paying special attention to growing points and the underside of leaves
- Consult manufacturer for guidance on efficacy before treating outdoor crops for the first time
- Product may be used in conjunction with biological control agents. Spray 24 h before they are introduced

Restrictions

- No limit on the number of treatments and no minimum interval between applications
- Before large scale use on a new crop, treat a few plants to check for crop safety

- Do not treat ornamental crops when in flower
- Avoid treating fruit or vegetables close to harvest as product may leave a sticky deposit

Crop-specific information

- HI zero. Product is not taken into plant

Environmental safety

- If bees are being used for pollination ensure hives are closed before treatment and do not re-open before spray has completely dried on the plants

Hazard classification and safety precautions

Operator protection U05a, U09a, U19, U20c

345 nicotine

A general purpose, non-persistent, contact, alkaloid insecticide

Products

1	Nico Soap	United Phosphorus	75 g/l	LI	07517
2	Nicotine 40% Shreds	Dow	40% w/w	FU	05725
3	No-Fid	Certis	75 g/l	LI	11183
4	XL-All Insecticide	Vitax	70 g/l	LI	02369
5	XL-All Nicotine 95%	Vitax	950 g/l	LI	07402

Uses

Aphids in ***all soft fruit, protected asparagus, protected celery, protected marrows, vegetables*** [4]; ***all top fruit, flower bulbs, leaf brassicas, navy beans, nursery stock, ornamental specimens, parsley, peas, protected carnations, protected chrysanthemums, protected crops, protected flowers, protected melons, protected roses*** [4, 5]; ***apples, artichokes, asparagus, broccoli, brussels sprouts, cabbages, calabrese, cauliflowers, celeriac, chicory, chinese cabbage, chives, combining peas, courgettes, cress, cucumbers, dwarf beans, edible podded peas, endives, fennel, frise, garlic, kale, lamb's lettuce, marrows, ornamental plant production, outdoor tomatoes, peppers, potatoes, pumpkins, radicchio, radishes, shallots, spring field beans, squashes, sweetcorn, vining peas, winter field beans*** [1, 3]; ***aubergines, protected herbs, protected potatoes, protected salad onions*** [2]; ***broad beans, carrots, celery, leaf spinach, lettuce, parsnips, red beet, runner beans, swedes, turnips*** [1, 3-5]; ***bush fruit, protected figs, protected grapevines, protected peaches*** [5]; ***leeks, onions*** [1, 3, 4]; ***protected courgettes, protected peppers*** [2, 4]; ***protected cucumbers, protected lettuce*** [2, 4, 5]; ***protected ornamentals, protected tomatoes*** [1-5]; ***strawberries*** [1, 3, 5]. Capsids in ***all soft fruit, vegetables*** [4]; ***all top fruit, flower bulbs, protected crops*** [4, 5]; ***apples*** [1, 3]; ***bush fruit*** [5]. Caterpillars in ***artichokes, asparagus, broad beans, broccoli, brussels sprouts, cabbages, calabrese, carrots, cauliflowers, celeriac, celery, chicory, chinese cabbage, chives, combining peas, courgettes, cress, cucumbers, dwarf beans, edible podded peas, endives, fennel, frise, garlic, kale, lamb's lettuce, leaf spinach, leeks, lettuce, marrows, onions, ornamental plant production, outdoor tomatoes, parsnips, peppers, potatoes, protected ornamentals, protected tomatoes, pumpkins, radicchio, radishes, red beet, runner beans, shallots, spring field beans, squashes, swedes, sweetcorn, turnips, vining peas, winter field beans*** [1, 3]. Glasshouse whitefly in ***aubergines, protected courgettes, protected cucumbers, protected herbs, protected lettuce, protected ornamentals, protected peppers, protected potatoes, protected salad onions, protected tomatoes*** [2]. Green leafhopper in ***aubergines, protected courgettes, protected cucumbers, protected herbs, protected lettuce, protected ornamentals, protected peppers, protected potatoes, protected salad onions, protected tomatoes*** [2]. Insect pests in ***french beans*** *(off-label)* [4]. Leaf miner in ***artichokes, asparagus, broad beans, broccoli, brussels sprouts, cabbages, calabrese, carrots, cauliflowers, celeriac, celery, chicory, chinese cabbage, chives, combining peas, courgettes, cress, cucumbers, dwarf beans, edible podded peas, endives, fennel, frise, garlic, kale, lamb's lettuce, leaf spinach, leeks, lettuce, marrows, onions, ornamental plant production, outdoor tomatoes, parsnips, peppers, potatoes, protected ornamentals, protected tomatoes, pumpkins, radicchio, radishes, red beet, runner beans, shallots, spring field beans, squashes, swedes, sweetcorn, turnips, vining peas, winter field beans*** [1, 3]; ***flower bulbs, leaf brassicas, protected crops*** [4, 5]. Leafhoppers in ***all soft fruit, vegetables*** [4]; ***bush fruit*** [5]; ***flower bulbs, protected crops*** [4, 5]. Peach-potato aphid in ***winter oilseed rape*** *(off-label)* [1]. Potato virus vectors in ***chitting***

potatoes [2]. Sawflies in ***all soft fruit, vegetables*** [4]; ***all top fruit, flower bulbs, protected crops*** [4, 5]; ***bush fruit*** [5]; ***gooseberries*** [1, 3]. Thrips in ***aubergines, protected courgettes, protected cucumbers, protected herbs, protected lettuce, protected ornamentals, protected peppers, protected potatoes, protected salad onions, protected tomatoes*** [2]; ***flower bulbs, protected crops*** [4, 5]; ***vegetables*** [4].

Specific Off-Label Approvals (SOLAs)

- ***french beans*** *(OLA 920080) Dec 2008* [4]
- ***winter oilseed rape*** *(OLA 012480) Dec 2008* [1]

Efficacy guidance

- Apply as foliar spray, taking care to cover undersides of leaves and repeat as necessary or dip young plants, cuttings or strawberry runners before planting out
- Best results achieved by spraying at air temperatures above 16°C but do not treat in bright sunlight or windy weather [2]
- Fumigate glasshouse crops at temperatures of at least 16°C. See label for details of recommended fumigation procedure [2]
- Potatoes may be fumigated in chitting houses to control virus-spreading aphids [2]. Field crops of potatoes may also be treated [1, 3]
- May be used in integrated control systems to give partial control of organophosphorus-, organochlorine- and pyrethroid-resistant whitefly

Restrictions

- Nicotine is subject to the Poisons Rules 1982 and the Poisons Act 1972. See notes in Section 1 [2, 5]
- Maximum number of treatments not specified
- On plants of unknown sensitivity test first on a small scale
- Keep unprotected persons out of treated glasshouses for at least 12 h
- Wear overall, hood, rubber gloves and respirator if entering glasshouse within 12 h of fumigating [2]

Crop-specific information

- HI 2 d for most crops - check labels

Environmental safety

- Highly flammable [2]
- Dangerous to livestock.
- Dangerous to game, wild birds and animals
- Harmful to bees. Do not apply to crops in flower or to those in which bees are actively foraging. Do not apply when flowering weeds are present
- Dangerous to fish or other aquatic life. Do not contaminate surface waters or ditches with chemical or used container
- Keep all livestock out of treated areas for at least 12 h

Hazard classification and safety precautions

Hazard R02 [5]; R03 [1-5]; R07c [2]
Risk phrases R02a, R02b, R02c [5]; R03a, R03b, R03c [1-5]
Operator protection A [1-5]; C [1, 3, 5]; D, J [2]; H [2, 5]; K, M [5]; U02, U04a, U05a, U13 [1, 3-5]; U05b [2]; U08, U10, U15 [4, 5]; U09a [1, 3]; U19, U20a [1-5]
Environmental protection E01, E10a, E12e, E13b [1-5], E02, E08a [1-3] (12 h), E02, E06b [4, 5]
Consumer protection C02 [2] (24 h); C02 [4, 5]; C03 [1-5]
Storage and disposal E30a, E31b [1, 3-5]; E30b [2, 5]; E31a [4, 5]; E32a, E34 [2, 4, 5]
Medical advice M03 [1-4]; M04 [5]

346 oxadiazon

A residual and contact oxadiazolone herbicide for fruit and ornamentals

See also *carbetamide + diflufenican + oxadiazon*
glyphosate + oxadiazon

Products

1	Ronstar 2G	Certis	2% w/w	GR	10011
2	Ronstar Liquid	Certis	250 g/l	EC	08974
3	Ronstar Liquid	Certis	250 g/l	EC	11215

FOR FULL CONDITIONS OF USE ALWAYS READ THE PRODUCT LABEL

SECTION 4

Uses

Annual dicotyledons in ***apple orchards, blackcurrants, gooseberries, grapevines, hops, pear orchards, raspberries, woody ornamentals*** [2, 3]; ***container-grown ornamentals*** [1]. Annual grasses in ***apple orchards, blackcurrants, gooseberries, grapevines, hops, pear orchards, raspberries, woody ornamentals*** [2, 3]; ***container-grown ornamentals*** [1]. Bindweeds in ***apple orchards, blackcurrants, gooseberries, grapevines, hops, pear orchards, raspberries, woody ornamentals*** [2, 3]. Cleavers in ***apple orchards, blackcurrants, gooseberries, grapevines, hops, pear orchards, raspberries, woody ornamentals*** [2, 3]. Knotgrass in ***apple orchards, blackcurrants, gooseberries, grapevines, hops, pear orchards, raspberries, woody ornamentals*** [2, 3].

Efficacy guidance

- Rain or overhead watering is needed soon after application for effective results
- Pre-emergence activity reduced on soils with more than 10% organic matter and, in these conditions, post-emergence treatment is more effective [2, 3]
- Best results on bindweed when first shoots are 10-15 cm long

Restrictions

- See label for list of ornamental species which may be treated with granules. Treat small numbers of other species to check safety. Do not treat hydrangea, spiraea or genista
- Do not cultivate after treatment [2, 3]
- Do not treat container stock under glass or use on plants rooted in media with high sand or non-organic content. Do not apply to plants with wet foliage
- Do not use more than 8 l/ha in any 12 mth period [2, 3]

Crop-specific information

- Latest use: Jul for apples, grapevines, hops, pears; Jun for raspberries, woody ornamentals
- Apply spray to apples and pears from Jan to Jul, avoiding young growth
- Treat bush fruit from Jan to bud-break, avoiding bushes, grapevines in Feb/Mar before start of new growth or in Jun/Jul avoiding foliage
- Treat hops cropped for at least 2 yr in Feb or in Jun/Jul after deleafing
- Treat woody ornamentals from Jan to Jun, avoiding young growth. Do not spray container stock overall

Environmental safety

- Flammable [2, 3]
- Dangerous to fish or other aquatic life. Do not contaminate surface waters or ditches with chemical or used container

Hazard classification and safety precautions

Hazard R04 [1-3]; R07d [2, 3]
Risk phrases R04a [1]; R04b [1-3]; R04d [2, 3]
Operator protection A, C; U05a, U08, U14, U15, U20b
Environmental protection E01, E13b
Consumer protection C03
Storage and disposal E30a, E32a, E34

347 oxamyl

A soil-applied, systemic oxime carbamate nematicide and insecticide

Products

Vydate 10G	DuPont	10% w/w	GR	02322

Uses

American serpentine leaf miner in ***aubergines*** *(off-label)*, ***ornamental plant production*** *(off-label)*, ***protected broad beans*** *(off-label)*, ***protected ornamentals*** *(off-label)*, ***protected peppers*** *(off-label)*, ***protected soya beans*** *(off-label)*, ***protected tomatoes*** *(off-label)*. Aphids in ***potatoes, sugar beet***. Docking disorder vectors in ***sugar beet***. Free-living nematodes in ***potatoes***. Leaf miner in ***aubergines*** *(off-label)*, ***ornamental plant production*** *(off-label)*, ***protected broad beans*** *(off-label)*, ***protected ornamentals*** *(off-label)*, ***protected peppers*** *(off-label)*, ***protected soya beans*** *(off-label)*, ***protected tomatoes*** *(off-label)*. Mangold fly in ***sugar beet***. Millipedes in ***sugar beet***. Potato cyst nematode in ***potatoes***. Pygmy beetle in ***sugar beet***. South American leaf miner in ***aubergines*** *(off-label)*, ***ornamental plant production*** *(off-label)*, ***protected broad***

beans *(off-label)*, ***protected ornamentals*** *(off-label)*, ***protected peppers*** *(off-label)*, ***protected soya beans*** *(off-label)*, ***protected tomatoes*** *(off-label)*. Spraing vectors in **potatoes**. Stem nematodes in ***garlic*** *(off-label)*, ***protected garlic*** *(off-label)*, ***protected onions*** *(off-label)*.

Specific Off-Label Approvals (SOLAs)

- ***aubergines***, ***ornamental plant production***, ***protected broad beans***, ***protected ornamentals***, ***protected peppers***, ***protected soya beans***, ***protected tomatoes*** *(OLA 930020) Dec 2008* [1]
- ***garlic*** *(OLA 920163) Dec 2008* [1]
- ***protected garlic***, ***protected onions*** *(OLA 940925) Dec 2008* [1]

Approval information

- Approvals for oxamyl allowed to continue, subject to label amendments, following the UK Review programme of cholinesterase compounds

Efficacy guidance

- Apply granules with suitable applicator before drilling or planting. See label for details of recommended machines
- In potatoes incorporate thoroughly to 10 cm and plant within 3-4 d
- In sugar beet apply in seed furrow at drilling

Restrictions

- This product contains an anticholinesterase carbamate compound. Do not use if under medical advice not to work with such compounds
- Maximum number of treatments 1 per crop or yr
- Keep in original container, tightly closed, in a safe place, under lock and key
- Wear protective gloves if handling treated compost or soil within 2 wk after treatment
- Allow at least 12 h, followed by at least 1 h ventilation, before entry of unprotected persons into treated glasshouses

Crop-specific information

- Latest use: at drilling/planting for vegetables; before drilling/planting for potatoes and peas.
- HI tomatoes, ornamentals 2 wk

Environmental safety

- Dangerous to fish or other aquatic life. Do not contaminate surface waters or ditches with chemical or used container
- Dangerous to game and wildlife. Cover granules completely. Bury spillages

Hazard classification and safety precautions

Hazard R03
Risk phrases R03b, R03c
Operator protection A, B, H, K, M [1]; C [1] (or D); U02, U04a, U05a, U09a, U13, U19, U20a
Environmental protection E01, E10a, E13b
Consumer protection C02, C03
Storage and disposal E30b, E32a, E34
Medical advice M02, M04

348 oxycarboxin

A protectant, eradicant and systemic carboxamide fungicide for use on ornamentals

Products

Plantvax 75	Fargro	75% w/w	WP	01601

Uses

Rust in ***ornamental specimens***, ***protected ornamentals***.

Approval information

- Products containing this active ingredient have been granted derogations for specified 'Essential Uses' for use until 31 December 2007. Sale and supply must cease by 30 June 2007 but growers have no guarantee that the products will continue to be available until then.
  For more information see 'The Review Programme' under 'Pesticide Legislation' in Section 1

Efficacy guidance

- Apply as a routine to established carnation beds from Sep to Feb at 7-10 d intervals. On rust-susceptible species or if rust already established add authorised wetting agent and repeat every 10-14 d until infection pressure ceases

- On stock plants or cuttings of carnations or geraniums apply after striking and repeat 3 times at 10-14 d intervals. Remove and burn any infected cuttings
- Can be used as a drench on carnations grown in peat bags

Restrictions

- This product must not be used on any crops other than those listed, including any extrapolations that would normally be permissible under the Long Term Arrangements for Extension of Use (see Section 1)

Crop-specific information

- Spray product alone and preferably not within 2 d of other sprays
- Can cause marginal leaf scorch when taken up by roots. Avoid spraying to run-off
- Product leaves visible deposit on plants so avoid treating when blooms are open

Environmental safety

- Dangerous to fish or other aquatic life. Do not contaminate surface waters or ditches with chemical or used container

Hazard classification and safety precautions

Hazard R04
Risk phrases R04a
Operator protection A, C; U02, U05a, U09a, U19, U20b
Environmental protection E01, E13b
Consumer protection C03
Storage and disposal E30a, E32a

349 paclobutrazol

A conazole plant growth regulator for ornamentals and fruit

Products

1	Bonzi	Syngenta Bioline	4 g/l	SC	10517
2	Cultar	Syngenta	250 g/l	SC	10523

Uses

Controlling vigour in ***apples***, ***pears*** [2]. Growth regulation in ***cherries*** *(off-label)*, ***plums*** *(off-label)* [2]. Improving colour in ***poinsettias*** [1]. Increasing flowering in ***azaleas***, ***bedding plants***, ***begonias***, ***kalanchoes***, ***lilies***, ***roses***, ***tulips*** [1]. Increasing fruit set in ***apples***, ***pears*** [2]. Stem shortening in ***azaleas***, ***bedding plants***, ***begonias***, ***kalanchoes***, ***lilies***, ***poinsettias***, ***roses***, ***tulips*** [1].

Specific Off-Label Approvals (SOLAs)

- ***cherries***, ***plums*** *(OLA 031235) Dec 2008* [2]

Efficacy guidance

- Chemical is active via both foliage and root uptake. For best results apply in dull weather when relative humidity not high
- Apply as spray to produce compact pot plants and to improve bract colour of poinsettias [1]
- Apply as compost drench to reduce flower stem length of potted tulips and Mid-Century hybrid lilies [1]
- Timing is critical and varies with species. See label for details [1]

Restrictions

- Maximum number of treatments 1 per specimen for some species [1]
- Maximum total dose equivalent to 3 full dose treatments for pears and 4 full dose treatments for apples [2]
- Some varietal restrictions apply in top fruit (see label for details) [2]
- Do not use on trees of low vigour or under stress [2]
- Do not use on trees from green cluster to 2 wk after full petal fall [2]
- Do not use in underplanted orchards or those recently interplanted in orchards which are required to make vigorous growth

Crop-specific information

- HI apples, pears 14 d [2]

- Apply to apple and pear trees under good growing conditions as pre-blossom spray (apples only) and post-blossom at 7-14 d intervals [2]
- Timing and dose of orchard treatments vary with species and cultivar. See label [2]

Following crops guidance

- Chemical has residual soil activity which can affect growth of following crops. Withhold treatment from orchards due for grubbing to allow the following intervals between the last treatment and planting the next crop: apples, pears 1 yr; beans, peas, onions 2 yr; stone fruit 3 yr; cereals, grass, oilseed rape, carrots 4 yr; market brassicas 6 yr; potatoes and other crops 7 yr [2]

Environmental safety

- Harmful to fish or other aquatic life. Do not contaminate surface waters or ditches with chemical or used container
- Do not use on food crops [1]
- Permissible in organic systems
- Keep livestock out of treated areas for at least 2 yr following treatment [2]

Hazard classification and safety precautions

Hazard R03 [2]
Risk phrases R03c [2]
Operator protection A, H, M [2]; U05a [2]; U08 [1]; U20a [1, 2]
Environmental protection E01 [2]; E07 [2] (2 yr); E13c [1, 2]; E15 [1]
Consumer protection C01 [1]; C02 [2] (2 wk); C03 [2]
Storage and disposal E26, E30a, E31b [1, 2]; E34 [2]
Medical advice M03 [2]

350 paraffin oil (commodity substance)

An agent for the control of birds by egg treatment

Products

paraffin oil	various	OL

Uses

Birds in ***miscellaneous pest control situations*** *(egg treatment).*

Approval information

- Approval for the use of paraffin oil as a commodity substance was granted on 25 April 1995 by Ministers under regulation 5 of the Control of Pesticides Regulations 1986
- Only to be used where a licence has been approved in accordance with Section 16(1) of the Wildlife and Countryside Act (1981)

Efficacy guidance

- Egg treatment should be undertaken as soon as clutch is complete
- Eggs should be treated by complete immersion in liquid paraffin

Restrictions

- Use to control eggs of birds covered by licences issued by the Agriculture and Environment Departments under Section 16(1) of the Wildlife and Countryside Act (1981)
- Treat eggs once only

Hazard classification and safety precautions

Operator protection A, C

351 paraquat

A non-selective, non-residual contact bipyridilium herbicide

See also diquat + paraquat
diuron + paraquat

Products

1	Agriguard Paraquat	AgriGuard	200 g/l	SL	09951
2	Barclay Total	Barclay	200 g/l	SL	11361
3	Dextrone X	n2n	200 g/l	SL	00687
4	Fernpath Graminite	AgriGuard	200 g/l	SL	11646
5	Gramoxone 100	Syngenta	200 g/l	SL	10526

Uses

Annual dicotyledons in **agastache spp.** *(off-label)*, **all top fruit**, **angelica** *(off-label)*, **aubergines** *(off-label - inter-row directed treatment)*, **balm** *(off-label)*, **basil** *(off-label)*, **bay** *(off-label)*, **borage** *(off-label)*, **camomile** *(off-label)*, **caraway** *(off-label)*, **catnip** *(off-label)*, **celery leaves** *(off-label)*, **chervil** *(off-label)*, **chives** *(off-label)*, **clary** *(off-label)*, **coriander** *(off-label)*, **curry plant** *(off-label)*, **dandelions** *(off-label)*, **dill** *(off-label)*, **dragonhead** *(off-label)*, **fennel** *(off-label)*, **fenugreek** *(off-label)*, **feverfew** *(off-label)*, **forest nursery beds**, **grapevines**, **grassland** *(sward destruction/direct drilling)*, **hardy ornamentals**, **hyssop** *(off-label)*, **land cress** *(off-label)*, **lavandin** *(off-label)*, **lavender** *(off-label)*, **lemon verbena** *(off-label)*, **lettuce** *(off-label - inter-row directed treatment)*, **lovage** *(off-label)*, **lucerne** *(off-label)*, **marigold** *(off-label)*, **marjoram** *(off-label)*, **mint** *(off-label)*, **nasturtium** *(off-label)*, **nettle** *(off-label)*, **oregano** *(off-label)*, **origanum heracleoticum** *(off-label)*, **parsley** *(off-label)*, **protected celery** *(off-label - inter-row directed treatment)*, **protected courgettes** *(off-label - inter-row directed treatment)*, **protected cucumbers** *(off-label - inter-row directed treatment)*, **protected gherkins** *(off-label - inter-row directed treatment)*, **protected lettuce** *(off-label - inter-row directed treatment)*, **protected marrows** *(off-label - inter-row directed treatment)*, **protected melons** *(off-label - inter-row directed treatment)*, **protected onions** *(off-label - inter-row directed treatment)*, **protected ornamentals** *(off-label - inter-row directed treatment)*, **protected peppers** *(off-label - inter-row directed treatment)*, **protected pumpkins** *(off-label - inter-row directed treatment)*, **protected squashes** *(off-label - inter-row directed treatment)*, **protected tomatoes** *(off-label - inter-row directed treatment)*, **rhubarb** *(off-label)*, **rocket** *(off-label)*, **rosemary** *(off-label)*, **row crops** *(stale seedbed/inter-row)*, **rue** *(off-label)*, **sage** *(off-label)*, **salad burnet** *(off-label)*, **savory** *(off-label)*, **sorrel** *(off-label)*, **spike lavender** *(off-label)*, **sweet ciceley** *(off-label)*, **tarragon** *(off-label)*, **thyme** *(off-label)*, **violet** *(off-label)*, **woodruff** *(off-label)* [5]; **blackcurrants**, **cultivated land/soil** *(minimum cultivation)*, **gooseberries**, **hops**, **potatoes**, **raspberries**, **stubbles**, **sugar beet** [1, 2, 4, 5]; **flower bulbs**, **forestry**, **forestry transplants**, **non-crop areas** [1-5]; **forest nursery beds** *(stale seedbed)* [1-4]; **orchards**, **redcurrants**, **row crops**, **whitecurrants** [1, 2, 4]; **woody ornamentals** [3]. Annual grasses in **agastache spp.** *(off-label)*, **all top fruit**, **angelica** *(off-label)*, **aubergines** *(off-label - inter-row directed treatment)*, **balm** *(off-label)*, **basil** *(off-label)*, **bay** *(off-label)*, **borage** *(off-label)*, **camomile** *(off-label)*, **caraway** *(off-label)*, **catnip** *(off-label)*, **celery leaves** *(off-label)*, **chervil** *(off-label)*, **chives** *(off-label)*, **clary** *(off-label)*, **coriander** *(off-label)*, **curry plant** *(off-label)*, **dandelions** *(off-label)*, **dill** *(off-label)*, **dragonhead** *(off-label)*, **fennel** *(off-label)*, **fenugreek** *(off-label)*, **feverfew** *(off-label)*, **grapevines**, **hardy ornamentals**, **hyssop** *(off-label)*, **land cress** *(off-label)*, **lavandin** *(off-label)*, **lavender** *(off-label)*, **lemon verbena** *(off-label)*, **lettuce** *(off-label - inter-row directed treatment)*, **lovage** *(off-label)*, **lucerne** *(off-label)*, **marigold** *(off-label)*, **marjoram** *(off-label)*, **mint** *(off-label)*, **nasturtium** *(off-label)*, **nettle** *(off-label)*, **oregano** *(off-label)*, **origanum heracleoticum** *(off-label)*, **parsley** *(off-label)*, **protected celery** *(off-label - inter-row directed treatment)*, **protected courgettes** *(off-label - inter-row directed treatment)*, **protected cucumbers** *(off-label - inter-row directed treatment)*, **protected gherkins** *(off-label - inter-row directed treatment)*, **protected lettuce** *(off-label - inter-row directed treatment)*, **protected marrows** *(off-label - inter-row directed treatment)*, **protected melons** *(off-label - inter-row directed treatment)*, **protected onions** *(off-label - inter-row directed treatment)*, **protected ornamentals** *(off-label - inter-row directed treatment)*, **protected peppers** *(off-label - inter-row directed treatment)*, **protected pumpkins** *(off-label - inter-row directed treatment)*, **protected squashes** *(off-label - inter-row directed treatment)*, **protected tomatoes** *(off-label - inter-row directed treatment)*, **rhubarb** *(off-label)*, **rocket** *(off-label)*, **rosemary** *(off-label)*, **row crops** *(stale seedbed/inter-row)*, **rue** *(off-label)*, **sage** *(off-label)*, **salad burnet** *(off-label)*, **savory** *(off-label)*, **sorrel** *(off-label)*, **spike lavender** *(off-*

label), ***sweet ciceley*** *(off-label)*, ***tarragon*** *(off-label)*, ***thyme*** *(off-label)*, ***violet*** *(off-label)*, ***woodruff*** *(off-label)* [5]; ***blackcurrants***, ***cultivated land/soil*** *(minimum cultivation)*, ***gooseberries***, ***grassland*** *(sward destruction/direct drilling)*, ***hops***, ***potatoes***, ***raspberries***, ***stubbles***, ***sugar beet*** [1, 2, 4, 5]; ***flower bulbs***, ***forest nursery beds*** *(stale seedbed)*, ***forestry***, ***forestry transplants***, ***non-crop areas*** [1-5]; ***orchards***, ***redcurrants***, ***row crops***, ***whitecurrants*** [1, 2, 4]; ***woody ornamentals*** [3]. Barren brome in ***field crops*** *(stubble treatment)* [1, 2, 4]; ***stubbles*** [5]. Chemical stripping in ***hops*** [1, 2, 4, 5]. Creeping bent in ***all top fruit***, ***hardy ornamentals*** [5]; ***blackcurrants***, ***cultivated land/soil*** *(minimum cultivation)*, ***gooseberries***, ***grassland*** *(sward destruction/direct drilling)*, ***hops***, ***raspberries***, ***stubbles*** [1, 2, 4, 5]; ***flower bulbs***, ***forestry***, ***forestry transplants***, ***non-crop areas*** [1-5]; ***orchards***, ***redcurrants***, ***whitecurrants*** [1, 2, 4]; ***woody ornamentals*** [3]. Firebreak desiccation in ***forestry*** [1-5]. Green cover in ***field margins*** [5]; ***land temporarily removed from production*** [1, 2, 4, 5]. Perennial ryegrass in ***all top fruit*** [5]; ***blackcurrants***, ***gooseberries***, ***grassland*** *(sward destruction/direct drilling)*, ***hops***, ***raspberries*** [1, 2, 4, 5]; ***non-crop areas*** [1-5]; ***orchards***, ***redcurrants***, ***whitecurrants*** [1, 2, 4]. Rough meadow grass in ***all top fruit*** [5]; ***blackcurrants***, ***gooseberries***, ***grassland*** *(sward destruction/direct drilling)*, ***hops***, ***raspberries*** [1, 2, 4, 5]; ***non-crop areas*** [1-5]; ***orchards***, ***redcurrants***, ***whitecurrants*** [1, 2, 4]. Runner desiccation in ***strawberries*** [1, 2, 4, 5]. Volunteer cereals in ***cultivated land/soil*** *(minimum cultivation)*, ***potatoes***, ***stubbles***, ***sugar beet*** [1, 2, 4, 5]; ***row crops*** [1, 2, 4]; ***row crops*** *(stale seedbed/inter-row)* [5]. Wild oats in ***stubbles*** [1, 2, 4, 5].

Specific Off-Label Approvals (SOLAs)

- ***agastache spp.***, ***angelica***, ***balm***, ***basil***, ***bay***, ***borage***, ***camomile***, ***caraway***, ***catnip***, ***celery leaves***, ***chervil***, ***chives***, ***clary***, ***coriander***, ***curry plant***, ***dandelions***, ***dill***, ***dragonhead***, ***fennel***, ***fenugreek***, ***feverfew***, ***hyssop***, ***land cress***, ***lavandin***, ***lavender***, ***lemon verbena***, ***lovage***, ***marigold***, ***marjoram***, ***mint***, ***nasturtium***, ***nettle***, ***oregano***, ***origanum heracleoticum***, ***parsley***, ***rocket***, ***rosemary***, ***rue***, ***sage***, ***salad burnet***, ***savory***, ***sorrel***, ***spike lavender***, ***sweet ciceley***, ***tarragon***, ***thyme***, ***violet***, ***woodruff*** *(OLA 021322) Apr 2005* [5]
- ***aubergines***, ***lettuce***, ***protected celery***, ***protected courgettes***, ***protected cucumbers***, ***protected gherkins***, ***protected lettuce***, ***protected marrows***, ***protected melons***, ***protected onions***, ***protected ornamentals***, ***protected peppers***, ***protected pumpkins***, ***protected squashes***, ***protected tomatoes*** *(inter-row directed treatment) (OLA 020225) Dec 2008* [5]
- ***lucerne***, ***rhubarb*** *(OLA 020225) Dec 2008* [5]

Efficacy guidance

- Apply to green weeds preferably less than 15 cm high. Spray is rainfast after 10 min
- Addition of wetter recommended with lower application rates
- Spray in autumn to suppress couch when shoots have 2 leaves, repeat as necessary and plough after last treatment
- For direct-drilling land should be free of perennial weeds and protection against slugs should be provided. Allow 7-10 d before drilling into sprayed grass
- Best time for use in top fruit is Nov-Apr
- For forestry fire-break use apply in Jul-Aug and fire 7-10 d later
- For interrow use in row crops apply with guarded, no-drift sprayer
- Use as a carefully directed spray in blackcurrants, gooseberries, grapevines and other fruit crops, in raspberries only when dormant
- For stawberry runner control use guarded sprayer, not when flowers or fruit present
- Apply to bulbs pre-emergence (at least 3 d pre emergence on sandy soils) or at end of season, provided no attached foliage and bulbs well covered (not on sandy soils)
- In forestry seedbed apply up to 3 d before seedling emergence

Restrictions

- Paraquat is subject to the Poisons Rules 1982 and the Poisons Act 1972. See notes in Section 1
- Products in this profile are for professional use only
- Do not put in a food or drinks container
- Maximum number of treatments 1 per yr for lucerne, outdoor lettuce, protected vegetables and ornamentals; 3 per yr for mint and leafy herbs
- Do not use on straw or other artificial growing media
- Do not use on potatoes under hot, dry conditions or on hops under drought conditions
- If using around glasshouses ensure vents and doors closed
- Only use clean water for mixing up spray
- Allow at least 4 h before cultivating, leave overnight if possible
- Observe label restrictions for maximum permitted concentration when spraying at low volume

SECTION 4

Crop-specific information

- Latest use: before shoots 15 cm high and before 10% of shoots emerged for early potatoes, before 40% of shoots emerged for maincrop potatoes [2, 5]; before 31 Mar for lucerne; see labels for other crops and situations
- HI mint, leafy herbs 8 wk
- Apply to lucerne in late Feb/early Mar when crop dormant

Following crops guidance

- Chemical is inactivated on contact with soil. Crops can be sown or planted soon after spraying on most soils but 3 d should be allowed on sandy or immature peat soils

Environmental safety

- Dangerous for the environment [5]
- Very toxic to aquatic organisms. May cause long-term adverse effects in the aquatic environment [5]
- Paraquat may be harmful to hares; stubbles must be sprayed early in the day [1-5]
- Harmful to fish or other aquatic life. Do not contaminate surface waters or ditches with chemical or used container [3]
- Keep in original container, tightly closed, in a safe place, under lock and key
- Harmful to livestock
- Keep all livestock out of treated areas for at least 24 h
- Do not contaminate surface waters or ditches with chemical or used container

Hazard classification and safety precautions

Hazard R02, R03, R04 [1-5]; R14 [5]
Risk phrases R02c [1-4]; R02d, R03c, R04c, R14c [5]; R03a, R04a, R04b [1-5]
Operator protection A, C; U02, U04a, U20a [1, 2, 4]; U04b, U20b [3, 5]; U05a, U19 [1-5]; U08 [1, 4]; U09a [2, 3, 5]
Environmental protection E01, E11 [1-5]; E06b [1-5] (24 h); E13c [3]; E15 [1, 2, 4, 5]; E38 [5]
Consumer protection C03
Storage and disposal E26, E30b, E34 [1-5]; E31a [1, 3, 4]; E31b [2]; E31c, E32d [5]
Medical advice M04

352 penconazole

A protectant conazole fungicide with antisporulant activity

See also *captan + penconazole*
dithianon + penconazole

Products

Topas	Syngenta	100 g/l	EC	09717

Uses

Powdery mildew in ***apples***, ***blackcurrants***, ***hops***, ***ornamental trees***. Scab in ***ornamental trees***.

Efficacy guidance

- Use as a protectant fungicide by treating at the earliest signs of disease
- Treat crops every 10-14 d (every 7-10 d in warm, humid weather) at first sign of infection or as a protective spray ensuring complete coverage. See label for details of timing
- Increase dose and volume with growth of hops but do not exceed 2000 l/ha. Little or no activity will be seen on established powdery mildew
- Antisporulant activity reduces development of secondary mildew in apples

Restrictions

- Maximum number of treatments 10 per yr for apples, 6 per yr for hops, 4 per yr for blackcurrants
- Check for varietal susceptibility in roses. Some defoliation may occur after repeat applications on Dearest

Crop-specific information

- HI apples, hops 14 d; blackcurrants 4 wk

Environmental safety

- Harmful to fish or other aquatic life. Do not contaminate surface waters or ditches with chemical or used container

Hazard classification and safety precautions

Hazard R04, R14

Risk phrases R04a, R14b

Operator protection A, C; U02, U05a, U08, U20b

Environmental protection E15, E22b, E38

Storage and disposal E26, E30a, E31c, E32d

353 pencycuron

A non-systemic urea fungicide for use on seed potatoes

See also imazalil + pencycuron

Products

1	Agriguard Pencycuron	AgriGuard	12.5% w/w	DP	10445
2	Me2 Penny	Me2	12.5% w/w	DS	10369
3	Monceren DS	Bayer	12.5% w/w	DS	04160
4	Monceren DS	Bayer CropScience	12.5% w/w	DS	11292
5	Standon Pencycuron DP	Standon	12.5% w/w	DS	08774

Uses

Black scurf in ***potatoes*** *(tuber treatment)*. Stem canker in ***potatoes*** *(tuber treatment)*.

Efficacy guidance

- Provides control of tuber-borne disease and gives some reduction of stem canker
- Apply to seed tubers in chitting trays, to bulk bins immediately before planting or in hopper at planting [2-5]
- Apply in accordance with detailed guidelines in manufacturer's literature [3]
- If rain interrupts planting cover tubers in hopper

Restrictions

- Maximum number of treatments 1 per batch of seed potatoes
- Treated tubers must be used only as seed and not for human or animal consumption
- Do not use on tubers previously treated with a dry powder seed treatment or hot water
- Use of suitable dust mask is mandatory when applying dust, filling the hopper or riding on planter

Crop-specific information

- Latest use: at planting

Environmental safety

- Do not use treated seed as food or feed
- Treated seed harmful to game and wildlife

Hazard classification and safety precautions

Operator protection A, F; U20b

Environmental protection E03, E15

Storage and disposal E30a, E32a

Treated seed S01, S02, S03, S04b, S05, S06

354 pendimethalin

A residual dinitroaniline herbicide for cereals and other crops

See also *bentazone + pendimethalin*
chlorotoluron + pendimethalin
cyanazine + pendimethalin
flufenacet + pendimethalin
isoproturon + pendimethalin

Products

1	Agriguard Pendimethalin	AgriGuard	400 g/l	SC	10864
2	Barclay Tremor	Barclay	400 g/l	SC	11342
3	Blazer	Headland	330 g/l	EC	11217
4	Claymore	BASF	400 g/l	SC	10214
5	Inter-Pendimethalin	I T Agro	330 g/l	EC	09841

Products – continued

6	Landgold Pendimethalin 400	Landgold	400 g/l	SC	10288
7	Plinth	Interfarm	330 g/l	EC	11094
8	Standon Pendimethalin 400 SC	Standon	400 g/l	SC	10729
9	Stomp 400 SC	BASF	400 g/l	SC	10229

Uses

Annual dicotyledons in ***angelica*** *(off-label)*, ***balm*** *(off-label)*, ***bay*** *(off-label)*, ***camomile*** *(off-label)*, ***chives*** *(off-label)*, ***hyssop*** *(off-label)*, ***land cress*** *(off-label)*, ***leeks*** *(off-label - drilled and transplanted)*, ***lemon verbena*** *(off-label)*, ***lovage*** *(off-label)*, ***marjoram*** *(off-label)*, ***mint*** *(off-label)*, ***nettle*** *(off-label)*, ***onions*** *(off-label - drilled and transplanted onion sets)*, ***oregano*** *(off-label)*, ***rosemary*** *(off-label)*, ***rue*** *(off-label)*, ***salad burnet*** *(off-label)*, ***sorrel*** *(off-label)*, ***sweetcorn*** *(off-label - grown under crop covers and mulches)*, ***tarragon*** *(off-label)*, ***thyme*** *(off-label)*, ***woodruff*** *(off-label)* [4]; ***apple orchards, blackcurrants, cherries, gooseberries, hops, pear orchards, plums, strawberries*** [1, 3-5, 8, 9]; ***basil*** *(off-label)*, ***borage*** *(off-label)*, ***caraway*** *(off-label)*, ***chervil*** *(off-label)*, ***clary*** *(off-label)*, ***coriander*** *(off-label)*, ***cress*** *(off-label)*, ***dill*** *(off-label)*, ***fennel*** *(off-label)*, ***fenugreek*** *(off-label)*, ***feverfew*** *(off-label)*, ***lamb's lettuce*** *(off-label)*, ***lettuce*** *(off-label)*, ***onions*** *(off-label)*, ***radicchio*** *(off-label)*, ***rocket*** *(off-label)*, ***runner beans*** *(off-label)*, ***savory*** *(off-label)*, ***scarole*** *(off-label)* [4, 9]; ***blackberries, loganberries, raspberries*** [1, 3, 4, 8, 9]; ***broccoli, brussels sprouts, cabbages, calabrese, cauliflowers, leeks*** [1, 3, 4, 7-9]; ***bulb onions***, ***cress*** *(off-label - under covers)*, ***lamb's lettuce*** *(off-label - under covers)*, ***lettuce*** *(off-label - under covers)*, ***radicchio*** *(off-label - under covers)*, ***scarole*** *(off-label - under covers)*, ***sweetcorn*** *(off-label - under crop covers and mulches)* [9]; ***carrots, parsley, parsnips*** [1, 3-5, 7-9]; ***combining peas, durum wheat, potatoes, spring barley, triticale, winter barley, winter rye, winter wheat*** [1-9]; ***forage maize*** [1-4, 8, 9]; ***onions*** [1, 3, 4, 7, 8]; ***rubus hybrids*** [1]; ***sunflowers*** [1-5, 8, 9]; ***tayberries*** [3, 4, 8, 9]. Annual grasses in ***apple orchards, blackcurrants, carrots, cherries, gooseberries, hops, parsley, parsnips, pear orchards, plums, strawberries*** [5]; ***basil*** *(off-label)*, ***borage*** *(off-label)*, ***caraway*** *(off-label)*, ***chervil*** *(off-label)*, ***clary*** *(off-label)*, ***coriander*** *(off-label)*, ***cress*** *(off-label)*, ***cress*** *(off-label - under covers)*, ***dill*** *(off-label)*, ***fennel*** *(off-label)*, ***fenugreek*** *(off-label)*, ***feverfew*** *(off-label)*, ***lamb's lettuce*** *(off-label)*, ***lamb's lettuce*** *(off-label - under covers)*, ***lettuce*** *(off-label)*, ***lettuce*** *(off-label - under covers)*, ***onions*** *(off-label)*, ***radicchio*** *(off-label)*, ***radicchio*** *(off-label - under covers)*, ***rocket*** *(off-label)*, ***runner beans*** *(off-label)*, ***savory*** *(off-label)*, ***scarole*** *(off-label)*, ***scarole*** *(off-label - under covers)*, ***sweetcorn*** *(off-label - under crop covers and mulches)* [9]; ***combining peas, durum wheat, potatoes, spring barley, triticale, winter barley, winter rye, winter wheat*** [2, 5, 6]; ***forage maize*** [2]; ***sunflowers*** [2, 5]. Annual meadow grass in ***apple orchards, blackberries, blackcurrants, cherries, forage maize, gooseberries, hops, loganberries, pear orchards, plums, raspberries, strawberries, sunflowers*** [1, 3, 4, 8, 9]; ***broccoli, brussels sprouts, cabbages, calabrese, carrots, cauliflowers, combining peas, durum wheat, leeks, parsley, parsnips, potatoes, spring barley, triticale, winter barley, winter rye, winter wheat*** [1, 3, 4, 7-9]; ***bulb onions*** [9]; ***onions*** [1, 3, 4, 7, 8]; ***rubus hybrids*** [1]; ***tayberries*** [3, 4, 8, 9]. Blackgrass in ***apple orchards, blackberries, blackcurrants, cherries, gooseberries, hops, loganberries, pear orchards, plums, raspberries, strawberries, sunflowers*** [1, 3, 4, 8, 9]; ***broccoli, brussels sprouts, cabbages, calabrese, carrots, cauliflowers, combining peas, durum wheat, leeks, parsley, parsnips, spring barley, triticale, winter barley, winter rye, winter wheat*** [1, 3, 4, 7-9]; ***bulb onions*** [9]; ***onions*** [1, 3, 4, 7, 8]; ***rubus hybrids*** [1]; ***tayberries*** [3, 4, 8, 9]. Cleavers in ***winter field beans*** *(off-label)* [4, 9]. Rough meadow grass in ***apple orchards, blackberries, blackcurrants, cherries, forage maize, gooseberries, hops, loganberries, pear orchards, plums, raspberries, strawberries, sunflowers*** [1, 3, 4, 8, 9]; ***broccoli, brussels sprouts, cabbages, calabrese, carrots, cauliflowers, combining peas, durum wheat, leeks, parsley, parsnips, potatoes, spring barley, triticale, winter barley, winter rye, winter wheat*** [1, 3, 4, 7-9]; ***bulb onions*** [9]; ***onions*** [1, 3, 4, 7, 8]; ***rubus hybrids*** [1]; ***tayberries*** [3, 4, 8, 9]. Volunteer oilseed rape in ***apple orchards, blackberries, blackcurrants, cherries, forage maize, gooseberries, hops, loganberries, pear orchards, plums, raspberries, strawberries, sunflowers*** [1, 3, 4, 8, 9]; ***broccoli, brussels sprouts, cabbages, calabrese, carrots, cauliflowers, combining peas, durum wheat, leeks, parsley, parsnips, potatoes, spring barley, triticale, winter barley, winter rye, winter wheat*** [1, 3, 4, 7-9]; ***bulb onions*** [9]; ***onions*** [1, 3, 4, 7, 8]; ***rubus hybrids*** [1]; ***tayberries*** [3, 4, 8, 9]. Wild oats in ***durum wheat, triticale, winter barley, winter rye, winter wheat*** [1, 3, 4, 7-9].

Specific Off-Label Approvals (SOLAs)

- ***angelica, balm, basil, bay, borage, camomile, caraway, chervil, chives, clary, coriander, cress, dill, fennel, fenugreek, feverfew, hyssop, lamb's lettuce, land cress, lemon verbena, lettuce, lovage, marjoram, mint, nettle, onions, oregano, radicchio, rocket, rosemary, rue, runner beans, salad burnet, savory, scarole, sorrel, tarragon, thyme, winter field beans, woodruff*** *(OLA 012354) Dec 2008* [4]
- ***basil, borage, caraway, chervil, clary, coriander, cress, dill, fennel, fenugreek, feverfew, lamb's lettuce, lettuce, radicchio, rocket, savory, scarole*** *(OLA 012356) Dec 2008* [9]
- ***cress, lamb's lettuce, lettuce, radicchio, scarole*** *(under covers) (OLA 012355) Dec 2008* [9]
- ***leeks*** *(drilled and transplanted) (OLA 012354) Dec 2008* [4]
- ***onions*** *(drilled and transplanted onion sets) (OLA 012354) Dec 2008* [4]
- ***onions*** *(OLA 012358) Dec 2008* [9]
- ***onions*** *(OLA 012360) Dec 2008* [9]
- ***runner beans*** *(OLA 012357) Dec 2008* [9]
- ***sweetcorn*** *(grown under crop covers and mulches) (OLA 012354) Dec 2008* [4]
- ***sweetcorn*** *(under crop covers and mulches) (OLA 012359) Dec 2008* [9]
- ***winter field beans*** *(OLA 012361) Dec 2008* [9]

Approval information

- Pendimethalin included in Annex I under EC Directive 91/414

Efficacy guidance

- Apply as soon as possible after drilling. Weeds are controlled as they germinate and emerged weeds will not be controlled by use of the product alone
- For effective blackgrass control apply not more than 2 d after final cultivation and before weed seeds germinate
- Tank mixes with approved formulations of isoproturon or chlorotoluron recommended for improved pre- and post-emergence control of blackgrass [9]
- Tank mixture with atrazine recommended for weed control in fodder maize
- Best results by application to fine firm, moist, clod-free seedbeds when rain follows treatment. Effectiveness reduced by prolonged dry weather after treatment
- Effectiveness reduced on soils with more than 6% organic matter. Do not use where organic matter exceeds 10%
- Any trash, ash or straw should be incorporated evenly during seedbed preparation
- Do not disturb soil after treatment
- On peas drilled after end Mar (mid-Apr in Scotland) tank-mix with cyanazine [9]
- Apply to potatoes as soon as possible after planting and ridging in tank-mix with cyanazine or metribuzin but note that the use of cyanazine or metribuzin will restrict the varieties that may be treated

Restrictions

- Maximum number of treatments 1 per crop or yr
- Maximum total dose equivalent to one full dose treatment on most crops; 2 full dose treatments on leeks
- May be applied pre-emergence of cereal crops sown before 30 Nov provided seed covered by at least 32 mm soil, or post-emergence to early tillering stage (GS 23)
- Do not undersow treated crops
- Do not use on crops suffering stress due to disease, drought, waterlogging, poor seedbed conditions or chemical treatment or on soils where water may accumulate

Crop-specific information

- Latest use: pre-emergence for spring barley, carrots, lettuce, fodder maize, parsnips, parsley, sage, peas, runner beans, potatoes, onions and leeks; before transplanting for brassicas; before leaf sheaths erect for winter cereals; before bud burst for blackcurrants, gooseberries, cane fruit, hops; before flower trusses emerge for strawberries; 14 d after transplanting for leaf herbs
- Do not use on spring barley after end Mar (mid-Apr in Scotland on some labels) because dry conditions likely. Do not apply to dry seedbeds in spring unless rain imminent
- Apply to combining peas as soon as possible after sowing. Do not spray if plumule less than 13 mm below soil surface
- Apply to potatoes up to 7 d before first shoot emerges
- Apply to drilled crops as soon as possible after drilling but before crop and weed emergence
- Apply in top fruit, bush fruit and hops from autumn to early spring when crop dormant

- In cane fruit apply to weed free soil from autumn to early spring, immediately after planting new crops and after cutting out canes in established crops
- Apply in strawberries from autumn to early spring (not before Oct on newly planted bed). Do not apply pre-planting or during flower initiation period (post-harvest to mid-Sep)
- Apply pre-emergence in drilled onions or leeks, not on Sands, Very Light, organic or peaty soils or when heavy rain forecast
- Apply to brassicas after final plant-bed cultivation but before transplanting. Avoid unnecessary soil disturbance after application and take care not to introduce treated soil into the root zone when transplanting. Follow transplanting with specified post-planting treatments - see label
- Do not use on protected crops or in greenhouses

Following crops guidance

- Before ryegrass is drilled after a very dry season plough or cultivate to at least 15 cm. If treated spring crops are to be followed by crops other than cereals, plough or cultivate to at least 15 cm
- In the event of crop failure land must be ploughed or thoroughly cultivated to at least 15 cm. See label for minimum intervals that should elapse between treatment and sowing a range of replacement crops

Environmental safety

- Flammable [3]
- Dangerous for the environment [4, 9]
- Very toxic to aquatic organisms. May cause long-term adverse effects in the aquatic environment [4, 9]
- Dangerous to fish or other aquatic life. Do not contaminate surface waters or ditches with chemical or used container

Hazard classification and safety precautions

Hazard R03, R04 [3, 5, 7]; R07d [3, 7]; R14 [4, 9]
Risk phrases R03a, R03b, R04a, R04b [5]; R04g [3, 7]; R14c [4, 9]
Operator protection A, C [5]; U02, U11 [5]; U05a [3-5, 7-9]; U08, U19 [1-9]; U13 [1, 2, 4-6, 8, 9]; U20a [8]; U20b [1, 2, 4-6, 9]
Environmental protection E01 [3-5, 7-9]; E13b [1-3, 5-8]; E15, E38 [4, 9]; E16a [7]
Consumer protection C03 [3-5, 7-9]
Storage and disposal E26 [1-7, 9]; E30a, E31b [1-9]; E32d [4, 9]; E34 [3, 5, 7, 8]
Medical advice M03, M05 [3, 7]

355 pendimethalin + picolinafen

A post-emergence broad-spectrum herbicide mixture for winter cereals

Products

1	PicoMax	BASF	320:16 g/l	SC	10739
2	Picona	BASF	320:16 g/l	SC	10715
3	PicoPro	BASF	320:16 g/l	SC	10740
4	Stomp Pico	BASF	320:16 g/l	SC	10738

Uses

Annual dicotyledons in ***winter barley***, ***winter wheat***. Annual meadow grass in ***winter barley***, ***winter wheat***. Chickweed in ***winter barley***, ***winter wheat***. Cleavers in ***winter barley***, ***winter wheat***. Rough meadow grass in ***winter barley***, ***winter wheat***. Speedwells in ***winter barley***, ***winter wheat***.

Approval information

- Pendimethalin and picolinafen included in Annex I under EC Directive 91/414

Efficacy guidance

- Best results obtained on crops growing in a fine, firm tilth and when rain falls within 7 d of application
- Loose or cloddy seed beds must be consolidated prior to application otherwise reduced weed control may occur
- Residual weed control may be reduced on soils with more than 6% organic matter or under prolonged dry conditions

Restrictions

- Maximum number of treatments 1 per crop

- Do not treat undersown cereals or those to be undersown
- Do not roll emerged crops before treatment nor autumn treated crops until the following spring
- Do not use on stony or gravelly soils, on soils that are waterlogged or prone to waterlogging, or on soils with more than 10% organic matter
- Do not treat crops under stress from any cause

Crop-specific information
- Latest use: before pseudo-stem erect stage (GS 30)
- Transient bleaching may occur after treatment but it does not lead to yield loss

Following crops guidance
- In the event of failure of a treated crop plough to at least 15 cm and allow at least 8 wk from the time of treatment before re-drilling either spring wheat or spring barley
- After a normally harvested treated crop there are no restrictions on following crops except that, in a very dry season, plough or cultivate to at least 15 cm before drilling rye-grass

Environmental safety
- Dangerous for the environment
- Very toxic to aquatic organisms. May cause long-term adverse effects in the aquatic environment
- Product binds strongly to soil minimising likelihood of movement into groundwater

Hazard classification and safety precautions

Hazard R14
Risk phrases R14c
Operator protection A, H; U02, U05a, U20c
Environmental protection E01, E15, E38
Consumer protection C03
Storage and disposal E26, E30a, E31c, E32d, E34
Medical advice M03

356 pendimethalin + simazine

A contact and residual herbicide for use in winter cereals

Products

Merit	BASF	300:100 g/l	SC	10384

Uses

Annual dicotyledons in ***winter barley***, ***winter wheat***. Meadow grasses in ***winter barley***, ***winter wheat***.

Approval information
- Pendimethalin included in Annex I under EC Directive 91/414

Efficacy guidance
- Apply from pre-emergence to early tillering stage of crop
- Best results achieved on fine, firm, moist seedbed
- Any straw residues should be incorporated before spraying

Restrictions
- Maximum number of treatments 1 per crop
- Do not use on durum wheat
- Do not use on Sands, Very Light, stony or gravelly soils
- Do not use on soils with more than 10% organic matter
- Do not apply to crops under stress or on waterlogged soils
- Avoid overlapping spray swathes

Crop-specific information
- Latest use: before main shoot and 2 tillers (GS 21)
- Ensure crop seed is evenly covered with at least 32 mm of settled soil

Environmental safety
- Dangerous to fish or other aquatic life. Do not contaminate surface waters or ditches with chemical or used container
- LERAP Category B

Hazard classification and safety precautions

Operator protection A, C, H; U05a, U08, U13, U19, U20b

Environmental protection E01, E13b, E16a, E16b

Consumer protection C03

Storage and disposal E26, E30a, E31b, E34

357 pentanochlor

A contact anilide herbicide

See also chlorpropham + pentanochlor

Products

1	Atlas Solan 40	Nufarm UK	400 g/l	EC	07726
2	Croptex Bronze	Certis	400 g/l	EC	04087
3	Croptex Bronze	Certis	400 g/l	EC	11329

Uses

Annual dicotyledons in ***anemones, carnations, celeriac, foxgloves, freesias, larkspur, roses, sweet williams, wallflowers*** [1]; ***carrots, celery, chrysanthemums, nursery stock, parsley, parsnips, sweet peas*** [1-3]. Annual meadow grass in ***anemones, carnations, celeriac, foxgloves, freesias, larkspur, roses, sweet williams, wallflowers*** [1]; ***carrots, celery, chrysanthemums, nursery stock, parsley, parsnips, sweet peas*** [1-3].

Approval information

- Products containing this active ingredient have been granted derogations for specified 'Essential Uses' for use until 31 December 2007. Sale and supply must cease by 30 June 2007 but growers have no guarantee that the products will continue to be available until then.
  For more information see 'The Review Programme' under 'Pesticide Legislation' in Section 1

Efficacy guidance

- Pentanochlor acts mainly through contact activity with some residual effect in moist soils. Best results achieved on young weed seedlings under warm, moist conditions
- Weeds most susceptible in cotyledon to 2-3 leaf stage although some important species controlled at later stages
- Effectiveness reduced by very cold weather or drought

Restrictions

- Maximum number of treatments 2 per crop for edible crops and sweet peas; 1 per yr for other horticultural ornamental crops
- Not recommended for flower crops on light sandy soils
- Do not treat crops suffering from stress, drought, mineral deficiency, pest attack or damage from previous herbicide applications
- Do not harvest crops for human or animal consumption for at least 28 d after last application
- These products must not be used on any crops other than those listed, including any extrapolations that would normally be permissible under the Long Term Arrangements for Extension of Use (see Section 1), except on herbs and other ornamentals

Crop-specific information

- HI 28 d for all edible crops
- Apply pre-emergence in anemones, freesias, foxgloves, larkspur

Following crops guidance

- In the event of failure of a treated crop only recommended crops should be re-planted
- Any crop may be planted 4-5 wk after application following mould-board ploughing or repeated cultivation

Hazard classification and safety precautions

Hazard R04

Risk phrases R04a, R04b [1-3]; R04c, R04g [1]

Operator protection A, C; U05a, U08, U19 [1-3]; U20a [1, 2]; U20b [3]

Environmental protection E01 [1-3]; E13b [3]; E15 [1, 2]

Consumer protection C02 [3] (28 d); C03 [1-3]
Storage and disposal E26 [3]; E30a, E31b [1-3]

358 petroleum oil

An insecticidal and acaricidal hydrocarbon oil

Products

Certis Spraying Oil	Certis	710 g/l	EC	00000

Uses

Mealybugs in ***protected cucumbers, protected grapevines, protected pot plants, protected tomatoes***. Red spider mites in ***protected cucumbers, protected grapevines, protected pot plants, protected tomatoes***. Scale insects in ***protected cucumbers, protected grapevines, protected pot plants, protected tomatoes***.

Approval information

- Product not controlled by Control of Pesticides Regulations because it acts by physical means only

Efficacy guidance

- Spray at 1% (0.5% on tender foliage) to wet plants thoroughly, particularly the underside of leaves, and repeat as necessary
- Apply under quick drying conditions but not in bright sun unless glass well shaded
- On plants of unknown sensitivity test first on a small scale
- Mixtures with certain pesticides may damage crop plants. If mixing, spray a few plants to test for tolerance before treating larger areas

Restrictions

- Do not mix with sulphur or use sulphur sprays within 28 d of treatment

Crop-specific information

- HI zero
- Treat grapevines before flowering

Hazard classification and safety precautions

Hazard R03, R04
Risk phrases R03c, R04a, R04c
Operator protection A, C; U05a, U08, U19
Environmental protection E01
Consumer protection C03
Storage and disposal E30a, E34
Medical advice M03

359 phenmedipham

A contact carbamate herbicide for beet crops and strawberries

See also *desmedipham + ethofumesate + phenmedipham*
desmedipham + phenmedipham
ethofumesate + phenmedipham
lenacil + phenmedipham

Products

1	Agriguard Phenmedipham	AgriGuard	114 g/l	EC	10712
2	Beetup	United Phosphorus	114 g/l	EC	07520
3	Beetup Flo 160	United Phosphorus	160 g/l	SC	10382
4	Betanal Flow	Aventis	160 g/l	SE	10997
5	Crotale	Sipcam	471 g/l	SC	11593
6	Dancer FL	Sipcam	160 g/l	SC	10048
7	Herbasan	Nufarm UK	160 g/l	SC	10734
8	Herbasan Flow	Nufarm UK	160 g/l	SC	11448
9	Mandolin Flow	Nufarm UK	160 g/l	SE	11162
10	Tripart Beta	Tripart	118 g/l	EC	03111

Uses

Annual dicotyledons in ***fodder beet, mangels, red beet, sugar beet*** [1-10]; ***strawberries*** [1, 2].

Approval information

- Approval expiry 31 Mar 2004 [1]

Efficacy guidance

- Best results achieved by application to young seedling weeds, preferably cotyledon stage, under good growing conditions when low doses are effective
- 2-3 repeat applications at 7-10 d intervals using a low dose are recommended on mineral soils, 3-5 applications may be needed on organic soils
- Addition of adjuvant oil may improve effectiveness on some weeds
- Various tank-mixtures with other beet herbicides recommended. See label for details
- Use of certain pre-emergence herbicides is recommended in combination with post-emergence treatment. See label for details

Restrictions

- Maximum number of treatments - see labels
- At high temperatures (above 21°C) reduce rate and spray after 5 pm
- Do not apply immediately after frost or if frost expected
- Do not spray wet foliage or if rain imminent
- Do not spray crops stressed by wind damage, nutrient deficiency, pest or disease attack etc. Do not roll or harrow for 7 d before or after treatment
- Do not use on strawberries under cloches or polythene tunnels
- Do not use after 31 Jul in yr of harvest [3-9]

Crop-specific information

- Latest use: before crop leaves meet between rows for beet crops; before flowering for strawberries
- Apply to beet crops at any stage as low dose/low volume spray or from fully developed cotyledon stage with full rate. Apply to red beet after fully developed cotyledon stage
- Apply to strawberries at any time when weeds in susceptible stage, except in period from start of flowering to picking

Environmental safety

- Harmful (dangerous [3, 6-8]) to fish or other aquatic life. Do not contaminate surface waters or ditches with chemical or used container [1, 2, 4, 5, 9, 10]

Hazard classification and safety precautions

Hazard R03 [2, 10]; R04 [1, 2, 5, 10]

Risk phrases R03c [2, 10]; R04a, R04b, R04c [1, 2, 10]; R04e [5]

Operator protection A [1, 2, 4, 5, 9, 10]; C [1, 2, 10]; H [5]; P [10]; U05a [1, 2, 5, 7, 8, 10]; U08 [1-3, 7, 8, 10]; U09a [6]; U13 [2]; U19 [1-3, 6-8, 10]; U20b [1-10]

Environmental protection E01 [1, 2, 5, 7, 8, 10]; E13b [3, 6-8]; E13c [1, 2, 4, 5, 9, 10]

Consumer protection C03 [1, 2, 5, 7, 8, 10]

Storage and disposal E26 [1-4, 6, 9]; E30a [1-10]; E31b [1, 2, 4, 5, 7, 8, 10]; E31c [3, 6, 9]; E34 [1, 2]

Medical advice M03 [1, 2, 10]

360 d-phenothrin

A non-systemic pyrethroid insecticide available only in mixtures

361 d-phenothrin + tetramethrin

A pyrethroid insecticide mixture for control of flying insects

Products

1	Killgerm ULV 500	Killgerm	36.8:18.4 g/l	UL	H4647
2	Sorex Super Fly Spray	Sorex	0.10:0.25% w/w	AE	H6297

Uses

Flies in ***agricultural premises*** [1, 2]. Grain storage mite in ***grain stores*** [1]. Mosquitoes in ***agricultural premises*** [1, 2]. Wasps in ***agricultural premises*** [1, 2].

Approval information

- Product approved for ULV application. See label for details [1]

Efficacy guidance

- Close doors and windows and spray in all directions for 3-5 sec. Keep room closed for at least 10 min
- Product synergised with piperonyl butoxide [2]

Restrictions

- Do not use space sprays containing pyrethrins or pyrethroid more than once per week in intensive or controlled environment animal houses in order to avoid development of resistance. If necessary, use a different control method or product [1]

Crop-specific information

- May be used in the presence of poultry and livestock

Environmental safety

- Extremely flammable [2]
- Do not apply directly to livestock/poultry
- Extremely dangerous to fish or other aquatic life. Do not contaminate surface waters or ditches with chemical or used container
- Remove exposed milk and collect eggs before application. Protect milk machinery and containers from contamination

Hazard classification and safety precautions

Hazard R03, R04 [1]; R07b [2]
Risk phrases R04g [1]
Operator protection A, H [1]; U02, U09a, U16, U17, U20b [1]; U14, U20c [2]; U19 [1, 2]
Environmental protection E01, E05, E13a [1, 2]; E02 [2]; E04, E15 [1]
Consumer protection C04, C05, C08, C10 [2]; C06, C09 [1]; C07 [1, 2]
Storage and disposal E30a [1, 2]; E32a [1]
Medical advice M03 [2]; M05 [1]

362 picloram

A persistent, translocated pyridine carboxylic acid herbicide for non-crop areas

See also 2,4-D + picloram

Products

Tordon 22K	n2n	240 g/l	SL	05790

Uses

Annual dicotyledons in ***non-crop areas***, ***non-crop grass***. Bracken in ***non-crop areas***, ***non-crop grass***. Japanese knotweed in ***non-crop areas***, ***non-crop grass***. Perennial dicotyledons in ***non-crop areas***, ***non-crop grass***. Woody weeds in ***non-crop areas***, ***non-crop grass***.

Efficacy guidance

- May be applied at any time of year. Best results achieved by application as foliage spray in late winter to early spring
- For bracken control apply 2-4 wk before frond emergence
- Clovers are highly sensitive and eliminated at very low doses
- Persists in soil for up to 2 yr

Restrictions

- Maximum number of treatments 1 per yr
- Do not apply around desirable trees or shrubs where roots may absorb chemical
- Do not apply on slopes where chemical may be leached onto areas of desirable plants

Environmental safety

- Harmful to fish or other aquatic life. Do not contaminate surface waters or ditches with chemical or used container
- Keep livestock out of treated areas until foliage of any poisonous weeds such as ragwort has died and become unpalatable

Hazard classification and safety precautions

Hazard R04
Risk phrases R04a
Operator protection U05a, U08, U20b

Environmental protection E01, E07, E13c
Consumer protection C03
Storage and disposal E30a, E31b

363 picolinafen

An aryloxypicolinamide herbicide for cereals available only in mixtures

See also pendimethalin + picolinafen

364 picoxystrobin

A broad spectrum strobilurin fungicide for cereals

Products

Acanto	Syngenta	250 g/l	SC	10978

Uses

Brown rust in ***spring barley, spring wheat, winter barley, winter wheat***. Crown rust in ***spring oats, winter oats***. Eyespot in ***spring wheat*** *(reduction)*, ***winter wheat*** *(reduction)*. Late ear diseases in ***spring wheat, winter wheat***. Net blotch in ***spring barley, winter barley***. Powdery mildew in ***spring barley, spring oats, winter barley, winter oats***. Rhynchosporium in ***spring barley, winter barley***. Septoria diseases in ***spring wheat, winter wheat***. Yellow rust in ***spring wheat, winter wheat***.

Efficacy guidance

- Best results achieved from protectant treatments made before disease establishes in the crop
- Use of tank mixtures is recommended where disease has become established at the time of treatment
- Persistence of yellow rust control is less than for other foliar diseases but can be improved by use of an appropriate tank mixture
- Eyespot control is not reliable and an appropriate tank mix should be used where crops are at risk
- Picoxystrobin is a member of the QoI cross resistance group. Product should be used preventatively and not relied on for its curative potential
- Use product as part of an Integrated Crop Management strategy incorporating other methods of control, including where appropriate other fungicides with a different mode of action. Do not apply more than two foliar applications of QoI containing products to any cereal crop
- There is a significant risk of widespread resistance occurring in *Septoria tritici* populations in UK. Failure to follow resistance management action may result in reduced levels of disease control
- On cereal crops product must always be used in mixture with another product, recommended for control of the same target disease, that contains a fungicide from a different cross resistance group and is applied at a dose that will give robust control
- Strains of barley powdery mildew resistant to QoI's are common in the UK

Restrictions

- Maximum number of treatments 2 per crop per yr

Crop-specific information

- Latest use: grain watery ripe (GS 71)

Environmental safety

- Dangerous for the environment
- Very toxic to aquatic organisms. May cause long-term adverse effects in the aquatic environment

Hazard classification and safety precautions

Hazard R14
Risk phrases R14c
Operator protection U05a, U09a, U14, U15, U20a
Environmental protection E01, E15, E38
Consumer protection C03
Storage and disposal E26, E30a, E31c, E32d

365 pirimicarb

A carbamate insecticide for aphid control

See also *deltamethrin + pirimicarb*
lambda-cyhalothrin + pirimicarb

Products

1	Agriguard Pirimicarb	AgriGuard	50% w/w	WG	09620
2	Aphox	Syngenta	50% w/w	WG	10515
3	Greencrop Glenroe	Greencrop	50% w/w	WG	09903
4	Landgold Pirimicarb 50	Landgold	50% w/w	WG	09018
5	Standon Pirimicarb 50	Standon	50% w/w	WG	08878

Uses

Aphids in ***apples, broad beans, broccoli, brussels sprouts, cabbages, calabrese, cauliflowers, durum wheat, pears, peas, potatoes, spring barley, spring field beans, spring oats, spring rye, spring wheat, strawberries, sugar beet, swedes, triticale, turnips, winter barley, winter field beans, winter oats, winter rye, winter wheat*** [1-5]; ***blackcurrants, carrots, chinese cabbage, collards, dwarf beans, gooseberries, maize, parsnips, raspberries, redcurrants, runner beans, spring oilseed rape, sweetcorn, winter oilseed rape*** [1-3, 5]; ***carnations, celeriac*** *(off-label)*, ***chicory*** *(off-label - for forcing)*, ***chrysanthemums, cinerarias, courgettes*** *(off-label)*, ***cyclamen, endives*** *(off-label)*, ***fennel*** *(off-label)*, ***gherkins*** *(off-label)*, ***honesty*** *(off-label)*, ***horseradish*** *(off-label)*, ***leaf spinach*** *(off-label)*, ***marrows*** *(off-label)*, ***parsley*** *(off-label)*, ***parsley root*** *(off-label)*, ***protected courgettes*** *(off-label)*, ***protected endives*** *(off-label)*, ***protected gherkins*** *(off-label)*, ***protected radishes*** *(off-label)*, ***protected spinach*** *(off-label)*, ***protected spinach beet*** *(off-label)*, ***radishes*** *(off-label)*, ***red beet*** *(off-label)*, ***roses, salad brassicas, salad brassicas*** *(off-label - for baby leaf production)*, ***spinach beet*** *(off-label)*, ***sweetcorn*** *(off-label)* [2]; ***cherries*** [1-3]; ***cucumbers, forest nurseries, lettuce*** *(outdoor crops)*, ***outdoor tomatoes, peppers, protected carnations, protected chrysanthemums, protected cinerarias, protected cyclamen, protected lettuce, protected roses, protected tomatoes*** [2, 3]; ***grassland*** [1]; ***kale*** [1, 2, 4, 5]; ***ornamental specimens*** [3]. Blackfly in ***cherries*** [5]. Leaf curling plum aphid in ***plums*** *(off-label)* [2]. Mealy plum aphid in ***plums*** *(off-label)* [2]. Thrips in ***endives*** *(off-label)*, ***parsley*** *(off-label)*, ***protected endives*** *(off-label)* [2].

Specific Off-Label Approvals (SOLAs)

- ***celeriac, fennel, red beet*** *(OLA 011292) Dec 2008* [2]
- ***chicory*** *(for forcing) (OLA 011283) Dec 2008* [2]
- ***courgettes, gherkins, horseradish, leaf spinach, marrows, parsley root, protected radishes, protected spinach, protected spinach beet, radishes, spinach beet, sweetcorn*** *(OLA 011298) Dec 2008* [2]
- ***endives, parsley, protected endives*** *(OLA 011286) Dec 2008* [2]
- ***honesty*** *(OLA 011289) Dec 2008* [2]
- ***plums*** *(OLA 012687) Dec 2008* [2]
- ***protected courgettes, protected gherkins*** *(OLA 011301) Dec 2008* [2]
- ***salad brassicas*** *(for baby leaf production) (OLA 011295) Dec 2008* [2]

Approval information

- Pirimicarb has been included in a comprehensive review of anticholinesterase compounds and considered by the ACP. As a result approvals for some SOLAs, and for uses on grassland and celery were revoked and approvals expire in Dec 2003
- Approved for aerial application on cereals [2, 3]. See notes in Section 1

Efficacy guidance

- Chemical has contact, fumigant and translaminar activity
- Best results achieved under warm, calm conditions when plants not wilting and spray does not dry too rapidly. Little vapour activity at temperatures below 15°C
- Apply as soon as aphids seen or warning issued and repeat as necessary
- Addition of non-ionic wetter recommended for use on brassicas
- On cucumbers and tomatoes a root drench is preferable to spraying when using predators in an integrated control programme
- Where aphids resistant to pirimicarb occur control is unlikely to be satisfactory

SECTION 4

Restrictions

- This product contains an anticholinesterase carbamate compound. Do not use if under medical advice not to work with such compounds
- Maximum number of treatments normally 2 per crop but not specified in some cases
- When treating ornamentals check safety by treating a small number of plants first
- Spray equipment must only be used where the operator's normal working position is within a closed cab on a tractor or self-propelled sprayer when making air-assisted applications to apples or other top fruit

Crop-specific information

- HI oilseed rape, cereals, maize, sweetcorn, lettuce under glass 14 d (sweetcorn off-label 3 d); grassland, chicory, plums 7 d; cucumbers, tomatoes and peppers under glass 2 d; protected courgettes and gherkins 24 h; other edible crops 3 d; flowers and ornamentals zero

Environmental safety

- Dangerous for the environment [2]
- Very toxic to aquatic organisms. May cause long-term adverse effects in the aquatic environment [2]
- Dangerous to fish or other aquatic life. Do not contaminate surface waters or ditches with chemical or used container [1, 3-5]
- Chemical has little effect on bees, ladybirds and other insects and is suitable for use in integrated control programmes on apples and pears
- Keep all livestock out of treated areas for at least 7 d. Bury or remove spillages

Hazard classification and safety precautions

Hazard R02, R04, R14 [2]; R03 [1, 3-5]
Risk phrases R02b, R02c, R04a, R14c [2]; R03c [1, 3-5]
Operator protection A, C, D, H, J, M; U05a, U08 [1-5]; U19, U20b [1, 2]; U20a [3-5]
Environmental protection E01 [1-5]; E06b [1-3] (7 d); E06b [4]; E06b [5] (7 d); E13b [1, 3-5]; E38 [2]
Consumer protection C03
Storage and disposal E26, E32d [2]; E30a, E32a, E34 [1-5]
Medical advice M02 [1-3, 5]; M03 [1-5]; M04 [2]

366 pirimiphos-methyl

A contact, fumigant and translaminar organophosphorus insecticide

Products

1	Actellic D	Syngenta	250 g/l	EC	10509
2	Actellic Smoke Generator No. 20	Syngenta	22.5% w/w	FU	10540
3	Fumite Pirimiphos Methyl Smoke	Certis	22.5% w/w	FU	00941

Uses

Ants in ***protected cucumbers***, ***protected ornamentals***, ***protected peppers***, ***protected tomatoes*** [3]. Aphids in ***protected cucumbers***, ***protected ornamentals***, ***protected peppers***, ***protected tomatoes*** [3]. Capsids in ***protected cucumbers***, ***protected ornamentals***, ***protected peppers***, ***protected tomatoes*** [3]. Earwigs in ***protected cucumbers***, ***protected ornamentals***, ***protected peppers***, ***protected tomatoes*** [3]. Flour beetle in ***stored grain*** [1]. Flour moth in ***stored grain*** [1]. Grain beetle in ***stored grain*** [1]. Grain storage mite in ***stored grain*** [1]. Grain storage pests in ***grain stores*** [1, 2]. Grain weevil in ***stored grain*** [1]. Leaf miner in ***protected cucumbers***, ***protected ornamentals***, ***protected peppers***, ***protected tomatoes*** [3]. Red spider mites in ***protected cucumbers***, ***protected ornamentals***, ***protected peppers***, ***protected tomatoes*** [3]. Sawflies in ***protected cucumbers***, ***protected ornamentals***, ***protected peppers***, ***protected tomatoes*** [3]. Thrips in ***protected cucumbers***, ***protected ornamentals***, ***protected peppers***, ***protected tomatoes*** [3]. Warehouse moth in ***stored grain*** [1]. Whitefly in ***protected cucumbers***, ***protected ornamentals***, ***protected peppers***, ***protected tomatoes*** [3].

Efficacy guidance

- Chemical acts rapidly and has short persistence in plants but persists for long periods on inert surfaces

- Best results for protection of stored grain achieved by cleaning store thoroughly before use and employing a combination of pre-harvest and grain/seed treatments
- Best results for admixture treatment obtained when grain stored at 15% moisture or less. Dry and cool moist grain coming into store but then treat as soon as possible, ideally as it is loaded
- Surface admixture can be highly effective on localised surface infestations but should not be relied on for long-term control unless application can be made to the full depth of the infestation
- For control of whitefly and other glasshouse pests apply as smoke or by thermal fogging. See label for details of techniques and suitable machines [3]
- Where insect pests resistant to pirimiphos-methyl occur control is unlikely to be satisfactory

Restrictions

- This product contains an organophosphorus anticholinesterase compound. Do not use if under medical advice not to work with such compounds
- Maximum number of treatments 2 per grain store; 1 per batch of stored grain
- Do not mix with grain store disinfectants because of risk of chemical interaction [1]
- Do not apply treated seed from the air [1]
- Do not fumigate glasshouses in bright sunshine or when foliage is wet or roots dry
- Do not fumigate young seedlings or plants being hardened off
- Do not fog open flowers of ornamentals without first consulting firm
- Do not fog mushrooms when wet as slight spotting may occur

Crop-specific information

- Latest use: before storing grain
- HI for edible crops zero [3]
- Disinfect empty grain stores by spraying surfaces and/or fumigation and treat grain by full or surface admixture. Treat well before harvest in late spring or early summer and repeat 6 wk later or just before harvest if heavily infested. See label for details of treatment and suitable application machinery
- Treatment volumes on structural surfaces should be adjusted according to surface porosity. See label for guidelines
- Treat inaccessible areas with smoke generating product used in conjunction with spray treatment of remainder of store

Environmental safety

- Highly flammable [2]; flammable [1, 3]
- Ventilate fumigated or fogged spaces thoroughly before re-entry
- Unprotected persons must be kept out of fumigated areas within 3 h of ignition and for 4 h after treatment
- Do not harvest for human or animal consumption for at least 5 d after last application
- Extremely dangerous to bees. Do not apply to crops in flower or to those in which bees are actively foraging. Do not apply when flowering weeds are present [3]
- Harmful (extremely dangerous [3]) to fish or other aquatic life. Do not contaminate surface waters or ditches with chemical or used container [1, 2]
- Wildlife must be excluded from buildings during treatment [1]
- Grain treated by admixture as specified may be consumed by humans and livestock

Hazard classification and safety precautions

Hazard R04 [1-3], R07c [2]; R07d [1, 3]
Risk phrases R04a [1-3]; R04c [2, 3]
Operator protection A, C [1]; D, H [1, 2]; U04a, U20b [1, 2]; U05a [1, 3]; U08, U20a [3]; U09a, U16, U17, U24 [1]; U10 [2]; U19 [1-3]
Environmental protection E01 [1-3]; E02 [2] (3 h); E02 [3] (4 h); E12b, E13a [3]; E13c [1, 2]
Consumer protection C02 [3] (5 d); C03 [1, 3]; C09 [1]
Storage and disposal E26, E27, E31c [1]; E30a [1-3]; E34 [3]
Treated seed S05, S07 [1]
Medical advice M01 [1-3]; M05 [1, 2]

367 prochloraz

A broad-spectrum protectant and eradicant conazole fungicide

See also *carbendazim + prochloraz*
cyproconazole + prochloraz
fenpropidin + prochloraz
fluquinconazole + prochloraz

Products

1	Barclay Eyetak 40	Barclay	400 g/l	EC	11352
2	Mirage 40 EC	Makhteshim	400 g/l	EC	06770
3	Panache 40	DAPT	400 g/l	EC	10632
4	Poraz	Aventis	450 g/l	EC	10474
5	Poraz	BASF	450 g/l	EC	11701
6	Prelude 20LF	Agrichem	200 g/l	LS	04371
7	Scotts Octave	Scotts	46% w/w	WP	09275
8	Sportak 45 EW	Aventis	450 g/l	EW	07996

Uses

Alternaria in ***spring oilseed rape*** [8]; ***winter oilseed rape*** [4, 5, 8]. Eyespot in ***spring wheat*** [1-3]; ***winter barley***, ***winter wheat*** [1-5, 8]; ***winter rye*** [4, 5, 8]. Fungus diseases in ***container-grown ornamentals***, ***hardy ornamentals***, ***woody ornamentals*** [7]. Glume blotch in ***winter wheat*** [4, 5, 8]. Grey mould in ***spring oilseed rape*** [8]; ***winter oilseed rape*** [4, 5, 8]. Light leaf spot in ***spring oilseed rape*** [2, 3, 8]; ***winter oilseed rape*** [2-5, 8]. Net blotch in ***spring barley***, ***winter barley*** [4, 5, 8]. Phoma in ***spring oilseed rape*** [2, 3, 8]; ***winter oilseed rape*** [2-5, 8]. Powdery mildew in ***spring barley***, ***winter barley***, ***winter rye*** [4, 5, 8]; ***spring wheat***, ***winter wheat*** [4, 5]; ***spring wheat*** *(protection)* [8]. Rhynchosporium in ***spring barley***, ***winter barley***, ***winter rye*** [4, 5, 8]. Sclerotinia stem rot in ***spring oilseed rape*** [8]; ***winter oilseed rape*** [4, 5, 8]. Seed-borne diseases in ***flax*** *(seed treatment)*, ***linseed*** *(seed treatment)* [6]. Septoria leaf spot in ***winter rye*** [4, 5]; ***winter wheat*** [4, 5, 8]. White leaf spot in ***spring oilseed rape*** [8]; ***winter oilseed rape*** [4, 5, 8].

Efficacy guidance

- Spray cereals at first signs of disease. Protection of winter crops through season usually requires at least 2 treatments. See label for details of rates and timing. Treatment active against strains of eyespot resistant to benzimidazole fungicides
- Tank mixes with other fungicides recommended to improve control of rusts in wheat and barley. See label for details
- A period of at least 3 h without rain should follow spraying
- Can be used through most seed treatment machines if good even seed coverage is obtained. Check drill calibration before drilling treated seed [6]
- Apply as drench against soil diseases, as a spray against aerial diseases, as a dip for cuttings, or as a drench at propagation. Spray applications may be repeated at 10-14 d intervals. Under mist propagation use 7 d intervals [7]

Restrictions

- Maximum number of treatments 1 per batch for flax, linseed [6]; varies with dose and crop for other uses - see labels for details
- Do not treat linseed varieties Linda, Bolas, Karen, Laura, Mikael, Norlin, Moonraker, Abbey, Agiace [6]
- Do not use treated seed as food or feed [6]

Crop-specific information

- Latest use: milky ripe stage (GS 77) for cereals; before drilling flax or linseed [6]
- HI 6 wk for oilseed rape and cereals; 21 d for herbs
- Application with other fungicides may cause cereal crop scorch [1-5]

Environmental safety

- Flammable [1-3]
- Dangerous to fish or other aquatic life. Do not contaminate surface waters or ditches with chemical or used container
- Treated seed harmful to game and wildlife [6]
- Product available in refillable container (see label for instructions) [6]

Hazard classification and safety precautions

Hazard R03 [4, 5]; R04 [1-7]; R07d [1-3]

Risk phrases R03a [4, 5]; R04a [1-7]; R04b [1-5, 7]; R04e [1-3]; R04g [2]

Operator protection A, C; U05a [1-8]; U07, U11 [6]; U08 [1-6]; U09a [7]; U20a [8]; U20b [1-7]

Environmental protection E01, E13b [1-8]; E03 [6]

Consumer protection C03 [1-5, 7]

Storage and disposal E26 [1-3, 6, 8]; E30a [1-8]; E31b [1-5, 7, 8]; E33, E36 [6]; E34 [4-6, 8]

Treated seed S01, S02, S03, S04b, S05, S06, S07 [6]

Medical advice M03 [4, 5]; M05 [2]

368 prochloraz + propiconazole

A broad spectrum fungicide mixture for wheat, barley and oilseed rape

Products

1	Bumper P	Makhteshim	400:90 g/l	EC	08548
2	Greencrop Twinstar	Greencrop	400:90 g/l	EC	09516

Uses

Eyespot in ***spring barley***, ***spring wheat*** [1]; ***winter barley***, ***winter wheat*** [1, 2]. Light leaf spot in ***spring oilseed rape***, ***winter oilseed rape*** [1]. Phoma in ***spring oilseed rape***, ***winter oilseed rape*** [1]. Rhynchosporium in ***spring barley*** [1]; ***winter barley*** [1, 2]. Septoria leaf spot in ***spring wheat*** [1]; ***winter wheat*** [1, 2].

Approval information

- Propiconazole included in Annex I under EC Directive 91/414

Efficacy guidance

- Best results obtained from treatment when disease is active but not well established
- Treat wheat normally from flag leaf ligule just visible stage (GS 39) but earlier if there is a high risk of Septoria
- Treat barley from the first node detectable stage (GS 31)
- If disease pressure persists a second application may be necessary

Restrictions

- Maximum number of treatments 2 per crop
- Propiconazole may produce an allergic reaction

Crop-specific information

- Latest use: before grain watery ripe (GS 71) for cereals; before most seeds green for winter oilseed rape
- HI 6 wk

Environmental safety

- Dangerous to fish or other aquatic life. Do not contaminate surface waters or ditches with chemical or used container

Hazard classification and safety precautions

Hazard R04

Risk phrases R04a

Operator protection A, C; U05a, U08, U20b

Environmental protection E01, E13b

Consumer protection C03

Storage and disposal E26, E30a, E31b [1, 2]; E34 [1]

369 prochloraz + tebuconazole

A broad spectrum systemic fungicide mixture for cereals and oilseed rape

Products

1	Agate	Bayer	267:133 g/l	EC	08826
2	Agate	Bayer CropScience	267:133 g/l	EC	11235

SECTION 4

Uses

Brown rust in ***spring barley, spring rye, spring wheat, winter barley, winter rye, winter wheat***. Eyespot in ***spring barley, spring rye, spring wheat, winter barley, winter rye, winter wheat***. Glume blotch in ***spring wheat, winter wheat***. Late ear diseases in ***spring wheat, winter wheat***. Net blotch in ***spring barley, winter barley***. Powdery mildew in ***spring barley, spring rye, spring wheat, winter barley, winter rye, winter wheat***. Rhynchosporium in ***spring barley, spring rye, winter barley, winter rye***. Sclerotinia stem rot in ***spring oilseed rape, winter oilseed rape***. Septoria leaf spot in ***spring wheat, winter wheat***. Yellow rust in ***spring barley, spring rye, spring wheat, winter barley, winter rye, winter wheat***.

Efficacy guidance

- Best results achieved from applications at an early stage of disease development before infection spreads to new crop growth
- Adequate protection of winter cereals will usually require a programme of at least two treatments
- Optimum application timing is normally when disease first seen but varies with main target disease - see label
- To minimise possibility of development of resistance repeated applications should not be made against the same pathogen on cereals. See notes on Resistance Management in Section 1

Restrictions

- Maximum total dose equivalent to two full dose treatments

Crop-specific information

- Latest use: up to and including inflorescence fully emerged (GS 59) for cereals; most seeds green for oilseed rape
- HI 6 wk
- Occasionally transient leaf speckling may occur after treating wheat. Yield responses should not be affected

Environmental safety

- Dangerous to fish or other aquatic life. Do not contaminate surface waters or ditches with chemical or used container

Hazard classification and safety precautions

Hazard R03, R04
Risk phrases R03a, R03c, R04d, R04e
Operator protection A, C, H; U05a, U09b, U20a
Environmental protection E01, E13b
Consumer protection C03
Storage and disposal E26, E30a, E31c, E34
Medical advice M03

370 prometryn

A contact and residual triazine herbicide

Products

1	Alpha Prometryne 50 WP	Makhteshim	50% w/w	WP	04871
2	Gesagard	Syngenta	50% w/w	WP	08410
3	Pennine 50	DAPT	50% w/w	WP	10638

Uses

Annual dicotyledons in ***basil*** *(off-label)*, ***bay*** *(off-label)*, ***borage*** *(off-label)*, ***bulb onions*** *(off-label)*, ***camomile*** *(off-label)*, ***caraway*** *(off-label)*, ***chervil*** *(off-label)*, ***chives*** *(off-label)*, ***coriander*** *(off-label)*, ***dill*** *(off-label)*, ***fennel*** *(off-label)*, ***feverfew*** *(off-label)*, ***lovage*** *(off-label)*, ***marjoram*** *(off-label)*, ***mint*** *(off-label)*, ***protected basil*** *(off-label)*, ***protected bay*** *(off-label)*, ***protected borage*** *(off-label)*, ***protected camomile*** *(off-label)*, ***protected caraway*** *(off-label)*, ***protected chervil*** *(off-label)*, ***protected chives*** *(off-label)*, ***protected coriander*** *(off-label)*, ***protected dill*** *(off-label)*, ***protected fennel*** *(off-label)*, ***protected feverfew*** *(off-label)*, ***protected lovage*** *(off-label)*, ***protected marjoram*** *(off-label)*, ***protected mint*** *(off-label)*, ***protected rosemary*** *(off-label)*, ***protected sage*** *(off-label)*, ***protected savory*** *(off-label)*, ***protected sorrel*** *(off-label)*, ***protected tarragon*** *(off-label)*, ***protected thyme*** *(off-label)*, ***rosemary*** *(off-label)*, ***sage*** *(off-label)*, ***salad onions*** *(off-label)*, ***savory*** *(off-label)*, ***sorrel*** *(off-label)*, ***tarragon*** *(off-label)*, ***thyme*** *(off-label)* [1, 2]; ***carrots, celery, parsley, protected parsley*** [1-3]; ***leeks*** *(off-label)*, ***transplanted leeks*** *(off-label)*

[2]; ***protected celery, transplanted leeks*** [1, 3]. Annual grasses in ***basil*** *(off-label)*, ***bay*** *(off-label)*, ***borage*** *(off-label)*, ***bulb onions*** *(off-label)*, ***camomile*** *(off-label)*, ***caraway*** *(off-label)*, ***chervil*** *(off-label)*, ***chives*** *(off-label)*, ***coriander*** *(off-label)*, ***dill*** *(off-label)*, ***fennel*** *(off-label)*, ***feverfew*** *(off-label)*, ***lovage*** *(off-label)*, ***marjoram*** *(off-label)*, ***mint*** *(off-label)*, ***protected basil*** *(off-label)*, ***protected bay*** *(off-label)*, ***protected borage*** *(off-label)*, ***protected camomile*** *(off-label)*, ***protected caraway*** *(off-label)*, ***protected chervil*** *(off-label)*, ***protected chives*** *(off-label)*, ***protected coriander*** *(off-label)*, ***protected dill*** *(off-label)*, ***protected fennel*** *(off-label)*, ***protected feverfew*** *(off-label)*, ***protected lovage*** *(off-label)*, ***protected marjoram*** *(off-label)*, ***protected mint*** *(off-label)*, ***protected rosemary*** *(off-label)*, ***protected sage*** *(off-label)*, ***protected savory*** *(off-label)*, ***protected sorrel*** *(off-label)*, ***protected tarragon*** *(off-label)*, ***protected thyme*** *(off-label)*, ***rosemary*** *(off-label)*, ***sage*** *(off-label)*, ***salad onions*** *(off-label)*, ***savory*** *(off-label)*, ***sorrel*** *(off-label)*, ***tarragon*** *(off-label)*, ***thyme*** *(off-label)* [1, 2]; ***carrots, celery, parsley, protected parsley*** [1-3]; ***leeks*** *(off-label)*, ***transplanted leeks*** *(off-label)* [2]; ***protected celery, transplanted leeks*** [1, 3].

Specific Off-Label Approvals (SOLAs)

- ***basil, bay, borage, camomile, caraway, chervil, chives, coriander, dill, fennel, feverfew, lovage, marjoram, mint, protected basil, protected bay, protected borage, protected camomile, protected caraway, protected chervil, protected chives, protected coriander, protected dill, protected fennel, protected feverfew, protected lovage, protected marjoram, protected mint, protected rosemary, protected sage, protected savory, protected sorrel, protected tarragon, protected thyme, rosemary, sage, savory, sorrel, tarragon, thyme*** *(OLA 023394) Dec 2007* [2]
- ***basil, bay, borage, bulb onions, camomile, caraway, chervil, chives, coriander, dill, fennel, feverfew, lovage, marjoram, mint, protected basil, protected bay, protected borage, protected camomile, protected caraway, protected chervil, protected chives, protected coriander, protected dill, protected fennel, protected feverfew, protected lovage, protected marjoram, protected mint, protected rosemary, protected sage, protected savory, protected sorrel, protected tarragon, protected thyme, rosemary, sage, salad onions, savory, sorrel, tarragon, thyme*** *(OLA 031794) Dec 2007* [1]
- ***bulb onions, salad onions*** *(OLA 011658) Dec 2008* [2]
- ***leeks, transplanted leeks*** *(OLA 011659) Dec 2008* [2]

Approval information

- Products containing this active ingredient have been granted derogations for specified 'Essential Uses' for use until 31 December 2007. Sale and supply must cease by 30 June 2007 but growers have no guarantee that the products will continue to be available until then.
  For more information see 'The Review Programme' under 'Pesticide Legislation' in Section 1

Efficacy guidance

- Best results achieved by application to young seedling weeds up to 5 cm high (cotyledon stage for knotgrass, mayweed and corn marigold) on fine, moist seedbed when rain falls afterwards. Do not use on very cloddy soils
- On organic soils only contact action effective and repeat application may be needed (certain crops only)
- Excessive rain after treatment may check crop

Restrictions

- Maximum number of treatments 1 per crop for transplanted leeks; 1 per crop for transplanted celery
- Maximum total dose equivalent to one full dose treatment on all crops
- This product must not be used on any crops other than those listed, including any extrapolations that would normally be permissible under the Long Term Arrangements for Extension of Use (see Section 1)

Crop-specific information

- Latest Use: 3 wk after planting out for garlic
- HI onions 10 wk; herbs 6 wk
- Apply to carrots, celery, parsley post-emergence after 2-rough leaf stage or after transplants established
- Apply to transplanted leeks or celery after transplants established. Do not use on drilled leeks

Following crops guidance

- In the event of crop failure within 8 wk of treatment only peas, potatoes, carrots, parsley or celery may be planted. After 8 wk the land should be cultivated or ploughed to 150 mm min before planting the next crop

Environmental safety

- Dangerous for the environment [2]
- Toxic to aquatic organisms. May cause long-term adverse effects to the aquatic environment [2]
- Harmful to fish or other aquatic life. Do not contaminate surface waters or ditches with chemical or used container [1, 3]

Hazard classification and safety precautions

Hazard R14 [2]
Risk phrases R14b [2]
Operator protection U20b
Environmental protection E01, E15, E38 [2]; E13c [1, 3]
Consumer protection C02 [1, 3] (6 wk)
Storage and disposal E29, E32d, E34 [2]; E30a, E32a [1-3]

371 propachlor

A pre-emergence chloroacetanilide herbicide for various horticultural crops

See also *chloridazon + propachlor*
chlorthal-dimethyl + propachlor

Products

1	Alpha Propachlor 50 SC	Makhteshim	500 g/l	SC	04873
2	Brasson	Monsanto	480 g/l	SC	10560
3	Ramrod Flowable	Monsanto	480 g/l	SC	10314
4	Tripart Sentinel	AgriGuard	480 g/l	SC	03250
5	Tripart Sentinel 2	Tripart	480 g/l	SC	05140

Uses

Annual dicotyledons in ***balm*** *(off-label)*, ***basil*** *(off-label)*, ***bay*** *(off-label)*, ***blackcurrants*** *(off-label)*, ***blueberries*** *(off-label)*, ***chinese cabbage*** *(off-label)*, ***chives*** *(off-label)*, ***cress*** *(off-label)*, ***frise*** *(off-label)*, ***gooseberries*** *(off-label)*, ***hyssop*** *(off-label)*, ***kohlrabi*** *(off-label)*, ***lamb's lettuce*** *(off-label)*, ***land cress*** *(off-label)*, ***lettuce*** *(off-label)*, ***marjoram*** *(off-label)*, ***mint*** *(off-label)*, ***oregano*** *(off-label)*, ***parsley*** *(off-label)*, ***radicchio*** *(off-label)*, ***redcurrants*** *(off-label)*, ***rocket*** *(off-label)*, ***rosemary*** *(off-label)*, ***rue*** *(off-label)*, ***savory*** *(off-label)*, ***scarole*** *(off-label)*, ***sorrel*** *(off-label)*, ***strawberries*** *(off-label)*, ***tarragon*** *(off-label)*, ***thyme*** *(off-label)*, ***whitecurrants*** *(off-label)*, ***woodruff*** *(off-label)* [1, 3]; ***broccoli***, ***brussels sprouts***, ***cabbages***, ***cauliflowers***, ***kale***, ***leeks***, ***onions***, ***spring oilseed rape***, ***swedes***, ***turnips*** [1-5]; ***brown mustard***, ***sage***, ***white mustard*** [2, 3]; ***calabrese*** [1-3, 5]; ***fodder rape*** [1-3]; ***frise*** *(off-label - under crop covers)*, ***lamb's lettuce*** *(off-label - under crop covers)*, ***lettuce*** *(off-label - under crop covers)*, ***peppermint*** *(off-label)*, ***radicchio*** *(off-label - under crop covers)*, ***ribes hybrids*** *(off-label)*, ***salad brassicas*** *(off-label - for baby leaf production)*, ***salad burnet*** *(off-label)*, ***scarole*** *(off-label - under crop covers)*, ***spearmint*** *(off-label)* [3]; ***mustard***, ***rubus hybrids*** *(off-label)*, ***salad burnet*** *(off-label - salad)*, ***salad onions*** [1]; ***ornamental plant production*** [1, 2]; ***strawberries*** [1, 4]; ***winter oilseed rape*** [1, 4, 5]. Annual grasses in ***balm*** *(off-label)*, ***basil*** *(off-label)*, ***bay*** *(off-label)*, ***blackcurrants*** *(off-label)*, ***blueberries*** *(off-label)*, ***chinese cabbage*** *(off-label)*, ***chives*** *(off-label)*, ***cress*** *(off-label)*, ***frise*** *(off-label)*, ***gooseberries*** *(off-label)*, ***hyssop*** *(off-label)*, ***kohlrabi*** *(off-label)*, ***lamb's lettuce*** *(off-label)*, ***land cress*** *(off-label)*, ***lettuce*** *(off-label)*, ***marjoram*** *(off-label)*, ***mint*** *(off-label)*, ***oregano*** *(off-label)*, ***parsley*** *(off-label)*, ***radicchio*** *(off-label)*, ***redcurrants*** *(off-label)*, ***rocket*** *(off-label)*, ***rosemary*** *(off-label)*, ***rue*** *(off-label)*, ***savory*** *(off-label)*, ***scarole*** *(off-label)*, ***sorrel*** *(off-label)*, ***strawberries*** *(off-label)*, ***tarragon*** *(off-label)*, ***thyme*** *(off-label)*, ***whitecurrants*** *(off-label)*, ***woodruff*** *(off-label)* [1, 3]; ***broccoli***, ***brussels sprouts***, ***cabbages***, ***cauliflowers***, ***kale***, ***leeks***, ***onions***, ***spring oilseed rape***, ***swedes***, ***turnips*** [1-5]; ***brown mustard***, ***sage***, ***white mustard*** [2, 3]; ***calabrese*** [1-3, 5]; ***fodder rape*** [1-3]; ***frise*** *(off-label - under crop covers)*, ***lamb's lettuce*** *(off-label - under crop covers)*, ***lettuce*** *(off-label - under crop covers)*, ***peppermint*** *(off-label)*, ***radicchio*** *(off-label - under crop covers)*, ***ribes hybrids*** *(off-label)*, ***salad burnet*** *(off-label)*, ***scarole*** *(off-label - under crop covers)*, ***spearmint*** *(off-label)* [3];

mustard, ***rubus hybrids*** *(off-label)*, ***salad burnet*** *(off-label - salad)*, ***salad onions*** [1]; ***ornamental plant production*** [1, 2]; ***strawberries*** [1, 4]; ***winter oilseed rape*** [1, 4, 5].

Specific Off-Label Approvals (SOLAs)

- ***balm***, ***basil***, ***bay***, ***blackcurrants***, ***blueberries***, ***chinese cabbage***, ***chives***, ***cress***, ***frise***, ***gooseberries***, ***hyssop***, ***kohlrabi***, ***lamb's lettuce***, ***land cress***, ***lettuce***, ***marjoram***, ***mint***, ***oregano***, ***parsley***, ***radicchio***, ***redcurrants***, ***rocket***, ***rosemary***, ***rubus hybrids***, ***rue***, ***savory***, ***scarole***, ***sorrel***, ***strawberries***, ***tarragon***, ***thyme***, ***whitecurrants***, ***woodruff*** *(OLA 002424) Dec 2008* [1]
- ***balm***, ***basil***, ***bay***, ***blackcurrants***, ***blueberries***, ***chinese cabbage***, ***chives***, ***cress***, ***frise***, ***gooseberries***, ***hyssop***, ***kohlrabi***, ***lamb's lettuce***, ***land cress***, ***lettuce***, ***marjoram***, ***mint***, ***oregano***, ***parsley***, ***peppermint***, ***radicchio***, ***redcurrants***, ***ribes hybrids***, ***rocket***, ***rosemary***, ***rue***, ***salad burnet***, ***savory***, ***scarole***, ***sorrel***, ***spearmint***, ***strawberries***, ***tarragon***, ***thyme***, ***whitecurrants***, ***woodruff*** *(OLA 021159) Dec 2008* [3]
- ***frise***, ***lamb's lettuce***, ***lettuce***, ***radicchio***, ***scarole*** *(under crop covers) (OLA 021159) Dec 2008* [3]
- ***salad brassicas*** *(for baby leaf production) (OLA 030436) Dec 2008* [3]
- ***salad burnet*** *(salad) (OLA 002424) Dec 2008* [1]

Efficacy guidance

- Controls germinating (not emerged) weeds for 6-8 wk
- Best results achieved by application to fine, firm, moist seedbed free of established weeds in spring, summer and early autumn
- Use higher rate on soils with more than 10% organic matter
- Recommended as tank-mix with glyphosate or chlorthal-dimethyl on onions, leeks, swedes, turnips. See label for details

Restrictions

- Maximum number of treatments 1 or 2 per crop, varies with product and crop - see label for details
- Maximum total dose equivalent to two full dose treatments on brassicas and leeks; one full dose treatment on most other crops [1]
- Do not use pre-emergence of drilled wallflower seed
- Do not use on crops under glass or polythene
- Do not use under extremely wet, dry or other adverse growth conditions

Crop-specific information

- Latest use: before weed emergence on brassicas, oilseed rape, swedes, turnips, leeks, onions strawberries; 14 d after transplanting salad vegetables; before flowering for leaf herbs
- HI salad vegetables, herbs 6 wk
- Apply in brassicas from drilling to time seed chits or after 3-4 true leaf stage but before weed emergence, in swedes and turnips pre-emergence only
- Apply in transplanted brassicas within 48 h of planting in warm weather. Plants must be hardened off and special care needed with block sown or modular propagated plants
- Apply in onions and leeks pre-emergence or from post-crook to young plant stage
- Apply to newly planted strawberries soon after transplanting, to weed-free soil in established crops in early spring before new weeds emerge
- Granules may be applied to onion, leek and brassica nurseries and to most flower crops after bedding out and hardening off

Following crops guidance

- In the event of crop failure only replant recommended crops in treated soil

Hazard classification and safety precautions

Hazard R03 [1-3]; R04 [1-5]

Risk phrases R03a [1]; R03c [1-3]; R04a, R04b [1-5]; R04c [5]; R04e [2, 3, 5]

Operator protection A, C; U02, U04a, U05a, U08, U13, U19 [1-5]; U20a [2, 5]; U20b [1, 3, 4]

Environmental protection E01, E15

Consumer protection C03

Storage and disposal E26, E30a, E31b, E34

Medical advice M03 [1-3]

372 propamocarb hydrochloride

A translocated protectant carbamate fungicide

See also *chlorothalonil + propamocarb hydrochloride*
mancozeb + propamocarb hydrochloride

Products

1	Filex	Scotts	722 g/l	SL	07631
2	Proplant	Fargro	722 g/l	SL	08572

Uses

Blight in ***outdoor tomatoes***, ***protected tomatoes*** [2]. Botrytis in ***angelica*** *(off-label)*, ***balm*** *(off-label)*, ***basil*** *(off-label)*, ***bay*** *(off-label)*, ***borage*** *(off-label)*, ***camomile*** *(off-label)*, ***caraway*** *(off-label)*, ***chervil*** *(off-label)*, ***chives*** *(off-label)*, ***clary*** *(off-label)*, ***coriander*** *(off-label)*, ***cress*** *(off-label)*, ***dill*** *(off-label)*, ***fennel*** *(off-label)*, ***fenugreek*** *(off-label)*, ***feverfew*** *(off-label)*, ***hyssop*** *(off-label)*, ***land cress*** *(off-label)*, ***lettuce*** *(off-label)*, ***lovage*** *(off-label)*, ***marigold*** *(off-label)*, ***marjoram*** *(off-label)*, ***mint*** *(off-label)*, ***nasturtium*** *(off-label)*, ***nettle*** *(off-label)*, ***oregano*** *(off-label)*, ***parsley*** *(off-label)*, ***rocket*** *(off-label)*, ***rosemary*** *(off-label)*, ***rue*** *(off-label)*, ***sage*** *(off-label)*, ***salad burnet*** *(off-label)*, ***savory*** *(off-label)*, ***scarole*** *(off-label)*, ***sorrel*** *(off-label)*, ***tarragon*** *(off-label)*, ***thyme*** *(off-label)* [1, 2]; ***endives*** *(off-label)*, ***protected angelica*** *(off-label)*, ***protected balm*** *(off-label)*, ***protected basil*** *(off-label)*, ***protected bay*** *(off-label)*, ***protected borage*** *(off-label)*, ***protected camomile*** *(off-label)*, ***protected caraway*** *(off-label)*, ***protected chervil*** *(off-label)*, ***protected chives*** *(off-label)*, ***protected clary*** *(off-label)*, ***protected coriander*** *(off-label)*, ***protected dill*** *(off-label)*, ***protected endives*** *(off-label)*, ***protected fennel*** *(off-label)*, ***protected fenugreek*** *(off-label)*, ***protected feverfew*** *(off-label)*, ***protected hyssop*** *(off-label)*, ***protected land cress*** *(off-label)*, ***protected lemon verbena*** *(off-label)*, ***protected lovage*** *(off-label)*, ***protected marigold*** *(off-label)*, ***protected marjoram*** *(off-label)*, ***protected mint*** *(off-label)*, ***protected nasturtium*** *(off-label)*, ***protected nettle*** *(off-label)*, ***protected oregano*** *(off-label)*, ***protected parsley*** *(off-label)*, ***protected plantain*** *(off-label)*, ***protected rocket*** *(off-label)*, ***protected rosemary*** *(off-label)*, ***protected rue*** *(off-label)*, ***protected sage*** *(off-label)*, ***protected salad burnet*** *(off-label)*, ***protected savory*** *(off-label)*, ***protected sorrel*** *(off-label)*, ***protected tarragon*** *(off-label)*, ***protected thyme*** *(off-label)*, ***verbena*** *(off-label)* [2]; ***lamb's lettuce*** *(off-label)*, ***lemon verbena*** *(off-label)*, ***protected herbs*** *(off-label)*, ***woodruff*** *(off-label)* [1]. Damping off in ***inert substrate cucumbers*** *(off-label)*, ***inert substrate peppers*** *(off-label)*, ***inert substrate tomatoes*** *(off-label)*, ***nft peppers*** *(off-label)*, ***nft tomatoes*** *(off-label)* [1, 2]; ***protected peppers*** *(off-label)*, ***protected tomatoes*** *(off-label)* [1]. Downy mildew in ***angelica*** *(off-label)*, ***balm*** *(off-label)*, ***basil*** *(off-label)*, ***bay*** *(off-label)*, ***borage*** *(off-label)*, ***broccoli***, ***brussels sprouts***, ***cabbages***, ***calabrese***, ***camomile*** *(off-label)*, ***caraway*** *(off-label)*, ***cauliflowers***, ***chervil*** *(off-label)*, ***chinese cabbage***, ***chives*** *(off-label)*, ***clary*** *(off-label)*, ***coriander*** *(off-label)*, ***cress*** *(off-label)*, ***dill*** *(off-label)*, ***fennel*** *(off-label)*, ***fenugreek*** *(off-label)*, ***feverfew*** *(off-label)*, ***hyssop*** *(off-label)*, ***land cress*** *(off-label)*, ***lettuce*** *(off-label)*, ***lovage*** *(off-label)*, ***marigold*** *(off-label)*, ***marjoram*** *(off-label)*, ***mint*** *(off-label)*, ***nasturtium*** *(off-label)*, ***nettle*** *(off-label)*, ***oregano*** *(off-label)*, ***parsley*** *(off-label)*, ***protected radishes*** *(off-label)*, ***radishes*** *(off-label)*, ***rocket*** *(off-label)*, ***rosemary*** *(off-label)*, ***rue*** *(off-label)*, ***sage*** *(off-label)*, ***salad burnet*** *(off-label)*, ***savory*** *(off-label)*, ***scarole*** *(off-label)*, ***sorrel*** *(off-label)*, ***tarragon*** *(off-label)*, ***thyme*** *(off-label)* [1, 2]; ***endives*** *(off-label)*, ***protected angelica*** *(off-label)*, ***protected balm*** *(off-label)*, ***protected basil*** *(off-label)*, ***protected bay*** *(off-label)*, ***protected borage*** *(off-label)*, ***protected camomile*** *(off-label)*, ***protected caraway*** *(off-label)*, ***protected chervil*** *(off-label)*, ***protected chives*** *(off-label)*, ***protected clary*** *(off-label)*, ***protected coriander*** *(off-label)*, ***protected dill*** *(off-label)*, ***protected endives*** *(off-label)*, ***protected fennel*** *(off-label)*, ***protected fenugreek*** *(off-label)*, ***protected feverfew*** *(off-label)*, ***protected hyssop*** *(off-label)*, ***protected land cress*** *(off-label)*, ***protected lemon verbena*** *(off-label)*, ***protected lovage*** *(off-label)*, ***protected marigold*** *(off-label)*, ***protected marjoram*** *(off-label)*, ***protected mint*** *(off-label)*, ***protected nasturtium*** *(off-label)*, ***protected nettle*** *(off-label)*, ***protected oregano*** *(off-label)*, ***protected parsley*** *(off-label)*, ***protected plantain*** *(off-label)*, ***protected rocket*** *(off-label)*, ***protected rosemary*** *(off-label)*, ***protected rue*** *(off-label)*, ***protected sage*** *(off-label)*, ***protected salad burnet*** *(off-label)*, ***protected savory*** *(off-label)*, ***protected sorrel*** *(off-label)*, ***protected tarragon*** *(off-label)*, ***protected thyme*** *(off-label)*, ***verbena*** *(off-label)*, ***watercress*** *(off-label - during propagation)* [2]; ***lamb's lettuce*** *(off-label)*, ***lemon verbena*** *(off-label)*, ***protected herbs*** *(off-label)*, ***watercress*** *(off-label - under protection)*, ***woodruff*** *(off-label)* [1]. Phytophthora in

aubergines, bedding plants, broccoli, brussels sprouts, cabbages, calabrese, cauliflowers, chinese cabbage, cucumbers, leeks, nursery stock, onions, peppers, pot plants, tulips [1, 2]; ***container-grown ornamentals, ornamental specimens, watercress*** *(off-label - under protection)* [1]; ***flower bulbs, outdoor tomatoes, protected tomatoes, rockwool aubergines, rockwool cucumbers, rockwool peppers, rockwool tomatoes, watercress*** *(off-label - during propagation)* [2]. Pythium in ***aubergines, bedding plants, broccoli, brussels sprouts, cabbages, calabrese, cauliflowers, chinese cabbage, cucumbers, flower bulbs, leeks, nursery stock, onions, outdoor tomatoes, peppers, pot plants, protected tomatoes, rockwool aubergines, rockwool cucumbers, rockwool peppers, rockwool tomatoes, tulips*** [1, 2]; ***ornamental specimens, watercress*** *(off-label - under protection)* [1]; ***watercress*** *(off-label - during propagation)* [2]. Root malformation disorder in ***red beet*** *(off-label)* [1, 2]. Root rot in ***inert substrate cucumbers*** *(off-label)*, ***inert substrate peppers*** *(off-label)*, ***inert substrate tomatoes*** *(off-label)*, ***nft peppers*** *(off-label)*, ***nft tomatoes*** *(off-label)* [2]. White blister in ***horseradish*** *(off-label)*, ***protected horseradish*** *(off-label)* [1, 2]. White tip in ***leeks*** *(off-label)* [1, 2].

Specific Off-Label Approvals (SOLAs)

- ***angelica, balm, basil, bay, borage, camomile, caraway, chervil, chives, clary, coriander, cress, dill, endives, fennel, fenugreek, feverfew, horseradish, hyssop, land cress, lettuce, lovage, marigold, marjoram, mint, nasturtium, nettle, oregano, parsley, protected angelica, protected balm, protected basil, protected bay, protected borage, protected camomile, protected caraway, protected chervil, protected chives, protected clary, protected coriander, protected dill, protected endives, protected fennel, protected fenugreek, protected feverfew, protected horseradish, protected hyssop, protected land cress, protected lemon verbena, protected lovage, protected marigold, protected marjoram, protected mint, protected nasturtium, protected nettle, protected oregano, protected parsley, protected plantain, protected radishes, protected rocket, protected rosemary, protected rue, protected sage, protected salad burnet, protected savory, protected sorrel, protected tarragon, protected thyme, radishes, rocket, rosemary, rue, sage, salad burnet, savory, scarole, sorrel, tarragon, thyme, verbena*** *(OLA 030784) Dec 2008* [2]
- ***angelica, balm, basil, bay, borage, camomile, caraway, chervil, chives, clary, coriander, cress, dill, fennel, fenugreek, feverfew, hyssop, lamb's lettuce, land cress, lemon verbena, lettuce, lovage, marigold, marjoram, mint, nasturtium, nettle, oregano, parsley, protected herbs, rocket, rosemary, rue, sage, salad burnet, savory, scarole, sorrel, tarragon, thyme, woodruff*** *(OLA 991971) Dec 2008* [1]
- ***horseradish, protected horseradish*** *(OLA 992036) Dec 2008* [1]
- ***inert substrate cucumbers, inert substrate peppers, inert substrate tomatoes, nft peppers, nft tomatoes*** *(OLA 002527) Dec 2008* [2]
- ***inert substrate cucumbers, inert substrate peppers, inert substrate tomatoes, nft peppers, nft tomatoes, protected peppers, protected tomatoes*** *(OLA 992032) Dec 2008* [1]
- ***leeks*** *(OLA 002975) Oct 2004* [1]
- ***leeks*** *(OLA 003029) Oct 2004* [2]
- ***protected radishes, radishes*** *(OLA 992030) Dec 2008* [1]
- ***red beet*** *(OLA 031247) Dec 2008* [1]
- ***red beet*** *(OLA 031453) Jul 2008* [2]
- ***watercress*** *(during propagation) (OLA 030784) Dec 2008* [2]
- ***watercress*** *(under protection) (OLA 010439) Dec 2008* [1]

Efficacy guidance

- Chemical is absorbed through roots and translocated throughout plant
- Incorporate in compost before use or drench moist compost or soil before sowing, pricking out, striking cuttings or potting up
- Drench treatment can be repeated at 3-6 wk intervals
- Concentrated solution is corrosive to all metals other than stainless steel
- May also be applied in trickle irrigation systems
- To prevent root rot in tulip bulbs apply as dip for 20 min but full protection only achieved by soil drench treatment as well

Restrictions

- Maximum number of treatments 4 per crop for cucumbers, tomatoes, peppers, aubergines; 1 per crop for listed brassicas; 1 compost incorporation and/or 1 drench treatment for leeks, onions; 1 dip and 1 soil drench for tulips

SECTION 4

- When applied over established seedlings rinse off foliage with water and do not apply under hot, dry conditions
- On plants of unknown tolerance test first on a small scale
- Do not apply in a recirculating system
- Store away from seeds and fertilizers

Crop-specific information

- Latest use: before transplanting for brassicas, herbs; before crop emergence for watercress
- HI 2 d for inert substrate and NFT vegetable crops; 14 d for cucumbers, tomatoes, peppers, aubergines, horseradish, radishes, herbs and protected herbs; 4 wk for calabrese, cauliflower, sprouting broccoli, Brussels sprouts, Chinese cabbage. watercress; 19 wk for leeks, onions

Hazard classification and safety precautions

Operator protection A, B, C, H, K, M; U08, U20b
Environmental protection E15
Storage and disposal E26, E29, E34 [2]; E30a, E32a [1, 2]

373 propaquizafop

A phenoxy alkanoic acid foliar acting grass herbicide

Products

1	Barclay Rebel	Barclay	100 g/l	EC	11648
2	Falcon	Makhteshim	100 g/l	EC	10585
3	Greencrop Satchmo	Greencrop	100 g/l	EC	10748
4	Landgold PQF 100	Landgold	100 g/l	EC	10763
5	Standon Propaquizafop	Standon	100 g/l	EC	11536

Uses

Annual grasses in ***bulb onions, carrots, combining peas, early potatoes, farm forestry, fodder beet, linseed, maincrop potatoes, parsnips, spring field beans, spring oilseed rape, sugar beet, swedes, turnips, winter field beans, winter oilseed rape*** [1-5]. Annual meadow grass in ***angelica*** *(off-label)*, ***balm*** *(off-label)*, ***basil*** *(off-label)*, ***bay*** *(off-label)*, ***borage*** *(off-label)*, ***camomile*** *(off-label)*, ***caraway*** *(off-label)*, ***chervil*** *(off-label)*, ***chives*** *(off-label)*, ***clary*** *(off-label)*, ***coriander*** *(off-label)*, ***dill*** *(off-label)*, ***fennel*** *(off-label)*, ***fenugreek*** *(off-label)*, ***feverfew*** *(off-label)*, ***hyssop*** *(off-label)*, ***land cress*** *(off-label)*, ***leaf spinach*** *(off-label)*, ***leeks*** *(off-label)*, ***lemon verbena*** *(off-label)*, ***lovage*** *(off-label)*, ***marigold*** *(off-label)*, ***marjoram*** *(off-label)*, ***mint*** *(off-label)*, ***nasturtium*** *(off-label)*, ***nettle*** *(off-label)*, ***oregano*** *(off-label)*, ***parsley*** *(off-label)*, ***red beet*** *(off-label)*, ***rocket*** *(off-label)*, ***rosemary*** *(off-label)*, ***rue*** *(off-label)*, ***sage*** *(off-label)*, ***salad burnet*** *(off-label)*, ***savory*** *(off-label)*, ***sorrel*** *(off-label)*, ***tarragon*** *(off-label)*, ***thyme*** *(off-label)*, ***woodruff*** *(off-label)* [2]. Couch in ***angelica*** *(off-label)*, ***balm*** *(off-label)*, ***basil*** *(off-label)*, ***bay*** *(off-label)*, ***borage*** *(off-label)*, ***camomile*** *(off-label)*, ***caraway*** *(off-label)*, ***chervil*** *(off-label)*, ***chives*** *(off-label)*, ***clary*** *(off-label)*, ***coriander*** *(off-label)*, ***dill*** *(off-label)*, ***fennel*** *(off-label)*, ***fenugreek*** *(off-label)*, ***feverfew*** *(off-label)*, ***hyssop*** *(off-label)*, ***land cress*** *(off-label)*, ***leaf spinach*** *(off-label)*, ***leeks*** *(off-label)*, ***lemon verbena*** *(off-label)*, ***lovage*** *(off-label)*, ***marigold*** *(off-label)*, ***marjoram*** *(off-label)*, ***mint*** *(off-label)*, ***nasturtium*** *(off-label)*, ***nettle*** *(off-label)*, ***oregano*** *(off-label)*, ***parsley*** *(off-label)*, ***red beet*** *(off-label)*, ***rocket*** *(off-label)*, ***rosemary*** *(off-label)*, ***rue*** *(off-label)*, ***sage*** *(off-label)*, ***salad burnet*** *(off-label)*, ***savory*** *(off-label)*, ***sorrel*** *(off-label)*, ***tarragon*** *(off-label)*, ***thyme*** *(off-label)*, ***woodruff*** *(off-label)* [2]. Perennial grasses in ***bulb onions, carrots, combining peas, early potatoes, farm forestry, fodder beet, linseed, maincrop potatoes, parsnips, spring field beans, spring oilseed rape, sugar beet, swedes, turnips, winter field beans, winter oilseed rape*** [1-5]; ***celery*** *(off-label)* [2]. Volunteer cereals in ***angelica*** *(off-label)*, ***balm*** *(off-label)*, ***basil*** *(off-label)*, ***bay*** *(off-label)*, ***borage*** *(off-label)*, ***camomile*** *(off-label)*, ***caraway*** *(off-label)*, ***celery*** *(off-label)*, ***chervil*** *(off-label)*, ***chives*** *(off-label)*, ***clary*** *(off-label)*, ***coriander*** *(off-label)*, ***dill*** *(off-label)*, ***fennel*** *(off-label)*, ***fenugreek*** *(off-label)*, ***feverfew*** *(off-label)*, ***hyssop*** *(off-label)*, ***land cress*** *(off-label)*, ***leaf spinach*** *(off-label)*, ***leeks*** *(off-label)*, ***lemon verbena*** *(off-label)*, ***lovage*** *(off-label)*, ***marigold*** *(off-label)*, ***marjoram*** *(off-label)*, ***mint*** *(off-label)*, ***nasturtium*** *(off-label)*, ***nettle*** *(off-label)*, ***oregano*** *(off-label)*, ***parsley*** *(off-label)*, ***red beet*** *(off-label)*, ***rocket*** *(off-label)*, ***rosemary*** *(off-label)*, ***rue*** *(off-label)*, ***sage*** *(off-label)*, ***salad burnet*** *(off-label)*, ***savory*** *(off-label)*, ***sorrel*** *(off-label)*, ***tarragon*** *(off-label)*, ***thyme*** *(off-label)*, ***woodruff*** *(off-label)* [2]. Wild oats in ***celery*** *(off-label)* [2].

Specific Off-Label Approvals (SOLAs)

- ***angelica, balm, basil, bay, borage, camomile, caraway, chervil, chives, clary, coriander, dill, fennel, fenugreek, feverfew, hyssop, land cress, leaf spinach, leeks, lemon verbena, lovage, marigold, marjoram, mint, nasturtium, nettle, oregano, parsley, rocket, rosemary, rue, sage, salad burnet, savory, sorrel, tarragon, thyme, woodruff*** *(OLA 010907) Dec 2008* [2]
- ***angelica, balm, basil, bay, borage, camomile, caraway, chervil, chives, clary, coriander, dill, fennel, fenugreek, feverfew, hyssop, land cress, leaf spinach, leeks, lemon verbena, lovage, marigold, marjoram, mint, nasturtium, nettle, oregano, parsley, red beet, rocket, rosemary, rue, sage, salad burnet, savory, sorrel, tarragon, thyme, woodruff*** *(OLA 020148) Dec 2008* [2]
- ***celery*** *(OLA 021502) Dec 2008* [2]
- ***red beet*** *(OLA 010908) Dec 2008* [2]

Efficacy guidance

- Apply to emerged weeds when they are growing actively with adequate soil moisture
- Activity is slower under cool conditions
- Broad-leaved weeds and any weeds germinating after treatment are not controlled
- Annual meadow grass up to 3 leaves checked at low doses and severely checked at highest dose
- Spray barley cover crops when risk of wind blow has passed and before there is serious competition with the crop
- Various tank mixtures and sequences recommended for broader spectrum weed control in oilseed rape, peas and sugar beet. See label for details
- Severe couch infestations may require a second application at reduced dose when regrowth has 3-4 leaves unfolded
- Products contain surfactants. Tank mixing with adjuvants not required or recommended

Restrictions

- Maximum total dose varies according to crop treated. See label for details
- See label for list of tolerant tree species
- Do not treat seed potatoes

Crop-specific information

- Latest use: before crop flower buds visible for winter oilseed rape, linseed, field beans; before 8 fully expanded leaf stage for spring oilseed rape, mustard; before weeds are covered by the crop for potatoes, sugar beet, fodder beet; when flower buds visible for peas
- HI herbs 3 wk; early potatoes, carrots, parsnips, bulb onions, celery 4 wk; peas 7 wk; sugar beet, fodder beet, maincrop potatoes, swedes, turnips 8 wk
- Application in high temperatures and/or low soil moisture content may cause chlorotic spotting especially on combining peas and field beans
- Overlaps at the highest dose can cause damage from early applications to carrots and parsnips

Following crops guidance

- In the event of a failed treated crop an interval of 2 wk must elapse between the last application and redrilling with winter wheat or winter barley. 4 wk must elapse before sowing oilseed rape, peas or field beans, and 16 wk before sowing ryegrass or oats

Environmental safety

- Harmful to fish or other aquatic life. Do not contaminate surface waters or ditches with chemical or used container
- Risk to certain non-target insects or other arthropods. See directions for use

Hazard classification and safety precautions

Hazard R04
Risk phrases R04a, R04b
Operator protection A, C; U02, U05a, U08, U19, U20b [1-5]; U13 [2]; U14, U15 [1, 2, 4, 5]; U23a [3]
Environmental protection E01, E13c, E22b
Consumer protection C03
Storage and disposal E26, E30a, E31b
Medical advice M05

374 propham

A pre-sowing carbamate herbicide, all approvals for which were revoked in 1997

FOR FULL CONDITIONS OF USE ALWAYS READ THE PRODUCT LABEL

375 propiconazole

A systemic, curative and protectant conazole fungicide

See also *cyproconazole + propiconazole*
prochloraz + propiconazole

Products

1	Barclay Bolt	Barclay	250 g/l	EC	11344
2	Bumper 250 EC	Makhteshim	250 g/l	EC	09039

Uses

Brown rust in ***spring barley, spring wheat, winter barley, winter rye, winter wheat*** [1, 2]; ***spring rye*** [1]. Crown rust in ***grass for ensiling, grass seed crops*** [1, 2]; ***spring oats, winter oats*** [2]. Drechslera leaf spot in ***grass for ensiling, grass seed crops*** [1, 2]. Eyespot in ***winter barley*** *(low levels only)*, ***winter wheat*** *(low levels only)* [1]. Leaf blotch in ***garlic*** *(off-label)*, ***onions*** *(off-label)*, ***shallots*** *(off-label)* [2]. Light leaf spot in ***spring oilseed rape, winter oilseed rape*** [1]; ***spring oilseed rape*** *(reduction)*, ***winter oilseed rape*** *(reduction)* [2]. Mildew in ***grass for ensiling, grass seed crops*** [1, 2]; ***sugar beet*** [1]. Net blotch in ***spring barley, winter barley*** [1]. Powdery mildew in ***spring barley, spring oats, spring wheat, winter barley, winter oats, winter rye, winter wheat*** [1, 2]; ***spring rye*** [1]. Ramularia leaf spots in ***sugar beet*** [1]; ***sugar beet*** *(reduction)* [2]. Rhynchosporium in ***grass for ensiling, grass seed crops, spring barley, winter barley, winter rye*** [1, 2]; ***spring rye*** [1]. Rust in ***garlic*** *(off-label)*, ***leeks, leeks*** *(qualified minor use)* [2]; ***sugar beet*** [1, 2]. Septoria in ***spring rye*** [1]; ***spring wheat, winter rye, winter wheat*** [1, 2]. Sooty moulds in ***spring wheat, winter wheat*** [2]. White blister in ***honesty*** *(off-label)* [2]. White rust in ***chrysanthemums*** *(off-label)*, ***protected chrysanthemums*** *(off-label)* [2]. Yellow rust in ***spring barley, spring wheat, winter barley, winter wheat*** [1, 2].

Specific Off-Label Approvals (SOLAs)

- ***chrysanthemums, garlic, honesty, onions, protected chrysanthemums, shallots*** *(OLA 012142) Dec 2008* [2]

Approval information

- Propiconazole included in Annex I under EC Directive 91/414
- Approved for aerial application on wheat, barley, rye, oats [1]. See notes in Section 1

Efficacy guidance

- Best results achieved by applying at early stage of disease. Recommended spray programmes vary with crop, disease, season, soil type and product. See label for details

Restrictions

- Maximum number of treatments 4 per crop for wheat (up to 3 in yr of harvest); 3 per crop for barley, oats, rye, triticale (up to 2 in yr of harvest), leeks; 2 per crop or yr for oilseed rape, sugar beet, grass seed crops, honesty; 1 per yr on grass for ensiling
- Propiconazole may produce an allergic reaction
- On oilseed rape do not apply during flowering. Apply first treatment before flowering, the second afterwards
- Grass seed crops must be treated in yr of harvest
- A minimum interval of 14 d must elapse between treatments on leeks and 21 d on sugar beet
- Avoid spraying crops under stress, eg during cold weather or periods of frost

Crop-specific information

- Latest use: before 4 pairs of true leaves for honesty
- HI cereals, grass seed crops, leeks 35 d; oilseed rape, sugar beet, garlic, onions, shallots, grass for ensiling 28 d

Environmental safety

- Dangerous to fish or other aquatic life. Do not contaminate surface waters or ditches with chemical or used container

Hazard classification and safety precautions

Hazard R04 [1]
Risk phrases R04a, R04b [1]
Operator protection A, C; U02, U05a, U19 [1, 2]; U04a, U09b, U20a [1]; U08 [2]
Environmental protection E01, E13b
Consumer protection C03

Storage and disposal E26, E34 [2]; E30a, E31b [1, 2]
Treated seed S06 [2]

376 propiconazole + trifloxystrobin

A mixture of strobilurin and conazole fungicides for cereals

Products

Rombus	Bayer CropScience	125:125 g/l	EC	11301

Uses

Brown rust in ***spring barley***, ***winter barley***, ***winter wheat***. Net blotch in ***spring barley***, ***winter barley***. Powdery mildew in ***spring barley***, ***winter barley***. Rhynchosporium in ***spring barley***, ***winter barley***. Septoria in ***winter wheat***. Yellow rust in ***spring barley***, ***winter barley***, ***winter wheat***.

Approval information

- Propiconazole included in Annex I under EC Directive 91/414

Efficacy guidance

- Best results obtained from treatment at early stages of disease development
- Trifloxystrobin is a member of the QoI cross resistance group. Product should be used preventatively and not relied on for its curative potential
- Use product as part of an Integrated Crop Management strategy incorporating other methods of control, including where appropriate other fungicides with a different mode of action. Do not apply more than two foliar applications of QoI containing products to any cereal crop
- There is a significant risk of widespread resistance occurring in *Septoria tritici* populations in UK. Failure to follow resistance management action may result in reduced levels of disease control

Restrictions

- Maximum number of treatments 2 per crop and maximum total dose of propiconazole (including other products containing propiconazole) equivalent to two full dose treatments

Crop-specific information

- HI 35 d for all crops

Environmental safety

- Dangerous to fish or other aquatic life. Do not contaminate surface waters or ditches with chemical or used container
- LERAP Category B

Hazard classification and safety precautions

Hazard H04
Risk phrases R04a
Operator protection A, C; U02, U05a, U09a, U20b
Environmental protection E01, E13b, E16a, E16b
Consumer protection C02 (35 d); C03
Storage and disposal E26, E29, E30a, E31b

377 propoxycarbazone-sodium

A sulfonylaminocarbonyltriazolinone residual grass weed herbicide for winter wheat

Products

1	Attribut	Bayer	70% w/w	SG	10671
2	Attribut	Bayer CropScience	70% w/w	SG	11221
3	Ethos	Bayer	70% w/w	SG	10736
4	Ethos	Bayer CropScience	70% w/w	SG	11223

Uses

Blackgrass in ***winter wheat***. Couch in ***winter wheat***.

Efficacy guidance

- Activity by root and foliar absorption but depends on presence of sufficient soil moisture to ensure root uptake by weeds. Control is enhanced by use of an adjuvant to encourage foliar uptake

SECTION 4

- Best results obtained from treatments applied when weed grasses are growing actively. Symptoms may not become apparent for 3-4 wk after application
- Ensure good even spray coverage
- To achieve best control of couch and reduction of infestation in subsequent crop treat between 2 true leaves and first node stage of the weed. A split dose programme may improve control
- Blackgrass should be treated after using specialist blackgrass herbicides with a different mode of action
- Proxycarbazone-sodium is a member of the ALS-inhibitor group of herbicides and products should be used in a planned Resistance Management strategy. See Section 1 for more information

Restrictions

- Maximum total dose equivalent to one full dose treatment
- Do not use in a programme with other aceto-lactase synthesis (ALS) inhibitors

Crop-specific information

- Latest use: flag leaf just visible (GS 37)

Following crops guidance

- Only winter wheat, field beans or winter barley may be sown in the autumn following spring treatment. Any crop may be grown in the spring on land treated during the previous calendar yr

Environmental safety

- Dangerous to fish or other aquatic life. Do not contaminate surface waters or ditches with chemical or used container
- LERAP Category B
- Take extreme care to avoid drift onto adjacent plants or land as this could result in severe damage

Hazard classification and safety precautions

Operator protection A; U05a, U20b
Environmental protection E01, E13b, E16a, E16b
Consumer protection C03
Storage and disposal E30a, E32a

378 propyzamide

A residual amide herbicide for use in a wide range of crops

See also clopyralid + propyzamide

Products

1	Agriguard Propyzamide 50 WP	AgriGuard	50% w/w	WP	10187
2	Agriguard Propyzamide Flo	AgriGuard	400 g/l	SC	10435
3	Barclay Piza 400FL	Barclay	400 g/l	SC	11495
4	Base 50 W	Interfarm	50% w/w	WP	10202
5	Base 50 W	Dow	50% w/w	WP	10765
6	Bulwark Flo	Interfarm	400 g/l	SC	10161
7	Flomide	Interfarm	400 g/l	SC	11171
8	Greencrop Saffron FL	Greencrop	400 g/l	SC	10244
9	Kerb 50 W	Interfarm	50% w/w	WP	10200
10	Kerb 50 W	Dow	50% w/w	WP	10771
11	Kerb 80 EDF	Dow	80% w/w	WG	10761
12	Kerb Flo	Interfarm	400 g/l	SC	10160
13	Kerb Flo	Dow	400 g/l	SC	10768
14	Kerb Granules	SumiAgro	4% w/w	GR	08917
15	Kerb Pro Flo	SumiAgro Amenity	400 g/l	SC	10158
16	Kerb Pro Flo	Dow	400 g/l	SC	10760
17	Kerb Pro Granules	SumiAgro Amenity	4% w/w	GR	09600
18	Menace 80 EDF	Dow	80% w/w	WG	10766
19	Mithras 80 EDF	Interfarm	80% w/w	WG	10164
20	Mithras 80 EDF	Dow	80% w/w	WG	10767
21	Precis	Dow	400 g/l	SC	10762
22	Propose	Interfarm	50% w/w	WP	11170
23	Propose	Interfarm	50% w/w	WP	11195
24	Quaver Flo	Interfarm	400 g/l	SC	10162
25	Quaver Flo	Dow	400 g/l	SC	10769
26	Redeem Flo	Interfarm	400 g/l	SC	10236
27	Redeem Flo	Dow	400 g/l	SC	10764

Products – continued

28	Standon Propyzamide 400 SC	Standon	400 g/l	SC	10255
29	Standon Propyzamide 50	Standon	50% w/w	WP	10814

Uses

Annual dicotyledons in ***apple orchards, blackberries, clover seed crops, loganberries, lucerne, plums, sugar beet seed crops, winter field beans, winter oilseed rape*** [1-5, 7-13, 18-29]; ***blackcurrants, gooseberries, redcurrants, strawberries*** [1, 3-5, 7-13, 18-29]; ***brassica seed crops, lettuce*** *(outdoor crops)*, ***rhubarb*** *(outdoor)* [11, 18-20, 27]; ***fodder rape seed crops, kale seed crops, lettuce, rhubarb, turnip seed crops*** [1-5, 7-10, 12, 13, 21-26, 28, 29]; ***forestry, trees and shrubs*** [1, 4-6, 9, 10, 14-17, 22, 23]; ***ornamental plant production, raspberries*** [29]; ***pear orchards*** [1-5, 7-13, 18, 19, 21-29]; ***pears*** [20]; ***raspberries*** *(England only)* [1-5, 7-13, 18-28]; ***roses*** [1, 4, 5, 9, 10, 22, 23, 29]; ***woody ornamentals*** [1, 4-6, 9-11, 14-20, 22, 23, 27]. Annual grasses in ***apple orchards, blackberries, clover seed crops, loganberries, lucerne, plums, sugar beet seed crops, winter field beans, winter oilseed rape*** [1-5, 7-13, 18-29]; ***blackcurrants, gooseberries, redcurrants, strawberries*** [1, 3-5, 7-13, 18-29]; ***brassica seed crops, lettuce*** *(outdoor crops)*, ***rhubarb*** *(outdoor)* [11, 18-20, 27]; ***camomile*** *(off-label)*, ***cherries*** *(off-label)*, ***chicory*** *(off-label)*, ***courgettes*** *(off-label)*, ***evening primrose*** *(off-label)*, ***fenugreek*** *(off-label)*, ***grapevines*** *(off-label)*, ***honesty*** *(off-label)*, ***hops*** *(off-label)*, ***marrows*** *(off-label)*, ***mirabelles*** *(off-label)*, ***protected agastache spp.*** *(off-label)*, ***protected angelica*** *(off-label)*, ***protected balm*** *(off-label)*, ***protected basil*** *(off-label)*, ***protected bay*** *(off-label)*, ***protected borage*** *(off-label)*, ***protected camomile*** *(off-label)*, ***protected caraway*** *(off-label)*, ***protected catnip*** *(off-label)*, ***protected chervil*** *(off-label)*, ***protected chives*** *(off-label)*, ***protected clary*** *(off-label)*, ***protected coriander*** *(off-label)*, ***protected courgettes*** *(off-label)*, ***protected curry plant*** *(off-label)*, ***protected dill*** *(off-label)*, ***protected dragonhead*** *(off-label)*, ***protected fennel*** *(off-label)*, ***protected fenugreek*** *(off-label)*, ***protected feverfew*** *(off-label)*, ***protected hyssop*** *(off-label)*, ***protected lamb's lettuce*** *(off-label)*, ***protected land cress*** *(off-label)*, ***protected lavandin*** *(off-label)*, ***protected lavender*** *(off-label)*, ***protected lemon verbena*** *(off-label)*, ***protected lettuce*** *(off-label)*, ***protected lovage*** *(off-label)*, ***protected marigold*** *(off-label)*, ***protected marjoram*** *(off-label)*, ***protected marrows*** *(off-label)*, ***protected mint*** *(off-label)*, ***protected nasturtium*** *(off-label)*, ***protected nettle*** *(off-label)*, ***protected oregano*** *(off-label)*, ***protected origanum heracleoticum*** *(off-label)*, ***protected parsley*** *(off-label)*, ***protected pumpkins*** *(off-label)*, ***protected radicchio*** *(off-label)*, ***protected rocket*** *(off-label)*, ***protected rosemary*** *(off-label)*, ***protected rue*** *(off-label)*, ***protected sage*** *(off-label)*, ***protected salad burnet*** *(off-label)*, ***protected savory*** *(off-label)*, ***protected scarole*** *(off-label)*, ***protected sorrel*** *(off-label)*, ***protected spike lavender*** *(off-label)*, ***protected squashes*** *(off-label)*, ***protected tarragon*** *(off-label)*, ***protected thyme*** *(off-label)*, ***protected violet*** *(off-label)*, ***protected woodruff*** *(off-label)*, ***pumpkins*** *(off-label)*, ***radicchio*** *(off-label)*, ***sage*** *(off-label)*, ***squashes*** *(off-label)*, ***tarragon*** *(off-label)* [10]; ***fodder rape seed crops, kale seed crops, lettuce, rhubarb, turnip seed crops*** [1-5, 7-10, 12, 13, 21-26, 28, 29]; ***forestry, trees and shrubs*** [1, 4-6, 9, 10, 14-17, 22, 23]; ***ornamental plant production, raspberries*** [29]; ***pear orchards*** [1-5, 7-13, 18, 19, 21-29]; ***pears*** [20]; ***raspberries*** *(England only)* [1-5, 7-13, 18-28]; ***roses*** [1, 4, 5, 9, 10, 22, 23, 29]; ***woody ornamentals*** [1, 4-6, 9-11, 14-20, 22, 23, 27]. Horsetails in ***forestry*** [2, 6, 11, 15, 16, 18-20, 27]; ***woody ornamentals*** [6, 15, 16]. Perennial grasses in ***apple orchards, blackberries, loganberries, plums*** [1-5, 7-13, 18-29]; ***blackcurrants, gooseberries, redcurrants*** [1, 3-5, 7-13, 18-29]; ***camomile*** *(off-label)*, ***cherries*** *(off-label)*, ***chicory*** *(off-label)*, ***courgettes*** *(off-label)*, ***evening primrose*** *(off-label)*, ***fenugreek*** *(off-label)*, ***grapevines*** *(off-label)*, ***honesty*** *(off-label)*, ***hops*** *(off-label)*, ***marrows*** *(off-label)*, ***mirabelles*** *(off-label)*, ***protected agastache spp.*** *(off-label)*, ***protected angelica*** *(off-label)*, ***protected balm*** *(off-label)*, ***protected basil*** *(off-label)*, ***protected bay*** *(off-label)*, ***protected borage*** *(off-label)*, ***protected camomile*** *(off-label)*, ***protected caraway*** *(off-label)*, ***protected catnip*** *(off-label)*, ***protected chervil*** *(off-label)*, ***protected chives*** *(off-label)*, ***protected clary*** *(off-label)*, ***protected coriander*** *(off-label)*, ***protected courgettes*** *(off-label)*, ***protected curry plant*** *(off-label)*, ***protected dill*** *(off-label)*, ***protected dragonhead*** *(off-label)*, ***protected fennel*** *(off-label)*, ***protected fenugreek*** *(off-label)*, ***protected feverfew*** *(off-label)*, ***protected hyssop*** *(off-label)*, ***protected lamb's lettuce*** *(off-label)*, ***protected land cress*** *(off-label)*, ***protected lavandin*** *(off-label)*, ***protected lavender*** *(off-label)*, ***protected lemon verbena*** *(off-label)*, ***protected lettuce*** *(off-label)*, ***protected lovage*** *(off-label)*, ***protected marigold*** *(off-label)*, ***protected marjoram*** *(off-label)*, ***protected marrows*** *(off-label)*, ***protected mint*** *(off-label)*, ***protected nasturtium*** *(off-label)*, ***protected nettle*** *(off-label)*, ***protected oregano*** *(off-label)*, ***protected origanum heracleoticum*** *(off-label)*, ***protected parsley*** *(off-label)*, ***protected***

pumpkins *(off-label)*, ***protected radicchio*** *(off-label)*, ***protected rocket*** *(off-label)*, ***protected rosemary*** *(off-label)*, ***protected rue*** *(off-label)*, ***protected sage*** *(off-label)*, ***protected salad burnet*** *(off-label)*, ***protected savory*** *(off-label)*, ***protected scarole*** *(off-label)*, ***protected sorrel*** *(off-label)*, ***protected spike lavender*** *(off-label)*, ***protected squashes*** *(off-label)*, ***protected tarragon*** *(off-label)*, ***protected thyme*** *(off-label)*, ***protected violet*** *(off-label)*, ***protected woodruff*** *(off-label)*, ***pumpkins*** *(off-label)*, ***radicchio*** *(off-label)*, ***sage*** *(off-label)*, ***squashes*** *(off-label)*, ***tarragon*** *(off-label)* [10]; ***clover seed crops***, ***fodder rape seed crops***, ***kale seed crops***, ***lettuce***, ***lucerne***, ***rhubarb***, ***sugar beet seed crops***, ***turnip seed crops***, ***winter field beans***, ***winter oilseed rape*** [1-5, 7-10, 12, 13, 21-26, 28, 29]; ***forestry*** [1, 2, 4-6, 9-11, 14-20, 22, 23, 27]; ***ornamental plant production***, ***raspberries*** [29]; ***pear orchards*** [1-5, 7-13, 18, 19, 21-29]; ***pears*** [20]; ***raspberries*** *(England only)* [1-5, 7-13, 18-28]; ***rhubarb*** *(outdoor)* [11, 18-20, 27]; ***roses*** [1, 4, 5, 9, 10, 22, 23, 29]; ***strawberries*** [1, 3-5, 7-10, 12, 13, 21-26, 28, 29]; ***trees and shrubs*** [1, 4-6, 9, 10, 14-17, 22, 23]; ***woody ornamentals*** [1, 4-6, 9-11, 14-20, 22, 23, 27]. Sedges in ***forestry*** [2, 6, 11, 15, 16, 18-20, 27]; ***woody ornamentals*** [6, 15, 16]. Volunteer cereals in ***sugar beet seed crops***, ***winter field beans***, ***winter oilseed rape*** [11, 18-20, 27]. Wild oats in ***sugar beet seed crops***, ***winter field beans***, ***winter oilseed rape*** [11, 18-20, 27].

Specific Off-Label Approvals (SOLAs)

- ***camomile, cherries, chicory, courgettes, evening primrose, fenugreek, grapevines, honesty, hops, marrows, mirabelles, protected agastache spp., protected angelica, protected balm, protected basil, protected bay, protected borage, protected camomile, protected caraway, protected catnip, protected chervil, protected chives, protected clary, protected coriander, protected courgettes, protected curry plant, protected dill, protected dragonhead, protected fennel, protected fenugreek, protected feverfew, protected hyssop, protected lamb's lettuce, protected land cress, protected lavandin, protected lavender, protected lemon verbena, protected lettuce, protected lovage, protected marigold, protected marjoram, protected marrows, protected mint, protected nasturtium, protected nettle, protected oregano, protected origanum heracleoticum, protected parsley, protected pumpkins, protected radicchio, protected rocket, protected rosemary, protected rue, protected sage, protected salad burnet, protected savory, protected scarole, protected sorrel, protected spike lavender, protected squashes, protected tarragon, protected thyme, protected violet, protected woodruff, pumpkins, radicchio, sage, squashes, tarragon*** *(OLA 021416) Dec 2008* [10]

Approval information

- Propyzamide included in Annex I under EC Directive 91/414
- Some products may be applied through CDA equipment. See labels for details

Efficacy guidance

- Active via root uptake. Weeds controlled from germination to young seedling stage, some species (including many grasses) also when established
- Best results achieved by winter application to fine, firm, moist soil. Rain is required after application if soil dry
- Excessive organic debris or ploughed-up turf may reduce efficacy
- For heavy couch infestations a repeat application may be needed in following winter
- In lettuce lightly incorporate in top 25 mm pre-drilling or irrigate on dry soil

Restrictions

- Maximum number of treatments 1 per crop or yr
- Maximum total dose equivalent to one full dose treatment for all crops
- Do not treat protected crops
- See label for lists of ornamental and forest species which may be treated
- Do not use on soils with more than 10% organic matter except in forestry

Crop-specific information

- Latest use: labels vary but normally before 31 Dec for rhubarb, lucerne, strawberries and winter field beans; before 31 Jan for other crops; pre-emergence for chicory
- HI normally 6 wk for edible crops and 3 mth for protected herbs but labels vary in detail
- May be applied to listed edible crops only between 1 Oct and the date specified as the latest time of application (except lettuce)
- Apply as soon as possible after 3-true leaf stage of oilseed rape (GS 1,3) and seed brassicas, after 4-leaf stage of sugar beet for seed, within 7 d after sowing but before emergence for field beans, after perennial crops established for at least 1 season, strawberries after 1 yr

- Only apply to strawberries on heavy soils. Use reduced dose on matted row crops
- Only apply to field beans on medium and heavy soils
- Only apply to established lucerne not less than 7 d after last cut

Following crops guidance

- Lettuce may be sown or planted immediately after treatment; with other crops period varies from 5 to 40 wk. See label for details
- Do not apply to the same land less than 9 mth after an earlier application

Environmental safety

- Do not harvest crops for human or animal consumption for at least 6 wk after last application

Hazard classification and safety precautions

Operator protection U20a [14, 17]; U20b [27, 29]; U20c [1-13, 15, 16, 18-26, 28]

Environmental protection E01 [2, 14, 17, 21, 29]; E15 [1-29]

Consumer protection C02 [1-5, 7-10] (6 wk); C02 [11]; C02 [12, 13] (6 wk); C02 [14, 17-20]; C02 [21-29] (6 wk)

Storage and disposal E26 [1, 6-8, 11-13, 15, 16, 18-21, 24-28]; E30a [1-29]; E31b [3, 8, 28]; E32a [1, 2, 4-7, 9-27, 29]; E34 [1, 2, 14, 17, 21]

379 prosulfuron

A contact and residual sulfonyl urea herbicide available only in mixture

See also *bromoxynil + prosulfuron*

380 pymetrozine

A novel azomethine insecticide

Products

1	Chess WG	Syngenta Bioline	50% w/w	WG	10651
2	Plenum WG	Syngenta	50% w/w	WG	10652

Uses

Aphids in ***angelica*** *(off-label)*, ***balm*** *(off-label)*, ***basil*** *(off-label)*, ***bay*** *(off-label)*, ***borage*** *(off-label)*, ***camomile*** *(off-label)*, ***caraway*** *(off-label)*, ***chervil*** *(off-label)*, ***chives*** *(off-label)*, ***clary*** *(off-label)*, ***coriander*** *(off-label)*, ***dill*** *(off-label)*, ***fennel*** *(off-label)*, ***fenugreek*** *(off-label)*, ***feverfew*** *(off-label)*, ***hyssop*** *(off-label)*, ***land cress*** *(off-label)*, ***lettuce*** *(off-label)*, ***lovage*** *(off-label)*, ***marigold*** *(off-label)*, ***marjoram*** *(off-label)*, ***mint*** *(off-label)*, ***nasturtium*** *(off-label)*, ***nettle*** *(off-label)*, ***oregano*** *(off-label)*, ***parsley*** *(off-label)*, ***potatoes***, ***rocket*** *(off-label)*, ***rosemary*** *(off-label)*, ***rue*** *(off-label)*, ***sage*** *(off-label)*, ***salad brassicas*** *(off-label - for baby leaf production)*, ***salad burnet*** *(off-label)*, ***savory*** *(off-label)*, ***sorrel*** *(off-label)*, ***tarragon*** *(off-label)*, ***thyme*** *(off-label)*, ***verbena*** *(off-label - lemon) and woodruff)* [2]; ***ornamental specimens***, ***protected cucumbers***, ***protected ornamentals*** [1]; ***protected blackberries*** *(off-label)*, ***protected blackcurrants*** *(off-label)*, ***protected gooseberries*** *(off-label)*, ***protected loganberries*** *(off-label)*, ***protected raspberries*** *(off-label)*, ***protected redcurrants*** *(off-label)*, ***protected ribes hybrids*** *(off-label)*, ***protected rubus hybrids*** *(off-label)*, ***protected strawberries*** *(off-label)* [1, 2]. Currant lettuce aphid in ***lettuce*** *(off-label)*, ***protected agastache spp.*** *(off-label)*, ***protected angelica*** *(off-label)*, ***protected balm*** *(off-label)*, ***protected basil*** *(off-label)*, ***protected bay*** *(off-label)*, ***protected borage*** *(off-label)*, ***protected camomile*** *(off-label)*, ***protected caraway*** *(off-label)*, ***protected catnip*** *(off-label)*, ***protected chervil*** *(off-label)*, ***protected chives*** *(off-label)*, ***protected clary*** *(off-label)*, ***protected coriander*** *(off-label)*, ***protected curry plant*** *(off-label)*, ***protected dill*** *(off-label)*, ***protected dragonhead*** *(off-label)*, ***protected endives*** *(off-label)*, ***protected fennel*** *(off-label)*, ***protected fenugreek*** *(off-label)*, ***protected feverfew*** *(off-label)*, ***protected horehound*** *(off-label)*, ***protected hyssop*** *(off-label)*, ***protected land cress*** *(off-label)*, ***protected lavandin*** *(off-label)*, ***protected lavender*** *(off-label)*, ***protected lemon verbena*** *(off-label)*, ***protected lettuce*** *(off-label)*, ***protected lovage*** *(off-label)*, ***protected marigold*** *(off-label)*, ***protected marjoram*** *(off-label)*, ***protected mint*** *(off-label)*, ***protected nasturtium*** *(off-label)*, ***protected nettle*** *(off-label)*, ***protected oregano*** *(off-label)*, ***protected origanum heracleoticum*** *(off-label)*, ***protected parsley*** *(off-label)*, ***protected plantain*** *(off-label)*, ***protected rocket*** *(off-label)*, ***protected rosemary*** *(off-label)*, ***protected rue*** *(off-label)*, ***protected sage*** *(off-label)*, ***protected salad brassicas*** *(off-label - for baby leaf*

SECTION 4

production), ***protected salad burnet*** *(off-label),* ***protected savory*** *(off-label),* ***protected sorrel*** *(off-label),* ***protected spike lavender*** *(off-label),* ***protected st john's wort*** *(off-label),* ***protected tarragon*** *(off-label),* ***protected thyme*** *(off-label),* ***protected violet*** *(off-label),* ***protected woodruff*** *(off-label),* ***protected yarrow*** *(off-label)* [2]. Glasshouse whitefly in ***protected tomatoes*** *(off-label)* [1]. Peach-potato aphid in ***aubergines*** *(off-label),* ***protected aubergines*** *(off-label),* ***protected peppers*** *(off-label)* [1]; ***broccoli*** *(off-label),* ***brussels sprouts*** *(off-label),* ***cabbages*** *(off-label),* ***calabrese*** *(off-label),* ***cauliflowers*** *(off-label)* [2]; ***lettuce*** *(off-label),* ***protected agastache spp.*** *(off-label),* ***protected angelica*** *(off-label),* ***protected balm*** *(off-label),* ***protected basil*** *(off-label),* ***protected bay*** *(off-label),* ***protected borage*** *(off-label),* ***protected camomile*** *(off-label),* ***protected caraway*** *(off-label),* ***protected catnip*** *(off-label),* ***protected chervil*** *(off-label),* ***protected chives*** *(off-label),* ***protected clary*** *(off-label),* ***protected coriander*** *(off-label),* ***protected curry plant*** *(off-label),* ***protected dill*** *(off-label),* ***protected dragonhead*** *(off-label),* ***protected endives*** *(off-label),* ***protected fennel*** *(off-label),* ***protected fenugreek*** *(off-label),* ***protected feverfew*** *(off-label),* ***protected horehound*** *(off-label),* ***protected hyssop*** *(off-label),* ***protected land cress*** *(off-label),* ***protected lavandin*** *(off-label),* ***protected lavender*** *(off-label),* ***protected lemon verbena*** *(off-label),* ***protected lettuce*** *(off-label),* ***protected lovage*** *(off-label),* ***protected marigold*** *(off-label),* ***protected marjoram*** *(off-label),* ***protected mint*** *(off-label),* ***protected nasturtium*** *(off-label),* ***protected nettle*** *(off-label),* ***protected oregano*** *(off-label),* ***protected origanum heracleoticum*** *(off-label),* ***protected parsley*** *(off-label),* ***protected plantain*** *(off-label),* ***protected rocket*** *(off-label),* ***protected rosemary*** *(off-label),* ***protected rue*** *(off-label),* ***protected sage*** *(off-label),* ***protected salad brassicas*** *(off-label - for baby leaf production),* ***protected salad burnet*** *(off-label),* ***protected savory*** *(off-label),* ***protected sorrel*** *(off-label),* ***protected spike lavender*** *(off-label),* ***protected st john's wort*** *(off-label),* ***protected tarragon*** *(off-label),* ***protected thyme*** *(off-label),* ***protected violet*** *(off-label),* ***protected woodruff*** *(off-label),* ***protected yarrow*** *(off-label)* [1, 2]. Pre-harvest desiccation in ***hops*** *(off-label)* [2]. Tobacco whitefly in ***protected tomatoes*** *(off-label)* [1].

Specific Off-Label Approvals (SOLAs)

- ***angelica, balm, basil, bay, borage, camomile, caraway, chervil, chives, clary, coriander, dill, fennel, fenugreek, feverfew, hyssop, land cress, lettuce, lovage, marigold, marjoram, mint, nasturtium, nettle, oregano, parsley, rocket, rosemary, rue, sage, salad burnet, savory, sorrel, tarragon, thyme*** *(OLA 021429) May 2004* [2]
- ***aubergines, protected peppers*** *(OLA 021428) May 2004* [1]
- ***broccoli, brussels sprouts, cabbages, calabrese, cauliflowers*** *(OLA 031478) Oct 2011* [2]
- ***hops*** *(OLA 031423) Oct 2011* [2]
- ***lettuce, protected agastache spp., protected angelica, protected balm, protected basil, protected bay, protected borage, protected camomile, protected caraway, protected catnip, protected chervil, protected chives, protected clary, protected coriander, protected curry plant, protected dill, protected dragonhead, protected endives, protected fennel, protected fenugreek, protected feverfew, protected horehound, protected hyssop, protected land cress, protected lavandin, protected lavender, protected lemon verbena, protected lettuce, protected lovage, protected marigold, protected marjoram, protected mint, protected nasturtium, protected nettle, protected oregano, protected origanum heracleoticum, protected parsley, protected plantain, protected rocket, protected rosemary, protected rue, protected sage, protected salad burnet, protected savory, protected sorrel, protected spike lavender, protected st john's wort, protected tarragon, protected thyme, protected violet, protected woodruff, protected yarrow*** *(OLA 030843) Oct 2011* [1]
- ***lettuce, protected agastache spp., protected angelica, protected balm, protected basil, protected bay, protected borage, protected camomile, protected caraway, protected catnip, protected chervil, protected chives, protected clary, protected coriander, protected curry plant, protected dill, protected dragonhead, protected endives, protected fennel, protected fenugreek, protected feverfew, protected horehound, protected hyssop, protected land cress, protected lavandin, protected lavender, protected lemon verbena, protected lettuce, protected lovage, protected marigold, protected marjoram, protected mint, protected nasturtium, protected nettle, protected oregano, protected origanum heracleoticum, protected parsley, protected plantain, protected rocket, protected rosemary, protected rue, protected sage, protected salad burnet, protected savory, protected sorrel, protected spike lavender, protected st john's wort, protected tarragon, protected thyme, protected violet, protected woodruff, protected yarrow*** *(OLA 030844) Oct 2011* [2]
- ***protected aubergines, protected peppers, protected tomatoes*** *(OLA 030845) Oct 2011* [1]

- ***protected blackberries*, *protected blackcurrants*, *protected gooseberries*, *protected loganberries*, *protected raspberries*, *protected redcurrants*, *protected ribes hybrids*, *protected rubus hybrids*, *protected strawberries*** *(OLA 031072) Oct 2011* [1]
- ***protected blackberries*, *protected blackcurrants*, *protected gooseberries*, *protected loganberries*, *protected raspberries*, *protected redcurrants*, *protected ribes hybrids*, *protected rubus hybrids*, *protected strawberries*** *(OLA 031073) Oct 2011* [2]
- ***protected salad brassicas*** *(for baby leaf production) (OLA 030843) Oct 2011* [1]
- ***protected salad brassicas*** *(for baby leaf production) (OLA 030844) Oct 2011* [2]
- ***protected tomatoes*** *(OLA 030170) Dec 2008* [1]
- ***salad brassicas*** *(for baby leaf production) (OLA 021429) May 2004* [2]
- ***verbena*** *(lemon) and woodruff) (OLA 021429) May 2004* [2]

Approval information

- Pymetrozine included in Annex I under EC Directive 91/414

Efficacy guidance

- Pymetrozine moves systemically in the plant and acts by preventing feeding leading to death by starvation in 1-4 d. There is no immediate knockdown
- Aphids controlled include those resistant to organophosphorus and carbamate insecticides
- To prevent development of resistance do not use continuously or as the sole method of control
- Best results achieved by starting spraying as soon as aphids seen in crop and repeating as necessary.
- To limit spread of persistent viruses such as potato leaf roll virus, apply from 90% crop emergence

Restrictions

- Maximum total dose equivalent to two full dose treatments on potatoes; three full dose treatments on edible crops; four full dose treatments on ornamentals
- Consult processors before use on potatoes for processing
- Check tolerance of ornamental species before large scale use. See label for list of species known to have been treated without damage. Visible spray deposits may be seen on leaves of some species [1]

Crop-specific information

- HI aubergines, peppers, cucumbers 3 d; potatoes 7 d , herbs 7 d

Environmental safety

- High risk to bees (outdoor use only). Do not apply to crops in flower or to those in which bees are actively foraging. Do not apply when flowering weeds are present
- Avoid spraying within 6 m of field boundaries to reduce effects on non-target insects or arthropods

Hazard classification and safety precautions

Hazard R03, R13 [2]
Operator protection A; U05a, U20c
Environmental protection E01, E15 [1, 2]; E12a [1, 2] (outdoor use only)
Consumer protection C03
Storage and disposal E26 [2]; E30a, E32a [1, 2]

381 pyraclostrobin

A protectant and curative strobilurin fungicide for cereals

See also *boscalid + pyraclostrobin*
epoxiconazole + fenpropimorph + pyraclostrobin
epoxiconazole + kresoxim-methyl + pyraclostrobin
epoxiconazole + pyraclostrobin

Products

1	Comet	BASF	250 g/l	EC	10875
2	Tucana	BASF	250 g/l	EC	10899
3	Vivid	BASF	250 g/l	EC	10898

Uses

Brown rust in ***spring barley*, *spring wheat*, *winter barley*, *winter wheat***. Crown rust in ***spring oats*, *winter oats***. Net blotch in ***spring barley*, *winter barley***. Rhynchosporium in ***spring barley*** (moderate), ***winter barley*** (moderate). Septoria diseases in ***spring wheat***, ***winter wheat***. Yellow rust in ***spring barley***, ***spring wheat***, ***winter barley***, ***winter wheat***.

Efficacy guidance

- For best results apply at the start of foliar disease attack
- Best results on Septoria glume blotch achieved when used as a protective treatment and against Septoria leaf blotch when treated in the latent phase
- Yield response may be obtained in the absence of visual disease symptoms
- Pyraclostrobin is a member of the QoI cross resistance group. Product should be used preventatively and not relied on for its curative potential
- Use product as part of an Integrated Crop Management strategy incorporating other methods of control, including where appropriate other fungicides with a different mode of action. Do not apply more than two foliar applications of QoI containing products to any cereal crop
- There is a significant risk of widespread resistance occurring in *Septoria tritici* populations in UK. Failure to follow resistance management action may result in reduced levels of disease control
- On cereal crops product must always be used in mixture with another product, recommended for control of the same target disease, that contains a fungicide from a different cross resistance group and is applied at a dose that will give robust control

Restrictions

- Maximum total dose equivalent to two full dose treatments

Crop-specific information

- Latest use: before grain watery ripe (GS 71) for wheat; up to and including emergence of ear just complete (GS 59) for barley and oats

Environmental safety

- Dangerous to fish or other aquatic life. Do not contaminate surface waters or ditches with chemical or used container
- LERAP Category B

Hazard classification and safety precautions

Hazard R03, R04
Risk phrases R03b, R03c, R04b
Operator protection A; U05a, U14, U20b
Environmental protection E01, E13b, E16a, E16b
Consumer protection C03
Storage and disposal E26, E30a, E31c, E34
Medical advice M03

382 pyrethrins

A non-persistent, contact acting insecticide extracted from Pyrethrum

Products

1	Alfadex	Novartis A H	0.75 g/l	AL	H7221
2	Dairy Fly Spray	B H & B	0.75 g/l	AL	H5579
3	Killgerm ULV 400	Killgerm	30 g/l	UL	H4838
4	Turbair Super Flydown	SumiAgro	2.5 g/l	RH	H7225

Uses

Flies in ***dairies***, ***farm buildings*** [1-3]; ***livestock houses***, ***poultry houses*** [1-4].

Approval information

- Products formulated for ULV application [2, 3]. May be applied through fogging machine or sprayer [2]. See label for details

Efficacy guidance

- For fly control close doors and windows and spray or apply fog as appropriate
- For best results outdoors spray during early morning or late afternoon and evening when conditions are still
- Ensure spray is directed to all areas where flies are seen to be numerous
- To avoid possibility of development of resistance do not spray more frequently than once per week

Restrictions

- Do not allow spray to contact open food products or food preparing equipment or utensils
- Remove exposed milk and collect eggs before application
- Do not treat plants
- Do not use space sprays containing pyrethrins or pyrethroid more than once per week in intensive or controlled environment animal houses in order to avoid development of resistance. If necessary, use a different control method or product [3]

Crop-specific information

- Product formulated for use through Turbair Flydowner machines [4]

Environmental safety

- Do not apply directly to livestock
- Harmful to fish or other aquatic life. Do not contaminate surface waters or ditches with chemical or used container
- Exclude all persons and animals during treatment [2]

Hazard classification and safety precautions

Hazard R03 [2, 4]; R04 [1-4]
Risk phrases R03c [2]; R04a, R04b [1-4]; R04c [1, 3, 4]; R04g [4]
Operator protection A [1-3]; B, E [2]; C, D, H [1, 3]; U05a, U19 [1-4]; U09a [2]; U20a [1, 3]; U20b [2, 4]
Environmental protection E01, E05, E13c [1-4]; E04 [2]
Consumer protection C03 [1-4]; C04, C08, C09, C10 [2]; C06, C07 [2, 4]
Storage and disposal E30a [1-4]; E31a [1, 3, 4]; E32a [2]; E34 [4]
Medical advice M05 [4]

383 pyridate

A contact pyridazine herbicide for cereals, maize and brassicas

Products

Lentagran WP	Syngenta	45% w/w	WB	08478

Uses

Annual dicotyledons in ***spring oilseed rape*** *(off-label)*, ***winter oilseed rape*** *(off-label)*. Black nightshade in ***brussels sprouts***, ***cabbages***, ***maize***, ***onions***. Cleavers in ***brussels sprouts***, ***cabbages***, ***maize***, ***onions***, ***spring oilseed rape*** *(off-label)*, ***winter oilseed rape*** *(off-label)*. Fat hen in ***brussels sprouts***, ***cabbages***, ***maize***, ***onions***.

Specific Off-Label Approvals (SOLAs)

- ***spring oilseed rape***, ***winter oilseed rape*** *(OLA 011663) Dec 2008* [1]

Approval information

- Pyridate included in Annex I under EC Directive 91/414

Efficacy guidance

- Best results achieved by application to actively growing weeds at 6-8 leaf stage when temperatures are above 8°C before crop foliage forms canopy

Restrictions

- Maximum number of treatments 1 per crop
- Do not apply in mixture with or within 14 d of any other product which may result in dewaxing of crop foliage
- Apply to maize after first-leaf stage. Do not use on cv. Meritos, Sunrise or Tainon 236
- Do not use on crops suffering stress from frost, drought, disease or pest attack

Crop-specific information

- Latest use: before 7 leaf stage for maize; before 5 true leaves for onions; before flower buds visible for oilseed rape
- HI 6 wk for Brussels sprouts, cabbages
- Apply to cabbages and Brussels sprouts after 4 fully expanded leaf stage. Allow 2 wk after transplanting before treating

Environmental safety

- Harmful to fish or other aquatic life. Do not contaminate surface waters or ditches with chemical or used container

Hazard classification and safety precautions

Hazard R04
Risk phrases R04a, R04b, R04e
Operator protection A, C; U05a, U08, U20b, U22a
Environmental protection E01, E13c
Consumer protection C03
Storage and disposal E26, E30a

384 pyrimethanil

An anilinopyrimidine fungicide for apples and strawberries

Products

1	Scala	Aventis	400 g/l	SC	07806
2	Scala	BASF	400 g/l	SC	11695
3	Standon Pyrimethanil	Standon	400 g/l	SC	10576

Uses

Botrytis in ***aubergines*** *(off-label)*, ***bilberries*** *(off-label)*, ***blackcurrants*** *(off-label)*, ***cranberries*** *(off-label)*, ***gooseberries*** *(off-label)*, ***grapevines*** *(off-label)*, ***outdoor tomatoes*** *(off-label)*, ***protected blackberries*** *(off-label)*, ***protected boysenberries*** *(off-label)*, ***protected lettuce*** *(off-label)*, ***protected loganberries*** *(off-label)*, ***protected marionberries*** *(off-label)*, ***protected raspberries*** *(off-label)*, ***protected sunberries*** *(off-label)*, ***protected tayberries*** *(off-label)*, ***protected tomatoes*** *(off-label)*, ***redcurrants*** *(off-label)*, ***whitecurrants*** *(off-label)* [1]; ***strawberries*** [1-3]. Scab in ***apples*** [1-3].

Specific Off-Label Approvals (SOLAs)

- ***aubergines, outdoor tomatoes, protected tomatoes*** *(OLA 021321) Dec 2008* [1]
- ***bilberries, blackcurrants, cranberries, gooseberries, redcurrants, whitecurrants*** *(OLA 001883) Dec 2008* [1]
- ***grapevines*** *(OLA 021414) Dec 2008* [1]
- ***protected blackberries, protected boysenberries, protected loganberries, protected marionberries, protected raspberries, protected sunberries, protected tayberries*** *(OLA 023411) Dec 2008* [1]
- ***protected lettuce*** *(OLA 001590) Dec 2008* [1]

Efficacy guidance

- On apples a programme of sprays will give early season control of scab. Season long control can be achieved by continuing programme with other approved fungicides
- In strawberries product should be used as part of a programme of disease control treatments which should alternate with other materials to prevent or limit development of less sensitive strains of grey mould

Restrictions

- Maximum number of treatments 5 per yr for apples; 4 per yr for protected crops; 3 per yr for grapevines; 2 per yr for cane fruit, bush fruit, strawberries, tomatoes
- Product does not taint apples. Processors should be consulted before use on strawberries

Crop-specific information

- Latest use: before end of flowering for apples
- HI cane fruit, strawberries 1 d; aubergines, tomatoes 3 d; protected lettuce 14 d; bush fruit, grapevines 21 d
- All varieties of apples and strawberries may be treated
- Treat apples from bud burst at 10-14 d intervals
- In strawberries start treatments at white bud to give maximum protection of flowers against grey mould and treat every 7-10 d. Product should not be used more than once in a 3 or 4 spray programme

Environmental safety

- Harmful to fish or other aquatic life. Do not contaminate surface waters or ditches with chemical or used container

- Product has negligible effect on hoverflies and lacewings. Limited evidence indicates some margin of safety to *Typhlodromus pyri*
- Broadcast air-assisted LERAP (20 m) [1, 2]

Hazard classification and safety precautions

Operator protection U08, U20b
Environmental protection E01, E13c [1-3]; E17a [1, 2] (20 m)
Consumer protection C03
Storage and disposal E26, E30a, E31b

385 quinmerac

A residual herbicide available only in mixtures

See also *chloridazon + quinmerac*
metazachlor + quinmerac

386 quinoxyfen

A systemic protectant fungicide for cereals

See also *cyproconazole + quinoxyfen*
fenpropimorph + quinoxyfen

Products

1	Apres	Dow	500 g/l	SC	08881
2	Erysto	Dow	250 g/l	SC	08697
3	Fortress	Dow	500 g/l	SC	08279
4	Me2 After	Me2	500 g/l	SC	10463

Uses

Powdery mildew in ***durum wheat***, ***spring oats***, ***spring rye***, ***sugar beet***, ***triticale***, ***winter oats***, ***winter rye*** [1-3]; ***spring barley***, ***spring wheat***, ***winter barley***, ***winter wheat*** [1-4].

Efficacy guidance

- For best results treat at early stage of disease development before infection spreads to new crop growth. Further treatment may be necessary if disease pressure remains high
- Product not curative and will not control latent or established disease infections
- For broad spectrum control in cereals use in tank mixtures - see label
- Product rainfast after 1 h
- Systemic activity may be reduced in severe drought

Restrictions

- Maximum total dose equivalent to two full dose treatments in cereals; see labels for split dose recommendation in sugar beet
- On cereals apply only in the spring from mid-tillering stage (GS 25)

Crop-specific information

- Latest use: first awns visible stage (GS 49) for cereals
- HI sugar beet 28 d

Environmental safety

- Dangerous for the environment [1-3]
- Very toxic to aquatic organisms. May cause long-term adverse effects in the aquatic environment [1-3]
- Dangerous to fish or other aquatic life. Do not contaminate surface waters or ditches with chemical or used container [4]

Hazard classification and safety precautions

Hazard H04 [1-4]; R14 [1-3]
Risk phrases R04a, R04b, R04c [4]; R04e [1-4]; R14c [1-3]
Operator protection A, C, H; U05a, U14
Environmental protection E01 [1-4]; E13b, E16a, E16b [4]; E15, E38 [1-3]

SECTION 4

Consumer protection C03
Storage and disposal E26 [4]; E32d [1-3]; E34 [1-4]

387 quizalofop-P-ethyl

An aryl phenoxypropionic acid post-emergence grass herbicide

Products

1	CoPilot	Bayer CropScience	100 g/l	EC	08042
2	Sceptre	Bayer CropScience	250 g/l	SC	08043

Uses

Annual grasses in ***combining peas, fodder beet, linseed, mangels, red beet, spring field beans, spring oilseed rape, sugar beet, vining peas, winter field beans, winter oilseed rape***. Couch in ***combining peas, fodder beet, linseed, mangels, red beet, spring field beans, spring oilseed rape, sugar beet, vining peas, winter field beans, winter oilseed rape***. Perennial grasses in ***combining peas, fodder beet, linseed, mangels, red beet, spring field beans, spring oilseed rape, sugar beet, vining peas, winter field beans, winter oilseed rape***. Volunteer cereals in ***combining peas, fodder beet, linseed, mangels, red beet, spring field beans, spring oilseed rape, sugar beet, vining peas, winter field beans, winter oilseed rape***.

Efficacy guidance

- Best results achieved by application to emerged weeds growing actively in warm conditions with adequate soil moisture. Use split treatment to extend period of control in oilseed rape
- Weed control may be reduced under conditions such as drought that limit uptake and translocation
- Must be used with a recommended adjuvant - see label for details [2]
- Annual meadow-grass is not controlled
- Various spray programmes and tank-mixtures recommended to control mixed dicotyledon/grass weed populations. See label for details
- For effective couch control do not hoe beet crops within 21 d after spraying
- At least 2 h rain free period required for effective results

Restrictions

- Maximum number of treatments 1 on peas and beans, 2 as split dose on oilseed rape, 2 for couch control in all other recommended crops. See labels for maximum total dose of individual products
- Do not spray crops under stress from any cause or in frosty weather
- Consult processor before use on peas or beans for processing
- An interval of at least 3 d must elapse between treatment and use of another herbicide on beet crops, 14 d on oilseed rape, 21 d on linseed and other recommended crops

Crop-specific information

- HI 16 wk for beet crops; 11 wk for oilseed rape, linseed; 8 wk for field beans; 5 wk for peas

Following crops guidance

- In the event of crop failure broad-leaved crops may be resown at any time, cereals, onions, leeks or maize may be sown after 2-6 wk depending on dose used. See label for details. Onions, leeks and maize are not recommended to follow a failed treated crop

Environmental safety

- Flammable [1]
- Dangerous to fish or other aquatic life. Do not contaminate surface waters or ditches with chemical or used container

Hazard classification and safety precautions

Hazard R04 [1, 2]; R07d [1]
Risk phrases R04a, R04b [2]; R04d [1]
Operator protection A, C; U05a, U11, U20b [1]; U09a, U20a [2]
Environmental protection E01 [1]; E13b [1, 2]
Consumer protection C03 [1]
Storage and disposal E26, E30a, E31b

388 resmethrin

A contact acting pyrethroid insecticide available only in mixtures

See also *pyrethrins + resmethrin*

389 resmethrin + tetramethrin

A contact acting pyrethroid insecticide mixture

Products

Sorex Wasp Nest Destroyer	Sorex	0.10:0.05% w/w	AE	H6294

Uses

Wasps in ***miscellaneous pest control situations***.

Efficacy guidance

- Product acts by lowering temperature in and around nest and producing a stupefying vapour in addition to its contact insecticidal effect. Spray only on surfaces
- Apply by directing jet at nest from up to 3 m away and approach nest gradually until jet can be directed into entrance hole. Continue spraying until nest saturated

Restrictions

- For use only by trained operators and owners of commercial and agricultural premises

Environmental safety

- Harmful to caged birds and pets
- Dangerous to bees. Do not apply to crops in flower or to those in which bees are actively foraging. Do not apply when flowering weeds are present
- Extremely dangerous to fish or other aquatic life. Remove fish tanks before spraying

Hazard classification and safety precautions

Hazard R03
Risk phrases R03b
Operator protection A, C, H; U14, U19, U20a
Environmental protection E01, E10b, E12c, E13a, E15
Consumer protection C03
Storage and disposal E30a

SECTION 4

390 rimsulfuron

A selective systemic sulfonylurea herbicide

Products

1	Landgold Rimsulfuron	Landgold	25% w/w	SG	08959
2	Me2 Aducksbackside	Me2	25% w/w	SG	10065
3	Standon Rimsulfuron	Standon	25% w/w	SG	09955
4	Titus	DuPont	25% w/w	SG	07908

Uses

Annual dicotyledons in ***forage maize***, ***potatoes***. Volunteer oilseed rape in ***forage maize***, ***potatoes***.

Efficacy guidance

- Product should be used with a suitable adjuvant or a suitable herbicide tank-mix partner. See label for details
- Product acts by foliar action. Best results obtained from good spray cover of small actively growing weeds. Effectiveness is reduced in very dry conditions
- Split application provides control over a longer period and improves control of fat hen and polygonums
- Weed spectrum can be broadened by tank mixture with other herbicides. See label for details
- Susceptible weeds cease growth immediately and symptoms can be seen 10 d later
- Rimsulfuron is a member of the ALS-inhibitor group of herbicides

Restrictions
- Maximum number of treatments 2 per crop
- Do not treat maize previously treated with organophosphorus insecticides
- Avoid high light intensity (full sunlight) and high temperatures on the day of spraying
- Do not treat during periods of substantial diurnal temperature fluctuation or when frost anticipated
- Do not apply to any crop stressed by drought, water-logging, low temperatures, pest or disease attack, nutrient or lime deficiency

Crop-specific information
- Latest use: before most advanced potato plants are 25 cm high; before 4-collar stage of fodder maize
- All varieties of ware potatoes and fodder maize may be treated, but variety restrictions of any tank-mix partner must be observed

Following crops guidance
- Only winter wheat should follow a treated crop in the same calendar yr. Only barley, wheat or maize should be sown in the spring of the yr following treatment. In the second autumn after treatment any crop except brassicas or oilseed rape may be drilled

Environmental safety
- Extremely dangerous to fish or other aquatic life. Do not contaminate surface waters or ditches with chemical or used container
- Herbicide is very active. Take particular care to avoid drift onto plants outside the target area
- Spraying equipment should not be drained or flushed onto land planted, or to be planted, with trees or crops other than potatoes or forage maize and should be thoroughly cleansed after use - see label for instructions

Hazard classification and safety precautions

Operator protection U08, U19 [1-4]; U20a [1, 2, 4]; U20b [3]
Environmental protection E01 [3]; E13a [1-4]
Storage and disposal E30a, E32a [1-4]; E34 [3]

391 rotenone

A natural, contact insecticide of low persistence

Products

FS Liquid Derris	Ford Smith	50 g/l	EC	01213

Uses

Aphids in ***all soft fruit, all top fruit, miscellaneous flowers, ornamental specimens, protected crops, vegetables***. Raspberry beetle in ***cane fruit***. Sawflies in ***gooseberries***. Slug sawfly in ***pears, roses***.

Efficacy guidance
- Apply as high volume spray when pest first seen and repeat as necessary
- Spray raspberries at first pink fruit, loganberries when most of blossom over, blackberries as first blossoms open. Repeat 2 wk later if necessary
- Spray to obtain thorough coverage, especially on undersurfaces of leaves

Crop-specific information
- HI 1 d

Environmental safety
- Flammable
- Dangerous to fish or other aquatic life. Do not contaminate surface waters or ditches with chemical or used container

Hazard classification and safety precautions

Hazard R07d
Operator protection U20a
Environmental protection E13b
Consumer protection C02
Storage and disposal E30a, E31a, E34

392 silthiofam

A benzamide fungicide seed dressing for cereals

Products

Latitude	Monsanto	125 g/l	FS	10695

Uses

Take-all in ***spring wheat*** *(seed treatment)*, ***winter barley*** *(seed treatment)*, ***winter wheat*** *(seed treatment)*.

Efficacy guidance

- Apply using approved seed treatment equipment which has been accurately calibrated
- Apply simultaneously (i.e. not in mixture) with a standard seed treatment
- Drill treated seed in the season of purchase. The viability of treated seed and fungicide activity may be reduced by physical storage
- Drill treated seed at 2.5-4 cm into a well prepared firm seedbed
- As precaution against possible development of disease resistance do not treat more than three consecutive susceptible cereal crops in any one rotation

Restrictions

- Maximum number of treatments one per batch of seed
- Do not use on seed with more than 16% moisture content, or on sprouted, cracked or skinned seed
- Test germination of all seed batches before treatment

Crop-specific information

- Latest use: immediately prior to drilling

Environmental safety

- Harmful to fish or other aquatic life. Do not contaminate surface waters or ditches with chemical or used container

Hazard classification and safety precautions

Operator protection A, H; U20b
Environmental protection E03, E13c
Storage and disposal E26, E30a, E31a, E34
Treated seed S01, S02, S04b, S05, S06, S07

SECTION 4

393 simazine

A soil-acting triazine herbicide with restricted permitted uses

See also *isoproturon + simazine*
pendimethalin + simazine

Products

1	Alpha Simazine 50 SC	Makhteshim	500 g/l	SC	04801
2	Gesatop	Syngenta	500 g/l	SC	08412
3	Sipcam Simazine Flowable	Sipcam	500 g/l	SC	07622

Uses

Annual dicotyledons in ***angelica*** *(off-label)*, ***asparagus*** *(off-label)*, ***balm*** *(off-label)*, ***basil*** *(off-label)*, ***bay*** *(off-label)*, ***borage*** *(off-label)*, ***camomile*** *(off-label)*, ***caraway*** *(off-label)*, ***chervil*** *(off-label)*, ***chives*** *(off-label)*, ***clary*** *(off-label)*, ***coriander*** *(off-label)*, ***dill*** *(off-label)*, ***edible podded peas*** *(off-label)*, ***fennel*** *(off-label)*, ***fenugreek*** *(off-label)*, ***feverfew*** *(off-label)*, ***hyssop*** *(off-label)*, ***land cress*** *(off-label)*, ***lemon verbena*** *(off-label)*, ***lovage*** *(off-label)*, ***marjoram*** *(off-label)*, ***mint*** *(off-label)*, ***nettle*** *(off-label)*, ***oregano*** *(off-label)*, ***parsley*** *(off-label)*, ***rhubarb*** *(off-label)*, ***rocket*** *(off-label)*, ***rosemary*** *(off-label)*, ***rue*** *(off-label)*, ***runner beans*** *(off-label)*, ***sage*** *(off-label)*, ***salad burnet*** *(off-label)*, ***savory*** *(off-label)*, ***sorrel*** *(off-label)*, ***tarragon*** *(off-label)*, ***thyme*** *(off-label)*, ***woodruff*** *(off-label)* [1, 2]; ***apple orchards, blackberries, blackcurrants, broad beans, forestry transplants, gooseberries, hops, loganberries, pear orchards, raspberries, redcurrants, spring field beans, strawberries, winter field beans*** [1-3]; ***asparagus, forestry plantations, sweetcorn*** [1, 3]; ***forage maize, ornamental plant production*** *(for resale only)*, ***rhubarb*** [1]; ***forest nursery beds*** [2, 3]; ***maize, trees and shrubs, woody ornamentals*** [3]; ***roses*** [2]. Annual grasses in

angelica *(off-label)*, ***asparagus*** *(off-label)*, ***balm*** *(off-label)*, ***basil*** *(off-label)*, ***bay*** *(off-label)*, ***borage*** *(off-label)*, ***camomile*** *(off-label)*, ***caraway*** *(off-label)*, ***chervil*** *(off-label)*, ***chives*** *(off-label)*, ***clary*** *(off-label)*, ***coriander*** *(off-label)*, ***dill*** *(off-label)*, ***edible podded peas*** *(off-label)*, ***fennel*** *(off-label)*, ***fenugreek*** *(off-label)*, ***feverfew*** *(off-label)*, ***hyssop*** *(off-label)*, ***land cress*** *(off-label)*, ***lemon verbena*** *(off-label)*, ***lovage*** *(off-label)*, ***marjoram*** *(off-label)*, ***mint*** *(off-label)*, ***nettle*** *(off-label)*, ***oregano*** *(off-label)*, ***parsley*** *(off-label)*, ***rhubarb*** *(off-label)*, ***rocket*** *(off-label)*, ***rosemary*** *(off-label)*, ***rue*** *(off-label)*, ***runner beans*** *(off-label)*, ***sage*** *(off-label)*, ***salad burnet*** *(off-label)*, ***savory*** *(off-label)*, ***sorrel*** *(off-label)*, ***tarragon*** *(off-label)*, ***thyme*** *(off-label)*, ***woodruff*** *(off-label)* [1, 2]; ***apple orchards, blackberries, blackcurrants, broad beans, forestry transplants, gooseberries, hops, loganberries, pear orchards, raspberries, redcurrants, spring field beans, strawberries, winter field beans*** [1-3]; ***asparagus, forestry plantations, sweetcorn*** [1, 3]; ***forage maize, ornamental plant production*** *(for resale only)*, ***rhubarb*** [1]; ***forest nursery beds*** [2, 3]; ***maize, trees and shrubs, woody ornamentals*** [3]; ***roses*** [2].

Specific Off-Label Approvals (SOLAs)

- ***angelica, balm, basil, bay, borage, camomile, caraway, chervil, chives, clary, coriander, dill, fennel, fenugreek, feverfew, hyssop, land cress, lemon verbena, lovage, marjoram, mint, nettle, oregano, parsley, rocket, rosemary, rue, sage, salad burnet, savory, sorrel, tarragon, thyme, woodruff*** *(OLA 011661) Dec 2008* [2]
- ***angelica, asparagus, balm, basil, bay, borage, camomile, caraway, chervil, chives, clary, coriander, dill, edible podded peas, fennel, fenugreek, feverfew, hyssop, land cress, lemon verbena, lovage, marjoram, mint, nettle, oregano, parsley, rhubarb, rocket, rosemary, rue, runner beans, sage, salad burnet, savory, sorrel, tarragon, thyme, woodruff*** *(OLA 021255) Dec 2008* [1]
- ***asparagus, rhubarb*** *(OLA 011660) Dec 2008* [2]
- ***edible podded peas, runner beans*** *(OLA 011662) Dec 2008* [2]

Approval information

- Approvals for sale, supply and use of simazine on non-crop land were revoked in 1992 and 1993

Efficacy guidance

- Active via root uptake. Best results achieved by application to fine, firm, moist soil, free of established weeds, when rain falls after treatment
- Do not use on highly organic soils as efficacy will be severely impaired
- Following repeated use of simazine or other triazine herbicides resistant strains of groundsel and some other annual weeds may develop

Restrictions

- Use must be restricted to one product containing simazine or atrazine applied either as a single application at the maximum approved dose or (subject to any existing maximum permitted number of treatments) to several applications at lower doses up to the maximum approved dose for a single application
- Do not spray beans on sandy or gravelly soils or cultivate after treatment. Do not treat varieties Beryl, Feligreen or Rowena
- Allow at least 7 mth before drilling or planting other crops, longer if weather dry
- Do not sow oats in autumn following spring application in maize

Crop-specific information

- Latest use: before emergence of spears for asparagus; 7 d after drilling for sweetcorn; before end of Feb (autumn planted), 10 d after drilling for spring planted field and broad beans; before end of Mar for top, bush and cane fruit; after harvest, before end of Nov for strawberries; after harvest, before 1 May for hops.
- HI mint and sage 4 mth; rhubarb 3 mth; herbs 14 wk; asparagus 5 wk [1, 2]
- Apply in top, bush and cane fruit and woody ornamentals established at least 12 mth in Feb-Mar. Roses may be sprayed immediately after planting. See label for lists of resistant and susceptible species
- Apply to strawberries in Jul-Dec, not in spring. Do not treat spring-planted crops established less than 6 mth, or winter-planted crops less than 9 mth. Do not spray varieties Huxley Giant, Madame Montal or Regina
- Apply to hops overall in Feb-Apr before weeds emerge. Newly planted sets must be covered with minimum of 50 mm soil

- Apply to forest nursery seedbed in second yr or to transplant lines after plants 5 cm tall. Do not treat Norway spruce (Christmas trees)
- On Sands, stony or gravelly soils there is risk of crop damage, especially with heavy rain

Environmental safety

- Very toxic to aquatic organisms. May cause long-term adverse effects in the aquatic environment [2]
- Dangerous to fish or other aquatic life. Do not contaminate surface waters or ditches with chemical or used container [1, 3]
- LERAP Category B
- Use must be restricted to one product containing atrazine or simazine, and either to a single application at the maximum approved rate or (subject to any existing maximum permitted number of treatments) to several applications at lower doses up to the maximum approved rate for a single application
- To reduce soil run-off on gradients especially in orchard and forest plantations, users are advised to plant grass strips or leave 6 m wide strips between treated areas and surface waters

Hazard classification and safety precautions

Hazard R04, R13, R14 [2]
Risk phrases R04e, R14c [2]
Operator protection A, C, H [1-3]; B [3]; D, M [1, 3]; U05a, U14, U23a [2]; U20b [1]; U20c [2, 3]
Environmental protection E13b, E16b [1, 3]; E16a [1-3]
Storage and disposal E26, E30a [1-3]; E31a [3]; E31b [1]; E31c [2]
Medical advice M05a [2]

394 sodium chlorate

A non-selective inorganic herbicide for total vegetation control

Products

1	Atlacide Soluble Powder	n2n	58.2% w/w	SP	00125
2	Doff Sodium Chlorate Weedkiller	Doff Portland	53% w/w	SP	06049

Uses

Total vegetation control in ***non-crop areas, paths and drives***.

Efficacy guidance

- Active through foliar and root uptake
- Apply as overall spray at any time during growing season. Best results obtained from application to moist soil in spring or early summer

Restrictions

- Maximum number of treatments 2 per yr for non-crop areas, paths and drives [2]
- Treated ground must not be replanted for at least 6 mth after treatment [2]
- Do not apply before heavy rain

Environmental safety

- Oxidizing - contact with combustible material may cause fire
- Clothing, paper, plant debris etc become highly inflammable when dry if contaminated with sodium chlorate
- Fire risk has been reduced by inclusion of fire depressant but product should not be used in areas of exceptionally high fire risk eg oil installations, timber yards
- Wash clothing thoroughly after use
- If clothes become contaminated do not stand near an open fire
- Must not be sold to persons under 18
- Keep livestock out of treated areas until foliage of any poisonous weeds such as ragwort has died and become unpalatable

Hazard classification and safety precautions

Hazard R03, R08
Risk phrases R03c
Operator protection A, H [2]; L [1]; M; U02 [1]; U05a, U20b [1, 2]
Environmental protection E01, E15 [1, 2]; E07 [1]
Consumer protection C03

Storage and disposal E30a, E32a, E34
Medical advice M03

395 sodium chloride (commodity substance)

An inorganic salt for use as a sugar beet herbicide

Products

sodium chloride	various	-	SL

Uses

Polygonums in ***sugar beet***. Volunteer potatoes in ***sugar beet***.

Approval information

- Approval for the use of sodium chloride as a commodity substance was granted on 11 September 1996 by Ministers under regulation 5 of the Control of Pesticides Regulations 1986

Efficacy guidance

- Best results obtained from good spray cover treatment in hot, humid weather
- Use as a follow-up treatment where earlier sprays have failed to control large volunteer potatoes or polygonums
- Acts by contact scorch and has no direct effect on daughter potato tubers
- Saturated spray solution should contain 0.1% w/w non-ionic wetter and be applied at 1000 l/ha
- Spray between three true leaf stage of crops and end Jul
- Crop scorch may occur after treatment

396 sodium hypochlorite (commodity substance)

An inorganic horticultural bactericide for use in mushrooms

Products

sodium hypochlorite	various	100% w/v	ZZ

Uses

Bacterial blotch in ***mushrooms***.

Approval information

- Approval for the use of sodium hypochlorite as a commodity substance was granted on 5 December 1996 by Ministers under regulation 5 of the Control of Pesticides Regulations 1986

Restrictions

- Maximum concentration 315 mg/litre of water
- Mixing and loading must only take place in a ventilated area
- Must only be used by suitably trained and competent operators

Crop-specific information

- HI 1 d

Environmental safety

- Harmful to fish or other aquatic life. Do not contaminate surface waters or ditches with chemical or used container

Hazard classification and safety precautions

Operator protection A, C, H
Environmental protection E13c

397 sodium monochloroacetate

A contact herbicide for various horticultural crops

Products

1	Atlas Somon	Nufarm UK	96% w/w	SP	07727
2	Croptex Steel	Certis	95% w/w	SP	02418

Uses

Annual dicotyledons in ***broccoli*** *(off-label)*, ***calabrese*** *(off-label)*, ***cauliflowers*** *(off-label)*, ***onions*** [2]; ***brussels sprouts***, ***cabbages***, ***kale***, ***leeks*** [1, 2]; ***bulb onions***, ***salad onions*** [1]. Basal defoliation in ***hops*** *(off-label)* [2]. Sucker control in ***blackberries*** *(off-label)*, ***loganberries*** *(off-label)*, ***raspberries*** *(off-label)*, ***rubus hybrids*** *(off-label)* [2].

Specific Off-Label Approvals (SOLAs)

- ***blackberries***, ***loganberries***, ***raspberries***, ***rubus hybrids*** *(OLA 930115) Dec 2008* [2]
- ***broccoli***, ***calabrese***, ***cauliflowers*** *(OLA 992602) Dec 2008* [2]
- ***hops*** *(OLA 930357) Dec 2008* [2]

Efficacy guidance

- Sodium monochloroacetate acts by contact action. Best results achieved by application to emerged weed seedlings up to young plant stage in warm humid weather when weeds are growing actively
- Effectiveness reduced by rain within 12 h

Restrictions

- Maximum number of treatments 1 per crop; 2 per yr for sucker control in cane fruit
- Do not spray cabbage that has begun to heart
- Do not treat transplanted onions or onion sets
- Safety on brassicas, onions and leeks depends on presence of adequate leaf wax, check by crystal violet wax test. Do not add wetters, pesticides or nutrients to spray
- These products must not be used on any crops other than those listed, including any extrapolations that would normally be permissible under the Long Term Arrangements for Extension of Use (see Section 1)
- Do not spray if frost likely or if temperature likely to exceed 27°C
- Do not apply any other pesticide within 7 d after use

Crop-specific information

- Latest use: before 4 true leaf stage for onions and leeks.
- HI 21 d for cabbage, kale, Brussels sprouts; 28 d for cane fruit; 14 wk for hops [2]
- Apply to brassicas from 2-4 leaf stage or after recovery from transplanting
- Apply to onions and leeks after crook stage but before 4-leaf stage
- Apply to cane fruit crops established for at least 1 yr as a directed spray
- Apply for sucker control in raspberries when canes 10-20 cm high with addition of Wayfarer adjuvant [2]

Environmental safety

- Corrosive. Causes burns
- Dangerous for the environment [2]
- Very toxic to aquatic organisms. May cause long-term adverse effects in the aquatic environment [2]
- Do not apply directly to livestock/poultry [1]
- Keep livestock, especially poultry, out of treated areas for at least 2 wk
- Harmful to bees. Do not apply to crops in flower or to those in which bees are actively foraging. Do not apply when flowering weeds are present

Hazard classification and safety precautions

Hazard R02, R14 [2]; R03, R04, R05 [1, 2]
Risk phrases R02c, R04b, R14c [2]; R03a, R03b, R04c, R04d, R05b [1, 2]; R03c [1]
Operator protection A, C, D, E; U05a, U10, U11, U19 [1, 2]; U08, U20a [1]; U20b [2]
Environmental protection E01, E05a, E07, E12e [1, 2]; E05 [1]
Consumer protection C03
Storage and disposal E30a, E34 [1, 2]; E31a [1]; E32a [2]
Medical advice M03, M04 [2]

398 spinosad

A selective insecticide derived from naturally occurring soil fungi

Products

Conserve	Fargro	120 g/l	SC	11011

Uses

Western flower thrips in ***protected ornamentals***.

Efficacy guidance

- Product enters insects by contact from a treated surface or ingestion of treated plant material therefore good spray coverage is essential
- Some plants, for example Fuchsia flowers, can provide effective refuges from spray deposits and control may be reduced
- Apply when nymphs or adults are first seen
- Monitor pest development carefully to see whether further applications are necessary. A 2 or 3 spray programme at 5-7 d intervals may be needed when conditions favour rapid pest development
- To reduce possibility of development of resistance, adopt resistance management measures. See label and Section 1

Restrictions

- Establish whether any incoming plants have been treated and apply no more than 3 consecutive sprays. Maximum number of treatments 6 per structure per yr
- Test for tolerance on a small number of plants before large scale treatment
- Avoid application in bright sunlight or into open flowers

Crop-specific information

- Some spotting of african violet flowers may occur

Environmental safety

- Harmful to fish or other aquatic life. Do not contaminate surface waters or ditches with chemical or used container
- Whenever possible use an Integrated Pest Management system
- Product has low impact on many insect and mite predators but is harmful to adults of most parasitic wasps. Most beneficials may be introduced to treated plants when spray deposits are dry but an interval of 2 wk should elapse before introduction of parasitic wasps. See label for details

Hazard classification and safety precautions

Operator protection A, H; U05a
Environmental protection E01, E13c
Consumer protection C03
Storage and disposal E26, E31b, E34

399 spiroxamine

A spiroketal amine fungicide for cereals

Products

1	Neon	Bayer	500 g/l	EW	08337
2	Neon	Bayer CropScience	500 g/l	EW	11226
3	Torch	Bayer	500 g/l	EW	08336
4	Torch	Bayer CropScience	500 g/l	EW	11258
5	Zenon	Bayer	800 g/l	EC	09193
6	Zenon	Bayer CropScience	800 g/l	EC	11232

Uses

Brown rust in ***spring barley***, ***spring rye***, ***spring wheat***, ***winter barley***, ***winter rye***, ***winter wheat*** [1-6]. Powdery mildew in ***spring barley***, ***spring rye***, ***spring wheat***, ***winter barley***, ***winter rye***, ***winter wheat*** [1-6]. Rhynchosporium in ***spring barley*** *(reduction)*, ***winter barley*** *(reduction)* [1-6]; ***spring rye*** *(reduction)*, ***winter rye*** *(reduction)* [1-4]. Yellow rust in ***spring barley***, ***spring rye***, ***spring wheat***, ***winter barley***, ***winter rye***, ***winter wheat*** [1-6].

Approval information

- Spiroxamine included in Annex I under EC Directive 91/414

Efficacy guidance

- For best results treat at an early stage of disease development before infection spreads to new growth
- Active against net blotch when applied at first signs of disease but better control achieved with mixtures. See label for details

- To reduce the risk of development of resistance avoid repeat treatments on diseases such as powdery mildew. If necessary tank mix or alternate with other non-morpholine fungicides

Restrictions

- Maximum total dose equivalent to two full dose treatments

Crop-specific information

- Latest use: before caryopsis watery ripe (GS 71) for spring wheat, winter rye, winter wheat; ear emergence complete (GS 59) for spring barley, winter barley

Environmental safety

- Dangerous to fish or other aquatic life. Do not contaminate surface waters or ditches with chemical or used container
- LERAP Category B

Hazard classification and safety precautions

Hazard R03, R04
Risk phrases R03a [5, 6]; R03c, R04d, R04e [1-6]; R04b [1-4]
Operator protection A, C, H; U05a, U11, U13, U14, U15, U20a
Environmental protection E01, E13b, E16a, E16b
Consumer protection C03
Storage and disposal E26, E30a, E31b, E34
Medical advice M03

400 spiroxamine + tebuconazole

A broad spectrum systemic fungicide mixture for cereals

Products

1	Beam	Bayer	250:133 g/l	EW	10411
2	Beam	Bayer CropScience	250:133 g/l	EW	11255
3	Sage	Bayer	250:133 g/l	EW	10413
4	Sage	Bayer CropScience	250:133 g/l	EW	11303

Uses

Brown rust in ***spring barley***, ***spring rye***, ***spring wheat***, ***winter barley***, ***winter rye***, ***winter wheat***. Fusarium ear blight in ***spring wheat***, ***winter wheat***. Glume blotch in ***spring wheat***, ***winter wheat***. Net blotch in ***spring barley***, ***winter barley***. Powdery mildew in ***spring barley***, ***spring rye***, ***spring wheat***, ***winter barley***, ***winter rye***, ***winter wheat***. Rhynchosporium in ***spring barley***, ***winter barley***. Septoria leaf spot in ***spring wheat***, ***winter wheat***. Sooty moulds in ***spring wheat***, ***winter wheat***. Yellow rust in ***spring barley***, ***spring rye***, ***spring wheat***, ***winter barley***, ***winter rye***, ***winter wheat***.

Approval information

- Spiroxamine included in Annex I under EC Directive 91/414

Efficacy guidance

- Best results achieved from treatment at early stage of disease development before infection spreads to new growth
- To protect flag leaf and ear from Septoria apply from flag leaf emergence to ear fully emerged (GS 37-59). Earlier treatment may be necessary where there is high disease risk
- Control of rusts, powdery mildew, leaf blotch and net blotch may require second treatment 2-3 wk later

Restrictions

- Maximum total dose equivalent to two full dose treatments
- Do not use on durum wheat

Crop-specific information

- Latest use: before caryopsis watery ripe (GS 71) for rye, wheat; ear emergence complete (GS 59) for barley
- Some transient leaf speckling may occur on wheat but this has not been shown to reduce yield reponse or disease control

SECTION 4

Environmental safety

- Dangerous to fish or other aquatic life. Do not contaminate surface waters or ditches with chemical or used container
- LERAP Category B

Hazard classification and safety precautions

Hazard R03, R04
Risk phrases R03c, R04b, R04d, R04e
Operator protection A, C, H; U05a, U11, U13, U14, U15, U20b
Environmental protection E01, E13b, E16a, E16b
Consumer protection C03
Storage and disposal E26, E30a, E31b, E34
Medical advice M03

401 strychnine hydrochloride (commodity substance)

A vertebrate control agent for destruction of moles underground

Products

strychnine	various	-	ZZ

Uses

Moles in ***grassland*** *(areas of restricted public access)*, ***miscellaneous pest control situations*** *(areas of restricted public access)*.

Approval information

- Approval for the use of strychnine hydrochloride as a commodity substance was granted on 19 June 1997 by Ministers under regulation 5 of the Control of Pesticides Regulations 1986

Efficacy guidance

- Only for use as poison bait against moles on commercial agricultural/horticultural land where public access restricted, on grassland associated with aircraft landing strips, horse paddocks, race and golf courses and other areas specifically approved by DEFRA in England and the appropriate authorities in Wales and Scotland

Restrictions

- Strychnine is subject to the Poisons Rules 1982 and the Poisons Act 1972. See notes in Section 1
- Must only be supplied to holders of a permit issued by DEFRA in England or the appropriate authorities in Wales and Scotland. Permits may only be issued to persons who satisfy the appropriate authority that they are trained and competent in its use
- Only to be supplied in original sealed pack in units up to 2 g. Quantities of more than 8 g may be supplied only to providers of a commercial service
- Other restrictions apply, see the PSD website at: www.pesticides.gov.uk/BlueBook/Annex_D/Strychnine%20Hydrochloride.htm

Environmental safety

- Store in original container under lock and key and only on premises under control of a permit holder or of a named individual who satisfies the designated requirements of competence
- A written COSHH assessment must be made before use
- Must be prepared for application with great care so that there is no contamination of the ground surface
- Any prepared bait remaining at the end of the day must be buried
- Operators must be provided with a Chemicals (Hazard Information and Supply) (CHIP) Safety Data Sheet, obtainable from the supplier, before commencing work

Hazard classification and safety precautions

Operator protection A

402 sulfosulfuron

A sulfonylurea herbicide for grass and broad-leaved weed control in winter wheat

Products

1	Landgold Sulfosulfuron	Landgold	80% w/w	WG	10908
2	Monitor	Monsanto	80% w/w	WG	10495

Uses

Brome grasses in ***winter wheat***. Chickweed in ***winter wheat***. Cleavers in ***winter wheat***. Loose silky bent in ***winter wheat***. Mayweeds in ***winter wheat***. Onion couch in ***winter wheat***.

Approval information

- Sulfosulfuron included in Annex I under EC Directive 91/414

Efficacy guidance

- For best results treat in early spring when annual weeds are small and growing actively. Avoid treatment when weeds are dormant for any reason
- Best control of onion couch is achieved when the weed has more than two leaves. Effects on bulbils or on growth in the following yr have not been examined
- An extended period of dry weather before or after treatment may result in reduced control
- The addition of a recommended surfactant is essential for full activity
- Specific follow-up treatments may be needed for complete control of some weeds
- Use only where competitively damaging weed populations have emerged otherwise yield may be reduced
- Sulfosulfuron is a member of the ALS-inhibitor group of herbicides and products should be used in a planned Resistance Management strategy. See Section 1 for more information

Restrictions

- Maximum total dose equivalent to one treatment at full dose
- Do not treat crops under stress
- Apply only after 1 Feb and from 3 expanded leaf stage
- Do not treat durum wheat or any undersown wheat crop
- Do not use in mixture or in sequence with any other sulfonyl urea herbicide on the same crop

Crop-specific information

- Latest use: flag leaf ligule just visible (GS 39)

Following crops guidance

- In the autumn following treatment winter wheat, winter rye, winter oats, triticale, winter oilseed rape, winter peas or winter field beans may be sown on any soil, and winter barley on soils with less than 60% sand
- In the next spring following application crops of wheat, barley, oats, maize, peas, beans, linseed, oilseed rape, potatoes or grass may be sown. Sugar beet or any other crop must not be drilled until the second spring following application

Environmental safety

- Extremely dangerous to fish or other aquatic life. Do not contaminate surface waters or ditches with chemical or used container
- LERAP Category B
- Take extreme care to avoid drift onto broad leaved plants or other crops, or onto ponds, waterways or ditches.
- Follow detailed label instructions for cleaning the sprayer to avoid damage to sensitive crops during subsequent use

Hazard classification and safety precautions

Operator protection U08 [1]; U20b [1, 2]
Environmental protection E13a, E16a, E16b
Storage and disposal E30a [1, 2], E31b [1], E32a, E34 [2]

403 sulphur

A broad-spectrum inorganic protectant fungicide, foliar feed and acaricide

Products

1	Cosavet DF	Headland	80% w/w	WG	11477
2	Headland Sulphur	Headland	800 g/l	SC	03711
3	Kumulus DF	BASF	80% w/w	SG	04707
4	Luxan Micro-Sulphur	Luxan	80% w/w	WG	06565
5	Sulfa	Nufarm UK	80% w/w	MG	06959
6	Sulphur Flowable	United Phosphorus	800 g/l	SC	07526
7	Thiovit Jet	Syngenta	80% w/w	WG	10928
8	Tripart Imber	Tripart	720 g/l	SC	04050

Uses

American gooseberry mildew in ***gooseberries*** [5]. Foliar feed in ***grassland*** [4, 5, 7]; ***spring barley, spring wheat, winter barley, winter wheat*** [1, 3, 4, 7, 8]; ***spring oats, winter oats*** [8]; ***spring oilseed rape, winter oilseed rape*** [1, 3-5, 7, 8]; ***sugar beet, swedes*** [4]. Gall mite in ***blackcurrants*** [1, 3, 5, 8]. Powdery mildew in ***apples, strawberries*** [1-3, 5, 6, 8]; ***aubergines*** *(off-label)*, ***parsnips*** *(off-label)*, ***protected agastache spp.*** *(off-label)*, ***protected angelica*** *(off-label)*, ***protected balm*** *(off-label)*, ***protected basil*** *(off-label)*, ***protected bay*** *(off-label)*, ***protected borage*** *(off-label)*, ***protected camomile*** *(off-label)*, ***protected caraway*** *(off-label)*, ***protected catnip*** *(off-label)*, ***protected chervil*** *(off-label)*, ***protected chives*** *(off-label)*, ***protected clary*** *(off-label)*, ***protected coriander*** *(off-label)*, ***protected cucumbers*** *(off-label)*, ***protected curry plant*** *(off-label)*, ***protected dill*** *(off-label)*, ***protected dragonhead*** *(off-label)*, ***protected fennel*** *(off-label)*, ***protected fenugreek*** *(off-label)*, ***protected feverfew*** *(off-label)*, ***protected horehound*** *(off-label)*, ***protected hyssop*** *(off-label)*, ***protected lamb's lettuce*** *(off-label)*, ***protected land cress*** *(off-label)*, ***protected lavandin*** *(off-label)*, ***protected lavender*** *(off-label)*, ***protected lemon verbena*** *(off-label)*, ***protected lovage*** *(off-label)*, ***protected marigold*** *(off-label)*, ***protected marjoram*** *(off-label)*, ***protected mint*** *(off-label)*, ***protected nasturtium*** *(off-label)*, ***protected nettle*** *(off-label)*, ***protected oregano*** *(off-label)*, ***protected origanum heracleoticum*** *(off-label)*, ***protected parsley*** *(off-label)*, ***protected peppers*** *(off-label)*, ***protected plantain*** *(off-label)*, ***protected rocket*** *(off-label)*, ***protected rosemary*** *(off-label)*, ***protected rue*** *(off-label)*, ***protected sage*** *(off-label)*, ***protected salad burnet*** *(off-label)*, ***protected savory*** *(off-label)*, ***protected sorrel*** *(off-label)*, ***protected spike lavender*** *(off-label)*, ***protected st john's wort*** *(off-label)*, ***protected tarragon*** *(off-label)*, ***protected thyme*** *(off-label)*, ***protected tomatoes*** *(off-label)*, ***protected violet*** *(off-label)*, ***protected woodruff*** *(off-label)*, ***protected yarrow*** *(off-label)* [7]; ***gooseberries*** [1, 3, 8]; ***grapevines*** [1, 3, 5, 8]; ***hops*** [1-3, 5-8]; ***pears*** [2, 6]; ***spring barley, spring wheat, winter barley, winter wheat*** [2, 4-6]; ***spring oats, winter oats*** [2, 5, 8]; ***spring oilseed rape, spring rye, winter oilseed rape, winter rye*** [2]; ***sugar beet*** [1-8]; ***swedes*** [2, 5-8]; ***turnips*** [5, 8]. Scab in ***apples*** [1-3, 6, 8]; ***pears*** [2, 6].

Specific Off-Label Approvals (SOLAs)

- ***aubergines, parsnips, protected basil, protected bay, protected borage, protected chervil, protected chives, protected coriander, protected cucumbers, protected dill, protected fenugreek, protected lamb's lettuce, protected lovage, protected marjoram, protected mint, protected parsley, protected peppers, protected rocket, protected rosemary, protected sage, protected savory, protected sorrel, protected tarragon, protected thyme, protected tomatoes*** *(OLA 021145) Dec 2008* [7]
- ***aubergines, protected agastache spp., protected angelica, protected balm, protected basil, protected bay, protected borage, protected camomile, protected caraway, protected catnip, protected chervil, protected chives, protected clary, protected coriander, protected cucumbers, protected curry plant, protected dill, protected dragonhead, protected fennel, protected fenugreek, protected feverfew, protected horehound, protected hyssop, protected land cress, protected lavandin, protected lavender, protected lemon verbena, protected lovage, protected marigold, protected marjoram, protected mint, protected nasturtium, protected nettle, protected oregano, protected origanum heracleoticum, protected parsley, protected peppers, protected plantain, protected rocket, protected rosemary, protected rue, protected sage, protected salad burnet, protected savory, protected sorrel, protected spike lavender, protected st john's wort, protected tarragon, protected thyme, protected tomatoes, protected violet, protected woodruff, protected yarrow*** *(OLA 023652) Dec 2008* [7]
- ***parsnips*** *(OLA 023654) Dec 2008* [7]

Efficacy guidance

- Apply when disease first appears and repeat 2-3 wk later. Details of application rates and timing vary with crop, disease and product. See label for information
- Sulphur acts as foliar feed as well as fungicide and with some crops product labels vary in whether treatment recommended for disease control or growth promotion
- In grassland best results obtained at least 2 wk before cutting for hay or silage, 3 wk before grazing
- Treatment unlikely to be effective if disease already established in crop

Restrictions

- Maximum number of treatments normally 2 per crop for grassland, sugar beet, parsnips, swedes, protected herbs; 3 per yr for blackcurrants, gooseberries; 4 per crop on apples, hops; variable on cereals. See labels

- Do not use on sulphur-shy apples (Beauty of Bath, Belle de Boskoop, Cox's Orange Pippin, Lanes Prince Albert, Lord Derby, Newton Wonder, Rival, Stirling Castle) or pears (Doyenne du Comice)
- Do not use on gooseberry cultivars Careless, Early Sulphur, Golden Drop, Leveller, Lord Derby, Roaring Lion, or Yellow Rough
- Do not use on apples or gooseberries when young, under stress or if frost imminent
- Do not use on fruit for processing, on grapevines during flowering or near harvest on grapes for wine-making
- Do not use on hops at or after burr stage
- Do not spray top or soft fruit with oil or within 30 d of an oil-containing spray

Crop-specific information

- Latest use: before burr stage for hops; before end Sep for parsnips, sugar beet, swedes; before end Sept in yr of harvest for sugar beet; fruit swell for gooseberries
- HI cutting grass for hay or silage 2 wk; grazing grassland 3 wk
- May be applied to cereals up to grain watery-ripe stage (GS 71) [2, 5, 8], up to milky ripe stage (GS 75) [7], to first node detectable (GS 31) [4], at any time [3]

Hazard classification and safety precautions

Hazard R04 [7]
Risk phrases R04e [7]
Operator protection A, H [7]; U02, U08, U14 [1, 7]; U05a [7]; U20a [2]; U20b [6, 8]; U20c [1, 3-5, 7]
Environmental protection E01 [4, 5, 7]; E15 [1-8]
Consumer protection C03 [7]
Storage and disposal E26 [6]; E30a [1-8]; E31b [2, 6]; E32a [1, 3-5, 7, 8]; E34 [4]

404 sulphuric acid (commodity substance)

A strong acid used as an agricultural desiccant

Products

sulphuric acid	various	77% w/w	SL

Uses

Haulm destruction in ***potatoes***. Pre-harvest desiccation in ***bulbs/corms***, ***peas***.

Approval information

- Approval for the use of sulphuric acid as a commodity substance was granted on 23 November 1995 by Ministers under regulation 5 of the Control of Pesticides Regulations 1986

Efficacy guidance

- Apply with suitable equipment between 1 Mar and 15 Nov

Restrictions

- Sulphuric acid is subject to the Poisons Rules 1982 and the Poisons Act 1972. See notes in Section 1
- Must only be used by suitably trained operators competent in use of equipment for applying sulphuric acid
- Not to be applied using hand-held or pedestrian controlled applicators
- A written COSHH assessment must be made before use. Operators should observe OES set out in HSE guidance note EH40/90 or subsequent issues
- Operators must have liquid suitable for eye irrigation immediately available at all times throughout spraying operation
- Maximum number of treatments 3 per crop for potatoes; 1 per crop for peas; 1 per yr for bulbs and corms
- Only 'sulphur burnt' sulphuric acid to be used

Crop-specific information

- Latest use: 15 Nov for bulbs, corms, peas, potatoes

Environmental safety

- Spray must not be deposited within 1 m of public footpaths
- Written notice of any intended spraying must be given to owners of neighbouring land and readable warning notices posted beforehand and left in place for 96 h afterwards

- Unprotected persons must be kept out of treated areas for at least 96 h after treatment
- Do not apply to crops in which bees are actively foraging. Do not apply when flowering weeds are present

Hazard classification and safety precautions

Hazard R05

Risk phrases R05b

Operator protection B, C, D, H, K, M

405 tar oils

Hydrocarbon and phenolic oils used as insecticidal and fungicidal winter washes

Products

1	Sterilite Tar Oil Winter Wash 60% Stock Emulsion	Coventry Chemicals	636 g/l	EC	05061
2	Sterilite Tar Oil Winter Wash 80% Miscible Quality	Coventry Chemicals	800 g/l	EC	05062

Uses

Aphids in ***apples***, ***blackberries***, ***blackcurrants***, ***cherries***, ***gooseberries***, ***loganberries***, ***nectarines***, ***peaches***, ***pears***, ***plums***, ***raspberries***, ***redcurrants***. Raspberry moth in ***blackberries***, ***loganberries***, ***raspberries***. Scale insects in ***apples***, ***blackberries***, ***blackcurrants***, ***cherries***, ***gooseberries***, ***grapevines***, ***loganberries***, ***nectarines***, ***peaches***, ***pears***, ***plums***, ***raspberries***, ***redcurrants***. Suckers in ***apples***. Winter moth in ***apples***, ***blackcurrants***, ***cherries***, ***gooseberries***, ***nectarines***, ***peaches***, ***pears***, ***plums***, ***redcurrants***.

Approval information

- Approval expiry 31 Dec 2004 [1, 2]

Efficacy guidance

- Spray kills hibernating insects and eggs as well as moss and lichens on trunk
- Spray winter wash over dormant branches and twigs. See labels for detailed timings
- Ensure that all parts of trees or bushes are completely wetted, especially cracks and crevices. Use higher concentration for cleaning up neglected orchard
- Treatment also partially controls winter and raspberry moths and reduces Botrytis on currants and powdery mildew on currants and gooseberries
- Combine with spring insecticide treatment for control of caterpillars
- When used in cold weather always add product to water and stir thoroughly before use

Restrictions

- Maximum number of treatments 2 per yr for hops
- Only spray fruit trees or bushes when fully dormant
- Do not spray hops if temperature is above 21°C or after cones have formed. Do not drench rootstocks
- Do not spray on windy, wet or frosty days

Crop-specific information

- Latest use: before any signs of bud swelling for fruit crops
- Apply hop defoliant by spraying when bines 1.5-3 m high and direct spray downward at 45°

Environmental safety

- Dangerous to fish or other aquatic life. Do not contaminate surface waters or ditches with chemical or used container

Hazard classification and safety precautions

Hazard R03, R04

Risk phrases R03c, R04a, R04b, R04c

Operator protection A, C, H; U05a, U08, U19, U20a

Environmental protection E01, E13b

Consumer protection C03

Storage and disposal E26, E27, E30a, E31a, E34

Medical advice M03

406 tau-fluvalinate

A contact pyrethroid insecticide for cereals and oilseed rape

Products

1	Greencrop Malin	Greencrop	240 g/l	EW	11787
2	Klartan	Makhteshim	240 g/l	EW	11074
3	Mavrik	Makhteshim	240 g/l	EW	10612

Uses

Aphids in ***spring barley***, ***spring oilseed rape***, ***spring wheat***, ***winter barley***, ***winter oilseed rape***, ***winter wheat***. Barley yellow dwarf virus vectors in ***winter barley***, ***winter wheat***. Pollen beetle in ***spring oilseed rape***, ***winter oilseed rape***.

Efficacy guidance

- For BYDV control on winter cereals follow local warnings or spray high risk crops in mid-Oct and make repeat application in late autumn/early winter if aphid activity persists
- For summer aphid control on cereals spray once when aphids present on two thirds of ears and increasing
- On oilseed rape treat peach potato aphids in autumn in response to local warning and repeat if necessary
- Best control of pollen beetle in oilseed rape obtained from treatment at green to yellow bud stage and repeat if necessary
- Good spray cover of target essential for best results

Restrictions

- Maximum total dose varies with crop. See label for details
- A minimum of 14 d must elapse between applications to cereals

Crop-specific information

- Latest use: before caryopsis watery ripe (GS 71) for barley; before flowering for oilseed rape; before kernel medium milk (GS 75) for wheat

Environmental safety

- Dangerous for the environment [1]
- Toxic to aquatic organisms. May cause long-term adverse effects to the aquatic environment [1]
- Extremely dangerous to fish or other aquatic life. Do not contaminate surface waters or ditches with chemical or used container [2, 3]
- High risk to non-target insects or other arthropods. Do not spray within 6 m of the field boundary [1-3]
- LERAP Category A
- Avoid spraying oilseed rape within 6 m of field boundary to reduce effects on certain non-target species or other arthropods
- Must not be applied to cereals if any product containing a pyrethroid insecticide or dimethoate has been sprayed after the start of ear emergence (GS 51)

Hazard classification and safety precautions

Hazard R04 [1-3]; R14 [1]
Risk phrases R04a, R04b [1-3]; R14b [1]
Operator protection A, C, H, U05a, U10, U11, U19, U20a
Environmental protection E01, E16c, E16d, E22a [1-3]; E13a [2, 3]; E15 [1]
Consumer protection C03
Storage and disposal E26, E31c [1-3]; E30a [2, 3]; E30b [1]

SECTION 4

407 tebuconazole

A systemic conazole fungicide for cereals and other field crops

See also *carbendazim + tebuconazole*
fenpropidin + tebuconazole
imidacloprid + tebuconazole + triazoxide
prochloraz + tebuconazole
spiroxamine + tebuconazole

Products

1	Barclay Busker	Barclay	250 g/l	EC	011345
2	Folicur	Bayer	250 g/l	EW	08691
3	Folicur	Bayer CropScience	250 g/l	EW	11278
4	Me2 Tebuconazole	Me2	250 g/l	EW	09751
5	Standon Tebuconazole	Standon	250 g/l	EC	09056

Uses

Alternaria in ***brussels sprouts*** [1]; ***cabbages*** [1-3]; ***carrots, horseradish*** [2, 3]; ***spring oilseed rape, winter oilseed rape*** [1-5]. Botrytis in ***linseed*** *(reduction)* [1-5]. Brown rust in ***spring barley, spring rye, spring wheat, winter barley, winter rye, winter wheat*** [1-5]. Chocolate spot in ***spring field beans, winter field beans*** [1-5]. Crown rust in ***spring oats*** *(reduction)*, ***winter oats*** *(reduction)* [2, 3]. Fungus diseases in ***kohlrabi*** *(off-label)* [2, 3]. Fusarium ear blight in ***spring wheat, winter wheat*** [2, 3]. Glume blotch in ***spring wheat, winter wheat*** [2, 3]. Light leaf spot in ***cabbages*** [2-5]; ***spring oilseed rape, winter oilseed rape*** [1-5]. Lodging control in ***spring oilseed rape, winter oilseed rape*** [2, 3]. Net blotch in ***spring barley, winter barley*** [1-5]. Phoma in ***spring oilseed rape, winter oilseed rape*** [2, 3]. Powdery mildew in ***angelica*** *(off-label)*, ***balsam*** *(off-label)*, ***basil*** *(off-label)*, ***bay*** *(off-label)*, ***borage*** *(off-label)*, ***camomile*** *(off-label)*, ***caraway*** *(off-label)*, ***carrots, chervil*** *(off-label)*, ***chives*** *(off-label)*, ***clary*** *(off-label)*, ***coriander*** *(off-label)*, ***dill*** *(off-label)*, ***fennel*** *(off-label)*, ***fenugreek*** *(off-label)*, ***feverfew*** *(off-label)*, ***hyssop*** *(off-label)*, ***land cress*** *(off-label)*, ***lemon verbena*** *(off-label)*, ***linseed, lovage*** *(off-label)*, ***marjoram*** *(off-label)*, ***mint*** *(off-label)*, ***nettle*** *(off-label)*, ***oregano*** *(off-label)*, ***parsley*** *(off-label)*, ***parsnips, rocket*** *(off-label)*, ***rosemary*** *(off-label)*, ***rue*** *(off-label)*, ***sage*** *(off-label)*, ***salad burnet*** *(off-label)*, ***savory*** *(off-label)*, ***sorrel*** *(off-label)*, ***tarragon*** *(off-label)*, ***thyme*** *(off-label)*, ***woodruff*** *(off-label)* [2, 3]; ***brussels sprouts*** [1]; ***cabbages, spring barley, spring oats, spring rye, spring wheat, swedes, turnips, winter barley, winter oats, winter rye, winter wheat*** [1-5]. Rhynchosporium in ***spring barley, spring rye, winter barley, winter rye*** [1-5]. Ring spot in ***broccoli*** *(off-label)*, ***calabrese*** *(off-label)*, ***cauliflowers*** *(off-label)*, ***chinese cabbage*** *(off-label)*, ***collards*** *(off-label)*, ***kale*** *(off-label)*, ***leaf brassicas*** *(off-label)*, ***spring oilseed rape*** *(reduction)*, ***winter oilseed rape*** *(reduction)* [2, 3]; ***brussels sprouts*** [1]; ***cabbages*** [1-5]. Rust in ***angelica*** *(off-label)*, ***balsam*** *(off-label)*, ***basil*** *(off-label)*, ***bay*** *(off-label)*, ***borage*** *(off-label)*, ***broad beans*** *(off-label)*, ***camomile*** *(off-label)*, ***caraway*** *(off-label)*, ***chervil*** *(off-label)*, ***chives*** *(off-label)*, ***clary*** *(off-label)*, ***coriander*** *(off-label)*, ***dill*** *(off-label)*, ***dwarf beans*** *(off-label)*, ***fennel*** *(off-label)*, ***fenugreek*** *(off-label)*, ***feverfew*** *(off-label)*, ***french beans*** *(off-label)*, ***hyssop*** *(off-label)*, ***land cress*** *(off-label)*, ***lemon verbena*** *(off-label)*, ***lovage*** *(off-label)*, ***marjoram*** *(off-label)*, ***mint*** *(off-label)*, ***nettle*** *(off-label)*, ***oregano*** *(off-label)*, ***parsley*** *(off-label)*, ***rocket*** *(off-label)*, ***rosemary*** *(off-label)*, ***rue*** *(off-label)*, ***runner beans*** *(off-label)*, ***sage*** *(off-label)*, ***salad burnet*** *(off-label)*, ***savory*** *(off-label)*, ***sorrel*** *(off-label)*, ***tarragon*** *(off-label)*, ***thyme*** *(off-label)*, ***woodruff*** *(off-label)* [2, 3]; ***leeks*** [2-4]; ***spring field beans, winter field beans*** [1-5]. Sclerotinia in ***carrots*** [2, 3]. Sclerotinia stem rot in ***spring oilseed rape, winter oilseed rape*** [1-5]. Septoria diseases in ***spring wheat, winter wheat*** [1, 4, 5]. Septoria leaf spot in ***spring wheat, winter wheat*** [2, 3]. Sooty moulds in ***spring wheat, winter wheat*** [2, 3]. Stem canker in ***spring oilseed rape, winter oilseed rape*** [2, 3]. White rot in ***bulb onions*** *(off-label)*, ***garlic*** *(off-label)*, ***salad onions*** *(off-label)*, ***shallots*** *(off-label)* [2, 3]. Yellow rust in ***spring barley, spring rye, spring wheat, winter barley, winter rye, winter wheat*** [1-5].

Specific Off-Label Approvals (SOLAs)

- ***angelica, balsam, basil, bay, borage, camomile, caraway, chervil, chives, clary, coriander, dill, fennel, fenugreek, feverfew, hyssop, land cress, lemon verbena, lovage, marjoram, mint, nettle, oregano, parsley, rocket, rosemary, rue, sage, salad burnet, savory, sorrel, tarragon, thyme, woodruff*** *(OLA 031873) Dec 2008* [3]

- ***angelica, balsam, basil, bay, borage, camomile, caraway, chervil, chives, clary, coriander, dill, fennel, fenugreek, feverfew, hyssop, land cress, lemon verbena, lovage, marjoram, mint, nettle, oregano, parsley, rocket, rosemary, rue, sage, salad burnet, savory, sorrel, tarragon, thyme, woodruff*** *(OLA 983020) Dec 2008* [2]
- ***broad beans*** *(OLA 011203) Dec 2008* [2]
- ***broad beans*** *(OLA 031878) Dec 2008* [3]
- ***broccoli, calabrese, cauliflowers, chinese cabbage*** *(OLA 031874) Dec 2008* [3]
- ***broccoli, calabrese, cauliflowers, chinese cabbage*** *(OLA 972048) Dec 2008* [2]
- ***bulb onions, garlic, shallots*** *(OLA 001831) Dec 2008* [2]
- ***bulb onions*** *(OLA 012846) Dec 2008* [2]
- ***bulb onions*** *(OLA 031877) Dec 2008* [3]
- ***bulb onions, garlic, shallots*** *(OLA 031879) Dec 2008* [3]
- ***collards, kale*** *(OLA 031872) Dec 2008* [3]
- ***collards, kale*** *(OLA 990510) Dec 2008* [2]
- ***dwarf beans, french beans, runner beans*** *(OLA 000073) Dec 2008* [2]
- ***dwarf beans, french beans, runner beans*** *(OLA 031880) Dec 2008* [3]
- ***kohlrabi*** *(OLA 000072) Dec 2008* [2]
- ***kohlrabi*** *(OLA 031881) Dec 2008* [3]
- ***leaf brassicas*** *(OLA 021624) Dec 2008* [2]
- ***leaf brassicas*** *(OLA 031876) Dec 2008* [3]
- ***salad onions*** *(OLA 021626) Jun 2007* [2]
- ***salad onions*** *(OLA 031875) Jun 2007* [3]

Efficacy guidance

- For best results apply at an early stage of disease development before infection spreads to new crop growth
- To protect flag leaf and ear from Septoria diseases apply from flag leaf emergence to ear fully emerged (GS 37-59). Earlier application may be necessary where there is a high risk of infection
- For light leaf spot control in oilseed rape apply in autumn/winter with a follow-up spray in spring/summer if required
- For control of most other diseases spray at first signs of infection with a follow-up spray 2-4 wk later if necessary. See label for details
- For disease control in cabbages a 3-spray programme at 21-28 d intervals will give good control

Restrictions

- Maximum total dose equivalent to 1 full dose treatment on linseed, green beans, runner beans; 2 full dose treatments on wheat, barley, rye, field beans, broad beans, swedes, turnips, onions, garlic, shallots; 2.25 full dose treatments on cabbages, herbs; 2.5 full dose treatments on oilseed rape; 3 full dose treatments on leeks, horseradish, parsnips, carrots
- Do not treat durum wheat
- Apply only to listed oat varieties (see label)
- Do not apply before swedes and turnips have a root diameter of 2.5 cm, or before heart formation in cabbages

Crop-specific information

- Latest use: before grain milky-ripe for cereals (GS 71); when most seed green-brown mottled for oilseed rape (GS 6,3); before brown capsule for linseed
- HI field beans, swedes, turnips 35 d; market brassicas, kale, carrots, parsnips, garlic, salad onions, salad brassicas, shallots 21 d; leeks, herbs 14 d; broad beans, green beans, runner beans 7 d
- Some transient leaf speckling on wheat or leaf reddening/scorch on oats may occur but this has not been shown to reduce yield response to disease control

Environmental safety

- Harmful to fish or other aquatic life and aquatic plants. Do not contaminate surface waters or ditches with chemical or used container

Hazard classification and safety precautions

Hazard R03, R04
Risk phrases R03c, R04b, R04d
Operator protection A, C, H; U05a, U11, U20a
Environmental protection E01, E13c
Consumer protection C03

Storage and disposal E26, E30a [1-5]; E31a [4]; E31b, E34 [1-3, 5]
Medical advice M03

408 tebuconazole + triadimenol

A broad spectrum systemic fungicide for cereals

Products

1	Silvacur	Bayer	250:125 g/l	EC	06387
2	Silvacur	Bayer CropScience	250:125 g/l	EC	11309
3	Veto F	Bayer	225:75 g/l	EC	08057
4	Veto F	Bayer CropScience	225:75 g/l	EC	11317

Uses

Brown rust in ***spring barley, spring rye, spring wheat, winter barley, winter rye, winter wheat*** [1-4]. Crown rust in ***spring oats, winter oats*** [1, 2]. Fusarium ear blight in ***spring wheat, winter wheat*** [1-4]. Glume blotch in ***spring wheat, winter wheat*** [1-4]. Net blotch in ***spring barley, winter barley*** [1-4]. Powdery mildew in ***spring barley, spring rye, spring wheat, winter barley, winter rye, winter wheat*** [1-4]; ***spring oats, winter oats*** [1, 2]. Rhynchosporium in ***spring barley, winter barley*** [1-4]. Septoria leaf spot in ***spring wheat, winter wheat*** [1-4]. Sooty moulds in ***spring wheat, winter wheat*** [1-4]. Yellow rust in ***spring barley, spring rye, spring wheat, winter barley, winter rye, winter wheat*** [1-4].

Efficacy guidance

- For best results apply at an early stage of disease development before infection spreads to new crop growth
- To protect flag leaf and ear from Septoria diseases apply from flag leaf emergence to ear fully emerged (GS 37-59). Earlier application may be necessary where there is a high risk of infection
- For control of rust, powdery mildew, leaf and net blotch apply at first signs of disease with a second application 2-3 wk later if necessary

Restrictions

- Maximum total dose equivalent to two full dose treatments
- Do not use on durum wheat
- Use only on listed varieties of oats. See label [1, 2]

Crop-specific information

- Latest use: before grain milky-ripe (GS 71)
- Some transient leaf speckling may occur on wheat or leaf reddening/scorch on oats but this has not been shown to reduce yield response or disease control

Environmental safety

- Harmful to fish or other aquatic life. Do not contaminate surface waters or ditches with chemical or used container

Hazard classification and safety precautions

Hazard R04
Risk phrases R04a
Operator protection A, C, H; U05a, U09b, U20a
Environmental protection E01, E13c
Consumer protection C03
Storage and disposal E30a, E31c, E34
Medical advice M03

409 tebuconazole + triazoxide

A triazole and benzotriazine fungicide seed treatment for use in barley

Products

1	Raxil S	Bayer	20:20 g/l	FS	06974
2	Raxil S	Bayer CropScience	20:20 g/l	LS	11296

Uses

Leaf stripe in ***spring barley*** *(seed treatment)*, ***winter barley*** *(seed treatment)*. Loose smut in ***spring barley*** *(seed treatment)*, ***winter barley*** *(seed treatment)*. Net blotch in ***spring barley*** *(seed treatment)*, ***winter barley*** *(seed treatment)*.

Efficacy guidance

- Best applied through recommended seed treatment machines
- Evenness of seed cover improved by simultaneous application of equal volumes of product and water or dilution of product with an equal volume of water
- Drill treated seed in the same season

Restrictions

- Maximum number of treatments 1 per batch of seed
- Do not use on seed with more than 16% moisture content, or on sprouted, cracked or skinned seed
- Diluted product must be used immediately
- Do not use treated seed as food or feed

Crop-specific information

- Latest use: before drilling
- Slightly delayed and reduced emergence may occur but this is normally outgrown
- Any delay in field emergence, for whatever reason, may be accentuated by treatment

Environmental safety

- Harmful to fish or other aquatic life. Do not contaminate surface waters or ditches with chemical or used container
- Treated seed harmful to game and wildlife
- Product also supplied in returnable containers. See label for guidance on handling, storage, protective clothing and precautions

Hazard classification and safety precautions

Operator protection A, H; U20b
Environmental protection E03, E13c
Storage and disposal E30a, E32a
Treated seed S01, S02, S03, S04b, S05, S06, S07

SECTION 4

410 tebufenpyrad

A pyrazole mitochondrial electron transport inhibitor aphicide and acaricide

Products

Masai	BASF	20% w/w	WB	10223

Uses

Damson-hop aphid in ***hops***, ***plums*** *(off-label)*. Gall mite in ***bilberries*** *(off-label)*, ***blackcurrants*** *(off-label)*, ***blueberries*** *(off-label)*, ***cranberries*** *(off-label)*, ***gooseberries*** *(off-label)*, ***redcurrants*** *(off-label)*, ***whitecurrants*** *(off-label)*. Red spider mites in ***apples***, ***pears***. Two-spotted spider mite in ***hops***, ***protected roses***, ***strawberries***.

Specific Off-Label Approvals (SOLAs)

- ***bilberries***, ***blackcurrants***, ***blueberries***, ***cranberries***, ***gooseberries***, ***redcurrants***, ***whitecurrants*** *(OLA 031741) Dec 2008* [1]
- ***plums*** *(OLA 031742) Dec 2008* [1]

Efficacy guidance

- Acts on eggs (except winter eggs) and all motile stages of spider mites up to adults
- Treat spider mites from 80% egg hatch but before mites become established
- For effective control total spray cover of the crop is required
- Product can be used in a programme to give season-long control of damson/hop aphids coupled with mite control
- Where aphids resistant to tebufenpyrad occur in hops control is unlikely to be satisfactory and repeat treatments may result in lower levels of control. Where possible use different active ingredients in a programme

Restrictions

- Maximum total dose equivalent to one full dose treatment on apples, pears, strawberries; 3 full dose treatments on hops
- Other mitochondrial electron transport inhibitor (METI) acaricides should not be applied to the same crop in the same calendar yr either separately or in mixture
- Do not treat apples before 90% petal fall
- Small-scale testing of rose varieties to establish tolerance recommended before use
- Inner liner of container must not be removed

Crop-specific information

- Latest use: end of burr stage for hops
- HI apples 7d
- Product has no effect on fruit quality or finish

Environmental safety

- Dangerous to fish or other aquatic life. Do not contaminate surface waters or ditches with chemical or used container
- High risk to bees. Do not apply to crops in flower or to those in which bees are actively foraging. Do not apply when flowering weeds are present
- LERAP Category B
- Broadcast air-assisted LERAP (18 m)

Hazard classification and safety precautions

Operator protection A; U02, U09a, U13, U14, U20b
Environmental protection E12a, E13b, E16a, E16b; E17a (18 m)
Storage and disposal E30a, E32a

411 teflubenzuron

A benzoylurea insecticide for use on ornamentals

Products

1	Nemolt	Fargro	150 g/l	SC	07012
2	Nemolt	Fargro	150 g/l	SC	10226

Uses

Browntail moth in ***ornamental plant production***. Caterpillars in ***ornamental plant production***. Whitefly in ***ornamental plant production***.

Efficacy guidance

- Product acts as larval stomach poison interfering with moulting process leading to cessation of feeding and larval death
- Apply as soon as first stage larvae seen. This will often coincide the peak of moth flight

Restrictions

- Maximum number of treatments 3 per yr
- Test specific varieties before carrying out extensive treatments

Environmental safety

- Limited evidence suggests some margin of safety to *Encarsia formosa*. Effects on other parasites and predators not fully tested

Hazard classification and safety precautions

Operator protection A, C; U20c
Environmental protection E15
Storage and disposal E26, E30a, E32a

412 tefluthrin

A soil acting pyrethroid insecticide seed treatment

Products

1	Evict	Bayer	100 g/l	CF	09150
2	Force ST	Bayer	200 g/l	CF	11085
3	Force ST	Bayer CropScience	200 g/l	CF	11671

Uses

Bean seed fly in ***bulb onions*** *(off-label)*, ***leeks*** *(off-label)*, ***salad onions*** *(off-label)* [2]. Carrot fly in ***carrots*** *(off-label)*, ***parsnips*** *(off-label)* [2]. Millipedes in ***fodder beet*** *(seed treatment)*, ***sugar beet*** *(seed treatment)* [2, 3]. Onion fly in ***bulb onions*** *(off-label)*, ***leeks*** *(off-label)*, ***salad onions*** *(off-label)* [2]. Pygmy beetle in ***fodder beet*** *(seed treatment)*, ***sugar beet*** *(seed treatment)* [2, 3]. Springtails in ***fodder beet*** *(seed treatment)*, ***sugar beet*** *(seed treatment)* [2, 3]. Symphylids in ***fodder beet*** *(seed treatment)*, ***sugar beet*** *(seed treatment)* [2, 3]. Wheat bulb fly in ***spring barley***, ***spring wheat***, ***winter barley***, ***winter wheat*** [1]. Wireworm in ***spring barley***, ***spring oats***, ***spring wheat***, ***winter barley***, ***winter oats***, ***winter wheat*** [1].

Specific Off-Label Approvals (SOLAs)

- ***bulb onions***, ***leeks***, ***salad onions*** *(OLA 001748) Dec 2008* [2]
- ***carrots***, ***parsnips*** *(OLA 000873) Dec 2008* [2]

Efficacy guidance

- Apply during process of pelleting beet seed. Consult manufacturer for details of specialist equipment required [2, 3]
- Apply to cereal seed through a suitable liquid seed treater calibrated to achieve even coverage [1]
- Micro-capsule formulation allows slow release to provide a protection zone around treated seed during establishment
- Product is non-systemic and may not protect against wireworm attack at the soil surface [1]

Restrictions

- Maximum number of treatments 1 per batch of seed
- Sow treated seed as soon as possible. Do not store treated seed from one drilling season to next
- Must be co-applied with suitable water-based fungicide seed treatment to protect against seed and soil borne diseases [1]
- Do not use on seed that is sprouted, cracked, damaged or over 16% moisture content [1]
- Powered visor respirator must be worn when bagging treated seed [1]
- Can cause a transient tingling or numbing sensation to exposed skin. Avoid skin contact with product, treated seed and dust throughout all operations in the seed treatment plant and at drilling
- Do not handle seed unnecessarily
- Do not use treated seed as food or feed

Crop-specific information

- Latest use: before drilling seed
- Co-application with other seed treatments may block or damage application equipment [1]
- Where wireworm levels exceed 1.25 million per ha, increased crop damage may occur

Environmental safety

- Extremely dangerous to fish or other aquatic life. Do not contaminate surface waters or ditches with chemical or used container
- Keep treated seed secure from people, domestic stock/pets and wildlife at all times during storage and use
- Treated seed harmful to game and wild life. Bury spillages
- Do not apply treated seed from the air
- Keep livestock out of areas drilled with treated seed for at least 80 d [2, 3]

Hazard classification and safety precautions

Hazard R03 [2, 3]; R04 [1-3]
Risk phrases R04e [1-3]; R04g [2, 3]
Operator protection A, D, H [1-3]; C [1]; E [2, 3]; U02, U08 [2, 3]; U04a, U05a, U20b [1-3]; U14 [1]
Environmental protection E01, E13a [1-3]; E03 [1, 3]; E07 [3] (80 d after sowing)
Consumer protection C03
Storage and disposal E26 [2, 3]; E30a [1, 3]; E32a [1-3]; E34 [1]
Treated seed S01, S02, S03, S04b, S05, S06, S07 [1-3]; S06a, S08 [1]
Medical advice M03 [1]; M05 [3]

413 tepraloxydim

A systemic post-emergence herbicide for control of annual grass weeds

Products

1	Aramo	BASF	50 g/l	EC	10280
2	Standon Tepraloxydim	Standon	50 g/l	EC	11119

Uses

Annual grasses in ***bulb onions, cabbages, carrots, cauliflowers, combining peas, flax, fodder beet, land temporarily removed from production, leeks, linseed, spring field beans, sugar beet, vining peas, winter field beans, winter oilseed rape*** [1]. Annual meadow grass in ***bulb onions, cabbages, carrots, cauliflowers, combining peas, fodder beet, leeks, linseed, spring field beans, spring oilseed rape, sugar beet, vining peas, winter field beans, winter oilseed rape*** [2]. Blackgrass in ***bulb onions, cabbages, carrots, cauliflowers, combining peas, fodder beet, leeks, linseed, spring field beans, spring oilseed rape, sugar beet, vining peas, winter field beans, winter oilseed rape*** [2]. Green cover in ***land temporarily removed from production*** [2]. Perennial grasses in ***bulb onions, cabbages, carrots, cauliflowers, combining peas, flax, fodder beet, land temporarily removed from production, leeks, linseed, spring field beans, sugar beet, vining peas, winter field beans, winter oilseed rape*** [1]. Volunteer barley in ***bulb onions, cabbages, carrots, cauliflowers, combining peas, fodder beet, leeks, linseed, spring field beans, spring oilseed rape, sugar beet, vining peas, winter field beans, winter oilseed rape*** [2]. Volunteer cereals in ***bulb onions, cabbages, carrots, cauliflowers, combining peas, flax, fodder beet, land temporarily removed from production, leeks, linseed, spring field beans, sugar beet, vining peas, winter field beans, winter oilseed rape*** [1]. Volunteer wheat in ***bulb onions, cabbages, carrots, cauliflowers, combining peas, fodder beet, leeks, linseed, spring field beans, spring oilseed rape, sugar beet, vining peas, winter field beans, winter oilseed rape*** [2].

Efficacy guidance

- Best results obtained from applications when weeds small and have not begun to compete with crop
- Only emerged weeds are controlled
- Cool conditions slow down activity and very dry conditions reduce activity by interfering with uptake and translocation
- Foliar death of susceptible weeds evident after 3-4 wks in warm conditions
- Reduced doses must not be used on resistant grass weed populations

Restrictions

- Maximum total dose equivalent to one full dose treatment on all crops
- Do not apply to crops damaged or stressed by factors such as previous herbicide treatments, pest or disease attack
- Do not spray if rain or frost expected, or if foliage is wet
- Do not treat oilseed rape with very low vigour and poor yield potential. Overlapping on oilseed rape may cause damage and reduce yields
- On peas a satisfactory crystal violet leaf wax test must be carried out before treatment. Winter varieties may be treated only in the spring
- For sugar beet, fodder beet, linseed and green cover on land temporarily removed from production, applications are prohibited between 1 Nov and 31 Mar
- For field beans, peas, leeks, bulb onions, carrots, cabbages and cauliflowers applications are prohibited between 1 Nov and 1 Mar

Crop-specific information

- Latest use : before end Nov or before crop has 9 true leaves, whichever is first for winter oilseed rape; before crop has 6 true leaves for spring oilseed rape [2]; before flower buds visible for flax [1], linseed; before head formation for cabbages, cauliflowers
- HI 3 wk for carrots; 4 wk for bulb onions, leeks; 5 wk for peas; 8 wk for fodder beet, field beans, sugar beet

Following crops guidance

- In the event of failure of a treated crop, wheat or barley may be drilled after 2 wk following normal seedbed cultivations, maize or Italian rye-grass may be planted after 8 wk following cultivation to 20 cm

- Graminaceous crops other than those mentioned above should not follow a treated crop in the rotation.
- Broad leaved crops may be planted at any time following a failed or normally harvested treated crop

Environmental safety

- Dangerous for the environment [1]
- Toxic to aquatic organisms. May cause long-term adverse effects to the aquatic environment [1]
- Harmful to fish or other aquatic life. Do not contaminate surface waters or ditches with chemical or used container [2]

Hazard classification and safety precautions

Hazard R03, R14 [1]; R04 [1, 2]
Risk phrases R04b, R04g [1, 2]; R10a, R12b, R14b [1]
Operator protection A [1, 2]; P [1]; U05a, U08, U20b
Environmental protection E01 [1, 2]; E13c [2]; E38 [1]
Consumer protection C03
Storage and disposal E26, E29, E30a, E31b [2]; E32d [1]
Medical advice M05

414 terbacil

A residual uracil herbicide for use in herbs

Products

Sinbar	DuPont	80% w/w	WP	01956

Uses

Annual dicotyledons in ***basil*** *(off-label)*, ***bay*** *(off-label)*, ***borage*** *(off-label)*, ***camomile*** *(off-label)*, ***caraway*** *(off-label)*, ***chervil*** *(off-label)*, ***chives*** *(off-label)*, ***coriander*** *(off-label)*, ***dill*** *(off-label)*, ***fennel*** *(off-label)*, ***fenugreek*** *(off-label)*, ***feverfew*** *(off-label)*, ***lovage*** *(off-label)*, ***marjoram*** *(off-label)*, ***mint*** *(off-label)*, ***parsley*** *(off-label)*, ***protected herbs*** *(off-label)*, ***rosemary*** *(off-label)*, ***sage*** *(off-label)*, ***savory*** *(off-label)*, ***sorrel*** *(off-label)*, ***tarragon*** *(off-label)*, ***thyme*** *(off-label)*. Annual grasses in ***basil*** *(off-label)*, ***bay*** *(off-label)*, ***borage*** *(off-label)*, ***camomile*** *(off-label)*, ***caraway*** *(off-label)*, ***chervil*** *(off-label)*, ***chives*** *(off-label)*, ***coriander*** *(off-label)*, ***dill*** *(off-label)*, ***fennel*** *(off-label)*, ***fenugreek*** *(off-label)*, ***feverfew*** *(off-label)*, ***lovage*** *(off-label)*, ***marjoram*** *(off-label)*, ***mint*** *(off-label)*, ***parsley*** *(off-label)*, ***protected herbs*** *(off-label)*, ***rosemary*** *(off-label)*, ***sage*** *(off-label)*, ***savory*** *(off-label)*, ***sorrel*** *(off-label)*, ***tarragon*** *(off-label)*, ***thyme*** *(off-label)*.

Specific Off Label Approvals (SOLAs)

- ***basil, bay, borage, camomile, caraway, chervil, chives, coriander, dill, fennel, fenugreek, feverfew, lovage, marjoram, mint, parsley, protected herbs, rosemary, sage, savory, sorrel, tarragon, thyme*** *(OLA 930082) Dec 2008* [1]

Approval information

- Products containing this active ingredient have been granted derogations for specified 'Essential Uses' for use until 31 December 2007. Sale and supply must cease by 30 June 2007 but growers have no guarantee that the products will continue to be available until then.
For more information see 'The Review Programme' under 'Pesticide Legislation' in Section 1

Efficacy guidance

- Has some foliar activity but uptake mainly through roots
- Adequate rainfall needed to ensure penetration to weed root zone
- Best results achieved by application in early spring, before mid-Apr

Restrictions

- Maximum number of treatments 1 per yr
- Do not use on Sands, Loamy Sand or gravels or soils with less than 1% organic matter
- This product must not be used on any crops other than those listed, including any extrapolations that would normally be permissible under the Long Term Arrangements for Extension of Use (see Section 1)

Crop-specific information

- Latest use: pre-emergence of new growth in spring for herbs

Hazard classification and safety precautions

Operator protection U08, U19, U20a

Environmental protection E15

Storage and disposal E30a, E31a

415 terbuthylazine

A triazine herbicide available only in mixtures

416 terbuthylazine + terbutryn

A pre-emergence herbicide for peas, beans and lupins

Products

1	Batallion	Makhteshim	150:350 g/l	SC	08305
2	Opogard	Syngenta	150:350 g/l	SC	08427

Uses

Annual dicotyledons in ***broad beans, combining peas, spring field beans, vining peas, winter field beans*** [1, 2]; ***edible podded peas*** [1]; ***lupins*** [2]. Annual grasses in ***broad beans, combining peas, spring field beans, vining peas, winter field beans*** [1, 2]; ***edible podded peas*** [1]; ***lupins*** [2].

Approval information

- Certain products containing terbuthylazine and terbutryn have been granted derogations for specified 'Essential Uses' for use until 31 December 2007. Sale and supply must cease by 30 June 2007 but growers have no guarantee that the products will continue to be available until then.

  For more information see 'The Review Programme' under 'Pesticide Legislation' in Section 1

Efficacy guidance

- Active via root uptake and with foliar activity on cotyledon stage weeds
- Best results achieved by application to fine, firm, moist seedbed, preferably at weed emergence, when rain falls after spraying
- Effectiveness may be reduced by excessive rain, drought or cold
- Residual control lasts for up to 8 wk on mineral soils. Effectiveness reduced on highly organic soils and subsequent use of post-emergence treatment recommended
- Do not use on soils which are very cloddy or have more than 10% organic matter otherwise weed control will be reduced

Restrictions

- Maximum number of treatments 1 per crop
- Vedette and Printana peas may be damaged by treatment. Do not use on forage peas
- Do not treat peas, beans or lupins on soils lighter than loamy fine sand, or lupins on silty clay soil
- Do not cultivate after treatment

Crop-specific information

- Latest use: 3 d before crop emergence
- Apply as soon as possible after drilling
- Crop seed must be covered by at least 25 mm of settled soil
- Heavy rain after application may cause damage to peas on light soils

Following crops guidance

- Treated land should be mould-board ploughed to 150 mm before any succeeding crop is planted
- A period of 12 wk (14 wk in dry conditions) since application must elapse before planting any succeeding crop

Environmental safety

- Dangerous for the environment [2]
- Toxic to aquatic organisms. May cause long-term adverse effects to the aquatic environment [2]
- Harmful to fish or other aquatic life. Do not contaminate surface waters or ditches with chemical or used container [1]

Hazard classification and safety precautions

Hazard R03 [1]; R04, R14 [2]

Risk phrases R03c [1]; R04e, R14b [2]
Operator protection U02, U08, U13, U14, U15, U19 [1]; U05a, U20a [1, 2]
Environmental protection E01 [1, 2]; E13c [1]; E15 [2]
Consumer protection C03
Storage and disposal E26, E31c [2]; E30a [1, 2]; E31b, E34 [1]
Medical advice M03 [1]

417 terbutryn

A residual triazine herbicide aquatic weed control

See also *fomesafen + terbutryn*
prometryn + terbutryn
terbuthylazine + terbutryn

Products

Clarosan	Scotts	1% w/w	GR	09394

Uses

Aquatic weeds in ***areas of water***.

Approval information

- Approved for aquatic weed control [1]. See notes in Section 1 on use of herbicides in or near water
- Certain products containing terbutryn have been granted derogations for specified 'Essential Uses' for use until 31 December 2007. Sale and supply must cease by 30 June 2007 but growers have no guarantee that the products will continue to be available until then.
  For more information see 'The Review Programme' under 'Pesticide Legislation' in Section 1

Efficacy guidance

- For aquatic weed control apply granules to water surface with suitable equipment, when growth active but before heavy infestation develops, usually Apr-May, sometimes to Aug
- Use only in static or sluggishly moving water. The flow in moving water should be stopped for at least 7 d after treatment otherwise weed control may be reduced. Treat small ponds before start of new growth, usually before end Apr
- Effectiveness reduced in water with peaty bottom

Restrictions

- This product must not be used for any purpose other than that listed, including any extrapolations that would normally be permissible under the Long Term Arrangements for Extension of Use (see Section 1)

Crop-specific information

- Latest use: Aug for areas of water

Environmental safety

- Do not use treated water for irrigation purposes within 7 d of treatment, or dump surplus herbicide in water or ditch bottoms
- For maximum concentration in water see label
- Harmful to fish or other aquatic life. Do not contaminate surface waters or ditches with chemical or used container
- If dense weed growth in watercourses to be controlled without de-oxygenation treat in sections of about 400 m at a time with intervals of at least 14 d

Hazard classification and safety precautions

Operator protection U20c
Environmental protection E19, E21
Storage and disposal E29, E30a, E32a

SECTION 4

418 tetraconazole

A systemic, protectant and curative triazole fungicide for cereals

See also chlorothalonil + tetraconazole

Products

Juggler	Sipcam	100 g/l	EC	09391

Uses

Brown rust in ***spring barley, spring wheat, winter barley, winter wheat***. Crown rust in ***spring oats, winter oats***. Powdery mildew in ***spring barley, spring wheat, winter barley, winter wheat***. Septoria leaf spot in ***spring wheat, winter wheat***. Sooty moulds in ***spring wheat, winter wheat***. Yellow rust in ***spring barley, spring wheat, winter barley, winter wheat***.

Efficacy guidance

- Apply at any time from the tillering stage up to the latest times indicated for each crop
- Best results achieved from applications before disease becomes established
- If disease pressure persists a second treatment may be required to prevent late season attacks
- Best control of ear diseases in wheat obtained from application at or after the ear completely emerged stage

Restrictions

- Maximum total dose equivalent to three full dose applications on wheat and two full dose applications on other cereals

Crop-specific information

- Latest use: before end of ear emergence (GS 59) for oats; before end of flowering (GS 69) for barley; before grain watery ripe (GS 71) for wheat

Environmental safety

- Harmful to fish or other aquatic life. Do not contaminate surface waters or ditches with chemical or used container
- LERAP Category B

Hazard classification and safety precautions

Hazard R03, R04
Risk phrases R03c, R04a, R04b
Operator protection A, C, H; U05a, U19, U20a
Environmental protection E01, E13c, E16a, E16b
Consumer protection C03
Storage and disposal E26, E30a, E31b
Medical advice M03

419 tetramethrin

A contact acting pyrethroid insecticide

See also d-phenothrin + tetramethrin
resmethrin + tetramethrin

Products

Killgerm Py-Kill W	Killgerm	10.2% w/v	EC	H4632

Uses

Flies in ***agricultural premises, livestock houses***.

Efficacy guidance

- Dilute in accordance with directions and apply as space or surface spray

Restrictions

- Do not apply directly on food or livestock
- Remove exposed milk before application. Protect milk machinery and containers from contamination
- Do not use space sprays containing pyrethrins or pyrethroid more than once per wk in intensive or controlled environment animal houses in order to avoid development of resistance. If necessary, use a different control method or product

Environmental safety

- Harmful to fish or other aquatic life. Do not contaminate surface waters or ditches with chemical or used container

Hazard classification and safety precautions

Hazard R07d
Operator protection U08, U19, U20c
Environmental protection E05, E13c
Consumer protection C06, C10
Storage and disposal E30a, E31a

420 thiabendazole

A systemic, curative and protectant benzimidazole (MBC) fungicide

See also imazalil + thiabendazole

Products

1	Hykeep	Agrichem	2% ww	DS	06744
2	Storite Clear Liquid	Banks Cargill	220 g/l	LS	08982
3	Storite Excel	Banks Cargill	500 g/l	SC	09542
4	Storite Flowable	Banks Cargill	450 g/l	FS	08703

Uses

Dry rot in ***seed potatoes*** *(tuber treatment - post-harvest)* [2, 4]; ***ware potatoes*** *(tuber treatment - post-harvest)* [1-4]. Fusarium basal rot in ***narcissi*** [2]. Gangrene in ***seed potatoes*** *(tuber treatment - post harvest)* [2, 4]; ***ware potatoes*** *(tuber treatment - post-harvest)* [1-4]. Silver scurf in ***seed potatoes*** *(tuber treatment - post-harvest)* [2, 4]; ***ware potatoes*** *(tuber treatment - post-harvest)* [1-4]. Skin spot in ***seed potatoes*** *(tuber treatment - post-harvest)* [2, 4]; ***ware potatoes*** *(tuber treatment - post-harvest)* [1-4].

Approval information

- Thiabendazole included in Annex I under EC Directive 91/414
- Product formulated for application with ULV equipment (see label for details) [2]
- Approval expiry 29 Feb 2004 [2]

Restrictions

- Maximum number of treatments 1 per batch for seed potato treatments; 2 per yr for bulbs [2]
- Treated seed potatoes must not be used for food or feed

Crop-specific information

- Latest use: before planting for seed potatoes; 21 d before removal from store for sale, processing or consumption for ware potatoes
- Apply to potatoes as soon as possible after harvest using suitable equipment and always within 2 wk of lifting. See label for details
- Potatoes should only be treated by systems that provide an accurate dose to tubers not carrying excessive quantities of soil
- Apply to bulbs as a dip. Ensure bulbs are clean [2]

Environmental safety

- Harmful to fish or other aquatic life. Do not contaminate surface waters or ditches with chemical or used container

Hazard classification and safety precautions

Operator protection A, C, D, H [1-3]; B, K [2]; M [1, 2]; U19, U20c
Environmental protection E01, E13c
Storage and disposal E26, E31a [2-4]; E29, E32a [1]; E30a [1-4]
Treated seed S01, S05 [2]

421 thiabendazole + thiram

A fungicide seed dressing mixture for field and vegetable crops

Products

1	Hy-TL	Agrichem	225:300 g/l	FS	06246
2	sHYlin	Agrichem	258:255 g/l	FS	10030

Uses

Ascochyta in ***broad beans*** *(seed treatment)*, ***peas*** *(seed treatment)*, ***spring field beans*** *(seed treatment)*, ***winter field beans*** *(seed treatment)* [1]. Botrytis root rot in ***flax*** *(seed treatment)*, ***linseed*** *(seed treatment)* [2]. Damping off in ***broad beans*** *(seed treatment)*, ***peas*** *(seed treatment)*, ***spring field beans*** *(seed treatment)*, ***winter field beans*** *(seed treatment)* [1]; ***carrots*** *(seed treatment)*, ***flax*** *(seed treatment)*, ***linseed*** *(seed treatment)*, ***lupins*** *(seed treatment)*, ***parsnips*** *(seed treatment)* [2]. Fusarium in ***bulb onions*** *(off-label - seed)* [1]. Fusarium root rot in ***flax*** *(seed treatment)*, ***linseed*** *(seed treatment)* [2]. Neck rot in ***bulb onions*** *(off-label - seed)* [1].

Specific Off-Label Approvals (SOLAs)

- ***bulb onions*** *(seed) (OLA 021299) Dec 2008* [1]

Approval information

- Thiabendazole included in Annex I under EC Directive 91/414

Efficacy guidance

- Dress seed as near to sowing as possible
- Dilution may be needed with particularly absorbent types of seed. If diluted material used, seed may require drying before storage

Restrictions

- Maximum number of treatments 1 per batch
- Seed to be treated should be of satisfactory quality and moisture content
- Do not use treated seed as food or feed

Crop-specific information

- Latest use: before drilling

Environmental safety

- Dangerous to fish or other aquatic life. Do not contaminate surface waters or ditches with chemical or used container
- Treated seed harmful to game and wildlife

Hazard classification and safety precautions

Hazard R03, R04
Risk phrases R03a, R03c, R04a, R04b, R04c
Operator protection A, C, D, H, M; U05a, U08, U20a [1, 2]; U20b [1]
Environmental protection E01, E03, E13b
Consumer protection C03
Storage and disposal E26, E30a, E31c, E34
Treated seed S01, S02, S03, S04b, S05, S06, S07
Medical advice M03

422 thiacloprid

A chloronicotinyl insecticide for use in apples

Products

1	Calypso	Bayer	480 g/l	SC	10342
2	Calypso	Bayer CropScience	480 g/l	SC	11257

Uses

Rosy apple aphid in ***apples***.

Efficacy guidance

- Best results obtained from a programme of sprays commencing pre-blossom at the first sign of aphids

- Minimise the possibility of development of resistance by alternating insecticides with different modes of action in the programme
- In dense canopies and on larger trees increase water volume to ensure full coverage

Restrictions

- Maximum number of treatments 2 per yr

Crop-specific information

- HI apples 14 d; protected vegetables 3 d

Environmental safety

- Harmful to fish or other aquatic life. Do not contaminate surface waters or ditches with chemical or used container
- Broadcast air-assisted LERAP (30 m)
- Risk to certain non-target insects or other arthropods. See directions for use

Hazard classification and safety precautions

Hazard R03, R04
Risk phrases R03c, R04e
Operator protection A, H; U05a, U14, U20b
Environmental protection E01, E13c, E22b; E17a (30 m)
Consumer protection C03
Storage and disposal E26, E30a, E31b, E34
Medical advice M03, M05

423 thifensulfuron-methyl

A translocated sulfonylurea herbicide

See also *carfentrazone-ethyl + thifensulfuron-methyl*
flupyrsulfuron-methyl + thifensulfuron-methyl
metsulfuron-methyl + thifensulfuron-methyl

Products

Prospect	Nufarm UK	75% w/w	WB	09389

Uses

Docks in ***established grassland*, *land temporarily removed from production*** *(in green cover)*.

Approval information

- Thifensulfuron-methyl included in Annex I under EC Directive 91/414

Efficacy guidance

- Best results achieved from application to young green dock foliage when growing actively
- Only broad-leaved docks (*Rumex obtusifolius*) are controlled; curled docks (*Rumex crispus*) are resistant
- Apply 7-10 d before grazing and do not graze for 7 d afterwards
- Docks with developing or mature seed heads should be topped and the regrowth treated later
- Established docks with large tap roots may require follow-up treatment
- High populations or poached grassland will require further treatment in following yr
- Ensure good spray coverage and apply to dry foliage
- Thifensulfuron-methyl is a member of the ALS-inhibitor group of herbicides

Restrictions

- Maximum number of treatments 1 per calender yr. Must only be applied from 1 Feb in yr of harvest
- Do not treat new leys in year of sowing
- Do not treat where nutrient imbalances, drought, waterlogging, low temperatures, lime deficiency, pest or disease attack have reduced sward vigour
- Do not roll or harrow within 7 d of spraying

Crop-specific information

- Latest use: before 1 Aug
- Product may cause a check to both sward and clover which is usually outgrown

Following crops guidance

- Only grass or cereals may be sown within 4 wk of treatment

Environmental safety

- Extremely dangerous to fish or other aquatic life. Do not contaminate surface waters or ditches with chemical or used container
- Extremely dangerous to aquatic higher plants. Do not contaminate surface waters or ditches with chemical or used container
- LERAP Category B
- Take extreme care to avoid drift onto broad-leaved plants outside the target area or onto surface waters or ditches, or land intended for cropping
- Spraying equipment should not be drained or flushed onto land planted, or to be planted, with trees or crops other than cereals and should be thoroughly cleansed after use - see label for instructions
- Keep livestock out of treated areas for at least 7 d following treatment

Hazard classification and safety precautions

Operator protection U08, U19, U20b
Environmental protection E07 (7 d); E13a, E14a, E16a, E16b
Storage and disposal E30a, E32a

424 thifensulfuron-methyl + tribenuron-methyl

A mixture of two sulfonylurea herbicides for cereals

Products

Calibre	DuPont	50:25% w/w	WG	07795

Uses

Annual dicotyledons in ***spring barley***, ***spring wheat***, ***winter barley***, ***winter wheat***. Charlock in ***spring barley***, ***spring wheat***, ***winter barley***, ***winter wheat***. Chickweed in ***spring barley***, ***spring wheat***, ***winter barley***, ***winter wheat***. Mayweeds in ***spring barley***, ***spring wheat***, ***winter barley***, ***winter wheat***.

Approval information

- Thifensulfuron-methyl included in Annex I under EC Directive 91/414

Efficacy guidance

- Apply after 1 Feb when weeds are small and actively growing
- Ensure good spray cover of the weeds
- Ensure that weeds present are those that are susceptible. See label
- Follow label mixing instructions
- Effectiveness reduced by rain within 4 h of treatment
- Various tank mixtures recommended to broaden weed control spectrum. Other mixtures are specifically excluded. See label
- Thifensulfuron-methyl and tribenuron-methyl are members of the ALS-inhibitor group of herbicides and products should be used in a planned Resistance Management strategy. See Section 1 for more information

Restrictions

- Maximum number of treatments 1 per crop
- Do not apply to cereals undersown with grass, clover or other legumes
- Do not apply within 7 d of rolling
- Do not apply in sequence or in tank mixture with a product containing any other sulfonyl urea except as directed on label
- Do not apply to any crop suffering from stress or not actively growing

Crop-specific information

- Latest use: before flag leaf ligule first visible (GS 39)
- May be sprayed on all varieties of wheat and barley from 3 leaf stage (GS 13) up to and including flag leaf fully emerged (GS 39)
- Igri winter barley may suffer damage during period of rapid growth and must not be treated before leaf sheath erect stage (GS 30)

Following crops guidance

- Only cereals, field beans or oilseed rape may be sown in the same calendar year as harvest of a treated crop. In the event of failure of a treated crop sow only a cereal crop within 3 mth of product application

Environmental safety

- Extremely dangerous to fish or other aquatic life. Do not contaminate surface waters or ditches with chemical or used container
- LERAP Category B
- Spraying equipment should not be drained or flushed onto land planted, or to be planted, with trees or crops other than cereals and should be thoroughly cleansed after use - see label for instructions
- Take particular care to avoid damage by drift onto broad-leaved plants outside the target area or onto surface waters or ditches

Hazard classification and safety precautions

Operator protection U08, U19, U20a
Environmental protection E13a, E16a, E16b
Storage and disposal E30a, E32a

425 thiodicarb

A carbamate insecticide with molluscicide uses

Products

1	Gonosis	Sipcam	4% w/w	RB	09993
2	Judge	Sipcam	4% w/w	RB	10000
3	Me2 Exodus	Me2	4.0% w/w	RB	09786

Uses

Slugs in ***brussels sprouts*** [1, 2]; ***durum wheat, potatoes, spring barley, spring oats, spring oilseed rape, spring wheat, triticale, winter barley, winter oats, winter oilseed rape, winter wheat*** [1-3].

Efficacy guidance

- Apply bait as broadcast treatment or admixed with seed
- Additional broadcast treatments may be needed after drilling admixed seed
- Sow admixed seed as soon as possible after treatment

Restrictions

- Product contains an anticholinesterase carbamate compound. Do not use if under medical advice not to work with such compounds
- Maximum number of treatments 1 per crop (admixture), 3 per crop (broadcast)
- Do not treat grain with more than 16% moisture content or allow treated seed to rise above this level
- May only be applied to oilseed rape and potatoes as broadcast application
- Must not be admixed with cereal seed already admixed with any other product containing thiodicarb
- Do not use treated seed as food or feed

Crop-specific information

- Latest use: before drilling when admixed with barley, durum wheat, oats, triticale, wheat; before second node detectable (GS 32) for barley, durum wheat, oats, triticale, wheat; before stem extension (GS 2,0) for oilseed rape
- HI potatoes, Brussels sprouts 3 wk

Environmental safety

- Harmful to game, wild birds and animals
- Dangerous to fish or other aquatic life. Do not contaminate surface waters or ditches with chemical or used container

Hazard classification and safety precautions

Hazard R03, R04
Risk phrases R03c, R04a, R04e
Operator protection A, H, J [1-3]; D [1, 3]; U05a, U13, U20b

SECTION 4

Environmental protection E01, E10b, E13b
Consumer protection C03
Storage and disposal E30a, E32a, E34
Treated seed S01, S02, S03, S04b, S05, S06, S07 [1, 3]
Medical advice M02 [1-3]; M03 [3]; M04 [1, 2]

426 thiophanate-methyl

A carbendazim precursor fungicide with protectant and curative activity

See also iprodione + thiophanate-methyl

Products

Mildothane Turf Liquid	Bayer Environ.	500 g/l	SC	09935

Uses

Dollar spot in ***managed amenity turf***. Fusarium patch in ***managed amenity turf***. Red thread in ***managed amenity turf***. Wormcast formation in ***managed amenity turf***.

Approval information

- Following implementation of Directive 98/82/EC, approval for use of thiophanate-methyl on several crops was revoked in 1999

Efficacy guidance

- Treat turf at the first sign of disease and repeat at monthly intervals as necessary
- Apply to turf during period of active growth and do not mow for 48 h
- Control of Red Thread and Dollar Spot will be helped by application of a nitrogenous fertilizer 7 d before fungicide treatment

Hazard classification and safety precautions

Operator protection A, C, H, M
Environmental protection E15
Storage and disposal E26, E30a, E31a, E31b

427 thiram

A protectant dithiocarbamate fungicide

See also carboxin + thiram
thiabendazole + thiram

Products

1	Agrichem Flowable Thiram	Agrichem	600 g/l	LS	10784
2	Thiraflo	Crompton	480 g/l	FS	11526
3	Thyram Plus	Agrichem	600 g/l	LS	10785
4	Unicrop Thianosan DG	Unicrop	80% w/w	WG	05454

Uses

Botrytis in ***chrysanthemums***, ***freesias***, ***lettuce*** *(outdoor crops)*, ***ornamental specimens*** *(except Hydrangea)*, ***outdoor tomatoes***, ***protected lettuce***, ***protected tomatoes***, ***raspberries***, ***strawberries*** [4]. Botrytis fruit rot in ***apples*** [4]. Cane spot in ***raspberries*** [4]. Damping off in ***broad beans*** *(seed treatment)*, ***dwarf beans*** *(seed treatment)*, ***grass seed*** *(seed treatment)*, ***maize*** *(seed treatment)*, ***runner beans*** *(seed treatment)*, ***spring field beans*** *(seed treatment)*, ***spring oilseed rape*** *(seed treatment)*, ***winter field beans*** *(seed treatment)*, ***winter oilseed rape*** *(seed treatment)* [1-3]; ***cabbages*** *(seed treatment)*, ***carrots*** *(seed treatment)*, ***cauliflowers*** *(seed treatment)*, ***leeks*** *(seed treatment)*, ***lettuce*** *(seed treatment)*, ***onions*** *(seed treatment)*, ***peas*** *(seed treatment)*, ***radishes*** *(seed treatment)*, ***salad onions*** *(seed treatment)*, ***turnips*** *(seed treatment)* [1, 3]; ***combining peas*** *(seed treatment)*, ***vining peas*** *(seed treatment)* [2]. Downy mildew in ***protected lettuce*** [4]. Fire in ***tulips*** [4]. Gloeosporium in ***apples*** [4]. Rust in ***blackcurrants***, ***carnations***, ***chrysanthemums*** [4]. Scab in ***apples***, ***pears*** [4]. Seed-borne diseases in ***carrots*** *(seed soak)*, ***celery*** *(seed soak)*, ***fodder beet*** *(seed soak)*, ***mangels*** *(seed soak)*, ***parsley*** *(seed soak)*, ***red beet*** *(seed soak)*, ***sugar beet*** *(seed soak)* [1]. Spur blight in ***raspberries*** [4].

Efficacy guidance

- Spray before onset of disease and repeat every 7-14 d. Spray interval varies with crop and disease. See label for details [4]
- Apply as seed treatment for protection against damping off. May be applied through most types of seed treatment machinery, from automated continuous flow machines to smaller batch treating apparatus [1, 3]
- Co-application of 175 ml water per 100 kg of seed likely to improve evenness of seed coverage [1-3]
- Do not spray when rain imminent [4]
- For use on tulips, chrysanthemums and carnations add non-ionic wetter [4]

Restrictions

- Maximum number of treatments 3 per crop for protected winter lettuce (thiram based products only [4]; 2 per crop if sequence of thiram and other EBDC fungicides used) [4]; 2 per crop for protected summer lettuce [4]; 1 per batch of seed for seed treatments
- Do not apply to hydrangeas [4]
- Notify processor before dusting or spraying crops for processing [4]
- Do not dip roots of forestry transplants [4]
- Do not treat seed of tomatoes, peppers or aubergines [1, 3]

Crop-specific information

- Latest use: 21 d after planting out or 21 d before harvest, whichever is earlier, for protected winter lettuce; 14 d after planting out or 21 d before harvest, whichever is earlier, for protected summer lettuce; before drilling for soya beans [1]
- HI protected lettuce 21 d; outdoor lettuce 14 d; apples, pears, blackcurrants, raspberries, strawberries, tomatoes 7 d
- Seed to be treated should be of satisfactory quality and moisture content [1-3]
- Follow label instructions for treating small quantities of seed [1, 3]

Environmental safety

- Dangerous to fish or other aquatic life. Do not contaminate surface waters or ditches with chemical or used container [4]
- Do not use treated seed as food or feed [2]
- Treated seed harmful to game and wildlife [2]
- A red dye is available from manufacturer to colour treated seed. See label for details [1, 3]
- Additional precautions apply if using small volume returnable container- see labels [1, 3]

Hazard classification and safety precautions

Hazard R03 [1, 3]; R04 [1, 3, 4]
Risk phrases R03a, R03c [1, 3]; R04a, R04b, R04c [1, 3, 4]
Operator protection A; U02, U09a [1, 3]; U05a, U20b [1-4]; U07 [1, 3] (SVR only); U08 [2, 4]; U19 [4]
Environmental protection E01 [1-4]; E13b [4]; E15 [1, 3]
Consumer protection C03
Storage and disposal E26 [1-3]; E30a [1-4]; E31a [2]; E31c, E34 [1, 3]; E32a [4]; E33, E36 [1, 3] (SVR only)
Treated seed S01, S02, S03, S04b, S05, S07 [1-3]; S06 [2]
Medical advice M03 [1-3]

428 tolclofos-methyl

A protectant organophosphorus fungicide for soil-borne diseases

Products

1	Basilex	Scotts	50% w/w	WP	07494
2	Me2 Cindy	Me2	500 g/l	SC	10634
3	Rizolex	Certis	10% w/w	DS	09673
4	Rizolex Flowable	Certis	500 g/l	FS	09358
5	Rizolex Flowable	Certis	500 g/l	FS	11399

Uses

Black scurf and stem canker in ***potatoes*** *(tuber treatment)* [2-5]. Bottom rot in ***protected lettuce*** [1]. Damping off in ***seedlings of ornamentals*** [1]. Damping off and wirestem in ***brussels sprouts,***

SECTION 4

cabbages, calabrese, cauliflowers, chinese cabbage [1]. Foot rot in ***ornamental specimens, seedlings of ornamentals*** [1]. Rhizoctonia in ***potatoes*** *(off-label - chitted seed treatment)* [4, 5]; ***protected celery*** *(off-label)*, ***protected radishes*** *(off-label)*, ***swedes*** *(off-label - with fleece covers)*, ***swedes*** *(off-label - without covers)*, ***turnips*** *(off-label - with fleece covers)*, ***turnips*** *(off-label - without covers)* [1]. Root rot in ***ornamental specimens, seedlings of ornamentals*** [1].

Specific Off-Label Approvals (SOLAs)

- ***potatoes*** *(chitted seed treatment) (OLA 020001) Dec 2008* [4, 5]
- ***protected celery*** *(OLA 992009) Dec 2008* [1]
- ***protected radishes*** *(OLA 002006) Dec 2008* [1]
- ***swedes, turnips*** *(with fleece covers) (OLA 023749) Dec 2006* [1]
- ***swedes, turnips*** *(without covers) (OLA 023749) Dec 2006* [1]

Approval information

- Tolclofos-methyl has been included in a comprehensive review of anticholinesterase compounds and considered by the ACP. As a result approvals have been allowed to continue subject to the imposition of additional operator protection requirements.
- May be applied by misting equipment mounted over roller table. See label for details [3-5]

Efficacy guidance

- Apply dust to seed potatoes during hopper loading [3]
- Apply flowable formulation to clean tubers with suitable misting equipment over a roller table. Spray as potatoes taken into store (first earlies) or as taken out of store (second earlies, maincrop, crops for seed) pre-chitting [2, 4, 5]
- Do not mix flowable formulation with any other product [2, 4, 5]
- To control Rhizoctonia in vegetables and ornamentals apply as drench before sowing, pricking out or planting [1]
- On established seedlings and pot plants apply as drench and rinse off foliage [1]

Restrictions

- Tolclofos-methyl is an atypical organophosphorus compound which has weak anticholinesterase activity. Do not use if under medical advice not to work with such compounds
- Maximum number of treatments 1 per batch for seed potatoes; 1 per crop for lettuce, brassicas, celery, radishes; 1 at each stage of growth (ie sowing, pricking out, potting) to a maximum of 3, for ornamentals
- Only to be used with automatic planters [3]
- Not recommended for use on seed potatoes where hot water treatment used or to be used [3]
- Do not apply as overhead drench to vegetables or ornamentals when hot and sunny [1]
- Do not use on heathers [1]
- Must not be used via hand-held equipment
- Application to protected crops must only be made where the operator is outside the structure at the time of treatment

Crop-specific information

- Latest use: at planting of seed potatoes [2-5]; before planting celery [1]; before transplanting for lettuce, brassicas, ornamental specimens [1]
- HI potatoes 12 wk [4, 5]

Environmental safety

- Treated tubers to be used as seed only, not for food or feed [2-4]

Hazard classification and safety precautions

Hazard R04

Risk phrases R04a, R04b, R04c

Operator protection A, D, H [1-5]; C [2, 4, 5]; E [2-5]; J, M [3]; U05a [3]; U16 [2, 4, 5]; U19, U20b [1-5]

Environmental protection E01 [3]; E13c [2-5]

Consumer protection C03 [3]

Storage and disposal E30a [1-5]; E31a [2, 4, 5]; E32a [1, 3]

Medical advice M01 [1]

429 tolylfluanid

A multi-site protectant fungicide

See also fenhexamid + tolylfluanid

Products

1	Elvaron Multi	Bayer	50.5% w/w	WG	10080
2	Elvaron Multi	Bayer CropScience	50.5% w/w	WG	11422

Uses

Botrytis in ***blackberries***, ***blackcurrants***, ***gooseberries***, ***loganberries***, ***raspberries***, ***redcurrants***, ***strawberries***, ***whitecurrants***.

Efficacy guidance

- Best results in soft fruit achieved by good spray coverage of flowers and young fruitlets
- To achieve required coverage of flowers and young fruitlets spray using a hand lance, drop arm attachments or a special strawberry boom with leaf rolling bar
- Best results using high volume directed over the crop rows

Restrictions

- Maximum total dose equivalent to 2.5 full dose treatments on currants, gooseberries, and four full dose treatments on other soft fruit crops

Crop-specific information

- HI blackberries, loganberries, raspberries, strawberries 14 d; currants, gooseberries 21 d

Environmental safety

- Dangerous to fish or other aquatic life. Do not contaminate surface waters or ditches with chemical or used container
- LERAP Category B
- Use best available application technique that minimises off-target drift to reduce effects on non-target insects or other arthropods

Hazard classification and safety precautions

Hazard R04
Risk phrases R04e
Operator protection A, H, M; U02, U05a, U20b
Environmental protection E01, E13b, E16a, E16b, E22b
Consumer protection C03
Storage and disposal E30a, E32a

430 tralkoxydim

A foliar applied oxime herbicide for grass weed control in cereals.

Products

1	Grasp	Syngenta	250 g/l	SC	10441
2	Greencrop Gweedore	Greencrop	250 g/l	SC	09882
3	Landgold Tralkoxydim	Landgold	250 g/l	SC	08604
4	Standon Tralkoxydim	Standon	250 g/l	SC	09579

Uses

Awned canary grass in ***spring barley*** *(qualified minor use)*, ***spring wheat*** *(qualified minor use)* [1]. Blackgrass in ***durum wheat***, ***spring barley***, ***spring wheat***, ***triticale***, ***winter barley***, ***winter rye***, ***winter wheat*** [1-4]; ***spring barley*** *(autumn sown)*, ***spring wheat*** *(autumn sown)* [1]. Rough meadow grass in ***durum wheat***, ***spring barley*** *(autumn sown)*, ***spring wheat***, ***spring wheat*** *(autumn sown)*, ***triticale***, ***winter barley***, ***winter rye***, ***winter wheat*** [1]. Ryegrass in ***durum wheat***, ***spring barley***, ***spring barley*** *(autumn sown)*, ***spring wheat***, ***spring wheat*** *(autumn sown)*, ***triticale***, ***winter barley***, ***winter rye***, ***winter wheat*** [1]. Wild oats in ***durum wheat***, ***spring barley***, ***spring wheat***, ***triticale***, ***winter barley***, ***winter rye***, ***winter wheat*** [1-4]; ***spring barley*** *(autumn sown)*, ***spring wheat*** *(autumn sown)* [1]. Yorkshire fog in ***durum wheat*** *(qualified minor use)*, ***spring barley*** *(autumn sown - qualified minor use)*, ***spring wheat*** *(autumn sown - qualified minor use)*, ***triticale*** *(qualified minor use)*, ***winter barley*** *(qualified minor use)*, ***winter rye*** *(qualified minor use)*, ***winter wheat*** *(qualified minor use)* [1].

Efficacy guidance

- Product leaf-absorbed and translocated rapidly to growing points. Best results achieved when weeds growing actively in competitive crops under warm humid conditions with adequate soil moisture
- Activity not dependent on soil type. Weeds germinating after application will not be controlled
- Best control of wild oats obtained from 2 leaf to 1st node detectable stage of weeds, and of blackgrass up to 3 tillers

Restrictions

- Maximum number of treatments 1 or 2 per crop. See label
- Authorised adjuvant must always be added. See label
- Do not spray undersown crops or crops to be undersown
- Do not spray when foliage wet or covered in ice or crop otherwise under stress
- Do not spray if a protracted period of cold weather forecast
- Do not spray crops under stress from chemical treatment, grazing, pest attack, mineral deficiency or low fertility
- Do not roll or harrow within 1 wk of spraying
- Do not tank-mix with phenoxy hormone or sulfonylurea herbicides or apply these chemicals within 15 d before and 5 d after treatment

Crop-specific information

- Latest use: before booting (GS 41)
- Apply to winter cereals from 2 leaves unfolded up to and including flag leaf sheath extending (GS 12-41). If necessary winter cereals may be sprayed twice: once in autumn and once in spring
- Apply to spring cereals from end of tillering up to and including flag leaf sheath extending (GS 29-41)

Environmental safety

- Harmful to aquatic organisms. May cause long-term adverse effects to the aquatic environment [1]

Hazard classification and safety precautions

Hazard R04 [1-4]; R14 [1]
Risk phrases R04a [2-4]; R04e [1-4]; R14a [1]
Operator protection A, C, H; U02, U04a, U05a, U08, U20b
Environmental protection E01, E15 [1-4]; E38 [1]
Consumer protection C03
Storage and disposal E26, E30a [1-4]; E31b [2-4]; E31c, E32d [1]; E34 [4]
Medical advice M03 [4]

431 triazamate

A carbamoyl triazole insecticide for apples and specified field crops

Products

Aztec	BASF	140 g/l	EW	10211

Uses

Aphids in ***brussels sprouts***, ***sugar beet*** *(including Myzus persicae)*. Green aphid in ***apples***. Pea aphid in ***combining peas***, ***vining peas***. Rosy apple aphid in ***apples***.

Efficacy guidance

- Treat as soon as aphids seen in crop or (on sugar beet) immediately after official warnings are issued. Treat vining peas when 15% of plants infested and combining peas when 20% of plants infested
- Products are systemic and act through contact and stomach action. Ensure foliage well covered with spray by using higher volumes in dense crops
- Controls aphids resistant to other chemical groups
- To reduce risk of development of resistance consider use of products with alternative modes of action in intensive pest control programmes. If tank mixtures used, do not reduce dose of either component

Restrictions

- Maximum number of treatments on sugar beet 3 per crop

- Maximum total dose on brassicas equivalent to three full dose treatments and on apples to two full dose treatments
- Product must be used with recommended adjuvant, except on apples. Consult manufacturer
- Consult processor before use on crops for processing

Crop-specific information
- HI apples, Brussels sprouts, sugar beet 28 d; peas 21 d
- Label warns that damage may result unless all recommendations carefully followed
- Treatment may cause increased russetting of apples

Environmental safety
- Dangerous for the environment
- Very toxic to aquatic organisms. May cause long-term adverse effects in the aquatic environment
- Product not harmful to bees when used as directed but spraying in late evening/early morning or in dull weather recommended to avoid unnecessary stress on foraging bees. Do not apply to apples during flowering
- Do not store above 40°C

Hazard classification and safety precautions

Hazard R03, R04, R14
Risk phrases R03b, R03c, R14c
Operator protection A, H; U02, U05a, U09a, U13, U14, U19, U20b
Environmental protection E01, E38
Consumer protection C03
Storage and disposal E26, E30a, E32a, E32d
Medical advice M03, M05a

432 triazoxide

A benzotriazine fungicide available only in mixtures

See also *imidacloprid + tebuconazole + triazoxide*
tebuconazole + triazoxide

433 tribenuron-methyl

A foliar acting sulfonylurea herbicide with some root activity for use in cereals

See also *thifensulfuron-methyl + tribenuron-methyl*

Products

1	Landgold Tribenuron 75	Landgold	75% w/w	WG	11203
2	Quantum	DuPont	50% w/w	TB	06270
3	Quantum 75 DF	DuPont	75% w/w	WG	09340

Uses

Annual dicotyledons in ***durum wheat, spring barley, spring oats, spring wheat, triticale, winter barley, winter oats, winter rye, winter wheat***. Charlock in ***durum wheat, spring barley, spring oats, spring wheat, triticale, winter barley, winter oats, winter rye, winter wheat***. Chickweed in ***durum wheat, spring barley, spring oats, spring wheat, triticale, winter barley, winter oats, winter rye, winter wheat***. Mayweeds in ***durum wheat, spring barley, spring oats, spring wheat, triticale, winter barley, winter oats, winter rye, winter wheat***.

Efficacy guidance
- Best control achieved when weeds small and actively growing
- Good spray cover must be achieved since larger weeds often become less susceptible
- Susceptible weeds cease growth almost immediately after treatment and symptoms can be seen in about 2 wk
- Products can be used on all soil types
- Weed control may be reduced when conditions very dry
- See label for details of technique to be used for dissolving tablets in spray tank
- When tank mixing ensure product fully dispersed before adding other products

- Tribenuron-methyl is a member of the ALS-inhibitor group of herbicides and products should be used in a planned Resistance Management strategy. See Section 1 for more information

Restrictions

- Maximum number of treatments 1 per crop
- Do not spray in tank mixture, or in sequence, with a product containing any other sulfonylurea except as directed on label
- Do not apply to crops undersown with grass, clover or other broad-leaved crops
- Do not apply to any crop suffering stress from any cause or not actively growing
- Do not apply within 7 d of rolling

Crop-specific information

- Latest use: up to and including flag leaf ligule/collar just visible (GS 39)
- Apply in autumn or in spring (after 1 Feb) from 3 leaf stage of crop up to and including flag leaf fully emerged (GS 13-39)

Following crops guidance

- In the event of crop failure sow only a cereal within 3 mth of application. See label for other restrictions on subsequent cropping

Environmental safety

- Extremely dangerous to fish or other aquatic life. Do not contaminate surface waters or ditches with chemical or used container [3]
- Special care must be taken to avoid damage by drift onto nearby broad-leaved crops, surface waters or ditches
- Spraying equipment should not be drained or flushed onto land planted, or to be planted, with trees or crops other than cereals and should be thoroughly cleansed after use - see label for instructions

Hazard classification and safety precautions

Hazard R04
Risk phrases R04e
Operator protection A; U05a, U08 [1-3]; U19 [1]; U20a [2]; U20b [1, 3]
Environmental protection E01 [1-3]; E13a [1, 3]; E15 [2]
Consumer protection C03
Storage and disposal E30a, E32a

434 triclopyr

An aryloxyalkanoic acid herbicide for perennial and woody weed control

See also 2,4-D + dicamba + triclopyr
clopyralid + fluroxypyr + triclopyr
clopyralid + triclopyr
fluroxypyr + triclopyr

Products

1	Agriguard Triclopyr	AgriGuard	480 g/l	EC	10679
2	Garlon 2	Syngenta	240 g/l	EC	10672
3	Garlon 4	Dow	480 g/l	EC	05090
4	n2n Garlon 4	n2n	480 g/l	EC	11806
5	Timbrel	Dow	480 g/l	EC	05815

Uses

Brambles in ***established grassland, land not intended for cropping, non-crop areas*** *(directed treatment)* [2]; ***forestry, land not intended to bear vegetation*** [1, 3-5]. Broom in ***established grassland, land not intended for cropping, non-crop areas*** *(directed treatment)* [2]; ***forestry, land not intended to bear vegetation*** [1, 3-5]. Brush clearance in ***industrial sites*** [3-5]. Docks in ***established grassland, land not intended for cropping, non-crop areas*** [2]; ***forestry, land not intended to bear vegetation*** [1, 3-5]. Gorse in ***established grassland, land not intended for cropping, non-crop areas*** *(directed treatment)* [2]; ***forestry, land not intended to bear vegetation*** [1, 3-5]. Hard rush in ***established grassland, land not intended for cropping, non-crop areas*** [2]. Perennial dicotyledons in ***established grassland, land not intended for cropping, non-crop areas*** [2]; ***forestry, land not intended to bear vegetation*** [1, 3-5];

industrial sites [3-5]. Rhododendrons in ***forestry, land not intended to bear vegetation*** [1, 3-5]. Scrub clearance in ***industrial sites*** [3-5]; ***land not intended for cropping, non-crop areas*** *(directed treatment)* [2]. Stinging nettle in ***established grassland, land not intended for cropping, non-crop areas*** [2]; ***forestry, land not intended to bear vegetation*** [1, 3-5]. Woody weeds in ***established grassland, land not intended for cropping, non-crop areas*** *(directed treatment)* [2]; ***forestry, land not intended to bear vegetation*** [1, 3-5]; ***industrial sites*** [3-5].

Efficacy guidance

- Apply in grassland as spot treatment or overall foliage spray when weeds in active growth in spring or summer. Details of dose and timing vary with species. See label
- Apply to woody weeds as summer foliage, winter shoot, basal bark, cut stump or tree injection treatment
- Apply foliage spray in water when leaves fully expanded but not senescent
- Do not spray in drought, in very hot or cold conditions
- Control may be reduced if rain falls within 2 h of application
- Control of rhododendron can be variable. If higher than 1.8 m cut stump treatment recommended. A follow-up shoot treatment may be required
- See label for maximum concentrations when applying in oil, water or via watering can

Restrictions

- Maximum number of treatments 1 per yr on non-crop land (as directed spray); 2 per yr on established grassland (including land not intended for cropping) and forestry
- Do not apply to grass leys less than 1 yr old
- Do not drill kale, swedes, turnips, grass or mixtures containing clover within 6 wk of treatment. Allow at least 6 wk before planting trees
- Not to be used on food crops
- Do not apply through hand held rotary atomisers
- Not to be applied in or near water

Crop-specific information

- Latest use: 6 wk before replanting; 7 d before grazing
- HI grass 6 wk
- Uses on land not intended for cropping include grassland of no agricultural interest such as roadside verges, railway and motorway embankments
- Apply winter shoot, basal bark or cut stump sprays in paraffin or diesel oil. Dose and timing vary with species. See label for details
- Inject undiluted or 1:1 dilution into cuts spaced every 7.5 cm round trunk
- Clover will be killed or severely checked by application in grassland

Environmental safety

- Flammable [2]
- Dangerous to fish or other aquatic life. Do not contaminate surface waters or ditches with chemical or used container
- LERAP Category B
- Do not allow spray to drift onto agricultural or horticultural crops, amenity plantings, gardens, ponds, lakes or water courses. Vapour drift may occur under hot conditions
- Keep livestock out of treated areas for at least 7 d and until foliage of any poisonous weeds such as buttercups or ragwort has died and become unpalatable

Hazard classification and safety precautions

Hazard R03, R04 [1-5]; R07d [2]
Risk phrases R03a, R03c [1, 3-5]; R04b, R04e [1-5]; R04g [2, 3, 5]
Operator protection A, C, M [1-5]; H [2-5]; U02, U05a, U08, U20b [1-5]; U19, U23b [2]
Environmental protection E01, E13b, E16a, E16b, E23; E07 (7 d)
Consumer protection C01 [1, 3-5]; C03 [1-5]
Storage and disposal E26 [2]; E30a, E31b, E34 [1-5]
Medical advice M05 [2, 3, 5]

435 trifloxystrobin

A protectant strobilurin fungicide for cereals

See also cyproconazole + trifloxystrobin

Products

1	Me2 Buy	Me2	125 g/l	EC	10655
2	Swift SC	Bayer	500 g/l	SC	10884
3	Swift SC	Bayer CropScience	500 g/l	SC	11227
4	Twist	Bayer	125 g/l	EC	10587
5	Twist	Bayer CropScience	125 g/l	EC	11230

Uses

Brown rust in ***spring barley, winter barley, winter wheat*** [1-5]. Net blotch in ***spring barley, winter barley*** [1-5]. Powdery mildew in ***spring barley, winter barley*** [1, 2, 4]. Rhynchosporium in ***spring barley, winter barley*** [1-5]. Septoria diseases in ***winter wheat*** [1-5].

Approval information

- Trifloxystrobin included in Annex I under EC Directive 91/414
- Approval expiry 13 Aug 2004 [2, 4]

Efficacy guidance

- Should be used protectively before disease is established in crop. Further treatment may be necessary if disease attack prolonged
- Trifloxystrobin is a member of the QoI cross resistance group. Product should be used preventatively and not relied on for its curative potential
- Use product as part of an Integrated Crop Management strategy incorporating other methods of control, including where appropriate other fungicides with a different mode of action. Do not apply more than two foliar applications of QoI containing products to any cereal crop
- There is a significant risk of widespread resistance occurring in *Septoria tritici* populations in UK. Failure to follow resistance management action may result in reduced levels of disease control
- On cereal crops product must always be used in mixture with another product, recommended for control of the same target disease, that contains a fungicide from a different cross resistance group and is applied at a dose that will give robust control
- Strains of barley powdery mildew resistant to QoI's are common in the UK

Restrictions

- Maximum number of treatments 2 per crop per yr

Crop-specific information

- HI barley, wheat 35 d

Environmental safety

- Do not harvest for human or animal consumption for at least 35 days after last application
- Dangerous to fish or other aquatic life. Do not contaminate surface waters or ditches with chemical or used container
- LERAP Category B

Hazard classification and safety precautions

Hazard R04

Risk phrases R04a, R04e

Operator protection A, H [1-5]; C [1, 4, 5]; U02 [1, 4, 5]; U05a, U09a, U19, U20b [1-5]

Environmental protection E01, E13b, E16a, E16b

Consumer protection C02 [1] (35 d); C02 [2, 3]; C02 [4, 5] (35 d); C03 [1-5]

Storage and disposal E26, E30a [1-5]; E29, E31b [1, 4, 5]; E31c [2, 3]

436 trifluralin

A soil-incorporated dinitroaniline herbicide for use in various crops

See also *clodinafop-propargyl + trifluralin*
diflufenican + trifluralin
isoxaben + trifluralin

Products

1	Alpha Trifluralin 48 EC	Makhteshim	480 g/l	EC	07406
2	Tandril 48	DAPT	480 g/l	EC	10643
3	Treflan	Dow	480 g/l	EC	05817
4	Trimaran	Nufarm UK	480 g/l	EC	11400
5	Tripart Trifluralin 48 EC	Tripart	480 g/l	EC	02215

Uses

Annual dicotyledons in ***angelica*** *(off-label)*, ***basil*** *(off-label)*, ***borage*** *(off-label)*, ***camomile*** *(off-label)*, ***celeriac*** *(off-label)*, ***chervil*** *(off-label)*, ***chives*** *(off-label)*, ***combining peas*** *(off-label)*, ***coriander*** *(off-label)*, ***dill*** *(off-label)*, ***evening primrose*** *(off-label)*, ***fennel*** *(off-label)*, ***fenugreek*** *(off-label)*, ***feverfew*** *(off-label)*, ***gold-of-pleasure*** *(off-label)*, ***hyssop*** *(off-label)*, ***kohlrabi*** *(off-label)*, ***lavender*** *(off-label)*, ***lemon barley*** *(off-label)*, ***linseed*** *(off-label)*, ***lovage*** *(off-label)*, ***marjoram*** *(off-label)*, ***mint*** *(off-label)*, ***nursery fruit trees and bushes*** *(off-label)*, ***oregano*** *(off-label)*, ***ornamental specimens*** *(off-label)*, ***parsley*** *(off-label)*, ***radish seed crops*** *(off-label)*, ***rocket*** *(off-label)*, ***rosemary*** *(off-label)*, ***sage*** *(off-label)*, ***savory*** *(off-label)*, ***soya beans*** *(off-label)*, ***sunflowers*** *(off-label)*, ***sweet bay*** *(off-label)*, ***tarragon*** *(off-label)*, ***thyme*** *(off-label)* [1, 3]; ***broad beans***, ***broccoli***, ***brussels sprouts***, ***cabbages***, ***carrots***, ***cauliflowers***, ***french beans***, ***kale***, ***lettuce***, ***parsnips***, ***raspberries***, ***runner beans***, ***strawberries***, ***swedes***, ***turnips***, ***winter barley***, ***winter wheat*** [1-5]; ***calabrese***, ***navy beans*** [3-5]; ***linseed*** [1]; ***mustard***, ***spring oilseed rape***, ***winter oilseed rape*** [1-3, 5]; ***ornamental plant production*** *(off-label)* [3]; ***parsley***, ***sugar beet*** [3, 5]; ***spring field beans***, ***winter field beans*** [1, 2]; ***spring linseed***, ***winter linseed*** [3, 4]. Annual grasses in ***angelica*** *(off-label)*, ***basil*** *(off-label)*, ***borage*** *(off-label)*, ***camomile*** *(off-label)*, ***celeriac*** *(off-label)*, ***chervil*** *(off-label)*, ***chives*** *(off-label)*, ***combining peas*** *(off-label)*, ***coriander*** *(off-label)*, ***dill*** *(off-label)*, ***evening primrose*** *(off-label)*, ***fennel*** *(off-label)*, ***fenugreek*** *(off-label)*, ***feverfew*** *(off-label)*, ***gold-of-pleasure*** *(off-label)*, ***hyssop*** *(off-label)*, ***kohlrabi*** *(off-label)*, ***lavender*** *(off-label)*, ***lemon barley*** *(off-label)*, ***linseed*** *(off-label)*, ***lovage*** *(off-label)*, ***marjoram*** *(off-label)*, ***mint*** *(off-label)*, ***nursery fruit trees and bushes*** *(off-label)*, ***oregano*** *(off-label)*, ***ornamental specimens*** *(off-label)*, ***parsley*** *(off-label)*, ***radish seed crops*** *(off-label)*, ***rocket*** *(off-label)*, ***rosemary*** *(off-label)*, ***sage*** *(off-label)*, ***savory*** *(off-label)*, ***soya beans*** *(off-label)*, ***sunflowers*** *(off-label)*, ***sweet bay*** *(off-label)*, ***tarragon*** *(off-label)*, ***thyme*** *(off-label)* [1, 3]; ***broad beans***, ***broccoli***, ***brussels sprouts***, ***cabbages***, ***carrots***, ***cauliflowers***, ***french beans***, ***kale***, ***lettuce***, ***parsnips***, ***raspberries***, ***runner beans***, ***strawberries***, ***swedes***, ***turnips***, ***winter barley***, ***winter wheat*** [1-5]; ***calabrese***, ***navy beans*** [3-5]; ***linseed*** [1]; ***mustard***, ***spring oilseed rape***, ***winter oilseed rape*** [1-3, 5]; ***ornamental plant production*** *(off-label)* [3]; ***parsley***, ***sugar beet*** [3, 5]; ***spring field beans***, ***winter field beans*** [1, 2]; ***spring linseed***, ***winter linseed*** [3, 4]. Fat hen in ***angelica*** *(off-label)*, ***basil*** *(off-label)*, ***borage*** *(off-label)*, ***camomile*** *(off-label)*, ***celeriac*** *(off-label)*, ***chervil*** *(off-label)*, ***chives*** *(off-label)*, ***combining peas*** *(off-label)*, ***coriander*** *(off-label)*, ***dill*** *(off-label)*, ***evening primrose*** *(off-label)*, ***fennel*** *(off-label)*, ***fenugreek*** *(off-label)*, ***feverfew*** *(off-label)*, ***gold-of-pleasure*** *(off-label)*, ***hyssop*** *(off-label)*, ***kohlrabi*** *(off-label)*, ***lavender*** *(off-label)*, ***lemon barley*** *(off-label)*, ***linseed*** *(off-label)*, ***lovage*** *(off-label)*, ***marjoram*** *(off-label)*, ***mint*** *(off-label)*, ***nursery fruit trees and bushes*** *(off-label)*, ***oregano*** *(off-label)*, ***ornamental specimens*** *(off-label)*, ***parsley*** *(off-label)*, ***radish seed crops*** *(off-label)*, ***rocket*** *(off-label)*, ***rosemary*** *(off-label)*, ***sage*** *(off-label)*, ***savory*** *(off-label)*, ***soya beans*** *(off-label)*, ***sunflowers*** *(off-label)*, ***sweet bay*** *(off-label)*, ***tarragon*** *(off-label)*, ***thyme*** *(off-label)* [1].

Specific Off-Label Approvals (SOLAs)

- ***angelica***, ***basil***, ***borage***, ***camomile***, ***celeriac***, ***chervil***, ***chives***, ***combining peas***, ***coriander***, ***dill***, ***evening primrose***, ***fennel***, ***fenugreek***, ***feverfew***, ***gold-of-pleasure***, ***hyssop***, ***kohlrabi***, ***lavender***, ***lemon barley***, ***linseed***, ***lovage***, ***marjoram***, ***mint***, ***nursery fruit trees and bushes***, ***oregano***, ***ornamental specimens***, ***parsley***, ***radish seed crops***, ***rocket***, ***rosemary***, ***sage***, ***savory***, ***soya beans***, ***sunflowers***, ***sweet bay***, ***tarragon***, ***thyme*** *(OLA 012763) Dec 2008* [1]
- ***angelica***, ***basil***, ***borage***, ***camomile***, ***chervil***, ***chives***, ***coriander***, ***dill***, ***fennel***, ***fenugreek***, ***feverfew***, ***gold-of-pleasure***, ***hyssop***, ***lavender***, ***lemon barley***, ***lovage***, ***marjoram***, ***mint***,

***nursery fruit trees and bushes**, **oregano**, **ornamental plant production**, **parsley**, **radish seed crops**, **rocket**, **rosemary**, **sage**, **savory**, **soya beans**, **sweet bay**, **tarragon**, **thyme** (OLA 930074) Dec 2008* [3]

- ***celeriac**, **combining peas**, **linseed**, **ornamental specimens**, **sunflowers** (OLA 981560) Dec 2008* [3]
- ***evening primrose** (OLA 921080) Dec 2008* [3]
- ***kohlrabi** (OLA 982341) Dec 2008* [3]

Efficacy guidance

- Acts on germinating weeds and requires soil incorporation to 5 cm (10 cm for crops to be grown on ridges) within 30 min of spraying (2 h [5], except in mixtures on cereals). See label for details of suitable application equipment
- Apply and incorporate at any time during 2 wk before sowing or planting
- Best results achieved by application to fine, firm seedbed, free of clods, crop residues and established weeds
- In winter cereals normally applied as surface treatment without incorporation in tank-mixture with other herbicides to increase spectrum of control. See label for details
- Follow-up herbicide treatment recommended with some crops. See label for details

Restrictions

- Maximum number of treatments 1 per crop or per yr
- Do not apply to brassica plant raising beds
- Minimum interval between application and drilling or planting may be up to 12 mth. See label for details
- Do not use on sand, fen soil or soils with more than 10% organic matter

Crop-specific information

- Latest use: varies between products; check labels. Normally before 4 leaves unfolded (GS 13) for cereals; up to 6, 8 or 10 leaves for sugar beet; pre-sowing/planting for other crops
- Transplants should be hardened off prior to transplanting
- Apply in sugar beet after plants 10 cm high with 4-8 leaves and harrow into soil
- Apply in cereals after drilling up to and including 3 leaf stage (GS 13)

Environmental safety

- Flammable
- Harmful to fish or other aquatic life. Do not contaminate surface waters or ditches with chemical or used container

Hazard classification and safety precautions

Hazard R03 [2, 4]; R04, R07d [1-5]
Risk phrases R04a, R04b [1, 2, 5]; R04g [2-4]
Operator protection A, C [1, 2]; U05a [1, 2, 5]; U08, U13 [1-5]; U20a [5]; U20b [1-4]
Environmental protection E01 [1, 2, 5]; E13c [1-5]
Consumer protection C03 [1, 2, 5]
Storage and disposal E26, E30a, E31b [1-5]; E27 [1-4]; E28 [3]; E29 [1, 2]; E34 [4]
Medical advice M05 [2-4]

437 triflusulfuron-methyl

A sulfonyl urea herbicide for sugar beet

Products

1	Debut	DuPont	50% w/w	WG	07804
2	Landgold TFS 50	Landgold	50% w/w	WG	08941
3	Standon Triflusulfuron	Standon	50% w/w	WG	09487

Uses

Annual dicotyledons in ***fodder beet*** [1, 3]; ***sugar beet*** [1-3]. Charlock in ***red beet*** *(off-label)* [1]. Cleavers in ***red beet*** *(off-label)* [1]. Flixweed in ***red beet*** *(off-label)* [1]. Fool's parsley in ***red beet*** *(off-label)* [1]. Nipplewort in ***red beet*** *(off-label)* [1]. Ox-eye daisy in ***red beet*** *(off-label)* [1].

Specific Off-Label Approvals (SOLAs)

- ***red beet** (OLA 023629) Dec 2008* [1]

Efficacy guidance

- Product should be used with a recommended adjuvant or a suitable herbicide tank-mix partner - see label for details
- Product acts by foliar action. Best results obtained from good spray cover of small actively growing weeds
- Susceptible weeds cease growth immediately and symptoms can be seen 5-10 d later
- Best results achieved from a programme of up to 4 treatments starting when first weeds have emerged with subsequent applications every 5-14 d when new weed flushes at cotyledon stage
- Weed spectrum can be broadened by tank mixture with other herbicides. See label for details
- Product may be applied overall or via band sprayer
- Triflusulfuron-methyl is a member of the ALS-inhibitor group of herbicides

Restrictions

- Maximum number of treatments 4 per crop
- Do not apply to any crop stressed by drought, water-logging, low temperatures, pest or disease attack, nutrient or lime deficiency

Crop-specific information

- Latest use: before crop leaves meet between rows
- All varieties of sugar beet (and fodder beet [1, 3]) may be treated from early cotyledon stage until the leaves begin to meet between the rows

Following crops guidance

- Only winter cereals should follow a treated crop in the same calendar yr. Any crop may be sown in the next calendar yr
- After failure of a treated crop, sow only spring barley, linseed or sugar beet within 4 mth of spraying unless prohibited by tank-mix partner

Environmental safety

- Irritant. May cause sensitization by skin contact
- Extremely dangerous to fish or other aquatic life. Do not contaminate surface waters or ditches with chemical or used container
- LERAP Category B
- Triflusulfuron-methyl is very active. Take particular care to avoid drift onto plants outside the target area
- Spraying equipment should not be drained or flushed onto land planted, or to be planted, with trees or crops other than sugar beet and should be thoroughly cleansed after use - see label for instructions

Hazard classification and safety precautions

Hazard H04

Risk phrases R04e

Operator protection A; U08, U19 [1-3]; U20a [1, 2]; U20b [3]

Environmental protection E13a, E16a

Storage and disposal E26 [1, 2]; E30a, E32a [1-3]

SECTION 4

438 trinexapac-ethyl

A novel cyclohexanecarboxylate plant growth regulator for cereals, turf and amenity grassland

Products

1	Moddus	Syngenta	250 g/l	EC	08801
2	Shortcut	Scotts	25% w/w	WB	09254

Uses

Growth retardation in ***amenity grass***, ***amenity turf*** [2]. Lodging control in ***durum wheat***, ***ryegrass seed crops***, ***spring barley***, ***spring oats***, ***spring rye***, ***triticale***, ***winter barley***, ***winter oats***, ***winter rye***, ***winter wheat*** [1].

Efficacy guidance

- Best results on turf achieved from application to actively growing weed free turf grass that is adequately fertilized and watered and is not under stress [2]
- Product rainfast after 12 h [2]

Restrictions
- Maximum total dose equivalent to one full dose on cereals, ryegrass seed crops [1]
- Maximum number of treatments on turf 5 per yr [2]
- Do not apply if rain or frost expected or if crop wet
- Only use on crops at risk of lodging [1]
- Turf under stress when treated may show signs of damage. Do not apply within 12 h of mowing [2]
- Not recommended for closely mown fine turf [2]
- Do not treat newly sown turf [2]
- Not to be used on food crops [2]
- Do not compost or mulch clippings [2]

Crop-specific information
- Latest use: before 2nd node detectable (GS 32) for oats, ryegrass seed crops; before 3rd node detectable (GS 33) for durum wheat, spring barley, triticale, rye; before flag leaf sheath extending (GS 41) for winter barley, winter wheat
- On wheat apply as single treatment between leaf sheath erect stage (GS 30) and flag leaf fully emerged (GS 39) [1]
- On barley, rye, triticale and durum wheat apply as single treatment between leaf sheath erect stage (GS 30) and second node detectable (GS 32), or on winter barley at higher dose between flag leaf just visible (GS 37) and flag leaf fully emerged (GS 39) [1]
- On oats and ryegrass seed crops apply between leaf sheath erect stage (GS 30) and first node detectable stage (GS 32) [1]
- Treatment may cause ears to remain erect through to harvest [1]

Environmental safety
- Dangerous for the environment [1]
- Toxic to aquatic organisms. May cause long-term adverse effects to the aquatic environment [1]
- Harmful to fish or other aquatic life. Do not contaminate surface waters or ditches with chemical or used container [1, 2]
- Avoid drift outside target area

Hazard classification and safety precautions

Hazard R04, R14 [1]
Risk phrases R04e, R14b [1]
Operator protection A [1, 2]; C [1]; U05a, U15, U20c [1]; U20b [2]
Environmental protection E01, E38 [1]; E13c [2]
Consumer protection C01 [2]; C03 [1]
Storage and disposal E26, E30a, E31a [1, 2]; E32d [1]

439 triticonazole

A conazole fungicide available only in mixtures

See also *guazatine + triticonazole*
imazalil + triticonazole

440 Verticillium lecanii

A fungal parasite of aphids and whitefly

Products

1	Mycotal	Koppert	16.1% w/w	WP	04782
2	Vertalec	Koppert	2.5% w/w	WP	04781

Uses

Aphids in ***aubergines***, ***protected beans***, ***protected chrysanthemums***, ***protected cucumbers***, ***protected lettuce***, ***protected ornamentals***, ***protected peppers***, ***protected roses***, ***protected tomatoes*** [2]. Whitefly in ***aubergines***, ***protected beans***, ***protected cucumbers***, ***protected lettuce***, ***protected ornamentals***, ***protected peppers***, ***protected tomatoes*** [1].

Efficacy guidance
- Verticillium lecanii is a pathogenic fungus that infects the target pests and destroys them

- Apply spore powder as spray as part of biological control programme keeping the spray liquid well agitated
- Pre-soak the product for 2-4 h before application to rehydrate the spores and assist in dispersion
- Treat before infestations build to high levels and repeat as directed on the label
- Spray during late afternoon and early evening directing spray onto underside of leaves and to growing points
- Best results require minimum 80% relative humidity and 18°C within the crop canopy
- Product highly infective to many aphid species except the chrysanthemum aphid. Follow specific label directions for this pest [2]

Restrictions

- Never use in tank mixture
- Do not use a fungicide within 3 d of treatment. Pesticides containing captan, chlorothalonil, fenarimol, dichlofluanid, imazalil, maneb, prochloraz, quinomethionate, thiram or tolylfluanid may not be used on the same crop
- Keep in a refrigerated store at 2-6°C but do not freeze

Environmental safety

- Products have negligible effects on commercially available natural predators or parasites but consult manufacturer before using with a particular biological control agent for the first time

Hazard classification and safety precautions

Operator protection U19, U20b
Environmental protection E15
Storage and disposal E30a, E32a

441 vinclozolin

A protectant dicarboximide fungicide

See also *carbendazim + vinclozolin*
chlorothalonil + vinclozolin

Products

1	Barclay Flotilla	Barclay	500 g/l	SC	11354
2	Landgold Vinclozolin SC	Landgold	500 g/l	SC	06459
3	Ronilan FL	BASF	500 g/l	SC	02960
4	Standon Vinclozolin	Standon	500 g/l	SC	07836

Uses

Alternaria in ***spring oilseed rape***, ***winter oilseed rape*** [1-4]. Ascochyta in ***combining peas*** [2-4]. Blossom wilt in ***apples*** [1, 3]. Botrytis in ***combining peas***, ***dwarf beans***, ***navy beans***, ***runner beans***, ***spring oilseed rape***, ***vining peas***, ***winter oilseed rape*** [1-4]. Chocolate spot in ***broad beans***, ***spring field beans***, ***winter field beans*** [1-4]. Mycosphaerella in ***combining peas*** [2-4]. Sclerotinia stem rot in ***spring oilseed rape***, ***winter oilseed rape*** [1-4].

Efficacy guidance

- Timing of sprays varies with crop and disease. See label for details
- May be used at reduced rate on peas and field beans in tank-mix with chlorothalonil
- Where dicarboximide resistant strains have developed product may not be effective

Restrictions

- Maximum number of treatments 2 per crop or per yr
- Do not treat mange-tout varieties of peas
- Do not spray if crop wet or if rain or frost expected
- Operator must use a vehicle fitted with a cab and forced air filtration unit with a pesticide filter complying with HSE Guidance Note PM 74 or to an equally effective standard

Crop-specific information

- Before end of petal fall for apples
- HI beans, peas 2 wk; oilseed rape 7 wk

Environmental safety

- Product presents a minimal hazard to bees when used as directed but consider informing local bee-keepers if intending to spray crops in flower
- Harmful to fish and aquatic life. Do not contaminate surface waters or ditches with chemical or used container

Hazard classification and safety precautions

Hazard R04
Risk phrases R04b [2]; R04e [1-4]
Operator protection A, C, H, K [1-4]; M [1-3]; U05a, U19 [1-4]; U08 [1-3]; U09a [4]; U20a [1, 2, 4]; U20b [3]
Environmental protection E01, E13c
Consumer protection C03
Storage and disposal E26 [2, 4]; E30a [1-4]; E31b [1, 2, 4]; E31c [3]; E34 [4]

442 warfarin

A hydroxycoumarin rodenticide

Products

1	Grey Squirrel Liquid Concentrate	Killgerm	0.5% w/w	CB	06455
2	Sakarat Ready-to-Use (Cut Wheat Base)	Killgerm	0.025% w/w	RB	H6807
3	Sakarat Ready-to-Use (Whole Wheat)	Killgerm	0.025% w/w	RB	H6808
4	Sakarat X	Killgerm	0.05% w/w	RB	H6809
5	Sewarin Extra	Killgerm	0.05% w/w	RB	H6810
6	Sewarin P	Killgerm	0.025% w/w	RB	H6811
7	Sewercide Cut Wheat Rat Bait	Killgerm	0.05% w/w	RB	H6805
8	Sewercide Whole Wheat Rat Bait	Killgerm	0.05% w/w	RB	H6806
9	Sorex Warfarin 250 ppm Rat Bait	Sorex	0.025% w/w	AB	H6812
10	Sorex Warfarin 500 ppm Rat Bait	Sorex	0.05% w/w	AB	H6813
11	Sorex Warfarin Sewer Bait	Sorex	0.05% w/w	AB	H6814
12	Warfarin 0.5% Concentrate	B H & B	0.5%	CB	H6815
13	Warfarin Ready Mixed Bait	B H & B	0.025%	RB	H6816

Uses

Grey squirrels in ***agricultural premises***, ***forestry***, ***industrial sites*** [1]. Mice in ***agricultural premises*** [9-11]. Rats in ***agricultural premises*** [2-13].

Approval information

- Product not approved for use in N Ireland [1]

Efficacy guidance

- For rodent control place ready-to-use or prepared baits at many points wherever rats active. Out of doors shelter bait from weather
- Inspect baits frequently and replace or top up as long as evidence of feeding. Do not underbait
- For grey squirrel control mix with whole wheat and leave to stand for 2-3 h before use
- Use bait in specially constructed hoppers and inspect every 2-3 d. Replace as necessary

Restrictions

- For use only by local authorities, professional operators providing a pest control service and persons occupying industrial, agricultural or horticultural premises
- For use only between 15 Mar and 15 Aug for tree protection [1]
- Must not be used outdoors at all in Scotland, nor in areas of England or Wales where pine martens occur naturally [1]

Environmental safety

- Prevent access to baits by children and animals, especially cats, dogs and pigs
- Rodent bodies must be searched for and burned or buried, not placed in refuse bins or rubbish tips. Remains of bait and containers must be removed after treatment and burned or buried

- Bait must not be used where food, feed or water could become contaminated
- The use of warfarin to control grey squirrels is illegal unless the provisions of the Grey Squirrels Order 1973 are observed. See label for list of counties in which bait may not be used [1]

Hazard classification and safety precautions

Operator protection A, C, D, E, H [2-8]; U13 [1-13]; U20a [1-6]; U20b [7-13]

Storage and disposal E30a [1-13]; E31a [12, 13]; E32a [1-11]

Vertebrate/rodent control products V01a, V03a [1, 4, 7-10, 12, 13]; V01b, V03b, V04b [2, 3, 5, 6]; V02 [1-10, 12, 13]; V04a [1, 4, 7-13]; V05 [1]

Medical advice M03 [2, 3, 5, 6]; M05 [9-11]

443 zeta-cypermethrin

A contact and stomach acting pyrethroid insecticide

Products

1	Fury 10 EW	FMC	100 g/l	EW	10718
2	Fury 10 EW	Belchim	100 g/l	EW	11608
3	Minuet EW	FMC	100 g/l	EW	10716
4	Minuet EW	Belchim	100 g/l	EW	11610

Uses

Aphids in ***spring barley, spring oats, spring wheat, winter barley, winter oats, winter wheat*** [1-4]. Barley yellow dwarf virus vectors in ***spring barley, spring wheat, winter barley, winter wheat*** [1-4]. Cabbage stem flea beetle in ***spring oilseed rape, winter oilseed rape*** [1-4]. Cutworms in ***potatoes, sugar beet*** [1-4]. Flax flea beetle in ***linseed*** [2, 4]. Flea beetle in ***spring oilseed rape, winter oilseed rape*** [2, 4]. Large flax flea beetle in ***linseed*** [2, 4]. Pea and bean weevil in ***combining peas, spring field beans, vining peas, winter field beans*** [1-4]. Pea aphid in ***combining peas, vining peas*** [1-4]. Pea moth in ***combining peas, vining peas*** [1-4]. Pod midge in ***spring oilseed rape, winter oilseed rape*** [1-4]. Pollen beetle in ***spring oilseed rape, winter oilseed rape*** [1-4]. Rape winter stem weevil in ***spring oilseed rape, winter oilseed rape*** [1-4]. Seed weevil in ***spring oilseed rape, winter oilseed rape*** [1-4].

Approval information

- Following implementation of Directive 98/82/EC, approval for use of zeta-cypermethrin on numerous crops was revoked in 1999

Efficacy guidance

- On winter cereals spray when aphids first found in the autumn for BYDV control. A second spray may be required on late drilled crops or in mild conditions
- For summer aphids on cereals spray when treatment threshold reached
- For listed pests in other crops spray when feeding damage first seen or when treatment threshold reached. Under high infestation pressure a second treatment may be necessary
- Best results for pod midge and seed weevil control in oilseed rape obtained from treatment after pod set but before 80% petal fall
- Pea moth treatments should be applied according to ADAS/PGRO warnings or when economic thresholds reached as indicated by pheromone traps
- Treatments for cutworms should be made at egg hatch and repeated no sooner than 10 d later

Restrictions

- Maximum number of treatments 2 per crop
- Consult processors before use on crops for processing

Crop-specific information

- Latest use: before end of flowering for oilseed rape; before flowering completed (GS 69) for cereals
- HI potatoes, field beans 14 d; sugar beet 60 d. No information for other crops

Environmental safety

- Extremely dangerous to fish or other aquatic life. Do not contaminate surface waters or ditches with chemical or used container
- High risk to non-target insects or other arthropods. Do not spray within 6 m of the field boundary
- Risk to certain non-target insects or other arthropods. See directions for use
- LERAP Category A

SECTION 4

Hazard classification and safety precautions

Hazard R03, R04
Risk phrases R03c, R04e
Operator protection A, C, H; U05a, U08, U14, U15, U19, U20b
Environmental protection E01, E13a, E16c, E16d, E22a, E22b
Consumer protection C03
Storage and disposal E30a, E31b, E34
Medical advice M03

444 zinc phosphide

A phosphine generating rodenticide

Products

1	Grovex Zinc Phosphide	Killgerm	100%	CB	H6800
2	RCR Zinc Phosphide	Killgerm	100%	CB	H6801
3	ZP Rodent Pellets	Antec	2.0% w/w	GB	07814

Uses

Mice in ***agricultural premises*** [1, 2]; ***farm buildings*** [3]. Rats in ***agricultural premises*** [1, 2]; ***farm buildings*** [3].

Efficacy guidance

- Use to prepare baits either by dry or wet baiting as directed

Restrictions

- Zinc phosphide is subject to the Poisons Rules 1982 and the Poisons Act 1972. See notes in Section 1
- For use only by local authorities, professional operators providing a pest control service and persons occupying industrial, agricultural or horticultural premises

Environmental safety

- Spontaneously inflammable in contact with acid
- Keep in original container, tightly closed, in a safe place, under lock and key
- Wear respirator if mixing baits in a confined space
- Prevent access to baits by children and domestic animals, especially cats, dogs and pigs
- Search for and burn or bury all rodent bodies. Do not place in refuse bins or on rubbish tips
- Do not prepare or use baits where food or water could be contaminated
- Remove all remains of baits and bait containers after treatment and burn or bury
- Wash out all mixing equipment thoroughly at the end of every operation

Hazard classification and safety precautions

Hazard R02 [1, 2]
Risk phrases R02c [1, 2]
Operator protection A [1-3]; D, H [1, 2]; U02, U04a, U05a [1, 2]; U13, U20a [1-3]
Environmental protection E01 [1, 2]; E15 [3]
Consumer protection C03 [1, 2]
Storage and disposal E30b, E32a [1-3]; E34 [1, 2]
Vertebrate/rodent control products V01a, V02, V04a [1-3]; V03a [1, 2]; V03b [3]
Medical advice M03 [3]

445 ziram

A dithiocarbamate bird and animal repellent

Products

AAprotect	Unicrop	32% w/w	PA	03784

Uses

Birds in ***all top fruit***, ***field crops***, ***forestry***, ***ornamental specimens***. Deer in ***all top fruit***, ***field crops***, ***forestry***, ***ornamental specimens***. Hares in ***all top fruit***, ***field crops***, ***forestry***, ***ornamental specimens***. Rabbits in ***all top fruit***, ***field crops***, ***forestry***, ***ornamental specimens***.

Efficacy guidance

- Apply undiluted to main stems up to knee height to protect against browsing animals at any time of yr or spray 1:1 dilution on stems and branches in dormant season
- Use dilute spray on fully dormant fruit buds to protect against bullfinches
- Only apply to dry stems, branches or buds
- Use of diluted spray can give limited protection to field crops in areas of high risk during establishment period

Restrictions

- Do not spray elongating shoots or buds about to open
- Do not apply concentrated spray to foliage, fruit buds or field crops

Crop-specific information

- HI edible crops 8 wk

Hazard classification and safety precautions

Hazard R04
Risk phrases R04a, R04b, R04c
Operator protection A, C; U05a, U08, U19, U20b
Environmental protection E01, E13c
Consumer protection C02 (8 wk); C03
Storage and disposal E30a, E31c

446 zoxamide

A substituted benzamide available only in mixtures

See also *mancozeb + zoxamide*

SECTION 5
PRODUCTS ALSO REGISTERED

Products also Registered

Products listed in the table below have not been notified for inclusion in Section 4 of this edition of the *Guide*. However they have extant approval until 31 December 2008 unless an earlier expiry date is shown. These products may legally be used in accordance with their label until their approval expires, but they may not still be available for purchase.

Product	Marketing Co.	Reg. No.	Expiry Date
acibenzolar-S-methyl			
Bion	Syngenta	09803	
aldicarb			
Aventis Temik 10G	Aventis	09749	
Marnoch Carbadil 10	Marnoch	10338	31-12-07
alpha-cypermethrin			
Alert	BASF	10272	
Alphaguard 100 EC	Gharda	10772	
Alphathrin	Nufarm UK	10025	31-03-04
Bestseller 100 EC	Chimac-Agriphar	10461	
Cleancrop Acymet	United Agri	10497	
Cleancrop Scope	United Agri	10273	
Contest	Cyanamid	09024	
Fastac	Cyanamid	07008	
Fastac Dry	BASF	09025	29-02-04
Fastac Dry	BASF	10221	
I T Alpha-cypermethrin	I T Agro	08274	
Standon Alpha-C10	Standon	08823	31-03-04
aluminium ammonium sulphate			
Curb Crop Spray Powder	Sphere	02480	
Rezist	Barrettine	08576	
aluminium phosphide			
Detia Gas Ex-P	Igrox	09802	
Detia Gas Ex-B	Detia Degesch	06927	
Detia Gas-Ex-T	Igrox	03792	
amidosulfuron			
Aventis Eagle	Aventis	09765	
Barclay Cleave	Barclay	09480	
Pursuit 50	Bayer CropScience	08716	
2-aminobutane			
Hortichem 2-Aminobutane	Certis	06147	31-12-07
amitraz			
Bye Bye 20 EC	Chimac-Agriphar	09346	
Cleancrop Bye Bye 20EC	United Agri	11530	

SECTION 5

Product	Marketing Co.	Reg. No.	Expiry Date
Ovasyn	Bayer CropScience	09190	
amitrole			
Aminotriazole Technical	Nufarm UK	10855	
Loft	Marks	06030	
MSS Aminotriazole Technical	Mirfield	04645	
Weedazol-TL	Marks	02349	
amitrole + 2,4-D + diuron			
Trik	Mirfield	07853	31-10-05
amitrole + simazine			
Alpha Simazol	Makhteshim	04799	30-06-04
asulam			
Asulox	RP Agric.	06124	
Cleancrop Asulam	United Agri	10465	
Inter Asulam	I T Agro	10929	
atrazine			
Alpha Atrazine 50 WP	Makhteshim	04793	
Atlas Atrazine	Nufarm UK	07702	
Atrazine 90 WG	Sipcam	09310	31-08-05
DAPT Atrazine 50 SC	DAPT	08031	
MSS Atrazine 50 FL	Nufarm UK	01398	
MSS Atrazine 80 WP	Mirfield	04360	
Unicrop Atrazine 50	Unicrop	02645	
Unicrop Atrazine FL	Unicrop	08045	
Unicrop Flowable Atrazine	Unicrop	05446	
azoxystrobin			
Amistar	Zeneca	08517	
Barclay ZX	Barclay	09570	
Clayton Stobik	Clayton	09440	01-07-08
Marnoch Binoxy	Marnoch	10622	01-07-08
Olympus	Syngenta	08520	01-10-05
Olympus	Syngenta	10541	01-07-08
Ortiva	Syngenta	09843	02-10-05
Ortiva	Syngenta	10542	01-07-08
Priori	Syngenta	08516	01-07-08
Priori	Syngenta	10543	01-07-08
azoxystrobin + fenpropimorph			
Amistar Pro	Zeneca	08871	
Aspect	Syngenta	08872	31-05-04
Aspect	Syngenta	10516	
azoxystrobin + flutriafol			
Amigo	Syngenta	08834	31-05-04
Amigo	Syngenta	10511	30-06-04

Product	Marketing Co.	Reg. No.	Expiry Date
Amistar Gem	Syngenta	08833	31-05-04
Amistar Gem	Syngenta	10512	30-06-04
benalaxyl + mancozeb			
Galben M	Sipcam	05092	
bentazone			
Basagran	BASF	00188	
I T Bentazone	I T Agro	08677	
I T Bentazone 48	I T Agro	09283	
bentazone + dichlorprop-P			
Quitt SL	BASF	08108	
bentazone + MCPA + MCPB			
Headland Archer	Headland	08814	
beta-cyfluthrin + imidacloprid			
Chinook	Bayer CropScience	11421	
Chinook	Bayer	11151	
Chinook	Bayer	10660	
Chinook Blue	Bayer	10912	
bifenox + chlorotoluron			
Dicurane Duo 446 SC	Syngenta	08404	
bifenox + isoproturon			
Banco	Feinchemie	10606	31-08-05
Banco	Portman	09027	30-04-04
Banco	Makhteshim	11191	
RP 4169	Bayer CropScience	05801	30-04-04
RP 4169	Feinchemie	10605	31-08-05
RP 4169	Makhteshim	11190	
bifenthrin			
Starion	FMC	10724	31-10-04
Talstar	FMC	10717	31-10-04
Talstar	FMC	06013	31-10-04
bitertanol + fuberidazole			
Sibutol CF	Bayer CropScience	11306	
Sibutol CF	Bayer	08174	31-05-05
Sibutol New Formula	Bayer CropScience	11307	
Sibutol New Formula	Bayer	08270	31-05-05
UK 743	Bayer CropScience	11315	
UK 743	Bayer	10707	
bitertanol + fuberidazole + imidacloprid + triadimenol			
Cereline Secur	Bayer CropScience	11261	
Cereline Secur	Bayer	09511	31-10-05

SECTION 5

Product	Marketing Co.	Reg. No.	Expiry Date
bitertanol + fuberidazole + triadimenol			
Cereline	Bayer CropScience	11260	
Cereline	Bayer	07239	31-10-05
bromoxynil			
Barclay Mutiny	Barclay	11136	
bromoxynil + ethofumesate + ioxynil			
Stefes Leyclene	Bayer CropScience	10086	31-10-05
bromoxynil + fluroxypyr			
Sickle	Dow	05187	
Tomahawk Plus	Makhteshim	09836	
bromoxynil + fluroxypyr + ioxynil			
Advance	Dow	05173	
bromoxynil + ioxynil			
Deloxil	Bayer CropScience	07405	
Deloxil	Bayer CropScience	09987	31-10-05
Deloxil	Bayer CropScience	07313	
Oxytril CM	RP Agric.	08667	
Percept	United Phosphorus	05481	
Status	Bayer CropScience	08668	31-10-05
Status	Bayer CropScience	10019	
Stellox	Nufarm UK	11551	
Stellox 380 EC	Syngenta	08451	
bromoxynil + ioxynil + trifluralin			
Masterspray	SumiAgro	02971	
Masterspray	SumiAgro	09603	
bromuconazole			
Granit	RP Agric.	08268	
bupirimate			
Nimrod	Syngenta	06686	31-03-04
Nimrod	Syngenta	10530	
buprofezin			
Applaud	Certis	11532	
Applaud	Syngenta	06900	31-07-04
Applaud	Syngenta	10678	
captan			
Alpha Captan 50 WP	Makhteshim	04797	
Alpha Captan 83 WP	Makhteshim	04806	
PP Captan 80 WG	Tomen	08971	28-02-05
PP Captan 83	Tomen	08768	

Product	Marketing Co.	Reg. No.	Expiry Date
carbendazim			
Ashlade Carbendazim Flowable	Nufarm UK	06213	
Barclay Shelter	Barclay	09000	
Bavistin	BASF	00217	
Carbate Flowable	SumiAgro	08957	
Clayton Chizm FL	Clayton	11050	
Cleancrop Curve	United Agri	11774	
Delsene 50 Flo	Nufarm UK	11452	
Derosal Liquid	Bayer CropScience	07315	
Derosal WDG	Aventis	07316	
Goldazim 500 SC	Chimac-Agriphar	10675	
Headland Addstem	Headland	06755	
Headland Addstem DF	Headland	08904	
Mascot Systemic	Rigby Taylor	11335	
Mascot Systemic	Scotts	09132	
MSS Mircarb	Mirfield	08788	
Occidor 500 SC	Chimac-Agriphar	10937	
Quadrangle Hinge	Quadrangle	08070	
Stefes Derosal Liquid	Bayer CropScience	10079	31-10-05
Stefes Derosal WDG	Bayer CropScience	10078	31-10-05
Supercarb	SumiAgro	09610	
Supercarb	SumiAgro	01560	
Top Farm Carbendazim 435	Top Farm	05307	
Turf Systemic Fungicide	SumiAgro	10821	
Turfclear	Scotts	07506	
Turfclear WDG	Scotts	07490	
carbendazim + chlorothalonil			
Bravocarb	Syngenta	10520	
Bravocarb	Zeneca	09105	
Greenshield	Scotts	07988	
carbendazim + iprodione			
Calidan	Aventis	09980	
Calidan	BASF	11687	
Calidan	RP Agric.	06536	
Vitesse	RP Amenity	06537	
carbendazim + mancozeb			
Headland Maple	Headland	09044	31-01-04
carbendazim + maneb			
Multi-W FL	SumiAgro	09447	
Protector	Procam	09448	
Tripart 147	Tripart	07978	
carbendazim + prochloraz			
Sportak Alpha HF	Aventis	07225	

SECTION 5

Product	Marketing Co.	Reg. No.	Expiry Date
carbendazim + tebuconazole			
Bayer UK 413	Bayer CropScience	11243	
Bayer UK 413	Bayer	08277	
Tricur	Bayer CropScience	11314	
Tricur	Bayer	10377	
carbetamide			
Carbetamex	Feinchemie	10265	
carbetamide + diflufenican + oxadiazon			
Buffalo G	Bayer Environ.	11088	
carbosulfan			
Marshal 10G	Belchim	11609	31-10-04
Marshal 10G	FMC	10722	31-10-04
Marshal 10G	FMC	09804	
Posse 10G	Belchim	11612	30-11-04
Posse 10G	FMC	09805	30-11-04
carfentrazone-ethyl			
Aurora	FMC	10719	18-02-04
Aurora 40 WG	Belchim	11614	
Aurora 40 WG	FMC	11132	18-02-04
Platform	Belchim	11615	
Platform	FMC	09661	18-02-04
Platform	FMC	10721	18-02-04
carfentrazone-ethyl + flupyrsulfuron-methyl			
Lexus Class	DuPont	08636	17-02-04
Lexus Class	DuPont	10809	17-02-04
Lexus Class WSB	DuPont	10810	13-05-04
carfentrazone-ethyl + isoproturon			
Affinity	FMC	10723	
Affinity	FMC	09745	
carfentrazone-ethyl + mecoprop-P			
Platform S	FMC	10726	
chloridazon			
Lidazone 65 WG	Globachem	10847	
Luxan Booster WG	Luxan	11709	
Luxan Chloridazon	Luxan	06304	
PG Chloridazon 430 SC	PG-Crop Protection	11517	
Portman Weedmaster	Portman	06018	
Pyramin FL	BASF	01661	
Questar	BASF	07955	
Starter Flowable	Truchem	03421	
Stefes Chloridazon	Bayer CropScience	10110	
Takron	BASF	11627	

Product	Marketing Co.	Reg. No.	Expiry Date
Tripart Gladiator	Tripart	00986	
Weedmaster SC	Portman	08793	
chloridazon + chlorpropham + metamitron			
Newtron	Nufarm UK	10943	31-08-05
chloridazon + ethofumesate			
Magnum	BASF	08635	
Spectron	Aventis	07284	
chloridazon + lenacil			
Advizor	DuPont	06571	
Advizor S	Griffin	09799	31-12-04
Advizor S	Griffin	10948	
chloridazon + metamitron			
Volcan Combi FL	Sipcam	11442	
chloridazon + quinmerac			
Fiesta T	BASF	11734	
chlormequat			
Agriguard Chlormequat 700	AgriGuard	09782	
Alpha Chlormequat 460	Makhteshim	04804	
Alpha Pentagan	Makhteshim	04794	
Alpha Pentagan Extra	Makhteshim	04796	
Ashlade 460 CCC	Ashlade	06474	
Ashlade 700 CCC	Ashlade	06473	
Ashlade Brevis	Nufarm UK	08119	30-11-05
Atlas 3C:645 Chlormequat	Nufarm UK	05710	
Atlas 3C:645 Chlormequat	Nufarm UK	07700	
Atlas Chlormequat 46	Nufarm UK	07704	
Atlas Chlormequat 460:46	Nufarm UK	06258	
Atlas Chlormequat 700	Nufarm UK	03402	30-06-04
Atlas Chlormequat 700	Nufarm UK	07708	30-06-04
Atlas Chlormequat 700	Nufarm UK	10706	
Atlas Quintacol	Nufarm UK	07706	30-06-05
Atlas Terbine	Nufarm UK	06523	
Atlas Terbine	Nufarm UK	07709	
Atlas Tricol	Nufarm UK	07707	
Atlas Tricol	Nufarm UK	07190	
Barclay Holdup	Barclay	06709	
Barclay Holdup 600	Barclay	08794	
Barclay Holdup 640	Barclay	08795	
Barclay Liffey	Barclay	11368	
Barclay Liffey	Barclay	09856	
Barclay Lucan	Barclay	09855	
Barclay Lucan	Barclay	11367	
Barleyquat B	Mandops	06001	

SECTION 5

Product	Marketing Co.	Reg. No.	Expiry Date
BASF 3C Chlormequat 750	BASF	06878	
Belcocel	Griffin	08248	
Bettaquat B	Mandops	06004	
Chlormequat 46	Nufarm UK	11504	
Ciba Chlormequat 460	Ciba Specialty	09525	
Ciba Chlormequat 5C 460:320	Ciba Specialty	09527	
Ciba Chlormequat 730	Ciba Specialty	09526	
Clayton CCC 750	Clayton	07952	
Clayton Manquat	Clayton	09916	
Clayton Standup	Clayton	10546	30-06-04
Clayton Standup	Clayton	08771	29-02-04
Cleancrop Chlormequat 700	United Agri	10143	
Cropsafe 5C Chlormequat	Certis	07897	
Greencrop Cong 750	Greencrop	09383	
Hive	Nufarm UK	11392	
Hyquat 70	Agrichem	03364	
Intracrop Balance	Intracrop	08037	
Intracrop MCCC	Intracrop	08506	
Larke	Nufarm UK	10045	30-11-05
Larke	Nufarm UK	11453	
Mandops Chlormequat 460	Mandops	06090	
Mirquat	Nufarm UK	11406	
MSS Chlormequat 40	Mirfield	01401	
MSS Chlormequat 460	Mirfield	03935	
MSS Chlormequat 60	Mirfield	03936	
MSS Chlormequat 70	Mirfield	03937	
MSS Mirquat	Mirfield	08166	
Portman Chlormequat 400	Portman	01523	
Portman Chlormequat 460	Portman	02549	
Portman Chlormequat 700	Portman	03465	
Quadrangle Chlormequat 700	Quadrangle	03401	
Renown	Bayer CropScience	10103	
Sigma PCT	Nufarm UK	10705	30-09-05
Sigma PCT	Nufarm UK	08663	30-06-04
Stabilan 460	Nufarm UK	09304	
Stabilan 5C	Nufarm UK	08144	
Stabilan 640	Nufarm UK	09401	
Stabilan 670	Nufarm UK	09402	
Stabilan 700	Nufarm UK	10579	31-10-05
Stabilan 700	Nufarm UK	09302	31-03-04
Stabilan 750	Nufarm UK	09303	
Stay Up	Nufarm UK	09444	
Stefes CCC	Bayer CropScience	10104	31-10-05
Stefes CCC 640	Bayer CropScience	10111	31-10-05
Stefes CCC 700	Bayer CropScience	10105	31-10-05
Stefes CCC 720	Bayer CropScience	10108	31-10-05
Supaquat 720	Portman	09381	

Product	Marketing Co.	Reg. No.	Expiry Date
Tripart 5C	Tripart	04726	
Tripart Brevis 2	Tripart	06612	
Tripart Chlormequat 460	Tripart	03685	
Whyte Chlormequat 700	Nufarm UK	09641	
chlormequat + 2-chloroethylphosphonic acid			
Barclay Banshee XL	Barclay	09201	
Nomad	Bayer CropScience	10004	
Terpal C	BASF	07062	
chlormequat + choline chloride + imazaquin			
Meteor	Cyanamid	06505	
Upright	BASF	10404	
Upright	BASF	08290	31-07-04
chlormequat with choline chloride			
Ashlade 5C	Ashlade	06227	
Ashlade 700 5C	Ashlade	07046	
Atlas 5C Chlormequat	Nufarm UK	03084	
Barclay Take 5	Barclay	08524	
MSS Mircell	Mirfield	06939	
New 5C Cycocel	BASF	01483	
Portman Supaquat	Portman	03466	
2-chloroethylphosphonic acid			
Aventis Cerone	Bayer CropScience	09748	31-10-05
Aventis Cerone	Aventis	09972	
Barclay Coolmore	Barclay	07917	
Cerone	RP Agric.	06185	
Charger	Bayer CropScience	08827	31-10-05
Charger	Aventis	09986	
Ethrel C	Hortichem	06995	31-10-05
EXPO 3149D	Bayer CropScience	08828	31-10-05
EXPO 3149D	Bayer CropScience	09989	
Pan Ethephon	Pan Agriculture	11020	
Pan Stiffen	Pan Agriculture	11044	
2-chloroethylphosphonic acid + mepiquat chloride			
Barclay Banshee	Barclay	08175	
CleanCrop Fonic M	United Agri	09553	
Cleancrop Fonic M	United Agri	09533	
Marnoch Phonot	Marnoch	09979	
Terpal	Clayton	07626	
chlorothalonil			
Barclay Corrib 500	Barclay	08981	
Baton 500 SC	Sipcam	09850	29-02-04
BB Chlorothalonil	Brown Butlin	03320	
Bombardier	Unicrop	02675	

SECTION 5

Product	Marketing Co.	Reg. No.	Expiry Date
Bravo 500	BASF	05637	
Bravo 500	Syngenta	09059	31-01-06
Bravo 720	Syngenta	10519	
Bravo 720	Zeneca	09104	
Clayton Turret	Clayton	09400	
Cleancrop Chlorothalonil 720	United Agri	10102	
Cleancrop Rover	United Agri	10500	
Clortosip	Sipcam	06126	
ISK 375	Syngenta	10527	31-07-04
ISK 375	Zeneca	09103	
Jupital	Syngenta	09109	
Jupital	Syngenta	10528	
Jupital DG	Syngenta	09181	
Jupital DG	Syngenta	10529	
Mainstay	Quadrangle	05625	
Marnoch Chlorothalonil	Marnoch	09763	
Miros DF	Sipcam	04966	
Mycoguard	Gharda	10945	
Repulse	Syngenta	06705	
Repulse	Syngenta	10535	28-02-05
Strada 500	Sipcam	09849	29-02-04
Visclor 500 SC	Sipcam	09404	
Visclor 75 DF	Sipcam	09361	
chlorothalonil + cymoxanil			
Gex 44	Griffin	10168	
chlorothalonil + cyproconazole			
Cleancrop Cyprothal	United Agri	09580	
Octolan	Syngenta	11675	
Octolan	Novartis	08480	
SAN 703	Syngenta	11676	
SAN 703	Syngenta	08487	31-10-04
chlorothalonil + fluquinconazole			
Trident	Bayer CropScience	09369	29-02-04
Vista CT	Aventis	09368	
chlorothalonil + flutriafol			
Prospa	Headland	11548	
chlorothalonil + mancozeb			
Dreadnought Flo	SumiAgro	07599	30-09-04
Dreadnought Flo	SumiAgro	09599	30-09-04
Dreadnought Flo	Rohm & Haas	07600	30-09-04
Guru	Interfarm	10056	
Sipcam Flo	Sipcam	07601	

Product	Marketing Co.	Reg. No.	Expiry Date
chlorothalonil + propamocarb hydrochloride			
Aventis Merlin	Aventis	09719	
Tattoo C	Bayer CropScience	07623	
chlorothalonil + tetraconazole			
Eminent Star	Isagro	10447	
TC Elite	Monsanto	09412	30-11-04
TC Elite	Monsanto	10506	30-11-04
chlorothalonil + vinclozolin			
Curalan CL	BASF	07174	
chlorotoluron			
Atol	Nufarm UK	10235	
Clayton Chloron	Clayton	08148	
Clayton Chloron 500 FL	Clayton	10791	
Dicurane	Novartis	08403	
MSS Chlorotoluron 500	Nufarm UK	07871	
NWA CTU 500	Nufarm UK	11140	
NWA CTU 500	Nufarm UK	10173	
Stefes Toluron	Bayer CropScience	10106	31-10-05
Talisman	FCC	03109	
Tolugan 700	Makhteshim	08064	
Top Farm Toluron 500	Top Farm	05986	
Tripart Culmus	Tripart	06619	
chlorotoluron + pendimethalin			
Totem	BASF	10485	
chlorpropham			
Aceto Chlorpropham 50M	Aceto	08929	
Atlas CIPC 40	Nufarm UK	07710	
Croptex Pewter	Certis	02507	
MSS CIPC 30M	Nufarm UK	10064	
MSS CIPC 50 M	Nufarm UK	11165	
MSS Sprout Nip	Whyte Agrochem.	11804	
Sprout Nip	Aceto	11786	
Spud-Nic 100 Briquette	Aceto	11633	28-02-05
Triherbicide CIPC	Cerexagri	06874	
Triherbicide CIPC	Atofina	06426	
chlorpropham + fenuron			
Atlas Red	Nufarm UK	07724	
chlorpropham + metamitron			
Buckler	Nufarm UK	11032	
chlorpyrifos			
Ballad	Headland	09775	31-10-04
Barclay Clinch II	Barclay	11346	

SECTION 5

Product	Marketing Co.	Reg. No.	Expiry Date
Choir	Nufarm UK	09778	
Cleancrop Pychlorex	United Agri	11681	
Crossfire 480	Dow	08141	
Dispatch	Dow	08139	
Dursban 4	Dow	07815	
Lorsban 480	Dow	08076	
Talon	SumiAgro	08746	31-01-04
Tripart Audax	Tripart	08745	31-01-04
chlorthal-dimethyl			
Dacthal W75	AMVAC	10289	
Dacthal W75	Certis	05500	
Dacthal W75	United Agri	10617	
Dacthal W-75	ISK Biosciences	05556	
clodinafop-propargyl			
Unite A	Bayer CropScience	07686	
clodinafop-propargyl + diflufenican			
Amazon	Novartis	08128	
Lucifer	Bayer CropScience	10001	31-08-04
clofentezine			
Apollo 50 SC	Aventis	07242	
clomazone			
Centium 360 CS	FMC	10591	31-07-04
Centium 360 CS	FMC	10720	
clopyralid			
Barclay Karaoke	Barclay	08185	
Cliophar	Chimac-Agriphar	09430	
Dow Shield	Dow	05578	28-02-05
Glopyr 200 SL	Globachem	10979	
I T Clopyralid	I T Agro	09808	
Lontrel 200	Dow	11558	
clopyralid + fluroxypyr + MCPA			
Greenor	Rigby Taylor	10909	
clopyralid + fluroxypyr + triclopyr			
Pastor	Dow	11168	
clopyralid + ioxynil			
Escort	Dow	05466	
clopyralid + propyzamide			
Matrikerb	Dow	10806	
Matrikerb	SumiAgro	01308	

Product	Marketing Co.	Reg. No.	Expiry Date
clopyralid + triclopyr			
Blaster	Headland	10571	
clothianidin			
Bayer UK 978	Bayer CropScience	11201	
copper ammonium carbonate			
Croptex Fungex	Certis	02888	31-03-05
cyanazine			
Barclay Canter	Barclay	07530	
Cleancrop Vectro	United Agri	10475	31-12-07
Fortrol	BASF	10222	30-06-04
Fortrol	Feinchemie	10666	
Fortrol	Cyanamid	07009	
Standon Cyanazine 50	Standon	09230	31-07-04
cyanazine + pendimethalin			
Bullet	Feinchemie	10654	31-12-07
Bullet	Cyanamid	08049	31-12-07
cyazofamid			
Ranman	BASF	10598	24-04-04
Ranman TwinPack	BASF	10628	
Ranman Twinpack	BASF	11396	
cycloxydim			
Marnoch Clodim	Marnoch	09852	31-12-05
Stratos	BASF	06891	
cyfluthrin			
Baythroid	Bayer CropScience	11254	31-10-04
Baythroid	Bayer	04273	31-10-04
cymoxanil			
Curzate 60DF	DuPont	11515	
cymoxanil + mancozeb			
Besiege	DuPont	08086	
Clayton Krypton	Clayton	00300	
Cleancrop Xanilite	United Agri	10050	
Fytosporo 68	DuPont	09827	
Marnoch Mancym	Marnoch	10130	
Systol M	Quadrangle	03480	
Systol M WSB	Quadrangle	08074	31-05-04
cypermethrin			
Afrisect 10	Mitchell Cotts	10849	
Afrisect 10	Stefes	09114	
Arrivo	FMC	10731	
Brazil	Nufarm UK	11135	

Product	Marketing Co.	Reg. No.	Expiry Date
Cleancrop Pyrimet	United Agri	10619	
Cyperguard 100 EC	Gharda	10775	
Cyperkill 25	Chiltern	09038	
Cypertox	FCC	05122	
Permasect C	Nufarm UK	09200	30-06-05
cyproconazole			
Alto 240 EC	Bayer CropScience	11236	
Alto 240 EC	Bayer	10462	
Alto 240 EC	Novartis	08354	
Alto Eco A	Syngenta	08466	
Fort	Bayer	11027	31-03-06
cyproconazole + prochloraz			
SAN 710 HF	Syngenta	08489	30-06-04
Tiptor	Syngenta	08495	30-06-04
cyproconazole + propiconazole			
Alto Extra	Syngenta	10850	
cyproconazole + quinoxyfen			
Choice	Dow	10680	
DOE 1762	Dow	08962	
cyproconazole + trifloxystrobin			
Fencer	Bayer CropScience	11277	
Fencer	Bayer	10887	
Sphere	Bayer	10263	31-01-04
Sphere	Bayer	10586	
Sphere	Bayer	10460	30-04-04
cyprodinil			
Barclay Amtrak	Barclay	09562	
Chieftain	Syngenta	09301	29-02-04
Cleancrop Cyprodinil	United Agri	09668	
Unix	Syngenta	11512	
2,4-D			
Agrichem 2,4-D	Agrichem	04098	
Agricorn 2,4-D	FCC	07349	01-04-05
Agricorn D	FCC	00056	01-04-05
Atlas 2,4-D	Atlas	07699	
Barclay Haybob II	Barclay	08532	
Depitox	Nufarm UK	08000	31-07-04
Depitox	Nufarm UK	10746	31-07-05
Dupont 24-D	DuPont	11737	
Easel	Nufarm UK	09878	
Forester	Vitax	00914	
GroWell 2,4-D Amine	Marks	08750	

Product	Marketing Co.	Reg. No.	Expiry Date
GroWell 2,4-D Amine	GroWell	11395	
Herboxone 60	Marks	09693	
Marks 2,4-D-A	Marks	01282	01-04-05
Maton	Headland	10366	
MSS 2,4-D Amine	Nufarm UK	11139	
MSS 2,4-D Amine	Nufarm UK	10183	
MSS 2,4-D Ester	Nufarm UK	01393	
Nomix 2,4-D Herbicide	n2n	06394	
Palormone D	Unicrop	01534	01-04-05
Ragox	Nufarm UK	10407	
Ritefeed 2,4-D Amine	Ritefeed	05309	01-04-05
Silvapron D	BP	01935	01-04-05
Syford	Vitax	02026	
2,4-D + dicamba			
Lawn Builder Plus Weed Control	Scotts	08499	
2,4-D + dicamba + triclopyr			
Cleancrop Broadshot	United Agri	11664	
Terbel Triple	Chimac-Agriphar	09138	31-12-04
2,4-D + MCPA			
Agroxone Combi	Headland	10277	31-03-05
Agroxone Combi	Headland	11025	
2,4-D + mecoprop-P			
Mascot Selective-P	Rigby Taylor	06105	01-04-05
2,4-D + picloram			
Tordon 101	Dow	05816	
daminozide			
B-Nine	Crompton	11465	
B-Nine	Certis	11057	30-11-05
B-Nine	Certis	07844	30-04-05
B-Nine	Certis	04468	30-11-05
Dazide WSG	Fine	11763	
dazomet			
Basamid	Certis	11324	
Basamid	BASF	00192	
2,4-DB			
Butoxone DB	Marks	10482	
Marks 2,4-DB	Marks	01283	29-04-04
2,4-DB + linuron + MCPA			
Alistell	Syngenta	06515	30-04-05
Alistell	United Phosphorus	11053	
Alistell	Syngenta	10510	

SECTION 5

Product	Marketing Co.	Reg. No.	Expiry Date
2,4-DB + MCPA			
Butoxone DB Extra	Marks	10455	
Headland Cedar	Headland	09948	29-02-04
Marks 2,4-DB Extra	Marks	01284	31-01-04
MSS 2,4-DB + MCPA	Nufarm UK	01392	
deltamethrin			
Aventis Decis	Aventis	09710	
Decis Micro	Bayer CropScience	08618	
Deleet	Rentokil	09312	
Delta-M 2.5 EC	O Endres	11334	
Deltamex 2.5 EC	O Endres	10291	30-09-05
Deltaplan	Chimac-Agriphar	11678	
deltamethrin + pirimicarb			
Best	Bayer CropScience	08988	
desmedipham + ethofumesate + phenmedipham			
Aventis Betanal Progress OF	Aventis	09722	
Betanal Progress OF	Aventis	07629	
desmedipham + phenmedipham			
Betanal Compact	Aventis	07247	
dicamba			
Banvel T	Syngenta	08470	
Cadence	Syngenta	08796	
Cadence	Barclay	09578	
SAN 845H	Syngenta	08797	
dicamba + dichlorprop-P + MCPA			
Intrepid	Miracle	07819	
dicamba + MCPA			
Banvel M	Syngenta	08469	
dicamba + MCPA + mecoprop-P			
Docklene Super	Bayer CropScience	10728	
Headland Relay Turf	Headland	08935	
Herrisol New	Bayer CropScience	11445	
Herrisol New	Bayer	07166	
Mascot Super Selective-P	Rigby Taylor	06106	
Mircam Plus	Nufarm UK	11525	
Mircam Plus	Nufarm UK	09884	28-02-05
MSS Mircam Plus	Mirfield	01416	
Outrun	SumiAgro	10426	
Pasturol D	FCC	07033	
Pasturol P	FCC	09099	
Pasturol Plus	Headland	10278	
Premier Amenity Selective	Amenity Land	10098	

Product	Marketing Co.	Reg. No.	Expiry Date
Quadrangle Quadban	Nufarm UK	07090	
Stefes Docklene Super	Bayer CropScience	09873	31-07-04
Super Selective Plus	Nufarm UK	11666	
Trireme	Nufarm UK	10425	31-01-06
UPL Amenity Grassland Herbicide	United Phosphorus	11713	
dicamba + mecoprop-P			
Condox	Syngenta	09429	31-10-04
Condox	Syngenta	10522	
Dockmaster	Nufarm UK	10033	31-10-04
Headland Swift	Headland	10437	
Mircam	Whyte Agrochem.	09888	31-12-04
MSS Mircam	Mirfield	01415	
Optica Forte	Marks	07432	
dichlobenil			
Casoron G	Crompton	09022	
Casoron G	Rigby Taylor	09326	
Casoron G	Zeneca	08065	
Casoron G4	Miracle	07927	
Casoron G-SR	Miracle	07925	
Sierraron G	Scotts	09263	
Sierraron G4	Scotts	10491	
Sierraron GSR	Scotts	10459	30-11-04
dichlorophen			
Halophen RE 49	Superfog	10863	31-01-06
Nomix-Chipman Mosskiller	n2n	06271	
Ritefeed Dichlorophen	Ritefeed	05265	
dichlorprop-P			
Headland Link	Headland	10603	31-05-05
Optica DP	Marks	07818	
Optica DP	Marks	11067	
dichlorprop-P + ferrous sulphate + MCPA			
Vitagrow Granular Feed, Weed & Mosskiller	Sinclair	10971	
dichlorprop-P + ioxynil			
Mextrol DP	Nufarm UK	11529	
dichlorprop-P + MCPA			
Optica Duo	Marks	10343	
dichlorprop-P + MCPA + mecoprop-P			
Optica Trio	Marks	09747	
difenacoum			
Neokil	Sorex	05564	

SECTION 5

Product	Marketing Co.	Reg. No.	Expiry Date
Neokil Bait Bags	Sorex	08949	31-12-04
Neosorexa	Sorex	07756	
Neosorexa Bait Blocks	Sorex	07355	31-12-04
Neosorexa Concentrate	Sorex	01475	31-12-04
Neosorexa Liquid Concentrate	Sorex	04640	31-12-04
Neosorexa Mixed Grain Bait	Sorex	08782	31-12-04
Neosorexa Ratpacks	Sorex	04653	
difenoconazole			
Plover	Syngenta	11763	
diflubenzuron			
Dimilin 25-WP	Hortichem	08902	
Dimilin Flo	Certis	10618	30-04-05
Dimilin WP	Syngenta	08870	
diflufenican			
EXP 4005	Bayer CropScience	10582	
diflufenican + flurtamone			
Buzzard	Bayer CropScience	10793	31-08-04
diflufenican + flurtamone + isoproturon			
EXP 31502A	Bayer CropScience	09991	31-01-05
diflufenican + isoproturon			
Clayton Fenican 550	Clayton	11160	
DFF + IPU WDG	Aventis	10204	
Gavel	Bayer CropScience	09992	31-10-05
Grenadier	Bayer CropScience	09996	31-08-04
Shire	Bayer CropScience	08117	
Shire	Bayer CropScience	10017	31-08-04
Tolkan Turbo	Bayer CropScience	10024	31-10-04
Unite B	Bayer CropScience	07686	
dimethomorph + mancozeb			
Invader	Cyanamid	06989	
Invader WDG	BASF	09556	31-05-04
Invader WDG	BASF	10349	
Saracen	BASF	07872	31-05-04
Saracen	BASF	10392	
Saracen WDG	BASF	09557	31-05-04
Saracen WDG	BASF	10393	
diphenylamine			
No-Scald DPA	Cerexagri	08312	
Shield DPA 15%	United Agri	09550	31-10-04
diquat			
Clayton Diquat	Clayton	10730	

Product	Marketing Co.	Reg. No.	Expiry Date
Cleancrop Diquat	United Agri	09687	
Diquanet	Hermoo	10846	01-07-04
Midstream	Scotts	09267	
Midstream	Miracle	07739	
Reglone	Zeneca	09646	
Standon Diquat	Standon	10674	
Waterloo	Alfa	10839	31-08-05
diquat + paraquat			
PDQ	Zeneca	10241	
Speedway 2	Scotts	10790	
dithianon			
Barclay Cluster	Barclay	08792	
Dithianon Flowable	Cyanamid	07007	
dithianon + penconazole			
Topas D 275 SC	Syngenta	08460	
diuron			
Atlas Diuron	Nufarm UK	08214	
Chipko Diuron 80	n2n	00497	
Chipman Diuron 80	n2n	08054	
Chipman Diuron Flowable	n2n	05701	
Diuron 50 FL	Staveley	02814	
Diuron 80% WP	Staveley	00730	
Freeway	RP Amenity	06047	
Karmex	Headland	11045	
Karmex	Nufarm UK	11647	
Karmex	Griffin	09475	30-04-05
Mascot Diuron Flow	Rigby Taylor	11212	
Mascot Diuron Flow	Rigby Taylor	10889	30-09-05
MSS Diuron 50 FL	Nufarm UK	07160	
MSS Diuron 500 FL	Mirfield	08171	31-01-06
MSS Diuron 80 WP	Nufarm UK	09787	31-01-06
Nomix Chipman Flowable Diuron	n2n	10837	
Rescind	RP Amenity	08036	
Sanuron	Headland	09230	
diuron + glyphosate			
NX 2075	Nomix-Chipman	11104	
NX 3083	Nomix-Chipman	08712	
NX 3083	Nomix-Chipman	11100	
Total	Nomix-Chipman	07932	
Total	Nomix-Chipman	11099	
Touche	Nomix-Chipman	06797	01-01-05
Touche	Nomix-Chipman	11097	

Product	Marketing Co.	Reg. No.	Expiry Date
dodine			
Barclay Dodex	Barclay	09055	
Dodifun 400 SC	Hermoo	11657	
Syllit 400 SC	Chimac-Agriphar	11079	
epoxiconazole			
Clayton Oust	Clayton	11588	
Cleancrop EPX	United Agri	10381	
epoxiconazole + fenpropimorph			
Barclay Riverdance	Barclay	09658	
Eclipse	BASF	11731	
Opus Team	BASF	11759	
epoxiconazole + fenpropimorph + kresoxim-methyl			
BAS 493F	BASF	10438	31-01-05
BAS 493F	BASF	11748	
Cleancrop Chant	United Agri	11746	
Cleancrop Chant	United Agri	10466	31-01-05
Mantra	BASF	08886	
Mastiff	BASF	10857	31-01-05
Mastiff	BASF	11747	
epoxiconazole + kresoxim-methyl			
Barclay Avalon	Barclay	09466	
Clayton Gantry	Clayton	09482	
Cleancrop Kresoxazole	United Agri	09698	
Landmark	BASF	08889	
Marnoch Kempo	Marnoch	09918	
epoxiconazole + kresoxim-methyl + pyraclostrobin			
Covershield	BASF	10900	
epoxiconazole + kresoxim-methyl + quinoxyfen			
BAS 541 00F	BASF	10924	
TPF003	Dow	10916	
esfenvalerate			
Clayton Estate	Clayton	11155	
Sumi-Alpha	Cyanamid	07207	
ethofumesate			
Atlas Thor	Nufarm UK	06966	01-09-04
Atlas Thor	Atlas	07732	
Barb	Sipcam	10378	01-09-04
Barclay Keeper	Barclay	08835	01-09-04
Barclay Keeper 500 FL	Barclay	09438	01-09-04
Cleancrop ET 200	United Agri	10555	01-09-04
Cleancrop ET 500	United Agri	10556	01-09-04
Fumesate 500 SC	Hermoo	10923	01-09-04

Product	Marketing Co.	Reg. No.	Expiry Date
I T Ethofumesate	I T Agro	11108	01-09-04
Kubist	Nufarm UK	11123	01-09-04
Kubist	Griffin	09454	31-08-05
Kubist 2	Griffin	09880	01-09-04
Linesman 200	Hutchinsons	10477	01-09-04
Linesman 500	Hutchinsons	10458	01-09-04
MSS Thor	Nufarm UK	08817	
Nortron	Bayer CropScience	07266	
Standon Ethofumesate 200	Standon	08726	
Stefes Fumat 2	Stefes	07856	01-09-04
Stefes Fumat 2	Bayer CropScience	10084	
Thor	Nufarm UK	11131	01-09-04
ethofumesate + metamitron			
Stefes ST/STE	Bayer CropScience	09111	31-07-04
Stefes ST/STE	Bayer CropScience	10109	31-07-04
Torero	Feinchemie	10250	
ethofumesate + metamitron + phenmedipham			
Bayer UK 540	Bayer CropScience	11244	28-02-06
Bayer UK 540	Bayer	07399	28-02-06
Betanal Trio WG	Aventis	07537	
ethofumesate + phenmedipham			
Barclay Goalpost	Barclay	09497	01-09-04
Barclay Goalpost	Barclay	09192	
Barclay Goalpost	Barclay	11372	
Betanal Tandem	Aventis	07254	
Betosip Combi	Sipcam	08030	
Fenlander	Hutchinsons	11706	01-09-04
Fenlander	Hutchinsons	11581	01-09-04
Powertwin	Feinchemie	10442	
Stefes Tandem	Bayer CropScience	10088	01-09-04
Thunder	Feinchemie	10958	31-08-05
Thunder	Makhteshim	11186	
Twin	Feinchemie	09875	
ethoprophos			
Aventis Mocap 10G	Bayer CropScience	09750	30-09-04
Aventis Mocap 10G	Aventis	09973	
Mocap 10G	RP Agric.	06773	
etridiazole			
Aaterra WP	Zeneca	06625	
Terrazole 35 WP	Crompton	09800	
famoxadone			
Midas	DuPont	10549	11-03-04

SECTION 5

Product	Marketing Co.	Reg. No.	Expiry Date
famoxadone + flusilazole			
Medley	DuPont	10933	
fatty acids			
Safers Insecticidal Soap	Safer	07197	
fenarimol			
Rimidin	Dow	07938	
Rubigan	Margarita	11069	
Rubigan	Dow	05489	
fenazaquin			
Matador 200 SC	Margarita	11058	
Matador 200 SC	Dow	07960	
fenbuconazole			
Indar 5EW	Whelehan	09644	
Kruga 5EC	Interfarm	09874	31-01-05
fenbuconazole + fenpropidin			
Accolade	SumiAgro	09950	
fenbuconazole + prochloraz			
Mirage Extra	Bayer CropScience	07958	31-03-04
Mirage Extra	Makhteshim	10558	
fenbuconazole + propiconazole			
Graphic	Interfarm	10987	
Graphic	Rohm & Haas	09907	
Graphic	Dow	10800	
Indar CG	Syngenta	08578	30-09-04
fenbutatin oxide			
Torque	BASF	07148	
Torque	BASF	10399	
fenhexamid			
Lattice	Bayer CropScience	11224	
Lattice	Bayer	09198	31-05-11
fenhexamid + tolylfluanid			
Talat	Bayer	09655	
fenoxaprop-P-ethyl			
Aventis Cheetah Super	Aventis	09730	
fenoxaprop-P-ethyl + isoproturon			
Pinion	Griffin	10287	31-03-05
Puma X	Aventis	08779	
fenpropathrin			
Meothrin	Cyanamid	07206	

Product	Marketing Co.	Reg. No.	Expiry Date
fenpropidin			
Mallard	Syngenta	08662	
Patrol	Syngenta	10531	
Patrol	Zeneca	08661	
fenpropidin + prochloraz			
SL 552A	Syngenta	08673	
Sponsor	Aventis	08674	
fenpropidin + propiconazole + tebuconazole			
Bayer UK 593	Bayer CropScience	11416	
Bayer UK 593	Bayer	08285	
Gladio	Novartis	08413	
fenpropidin + tebuconazole			
Monicle	Bayer CropScience	11428	
Monicle	Bayer	07375	
SL 556 500 EC	Syngenta	08449	
fenpropimorph			
BAS 421F	BASF	06127	
Clayton Spigot	Clayton	11560	
Cleancrop Fenpro	United Agri	09885	
Cleancrop Fenpropimorph	United Agri	09445	
Keetak	BASF	06950	
Marnoch Phorm	Marnoch	10773	31-05-05
fenpropimorph + flusilazole			
BAS 48500F	BASF	06784	
Pluton	DuPont	10957	
fenpropimorph + kresoxim-methyl			
Ensign	BASF	08362	
fenpropimorph + quinoxyfen			
EF-1288	Dow	08241	
fenpyroximate			
NNI 850 5SC	Nihon Nohyaku	09887	
Sequel	Hortichem	09888	31-10-05
fentin hydroxide			
Barclay Fentin Flow	Barclay	07914	
Barclay Fentin Flow 532	Barclay	09434	
Super-Tin 4l	Griffin	09028	
Super-Tin 4L	Griffin	09559	
Super-Tin 80 WP	Griffin	09560	
ferrous sulphate			
Fisons Greenmaster Autumn	Fisons	03211	28-02-06
Fisons Greenmaster Mosskiller	Fisons	00881	28-02-06

SECTION 5

Product	Marketing Co.	Reg. No.	Expiry Date
Greenmaster Autumn	Scotts	11185	
Greenmaster Mosskiller	Scotts	11576	
Maxicrop Moss Killer & Conditioner	Maxicrop	04635	
Moss Control Plus Lawn Fertilizer	Miracle	07912	
No More Moss Lawn Feed	Wolf	10754	
Pentagon Prestige Lawn Sand	Sinclair	10456	
Vitagrow Lawn Sand	Vitagrow	05097	
fipronil			
Regent 1 GR	Bayer CropScience	11078	
florasulam			
Primus 25 SC	Dow	10175	
florasulam + fluroxypyr			
GF 184	Dow	10878	
Hiker	Dow	11451	
Starane Vantage	Dow	10922	18-11-04
fluazifop-P-butyl			
Fusilade 250 EW	Zeneca	06531	
PP 007	Syngenta	06533	
PP 007	Syngenta	10533	
Wizzard	Syngenta	06521	
Wizzard	Syngenta	10539	
fluazinam			
Barclay Cobbler	Barclay	08349	
Legacy	ISK Biosciences	09966	
Shirlan	Zeneca	07091	
Shirlan Programme	Syngenta	08761	
Shirlan Programme	Syngenta	10574	
Top Farm Fluazinam	Top Farm	07683	
fludioxonil			
Beret Gold	Syngenta	11635	
flufenacet + pendimethalin			
210	BASF	10683	
fluoroglycofen-ethyl + isoproturon			
Competitor	Bayer CropScience	07310	
flupyrsulfuron-methyl			
DPX-KE459 WSB	DuPont	08540	
KE 459 DF	DuPont	09835	
KE 459 PX	DuPont	11189	06-08-05
Lexus 50 PX	DuPont	11188	06-08-05
flupyrsulfuron-methyl + metsulfuron-methyl			
Lexus XPE	DuPont	08541	

Product	Marketing Co.	Reg. No.	Expiry Date
flupyrsulfuron-methyl + picolinafen			
BUK 960	BASF	11152	18-02-04
DP 944	DuPont	11120	18-02-04
DP 945	DuPont	11447	
flupyrsulfuron-methyl + thifensulfuron-methyl			
Lexus Millenium WSB	DuPont	09207	
fluquinconazole			
Diablo	Bayer CropScience	09914	
Jockey Flexi	Bayer CropScience	10073	
fluquinconazole + prochloraz			
Aventis Foil	Bayer CropScience	10906	31-03-04
Aventis Foil	Aventis	09709	
Baron	Bayer CropScience	09364	30-11-04
Baron	Bayer CropScience	10904	
Flamenco Plus	Bayer CropScience	09362	30-11-04
Flamenco Plus	Bayer CropScience	10862	31-03-04
Foil	Bayer CropScience	09363	30-11-04
Jockey Plus	Bayer CropScience	10075	
fluroxypyr			
Barclay Hurler	Barclay	08791	
BAS 91092H	BASF	11044	
Clayton Fluroxypyr	Clayton	06356	
Emrald Chopper	Emrald	11716	
Gala	Dow	09793	
Gex 353 B	Griffin	10185	30-11-05
Tomahawk 2000	Makhteshim	00307	
fluroxypyr + MCPA + mecoprop-P			
EM1658/01	SumiAgro	10892	
fluroxypyr + mecoprop-P			
EF-1374	Dow	10621	30-11-04
fluroxypyr + thifensulfuron-methyl + tribenuron-methyl			
GEX 353	DuPont	11474	
fluroxypyr + tribenuron-methyl			
DP953 PX	DuPont	11321	
fluroxypyr + triclopyr			
Doxstar	Dow	11003	
flusilazole			
Capitan 40	DuPont	10914	
Genie	DuPont	08238	
Sanction	DuPont	08237	

Product	Marketing Co.	Reg. No.	Expiry Date
flutolanil			
EXP 10066A	Bayer CropScience	11584	
NNF-136	Nihon Nohyaku	11585	
flutriafol			
Consul	Headland	11523	
Impact	Headland	11521	
fosetyl-aluminium			
Aliette	Certis	02484	
Aliette	Hortichem	05648	
I.T. Fosetyl-AL	I T Agro	11717	
fosthiazate			
Nemathorin 10G	Syngenta	11003	17-02-04
fuberidazole + imazalil + triadimenol			
Baytan IM	Bayer CropScience	11418	
Baytan IM	Bayer	00226	
fuberidazole + triadimenol			
Baytan	Bayer CropScience	11250	
Baytan	Bayer	00225	
Baytan CF	Bayer CropScience	11251	
Baytan CF	Bayer	08193	
Baytan Flowable	Bayer CropScience	11252	31-12-04
gibberellins			
Berelex	Nufarm UK	08903	
Gistar	Hermoo	11397	
Regulex	Sumitomo	10147	
glufosinate-ammonium			
Aventis Challenge	Aventis	09721	
Aventis Harvest	Aventis	09810	
Dash	n2n	05177	
Headland Sword	Headland	07676	
Kaspar	Certis	11214	
Nomix Touchweed	n2n	09596	
glyphosate			
Acrion	Bayer Environ.	09965	
Agriguard Glyphosate 180	AgriGuard	09806	01-01-05
Apache	Syngenta	06748	01-01-05
Azural	Monsanto	09582	01-01-05
Azural	Monsanto	10300	31-10-04
Barbarian	Barclay	07625	
Barclay Trustee	Barclay	09177	30-10-04
Biactive 270	Monsanto	10031	
Biactive 270	Monsanto	10301	

Product	Marketing Co.	Reg. No.	Expiry Date
Bioglyce	Applyworld	10054	30-11-05
Bioglyce	Austrital	11459	
Cardel Egret	Cardel	10252	30-06-04
Cardel Glyphosate	Cardel	09581	01-01-05
Cardel Glyphosate	Cardel	10253	01-01-05
Clarion	Syngenta	10521	
Clarion	Syngenta	08272	01-01-05
Clayton Glyphosate	Clayton	06608	
Clayton Rhizeup	Clayton	08920	
Clayton Swath	Clayton	06715	
Cleancrop Hoedown	Loveland	11194	
Cleancrop Hoedown	United Agri	10852	
Clean-Up-360	Top Farm	05076	
Do-Away	Barclay	11380	
Do-Away	NCH	10112	31-03-06
Dow Agrosciences Glyphosate 360	Dow	11552	
Economix	n2n	08000	
Fernpath Opsen	AgriGuard	10687	31-01-05
Fernpath Statis	AgriGuard	10663	01-01-05
First Line	Linemark	11386	
Glister	Sinon EU	11413	
Glyfonex 480	Danagri	07464	01-01-05
Glyfos 480	Cheminova	08014	31-03-05
Glyfos 480	Headland	10996	
Glyphosate 120B	Monsanto	10302	01-01-05
Glyphosate 120B	Monsanto	08292	01-01-05
Glyphosate 360	Danagri	09233	28-02-06
Glyphosate 360A	Danagri	09234	01-01-05
Glyphosate 360B	Danagri	09235	01-01-05
Glyphosate Biactive	Monsanto	10303	01-01-05
Gorgon	Portman	09182	01-01-05
Initial Line	Linemark	11384	
Marnoch Glyphosate	Marnoch	09744	
Mogul	Monsanto	10306	01-01-05
Mogul	Monsanto	07076	01-01-05
MON 240	Monsanto	10307	01-01-05
MON 44068 Pro	Monsanto	06815	31-07-04
MON 52276	Monsanto	10310	01-01-05
MON 52276	Monsanto	06949	28-02-05
Monty	Quadrangle	07796	01-01-05
MSS Glyfield	Nufarm UK	08009	31-07-05
Muster	Syngenta	06685	01-01-05
Muster LA	Zeneca Prof.	05762	01-01-05
New Total Weedkiller	Headland	11207	
Nomix G	Monsanto	08781	
Nomix Nova	n2n	08647	
Nomix Revenge	n2n	09483	

Product	Marketing Co.	Reg. No.	Expiry Date
Nomix Supernova	n2n	09473	
NX 2031	n2n	11105	
NX 2033	n2n	11106	
NX 8049	n2n	10099	
PMG	AgriGuard	10153	01-01-05
Pontil 360	Mastra	11433	
Preline	Linemark	10977	
Preline Plus	Linemark	11385	
Reliance	Monsanto	09954	01-01-05
Reliance	Monsanto	10315	31-10-04
Rival	Monsanto	09220	
Roundup	Monsanto	01828	
Roundup Biactive	Monsanto	06941	28-02-05
Roundup Biactive Dry	Monsanto	10321	
Roundup CF	Monsanto	10323	01-01-05
Roundup CF	Monsanto	09547	01-01-05
Roundup Four 80	Monsanto	10325	
Roundup GT	Monsanto	10327	
Roundup MX	Monsanto	10328	01-01-05
Roundup MX	Monsanto	09631	01-01-05
Roundup Pro	Monsanto	10329	
Roundup Provide	Monsanto	11721	
Roundup Rapide	Monsanto	08067	30-09-05
Samurai	Monsanto	09952	
Scorpion	Monsanto	10335	30-10-04
Scorpion	Monsanto	09953	01-01-05
Smart 360	Mastra	09476	
Stampede	Syngenta	10536	
Sting	Monsanto	02789	01-01-05
Sting	Monsanto	10439	01-01-05
Sting CT	Monsanto	10336	01-01-05
Sting Eco	Monsanto	08291	31-08-04
Stride	Syngenta	08282	01-01-05
Trustee Elite	Dalgety	11556	
Typhoon 360	Makhteshim	11175	
Typhoon 360	Feinchemie	09792	31-07-05
Unistar Glyfosate 360	Unistar	05928	
Unistar Glyphosate 360	Unistar	06332	
glyphosate (agricultural uses)			
AgriGuard Glyphosate 360	AgriGuard	09184	
Apache	Syngenta	10514	
Barclay Cleanup	Barclay	09179	
Barclay Dart	Barclay	05129	
Barclay Gallup Biograde 450	Dalgety	10151	31-03-06
Cardel Egret	Cardel	09703	
Glyfosate-360	Top Farm	05319	

Product	Marketing Co.	Reg. No.	Expiry Date
Glyper	PBI	07968	
Helosate	Helm	06499	
MON 240	Monsanto	04538	
Poise	Unicrop	08276	
Portman Glider	Portman	04695	
Portman Glyphosate	Portman	05891	
Portman Glyphosate 360	Portman	04699	
Roundup	Monsanto	10317	
Roundup 2000	Monsanto	08069	
Roundup Biactive Dry	Monsanto	06942	
Roundup Express	Monsanto	10318	
Roundup Four 80	Monsanto	03176	
Roundup GT	Monsanto	08068	
Roundup Pro	Monsanto	04146	
Roundup Pro Biactive	Monsanto	06954	
Roundup Rapide	Monsanto	10331	
Stampede	Zeneca	06327	
Sting CT	Monsanto	04754	
Stirrup	n2n	06132	
Stride	Syngenta	10537	
glyphosate (fruit, horticulture, forestry, amenity etc.)			
Barclay Barbarian	Barclay	09865	31-03-06
Barclay Gallup 360	Barclay	09127	28-02-06
CleanCrop Egret	United Agri	10689	
Glyphosate 360	Monsanto	11726	
Glyphosate 360	Austrital	11555	
MSS Glyfield	Nufarm UK	11117	
Rival	Monsanto	10316	
Roundup Amenity	Monsanto	08721	
Roundup Greenscape	Monsanto	10599	
Spasor	RP Amenity	07211	
Touchdown	Zeneca	06326	
Touchdown LA	Scotts	09270	
guazatine			
Panoctine	Makhteshim	11404	
hymexazol			
Tachigaren 70 WP	Sumitomo	02649	
imazalil			
Fungaflor	Hortichem	05967	31-12-04
Fungaflor	Brinkman	05968	31-12-04
Fungaflor Smoke	Hortichem	05969	31-12-04
Fungaflor Smoke	Brinkman	06009	
Fungazil 100 SL	Aventis	10270	
Sphinx	Aventis	10561	

SECTION 5

Product	Marketing Co.	Reg. No.	Expiry Date
Sphinx	Aventis	07607	
Sphinx	BASF	11764	
Stryper	Crompton	08310	
imazalil + pencycuron			
Monceren IM Flowable	Bayer CropScience	11427	
Monceren IM Flowable	Bayer	06731	
imazalil + thiabendazole			
Extratect Flowable	Banks Cargill	09157	29-02-04
imazamethabenz-methyl			
Assert	BASF	08656	31-01-05
Assert	BASF	10209	
Dagger	Cyanamid	03737	
imidacloprid			
Bayer UK 368	Bayer CropScience	11414	31-08-04
Bayer UK 368	Bayer	08125	31-12-05
Bayer UK 397	Bayer CropScience	11242	
Bayer UK 397	Bayer	08930	
Bayer UK 479	Bayer CropScience	11415	31-08-04
Bayer UK 479	Bayer	08584	31-08-04
Gaucho FS	Bayer CropScience	11282	
Gaucho FS	Bayer	08496	
Levington Professional Plus Intercept	BASF	08569	
Rentokil Desyst	Rentokil	09446	
indol-3-ylacetic acid			
Rhizopon A Powder	Fargro	07131	
Rhizopon A Tablets	Fargro	07132	
4-indol-3-yl-butyric acid			
Chryzoplus Grey	Fargro	07984	
Chryzopon Rose	Fargro	07982	
Chryzosan White	Fargro	07983	
Chryzotek Beige	Fargro	07125	
Chryzotop Green	Fargro	07129	
Rhizopon AA Powder (0.5%)	Fargro	07126	
Rhizopon AA Powder (1%)	Fargro	07127	
Rhizopon AA Powder (2%)	Fargro	07128	
Rhizopon AA Tablets	Fargro	07130	
ioxynil			
Totril	RP Agric.	06116	
iprodione			
Aventis Rovral Flo	Aventis	09974	
CDA Rovral	RP Amenity	04679	
Ipromex 50% WP	O Endres	11704	

Product	Marketing Co.	Reg. No.	Expiry Date
Rovral Green	RP Amenity	05702	
Rovral Liquid FS	Bayer CropScience	10014	31-03-04
Rovral Liquid FS	Bayer CropScience	09776	31-03-04
Rovral WP	RP Agric.	06091	
iprodione + thiophanate-methyl			
Aventis Compass	Aventis	10040	
isoproturon			
Agriguard IPU	AgriGuard	10818	
Aligran WDG	Agform	10711	30-06-04
Aligran WDG	Portman	08931	31-07-04
Arelon 2	Griffin	10276	03-10-04
Arelon 2	Nufarm UK	11670	
Arelon 500	DuPont	11110	30-06-05
Arelon 500	Aventis	08100	
Arelon 500	Griffin	10259	31-10-04
Arelon 700	Nufarm UK	11542	
Atlas Fieldgard	Nufarm UK	08582	31-07-04
Atum	Bayer CropScience	10897	
Atum	Bayer CropScience	07379	31-10-04
Atum	Bayer CropScience	09970	31-10-04
Atum WDG	Bayer CropScience	09971	
Atum WDG	RP Agric.	07778	
Auger	Headland	10896	
Auger	RP Agric.	06581	
Auger	Headland	10786	31-10-04
Bison	Gharda	08699	
Bison 83 WG	Nufarm UK	10063	28-02-06
Bison 83 WG	Gharda	10062	
Clayton Siptu 50 FL	Clayton	08751	
Cordelia	Griffin	09780	30-06-04
Cordelia 2	Griffin	10238	30-04-06
Cordelia 2	Nufarm UK	11591	
Cordelia 2	DuPont	11111	30-06-05
DAPT Isoproturon 500 FL	DAPT	08092	
Fieldgard	Nufarm UK	11770	
Fieldgard	Nufarm UK	10710	31-03-06
Greencrop Proturon 500	Greencrop	11394	30-06-04
Griffin IPU 700	Griffin	10409	31-03-06
Hytane 500 SC	Syngenta	08419	30-06-04
I.P.U. SUN	Phytorus	08798	30-06-04
Isoguard	Gharda	10817	
Isoguard 83 WG	Gharda	10061	
Isoguard SVR	BASF	10402	
Isoguard SVR	BASF	08696	29-02-04
Isoproturon 500	Nufarm UK	11603	

SECTION 5

Product	Marketing Co.	Reg. No.	Expiry Date
Isoproturon 500	Griffin	10261	31-03-06
Mysen	Portman	07695	
Mysen WDG	Portman	09141	
Portman Isotop SC	Portman	07663	
Quintil 500	Phytorus	06800	30-06-04
Standon IPU	Standon	08671	30-11-04
Stefes IPU	Bayer CropScience	10107	31-10-05
Tolkan	Bayer CropScience	10022	
Tolkan	Bayer CropScience	07365	31-10-05
Tripart Pugil 5	Tripart	08606	30-06-04
isoproturon + isoxaben			
Ipso	Dow	05736	31-01-05
isoproturon + pendimethalin			
Artillery	BASF	10372	
Artillery	BASF	07768	
Artillery SC	BASF	10385	
Artillery SC	BASF	08504	
Clayton Pendalin-IPU 37	Clayton	09154	30-06-04
Clayton Pendalin-IPU 47	Clayton	09170	30-06-04
Encore	Cyanamid	04737	
Encore SC	BASF	08502	29-02-04
Encore SC	BASF	10386	
Jolt	BASF	10373	
Jolt SC	BASF	10387	
Jolt SC	BASF	08503	
Orient	BASF	10374	
Orient	BASF	07798	
Orient SC	BASF	10388	
Orient SC	BASF	08501	
Trump	Cyanamid	03687	
Trump SC	BASF	08500	29-02-04
Trump SC	BASF	10389	
isoproturon + simazine			
Cleancrop Nevada	United Agri	11771	
isoproturon + trifluralin			
Autumn Kite	Nufarm UK	11549	
Autumn Kite	Griffin	10294	
Autumn Kite	Aventis	07119	
isoxaben			
Flexidor	Dow	05121	
Flexidor 125	Landseer	10946	
Gallery 125	Rigby Taylor	06889	

Product	Marketing Co.	Reg. No.	Expiry Date
isoxaben + methabenzthiazuron			
Glytex	Bayer CropScience	11424	
Glytex	Bayer	04230	
isoxaben + trifluralin			
Premiere	Scotts	10866	
lambda-cyhalothrin			
Hallmark with Zeon technology	Zeneca	09809	
I T Lambda	I T Agro	10833	31-07-04
lambda-cyhalothrin + pirimicarb			
Dovetail	Zeneca	07973	
lenacil			
Clayton Lenaflo	Clayton	11031	
Cleancrop Lenflow	United Agri	10059	
Lenazar Flowable	Hermoo	11068	
Lenazar WP	Hermoo	10792	
Lens Plus	Agricola	11587	
Venzar 80 WP	DuPont	09981	
Vizor	DuPont	06572	31-05-04
lenacil + phenmedipham			
DUK 880	DuPont	04121	
linuron			
Afalon	Makhteshim	11665	
Afalon	Aventis	08186	
Ashlade Linuron FL	Nufarm UK	06221	
DAPT Linuron 50 SC	DAPT	08058	
Linurex 50 SC	Makhteshim	07950	
Linuron Flowable	SumiAgro	09602	
MSS Linuron 500	Nufarm UK	08893	
linuron + trifluralin			
Chandor	Dow	05631	31-01-05
Linnet	SumiAgro	09601	
Neminfest	Montedison	02546	
Neminfest	Sipcam	07219	
magnesium phosphide			
Degesch Fumigation Pellets	Rentokil	11436	
Detia Gas-Ex-B Forte	Detia Degesch	10661	
malathion			
Fyfanon	Headland	11014	
Malathion 60	United Phosphorus	08018	
maleic hydrazide			
Bos MH 180	Crompton	06502	

Product	Marketing Co.	Reg. No.	Expiry Date
Cleancrop Malahide	United Agri	11066	
Fazor	Crompton	05461	
Royal MH 180	Mirfield	07840	
Royal MH 180	Crompton	07043	
mancozeb			
Agrichem Mancozeb 80	Agrichem	06354	
Agrizeb	Chimac-Agriphar	10980	
Alto Eco B	Syngenta	08466	
Ashlade Mancozeb FL	Nufarm UK	06226	
Barclay Manzeb 455	Barclay	07990	
Cleancrop Mancozeb	United Agri	11193	
Deny DF	Interfarm	10144	
Deny DF	Rohm & Haas	10797	
Deny WP	Interfarm	09899	30-09-04
Deny WP	Interfarm	10798	
Deny WP	Rohm & Haas	10140	
Dequiman MZ	Elf	06870	
Dithane 945	PBI	00719	
Dithane 945	SumiAgro	09889	
Dithane Dry Flowable Newtec	SumiAgro	09892	31-01-05
Dithane NT Dry Flowable	Interfarm	09868	
Dithane Superflo	Rohm & Haas	10142	
Headland Kor Flo	Headland	08019	
Headland Quell	Headland	08317	
Headland Zebra WP	Headland	07441	
Karamate Dry Flo	Landseer	09259	
Karamate N	Rohm & Haas	01125	
Karamate N	Dow	10802	31-01-05
Kor Flo	Interfarm	09895	
Kor NT Dry Flowable	Interfarm	09893	
Manconex	Griffin	09555	
Mancozeb 80	Dow	10804	
Mancozeb 80	Rohm & Haas	09896	
Manex II	Agrichem	07637	
Manfil	Indofil	11093	
Manzate 200 DF	Griffin	10081	29-02-04
Manzate 200 PI	Griffin	09480	
Manzate 75 WG	Headland	11046	
Manzate 75 WG	Griffin	10547	30-04-05
Micene 80	Sipcam	08560	
Mortar Flo	Griffin	09592	31-01-05
Nemispor	Sipcam	07348	
Opie 80 WP	SumiAgro	09890	31-01-05
Opie 80 WP	SumiAgro	08301	
Penncozeb	Cerexagri	07820	31-10-04
Penncozeb WDG	Nufarm UK	09690	31-07-05

Product	Marketing Co.	Reg. No.	Expiry Date
Penncozeb WDG	Nufarm UK	11065	
Restraint DF Newtec	Griffin	09755	31-01-05
Stefes Restraint	Bayer CropScience	08945	
Tariff 75 WG Newtec	SumiAgro	09891	31-01-05
Tariff 75 WG Newtec	SumiAgro	09756	30-09-04
Unicrop Mancozeb	Unicrop	05467	
mancozeb + metalaxyl			
Fubol 75 WP	Novartis	08409	
Osprey 58 WP	Novartis	08428	
Ridomil MZ 75	Syngenta	07640	
Ridomil MZ 75WP	Novartis	08438	02-11-04
mancozeb + metalaxyl-M			
Fubol Gold WP	Syngenta	08812	
mancozeb + propamocarb hydrochloride			
Aventis Tattoo	Aventis	09811	
mancozeb + zoxamide			
Electis 75 WG	Dow	11013	
RH 7281/Mancozeb 75 WG	Rohm & Haas	10564	30-09-05
RH 7281/Mancozeb 75 WG	Dow	10807	
Roxam 75 WG	Dow	11017	
Unikat 75 WG	Dow	11018	
maneb			
Agrichem Maneb 80	Agrichem	05474	
Ashlade Maneb Flowable	Ashlade	06477	
Luxan Maneb 80	Luxan	06570	
Maneb 80	Dow	10805	31-01-05
Manex	Griffin	09554	
Manex	Agrichem	07935	
RH Maneb 80	Rohm & Haas	01796	30-09-04
Trimangol 80	Cerexagri	06070	
Trimangol 80	Cerexagri	06871	
Trimangol WDG	Cerexagri	06992	
Unicrop Maneb 80	Unicrop	00920	
Unicrop Maneb FL	Unicrop	08025	
MCPA			
Agrichem MCPA-50	Agrichem	04097	
Agricorn 500	FCC	00055	
Agroxone 40	Marks	10264	
Agroxone 50	Mirfield	08345	
Agroxone 75	Marks	09208	
Atlas MCPA	Atlas	07717	
Atlas MCPA	Nufarm UK	03055	
Barclay Meadowman	Barclay	07639	

SECTION 5

Product	Marketing Co.	Reg. No.	Expiry Date
Barclay Meadowman II	Barclay	08525	
BH MCPA 75	RP Amenity	05395	
Circium II	Whyte Agrochem.	10293	
Dupont MCPA	DuPont	11738	
Empal	Unicrop	00795	
FCC Agricorn 50M	FCC	09032	
Marks MCPA P30	Marks	01292	
Marks MCPA S25	Marks	01294	
Marks MCPA SP	Marks	01293	
MCPA 25%	Nufarm UK	07998	
MCPA 25%	Nufarm UK	02766	
MCPA 500	Nufarm UK	08655	
MSS MCPA 50	Nufarm UK	10240	
MSS MCPA 50	Nufarm UK	11144	
Nufarm MCPA 750	Nufarm UK	11768	
Nufarm MCPA DMA 480	Nufarm UK	09686	
Nufarm MCPA DMA 500	Nufarm UK	09685	
Quadrangle MCPA 50	Quadrangle	10493	
Quadrangle MCPA 50	Quadrangle	06935	29-02-04
Tripart MCPA 50	Tripart	02206	
MCPA + MCPB			
Butoxone Plus	Headland	11080	
Butoxone Plus	Headland	10499	
MSS MCPB + MCPA	Nufarm UK	01413	
Trifolex-Tra	Cyanamid	07147	
Tropotox Plus	Nufarm UK	10701	
MCPA + mecoprop-P			
No More Weeds Lawn Feed	Wolf	10747	
Optica Combi	Marks	10118	
MCPB			
Marks MCPB	Marks	03497	29-02-04
Tropotox	Nufarm UK	10697	
mecoprop-P			
Astix	Aventis	06174	
Astix K	Bayer CropScience	06904	
BAS 03729H	Nufarm UK	05912	
Duplosan 500	BASF	07889	
Duplosan KV 500	BASF	08027	
Duplosan New System CMPP	BASF	04481	
Dupont Mecoprop-p	DuPont	11739	
Headland Charge	Headland	10981	
Isomec	Nufarm UK	11156	
Landgold Mecoprop-p	Landgold	06052	
MSS Mirprop	Nufarm UK	05911	

Product	Marketing Co.	Reg. No.	Expiry Date
MSS Optica	Mirfield	04973	
Optica	Marks	05814	
Optica 50	Marks	08283	
Raven B	Syngenta	08434	
mecoprop-P + metribuzin			
Centra	Bayer CropScience	11420	
Centra	Bayer	08112	
Optica Plus	Marks	09325	
mefluidide			
Check Turf II	Certified Laboratories	04463	
Embark	Gordon International	04810	
Gro-Tard II	Chemsearch	04462	
metalaxyl			
Polycote Universal	Syngenta	08431	02-11-04
metalaxyl + thiabendazole			
Apron T 69 WS	Syngenta	08387	02-11-04
Polycote Select	Germains	08430	29-02-04
Polycote Select	Germains	09718	02-11-04
metalaxyl-M			
SL 567A	Syngenta	10594	
metaldehyde			
Antares	CDP-Clartex	11654	
Aristo	De Sangosse	09824	31-12-04
Aristo	De Sangosse	11484	
Aristo M	De Sangosse	11599	
Bristol Blues	Doff Portland	11116	31-12-04
Bristol Blues	Doff Portland	11492	
Brits	Doff Portland	11435	
Clartex	CDP-Clartex	10942	
Clartex	SumiAgro	09213	31-03-05
Cleancrop Jekyll	United Agri	11724	
Condor	Doff Portland	11490	
Condor	Doff Portland	11113	31-12-04
Dixie 6	Greencrop	11488	
Dixie 6 M	De Sangosse	11597	
Doff Horticultural Slug Killer Blue Mini Pellets	Doff Portland	11463	
EM 1617/01	SumiAgro	09344	
Entice	CDP-Clartex	11090	
ESP	CDP-Clartex	10938	
ESP	SumiAgro	09428	31-05-05
FP 107	SumiAgro	09060	
Helimax	De Sangosse	07350	

Product	Marketing Co.	Reg. No.	Expiry Date
Helimax	De Sangosse	11720	
Lincoln VI	Doff Portland	11793	
Lincoln VI	Doff Portland	11114	31-12-04
Lincoln VI	Doff Portland	11491	28-02-05
Luxan 9363 Red	Luxan	11480	
Lynx	De Sangosse	09770	31-12-04
Lynx	De Sangosse	11482	
Lynx M	De Sangosse	11598	
Metasec M	De Sangosse	11600	
Murphy Slugits	Scotts	09576	
Optimol	CDP-Clartex	10939	
Optimol	SumiAgro	09061	
Optimol XL	SumiAgro	10502	
Optimol XL	CDP-Clartex	10940	
Pastel M	De Sangosse	11736	
Pesta	De Sangosse	11796	
Pesta	De Sangosse	11483	
Pesta	De Sangosse	10070	31-12-04
Pesta M	De Sangosse	11596	
Slug Pellets	CDP-Clartex	10941	
Slug Pellets	SumiAgro	10150	
Super Six	Doff Portland	11115	31-12-04
Super Six	Doff Portland	11489	28-02-05
Super Six	Doff Portland	11794	
Super-flor 6% Metaldehyde Slug Killer Mini Pellets	CMI	11487	28-02-05
Super-Flor 6% Metaldehyde Slug Killer Mini Pellets	De Sangosse	11789	
Unicrop 6% Mini Slug Pellets	Unicrop	09771	31-12-04
Unicrop 6% Mini Slug Pellets	Unicrop	11486	28-02-05
Unicrop 6% Mini Slug Pellets	Unicrop	11795	
metamitron			
Alpha Metamitron	Makhteshim	11081	
Barclay Seismic	Barclay	09323	
Barclay Seismic	Barclay	11377	
Barclay Seismic XL	Barclay	09328	
Bettix WG	United Phosphorus	11154	
Cobra	Interfarm	10578	
Goldbeet	Makhteshim	11538	
Goltix 90	Bayer CropScience	11283	28-02-06
Goltix Flowable	Bayer CropScience	11284	
Goltix WG	Bayer CropScience	11285	28-02-06
Homer	Bayer CropScience	11287	28-02-06
Homer	Makhteshim	11544	
Homer	Griffin	09834	
Homer	Bayer	10867	28-02-06

Product	Marketing Co.	Reg. No.	Expiry Date
Lektan	Bayer	06111	28-02-06
Lektan	Bayer CropScience	11290	28-02-06
Mitron 70 WG	Hermoo	11516	
Mitron 90 WG	Hermoo	10888	
Mitron SC	Hermoo	11643	
Mitron WG	Hermoo	10171	31-08-04
MM70	Gharda	09490	
Quartz	Quadrangle	05167	30-06-04
Quartz	Bayer CropScience	11431	30-06-04
Skater	Feinchemie	09790	
metam-sodium			
Fumetham	Chemical Nutrition	10047	
Metam 510	UCB	09796	
Sistan	Unicrop	01957	
Sistan 38	Unicrop	08646	
Vapam	United Agri	09194	
metazachlor			
Barclay Metaza	Barclay	09169	
Butisan S	BASF	11733	
Clayton Metazachlor	Clayton	09688	
Clayton Metazachlor 50 SC	Clayton	11719	
Clean Crop MTZ 500	United Agri	09222	
Fuego	Makhteshim	11177	
Fuego	Feinchemie	10296	
Fuego 50	Makhteshim	11473	
Marnoch Metazachlor	Marnoch	09761	
Rapsan 500 SC	Globachem	11468	
metazachlor + quinmerac			
Clayton Mazarac	Clayton	11161	
Katamaran	BASF	11732	
metconazole			
Caramba	Cyanamid	09864	
Clayton Tunik	Clayton	10545	
Pan Metconazole	Pan Agriculture	10470	
methabenzthiazuron			
Clayton Benson	Clayton	08687	
Tribunil	Bayer CropScience	11312	
Tribunil	Bayer	02169	
Tribunil WG	Bayer CropScience	11313	
Tribunil WG	Bayer	03260	
methiocarb			
Barclay Poacher	Barclay	09031	
Bayer UK 808	Bayer CropScience	11246	31-01-05

Product	Marketing Co.	Reg. No.	Expiry Date
Bayer UK 808	Bayer	09513	31-01-05
Bayer UK 809	Bayer	09514	
Bayer UK 809	Bayer CropScience	11247	
Bayer UK 892	Bayer	09540	31-01-05
Bayer UK 892	Bayer CropScience	11248	31-01-05
Bayer UK 935	Bayer	09541	28-02-05
Club	Syngenta	07176	31-01-05
Club	Bayer CropScience	11263	31-01-05
Decoy	Bayer	06535	
Decoy	Bayer CropScience	11264	
Decoy Plus	Bayer	07615	31-01-05
Decoy Plus	Bayer CropScience	11265	31-01-05
Draza	Bayer CropScience	11268	31-01-06
Draza	Bayer	00765	
Draza 2	Bayer	04748	31-01-05
Draza 2	Bayer CropScience	11269	
Draza Plus	Bayer	06553	31-01-05
Draza Plus	Bayer CropScience	11270	31-01-05
Draza ST	Bayer CropScience	11320	31-01-05
Draza ST	Bayer	05315	
Draza Wetex	Bayer CropScience	11271	
Draza Wetex	Bayer	09704	31-01-05
Elvitox	Bayer CropScience	11273	31-01-05
Elvitox	Bayer	06738	31-01-05
Epox	Bayer	06737	31-01-05
Epox	Bayer CropScience	11274	31-01-05
Exit	Bayer	07632	
Exit	Bayer CropScience	11275	31-01-05
Rescur	Bayer CropScience	11299	31-01-05
Rescur	Bayer	07942	31-01-05
methoxyfenozide			
RH-2485 240 SC	Bayer CropScience	11322	21-10-05
methyl bromide			
Bromomethane 100%	Brian Jones	09244	
Mebrom 100	Mebrom	04869	
Methyl Bromide	Rentokil	09316	
Methyl Bromide 100%	Bromine & Chem.	01336	
Rentokil Methyl Bromide	Rentokil	05646	
methyl bromide with chloropicrin			
Methyl Bromide 98	Bromine & Chem.	01335	
metosulam			
EF 1077	Bayer CropScience	11479	
EF 1077	Dow	07965	31-12-05

Product	Marketing Co.	Reg. No.	Expiry Date
metoxuron			
Dosaflo	Syngenta	09351	31-12-07
metribuzin			
Agriguard Metribuzin	AgriGuard	09853	
Bayer UK 093	Bayer CropScience	11240	
Cleancrop Metribuzin	United Agri	09621	
Cleancrop Solmet	United Agri	10865	
Python	Feinchemie	09791	
Tuberon	Unicrop	11061	
metsulfuron-methyl			
Agform MTS 20	Agform	10685	
Ally PX	DuPont	10844	
Barclay Flumen	Barclay	08752	
Clayton Metsulfuron	Clayton	06734	
Gem 690	DuPont	09436	31-05-04
Simba 20 DF	Griffin	09437	
metsulfuron-methyl + thifensulfuron-methyl			
Concert	DuPont	10984	
DP 928	DuPont	09632	
DP 928 PX	DuPont	10991	
Finish PX	DuPont	10989	
metsulfuron-methyl + tribenuron-methyl			
Biplay PX	DuPont	10990	
DP 911 PX	DuPont	10992	
DP 911 WSB	DuPont	09867	
myclobutanil			
Systhane 20 EW	Whelehan	09397	
Systhane 6 Flo	Whelehan	0000	
Systhane 6 W	Rohm & Haas	04570	
Systhane 6 W	Dow	09611	
Systhane 6W	SumiAgro	04571	
2-(1-naphthyl)acetic acid			
Rhizopon B Powder (0.1%)	Fargro	07133	
Rhizopon B Powder (0.2%)	Fargro	07134	
Rhizopon B Tablets	Fargro	07135	
Tipoff	Unicrop	05878	
napropamide			
AC 650	United Phosphorus	11102	
Devrinol	United Phosphorus	09375	
nicosulfuron			
Samson	Syngenta	10433	

SECTION 5

Product	Marketing Co.	Reg. No.	Expiry Date
nicotine			
No-FID	Certis	07959	
oxamyl			
Fielder	DuPont	05279	31-05-04
paclobutrazol			
Bonzi	Syngenta	06640	
Cultar	Zeneca	06649	
paraquat			
Barclay Total	Barclay	11122	
Barclay Total	Barclay	08822	31-10-04
Clayton Paragon	Clayton	11210	
Gramoxone 100	Zeneca	06674	
Speedway Liquid	Scotts	07744	
pencycuron			
Marnoch Penor D	Marnoch	09871	
Monceren Flowable	Bayer CropScience	11425	
Monceren Flowable	Bayer	04907	
Pan Pencycuron P	Pan Agriculture	10494	
Tubercare 12.5 DS	PG-Crop Protection	11147	
pendimethalin			
Barclay Tremor	Barclay	09188	
Blazer	Cheminova	10258	
Claymore	BASF	09232	31-07-04
Clayton Pendalin	Clayton	09708	
Cleancrop Stomp	United Agri	10696	
Ipimethalin	I Pi Ci	10664	
Ipimethalin	SumiAgro	09701	30-06-04
Parapet	Dow	10700	
PDM 330 EC	BASF	11084	
Plinth	SumiAgro	09702	30-06-04
Plinth	SumiAgro	10665	30-06-05
Sidewinder	Dow	10699	
Sovereign	BASF	08151	29-02-04
Sovereign	Novartis	08533	
Sovereign	BASF	10228	
Stomp 400 SC	BASF	11777	
Stomp 400 SC	Cyanamid	04183	
pendimethalin + simazine			
Deuce	BASF	06746	
Deuce	BASF	10383	
Merit	Cyanamid	04976	

Product	Marketing Co.	Reg. No.	Expiry Date
Peniophora gigantea			
PG Suspension	Ecological	08975	28-02-05
PG Suspension	Forest Research	11772	
permethrin			
Permit	SumiAgro	01577	
Permit	SumiAgro	09609	
phenmedipham			
Atlas Protrum K	Nufarm UK	03089	
Atlas Protrum K	Atlas	07723	
Barclay Punter XL	Barclay	08047	
Betanal E	Aventis	07248	
Betanal Flo	Aventis	08898	
Betosip	Sipcam	06787	
Betosip 114	Sipcam	05910	
Cleancrop Phenmedipham	United Agri	11586	
Dancer Flow	Sipcam	11405	
Hickson Phenmedipham	Griffin	09961	
Hickson Phenmedipham	Nufarm UK	11124	
Kemifam E	Bayer CropScience	08978	31-10-05
Kemifam E	Kemira	06104	
Luxan Phenmedipham	Luxan	06933	
Mandolin	Nufarm UK	11127	
Mandolin	Griffin	09456	
Mandolin Flo	Nufarm UK	11125	
Mandolin Flo	Griffin	10368	
MSS Protrum G	Nufarm UK	08342	
Protrum G	Nufarm UK	11133	
Stefes Forte 2	Stefes	08204	
Stefes Forte 2	Bayer CropScience	10083	31-03-04
Stefes Medipham 2	Stefes	08203	
Stefes Medipham 2	Bayer CropScience	10087	31-03-04
picloram			
Tordon 22K	Dow	05083	
picolinafen			
AC 900001	BASF	10714	12-08-04
pirimicarb			
Aphox	Zeneca	06633	
Barclay Pirimisect	Barclay	11300	
Barclay Pirimisect	Barclay	09057	
Clayton Pirimicarb 50 SG	Clayton	09221	
Cleancrop Miricide	United Agri	11776	
Pirimate	Portman	09568	

Product	Marketing Co.	Reg. No.	Expiry Date
pirimiphos-methyl			
Actellic Smoke Generator No. 10	Syngenta	07900	31-01-04
Actellic Smoke Generator No. 10	Syngenta	10448	
prochloraz			
Alpha Prochloraz 40 EC	Makhteshim	11002	
Barclay Eyetak 40	Barclay	07843	
Barclay Eyetak 450	Barclay	09484	
Mirage 45 SC	Makhteshim	10894	
Octave	Aventis	07267	
Octave	BASF	11741	
Prelude 20 LF	Bayer CropScience	07269	
Sporgon 50WP	Sylvan	08802	
Sportak 45 HF	Bayer CropScience	07287	
Stefes Poraz	Bayer CropScience	10090	31-01-04
Stefes Poraz	Stefes	07528	
prochloraz + propiconazole			
Bumper Excell	Makhteshim	11015	
prometryn			
Alpha Prometryne 80 WP	Makhteshim	04795	31-12-07
prometryn + terbutryn			
P-Weed	SumiAgro	08574	31-12-07
P-Weed	SumiAgro	09608	31-12-07
propachlor			
Alpha Propachlor 65 WP	Makhteshim	04807	
Atlas Orange	Atlas	03096	30-09-04
Atlas Propachlor	Atlas	06462	
Portman Brasson	Portman	08158	
Portman Propachlor 50 FL	Portman	06892	
Portman Propachlor SC	Portman	08159	
Ramrod 20 Granular	Monsanto	10313	
Ramrod 20 Granular	Monsanto	01687	
Ramrod Flowable	Monsanto	01688	
Ramrod Granular	Hortichem	05806	
propamocarb hydrochloride			
Previcur N	Bayer CropScience	08575	
propaquizafop			
Agil	Makhteshim	11048	
Bulldog	Makhteshim	11723	
Cleancrop GYR	United Agri	10646	
Falcon	Novartis	09384	
Headland Rocket	Headland	10998	
Raptor	Makhteshim	11092	

Product	Marketing Co.	Reg. No.	Expiry Date
Shogun	Syngenta	08443	31-03-04
Shogun	Makhteshim	10584	
propiconazole			
Barclay Bolt	Barclay	08341	
Controller Fungicide	Ritefeed	10340	
Mantis	Syngenta	08423	
Radar	Syngenta	09168	
Tilt	Syngenta	08456	
Tumbleblite	Scotts	07567	
propiconazole + trifloxystrobin			
Rombus	Syngenta	10262	31-01-04
Rombus	Bayer	10464	
propineb			
Antracol	Bayer CropScience	11237	
Antracol	Bayer	00104	
propyzamide			
Bulwark Flo	Interfarm	10755	
Clayton Propel	Clayton	09783	
Clayton Propel 80 WG	Clayton	11216	
Flomide	Alfa	11103	
Kerb 50 W	SumiAgro	10166	
Kerb 50W	SumiAgro	01133	
Kerb 80 EDF	Interfarm	10201	31-01-05
Kerb 80 EDF	SumiAgro	10163	31-01-05
Kerb Flo	SumiAgro	10167	31-01-05
Kerb Granules	Dow	10803	
Kerb Granules	Rohm & Haas	01136	
Kerb Granules	SumiAgro	01135	
Kerb Pro Granules	SumiAgro	08698	
Menace 80 EDF	Interfarm	10165	31-01-05
Precis	SumiAgro	10159	31-07-04
Precis	Interfarm	10737	31-01-05
Precis Flo	SumiAgro	08530	30-09-04
Propose	Alfa	10890	31-08-05
Rapior	MTM Agrochem.	05314	
Stefes Pride	Bayer CropScience	08899	30-09-04
Top Farm Propyzamide 500	Top Farm	05484	
prosulfuron			
SL 600	Syngenta	08633	17-02-04
pymetrozine			
Chess	Syngenta Bioline	09817	
Plenum	Syngenta	09816	

Product	Marketing Co.	Reg. No.	Expiry Date
pyraclostrobin			
Clayton Raucous	Clayton	11636	20-11-04
pyrethrins			
Py Spray Garden Insect Killer	Vitax	06085	
pyrethrins + resmethrin			
Pynosect 30 Water Miscible	Mitchell Cotts	01653	
quinoxyfen			
DOE 3000	Dow	11009	17-02-04
quizalofop-P-ethyl			
Pilot D	Aventis	08041	
Poseidon	Nissan	11040	
rotenone			
Devcol Liquid Derris	Nehra	06063	
simazine			
Alpha Simazine 50 WP	Makhteshim	04879	
Alpha Simazine 80 WP	Makhteshim	04800	
Atlas Simazine	Nufarm UK	05610	
Atlas Simazine	Nufarm UK	07725	
MSS Simazine 50 FL	Mirfield	01418	
Simazine 90WG	Sipcam	10641	
Unicrop Flowable Simazine	Unicrop	05447	
Unicrop Simazine 50	Unicrop	02646	
Unicrop Simazine FL	Unicrop	08032	
sodium chlorate			
Deosan Chlorate Weedkiller (Fire Suppressed)	JohnsonDiversey	08521	
Gem Sodium Chlorate Weedkiller	Joseph Metcalf	04276	
Growing Success Sodium Chlorate Weedkiller	Growing Success	10787	
Morgan's Sodium Chlorate Weedkiller	David Morgan	08927	30-09-04
Sodium Chlorate	Marlow Chemical	06294	
Strathclyde Sodium Chlorate Weedclear	Strathclyde	07420	
TWK Total Weedkiller	Reabrook	06393	
spiroxamine			
Accrue	Bayer CropScience	11219	
Accrue	Bayer	08335	01-09-09
Bayer UK 477	Bayer CropScience	11222	
Bayer UK 477	Bayer	08330	01-09-09
Talvin	Bayer CropScience	11228	
Talvin	Bayer	09253	01-09-09

Product	Marketing Co.	Reg. No.	Expiry Date
spiroxamine + tebuconazole			
Array	Bayer CropScience	11238	
Array	Bayer	09352	
Bayer UK 552	Bayer CropScience	11245	
Bayer UK 552	Bayer	10410	
Bronze	Bayer CropScience	11419	
Bronze	Bayer	10412	
Draco	Bayer CropScience	11267	
Draco	Bayer	09353	
Folicur Star	Bayer CropScience	11531	
sulfosulfuron			
Monitor C	Monsanto	10602	19-02-04
Monitor Sachet	Monsanto	10600	19-02-04
sulphur			
Ashlade Sulphur FL	Nufarm UK	06478	
Headland Venus	Headland	10611	
Headland Venus	Headland	09572	
Kumulus S	BASF	01170	
Microthiol Special	Cerexagri	06268	
MSS Sulphur 80	Mirfield	05752	
MSS Sulphur 80 WP	Mirfield	03225	31-01-05
Solfa	Nufarm UK	11390	31-03-06
Solfa WG	Nufarm UK	11602	
Thiomex	Omex	11176	
Thiovit	Syngenta	10589	31-12-04
Thiovit	Syngenta	08493	
tar oils			
Jeyes Fluid	Jeyes	04606	
tebuconazole			
Barclay Busker	Barclay	08994	
Barclay Busker	Barclay	11345	
Bayer UK 226	Bayer CropScience	11241	
Bayer UK 226	Bayer	09504	
Clayton Tebucon	Clayton	08707	
Gainer	Bayer CropScience	11279	
Gainer	Bayer	08692	
Halt	Bayer CropScience	11286	
Halt	Bayer	08693	
Raxil	Bayer CropScience	11295	
Raxil	Bayer	08400	
UK 200	Bayer CropScience	11316	
UK 200	Bayer	08246	
tebuconazole + triadimenol			
Garnet	Bayer CropScience	11280	

SECTION 5

Product	Marketing Co.	Reg. No.	Expiry Date
Garnet	Bayer	06391	
Ruby	Bayer CropScience	11302	
Ruby	Bayer	06389	
Silvacur 300	Bayer CropScience	11310	
Silvacur 300	Bayer	08056	
tebuconazole + triazoxide			
Raxil S CF	Bayer CropScience	11297	
Raxil S CF	Bayer	08192	
tebufenpyrad			
Masai	Cyanamid	07452	
Masai G	BASF	07453	31-03-05
Masai G	BASF	10224	
tefluthrin			
Evict	Syngenta	11735	
Evict	Bayer CropScience	11673	
Force ST	Syngenta	11752	
Force ST	Bayer	09713	31-05-05
terbutryn			
Alpha Terbutryn 50 SC	Makhteshim	04809	
Prebane	Novartis	08432	
tetraconazole			
Digit	Monsanto	09409	
Digit	Monsanto	10503	
Domark	Isagro	10452	
Emerald	Isagro	10449	
Eminent	Monsanto	09410	
Eminent	Monsanto	10504	
MON 10 EC	Monsanto	09347	
MON 10 EC	Monsanto	10505	
Omen	Sipcam	09413	
Tetra 5634	Monsanto	09411	
Tetra 5634	Monsanto	10507	
thiabendazole			
Storite Clear Liquid	Syngenta	09503	
Storite Excel	Syngenta	08993	29-02-04
Tecto Flowable Turf Fungicide	Vitax	06273	
Tecto Superflowable Turf Fungicide	Vitax	09258	
thiabendazole + thiram			
Hy-Vic	Agrichem	06247	
thifensulfuron-methyl			
DUK 118	DuPont	04596	
Harmony PX	DuPont	10845	

Product	Marketing Co.	Reg. No.	Expiry Date
thifensulfuron-methyl + tribenuron-methyl			
DUK 110	DuPont	09189	
Gex 353 A	Griffin	10185	30-11-05
thiodicarb			
Genesis	Bayer CropScience	11712	
Genesis	RP Agric.	06168	
Genesis ST	Bayer CropScience	11710	
Genesis ST	Bayer CropScience	09994	31-12-04
Judge	Bayer CropScience	11711	
thiophanate-methyl			
Cercobin Liquid	Bayer CropScience	06188	31-10-05
Cercobin Liquid	Aventis	09984	
Mildothane Liquid	Aventis	10002	
Mildothane Liquid	Hortichem	06211	
thiram			
Agrichem Flowable Thiram	Agrichem	06245	30-09-04
Hy-Flo	Agrichem	04637	
Robinson's Thiram 60	Agrichem	04638	
Thiraflo	Crompton	11225	
Thiraflo	Uniroyal	09496	31-10-05
Thyram Plus	Agrichem	10421	30-09-04
tolclofos-methyl			
Rizolex 50 WP	Bayer CropScience	07272	
tralkoxydim			
Grasp	Zeneca	06676	
Standon Tralkoxydim	Standon	08326	
triadimenol			
Bayfidan	Bayer CropScience	11417	
Bayfidan	Bayer	02672	
Hi-Shot	BASF	06508	
Hi-Shot	BASF	10397	
Spinnaker	BASF	10398	
Spinnaker	Cyanamid	07023	
tri-allate			
Avadex 15G	Monsanto	10167	31-07-04
Avadex 15G	Monsanto	10297	
Avadex BW Granular	Monsanto	10298	
Avadex BW Granular	Monsanto	00174	
Avadex Excel 15G	Monsanto	07117	31-07-04
Avadex Excel 15G	Monsanto	10299	
triasulfuron			
Banvel T	Syngenta	08470	

Product	Marketing Co.	Reg. No.	Expiry Date
Lo-Gran	Novartis	08421	
Raven A	Syngenta	08434	
triazamate			
Aztec	Cyanamid	07817	
tribenuron-methyl			
Quantum PX	DuPont	10843	
triclopyr			
Altix 240 EC	Chimac-Agriphar	10648	
Chipman Garlon 4	n2n	06016	
Cleancrop Triptic 48 EC	United Agri	11568	
Cleancrop Unival	United Agri	11388	
Garlon 2	Dow	05682	
Garlon 2	Zeneca	06616	
Triptic 48 EC	Chimac-Agriphar	10891	31-10-04
Triptic 48 EC	United Phosphorus	09294	
Unival	Chimac-Agriphar	10636	31-10-05
trifloxystrobin			
Aprix	Bayer CropScience	11220	
Flint	Bayer CropScience	11259	
Twist 500 SC	Bayer CropScience	11231	
Twist 500 SC	Bayer	10880	19-12-04
Zest SC	Bayer CropScience	11233	
Zest SC	Bayer	10885	19-12-04
trifluralin			
Ashlade Trifluralin	Nufarm UK	08303	
Ashlade Trimaran	Nufarm UK	06228	
Atlas Trifluralin	Nufarm UK	08498	
DAPT Trifluralin 48 EC	DAPT	07906	
Digermin	Sipcam	07221	
Digermin	Montedison	00701	
FCC Trigard	FCC	09030	
Ipifluor	I Pi Ci	04692	
MSS Trifluralin 48 EC	Nufarm UK	10749	31-12-05
MSS Trifluralin 48 EC	Nufarm UK	07753	
Portman Trifluralin	Portman	05751	
Triflur	Nufarm UK	09670	31-08-04
Triflur	Nufarm UK	10750	
Triflurex 48 EC	Makhteshim	07947	
Trifsan	Dow	09237	
Trigard	SumiAgro	09699	
Triplen	Sipcam	05897	
Tristar	SumiAgro	09612	
Whyte Trifluralin	Nufarm UK	09286	

Product	Marketing Co.	Reg. No.	Expiry Date
triflusulfuron-methyl			
Debut WSB	DuPont	07809	
vinclozolin			
Barclay Flotilla	Barclay	07905	
Landgold Vinclozolin DG	Landgold	06500	
Marnoch Vinol	Marnoch	09815	
Ronilan DF	BASF	04456	
zeta-cypermethrin			
Fury 10 EW	FMC	10268	31-01-05
Fury 10 EW	FMC	10718	
Minuet EW	FMC	10269	31-01-05
Minuet EW	FMC	10716	

SECTION 6
APPENDICES

Appendix 1
Suppliers of Pesticides and Adjuvants

Agrichem: Agrichem (International) Ltd
Industrial Estate
Station Road
Whittlesey
Cambs.
PE7 2EY
Tel: (01733) 204019
Fax: (01733) 204162
Email: info@agrichem.co.uk
Web: www.agrichem.co.uk

Agricola: Agricola Ltd
Abbey Farm
Snape
Saxmundham
Suffolk
IP17 1RQ
Tel: (01728) 688075
Fax: (01728) 688989

AgriGuard: AgriGuard Ltd
Unit 3
Tally House
Broomfield Business Park
Malahide
Co. Dublin
Ireland
Tel: (+353) 1 845 1599
Fax: (+353) 1 845 1620
Email: info@agriguard.ie
Web: www.agriguard.ie

Amega: Amega Sciences
Lanchester Way
Royal Oak Industrial Estate
Daventry
Northants.
NN11 5PU
Tel: (01327) 704444
Fax: (01327) 71154
Email: admin@amega-sciences.com
Web: www.amega-sciences.com

Antec: Antec International Ltd
Windham Road
Chilton Industrial Estate
Sudbury
Suffolk
CO10 2XD
Tel: (01787) 377305
Fax: (01787) 310846
Email: biosecurity@antecint.com
Web: www.antecint.com

Aquaspersions: Aquaspersions Ltd
Beacon Hill Road
Halifax
W. Yorks.
HX3 6AQ
Tel: (01422) 386200
Fax: (01422) 386239
Email: info@aquaspersions.co.uk
Web: www.aquaspersions.com

Aventis: Aventis CropScience UK Ltd
See Bayer CropScience Limited

B H & B: Battle Hayward & Bower Ltd
Victoria Chemical Works
Crofton Drive
Allenby Road Industrial Estate
Lincoln
LN3 4NP
Tel: (01522) 529206
Fax: (01522) 538960

Banks Cargill: Banks Cargill Agriculture Ltd
Cattle Market Chase
Wisbech
Cambs.
PE13 1RE
Tel: (01945) 582901
Email: david_turner@bankscargill.co.uk
Web: www.bankscargill.co.uk

Barclay: Barclay Chemicals Manufacturing Ltd
Damastown Way
Damastown Industrial Park
Mulhuddart
Dublin 15
Ireland
Tel: (+353) 1 822 4555
Fax: (+353) 1 822 4678
Email: info@barclay.ie
Web: www.barclay.ie

Barrier: Barrier BioTech Ltd
36/37 Haverscroft Industrial Estate
New Road
Attleborough
Norfolk
NR17 1YE
Tel: (01953) 456363
Fax: (01953) 455594
Email: sales@barrier-biotech.com
Web: www.barrier-biotech.com

BASF: BASF plc.
Agricultural Divison
PO Box 4, Earl Road
Cheadle Hulme
Cheadle
Cheshire
SK8 6QG
Tel: (0161) 485 6222
Fax: (0161) 485 2229
Web: www.agricentre.co.uk

Bayer: Bayer plc.
See Bayer CropScience Limited

Bayer CropScience: Bayer CropScience Limited
Hauxton
Cambridge
CB2 5HU
Tel: (01223) 870312
Fax: (01223) 872142
Email: tech@bayercropscience.co.uk
Web: www.bayercropscience.co.uk

Bayer Environ.: Bayer Environmental Science
Durkan House
214-224 High Street
Waltham Cross
Herts.
EN8 7DP
Tel: (01992) 784270
Fax: (01992) 784276
Email: john.hall1@bayercropscience.com
Web: www.bayer-escience.co.uk

Belchim: Belchim Crop Protection Ltd
Suite 2, Unit 3
Phoenix Park
Eaton Socon
St Neots
Cambs.
PE19 8EP
Tel: (01480) 403333
Fax: (01480) 403444
Email: info@belchim.com
Web: www.belchim.com

Brian Jones: Brian Jones and Associates Ltd
Fluorocarbon Building
Caxton Hill
Hertford
Herts.
SG13 7NH
Tel: (01992) 553065
Fax: (01992) 551873
Email: info@brian-jones.co.uk
Web: www.brian-jones.co.uk

Cardel: Cardel Agro SPRL
Avenue de Tervuren 270-272
11, Rue Pascal
B-1150 Brussels
Belgium
Tel: (+32) 2776 4157

Certis: Certis
1b Mills Way
Boscombe Down Business Park
Amesbury
Wilts.
SP4 7RX
Tel: (01980) 676500
Fax: (01980) 626555
Email: certis@certiseurope.co.uk
Web: www.certiseurope.co.uk

Cheminova: Cheminova Agro (UK) Ltd
Norfolk House
Gt. Chesterford Court
Gt. Chesterford
Saffron Walden
Essex
CB10 1PF
Tel: (01799) 530146
Fax: (01799) 530229

Chiltern: Chiltern Farm Chemicals Ltd
East Mellwaters
Stainmore
Bowes
Barnard Castle
Co. Durham
DL12 9RH
Tel: (01833) 628282
Fax: (01833) 628020
Web: www.chilternfarm.com

Ciba Specialty: Ciba Specialty Chemicals
Water Treatments Limited
P O Box 38
Low Moor
Bradford
W. Yorks.
BD12 0JZ
Tel: (01274) 417549
Fax: (01274) 417305
Email: soil.additives@cibasc.com
Web: www.cibasc.com

CMI: CMI Ltd (incorporating Collingham Marketing)
United House
113 High Street
Collingham
Newark
Notts.
NG23 7NG
Tel: (01636) 892078
Fax: (01636) 893037
Email: info@cmi.ltd.uk
Web: http:www.cmi.ltd.uk

Coalite: Coalite Chemicals
PO Box 152
Buttermilk Lane
Bolsover
Chesterfield
Derbyshire
S44 6AZ
Tel: (01246) 826816
Fax: (01246) 240309
Email: sales@coalitechemicals.com
Web: www.coalitechemicals.com

Coventry Chemicals: Coventry Chemicals Ltd
Woodhams Road
Siskin Drive
Coventry
CV3 4FX
Tel: (024) 7663 9739
Fax: (024) 7663 9717

Crompton: Crompton Europe Ltd
Kennet House
4 Langley Quay
Slough
Berks.
SL3 6EH
Tel: (01753) 603000
Fax: (01753) 603077

DAPT: DAPT Agrochemicals Limited
14 Monks Walk
Southfleet
Gravesend
Kent
DA13 9NZ
Tel: (01474) 834448
Fax: (01474) 834449
Email: rkjltd@supanet.com

Dax: Dax Products Ltd
P O Box 119
Nottingham
NG5 4WL
Tel: (0115) 926 9996
Fax: (0115) 966 1173
Web: www.daxproducts.co.uk

De Sangosse: De Sangosse UK
PO Box 191
Romsey
Hants
SO51 5YP
Tel: (023) 8064 4567
Fax: (023) 8064 2121
Email: Eric.Gussin@farmline.com
Web: www.desangosse.co.uk

Dewco-Lloyd: Dewco-Lloyd Ltd
Cyder House
Ixworth
Suffolk
IP31 2HT
Tel: (01359) 230555
Fax: (01359) 232553

Doff Portland: Doff Portland Ltd
Aerial Way
Hucknall
Nottingham
NG15 6DW
Tel: (0115) 963 2842
Fax: (0115) 963 8657
Email: info@doff.co.uk
Web: www.doff.co.uk

Dow: Dow AgroSciences
Latchmore Court
Brand Street
Hitchin
Herts.
SG5 1NH
Tel: (01462) 457272
Fax: (01462) 426605
Email: fhibotl@dow.com
Web: www.dowagro.com/uk

DuPont: DuPont (UK) Ltd
Agricultural Products Department
Wedgwood Way
Stevenage
Herts.
SG1 4QN
Tel: (01438) 734000
Fax: (01438) 734452

Elliott: Thomas Elliott Ltd
143A High Street
Edenbridge
Kent
TN8 5AX
Tel: (01732) 866566
Fax: (01732) 864709

SECTION 6

Euroagkem: Euroagkem Ltd
Byemoor Farm
Melmerby
Leyburn
N. Yorks.
DL8 4TW
Tel: (01969) 640655
Fax: (01969) 640633

Fargro: Fargro Ltd
Toddington Lane
Littlehampton
Sussex
BN17 7PP
Tel: (01903) 721591
Fax: (01903) 730737
Email: promos@fargro.co.uk
Web: www.fargro.co.uk

FCC: Farmers Crop Chemicals Ltd
Carmichael House
The Green
Inkberrow
Worcs.
WR7 4DZ
Tel: (01386) 793401
Fax: (01386) 793184
enquiries@farmerscropchemicals.com
Web: www.farmerscropchemicals.com

Fine: Fine Agrochemicals Ltd
Hill End House
Whittington
Worcester
WR5 2RQ
Tel: (01905) 361800
Fax: (01905) 361810
Email: enquire@fine-agrochemicals.com
Web: www.fine-agrochemicals.com

FMC: FMC Chemicals Ltd, Agricultural Group
See Belchim Crop Protection Ltd

Ford Smith: Ford Smith & Co. Ltd
Lyndean Industrial Estate
Felixstowe Road
Abbey Wood
London
SE2 9SG
Tel: (020) 8310 8127
Fax: (020) 8310 9563
Email: fordsmithltd@aol.com

Greenaway: Greenaway Amenity Ltd
7 Browntoft Lane
Donington
Spalding
Lincs
PE11 4TQ
Tel: (01775) 821031
Fax: (01775) 821034
Email: greenawayamenity@aol.com

Greencrop: Greencrop Technology Ltd
Burren House
2 Cowbrook Court
Glossop
Derbyshire
SK13 8SL
Tel: (01457) 856001
Fax: (01457) 857137
Email: mail@greencrop-technology.co.uk
Web: www.greencrop-technology.co.uk

Greenhill: Greenhill
Rawdon Works
Rawdon Road
Moira
Swadlincote
Derbyshire
DE12 6DA
Tel: (01283) 551500
Fax: (01283) 550413

Headland: Headland Agrochemicals Ltd
Norfolk House
Gt. Chesterford Court
Gt. Chesterford
Saffron Walden
Essex
CB10 1PF
Tel: (01799) 530146
Fax: (01799) 530229
Email: enquiry@headlandgroup.com
Web: www.headland-ag.co.uk

I T Agro: I T Agro Ltd
805 Salisbury House
31 Finsbury Circus
London
EC2M 5SQ
Tel: (+45) 98 93 89 77
Fax: (+45) 98 93 80 01
Email: kiil@inter-trade.dk
Web: www.inter-trade.dk

Interagro: Interagro (UK) Ltd
Sworders Barn
Sworders Yard
North Street
Bishop's Stortford
Herts.
CM23 2LD
Tel: (01279) 501995
Fax: (01279) 501996
Email: info@interagro.co.uk
Web: www.interagro.co.uk

Interfarm: Interfarm (UK) Ltd
Kinghams's Place
36 Newgate Street
Doddington
Cambs.
PE15 0SR
Tel: (01354) 741414
Fax: (01354) 741004
Email: technical@interfarm.co.uk
Web: www.interfarm.co.uk

Intracrop: Intracrop
Byemoor Farm
Melmerby
Leyburn
N. Yorks.
DL8 4TW
Tel: (01969) 640655
Fax: (01969) 640633

Irish Drugs: Irish Drugs Ltd
Burnfoot
Lifford
Co. Donegal
Ireland
Tel: (+353) 74 9368104
Fax: (+353) 74 9368311
Email: idl@eircom.net

Isagro: Isagro Sp.A
See Sipcam UK Ltd

JohnsonDiversey: JohnsonDiversey Ltd
Weston Favell Centre
Northampton
NN3 8PD
Tel: (01604) 783505
Fax: (01604) 783506

K & S Fumigation: K & S Fumigation Services Ltd
Shirley Farmhouse
Moor Lane
Woodchurch
Ashford
Kent
TN26 3SS
Tel: (01233) 758252
Fax: (01233) 758343
Email: k.s.treatments@talk21.com

Killgerm: Killgerm Chemicals Ltd
115 Wakefield Road
Flushdyke
Ossett
W. Yorks.
WF5 9AR
Tel: (01924) 268400
Fax: (01924) 264757
Email: info@killgerm.com
Web: www.killgerm.com

Koppert: Koppert (UK) Ltd
Homefield Road
Haverhill
Suffolk
CB9 8QP
Tel: (01440) 704488
Fax: (01440) 704487
Web: www.koppert.com

Landgold: Landgold & Co. Ltd
PO Box 829
Charles House
Charles Street
St. Helier
Jersey
JE4 0UE
Tel: (01534) 768446
Fax: (01534) 732843

Landseer: Landseer Ltd
Lodge Farm
Goat Hall Lane
Galleywood
Chelmsford
Essex
CM2 8PH
Tel: (01245) 357109
Fax: (01245) 494165

Loveland: Loveland Industries Ltd
Swaffham Bulbeck
Cambridge
CB5 0LU
Tel: (01223) 811215
Fax: (01223) 812725
Email: info@loveland.co.uk
Web: www.loveland.co.uk

SECTION 6

Luxan: Luxan (UK) Ltd
Sysonby Lodge
Nottingham Road
Melton Mowbray
Leics.
LE13 0NU
Tel: (01664) 566372
Fax: (01664) 480137
Email: enquiry@luxan.co.uk
Web: www.luxan.co.uk

Makhteshim: Makhteshim-Agan (UK) Ltd
Unit 16
Thatcham Business Village
Colthrop Way
Thatcham
Berks.
RG19 4LW
Tel: (01635) 860555
Fax: (01635) 861555
Email: admin@mauk.co.uk
Web: www.mauk.co.uk

Mandops: Mandops (UK) Ltd
36 Leigh Road
Eastleigh
Hants.
SO50 9DT
Tel: (023) 8064 1826
Fax: (023) 8062 9106
Email: enquiries@mandops.co.uk
Web: www.mandops.co.uk

Me2: Me2 Crop Protection Ltd
Tower House
Fishergate
York
YO10 4HA
Tel: (01904) 567331
Fax: (01904) 720094
Email: sales@me2cpl.co.uk
Web: www.me2cpl.co.uk

Microcide: Microcide Ltd
Shepherds Grove
Stanton
Bury St. Edmunds
Suffolk
IP31 2AR
Tel: (01359) 251077
Fax: (01359) 251545
Email: microcide@microcide co.uk
Web: www.microcide.co.uk

Mitchell Cotts: Mitchell Cotts Chemicals Ltd
PO Box 6
Steanard Lane
Mirfield
W. Yorks.
WF14 8QB
Tel: (01924) 493861
Fax: (01924) 490972

Monsanto: Monsanto (UK) Ltd
The Maris Centre
Hauxton Road
Trumpington
Cambridge
CB2 2LQ
Tel: (01223) 849200
Fax: (01223) 849414
Email:
technical.helpline.uk@monsanto.com
Web: www.monsanto-ag.co.uk

n2n: n2n enviro Ltd
Portland Building
Portland Street
Staple Hill
Bristol
BS16 4PS
Tel: (0117) 957 4574
Fax: (0117) 956 3461
Email: weedkiller@n2nenviro.co.uk
Web: www.n2nenviro.co.uk

Novartis: Novartis Crop Protection UK Ltd
see Syngenta Crop Protection UK Limited

Novartis A H: Novartis Animal Health UK Ltd
New Cambridge House
Litlington
Royston
Herts.
SG8 0SS
Tel: (01763) 850500
Fax: (01763) 850490

Nufarm UK: Nufarm UK Ltd
Crabtree Manorway North
Belvedere
Kent
DA17 6BQ
Tel: (020) 8319 7222
Fax: (020) 8319 7280
Email: infouk@uk.nufarm.com
Web: www.ag.nufarm.co.uk

Pan Agriculture: Pan Agriculture Ltd
8 Cromwell Mews
Station Road
St Ives
Huntingdon
Cambs.
PE27 5HJ
Tel: (01480) 467790
Fax: (01480) 467041
Email: info@panagriculture.co.uk

Reabrook: Reabrook Ltd
Stanhope Road
Swadlincote
Burton-on-Trent
Staffs.
DE11 9BE
Tel: (01283) 221044
Fax: (01283) 225731

Rentokil: Rentokil Initial plc
Felcourt
East Grinstead
W. Sussex
RH19 2JY
Tel: (01342) 833022
Fax: (01342) 326229
Web: www.ri-research.com

Rhizopon: Rhizopon UK Ltd
Croda Rosa
12 Bixley Road
Ipswich
Suffolk
IP3 8PL
Tel: (01473) 712666
Fax: (01473) 712666

Rigby Taylor: Rigby Taylor Ltd
Rigby Taylor House
Crown Lane
Horwich
Bolton
Lancs.
BL6 5HP
Tel: (01204) 677777
Fax: (01204) 677765
Email: info@rigbytaylor.com
Web: www.rigbytaylor.com

Roebuck Eyot: Roebuck Eyot Ltd
7a Hatfield Way
South Church Enterprise Park
Bishop Auckland
Co. Durham
DL14 6XF
Tel: (01388) 772233
Fax: (01388) 775233
Email: sales@roebuck-eyot.co.uk
Web: www.roebuck-eyot.co.uk

Rohm & Haas: Rohm & Haas (UK) Ltd
See Dow AgroSciences

Scotts: The Scotts Company (UK) Ltd
Paper Mill Lane
Bramford
Ipswich
Suffolk
IP8 4BZ
Tel: (01473) 830492
Fax: (01473) 830386

Sinclair: William Sinclair Horticulture Ltd
Firth Road
Lincoln
LN6 7AH
Tel: (01522) 537561
Fax: (01522) 513609

Sipcam: Sipcam UK Ltd
3 The Barn
27 Kneesworth Street
Royston
Herts.
SG8 5AB
Tel: (01763) 212100
Fax: (01763) 212101

Sorex: Sorex Ltd
St Michael's Industrial Estate
Widnes
Cheshire
WA8 8TJ
Tel: (0151) 420 7151
Fax: (0151) 495 1163
Email: enquiries@sorex.com
Web: www.sorex.com

Sphere: Sphere Laboratories (London) Ltd
The Yews
Main Street
Chilton
Oxon.
OX11 0RZ
Tel. (01235) 831802
Fax: (01235) 833896

Standon: Standon Chemicals Ltd
48 Grosvenor Square
London
W1K 2HT
Tel: (020) 7493 8648
Fax: (020) 7493 4219

SECTION 6

SumiAgro: SumiAgro (UK) Ltd
Merlin House
Falconry Court
Bakers Lane
Epping
Essex
CM16 5DQ
Tel: (01992) 563700
Fax: (01992) 563800
Email: sumiagro@sumiagro.co.uk
Web: www.sumiagro.co.uk

SumiAgro Amenity: SumiAgro Amenity, a division of SumiAgro (UK) Ltd
See SumiAgro (UK) Ltd

Syngenta: Syngenta Crop Protection UK Limited
Whittlesford
Cambridge
CB2 4QT
Tel: (0800) 169 6058
Fax: (01223) 493700
Web: www.syngenta-crop.co.uk

Syngenta Bioline: Syngenta Bioline
Telstar Nursery
Holland Road
Little Clacton
Essex
CO16 9QG
Tel: (01255) 863200
Fax: (01255) 863206
Email: syngenta.bioline@syngenta.com
Web: www.syngenta-bioline.co.uk

Tomen: Tomen (UK) plc
Tomen House
13 Charles II Street
London
SW1Y 4QT
Tel: (020) 7321 6621
Fax: (020) 7321 6624
Email: andy_meads@ov.tomen.com
Web: www.arvesta.com

Tripart: Tripart Farm Chemicals Ltd
See AgriGuard Ltd

Truchem: Truchem Ltd
The Knoll
The Cross
Horsley
Stroud
Gloucestershire
GL6 0PR
Tel: (01453) 833293
Fax: (01453) 833293
Email: truchem@btopenworld.com

Unicrop: Universal Crop Protection Ltd
Park House
Maidenhead Road
Cookham
Berks.
SL6 9DS
Tel: (01628) 526083
Fax: (01628) 810457
Email: enquiries@unicrop.com

United Phosphorus: United Phosphorus Ltd
Chadwick House
Birchwood Park
Warrington
Cheshire
WA3 6AE
Tel: (01925) 819999
Fax: (01925) 817425
Email: chris@upluk6.demon.co.uk

Vitax: Vitax Ltd
Owen Street
Coalville
Leicester
LE67 3DE
Tel: (01530) 510060
Fax: (01530) 510299
Email: info@vitax.co.uk
Web: www.vitax.co.uk

Whyte Agrochemicals: Whyte Agrochemicals Ltd
Marlborough House
298 Regents Park Road
Finchley
London
N3 2UA
Tel: (020) 8346 5946
Fax: (020) 8349 4589

Appendix 2
Useful Contacts

BASIS Ltd
Bank Chambers
34 St John Street
Ashbourne
Derbyshire DE6 1GH
Tel: (01335) 343945/346138
Fax: (01335) 346488
Web: www.basis-reg.co.uk

British Beekeepers' Association
National Agricultural Centre
Stoneleigh
Kenilworth
Warwickshire CV8 2LZ
Tel: (024) 7669 6679
Fax: (024) 7669 0682

British Crop Protection Council (BCPC)
7 Omni Business Centre
Omega Park
Alton
Hampshire GU34 2QD
Tel: (01420) 593200
Fax: (01420) 593209
Web: www.bcpc.org

British Crop Protection Council Publications Sales
7 Omni Business Centre
Omega Park
Alton
Hampshire GU34 2QD
Tel: (01420) 593200
Fax: (01420) 593209
Web: www.bcpc.org/bookshop
e-mail: publications@bcpc.org

British Pest Control Association
1 Gleneagles House
Vernon Gate
South Street
Derby DE1 1UP
Tel: (01332) 294288
Fax: (01332) 295904
Web: www.bpca.org.uk

Crop Protection Association Ltd
4 Lincoln Court
Lincoln Road
Peterborough
Cambs. PE1 2RP
Tel: (01733) 349225
Fax: (01733) 562523
Web: www.cropprotection.org.uk

CropLife International (previously the Global Crop Protection Federation)
Avenue Louise 143
B-1050 Brussels
Belgium
Tel: (+32) 2 542 0410
Fax: (+32) 2 542 0419
Web: www.gcpf.org

Department of Agriculture and Rural Development (Northern Ireland)
Pesticides Section
Dundonald House
Upper Newtownards Road
Belfast BT4 3SB
Tel: (028) 9052 4704
Fax: (028) 9052 4266

Department of Environment, Food and Rural Affairs (Defra)
Nobel House
17 Smith Square
London SW1P 3JR
Tel: (020) 7238 6000
Fax: (020) 7238 6591
Web: www.defra.gov.uk

Environment Agency
Rio House
Waterside Drive
Aztec West
Almondsbury
Bristol BS12 4UD
Tel: (01454) 624400
Fax: (01454) 624409
Web: www.environment-agency.gov.uk

European Crop Protection Association (ECPA)
Avenue E van Nieuwenhuyse 6
B-1160 Brussels
Belgium
Tel: (+32) 2 663 1550
Fax: (+32) 2 663 1560
Web: www.ecpa.be

Farmers' Union of Wales
Llys Amaeth
Queen's Square
Aberystwyth
Dyfed SY23 2EA
Tel: (01970) 612755
Fax: (01970) 624369

Forestry Commission
231 Corstorphine Road
Edinburgh EH12 7AT
Tel: (0131) 334 0303
Fax: (0131) 334 3047
Web: www.forestry.gov.uk

Health and Safety Executive
Information Services
Room 318, Daniel House
Stanley Precinct
Bootle
Merseyside L20 3TW
Tel: (0151) 951 3191
Fax: (0151) 951 3467

Health and Safety Executive
Biocides & Pesticides Assessment Unit
Room 123, Magdalen House
Bootle
Merseyside L20 3QZ
Tel: (0151) 951 3535
Fax: (0151) 951 3317

Health and Safety Executive - Books
PO Box 1999
Sudbury
Suffolk CO10 2WA
Tel: (01787) 881165
Fax: (01787) 313995

Lantra National Training Organisation Ltd
National Agricultural Centre
Stoneleigh
Kenilworth
Warwickshire CV8 2 LG
Tel: (024) 7669 6996
Fax: (024) 7669 6732
Web: www.lantra.co.uk/nto

National Association of Agricultural Contractors (NAAC)
Samuelson House
Paxton Road
Orton Centre
Peterborough
Cambs. PE2 5LT
Tel: (01733) 362920
Fax: (01733) 362921

National Farmers' Union
Agriculture House
164 Shaftesbury Avenue
London WC2H 8HL
Tel: (020) 7331 7200
Fax: (020) 7331 7313
Web: www.nfu.org.uk

National Poisons Information Service
Guys' and St Thomas' Hospital Trust
London SE14 5ER
Tel: (020) 7635 9191

The Royal Hospitals
Belfast BT12 6BA
Tel: (028) 9024 0503

City Hospital NHS Trust
Birmingham B18 7QH
Tel: (0121) 507 5588/5589

Llandough Hospital
Penarth CF64 2XX
Tel: (029) 2070 9901

Royal Infirmary
Edinburgh
Tel: (0131) 536 2300

The General Infirmary
Leeds LS1 3EX
Tel: (0113) 243 0715

Royal Victoria Infirmary
Newcastle NE1 4LP
Tel: (0191) 232 5131

National Proficiency Tests Council (NPTC)
National Agricultural Centre
Stoneleigh
Kenilworth
Warwickshire CV8 2LG
Tel: (024) 7669 6553
Fax: (024) 7669 6128
Web: www.nptc.org.uk

National Turfgrass Council
Hunter's Lodge
Dr Brown's Road
Minchinhampton
Glos. GL6 9BT
Tel: (01453) 883588
Fax: (01453) 731449

Pesticides Safety Directorate
Mallard House
King's Pool
3 Peasholme Green
York YO1 2PX
Tel: (01904) 640500
Fax: (01904) 455733
Web: www.pesticides.gov.uk

Processors and Growers Research Organisation
The Research Station
Great North Road
Thornhaugh
Peterborough
Cambs. PE8 6HJ
Tel: (01780) 782585
Fax: (01780) 783993
Web: www.pgro.co.uk

Scottish Beekeepers' Association
North Trinity House
114 Trinity Road
Edinburgh EH5 3JZ
Tel: (0131) 552 5341

Scottish Environment Protection Agency (SEPA)
Erskine Court
The Castle Business Park
Stirling FK9 4TR
Tel: (01786) 457 700
Fax: (01786) 446 885
Web: www.sepa.org.uk

Stationery Office
Publications Centre
PO Box 276
London SW8 5DT
Tel: (020) 7873 9090 (orders)
(020) 7873 0011 (enquiries)
Fax: (020) 7873 8200
Web: www.thestationeryoffice.com

Ulster Beekeepers' Association
57 Liberty Road
Carrickfergus
Co. Antrim BT38 9DJ
Tel: (01960) 362998

Welsh Beekeepers' Association
Trem y Clawdd
Fron Isaf
Chirk
Wrexham
Clwyd LL14 5AH
Tel/Fax: (01691) 773300

SECTION 6

Appendix 3
Keys to Crop and Weed Growth Stages

Decimal Code for the Growth Stages of Cereals

Illustrations of these growth stages can be found in the reference indicated below and in some company product manuals.

0 Germination
- 00 Dryseed
- 03 Imbibition complete
- 05 Radicle emerged from caryopsis
- 07 Coleoptile emerged from caryopsis
- 09 Leaf at coleoptile tip

1 Seedling growth
- 10 First leaf through coleoptile
- 11 First leaf unfolded
- 12 2 leaves unfolded
- 13 3 leaves unfolded
- 14 4 leaves unfolded
- 15 5 leaves unfolded
- 16 6 leaves unfolded
- 17 7 leaves unfolded
- 18 8 leaves unfolded
- 19 9 or more leaves unfolded

2 Tillering
- 20 Main shoot only
- 21 Main shoot and 1 tiller
- 22 Main shoot and 2 tillers
- 23 Main shoot and 3 tillers
- 24 Main shoot and 4 tillers
- 25 Main shoot and 5 tillers
- 26 Main shoot and 6 tillers
- 27 Main shoot and 7 tillers
- 28 Main shoot and 8 tillers
- 29 Main shoot and 9 or more tillers

3 Stem elongation
- 30 ear at 1 cm
- 31 1st node detectable
- 32 2nd node detectable
- 33 3rd node detectable
- 34 4th node detectable
- 35 5th node detectable
- 36 6th node detectable
- 37 Flag leaf just visible
- 39 Flag leaf ligule/collar just visible

4 Booting
- 41 Flag leaf sheath extending
- 43 Boots just visibly swollen
- 45 Boots swollen
- 47 Flag leaf sheath opening
- 49 First awns visible

5 Inflorescence
- 51 First spikelet of inflorescence just visible
- 52 1/4 of inflorescence emerged
- 55 1/2 of inflorescence emerged
- 57 3/4 of inflorescence emerged
- 59 Emergence of inflorescence completed

6 Anthesis
- 60
- 61 } Beginning of anthesis
- 64
- 65 } Anthesis half way
- 68
- 69 } Anthesis complete

7 Milk development
- 71 Caryopsis watery ripe
- 73 Early milk
- 75 Medium milk
- 77 Late milk

8 Dough development
- 83 Early dough
- 85 Soft dough
- 87 Hard dough

9 Ripening
- 91 Caryopsis hard (difficult to divide by thumb-nail)
- 92 Caryopsis hard (can no longer be dented by thumb-nail)
- 93 Caryopsis loosening in daytime

(From Tottman, 1987. *Annals of Applied Biology*, **110**, 441–454)

Stages in Development of Oilseed Rape

Illustrations of these growth stages can be found in the reference indicated below and in some company product manuals.

0 Germination and emergence

1 Leaf production

- 1,0 Both cotyledons unfolded and green
- 1,1 First true leaf
- 1,2 Second true leaf
- 1,3 Third true leaf
- 1,4 Fourth true leaf
- 1,5 Fifth true leaf
- 1,10 About tenth true leaf
- 1,15 About fifteenth true leaf

2 Stem extension

- 2,0 No internodes ('rosette')
- 2,5 About five internodes

3 Flower bud development

- 3,0 Only leaf buds present
- 3,1 Flower buds present but enclosed by leaves
- 3,3 Flower buds visible from above ('green bud')
- 3,5 Flower buds raised above leaves
- 3,6 First flower stalks extending
- 3,7 First flower buds yellow ('yellow bud')

4 Flowering

- 4,0 First flower opened
- 4,1 10% all buds opened
- 4,3 30% all buds opened
- 4,5 50% all buds opened

5 Pod development

- 5,3 30% potential pods
- 5,5 50% potential pods
- 5,7 70% potential pods
- 5,9 All potential pods

6 Seed development

- 6,1 Seeds expanding
- 6,2 Most seeds translucent but full size
- 6,3 Most seeds green
- 6,4 Most seeds green-brown mottled
- 6,5 Most seeds brown
- 6,6 Most seeds dark brown
- 6,7 Most seeds black but soft
- 6,8 Most seeds black and hard
- 6,9 All seeds black and hard

7 Leaf senescence

8 Stem senescence

- 8,1 Most stem green
- 8,5 Half stem green
- 8,9 Little stem green

9 Pod senescence

- 9,1 Most pods green
- 9,5 Half pods green
- 9.9 Few pods green

(From Sylvester-Bradley, 1985. *Aspects of Applied Biology*, **10**, 395–400)

SECTION 6

Stages in Development of Peas

Illustrations of these growth stages can be found in the reference indicated below and in some company product manuals.

0 Germination and emergence

000 Dry seed
001 Imbibed seed
002 Radicle apparent
003 Plumule and radicle apparent
004 Emergence

1 Vegetative stage

101 First node (leaf with one pair leaflets, no tendril)
102 Second node (leaf with one pair leaflets, simple tendril)
103 Third node (leaf with one pair leaflets, complex tendril)
•
•
l0x X nodes (leaf with more than one pair leaflets, complex tendril)
•
•
10n Last recorded node

2 Reproductive stage (main stem)

201 Enclosed buds
202 Visible buds
203 First open flower
204 Pod set (small immature pod)
205 Flat pod
206 Pod swell (seeds small, immature)
207 Podfill
208 Pod green, wrinkled
209 Pod yellow, wrinkled (seeds rubbery)
210 Dry seed

3 Senescence stage

301 Desiccant application stage. Lower pods dry and brown, middle yellow, upper green. Overall moisture content of seed less than 45%
302 Pre-harvest stage. Lower and middle pods dry and brown, upper yellow. Overall moisture content of seed less than 30%
303 Dry harvest stage. All pods dry and brown, seed dry

(From Knott, 1987. *Annals of Applied Biology*, **111**, 233–244)

Stages in Development of Faba Beans

Illustrations of these growth stages can be found in the reference indicated below and in some company product manuals.

0 Germination and emergence

000 Dry seed
001 Imbibed seed
002 Radicle apparent
003 Plumule and radicle apparent
004 Emergence
005 First leaf unfolding
006 First leaf unfolded

1 Vegetative stage

101 First node
102 Second node
103 Third node
•
•
l0x X nodes
•
•
10n N, last reoorded node

2 Reproductive stage (main stem)

201 Flower buds visible
203 First open flowers
204 First pod set
205 Pods fully formed, green
207 Pod fill, pods green
209 Seed rubbery, pods pliable, turning black
210 Seed dry and hard, pods dry and black

3 Pod senescence

301 10% pods dry and black
•
•
305 50% pods dry and black
•
•
308 80% pods dry and black, some upper pods green
309 90% pods dry and black, most seed dry. Desiccation stage.
310 All pods dry and black, seed hard. Pre-harvest (glyphosate application stage)

4 Stem senescence

401 10% stem brown/black
•
•
405 50% stem brown/black
•
•
409 90% stem brown/black
410 All stems brown/black. All pods dry and black, seed hard.

(From Knott, 1990 *Annals of Applied Biology*, **116**, 391–404)

SECTION 6

Stages in Development of Potato

Illustrations of these growth stages can be found in the reference indicated below and in some company product manuals.

0 Seed germination and seedling emergence

000 Dry seed
001 Imbibed seed
002 Radicle apparent
003 Elongation of hypocotyl
004 Seedling emergence
005 Cotyledons unfolded

1 Tuber dormancy

100 Innate dormancy (no sprout development under favourable conditions)
150 Enforced dormancy (sprout development inhibited by environmental conditions)

2 Tuber sprouting

200 Dormancy break, sprout development visible
21x Sprout with 1 node
22x Sprout with 2 nodes
•
•
29x Sprout with 9 nodes
21x(2) Second generation sprout with 1 node
22x(2) Second generation sprout with 2 nodes
•
•
29x(2) Second generation sprout with 9 nodes

Where x = 1, sprout <2 mm;
2, 2-5 mm; 3, 5-20 mm;
4, 20-30 mm; 5, 50-100 mm;
6, 100-150 mm long

3 Emergence and shoot expansion

300 Main stem emergence
301 Node 1
302 Node 2
•
•
319 Node 19
Second order branch
321 Node 1
•
•
Nth order branch
3N1 Node 1
•
•
3N9 Node 9

4 Flowering

Primary flower
400 No flowers
410 Appearance of flower bud
420 Flower unopen
430 Flower open
440 Flower closed
450 Berry swelling
460 Mature berry
Second order flowers
410(2) Appearance of flower bud
420(2) Flower unopen
430(2) Flower open
440(2) Flower closed
450(2) Berry swelling
460(2) Mature berry

5 Tuber development

500 No stolons
510 Stolon initials
520 Stolon elongation
530 Tuber initiation
540 Tuber bulking (>10 mm diam)
550 Skin set
560 Stolon development

6 Senescence

600 Onset of yellowing
650 Half leaves yellow
670 Yellowing of stems
690 Completely dead

(From Jefferies & Lawson, 1991. *Annals of Applied Biology*, **119**, 387–389)

Stages in Development of Linseed

Illustrations of these growth stages can be found in the reference indicated below and in some company product manuals.

0 Germination and emergence

- 00 Dry seed
- 01 Imbibed seed
- 02 Radicle apparent
- 04 Hypocotyl extending
- 05 Emergence
- 07 Cotyledon unfolding from seed case
- 09 Cotyledons unfolded and fully expanded

1 Vegetative stage (of main stem)

- 10 True leaves visible
- 12 First pair of true leaves fully expanded
- 13 Third pair of true leaves fully expanded
- 1*n* *n* leaf fully expanded

2 Basal branching

- 21 One branch
- 22 Two branches
- 23 Three branches
- 2*n* *n* branches

3 Flower bud development (on main stem)

- 31 Enclosed bud visible in leaf axils
- 33 Bud extending from axil
- 35 Corymb formed
- 37 Buds enclosed but petals visible
- 39 First flower open

4 Flowering (whole plant)

- 41 10% of flowers open
- 43 30% of flowers open
- 45 50% of flowers open
- 49 End of flowering

5 Capsule formation (whole plant)

- 51 10% of capsules formed
- 53 30% Of capsules formed
- 55 50% of capsules formed
- 59 End of capsule formation

6 Capsule senescence (on most advanced plant)

- 61 Capsules expanding
- 63 Capsules green and full size
- 65 Capsules turning yellow
- 67 Capsules all yellow brown but soft
- 69 Capsules brown, dry and senesced

7 Stem senescence (whole plant)

- 71 Stems mostly green below panicle
- 73 Most stems 30% brown
- 75 Most stems 50% brown
- 77 Stems 75% brown
- 79 Stems completely brown

8 Stems rotting (retting)

- 81 Outer tissue rotting
- 85 Vascular tissue easily removed
- 89 Stems completely collapsed

9 Seed development (whole plant)

- 91 Seeds expanding
- 92 Seeds white but full size
- 93 Most seeds turning ivory yellow
- 94 Most seeds turning brown
- 95 All seeds brown and hard
- 98 Some seeds shed from capsule
- 99 Most seeds shed from capsule

(From Freer, 1991. *Aspects of Applied Biology*, **28**, 33–40)

Stages in Development of Annual Grass Weeds

Illustrations of these growth stages can be found in the reference indicated below and in some company product manuals.

0 Germination and emergence

- 00 Dry seed
- 01 Start of imbibition
- 03 Imbibition complete
- 05 Radicle emerged from caryopsis
- 07 Coleoptile emerged from caryopsis
- 09 Leaf just at coleoptile tip

1 Seedling growth

- 10 First leaf through coleoptile
- 11 First leaf unfolded
- 12 2 leaves unfolded
- 13 3 leaves unfolded
- 14 4 leaves unfolded
- 15 5 leaves unfolded
- 16 6 leaves unfolded
- 17 7 leaves unfolded
- 18 8 leaves unfolded
- 19 9 or more leaves unfolded

2 Tillering

- 20 Main shoot only
- 21 Main shoot and 1 tiller
- 22 Main shoot and 2 tillers
- 23 Main shoot and 3 tillers
- 24 Main shoot and 4 tillers
- 25 Main shoot and 5 tillers
- 26 Main shoot and 6 tillers
- 27 Main shoot and 7 tillers
- 28 Main shoot and 8 tillers
- 29 Main shoot and 9 or more tillers

3 Stem elongation

- 31 First node detectable
- 32 2nd node detectable
- 33 3rd node detectable
- 34 4th node detectable
- 35 5th node detectable
- 36 6th node detectable
- 37 Flag leaf just visible
- 39 Flag leaf ligule just visible

4 Booting

- 41 Flag leaf sheath extending
- 43 Boots just visibly swollen
- 45 Boots swollen
- 47 Flag leaf sheath opening
- 49 First awns visible

5 Inflorescence emergence

- 51 First spikelet of inflorescence just visible
- 53 1/4 of inflorescence emerged
- 55 1/2 of inflorescence emerged
- 57 3/4 of inflorescence emerged
- 59 Emergence of inflorescence completed

6 Anthesis

- 61 Beginning of anthesis
- 65 Anthesis half-way
- 69 Anthesis complete

(From Lawson & Read, 1992. *Annals of Applied Biology*, **12**, 211–214)

Growth Stages of Annual Broad-leaved Weeds

Preferred Descriptive Phrases

Illustrations of these growth stages can be found in the reference indicated below and in some company product manuals.

Pre-emergence
Early cotyledons
Expanded cotyledons
One expanded true leaf
Two expanded true leaves
Four expanded true leaves
Six expanded true leaves
Plants up to 25 mm across/high
Plants up to 50 mm across/high
Plants up to 100 mm across/high
Plants up to 150 mm across/high
Plants up to 250 mm across/high
Flower buds visible
Plant flowering
Plant senescent

(From Lutman & Tucker, 1987. *Annals of Applied Biology*, **110**, 683–687)

Appendix 4
Key to Hazard Classifications and Safety Precautions

Every product label contains information to warn users of the risks from using the product, together with precautions that must be followed in order to minimise the risks. A hazard classification (if any) and symbol is shown with associated risk phrases, followed by a series of safety precautions. These are represented in the pesticide profiles in Section 4 by code letters and numbers under the heading **Hazard classification and safety precautions**.

The codes are defined below, under the same sub-headings as they appear in the pesticide profiles.

Where a product label specifies the use of personal protective equipment (PPE), the requirements are listed under the sub-heading **Operator protection**, using letter codes to denote the protective items, according to the list below. Often PPE requirements are different for specified operations, e.g. handling the concentrate, cleaning equipment etc., but it is not possible to list them separately. The lists of PPE are therefore an indication of what the user may require to have available to use the product in different ways. **When making a COSHH assessment it is therefore essential that the product label is consulted for information on the particular use that is being assessed**.

Where the generalised wording includes a phrase such as '... for *xx* days', the specific requirement for each pesticide is shown in brackets after the code.

Hazard

R01	Very toxic
R02	Toxic
R03	Harmful
R04	Irritant
R05	Corrosive
R06	Danger of serious damage to health by prolonged exposure
R07a	Extremely flammable liquefied gas
R07b	Extremely flammable
R07c	Highly flammable
R07d	Flammable
R08	Oxidising agent. Contact with combustible material may cause fire
R09	Explosive when mixed with oxidising substances
R11	May cause harm to breast-fed babies
R13	Limited evidence of a carcinogenic effect
R14	Dangerous for the environment
R15	Toxic to flora. May cause long-term adverse effects in the environment
R16	Toxic to fauna. May cause long-term adverse effects in the environment
R17	Toxic to soil organisms. May cause long-term adverse effects in the environment
R18	Toxic to bees. May cause long-term adverse effects in the environment

Risk phrases

R01a	Very toxic in contact with skin
R01b	Very toxic by inhalation
R01c	Very toxic if swallowed
R02a	Toxic in contact with skin
R02b	Toxic by inhalation
R02c	Toxic if swallowed
R02d	Toxic: Danger of serious damage to health by prolonged exposure if swallowed
R03a	Harmful in contact with skin
R03b	Harmful by inhalation

R03c Harmful if swallowed
R04a Irritating to eyes
R04b Irritating to skin
R04c Irritating to respiratory system
R04d Risk of serious damage to eyes
R04e May cause sensitization by skin contact
R04f May cause sensitization by inhalation
R04g May cause lung damage if swallowed
R05a Causes severe burns
R05b Causes burns
R10a Possible risk of harm to the unborn child
R10b May cause harm to the unborn child
R10c Possible risk of impaired fertility
R10d May impair fertility
R10e Possible risk of irreversible effects
R12a Vapours may cause drowsiness and dizziness
R12b Repeated exposure may cause skin dryness or cracking
R14a Harmful to aquatic organisms. May cause long-term adverse effects to the aquatic environment
R14b Toxic to aquatic organisms. May cause long-term adverse effects to the aquatic environment
R14c Very toxic to aquatic organisms. May cause long-term adverse effects in the aquatic environment
R14d Harmful to aquatic organisms

Operator protection

A Suitable protective gloves (the product label should be consulted for any specific requirements about the material of which the gloves should be made)
B Rubber gauntlet gloves
C Face-shield
D Approved respiratory protective equipment
E Goggles
F Dust mask
G Full face-piece respirator
H Coverall
J Hood
K Rubber apron
L Waterproof coat
M Rubber boots
N Waterproof jacket and trousers
P Suitable protective clothing
U01 To be used only by operators instructed or trained in the use of chemical/product/type of produce and familiar with the precautionary measures to be observed
U02 Wash all protective clothing thoroughly after use, especially the inside of gloves/Avoid excessive contamination of coveralls and launder regularly
U03 Wash splashes off gloves immediately
U04a Take off immediately all contaminated clothing
U04b Take off immediately all contaminated clothing and wash underlying skin. Wash clothes before re-use
U04c Wash clothes before re-use
U05a When using do not eat, drink or smoke
U05b When using do not eat, drink, smoke or use naked lights
U06 Handle with care and mix only in a closed container
U07 Open container only as directed
U08 Wash concentrate/dust from skin or eyes immediately
U09a Wash any contamination/splashes/dust/powder/concentrate from skin or eyes immediately
U09b Wash any contamination/splashes/dust/powder/concentrate from eyes immediately
U10 After contact with skin or eyes wash immediately with plenty of water
U11 In case of contact with eyes rinse immediately with plenty of water and seek medical advice
U12 In case of contact with skin rinse immediately with plenty of water and seek medical advice

SECTION 6

U13 Avoid all contact by mouth
U14 Avoid all contact with skin
U15 Avoid all contact with eyes
U16 Ensure adequate ventilation in confined spaces
U17 Ventilate treated areas thoroughly when smoke has cleared/Ventilate treated rooms thoroughly before occupying
U18 Extinguish all naked flames, including pilot lights, when applying the fumigant/dust/liquid/product
U19 Do not breathe dust/fog/fumes/gas/smoke/spray mist/vapour. Avoid working in spray mist
U20a Wash hands and exposed skin before eating, drinking or smoking and after work
U20b Wash hands and exposed skin before meals and after work
U20c Wash hands before meals and after work
U21 Before entering treated crops, cover exposed skin areas, particularly arms and legs
U22a Do not touch sachet with wet hands or gloves/Do not touch water soluble bag directly
U22b Protect sachets from rain or water
U23a This product must not be applied using knapsack sprayers/This product must not be applied by hand-held equipment
U23b Do not apply through hand held rotary atomisers
U24 Do not handle grain unnecessarily

Environmental protection

E01 Keep out of reach of children
E02 Keep unprotected persons out of treated areas for at least xx hours/days
E03 Label treated seed with the appropriate precautions, using the printed sacks, labels or bag tags supplied
E04 Remove all pets/livestock/fish tanks before treatment/spraying
E05 Do not apply directly to livestock/poultry
E05a Keep poultry out of treated areas for at least xx days/weeks
E06a Dangerous to livestock. Keep all livestock out of treated areas/away from treated water for at least xx days/weeks. Bury or remove spillages
E06b Harmful to livestock. Keep all livestock out of treated areas/away from treated water for at least xx days/weeks. Bury or remove spillages
E07 Keep livestock out of treated areas/Keep livestock out of treated areas for at least xx weeks and until foliage of any poisonous weeds such as ragwort has died and become unpalatable
E08 Do not feed treated straw or haulm to livestock within x days of spraying
E09 Do not use on crops if the straw is to be used as animal feed/bedding
E10a Dangerous to game, wild birds and animals
E10b Harmful to game, wild birds and animals
E11 Paraquat may be harmful to hares; stubbles must be sprayed early in the day
E12a High risk to bees. Do not apply to crops in flower or to those in which bees are actively foraging. Do not apply when flowering weeds are present
E12b Extremely dangerous to bees. Do not apply to crops in flower or to those in which bees are actively foraging. Do not apply when flowering weeds are present
E12c Dangerous to bees. Do not apply to crops in flower or to those in which bees are actively foraging. Do not apply when flowering weeds are present
E12d Dangerous to bees. Do not apply to crops in flower or to those in which bees are actively foraging except as directed on [crop]. Do not apply when flowering weeds are present
E12e Harmful to bees. Do not apply to crops in flower or to those in which bees are actively foraging. Do not apply when flowering weeds are present
E12f Apply away from bees
E13a Extremely dangerous to fish or other aquatic life. Do not contaminate surface waters or ditches with chemical or used container
E13b Dangerous to fish or other aquatic life. Do not contaminate surface waters or ditches with chemical or used container
E13c Harmful to fish or other aquatic life. Do not contaminate surface waters or ditches with chemical or used container
E13d Apply away from fish
E14a Extremely dangerous to aquatic higher plants. Do not contaminate surface waters or ditches with chemical or used container

E14b Dangerous to aquatic higher plants. Do not contaminate surface waters or ditches with chemical or used container

E15 Do not contaminate ponds, waterways or ditches with chemical or used container

E16 Do not allow direct spray/granule applications from vehicle mounted/drawn hydraulic sprayers/applicators to fall within 6 m of surface waters or ditches/Do not allow direct spray/granule applications from hand-held sprayers/applicators to fall within 2 m of surface waters or ditches. Direct spray/applications away from water

E16a Do not allow direct spray from horizontal boom sprayers to fall within 5 m of the top of the bank of a static or flowing waterbody, unless a Local Environment Risk Assessment for Pesticides (LERAP) permits a narrower buffer zone, or within 1 m of the top of a ditch which is dry at the time of application. Aim spray away from water

E16b Do not allow direct spray from hand-held sprayers to fall within 1 m of the top of the bank of a static or flowing waterbody. Aim spray away from water

E16c Do not allow direct spray from horizontal boom sprayers to fall within 5 m of the top of the bank of a static or flowing waterbody, or within 1m of the top of a ditch which is dry at the time of application. Aim spray away from water. This product is not eligible for buffer zone reduction under the LERAP horizontal boom sprayers scheme.

E16d Do not allow direct spray from hand-held sprayers to fall within 1 m of the top of the bank of a static or flowing waterbody. Aim spray away from water. This product is not eligible for buffer zone reduction under the LERAP horizontal boom sprayers scheme.

E16e Do not allow direct spray from horizontal boom sprayers to fall within 5 m of the top of the bank of a static or flowing water body or within 1 m from the top of any ditch which is dry at the time of application. Spray from hand held sprayers must not in any case be allowed to fall within 1 m of the top of the bank of a static or flowing water body. Always direct spray away from water. The LERAP scheme does not extend to adjuvants. This product is therefore not eligible for a reduced buffer zone under the LERAP scheme.

E17 Do not allow direct spray from broadcast air-assisted sprayers to fall within xx m of surface waters or ditches. Direct spray away from water

E17a Do not allow direct spray from broadcast air-assisted sprayers to fall within xx m of the bank of a static or flowing waterbody, unless a Local Environmental Risk Assessment for Pesticides (LERAP) permits a narrower buffer zone, or within 5 m of the top of a ditch which is dry at the time of application. Aim spray away from water

E18 Do not spray from the air within 250 m horizontal distance of surface waters or ditches

E19 Do not dump surplus herbicide in water or ditch bottoms

E20 Prevent any surface run-off from entering storm drains

E21 Do not use treated water for irrigation purposes within xx days/weeks of treatment

E22a High risk to non-target insects or other arthropods. Do not spray within 6 m of the field boundary

E22b Risk to certain non-target insects or other arthropods. See directions for use

E23 Avoid damage by drift onto susceptible crops or water courses

E37 Do not use with any pesticide which is to be applied in or near water

E38 Use appropriate containment to avoid environmental contamination

Consumer protection

C01 Not to be used on food crops

C02 Do not harvest for human or animal consumption for at least xx days/weeks after last application

C03 Keep away from food, drink and animal feeding stuffs

C04 Do not apply to surfaces on which food/feed is stored, prepared or eaten

C05 Remove/cover all foodstuffs before application

C06 Remove exposed milk before application

C07 Collect eggs before application

C08 Protect food preparing equipment and eating utensils from contamination during application

C09 Cover water storage tanks before application

C10 Protect exposed water/food/milk machinery/milk containers from contamination

Storage and disposal

E24 Store away from seeds, fertilizers, fungicides and insecticides

E25 Store well away from corms, bulbs, tubers and seeds

E26 Store away from frost/Protect from frost
E27 Store away from heat
E28 Flammable. Do not store near heat or open flame
E29 Store under cool, dry conditions
E30a Store/keep in original container, tightly closed, in a safe place
E30b Store/keep in original container, tightly closed, in a safe place, under lock and key
E31a Wash out container thoroughly and dispose of safely
E31b Wash out container thoroughly, empty washings into spray tank and dispose of safely
E31c Rinse container thoroughly by using an integrated pressure rinsing device or manually rinsing three times. Add washings to sprayer at time of filling and dispose of container safely
E32a Empty container completely and dispose of safely
E32b Empty container completely and dispose of it in the specified manner
E32c Treat used container as if it contained pesticide
E32d This material and its container must be disposed of in a safe way
E32e This material and its container must be disposed of as hazardous waste
E33 Return empty container as instructed by supplier
E34 Do not re-use this container for any purpose/Do not re-use this container for any other purpose
E35 Do not burn this container
E36 Do not rinse out the container

Treated Seed

S01 Do not handle treated seed unnecessarily
S02 Do not use treated seed as food or feed
S03 Keep treated seed secure from people, domestic stock/pets and wildlife at all times during storage and use
S04a Bury or remove spillages
S04b Harmful to game and wild life. Bury spillages/Bury or remove spillages
S04c Dangerous to birds, game and other wildlife. Treated seed should not be left on the soil surface. Bury spillages
S05 Do not re-use sack for food or feed/Do not reuse sacks or containers that have been used for treated seed for food or feed
S06 Wash hands and exposed skin before meals and after work
S06a Wash hands and exposed skin after cleaning and re-calibrating equipment
S07 Do not apply treated seed from the air
S08 Treated seed must not be applied as a broadcast application

Vertebrate/Rodent control products

V01a Prevent access to baits/powder by children, birds and other animals, particularly cats, dogs, pigs and poultry
V01b Prevent access to bait/gel/dust by children, birds and non-target animals, particularly dogs, cats, pigs, poultry
V02 Do not prepare/use/lay baits/dust/spray where food/feed/water could become contaminated
V03a Remove all remains of bait, tracking powder or bait containers after use and burn or bury
V03b Remove all remains of bait and bait containers/exposed dust/after treatment (except where used in sewers) and dispose of safely (e.g. burn/bury). Do not dispose of in refuse sacks or on open rubbish tips.
V04a Search for and burn or bury all rodent bodies. Do not place in refuse bins or on rubbish tips
V04b Search for rodent bodies (except where used in sewers) and dispose of safely (e.g. burn/bury). Do not dispose of in refuse sacks or on open rubbish tips
V04c Dispose of safely any rodent bodies and remains of bait and bait containers that are recovered after treatment (e.g. burn/bury). Do not dispose of in refuse sacks or on open rubbish tips
V05 Use bait containers clearly marked POISON at all surface baiting points

Medical advice

M01 This product contains an anticholinesterase organophosphorus compound. DO NOT USE if under medical advice NOT to work with such compounds

M02 This product contains an anticholinesterase carbamate compound. DO NOT USE if under medical advice NOT to work with such compounds

M03 If you feel unwell, seek medical advice (show the label where possible)

M04 In case of accident or if you feel unwell, seek medical advice immediately (show the label where possible)

M05 If swallowed, do not induce vomiting: seek medical advice immediately and show this container or label

M05a If swallowed, seek medical advice immediately and show this container or label

M05c If swallowed induce vomiting if not already occurring and take patient to hospital immediately

M06 This product contains an anticholinesterase carbamoyl triazole compound. DO NOT USE if under medical advice NOT to work with such compounds

Appendix 5
Key to Abbreviations and Acronyms

The abbreviations of formulation types in the following list are used in Section 4 (Pesticide Profiles) and are derived from the Catalogue of Pesticide Formulation Types and International Coding System (CropLife International Technical Monograph 2, 4th Edition, April 1999)

1 Formulation Types

AB	Grain bait
AE	Aerosol generator
AL	Other liquids to be applied undiluted
AP	Any other powder
BB	Block bait
BR	Briquette
CB	Bait concentrate
CF	Capsule suspension for seed treatment
CG	Encapsulated granule (controlled release)
CL	Contact liquid or gel (for direct application)
CP	Contact powder (for direct application)
CR	Crystals
CS	Capsule suspension
DC	Dispersible concentrate
DP	Dustable powder
DS	Powder for dry seed treatment
EC	Emulsifiable concentrate
EG	Emulsifiable granule
ES	Emulsion for seed treatment
EW	Oil in water emulsion
FG	Fine granules
FP	Smoke cartridge
FS	Flowable concentrate for seed treatment
FT	Smoke tablet
FU	Smoke generator
FW	Smoke pellets
GA	Gas
GB	Granular bait
GE	Gas-generating product
GG	Macrogranules
GL	Emulsifiable gel
GP	Flo-dust (for pneumatic application)
GR	Granules
GS	Grease
GW	Water soluble gel
HN	Hot fogging concentrate
KK	Combi-pack (solid/liquid)
KL	Combi-pack (liquid/liquid)
KN	Cold-fogging concentrate
KP	Combi-pack (solid/solid)
LA	Lacquer
LI	Liquid, unspecified
LS	Solution for seed treatment
ME	Microemulsion
MG	Microgranules
OL	Oil miscible liquid
PA	Paste
PC	Gel or paste concentrate
PS	Seed coated with a pesticide
PT	Pellet
RB	Ready-to-use bait

RH	Ready-to-use spray in hand-operated sprayer
SA	Sand
SC	Suspension concentrate (= flowable)
SE	Suspo-emulsion
SG	Water soluble granules
SL	Soluble concentrate
SP	Water soluble powder
SS	Water soluble powder for seed treatment
ST	Water soluble tablet
SU	Ultra low-volume suspension
TB	Tablets
TC	Technical material
TP	Tracking powder
UL	Ultra low-volume liquid
VP	Vapour releasing product
WB	Water soluble bags
WG	Water dispersible granules
WP	Wettable powder
WS	Water dispersible powder for slurry treatment of seed
WT	Water dispersible tablet
XX	Other formulations
ZZ	Not Applicable

2 Other abbreviations and acronyms

ACP	Advisory Committee on Pesticides
ACTS	Advisory Committee on Toxic Substances
ADAS	Agricultural Development and Advisory Service
a.i.	active ingredient
CDA	controlled droplet application
CPA	Crop Protection Association
cm	centimetre(s)
COPR	Control of Pesticides Regulations 1986
COSHH	Control of Substances Hazardous to Health Regulations
d	day(s)
Defra	Department for Environment, Food & Rural Affairs
EA	Environment Agency
EBDC	ethylene-bis-dithiocarbamate fungicide
FEPA	Food and Environment Protection Act 1985
g	gram(s)
GCPF	Global Crop Protection Federation
GS	growth stage (unless in formulation column)
h	hour(s)
ha	hectare(s)
HBN	hydroxybenzonitrile herbicide
HI	harvest interval
HSE	Health and Safety Executive
ICM	Integrated Crop Management
IPM	Integrated Pest Management
kg	kilogram(s)
l	litre(s)
LERAP	Local Environmental Risk Assessments for Pesticides
m	metre(s)
MAFF	Ministry of Agriculture, Fisheries and Food
MBC	methyl benzimidazole carbamate fungicide
MEL	Maximum Exposure Limit
min	minute(s)
mm	millimetre(s)
MRL	Maximum Residue Level
mth	month(s)
NA	notice of approval
NFU	National Farmers Union
OES	Occupational Exposure Standard

OLA	off-label approval
PPE	personal protective equipment
PPPR	Plant Protection Products Regulations
PSD	Pesticides Safety Directorate
SOLA	specific off-label approval
ULV	ultra-low volume
w/v	weight/volume
w/w	weight/weight
wk	week(s)
yr	year(s)

SECTION 7
INDEX

Index of Proprietary Names of Products

The references are to entry numbers, not to pages. Adjuvant names are referred to as 'Adj' and are listed separately in Section 2

The references are to entry numbers, not to pages

The references are to entry numbers, not to pages

INDEX

The references are to entry numbers, not to pages

The references are to entry numbers, not to pages

INDEX

The references are to entry numbers, not to pages

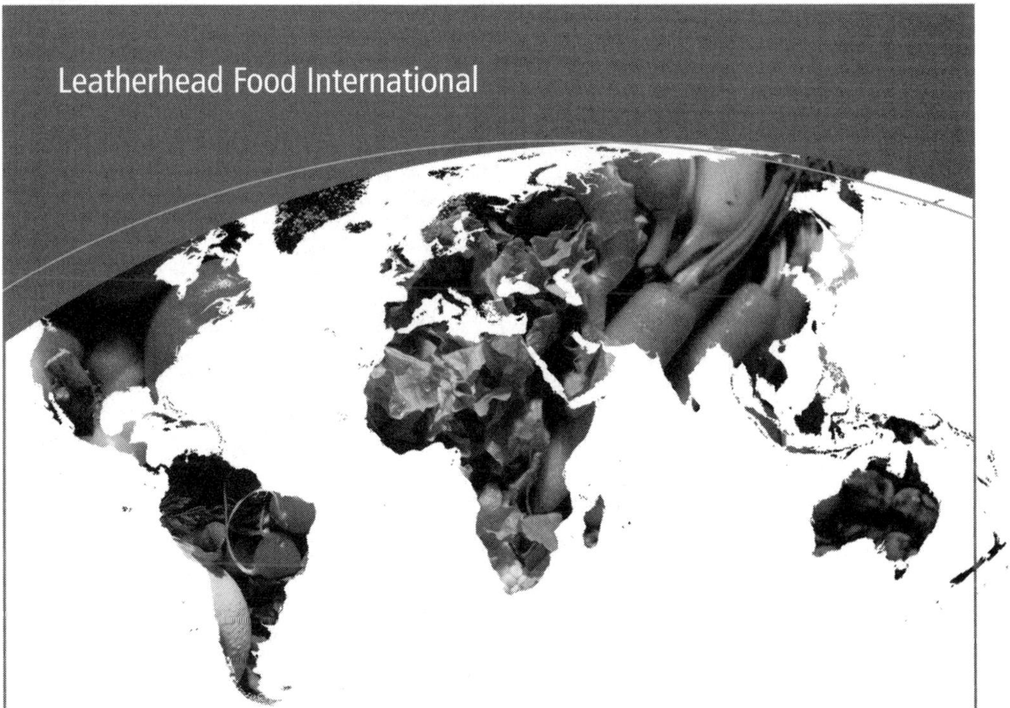

Pesticide Residue Analysis; let us be your partner

With Leatherhead Food International you can:

Keep abreast of changing legislation and regulations
We actively monitor legislation for changes to pesticide approval statuses and advise our customers via a free monthly bulletin (also available in Spanish) and a regularly updated MRL database.

Have access to a wide range of UKAS accredited analyses
Our multi-residue screen covers over 120 pesticides and their residues. The laboratory is also set up to carry out many more specialised analyses all using the latest technology and methodologies available.

Target your analysis
We offer targeted analysis for specific commodities and residues, as well as 24-48 hr turnaround for critical samples.

Obtain assistance in managing due diligence programmes
Our technical experts are available for preparation of due diligence programs with reference to the Pesticide Safety Directorates' monitoring programs and to deal with any other enquiries you may have.

In a climate of rapidly changing Food regulation we offer a high quality, confidential and independent analytical service for our clients, along with the unique advantages that come from having large legislation and information departments onsite. Our experts are available to answer any queries you may have and keep you informed of changes that may affect your business.

For further information on our comprehensive pesticides residue analyses, please contact our Customer Services Manager, Kath Rowles, on + 44 (0) 1372 822373 or E-mail krowles@leatherheadfood.com

Leatherhead Food International

Your food, our world

Leatherhead Food International, Randalls Road, Leatherhead, Surrey KT22 7RY, UK
Telephone: +44 (0) 1372 376761 Facsimile: +44 (0) 1372 386228 Web site: www.leatherheadfood.com

The UK Pesticide Guide
is also available on CD-ROM as

The e-UK Pesticide Guide

The e-UK Pesticide Guide contains all the information found in this book and much more, making searching for information quick and easy.

What's in the e-UK Pesticide Guide?

The CD-ROM contains everything contained in this book and more, with powerful search facilities, harvest interval data, product LERAP classification and on/off label approvals available as searchable fields.

The CD-ROM contains SOLA data which link directly to the PSD website for access to the SOLA recommendation sheets. There are embedded links to access* the regularly updated Leatherhead Food International (LFI) online database of Maximum Residue Levels (MRLs), plant growth stage illustrations and a glossary, Internet links and background information.
Networking and downloading the CD-ROM to portable hand-held devices is possible.

*Access will expire at the end of 2004

Free demonstration available via the web at:
www.ukpesticideguide.co.uk

To order your copy today,
contact BCPC Publishing sales on
01420 593 200

What's new in the 2004 Guide?

- Numerous products lost – Dozens of products were lost in 2003 as a result of the EU Review, whilst others have been withdrawn by manufacturers and suppliers. Check in the Guide to ensure the product you want to use is still available and legal
- Temporary reprieve for 'Essential Use' Products
- New products and actives
- FREE online news updates throughout 2004

The e-UK Pesticide Guide (CD-ROM)

Edited by R Whitehead
Release date January 2004
ISBN 0 85199 824 0

BCPC

CABI Publishing
A division of CAB International

Crop Protection Compendium

2003 Edition

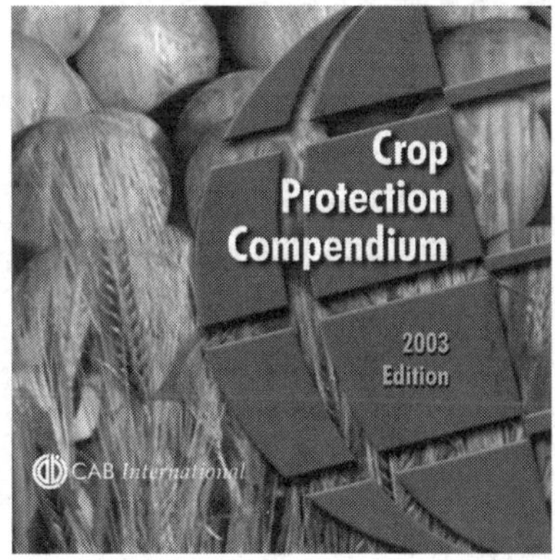

The Crop Protection Compendium is a global compilation of knowledge, which provides practical answers to tough crop protection health questions. Available on CD-ROM or via the Internet, it includes information on pests, diseases and weeds and their natural enemies, as well as on the crops that are their hosts and the countries where they occur.

The Crop Protection Compendium contains:

Interactive encyclopaedia of pests, diseases, weeds and crops

- The Crop Protection Compendium is an encyclopaedic, mixed media tool that brings together a wide range of different types of science-based information on all aspects of crop protection.

Photos, text, maps and illustrations

- The CPC contains information on 10,000 pests, diseases, weeds and natural enemies. Full datasheets are available for over 1,900 of these species, each with text, photos / illustrations and distribution map. The CPC also contains detailed data on over 200 different crops and a database of seedborne diseases.

Taxonomic framework

- The CPC is built around an intelligent taxonomic framework and an advanced cross-reference facility, which provide fast, effective navigation around vast amounts of data.

Sourced and developed by experts

- The CPC comprises information sourced from experts, edited and compiled by an independent scientific organization, and resourced by a diverse international Development Consortium. It is published on CD-ROM and the Internet.

Potential uses:

Identification and diagnostic tool

- The CPC provides solutions for identifying, preventing and solving crop health problems. It contains a step-by-step diagnostic tool and over 6,800 images, for identifying pests and diseases. It also provides information on pesticides, natural predators, parasites of pests, cultural practices and host-plant resistance.

Statistical analysis

- Advanced software makes statistical, economic, climatic and regional analysis by crop possible. Information on trade production statistics, pest distribution, population and GDP is supplied by FAOStat, USDA-FAS, EPPO-PQR and the world bank.

Risk assessment

- A decision-support tool for pest-risk analysis; allows users to assess the quarantine risks of importing and exporting plants and the raw commodities.

Create reports

- A copy and paste facility allows reports and lecture notes – containing text, pictures, maps, graphs and reference – to be created at the touch of a button.

CAB *International*

For more information, visit **www.cabicompendium.org/cpc**

New Edition

The Pesticide Manual

The world reference for:

chemical structure
nomenclature
physical chemistry
mode of action
commercialisation
manufacturers & products
toxicology
ecotoxicology
environmental fate

Sample entries downloadable from www.pesticidemanual.com

Available from:
BCPC Publications Sales,
7 Omni Business Centre, Omega Park,
Alton, Hampshire GU34 2QD, UK

Tel: +44 (0)1420 593 200
Fax: +44 (0)1420 593 209
Email: publications@bcpc.org
Web: www.bcpc.org/bookshop (secure orders)

BCPC

The Pesticide Manual includes 1,469 profiles –
858 detailed chemical entries plus abbreviated details on 611 superseded products.
1,344 + XXVI pages. Published November 2003. Price £195.

READ IT!
SEARCH IT!

The PC CD-ROM contains all the data from the book, plus a larger set of product names – over 8,000 current and 2,000 discontinued. Custom-built software enables speedy and complex searches.
Published November 2003 (Version 3.0). Price £250 + VAT.

Buy Version 3.0 with 3.1 (November 2004) and 3.2 (November 2005) and save £250. Price £500 + VAT.

THE UK PESTICIDE GUIDE 2004

RE-ORDERS

☐ Please send me ____ more copies of *The UK Pesticide Guide 2004* at £33.95 each

☐ Please send me —— copies of *The e-UK Pesticide Guide 2004* (PC CD-ROM, single-user version) at £49.95 + VAT (£58.69 incl. VAT)

Name ______________________ Position ______________________

Institution ______________________ Department ______________________

Address ____

City ______________ Region ______________ Postcode ______________

Country ______________________

Tel ______________ Fax ______________ E-mail ______________

EU countries except UK – VAT No: ______________________

Payment (pre-payment is required)

☐ I enclose a cheque/draft for £________ payable to BCPE Ltd. Please send me a receipt.

☐ I wish to pay by credit card: ☐ Visa ☐ Mastercard ☐ Amex ☐ Switch

Please charge to my card £________ and send me a receipt. Name of issuing bank ______________

Card no. ☐☐☐☐ ☐☐☐☐ ☐☐☐☐ ☐☐☐☐

Expiry date ☐☐/☐☐ Security code ☐☐/☐☐ Switch cards only: Start date ☐☐/☐☐ Issue no. ☐

Signature ______________________________ Date ______________

Name and address of cardholder if different from above: ______________________

____

Please photocopy and return to:
BCPC Publications Sales, 7 Omni Business Centre, Omega Park, Alton, Hants GU34 2QD, UK
Tel: 01420 593 200, Fax: 01420 593 209, Email: publications@bcpc.org, Web: www.bcpc.org

BCPC

FUTURE EDITIONS

☐ I wish to take out an annual order for ____ copies of each new edition of *The UK Pesticide Guide*

☐ Please send me advance price details for the 2005 edition of *The UK Pesticide Guide* when available

Order by phone: 01420 593 200 or online: www.bcpc.org/bookshop

Bookshop orders to:
Customer Services, CABI Publishing, CAB International,
Nosworthy Way, Wallingford, Oxfordshire OX10 8DE, UK
Tel: 01491 832111, Fax: 01491 829292,
Email: publishing@cabi.org, Web: www.cabi-publishing.org

Bulk discount:

100+ copies	30%
50–99	25%
10–49	15%
List price £33.95	

Thank you for your order